Lecture Notes in Computer Science 4992

Commenced Publication in 1973
Founding and Former Series Editors:
Gerhard Goos, Juris Hartmanis, and Jan van ?

David Coeurjolly   Isabelle Sivignon
Laure Tougne   Florent Dupont (Eds.)

Discrete Geometry for Computer Imagery

14th IAPR International Conference, DGCI 2008
Lyon, France, April 16-18, 2008
Proceedings

 Springer

Volume Editors

David Coeurjolly
Isabelle Sivignon
Florent Dupont
Université Claude Bernard Lyon 1
Laboratoire LIRIS, Equipe M2DisCo
Bâtiment 710, 43 boulevard du 11 novembre 1918
69622 Villeurbanne cedex, France
E-mail: {david.coeurjolly,isabelle.sivignon,florent.dupont}@liris.cnrs.fr

Laure Tougne
Université Lumière Lyon 2
Laboratoire LIRIS, Equipe M2DisCo
5, avenue Pierre Mendès-France, 69676 Bron cedex, France
E-mail: laure.tougne@liris.cnrs.fr

Library of Congress Control Number: 2008924376

CR Subject Classification (1998): I.4, I.3.5, G.2, I.6.8, F.2.1

LNCS Sublibrary: SL 6 – Image Processing, Computer Vision, Pattern Recognition, and Graphics

ISSN   0302-9743
ISBN-10   3-540-79125-6 Springer Berlin Heidelberg New York
ISBN-13   978-3-540-79125-6 Springer Berlin Heidelberg New York

Springer is a part of Springer Science+Business Media

springer.com

© Springer-Verlag Berlin Heidelberg 2008
Printed in Germany

Typesetting: Camera-ready by author, data conversion by Scientific Publishing Services, Chennai, India
Printed on acid-free paper   SPIN: 12255115   06/3180   5 4 3 2 1 0

Preface

The 14th edition of the International Conference on Discrete Geometry for Computer Imagery was held in Lyon, France, April 16–18, 2008. DGCI 2008 attracted many researchers from all around the world. Indeed, 76 papers were submitted, from 24 different countries (13 European and 11 non European), confirming the international status of the conference. Once reviewed, 45 papers were accepted for publication in the present LNCS volume. In all, 23 papers were scheduled for oral presentation in single-track sessions, and 22 papers were presented as posters, with preliminary plenary sessions with very short presentations of these posters. Three invited speakers gave lectures on topics ranging from connected fields to the theoretical foundations of discrete geometry: Dinesh Manocha (University of North Carolina at Chapel Hill, USA), Ullrich Köthe (University of Heidelberg, Germany) and Jean-Pierre Reveilles (University of Auvergne, France).

Building on the experience of the previous editions, this edition was the occasion to remodel the paper selection process in order to improve the overall quality of the conference. These changes were based on an update and redefinition of the topics covered by the conference, in such a way that most of the present-day works in discrete geometry naturally fit in one topic. The members of the Program Committee were then chosen for their expertise in these different topics. They supervised the reviewing process of the papers related to their topic. Hereby, we would like to thank them for the important part they have played in the careful reviews. Furthermore, this would not have been possible without the support and advice of the Steering Committee members who guided us in this renewal process. Last but not least, we are indebted to the reviewers for the remarkable work they have accomplished. We are also grateful to all the authors who answered to the call for papers of DGCI, and to the invited speakers who kindly accepted our invitation.

DGCI 2008 was supported by the International Association for Pattern Recognition (IAPR), and is the main conference associated with the Technical Committee 18 on Discrete Geometry of the IAPR. We also would like to express our gratefulness to the Ecole Supérieure Chimie Physique Electronique de Lyon (CPE) for hosting this event and providing all the necessary infrastructural support. We finally thank our sponsoring institutions for providing the financial support essential for a successful event.

We dedicate this book to our late friend and colleague Oscar Figueiredo, who should have been with us in the Organizing Committee.

April 2008

David Coeurjolly
Isabelle Sivignon
Laure Tougne
Florent Dupont

Organization

DGCI 2008 was organized by the Multiresolution, Discrete and Combinatorial Models Team (M2Disco, Laboratoire LIRIS) of the University of Lyon.

Organizing Committee

David Coeurjolly   LIRIS, CNRS, France (General Co-chair)
Isabelle Sivignon   LIRIS, CNRS, France (General Co-chair)
Laure Tougne   LIRIS, Université Lumière Lyon 2, France
Florent Dupont   LIRIS, Université Claude Bernard Lyon 1, France
Jean-Marie Becker   Ecole Supérieure Chimie Physique Electronique de Lyon, France
Serge Miguet   LIRIS, Université Lumière Lyon 2, France

Steering Committee

Eric Andres   XLIM-SIC, Université de Poitiers, France
Gunilla Borgefors   Centre for Image Analysis, Uppsala University, Sweden
Achille Braquelaire   LABRI, Université Bordeaux 1, France
Jean-Marc Chassery   Gipsa-Lab, CNRS, Grenoble, France
Annick Montanvert   Gipsa-Lab, Université Pierre-Mendès France, Grenoble France
Kalman Palagyi   Department of Image Processing and Computer Graphics, University of Szeged, Hungary
Gabriella Sanniti di Baja   Istituto di Cibernetica "Eduardo Caianiello", CNR, Italy
Stina Svensson   Centre for Image Analysis, Uppsala University, Sweden

Program Committee

Reneta Barneva   Department of Computer Science, SUNY Fredonia, USA
Gilles Bertrand   Laboratoire A2SI, ESIEE Paris, France
Isabelle Debled-Rennesson   LORIA, France
Atsushi Imiya   Institute of Media and Information Technology, Chiba University, Japan

Pieter Jonker	Department of Imaging Science and Technology, Delft University of Technology, The Netherlands
Walter G. Kropatsch	Institute of Computer Aided Automation, Vienna University of Technology, Austria
Jacques-Olivier Lachaud	LAMA, Université de Savoie, France
Remy Malgouyres	LAIC, Université d'Auvergne Clermont 1, France
Ingela Nyström	Centre for Image Analysis, Uppsala University, Sweden
Pierre Soille	Joint Research Centre of the European Commission, Italy
Edouard Thiel	LIF, Université de la Méditerranée Aix-Marseille 2, France
Alain Daurat	LSIIT, Université Louis Pasteur Strasbourg, France
Peter Veelaert	University College Ghent, Belgium
Alexandru Telea	Eindhoven University of Technology, The Netherlands

Reviewing Committee

Sylvie Alayrangues	Ulrich Eckhardt	Yung Kong
Dominique Attali	Fabien Feschet	Ullrich Köthe
Joost Batenburg	Christophe Fiorio	Gaëlle Largeteau-Skapin
Valérie Berthé	Céline Fouard	Joakim Lindblad
Isabelle Bloch	Sebastien Fourey	Longin Jan Latecki
Valentin Brimkov	Yan Gerard	Robert Melter
Alfred Bruckstein	Jean-Pierre Guédon	Nicolas Normand
Luc Brun	Yll Haxhimusa	László G. Nyúl
Sara Brunetti	Jérôme Hulin	Eric Remy
Michel Couprie	Damien Jamet	Ralf Reulke
Jose Crespo	Zoltan Kato	Christian Ronse
Guillaume Damiand	Yukiko Kenmochi	Mohamed Tajine
Olivier Devillers	Bertrand Kerautret	Hugues Talbot
Eric Domenjoud	Christer Kiselman	Laurent Vuillon

Sponsoring Institutions

DGCI 2008 was sponsored by the following institutions:

- The International Association for Pattern Recognition (IAPR)
- Université Claude Bernard Lyon 1
- Ecole Supérieure Chimie Physique Electronique de Lyon (CPE)
- GdR Informatique Mathématique
- Conseil Général du Rhône

Table of Contents

Discrete and Combinational Topology

Geometric Transforms

Discrete Shape Representation, Recognition and Analysis

Discrete Tomography

Morphological Analysis

Discrete Modelling and Visualization

Discrete and Combinational Tools for Image Segmentation and Analysis

Digital Geometry Processing with Topological Guarantees

Dinesh Manocha

Department of Computer Science
University of North Carolina at Chapel Hill
dm@cs.unc.edu
http://gamma.cs.unc.edu/recons

1 Introduction

We describe novel approaches to compute reliable solutions for many non-linear geometric problems that arise in geometric modeling, computer graphics and robotics. Specifically, we focus on problems that can be formulated as *surface extraction problems*. These include Boolean operations and Minkowski sums of polyhedral or higher models as well as reliable polygonization of general implicit surfaces. All these problems reduce to computing a topology preserving isosurface from a volumetric grid, i.e. the zero set of a scalar field. A common way of representing a scalar field is to discretize the continuous scalar field into discrete samples – to compute the value of the scalar field at the vertices of a volumetric grid. We refer to this step as a sampling of the scalar field. The grid is an approximate representation of the scalar field; the accuracy of the approximate representation depends on the rate of sampling – the resolution of the grid. An explicit boundary representation of the implicit surface can be obtained by extracting the zero-level isosurface using Marching Cubes or any of its variants. We refer to these isosurface extraction algorithms collectively as *MC-like algorithms*. The output of an MC-like algorithm is an approximation – usually a polygonal approximation – of the implicit surface. We refer to this step as reconstruction of the implicit surface.

Our goal is to exploit the desirable properties of implicit surface representations for geometric computations such as Boolean operations (i.e. union, intersection and difference), Minkowski sum computation, simplification, configuration space boundary computation and remeshing. In each case, we wish to obtain an accurate polygonal approximation of the boundary of the final solid. Let E denote this boundary. We represent E implicitly – as an isosurface of a scalar field. This scalar field is obtained by performing minimum/maximum operations over the scalar fields defined for the primitives. At a broad level, our approach performs three main steps.

1. *Sampling*: Generate a volumetric grid and compute a scalar field (e.g, a signed distance field) at its corner grid points.
2. *Geometric operation*: For each geometric operation (union or intersection), perform an analogous operation (e.g., min/max) on the scalar fields of the primitives. At the end of this step, the scalar values at the grid points define a sampled scalar field for E.

D. Coeurjolly et al. (Eds.): DGCI 2008, LNCS 4992, pp. 1–3, 2008.

3. *Reconstruction*: Perform isosurface extraction using an MC-like algorithm to obtain a topologically accurate polygonal approximation E. In fact, we guarantee that the reconstructed surface is homeomorphic to the exact surface without explicitly computing the exact surface representation.

Preserving topology is also important in many applications. In CAD, topological features such as tunnels often correspond to distinguishing characteristics of the model. The geometric models used to represent the organs in medical datasets often consist of handles. Retaining these topological features can be necessary in order to preserve the anatomical structure of the organ, which can be crucial for visualization and analysis. Apart from capturing important features present in E, guaranteeing topology is important for another reason. An algorithm that preserves topology avoids the introduction of extraneous topology; its output does not have unwanted additional components or handles.

We present a novel approach to compute a topology preserving isosurface using an MC-like algorithm for geometry processing applications. We present conservative sampling criteria such that if every cell in the volumetric grid satisfies the criteria, then the extracted isosurface will have the same topology as the exact isosurface. We present an adaptive subdivision algorithm to generate a volumetric grid such that every grid cell satisfies the sampling criteria. We present efficient computational techniques to verify the sampling criteria during grid generation. Our algorithm can easily perform these computations on polyhedra, algebraic or parametric primitives and their Boolean com- binations. Furthermore, we extend the adaptive subdivision algorithm to also bound the Hausdorff distance between the exact isosurface and the extracted isosurface. This ensures that the extracted isosurface is geometrically close to the exact isosurface. We have used our algorithm to perform accurate boundary evaluation of Boolean combinations of polyhedral and low degree algebraic primitives, model simplification, and remeshing of complex models. In each case, we compute a topology preserving polygonal app- roximation of the boundary of the final solid. The running time of our algorithm varies between a few seconds for simple models consisting of thousands of triangles and tens of seconds on complex primitives represented using hundreds of thousands of triangles on a desktop PC.

Some of the main benefits of our approach include:

- Conservative sampling criteria for the volumetric grid such that the topology of the isosurface is preserved [1,4].
- An efficient adaptive subdivision algorithm to generate an octree satisfying the sampling criteria [1].
- Efficient and accurate algorithms for boundary evaluation of solids defined by Boolean operations. A fast algorithm to compute topology preserving simplification and remeshing of a complex polygonal model [1].
- An accurate algorithm to compute the Minkowski sum of 3D polyhedral models [2].
- A reliable algorithm to polygonize implicit surfaces based on star-shaped decomposition [6].
- Accurate algorithm to compute the free-space of low degree of freedom rigid and articulate models among polygonal obstacles [5].
- Exact motion planning for low degree of freedom robots, including first practical and reliable algorithms for path non-existence [3,4].

We will give an overview of our approach and demonstrate the applications. Overall, this approach provides a powerful framework to perform topologically reliable computations on a discretized geometric grid for a variety of applications.

Acknowledgements

This is joint work with my current and former students and postdocs including Gokul Varadhan, Shankar Krishnan, Young Kim, TVN Sriram and Liangjun Zhang. This work was supported in part by grants from National Science Foundation, Army Research Office, Office of Naval Research, DARPA, RDECOM and Intel.

References

1. Varadhan, G., Krishnan, S., Sriram, T.V.N., Manocha, D.: Topology Preserving Surface Extraction using Adaptive Subdivision. In: Eurographics Symposium on Geometry Processing (2004)
2. Varadhan, G., Manocha, D.: Accurate Minkowski Sum Approximation of Polyhedral Surfaces. In: Proceedings of Pacific Graphics (2004)
3. Varadhan, G., Krishnan, S., Sriram, T.V.N., Manocha, D.: A Simple Algorithm for Motion Planning of Translating Polyhedral Robots. In: Workshop on Algorithmic Foundations of Robotics (2004)
4. Varadhan, G., Manocha, D.: Star-shaped Roadmaps – A Deterministic Sampling Approach for Complete Motion Planning. In: Proceedings of Robotics: Science and Systems (2005)
5. Varadhan, G., Kim, Y., Krishnan, S., Manocha, D.: Topology Preserving Free Configuration Space Approximation. In: Proceedings of IEEE International Conference on Robotics and Automation (2006)
6. Varadhan, G., Krishnan, S., Zhang, L., Manocha, D.: Reliable Implicit Surface Polygonization using Visibility Mapping. In: Proceedings of Eurographics Symposium on Geometry Processing (2006)

What Can We Learn from Discrete Images about the Continuous World?*

Ullrich Köthe

Multi-Dimensional Image Processing Group, University of Heidelberg, Germany
ullrich.koethe@iwr.uni-heidelberg.de

Abstract. Image analysis attempts to perceive properties of the continuous real world by means of digital algorithms. Since discretization discards an infinite amount of information, it is difficult to predict if and when digital methods will produce reliable results. This paper reviews theories which establish explicit connections between the continuous and digital domains (such as Shannon's sampling theorem and a recent geometric sampling theorem) and describes some of their consequences for image analysis. Although many problems are still open, we can already conclude that adherence to these theories leads to significantly more stable and accurate algorithms.

1   Introduction

As far as computer vision is concerned, the real world can be considered as a continuous domain.[1] With few exceptions, discrete atoms and molecules are below the relevant scales of image analysis. In contrast, computer-based reasoning is always discrete – an infinite amount of information is inevitably lost before algorithms produce answers. Experience tells us that these answers are nevertheless useful, so enough information is apparently preserved in spite of the loss. But failures in automatic image analysis are not infrequent, and the fundamental question under which conditions discrete methods provide valid conclusions about the analog real world is still largely unsolved.

This problem can be approached in different ways. In one approach (which is, for example, popular in physics), theories and models are formulated in the analog domain, e.g. by means of differential equations. Discretisation is then considered as an implementation issue that doesn't affect the theory itself. The correctness of discrete solutions is ensured by asymptotic convergence theorems: Discretization errors can be made as small as desired by choosing sufficiently fine

---

* The author gratefully acknowledges essential contributions by Hans Meine and Peer Stelldinger and many fruitful discussions with Bernd Neumann, Hans-Siegfried Stiehl, and Fred Hamprecht.

[1] To avoid ambiguity, we shall use the term *analog* domain in the sequel. Otherwise, it might remain unclear whether a "continuous image" is an image function without discontinuities, or an image defined on a non-discrete domain. We will call the latter "analog image" instead.

D. Coeurjolly et al. (Eds.): DGCI 2008, LNCS 4992, pp. 4–19, 2008.

discretizations, and automatic or interactive refinement mechanisms adaptively adjust the resolution as required.

Unfortunately, this approach doesn't work well in image analysis: Usually, we don't have the opportunity to take refined images if the original resolution turns out to be insufficient. Thus, we need *absolute* accuracy bounds (rather than asymptotic ones) to determine the correct resolution beforehand, or to decide what kind of additional information can be utilized if the desired resolution cannot be reached. Moreover, new discretisation schemes may be required because conventional ones are often unable to arrive at optimal results when the resolution is fixed. For example, [14] show that orientation artifacts in non-linear diffusion on regular grids can be significantly reduced by means of a better discretization of the diffusion operator.

A different approach has been taken by purely discrete methods such as digital geometry. Here, theories are formulated directly in the discrete domain, and implementations impose no further information loss. Algorithms of this type are automatically reliable within the scope of the underlying theories and models. However, the fundamental question of discretization remains unanswered: Digital theories take their (digital) input for granted and don't consider where it comes from and how it relates to the real world. This is no problem as long as one refers to naturally discrete entities of the real world (e.g. well distinguished objects). But pixels in digital images are not natural: They only arise from a particular image acquisition method, and their relation to the real world is largely coincidental. When the theoretical treatment only starts after discretization, one can never be sure that the desired information was present in the original pixel values to begin with.

A satisfactory discretization theory should therefore refer to the analog and digital domains simultaneously. It should establish connections between analog models and their digital representations which lead to realistic, yet tractable problem formulations and provable solutions. Well-defined transformations between the two domains should be possible, so that one can perform each task in the most suitable domain, without loosing correctness guaranties. This paper attempts to give an overview over important results in this area.

2   Consequences of Shannon's Sampling Theorem

The information loss from the analog to the digital domain occurs in three stages (cf. figure 1): First, the 3-dimensional world is projected onto a 2-dimensional surface. Regardless of the particular projection method (perspective, fish-eye, catadioptric etc.), the depth dimension is lost in this transformation, but the resulting representation is still an *analog image* with infinite resolution. We refer to the mathematical projection of a real-world scene as the *ideal geometric image*. However, this geometric image is an idealization – any real optical device (e.g. a camera lens) imposes certain imprecisions such as geometric distortions and blurring. The second stage models the effect of these errors and results in the *real analog image*. It still resides in the analog domain (i.e. contains an infinite amount of information), and is not observable.

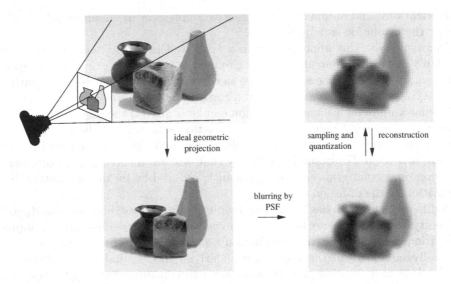

Fig. 1. The digitization model employed in this paper: the transformation from the real world to the digital image (top right) involves several stages of information loss

The third stage, discretization plus addition of noise (including photon counting noise, electronic noise, and quantization round-off noise), finally produces an observable *digital image* that only contains a finite amount of data. In the sense of set cardinality, the information loss during the last stage (from uncountable to finite) is very dramatic. Yet, much less effort has been devoted to understanding the consequences of discretization than to those of projection.

In this paper, we will not deal with projection but assume that the ideal geometric image is given and geometric distortions can be neglected or have been accounted for by camera calibration. The ideal image will be defined in terms of partitions of the plane that represent a given set of ground-truth boundaries:

Definition 1. *A* partition of the plane *is defined by a tuple* $[V, A, R]$, *where* $V = \{v_i \in \mathbb{R}^2\}$ *is a finite set of points and* $A = \{a_i \subset \mathbb{R}^2\}$ *a finite set of pairwise disjoint arcs such that each* a_i *is a bijective mapping of the open interval* $(0, 1)$ *into the plane, the end points* $\lim_{t \to \{0,1\}} a_i(t)$ *of each arc are in* V, *but no point of* V *belongs to an arc.* R *is a set of regions* r_i *which are defined as maximally connected components of the points not in* V *or any of the* a_i. *The union* $B = \{\boldsymbol{x} : \boldsymbol{x} \in V \lor \boldsymbol{x} \in a_i\}$ *of points and arcs determines the boundaries of the partition.*

Regions can be described by indicator functions χ_{r_k} taking the value 1 in the interior of region r_k and 0 outside. An ideal geometric image is now the superposition of a set of continuous functions f_{r_k} describing the intensities in every region

$$f_{\text{ideal}}(\boldsymbol{x}) = \sum_{r_k} \chi_{r_k}(\boldsymbol{x}) \, f_{r_k}(\boldsymbol{x})$$

such that f_{ideal} is discontinuous across region boundaries, i.e. at almost every boundary point $x \in B$.[2] An important special case is the well-known *step-edge model*, where all f_{r_k} are constant functions. When ideal geometric images were observable, image segmentation would be trivial: one could simply detect image discontinuities. Unfortunately, these discontinuities are lost in the transformation to the observed digital image, so that segmentation is considerably more complicated. Therefore, the quality of a low-level segmentation method can be judged by *how well it recovers the boundaries of the ideal image from the actual digital image*.

To keep the transition from the ideal to the real analog image tractable, we assume that the amount of blurring does not vary within the field of view, so that it can be modeled by a convolution of the ideal image. Additional blurring is introduced by the sensitivity profile of the CCD sensor elements (which necessarily have non-zero area). All sources of blurring are conveniently combined into a convolution with a single *point spread function* (PSF) of the image acquisition device:

$$f_{\text{real}}(x) = f_{\text{ideal}}(x) \star \text{psf}(x)$$

The Fourier transform of the PSF is called the *optical transfer function* (OTF), and its magnitude is the *magnitude transfer function* (MTF). An upper bound for all possible OTFs is defined by the OTF of the ideal *diffraction-limited system*: Since any real lens has finite extent, some blurring is inevitably caused by diffraction at the lens' exit pupil. If there exist no other aberrations, the OTF is

$$\text{OTF}_{\text{ideal}}(\nu, \nu_0) = \begin{cases} \frac{2}{\pi}\left(\cos^{-1}\left(\frac{\nu}{\nu_0}\right) - \frac{\nu}{\nu_0}\sqrt{1 - \left(\frac{\nu}{\nu_0}\right)^2}\right) & \text{if } \nu < \nu_0 \\ 0 & \text{otherwise} \end{cases}$$

where ν is the radial coordinate in the Fourier domain, and $\nu_0 = \frac{d}{z\lambda}$ is the *band-limit* of the OTF (with d - diameter of the exit pupil, z - distance between exit pupil and image, λ- wave length of the light). No real lens can perform better than the diffraction limited lens with the same exit pupil and magnification: $|\text{OTF}_{\text{real}}(\nu)| \leq |\text{OTF}_{\text{ideal}}(\nu, \nu_0)|$. Experimentally one finds that many real PSFs are well approximated by Gaussian kernels

$$\text{psf}_{\text{real}}(x) \approx \frac{1}{2\pi\sigma_{\text{PSF}}^2} \exp\left(-\frac{\|x\|^2}{2\sigma_{\text{PSF}}^2}\right)$$

with $0.45 < \sigma_{\text{PSF}} < 0.9$ (measured in pixel distances). The practical band-limit of this PSF is even lower than that of the diffraction-limited system because the corresponding Gaussian OTF rapidly converges to zero, see figure 2.

The fact that real OTFs are band-limited has an important consequence: Since Shannon's sampling theorem states that band-limited functions can be discretized and reconstructed without information loss (see e.g. [2]), this applies to

---

[2] Since B is set of measure zero, the exact values of f_{ideal} on B can be arbitrary as long as they remain finite.

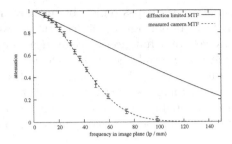

Fig. 2. The magnitude of the OTF of the ideal diffraction limited system and a real digital SLR camera. The sampling frequency of the latter's sensor is about 120 pixels per mm.

real images as well, provided the sampling frequency ν_S is high enough: $\nu_S \geq 2\nu_0$ (Nyquist limit). In practice, the reconstruction will never be perfect (mainly due to noise, but also because camera designers increase subjective image sharpness – at the price of some aliasing artifacts – by slight under-sampling), but it is good enough to consider the reconstructed analog image a close approximation of the real analog image

$$f_{\text{reconstructed}}(\boldsymbol{x}) = f_{\text{real}}(\boldsymbol{x}) + n(f_{\text{real}}(\boldsymbol{x}), \boldsymbol{x}) \tag{1}$$

where $n(f_{\text{real}}(\boldsymbol{x}), \boldsymbol{x})$ is a noise process with band-limit $\nu_S/2$ that may depend on image position and intensity and subsumes all sources of statistical reconstruction error. In many cases, the noise is or can be transformed into homogeneous, intensity-independent Gaussian noise, which is the easiest to handle. Performing this reconstruction is not always necessary in practice, although it can be done efficiently and accurately by means of cubic or quintic splines

$$f_{\text{reconstructed}}(\boldsymbol{x}) \approx \sum_{k,l} f_{kl}\, s_m(x_1 - k)\, s_m(x_2 - l) \tag{2}$$

where $f_{kl} = f_{\text{real}}(k, l) + n_{kl}$ is the digital image, and $s_m(.)$ is an mth-order cardinal spline reconstruction kernel [13]. Formulas (1) and (2) are key ingredients for the desired correspondence between analog and discrete domains: they ensure that an *analog theory* derived for f_{real} can actually be implemented with high accuracy by *discrete algorithms*, provided the noise is handled in a statistically sound way.

Due to convolution with the PSF, the real analog image does no longer exhibit any discontinuities, so that edge detection must be based on an alternative boundary definition. Many popular definitions involve the computation of some non-linear filter, for example the squared magnitude of the Gaussian gradient. Here we find another important consequence of Shannon's sampling theorem: When the image analysis algorithm includes *non-linear steps*, it is usually necessary to perform the analysis at a higher resolution than that of the original image in order to avoid further information loss. In other words, the image must be interpolated to a higher resolution *before* image analysis starts [7].

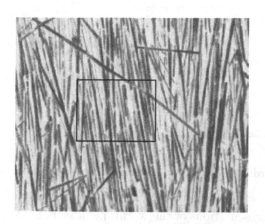 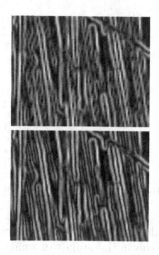

Fig. 3. Example image (Brodatz texture D15, left) where aliasing of the gradient magnitude at the original resolution is clearly visible (top right), whereas no aliasing occurs when the image is interpolated two-fold before gradient computation (bottom right)

We use the gradient to explain this seemingly counter-intuitive fact. The local maxima of the gradient squared magnitude along the gradient direction can be used as a boundary definition that replaces the (unobservable) discontinuities of the ideal image. Now suppose that the real analog image is band-limited according to the sampling rate of the sensor. When we apply any linear filter (e.g. a smoothing filter in order to reduce noise, or a derivative filter), the band-limit does not change, and the original resolution is still sufficient. However, the computation of the gradient squared magnitude involves the *square* of two linearly filtered images:

$$|\nabla f|^2 = f_x^2 + f_y^2$$

where f_x and f_y are derivatives along the horizontal and vertical directions respectively. As is generally known, a multiplication in the spatial domain corresponds to a convolution in the Fourier domain. If a band-limited spectrum is convolved with itself, the *band-limit is doubled*. Consequently, the gradient squared magnitude must be represented with doubled resolution in every direction (i.e. four times as many pixels). Figure 3 shows an example where severe aliasing artifacts occur when this is not done. Similar conclusions can be drawn for other non-linear methods, and it can generally be observed that image analysis results improve when the image is interpolated at the beginning.

3 Geometric Sampling Theorems

Shannon's sampling theorem guarantees a close relationship between a band-limited analog image function and its digitization. It is therefore a fundamental justification for digital image analysis in a continuous world. However, it only

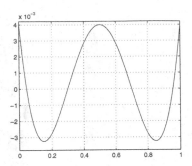

Fig. 4. A band-limited function according to (3) with 4 zero crossings in a single pixel

makes statements about iconic representations of the world, i.e. about the data at the initial stages of image analysis. But image analysis ultimately aims for descriptions at much higher levels of abstraction which are outside the scope of the sampling theorem.

We can demonstrate the limitations of Shannon's theorem by looking at basic geometric properties of band-limited 1-dimensional functions. Suppose we are interested in object edges defined by a level line of the image function. For simplicity, we choose level zero. When the image function is band-limited according to the grid's Nyquist limit, there will be at most one zero crossing per pixel *on average* (corresponding to the highest frequency component of the data). But the actual distribution of zero crossings can be rather uneven – there may be large regions without zero crossings, but there can also be many zeros in a very small interval. In fact, we can construct band-limited functions with as many zero-crossings in a given interval as we desire by means of a superposition of integer translates of the function $\operatorname{sinc}(x - k) = \frac{\sin(\pi(x-k))}{\pi(x-k)}$ (I'm indebted to P. Stelldinger for suggesting this construction). Since sinc-functions are band-limited, their superpositions are band-limited as well. Now consider, for example, the interval $[0, 1]$. A pair of two sinc-functions

$$s\,(x, k) = \operatorname{sinc}(x - k - 1) + \operatorname{sinc}(x + k)$$

has zeros at 0 and 1. Using two such pairs, we arrive at the function

$$f\,(x) = s(x, 1) + \frac{5}{3}s(x, 2) + 0.004 \tag{3}$$

which has four zeros in the interval $[0, 1]$ (approximately at 0.045, 0.3, 0.7, and 0.955, see figure 4). Obviously, more pairs can be added to this function to achieve arbitrary complicated behavior in the interval $[0, 1]$. In other words, even if the image function is band-limited, regions can become arbitrary narrow. Thus, the property of band-limitation is not sufficient to derive resolution requirements for the representation of *geometric features* such as region boundaries.

To overcome this limitation, it is necessary to complement the signal-theoretic sampling theorem with dedicated *geometric sampling theorems*. Such theorems must be based on certain assumptions on the geometry of the regions in the ideal geometric image, e.g. sufficient region size and limited boundary curvature. Necessary resolution requirements are then derived from these assumptions in conjunction with an adequate model of the image acquisition process. The challenge in developing these theories is to make them realistic while maintaining mathematical and algorithmic tractability – many existing attempts employ simplifications that exclude most of the practically relevant images.

One typical simplification is the notion of *r-regular shapes* first introduced in *[9,11]*. Here it is assumed that all objects in the ideal geometric image are invariant under morphologically opening and closing with a disc of radius r. This definition has a very undesirable consequence: Since the boundary curvature of r-regular shapes cannot exceed $1/r$, shapes are not allowed to have corners. This may not look like a very severe restriction at first because one can approximate corners as accurately as desired by choosing a small r. But this refinement strategy does not always work. Most importantly, it fails at occlusions: Occlusions give raise to configurations where the boundaries of three objects (the background, the occluding object, and the occluded object) meet in a single point, and it is impossible to arrange the configuration without a sharp corner in at least one of the three regions. Unfortunately, occlusions abound in many real images due to the very nature of projections from 3D to 2D.

Another unrealistic simplification is the direct use of the ideal geometric image in sampling analysis. For example, digital straight line algorithms are usually motivated by the direct digitization of a Euclidean straight line, for example by means of grid intersection or subset digitization, cf. [6]. That is, pixels (or interpixel cracks) are assigned to a digital line according to rules describing the relationship between the grid and the ideal Euclidean shape. However, this approach is far from what is actually happening in a camera: In reality, the ideal geometric image cannot be observed, and the intrinsic blurring and noise of the image acquisition process must be taken into account. We already mentioned the most important consequence of this: Convenient boundary definitions on the basis of set intersections and discontinuities are not applicable to the real camera image and must be replaced.

A third shortcoming in traditional geometric sampling analysis is the introduction of unnecessarily hard requirements on the quality of the digitization. In particular, it is often required that the digital image be topologically equivalent, i.e. homeomorphic, to the ideal geometric image. This excludes certain digitizations that we would intuitively consider as correct. For example, a junction of degree 4 in the ideal image is often digitized into a pair of junctions of degree 3 which cannot be continuously transformed into a 4-junction as required by topological equivalence (figure 5 left). Similarly, when a closed contour in the ideal image (i.e. a 1-dimensional manifold) is mapped onto a closed chain of 8-connected pixels (i.e. a 2-dimensional union of squares), this mapping cannot be continuous (figure 5 right). On the other hand, purely topological quality criteria are insufficient because they disregard geometric similarity between the ideal and digital shapes – since homeomorphisms can be arbitrary continuous

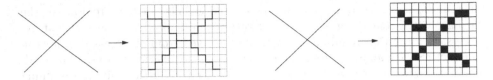

Fig. 5. Two cases where an intuitively correct boundary digitization is not homeomorphic to the ideal geometric image

distortions, topologically "correct" digitizations may actually fail to be close to the original shape by any intuitively plausible criteria. For example, the letters "L" and "G" are very different, yet topologically equivalent. In other words, a notion of geometric closeness should also be part of digitization analysis.

We addressed these shortcomings in [12]: There, we relaxed the requirement of topological equivalence to homotopy equivalence. In 2D, this is easily captured by the notion of *homotopy trees*:

Definition 2. *The* homotopy tree *of a partition P of the plane is defined as follows: The infinite region (containing the point at infinity) is the root of the tree. All connected boundary components adjacent to the infinite region are its children. In turn, the regions adjacent to these boundary components (if not already in the tree) become their children. This process is repeated recursively, alternating between boundary and regions levels, until all elements of P have been added to the tree.*

It has been shown by Serra [11] that two partitions of the plane are of the *same homotopy type* if and only if their homotopy trees are isomorphic, which is relatively easy to check. The definition ensures that the regions of a partition are correctly reconstructed, whereas certain errors in the boundary are permitted. For example, the digitizations in figure 5 are correct (have the same homotopy type as the original) under the new requirement. Homotopy allows us to relax the strict requirement of r-regular shapes into a much more realistic shape class [8,12]:

Definition 3. *A partition P of the plane is called r-stable when the boundary B of P can be morphologically dilated with an open disc of radius r without changing the homotopy tree of the partition.*

This definition essentially means two things: First, regions must be large enough for an r-disc to fit into them. Second, regions may not have narrow waists where boundary dilation would lead to splitting of a region into two or more disconnected parts. On the other hand, shapes with corners and junctions are allowed. Dilation extends the true boundary into a band of width $2r$ wherein the detected boundary may be distorted without changing the homotopy type of the partition. The key effect of this possibility is that segmentation needs no longer be error free: As long as errors are confined to the band, correct reconstruction is still possible. In fact, we have been able to prove the following geometric sampling theorem:

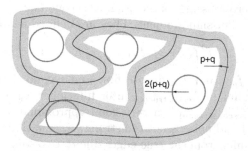

Fig. 6. Any r-regular shape (black lines) can be correctly segmented when boundary detection errors are confined to the gray strip around the true boundary, and every region contains a sufficiently large circle

Theorem 1. *Let S be a finite set of points approximating the true boundary B of a partition P of the plane such that the distance of any point on the true boundary from S is at most p, and the distance of any point in S from the true boundary is at most q. Then it is possible to link the points of S into the boundary B' of a reconstructed partition P' which has the same homotopy type as P, provided the following conditions are fulfilled: (i) P is r-stable with $r > 2q$ and $r > p+q$, and (ii) every region of P contains a disc with radius $r' > 2(p+q)$.*

The theorem is illustrated in figure 6. Its proof, as well as a detailed description of the reconstruction algorithm (based on Delaunay triangulation and α-shapes) can be found in [8,12]. The theorem does in no way depend on the method applied to compute the points in S, as long as the errors do not exceed the given bounds. Or, putting it the other way, correct segmentation is possible and indeed guaranteed when the regions in the true partition are, in a very precise sense, large enough with respect to the errors of the segmentation algorithm.

4 Error Analysis of Segmentation Methods

The geometric sampling theorem outlined in the last section poses the challenge of estimating the errors of a segmentation method. We must distinguish two kinds of error: systematic and statistical ones. The former arise mainly from the fact that the ideal, discontinuity-based boundary definition has to be replaced with some alternative definition (cf. section 2). Due to the unavoidable information loss incurred by a band-limiting pre-filter (i.e. the blurring effect of the PSF), these alternative definitions cannot in general reproduce the true boundaries exactly. Statistical errors are, on the one hand, a consequence of the inherent noise of any real image acquisition system. On the other hand, aliasing and round-off errors arising due to digitization (spatial sampling and intensity quantization) can also be treated as noise to good approximation.

Let us first look at systematic errors. Obviously, their nature and magnitude depend on the boundary definition applied. One particularly well understood possibility are the zero-crossings of the oriented second Gaussian derivative along the direction of the Gaussian gradient. They are defined as follows

$$f_x^2 f_{xx} + 2 f_x f_y f_{xy} + f_y^2 f_{yy} = 0 \tag{4}$$

where f_x, f_y, f_{xx}, f_{yy}, and f_{xy} denote first and second derivatives of the reconstructed analog image $f_{reconstructed}$ after filtering with a Gaussian at scale σ_{filter}. Recall that $f_{reconstructed}$ is a very good approximation of f_{real} when the PSF is band-limited and noise is not too strong, cf. equation (1). Also note that the zero-crossings differ from the true edges due to two sources of blurring: first by the PSF at scale σ_{PSF}, and then by the noise reduction filter at scale σ_{filter}. Since both blurring kernels are assumed to be Gaussians, the total blur can be expressed by a single scale parameter $\sigma = \sqrt{\sigma_{PSF}^2 + \sigma_{filter}^2}$. It can be shown that this blurring causes the following errors:

1. When the true edge is perfectly straight, and the intensities conform to the step-edge model (i.e. are constant within each region), no systematic errors occur, i.e. the true edge position can be estimated without bias.
2. When the true edge is circular with radius R and the regions still conform to the step edge model, the detected zero-crossing edge is located at the position $r = r_0$ which is the solution to the implicit equation

$$R \left(I_0 \left(\frac{rR}{\sigma^2} \right) + I_2 \left(\frac{rR}{\sigma^2} \right) \right) - 2r\, I_1 \left(\frac{rR}{\sigma^2} \right) = 0$$

where I_k are modified Bessel functions of order k [1]. This is the essential part of the second derivative of the image along the gradient direction, expressed in polar coordinates with origin at the center of curvature. It turns out that the detected edge is shifted toward the concave side of the true edge, and the bias can be well approximated by

$$\frac{r_0 - R}{\sigma} = 0.52 \sqrt{0.12^2 + \left(\frac{\sigma}{R} - 0.476 \right)^2} - 0.255$$

when $\sigma < R/2$ (which is true in most practical situations). For $\sigma < R/5$, the even simpler approximation $\frac{r_0 - R}{\sigma} \approx -\frac{\sigma}{2R}$ holds. The absolute bias $\Delta x = r_0 - R$ is thus essentially proportional to σ^2, the *square of the blurring scale*.
3. When the edge is straight, but the adjacent regions are shaded (i.e. do not conform to the step edge model), the bias depends on the orientation and strength of the shading gradient. When shading occurs perpendicular to the edge, and the resulting intensity profile can be described by two linear equations separated by a discontinuity of height b, the bias is

$$\Delta x = \frac{a}{b} \sigma^2$$

where a is the slope difference between the shading in the two regions (measured in the ideal geometric image). Again, the bias is proportional to the square of the scale σ.

4. At a sharp corner with opening angle β, the bias is maximal along the bisector of the corner. [10] derived the formula

$$\Delta x = \sigma\, x' \sqrt{1 + \left(\tan \frac{\beta}{2}\right)^2}$$

where x' is the solution of the implicit equation

$$\frac{1}{\sqrt{2\pi}}\, e^{-x'^2/2} - \left(\tan \frac{\beta}{2}\right)^2 \frac{x'}{2}\left(1 + \operatorname{erf}\left(\frac{x'}{\sqrt{2}}\right)\right) = 0 \qquad (5)$$

It can be seen that the displacement increases as the corner gets sharper. For $\beta = 90°$, the bias is about 0.7σ, but it reaches 2.2σ for $\beta = 15°$ and increases without bound for $\beta \to 0$.

5. At junctions of degree 3 and higher, the bias depends on the angles and intensities of the adjacent regions in very complicated ways, see e.g. [4]. Maximum errors can be determined experimentally and may be as high as 4σ. This mainly reflects the fact that equation (4) is only a very coarse approximation of the ideal geometric image near junctions. However, it is not at all easy to come up with an alternative boundary definition that produces consistently superior results, unless one is willing to give up generality and make heavy use of application-specific knowledge. A general definition is an unsolved problem of low-level image analysis.

The above theoretical predictions of the systematic errors of Gaussian zero-crossings are well confirmed by experiment, provided that edge positions are computed with floating-point accuracy and not rounded to integer coordinates. It should be stressed that errors frequently increase linearly or even quadratically with scale. This suggests that noise reduction filters should be chosen as small as possible in order to keep the induced bias to a minimum, unless we are certain that the image does not contain boundaries of the critical types.

Expected statistical errors can be derived from a suitable noise model. Photon counting sensors usually have a Poisson noise characteristic, i.e. the noise variance increases linearly with intensity I: $\sigma^2_{\text{noise}} = s\, I + t$, where the parameters s and t are determined by camera calibration. Since it is very inconvenient to handle intensity-dependent noise in algorithms, it is preferable to transform Poisson noise into unit-variance Gaussian noise by means of a noise normalization transformation

$$I' = \frac{2}{s}\sqrt{s\, I + t} - \frac{2\sqrt{t}}{s} \qquad (6)$$

where I' is the intensity after the transformation [5].[3] Other noise sources (electronic, quantization) are typically of much smaller magnitude. We can therefore

---

[3] Interestingly, most consumer cameras are internally performing a similar transformation, called γ-correction. Its main purpose is the creation of a perceptually more uniform color space (usually sRGB), but approximate noise normalization is achieved as a side effect since γ-correction curves are very similar to (6).

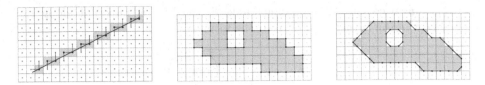

Fig. 7. Grid intersection digitization, subset digitization, mid-crack digitization of the reconstructed boundary

assume that the image is corrupted by additive white Gaussian noise. The noise standard deviation matters only in relation to the step height of the relevant discontinuities. We define the signal-to-noise ratio $SNR = b/\sigma_{\text{noise}}$ as the quotient of the step height and the noise standard deviation.

The effect of intensity noise on the localization accuracy of zero-crossing edges has, for example, been studied by [3]. Without loss of generality, one can assume that the true edge runs vertically through the coordinate origin. Than all y-derivatives in (4) vanish, and the estimated zero-crossing will be displaced along the x-axis by Δx. Formula (4) can be expanded into a Taylor series around $x = 0$, so that we get the estimated zero-crossing as the solution of

$$f_{xxx}(x)|_{x=0} \, \Delta x + n_{xx}(\Delta x) = 0$$

where $f_{xxx}(x)|_{x=0} = -\frac{b}{\sqrt{2\pi}\sigma^3}$ is the third derivative of the noise-free image blurred by both the PSF and noise reduction filters (i.e. with a total scale of $\sigma = \sqrt{\sigma_{\text{PSF}}^2 + \sigma_{\text{filter}}^2}$), and n_{xx} is the second derivative of the filtered noise (with only σ_{filter}, because the PSF acts before noise is added). The expected value of $|\Delta x|^2$ can be obtained as the quotient between the variance $N_{xx}^2 = \frac{3\sigma_{\text{noise}}^2}{16\pi\sigma_{\text{filter}}^6}$ of the second noise derivative, and the square of the third signal derivative. The standard deviation of the statistical edge displacement is therefore

$$\text{StdDev}\,[\Delta x] = \frac{\sigma_{\text{noise}}}{b} \frac{\sqrt{6}}{4} \left(1 + \frac{\sigma_{\text{PSF}}^2}{\sigma_{\text{filter}}^2}\right)^{3/2} \tag{7}$$

In typical situations, $\sigma_{\text{PSF}} < \sigma_{\text{filter}}$, so that $\text{StdDev}\,[\Delta x] \approx 1/SNR$. Typical signal-to-noise ratios in images of standard quality are above 10, so that $\text{StdDev}\,[\Delta x] < 0.1$ pixels is perfectly possible. The maximum error is below $3\,\text{StdDev}\,[\Delta x]$ in 99.7% of the cases. Therefore, the maximum noise-induced edge displacement is of the order of $3/SNR$, provided that edges are represented as polygons with floating-point knot coordinates.

When edges are represented on the grid instead, coordinate round-off errors must be accounted for as well. The magnitude of these errors depends on the particular method of rounding. We distinguish three common cases, cf. figure 7:

1. Grid intersection digitization: Whenever the polygonal edge crosses a grid line (i.e. a line connecting two directly adjacent pixel centers), the edge is rounded to the nearest pixel coordinate (both coordinates are integers).

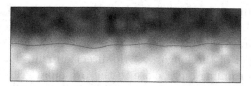

Fig. 8. Error correlation along a noisy straight edge at $SNR = 10$ and $\sigma_{\text{filter}} = 1$. The true edge runs in the image center. On average, the error of the detected edge has the same sign over about 4 pixels.

2. Subset digitization: Pixels are marked as being either inside or outside of an object of interest. The resulting boundary is therefore an interpixel boundary, and the edge coordinates are rounded to pixel corners (both coordinates are half-integers).
3. Mid-crack digitization: The boundary is defined as in subset digitization, but the coordinates are rounded to the midpoint between two directly adjacent pixels (one coordinate – either x or y – is integer, the other one half-integer).

In either case, round-off errors are uniformly distributed between the maximum possible displacements to the left and right of the polygonal boundary. It is easily verified that the standard deviation of a uniform distribution is $\text{StdDev}\,[\Delta x] = l/\sqrt{12}$, where l is the length of the permissible interval, and the maximum error is $l/2$. In cases 1 and 2, this interval is the pixel diagonal, i.e. $l = \sqrt{2}$, whereas in case 3 the interval is the pixel side length, i.e. $l = 1$. Since the round-off error is independent of the noise error, the total maximum error is the sum of the two maximum errors. In other words, the round-off error is significantly larger than the noise error.[4] We conclude once again that it is a good idea to perform image analysis at a higher resolution than the original image, either by representing edges as polygons with floating-point accuracy, or by interpolation to a denser grid. If the latter approach is chosen, mid-crack edges are to be preferred over other possibilities because they minimize the error at a given resolution.

Another consequence of the above noise analysis is the fact that the displacements of neighboring edge points are highly correlated. In case of noise-induced displacements according to (7), the error auto-correlation along the edge, or equivalently the noise power spectrum can be explicitly computed

$$\|\mathcal{F}\,[\Delta x]\,(u)\|^2 = 2\sqrt{\pi}\sigma_{\text{filter}}\,\text{StdDev}\,[\Delta x]^2\ e^{-4\pi^2\nu^2\sigma_{\text{filter}}^2}$$

where $\mathcal{F}\,[\Delta x]$ denotes the Fourier transform of the error. An immediate conclusion from this formula is that the correlation length along the edge is rather high: On average, the error Δx keeps its sign over an interval of $\sqrt{2}\pi\sigma_{\text{filter}}$ pixels along the boundary. That is, even for small filters with $\sigma_{\text{filter}} = 1$, the edge is

---

[4] This is in contrast to the situation for intensity quantization. Here, the noise standard deviation (usually about 3 gray levels) is much larger than the quantization standard deviation of $1/\sqrt{12}$ gray levels.

displaced to the same side of the true boundary over more than 4 pixels, see figure 8. The situation with round-off error is similar: As is well known, a digital straight line contains only two different directions, and the error sign remains equal within one half of the longest single-direction run. Thus, the correlation is highest for edges running on or close to the principal grid directions. In fact, the error of a perfectly horizontal or vertical digital line will have the same sign throughout.

This error correlation has important consequences for certain methods that attempt to improve an initial boundary (or properties derived from it) by filtering the edge coordinates *along* the boundary. This includes, in particular, methods that estimate the boundary tangent direction by means of derivative filters or geometric fits applied to the points of the boundary: Due to correlation, small filtering or fitting windows will not lead to significant improvements of the estimates from the raw boundary – correlated errors can only be averaged out when the windows are significantly larger than the correlation length. Sufficiently large windows will only be available for rather large objects. Consequently, it can be shown both theoretically and experimentally that tangent angle estimates derived from the direction $\phi_{\text{tangent}} = \arctan\left(-f_x/f_y\right)$ perpendicular to the simple Gaussian gradient of the gray levels are often vastly superior to much more complicated estimates derived from the boundary itself.

5   Conclusions

This contribution demonstrated that image analysis theories can and, in fact, should be constructed so that they apply to both the analog and digital domains. This approach offers the possibility to proof that computed properties actually correspond to properties of the analog real world. In contrast, no such guarantees can be established when the two domains are considered in isolation. We reported a number of useful and encouraging results, but the existing body of knowledge barely scratches the surface of this important area, and much remains to be desired.

First, the realism of the theories has still to be enhanced in order to be of universal practical value. For example, the geometric sampling theorem 1 is based on bounds for maximum edge detection errors. However, these bounds are quite large for certain boundary configurations such as junctions, and may become arbitrarily big if the edge detector produced some artifact edges (outliers). We would like to reduce these errors by using a more accurate and more robust boundary definition for blurred noisy images, but we would also like to generalize the theorem so that it gives probabilistic guarantees when the requirements can only be fulfilled with certain probabilities. Moreover, it would be highly desirable to proof similar theorems for higher dimensions and other image analysis tasks like object recognition – these bounds may differ in interesting ways from the bounds for segmentation.

Another important achievement would be the development of a truly multidimensional signal theory. Up to now, Fourier transforms, filters and related techniques are often generalized to 2- and higher dimensional spaces by simple outer products of 1-dimensional spaces. In practice, this often means that the desirable rotational invariance of the Euclidean space is lost during digitization, so that digital algorithms exhibit significant anisotropy. Solutions to this problem might eventually be built into dedicated image analysis cameras that are optimized for the requirements of digital computers, and not for that of human viewers like most current cameras.

References

1. Bouma, H., Vilanova, A., van Vliet, L.J., Gerritsen, F.A.: Correction for the Dislocation of Curved Surfaces Caused by the PSF in 2D and 3D CT Images. IEEE Trans. Pattern Analysis and Machine Intelligence 27(9), 1501–1507 (2005)
2. Bracewell, R.N.: The Fourier Transform and its Applications. McGraw-Hill, New York (1978)
3. Canny, J.: A Computational Approach to Edge Detection. IEEE Trans. Pattern Analysis and Machine Intelligence 8(6), 679–698 (1986)
4. Deriche, R., Giraudon, G.: A computational approach for corner and vertex detection. Intl. Journal of Computer Vision 10(2), 101–124 (1993)
5. Förstner, W.: Image Preprocessing for Feature Extraction in Digital Intensity, Color and Range Images. In: Proc. Summer School on Data Analysis and the Statistical Foundations of Geomatics. Lecture Notes in Earth Science, Springer, Berlin (1999)
6. Klette, R., Rosenfeld, A.: Digital Geometry. Elsevier, Amsterdam (2004)
7. Köthe, U.: Edge and Junction Detection with an Improved Structure Tensor. In: Michaelis, B., Krell, G. (eds.) DAGM 2003. LNCS, vol. 2781, pp. 25–32. Springer, Heidelberg (2003)
8. Köthe, U., Stelldinger, P., Meine, H.: Provably Correct Edgel Linking and Subpixel Boundary Reconstruction. In: Franke, K., Müller, K.-R., Nickolay, B., Schäfer, R. (eds.) DAGM 2006. LNCS, vol. 4174, pp. 81–90. Springer, Heidelberg (2006)
9. Pavlidis, T.: Algorithms for Graphics and Image Processing. Computer Science Press, Rockville (1982)
10. Rohr, K.: Localization Properties of Direct Corner Detectors. Journal of Mathematical Imaging and Vision 4, 139–150 (1994)
11. Serra, J.: Image Analysis and Mathematical Morphology. Academic Press, New York (1982)
12. Stelldinger, P., Köthe, U., Meine, H.: Topologically Correct Image Segmentation Using Alpha Shapes. In: Kuba, A., Nyúl, L.G., Palágyi, K. (eds.) DGCI 2006. LNCS, vol. 4245, pp. 542–554. Springer, Heidelberg (2006)
13. Unser, M., Aldroubi, A., Eden, M.: B-Spline Signal Processing. IEEE Trans. Signal Processing 41(2), 821–833 (part I), 834–848 (part II) (1993)
14. Weickert, J., Scharr, H.: A scheme for coherence-enhancing diffusion filtering with optimized rotation invariance. J. Visual Comm. Image Repr. 13(1/2), 103–118 (2002)

Weak Rational Computing for Digital Geometry

Jean-Pierre Reveillès

LAIC, Auvergne University,Clermont-Ferrand

Abstract. Since several centuries Mathematics award the prominent position to continuous concepts and a secondary one to discrete objects and integers. On the contrary Computer Science and digital technologies lay integers and discrete structures at the heart of their concerns and recover continuous notions from them.

During the eighties some Strasbourg's mathematicians (mainly G. Reeb and J. Harthong) showed, relying on Non Standard Analysis, how integers could be substituted to real numbers in areas like Analysis and Geometry.

Even if Strasbourg's NSA researchers were not, at first, motivated by relationships between Mathematics and Computer Science, they soon realized the interest of this issue and started some work in that direction, mainly on the use of *all integers methods* to integrate differential equations.

It was only from 1987 that convergence with Digital Geometry arose (A. Troesch, J.-P. Reveills) resulting in original definitions for discrete objects (lines, planes, circles...)

Work independently done since that time by digital geometers produced many results but, also, the need for new tools as one to treat *multiscale digital objects*.

We will briefly explain why mathematicians can be interested in Non Standard Analysis and some of the consequences this had in Strasbourg's mathematics department, mainly Harthong's Moir theory and Reeb's integration of equation $y' = y$ by an *all integer* method.

This last one, called *Weak Rational Computing*, is a kind of abstract *Multiscale System* which will be detailed with the help of simple linear and non linear differential equations and iteration systems applied to geometric entities.

D. Coeurjolly et al. (Eds.): DGCI 2008, LNCS 4992, p. 20, 2008.
© Springer-Verlag Berlin Heidelberg 2008

A First Look into a Formal and Constructive Approach for Discrete Geometry Using Nonstandard Analysis*

Laurent Fuchs[1], Gaëlle Largeteau-Skapin[1], Guy Wallet[2], Eric Andres[1], and Agathe Chollet[2]

[1] Laboratoire SIC,
Université de Poitiers,
BP 30179 86962 Futuroscope Chasseneuil cédex, France
{fuchs, glargeteau,andres}@sic.univ-poitiers.fr
[2] Laboratoire LMA,
Université de La Rochelle,
Avenue Michel Crépeau 17042 La Rochelle cedex, France
Guy.Wallet@univ-lr.fr, achollet@etudiant.univ-lr.fr

Abstract. In this paper, we recall the origins of discrete analytical geometry developed by J-P. Reveillès [1] in the nonstandard model of the continuum based on integers proposed by Harthong and Reeb [2,3]. We present some basis on constructive mathematics [4] and its link with programming [5,6]. We show that a suitable version of this new model of the continuum partly fits with the constructive axiomatic of $\mathbb{R}$ proposed by Bridges [7]. The aim of this paper is to take a first look at a possible formal and constructive approach to discrete geometry. This would open the way to better algorithmic definition of discrete differential concepts.

Keywords: discrete geometry, nonstandard analysis, constructive mathematics.

1   Introduction

In the last twenty years Reveillès' approach to discrete geometry, namely Discrete Analytic Geometry (DAG), has become a very successful approach. DAG is based on the development of a powerful arithmetical framework which was originally founded on a special view of calculus: nonstandard analysis. The goal of this paper is to revisit some of these results and relate them to recent results on constructive mathematics [7].

Calculus, as initiated by Leibniz and Newton, deals with the concept of infinitesimals that are very small non-zero quantities. These infinitesimals have been used to define the notion of derivatives. However, even if powerful methods were developed by Leibniz and Newton, the notion of infinitesimal numbers wasn't well defined. These numbers, that are smaller than any positive number

---

* Partially supported by the ANR program ANR-06-MDCA-008-05/FOGRIMMI.

D. Coeurjolly et al. (Eds.): DGCI 2008, LNCS 4992, pp. 21–32, 2008.

but still different from zero, didn't satisfy usual properties of real numbers. For example, any multiple of an infinitesimal number is still an infinitesimal number. This does not satisfy the Archimedean property: if x and y are two numbers such that $x < y$ then there exists an integer n such that $y < n.x$. Some paradoxes, such Zeno's paradox [8], also questioned the foundations of calculus. In the 19th century, this led to development of the, now, classical approach to calculus, based on the notion of limits defined on the continuum of real numbers. Later, in the mid of the 20th century an alternative approach, the nonstandard analysis, was proposed which adds infinitesimals and infinitely large numbers to the real numbers.

At the end of the eighties, at Strasbourg, Reeb and Harthong developed a nonstandard model of the continuum based on integers, the discrete-continuous Harthong-Reeb model [3]. This arithmetical description of the continuum was firstly tested on the numerical resolution of differential equations with integer numbers. We recall in this paper how the simple equation $y' = \alpha$, led Reveillès to his well known discrete analytical line definition and thus to the Discrete Analytic Geometry theory [1]. This study is mainly based on J. Harthong [2,3], F. Diener and G. Reeb [9], M. Diener [10] and J-P. Reveillès and D. Richard [1] works. Since part of these works are in french, in our paper we have tried to summarize them to be self contained.

Interestingly, one of the difficulties that the development of discrete geometry faces today is the difficulty of correctly defining and using differential concepts. Our claim is that these difficulties come from a lack of theoretical foundations and effective methods (algorithms) to compute them. This is our motivation to reinvestigate the original nonstandard analysis point of view of discrete geometry. As Georges Reeb himself noted [9,11], his model can be looked at from the constructivist (intuitionist) point of view. This has however never been really fully investigated although it represents a way to integrate the algorithmic point of view to the continuum theory [12]. In this paper, we take a first look into a formal and constructive approach for discrete geometry using nonstandard analysis.

For that, we show that a suitable version of the Harthong-Reeb model of the continuum partly fits with the constructive axiomatic of $\mathbb{R}$ proposed by Bridges [7]. Thus, this Harthong-Reeb model can be viewed as a constructive discrete-continuous model of the real line (called the Harthong-Reeb line). This is the first step in the project of giving theoretical and algorithmic definitions of discrete differential notions such as, for instance, the curvature of a curve.

2   Theoretical Framework

In this part, we start with the origin of the Reveillès line to illustrate the strong link between the Harthong-Reeb approach to nonstandard analysis and discrete geometry. We explain the link between the integer nonstandard set and the real line $\mathbb{R}$. We also present an abstract on constructive mathematics and its link with programming.

2.1   Origin of the Reveillès Discrete Analytical Line

The Reveillès definition of a discrete naive straight line is classically given by [1]:

Definition 1. *A discrete analytical line of Reveillès is defined by*

$$D(a, b, \gamma, \tau) = \{(x, y) \in \mathbb{Z}^2, \gamma \le ax - by < \gamma + \tau\}$$

where a, b, γ and τ are integers with $0 \le a \le b$, $b \ne 0$ and $\tau > 0$. In the case where $b = \tau$, this definition is equivalent to $y = \lfloor \frac{ax - \gamma}{b} \rfloor$.

In this definition, the *integer part* $\lfloor x \rfloor$ of a real number x is the largest integer less than or equal to x and the *fractional part* of x is the real number $\{x\} \in [0, 1[$ such that $x = \lfloor x \rfloor + \{x\}$.

Originally, the definition of the discrete analytical line comes from the use of the Euler method (that numerically resolve ordinary differential equations) to the resolution of the differential equation $y'(x) = a$ such that $y(0) = b$. Solution of this equation is the straight line with equation $y(x) = ax + b$.

The Euler method gives the system

$$\left\{ x_0 = 0, \ y_0 = b, \ x_{n+1} = x_n + \frac{1}{\beta}, \ y_{n+1} = y_n + \frac{a}{\beta} \right\}$$

where $\frac{1}{\beta}$ is an **infinitely small** integration step.

This system is arithmetized (i.e. transformed in an integer system) using an integer ω that can be viewed as a scale factor that allows to adjust the level of precision.

Everything works as if we move the coma in the usual real number representation: consider $\omega = 100$, then the real 3.12 becomes the integer 312 and no error is done on the two first digits after the coma. Hence, with an **infinitely large integer** ω, infinite precision is obtained. In practice, we work with an arithmetical analogous of the previous system:

$$\left\{ X_0 = 0, \ Y_0 = \lfloor \omega b \rfloor, \ X_{n+1} = X_n + \beta, \ Y_{n+1} = Y_n + \left\lfloor \frac{\lfloor \omega a \rfloor}{\beta} \right\rfloor \right\}$$

Note that, even if ω and β are independent, in practice it is useful to define $\omega = \beta^2$. For an integer T, we define $\mathbb{T}$ by the Euclidean division $T = \mathbb{T}\beta + r$. With this decomposition we have:

$$Y_{n+1} = \mathbb{Y}_{n+1}\beta + r_{n+1} = \mathbb{Y}_n\beta + \left\lfloor \frac{r_n + \left\lfloor \frac{\lfloor \omega a \rfloor}{\beta} \right\rfloor}{\beta} \right\rfloor \beta + \left\{ \frac{r_n + \left\lfloor \frac{\lfloor \omega a \rfloor}{\beta} \right\rfloor}{\beta} \right\}$$

$$= \left(\mathbb{Y}_n + \left\lfloor \frac{r_n + \left\lfloor \frac{\lfloor \omega a \rfloor}{\beta} \right\rfloor}{\beta} \right\rfloor \right) \beta + \left\{ \frac{r_n + \left\lfloor \frac{\lfloor \omega a \rfloor}{\beta} \right\rfloor}{\beta} \right\}$$

Since the decomposition in a base is unique, the following system is obtained:

$$\begin{cases} \mathbb{X}_0 = 0, \ \mathbb{X}_{n+1} = \mathbb{X}_n + 1, \mathbb{Y}_0 = B, \ r_0 = \left\{ \frac{\lfloor \omega b \rfloor}{\beta} \right\}, \\ \mathbb{Y}_{n+1} = \mathbb{Y}_n + \left\lfloor \frac{r_n + A}{\beta} \right\rfloor, \ r_{n+1} = \left\{ \frac{r_n + A}{\beta} \right\}. \end{cases}$$

where $A = \left\lfloor \frac{\lfloor \omega a \rfloor}{\beta} \right\rfloor$ and $B = \left\lfloor \frac{\lfloor \omega b \rfloor}{\beta} \right\rfloor$. This leads directly to the Reveillès algorithm [1] to draw the digitized straight line : $\mathbb{Y} = \frac{A}{\beta}\mathbb{X} + B$. The slope of this discrete line $\frac{A}{\beta}$ is an infinitely good approximation of a.

Note that to obtain the arithmetized Euler scheme, we have used an "infinitely small" value as integration step and an "infinitely large" value as scaling (precision) factor. Moreover, by choosing nonstandard values such that $\omega = \beta^2$ and adding the predicate "standard", denoted by st (which allows us to determine standard integers) to the usual Peano's axioms, we have the basis of the Harthong-Reeb model of nonstandard analysis (see below).

As usual, in order to deal with a model, it is useful to define a set of rules (axioms) that determine authorized expressions. Next section gives a minimal set of such rules that can be viewed as an approach of nonstandard analysis which is simpler and better adapted to our purpose than the usual theory [13,14].

2.2   Bases of Nonstandard Analysis on $\mathbb{N}$ and $\mathbb{Z}$

In this section we show that a continuum theory of the real line can be developed using only integers [2,3,9,10,1]. The ground idea of Harthong-Reeb model is that it suffices to introduce a scale over the usual set of integers to obtain a space that is both discrete and continuous. Nonstandard analysis is the paradigm that can be used to define such scales.

Even if axiomatic theories of nonstandard analysis, such as IST [13], are available, we present here axioms that are well suited for our purpose.

First we introduce a new predicate st over integer numbers: $st(x)$ "means" that the integer x is standard. This predicate is external to the classical integer theory and its meaning directly derives from the following axioms ANS1, ANS2, ANS3, ANS4 (and ANS5 which will be introduced later):

ANS1. *The number 1 is standard.*
ANS2. *The sum and the product of two standard numbers are standard.*
ANS3. *Nonstandard integer numbers exist.*
ANS4. *For all $(x,y) \in \mathbb{Z}^2$ such that x is standard and $|y| \leq |x|$, the number y is standard.*

For reading conveniences, we introduce the following notations:

- $\forall^{st}x \ F(x)$ is an abbreviation for $\forall x \ (st(x) \Rightarrow F(x))$ and can be read as "for all standard x, $F(x)$ stands".
- $\exists^{st}x \ F(x)$ is an abbreviation for $\exists x \ (st(x) \wedge F(x))$ and can be read as "exists a standard x such that $F(x)$".

Here we have to insist on the fact that these rules are added to every classical property (axioms or theorems) over integer numbers. Everything that was classically true remains true. We simply improve the language by a syntactic provision. These first rules imply that $\mathbb{N}$ is split into two classes, the class $\mathbb{N}_{st} := \{0, 1, \ldots\}$ of natural standard integers (closed by arithmetical operations), and the class of natural nonstandard integers (a nonstandard integer is bigger than every standard integer). These nonstandard integers are said *infinitely large*. These first axioms allow the development of an explicit and rigorous calculus on the different scales.

Let us add some technical but important remarks. A formula $\mathcal{P}$ is said *internal* if it does not bring in elements of the greatness scale. For example, the formula $x + 1 > x$ is internal. An internal formula is therefore a classical formula on numbers. In contrast, an *external* formula uses explicitly the greatness scale ; for example, $st(x)$ or $\forall^{st} x,\ y < x$ are external formulae. Since everything that was true remains true, for all internal formula $\mathcal{P}(x)$, we can build the set $P = \{x \in \mathbb{N}\ ;\ \mathcal{P}(x)\}$ which possesses the classic properties of subsets of $\mathbb{N}$; for example, if P is not empty and is bounded, then P possesses a bigger element which is not necessarily valid for an external property. For instance, if we consider the external property $st(x)$, the class $\mathbb{N}_{st} = \{x \in \mathbb{N}\ ;\ st(x)\}$ of standard integers is a non empty bounded part which cannot have a bigger element since $x + 1$ is standard for all standard x. A class of numbers defined by an external property which cannot be a set of numbers in the classical meaning is called *external set*. Hence, $\mathbb{N}_{st}$ is an external part of $\mathbb{N}$. Dealing with external sets that are not classical (internal) sets gives birth to a new process of demonstration called the *overspill principle*.

Proposition 1. *(Overspill principle) Let $\mathcal{P}(x)$ be an internal formula such that $\mathcal{P}(n)$ is true for all $n \in \mathbb{N}_{st}$. Then, there exists an infinitely large $\nu \in \mathbb{N}$ such that $\mathcal{P}(m)$ is true for all integers m such that $0 \leq m \leq \nu$.*

Proof. The class $A = \{x \in \mathbb{N}\ ;\ \forall y \in [0, x]\ \mathcal{P}(y)\}$ is an internal set (i.e. a classical set) containing $\mathbb{N}_{st}$. Since $\mathbb{N}_{st}$ is an external set, the inclusion $\mathbb{N}_{st} \subset A$ is strict and leads to the result. $\square$

In the same way, the application of an inductive reasoning on an external formula can be illegitimate. For example, number 0 is standard, $x + 1$ is standard for all standard x. Nevertheless not all integers are standard. To improve the power of our nonstandard tool, we have to add a special induction that fits with external formulae. In the following principle which is our last axiom, $\mathcal{P}$ denotes an internal or external formula:

ANS5. *(External inductive defining principle): We suppose that*

1. *there is $x_0 \in \mathbb{Z}^p$ such that $\mathcal{P}((x_0))$;*
2. *for all $n \in \mathbb{N}_{st}$ and all sequence $(x_k)_{0 \leq k \leq n}$ in $\mathbb{Z}^p$ such that $\mathcal{P}((x_k)_{0 \leq k \leq n})$ there is $x_{n+1} \in \mathbb{Z}^p$ such that $\mathcal{P}((x_k)_{0 \leq k \leq n+1})$.*

Therefore, there exists an internal sequence $(x_k)_{k\in\mathbb{N}}$ *in* $\mathbb{Z}^p$ *such that, for all* $n \in \mathbb{N}_{st}$, *we have* $\mathcal{P}((x_k)_{0\leq k\leq n})$.

This principle means that the sequence of values x_k for k standard can be prolonged in an infinite sequence $(x_k)_{k\in\mathbb{N}}$ defined for all integers. Saying that this sequence is internal means that this is a classical sequence of the number theory. Particularly, if $\mathcal{Q}(x)$ is an internal formula, then the class $\{k \in \mathbb{N} \; ; \; \mathcal{Q}(x_k)\}$ is an internal part of $\mathbb{N}$.

2.3   The System $\mathcal{A}_\omega$

Now we are going to give the definition of the system $\mathcal{A}_\omega$. Introduced by M. Diener [10], this system is the formal version of the so-called Harthong-Reeb line. The underlying set depends on a parameter ω which is an infinitely large integer. In the next section (cf. section 3) we prove that this system can be viewed as a constructive model of the real line.

Accordingly to axiom ANS3, the construction starts by considering $\omega \in \mathbb{N}$ is an infinitely large (nonstandard) integer. We then introduce the set:

Definition 2. *The set* $\mathcal{A}_\omega$ *of the admissible integers considering the scale* ω *is defined by:* $\mathcal{A}_\omega = \{x \in \mathbb{Z} \; ; \; \exists^{st}n \in \mathbb{N} \; |x| < n\omega\}$.

The set $\mathcal{A}_\omega$ is an external set. Moreover, it is an additive sub-group of $\mathbb{Z}$. We provide $\mathcal{A}_\omega$ with the operations $+_\omega$ and $*_\omega$, the ω-scale equality, the ω-scale inequality relations (noted $=_\omega$ and $\neq_\omega$) and the order relation $>_\omega$:
We note $+, -, ., /, >$ the usual operations and order relation in $\mathbb{Z}$.

Definition 3. *Let* X *and* Y *be any elements of* $\mathcal{A}_\omega$.

- *X and Y are equal at the scale ω and we write $X =_\omega Y$ when*
  $\forall^{st}n \in \mathbb{N} \quad n|X - Y| \leq \omega$.
- *Y is strictly greater than X at the scale ω and we write $Y >_\omega X$ when*
  $\exists^{st}n \in \mathbb{N} \quad n(Y - X) \geq \omega$.
- *X is different from Y at the scale ω and we write $X \neq_\omega Y$ when*
  $(X >_\omega Y$ or $Y >_\omega X)$
- *The sum of X and Y at the scale ω is $X +_\omega Y := X + Y$ (like the usual sum). For this operation, the neutral element is $0_\omega = 0$ and the opposite of each element $Z \in \mathcal{A}_\omega$ is $-_\omega Z := -Z$.*
- *The product of X and Y at the scale ω is $X \times_\omega Y := \lfloor \frac{X.Y}{\omega} \rfloor$ (different from the usual one). The neutral element is $1_\omega := \omega$, and the inverse of each element $Z \in \mathcal{A}_\omega$ such that $Z \neq_\omega 0_\omega$ is $Z^{(-1)\omega} := \lfloor \frac{\omega^2}{Z} \rfloor$.*

Let us give an informal description of $\mathcal{A}_\omega$. It is easy to see that $X = Y$ implies $X =_\omega Y$ but that the reverse is not true. It is a little less obvious to see that we have $\forall X \in \mathcal{A}_\omega$, $st(X)$ implies $X =_\omega 0$ but not the reverse. Indeed, for instance, $\lfloor \sqrt{\omega} \rfloor =_\omega 0$ (because $\forall^{st}n \in \mathbb{N}, n.\lfloor \sqrt{\omega} \rfloor < \omega$) but $\lfloor \sqrt{\omega} \rfloor$ isn't a standard integer or else ω would also be.

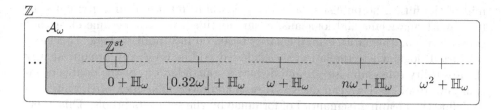

Fig. 1. The integer set $\mathbb{Z}$ and the set $\mathcal{A}_\omega$ (in grey)

Figure 1 illustrates how a representation of $\mathcal{A}_\omega$ could look like. Let us define the set $\mathbb{H}_\omega := \{X \in \mathcal{A}_\omega \; ; \; X =_\omega 0\}$. We have a strict inclusion of $\mathbb{Z}^{st}$ into $\mathbb{H}_\omega$. Let us consider the classical real value 0.32. In $\mathcal{A}_\omega$, the integer $\lfloor 0.32\omega \rfloor$ will be a representation of the classical real value 0.32 with an infinite precision. Of course, so does $\lfloor 0.32\omega \rfloor + 150$ or any integer x belonging to $\lfloor 0.32\omega \rfloor + \mathbb{H}_\omega$ since they all verify $x =_\omega \lfloor 0.32\omega \rfloor$. It is easy to see that $\lfloor 0.32\omega \rfloor$ is neither in $0 + \mathbb{H}_\omega$ nor in $\omega + \mathbb{H}_\omega$ (nor in $\lfloor 0.319\omega \rfloor + \mathbb{H}_\omega$ for that matter). The set $\lfloor 0.32\omega \rfloor + \mathbb{H}_\omega$ is sometimes called the halo of 0.32. The set $\mathcal{A}_\omega$ is in grey on the figure.

As we can see, $\mathcal{A}_\omega$ doesn't extend to $\omega^2 + \mathbb{H}_\omega$. The set $\mathcal{A}_\omega$ is a subset of $\mathbb{Z}$. It contains all integers from $-n\omega + \mathbb{H}_\omega$ to $n\omega + \mathbb{H}_\omega$ with n **standard**. Whereas the set $\mathbb{Z} \setminus \mathcal{A}_\omega$ contains all the integers $N\omega + \mathbb{H}_\omega$ with N **nonstandard**. Particularly $\mathcal{A}_\omega$ contains all $\lfloor k.\omega \rfloor + \mathbb{H}_\omega$ for k limited in $\mathbb{R}$ (i.e. $k \in \mathbb{R}$ such that $\exists^{st} n \in \mathbb{N}$ with $|k| \leq n$) and thus representations of all the classical real numbers.

2.4   Constructive Mathematics, Proofs and Programs

In this section we will briefly introduce constructive mathematics and shortly draw the links with programming. For interested readers more details can be found in [7,15,16,5,17].

As explained by P. Martin-Löf in [5] :

> The difference between constructive mathematics and programming does not concern the primitive notions [...] they are essentially the same, but lies in the programmer's insistence that his programs be written in a formal notation [...] whereas, in constructive mathematics [...] the computational procedures are normally left implicit in the proofs [...].

Constructive mathematics has its origins, at the beginning of 19th century, with the criticisms of the formalist mathematical point of view developed by Hilbert which led to what we now call "classical mathematics". Brouwer was the most radical opponent to formal mathematics in which one can prove the existence of a mathematical object without providing a way (an algorithm) to construct it [18]. But his metaphysical approach to constructivism (intuitionism) was not successful. Around 1930, the first who tried to define an axiomatization of constructive mathematics was Arend Heyting, a student of Brouwer. In the mid of the fifties, he published a treaty [19] where intuitionism is presented to both mathematicians and logicians. From Heyting's work it became clear that constructive mathematics is mathematics based on intuitionistic logic, i.e. classical (usual) logic where the law of the excluded middle $(A \vee \neg A)$, or equivalently the absurdity rule (suppose $\neg A$ and deduce a contradiction) or the double negation law (from $\neg\neg A$ we can derive A) aren't allowed. The idea of Heyting was to define the meaning (semantic) of formulae by the set of its proofs. This interpretation of formulae have in its sequels the rejection of the law of the excluded middle otherwise we would have a universal method for obtaining a proof of A or a proof of $\neg A$ for any proposition A. This idea, referred in the literature as BHK-interpretation [18], gives the way to link constructive mathematics to programming by the equivalence:

$$\text{proof} \ = \text{term} = \quad \text{program}$$
$$\text{theorem} = \text{type} = \text{specification}$$

This is the Curry-Howard correspondence which leads [6], via typed lambda-calculus, to a new programming paradigm [5,16,20,7]; Rather than write a program to compute a function one would instead prove a corresponding theorem and then extract the function from the proof. Examples of such systems are Nuprl [20] and Coq [16].

From the constructive mathematical point of view, as developed by Bishop [12], the algorithmic processes are usually left implicit in the proofs. This practice is more flexible but requires some work to obtain a form of the proof that is computer-readable.

3 A Discrete Nonstandard Constructive Model of $\mathbb{R}$

One of the common remarks about nonstandard analysis is that this theory is deeply nonconstructive. However, from the practical point of view, nonstandard analysis has undeniable constructive aspects. This is particularly true for the Harthong-Reeb line as Reeb himself explained [9]. In this work, we will consolidate this impression by showing that the system $\mathcal{A}_\omega$ verifies the constructive axiomatic proposed by Bridges [7]. First, let us note that, $\mathcal{A}_\omega$ comes (by construction) with a binary equality relation $=_\omega$, a binary relation $>_\omega$(greater than), a corresponding inequality relation $\neq_\omega$, two binary operations $+_\omega$ and $*_\omega$ with respectively neutral elements 0_ω and 1_ω (where $0_\omega \neq_\omega 1_\omega$) and two unary operation $-_\omega$ and $x \mapsto x^{(-1)_\omega}$. Let us note also that all the foregoing relations and operations are extensional. An important point in our treatment of the relations $=_\omega$ and $>_\omega$ is that our definitions and proofs comply with the constructive rules. Another important point is that we identify the standard integers (the elements of $\mathbb{Z}^{\text{st}}$) with the usual (constructive) integers. Nevertheless, we treat the relations $=$ and $>$ on the whole set $\mathbb{Z}$ with the usual rules of classical logic. Let us now prove that $(\mathcal{A}_\omega, +_\omega, *_\omega, =_\omega, >_\omega)$ satisfies a first group of axioms which deals with the basic algebraic properties of $\mathcal{A}_\omega$.

R1. $\mathcal{A}_\omega$ is a **Heyting field:** $\forall X, Y, Z \in \mathcal{A}_\omega$,

1. $X +_\omega Y =_\omega Y +_\omega X$,
2. $(X +_\omega Y) +_\omega Z =_\omega X +_\omega (Y +_\omega Z)$,
3. $0_\omega +_\omega X =_\omega X$,
4. $X +_\omega (-_\omega X) =_\omega 0_\omega$,
5. $X \times_\omega Y =_\omega Y \times_\omega X$,
6. $(X \times_\omega Y) \times_\omega Z =_\omega X \times_\omega (Y \times_\omega Z)$,
7. $1_\omega \times_\omega X =_\omega X$,
8. $X \times_\omega X^{(-1)_\omega} =_\omega 1_\omega$ if $X \neq_\omega 0_\omega$,
9. $X \times_\omega (Y +_\omega Z) =_\omega X \times_\omega Y +_\omega X \times_\omega Z$.

Proof. Since $+_\omega$ is the same as the classical $+$, the properties (1.), (2.), (3.) and (4.) are verified.

(5.) $X \times_\omega Y = \lfloor \frac{XY}{\omega} \rfloor = \lfloor \frac{YX}{\omega} \rfloor = Y \times_\omega X =_\omega Y \times_\omega X$.

(6.) From the definition, we get $(X \times_\omega Y) \times_\omega Z = \lfloor \lfloor \frac{X.Y}{\omega} \rfloor \frac{Z}{\omega} \rfloor$. Using several times the decomposition $U = \lfloor U \rfloor - \{U\}$ with $0 \le \lfloor U \rfloor < 1$, we obtain

$$(X \times_\omega Y) \times_\omega Z = \left\lfloor \frac{XYZ}{\omega^2} \right\rfloor + \left\{ \frac{XYZ}{\omega^2} \right\} - \left\{ \frac{XY}{\omega} \right\} \frac{Z}{\omega} - \left\{ \left\lfloor \frac{X.Y}{\omega} \right\rfloor \frac{Z}{\omega} \right\}$$

Since $Z \in \mathcal{A}_\omega$, there is a standard $n \in \mathbb{N}$ such that $|Z| \le n\omega$. Hence, we have

$$\left| \left\{ \frac{XYZ}{\omega^2} \right\} - \left\{ \frac{XY}{\omega} \right\} \frac{Z}{\omega} - \left\{ \left\lfloor \frac{X.Y}{\omega} \right\rfloor \frac{Z}{\omega} \right\} \right| \le n + 2$$

and thus, $(X \times_\omega Y) \times_\omega Z) =_\omega \left\lfloor \frac{XYZ}{\omega^2} \right\rfloor$.

A similar treatment gives $X \times_\omega (Y \times_\omega Z) =_\omega \left\lfloor \frac{XYZ}{\omega^2} \right\rfloor$.

(7.) $1_\omega \times_\omega X = \omega \times_\omega X = \lfloor \frac{\omega X}{\omega} \rfloor = \lfloor X \rfloor = X =_\omega X$.

(8.) $X \times_\omega X^{(-1)_\omega} = \lfloor \frac{X \frac{\omega^2}{X}}{\omega} \rfloor = \lfloor \omega \rfloor = \omega = 1_\omega$.

(9.) The definitions lead to $X \times_\omega (Y +_\omega Z) = \lfloor \frac{X.Y + X.Z}{\omega} \rfloor$ and also to

$$X \times_\omega Y +_\omega X \times_\omega Z = \left\lfloor \frac{XY}{\omega} + \frac{XZ}{\omega} \right\rfloor + \left\{ \frac{XY}{\omega} + \frac{XZ}{\omega} \right\} - \left\{ \frac{XY}{\omega} \right\} - \left\{ \frac{XZ}{\omega} \right\}$$

Since $\left| \left\{ \frac{XY}{\omega} + \frac{XZ}{\omega} \right\} - \left\{ \frac{XY}{\omega} \right\} - \left\{ \frac{XZ}{\omega} \right\} \right| \le 3$, we get the result. $\square$

R2. Basic properties of $>_\omega$: $\forall X, Y, Z \in \mathcal{A}_\omega$,

1. $\neg (X >_\omega Y \text{ and } Y >_\omega X)$,
2. $(X >_\omega Y) \Rightarrow \forall Z (X >_\omega Z \text{ or } Z >_\omega Y)$,
3. $\neg (X \neq_\omega Y) \Rightarrow X =_\omega Y$,
4. $(X >_\omega Y) \Rightarrow \forall Z (X +_\omega Z >_\omega Y +_\omega Z)$,
5. $(X >_\omega 0_\omega \text{ and } Y >_\omega 0_\omega) \Rightarrow X \times_\omega Y >_\omega 0_\omega$.

Proof. (1.) The definition of $X >_\omega Y$ implies $X > Y$. Thus, starting with $X >_\omega Y$ and $Y >_\omega X$, we get ($X > Y$ *and* $Y > X$) which is a contradiction for the usual rules on the integers.

(2.) We know that there is a standard $n \in \mathbb{N}$ such that $n(X - Y) \geq \omega$. Thus, for $Z \in \mathcal{A}_\omega$, we get $n(X - Z) + n(Z - Y) \geq \omega$. Hence, $2n(X - Z) \geq \omega$ or $2n(Z - Y) \geq \omega$ which gives the result.

(3.) Let us recall that $\neg(X \neq_\omega Y)$ is equivalent to $\neg((X >_\omega Y) \vee (Y >_\omega X))$. We suppose that the existence of a standard $n \in \mathbb{N}$ such that $n(X - Y) \geq \omega$ or $n(Y - X) \geq \omega$ leads to a contradiction. Let $k \in \mathbb{N}$ be an arbitrary standard number; since $(k|X - Y| < \omega) \vee (k|X - Y| \geq \omega)$, we get $k|X - Y| < \omega$.

(4.) We suppose that there exists a standard $n \in \mathbb{N}$ such that $n(X - Y) \geq \omega$. Hence, for every $Z \in \mathcal{A}_\omega$ we have $n((X + Z) - (Y + Z)) \geq \omega$.

(5.) We suppose that there exists a standard $(n, m) \in \mathbb{N}^2$ such that $nX \geq \omega$ and $mY \geq \omega$. Hence, $mnX \times_\omega Y = mn\lfloor \frac{XY}{\omega} \rfloor = mn\frac{XY}{\omega} - mn\{\frac{XY}{\omega}\} \geq \omega - mn \geq \frac{\omega}{2}$. Thus, $2mnX \times_\omega Y \geq \omega$. $\qquad\square$

Before we deal with the third group of axioms, let us just recall that we identify the constructive integers with the standard ones. As usual in a Heyting field, we embed the constructive integers in our system by the map $n \mapsto n *_\omega 1_\omega = n\omega$.

R3. Special properties of $>_\omega$

1. **Axiom of Archimedes:** For each $X \in \mathcal{A}_\omega$ there exists a constructive $n \in \mathbb{Z}$ such that $X < n$.

2. **The constructive least-upper-bound principle:** Let S be a nonempty subset of $\mathcal{A}_\omega$ that is bounded above relative to the relation $\geq_\omega$, such that for all $\alpha, \beta \in \mathcal{A}_\omega$ with $\beta >_\omega \alpha$, either β is an upper bound of S or else there exists $s \in S$ with $s >_\omega \alpha$; then S has a least upper bound.

Proof. (1.) Since the definition of $\mathcal{A}_\omega$ is $\{x \in \mathbb{Z} ; \exists^{st}n \in \mathbb{N} |x| < n\omega\}$, the property R3.1. is immediately satisfied.

(2.) The pattern or our proof follows the heuristic motivation given by Bridges in [21]. We choose an element s_0 of S and an upper bound b'_0 of S in $\mathcal{A}_\omega$. Then, we consider the new upper bound $b_0 := b'_0 + 1_\omega$ of S so that $s_0 <_\omega b_0$. We define $\alpha_0 := \frac{2}{3}s_0 + \frac{1}{3}b_0$ and $\beta_0 := \frac{1}{3}s_0 + \frac{2}{3}b_0$. Since $s_0 <_\omega b_0$, we also have $\alpha_0 <_\omega \beta_0$. According to the hypothesis relative to the set S, two cases occur.

- *First case: β_0 is an upper bound of S.* Therefore we define $s_1 := s_0$ and $b_1 := \beta_0$.
- *Second case: there is $s \in S$ such that $\alpha_0 <_\omega s$.* Then, we define $s_1 := s$ and $b_1 := b_0 + s - \alpha_0$.

In each case, we get an element s_1 of S and an upper bound b_1 of S such that $\min_{0 \leq k \leq 1} b_k \geq s_1 \geq s_0$ and $b_1 - s_1 =_\omega \frac{2}{3}(b_0 - s_0)$. According to the external inductive defining principle, there is an internal sequence $(s_k, b_k)_{k \in \mathbb{N}}$ in $\mathbb{Z}^2$ such that, for all standard $n \in \mathbb{N}$, we know that $s_n \in S$, b_n is an upper bound of S and

$$\min_{0 \leq k \leq n} b_k \geq s_n \geq \ldots \geq s_1 \geq s_0 \quad \text{and} \quad b_n - s_n =_\omega \left(\frac{2}{3}\right)^n (b_0 - s_0)$$

where the function min is relative to the usual order relation $\leq$ on $\mathbb{Z}$. Hence, from the overspill principle we can deduce the existence of an infinitely large number $\nu \in \mathbb{N}$, such that

$$\min_{0 \leq k \leq \nu} b_k \geq s_\nu \geq \ldots \geq s_1 \geq s_0$$

Then, we consider the element $b := \min_{0 \leq k \leq \nu} b_k$ of $\mathcal{A}_\omega$ and we want to show that b is a least upper bound of S.

- Given any element $s \in S$, we know that the property $b \geq_\omega s$ is constructively equivalent to $\neg(s >_\omega b)$. If we suppose that $s >_\omega b$, we can find a standard $n \in \mathbb{N}$ such that $s - b >_\omega b_n - s_n$. Since $b_n \geq b \geq s_n$, we have $s - b >_\omega b_n - s_n \geq b_n - b$ and thus $s - b >_\omega b_n - b$ which leads to the contradiction $s >_\omega b_n$. Hence, $b \geq_\omega s$.
- Given $b >_\omega b'$, we can choose a standard $n \in \mathbb{N}$ such that $b - b' >_\omega b_n - s_n$. Thus, we have also $b - b' > b_n - s_n$ and $b_n \geq b \geq s_n \geq b'$. As a consequence, $(b - s_n) + (s_n - b') >_\omega b_n - s_n \geq b - s_n$ so that $s_n >_\omega b'$. Hence, we have find an element s of S such that $s >_\omega b'$. $\square$

4 Conclusion

In this paper, we have proposed a first look into a formal and constructive approach to discrete geometry based on nonstandard analysis. One of the common remarks about nonstandard analysis is that this theory is deeply nonconstructive. However, nonstandard practice has undeniable constructive aspects. This is particularly true for the Harthong-Reeb line as Reeb himself wrote [9,11]. The arithmetization of the Euler scheme, that led to the Reveillès discrete straight line definition, is a good illustration of this. We tried to consolidate this constructive impression by showing that the system $\mathcal{A}_\omega$ verifies the constructive axiomatic proposed by Bridges [7]. The model $\mathcal{A}_\omega$ is defined with a weaker axiomatic than the one usually used for nonstandard analysis such as IST [13]. Indeed, IST allows all sorts of ideal (non constructive) objects to appear. Our weaker, restraint, axiomatic only induces a non trivial scale on the set of integers.

Of course, we are not the first to explore the relationship between constructive mathematics and nonstandard analysis. For instance, there are the deep works of Palmgren [22,23] who introduced some new constructive approaches to nonstandard analysis. Actually, our study is completely independent of these developments, mainly because we remain within the framework of an usual axiomatic which is just a weakening of the theory IST of Nelson [13].

Our long term goal is to show that $\mathcal{A}_\omega$ represents a "good" discrete constructive model of the continuum. This work represents only the very first step towards this goal. If this succeeds, every constructive proof done in this set and based on the constructive axiomatic can be translated into an algorithm (with some work). Future works includes the production of such proofs in the discrete geometry context. We hope that it may be possible, following ideas such as the arithmetization of the Euler scheme, to compute differential properties on discrete object such as normal, curvature and so by using constructive mathematics.

References

1. Reveillès, J.P., Richard, D.: Back and forth between continuous and discrete for the working computer scientist. Annals of Mathematics and Artificial Intelligence, Mathematics and Informatic 16, 89–152 (1996)
2. Harthong, J.: Éléments pour une théorie du continu. Astérisque 109/110, 235–244 (1983)
3. Harthong, J.: Une théorie du continu. In: Barreau, H., Harthong, J. (eds.) La mathématiques non standard, Paris, Editions du CNRS, pp. 307–329 (1989)
4. Bishop, E., Bridges, D.: Constructive Analysis. Springer, Heidelberg (1985)
5. Martin-Lof, P.: Constructive mathematics and computer programming. Logic, Methodology and Philosophy of Science VI, 153–175 (1980)
6. Howard, W.A.: The formulae-as-types notion of construction. H.B. Curry: Essays on Combinatory Logic, Lambda-calculus and Formalism, 479–490 (1980)
7. Bridges, D.S.: Constructive mathematics: A foundation for computable analysis. Theor. Comput. Sci. 219, 95–109 (1999)
8. Mc Laughlin, W.I.: Resolving Zeno's paradoxes. Scientific American, 84–89 (1994)
9. Diener, F., Reeb, G.: Analyse Non Standard. Hermann, Paris (1989)
10. Diener, M.: Application du calcul de Harthong-Reeb aux routines graphiques. In: [11], pp. 424–435
11. Salanski, J.M., Sinaceurs, H. (eds.): Le Labyrinthe du Continu. Springer, Heidelberg (1992)
12. Bishop, E.: Foundations of Constructive Analysis. McGraw-Hill, New York (1967)
13. Nelson, E.: Internal set theory: A new approach to nonstandard analysis. Bulletin of the American Mathematical Society 83, 1165–1198 (1977)
14. Robinson, A.: Non-standard analysis, 2nd edn. American Elsevier, New York (1974)
15. Coquand, T., Huet, G.P.: A selected bibliography on constructive mathematics, intuitionistic type theory and higher order deduction. j-J-SYMBOLIC-COMP 3, 323–328 (1986)
16. Coquand, T., Huet, G.P.: The calculus of constructions. Inf. Comput. 76, 95–120 (1988)
17. Beeson, M.J.: Foundations of constructive Mathematics. A Series of Modern Surveys in Mathematics. Springer, Heidelberg (1985)
18. Troelstra, A.S., van Dalen, D.: Constructivism in Mathematics: An Introduction, vol. I. North-Holland, Amsterdam (1988)
19. Heyting, A.: Intuitionism an introduction. Studies in Logic and Foundations of Mathematics (1956)
20. Constable, R.L., Allen, S.F., Bromley, H.M., Cleaveland, W.R., Cremer, J.F., Harper, R.W., Howe, D.J., Knoblock, T.B., Mendler, N.P., Panangaden, P., Sasaki, J.T., Smith, S.F.: Implementing Mathematics with the Nuprl Development System. Prentice-Hall, Englewood Cliffs (1986)
21. Bridges, D., Reeves, S.: Constructive mathematics, in theory and programming practice. Technical Report CDMTCS-068, Centre for Discrete Mathematics and Theorical Computer Science (1997)
22. Palmgren, E.: A constructive approach to nonstandard analysis. Annals of Pure and Applied Logic 73, 297–325 (1995)
23. Palmgren, E.: Developments in constructive nonstandard analysis. The Bulletin of Symbolic Logic 4, 233–272 (1998)

Generation and Recognition of Digital Planes Using Multi-dimensional Continued Fractions

Thomas Fernique

LIRMM, Univ. Montpellier 2, CNRS
161 rue Ada 34392 Montpellier - France
fernique@lirmm.fr

Abstract. This paper extends, in a multi-dimensional framework, pattern recognition technics for generation or recognition of digital lines. More precisely, we show how the connection between chain codes of digital lines and continued fractions can be generalized by a connection between tilings and multi-dimensional continued fractions. This leads to a new approach for generating and recognizing digital hyperplanes.

Introduction

Discrete (or digital) geometry deals with discrete sets considered to be digitized objects of the Euclidean space. A challenging problem is to decompose a huge complicated discrete set into elementary ones, which could be easily stored and from which one can easily reconstruct the original discrete set. Good candidates for such elementary discrete sets are digitizations of Euclidean hyperplanes, in particular *arithmetic discrete hyperplanes* (see [1,7,9]). We thus need efficient algorithms which generate arbitrarily big patches of such digitizations from given parameters and, conversely, recognize parameters from given digitizations.

In the particular case of digitizations of lines, among other technics, so-called *linguistic technics* provide a nice connection with words theory and continued fractions. Let us briefly detail this. A digital line made of horizontal or vertical unit segments can be coded by a two-letter word, called *chain code* or *Freeman code*. For example, if a horizontal (resp. vertical) unit segment is coded by 0 (resp. 1), then a segment of slope 1 can be coded by a word of the form $10 \ldots 10 = (10)^k$. Then, basic transformations on words correspond to basic operations on slopes of the segments they code. For example, replacing each 0 by 01 and each 1 by 0 in the previous word leads to the word $(001)^k$, which codes a segment of slope $1/2$. Many algorithms use this approach for both recognition and generation of digital lines, and continued fraction expansions of slopes of segments turn out to play a central role there (see *e.g.* [11] or references in [8]).

In higher dimensions, there are also various technics for generation or recognition of digital hyperplane as, for example, linear programming, computational geometry or preimage technics (see *e.g.* [4] and references therein). However, these approaches do not extend the connection between words theory and continued fractions. The aim of this paper is to introduce an approach which does

D. Coeurjolly et al. (Eds.): DGCI 2008, LNCS 4992, pp. 33–44, 2008.
© Springer-Verlag Berlin Heidelberg 2008

it. Such an approach extends the case of so-called stepped surfaces (which are particular infinite digitizations), studied in [3].

The paper is organized as follows. In Sec. 1, we introduce *binary functions*, which can be seen as unions of faces of unit hypercubes. Among them, the ones called *stepped planes* ([12]) play for Euclidean hyperplanes the role played by chain codes for Euclidean lines. We also introduce *dual maps* ([2,6]), which generalize the basic transformations on chain codes mentioned above. Then, in Sec. 2, we briefly describe the *Brun algorithm*, which is one of the existing multi-dimensional continued fraction algorithms (see [10]). The Brun algorithm computes so-called *Brun expansions* of real vectors. We also introduce particular dual maps which allow the Brun algorithm to act over stepped planes. This leads, in Sec. 3, to a method for obtaining a *fundamental domain* of a stepped plane, that is, a binary function which suffices to generate by periodicity the whole stepped plane (Th. 2). In Sec. 4, we describe a method to compute so-called *Brun expansions of stepped planes*, by grabing information from local configurations (namely *runs*). Actually, the Brun expansion of a stepped plane is nothing but the Brun expansion of its normal vector. So, the interest of this method is that it can be naturally extended to binary functions, leading to define, in Sec. 5, *Brun expansions of binary functions*. We finally use this extended notion of Brun expansion, in Sec. 6, to describe a recognition algorithm which decides whether a given binary function is a stepped plane or not (Th. 3).

1   Stepped Planes and Dual Maps

We here first introduce our basic digital objects, namely *binary functions* and *stepped planes*. Formally, it is convenient to consider the set of functions from $\mathbb{Z}^d \times \{1, \ldots, d\}$ to $\mathbb{Z}$, denoted by $\mathfrak{F}_d$. Then, we define:

Definition 1. *A* binary function *is a function in $\mathfrak{F}_d$ which takes values in $\{0, 1\}$. The* size *of a binary function $\mathcal{B}$, denoted by $|\mathcal{B}|$, is the cardinality of its support, that is, the subset of $\mathbb{Z}^d \times \{1, \ldots, d\}$ where $\mathcal{B}$ takes value one. We denote by $\mathfrak{B}_d$ the set of binary functions. For $\boldsymbol{x} \in \mathbb{Z}^d$ and $i \in \{1, \ldots, d\}$, we call* face of type *i located in $\boldsymbol{x}$ the binary function denoted by $(\boldsymbol{x}, i^*)$ whose support is $\{(\boldsymbol{x}, i)\}$.*

Note that binary functions (resp. functions of $\mathfrak{F}_d$) can be seen as sums of faces (resp. weighted sums of faces). Let us now provide a geometric interpretation of binary functions. Let $(\boldsymbol{e}_1, \ldots, \boldsymbol{e}_d)$ denote the canonical basis of $\mathbb{R}^d$. The geometric interpretation of a face $(\boldsymbol{x}, i^*)$ is the closed subset of $\mathbb{R}^d$ defined by (see Fig. 1):

$$\{\boldsymbol{x} + \boldsymbol{e}_i + \sum_{j \neq i} \lambda_j \boldsymbol{e}_j \mid 0 \leq \lambda_j \leq 1\}.$$

This subset is a hyperface of the unit cube of $\mathbb{R}^d$ whose lowest vertex is $\boldsymbol{x}$. Then, the geometric interpretation of a binary function, that is, of a sum of faces, is the union of the geometrical interpretations of these faces (see Fig. 3).

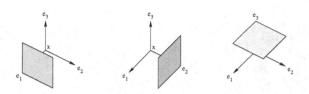

Fig. 1. Geometrical interpretations of faces (x, i^*), for $i = 1, 2, 3$ (from left to right)

Among binary functions, we are especially interested in so-called *stepped planes*:

Definition 2. *Let $\alpha \in \mathbb{R}_+^d \setminus \{0\}$ and $\rho \in \mathbb{R}$. The* stepped plane *of normal vector α and intercept $\rho \in \mathbb{R}$, denoted by $\mathcal{P}_{\alpha,\rho}$, is the binary function defined by:*

$$\mathcal{P}_{\alpha,\rho}(x, i) = 1 \iff \langle x | \alpha \rangle < \rho \leq \langle x + e_i | \alpha \rangle,$$

where $\langle | \rangle$ is the canonical dot product. We denote by $\mathfrak{P}_d$ the set of stepped planes.

Fig. 2 depicts the geometrical interpretation of a stepped plane. It is not hard to check that the vertices of a stepped plane $\mathcal{P}_{\alpha,\rho}$, that is, the integers vectors which belong to its geometrical interpretation, form a *standard arithmetic discrete plane of parameters* (α, ρ) (see [1,7,9]). Moreover, one checks that the orthogonal projection along $e_1 + \ldots + e_d$ maps the geometrical representation of a stepped plane onto a tiling of $\mathbb{R}^{d-1}$ whose tiles are projections of geometrical representations of faces (see also Fig. 2).

Let us now introduce the main tool of this paper, namely *dual maps*, which act over binary functions and stepped planes. First, let us recall some basic definitions and notations. We denote by F_d the free group generated by the alphabet $\{1, \ldots, d\}$, with the concatenation as a composition rule and the empty word as unit. An endomorphism of F_d is a *substitution* if it maps any letter to a non-empty concatenation of letters with non-negative powers. The *parikh mapping* is the map f from F_d to $\mathbb{Z}^d$ defined on $w \in F_d$ by:

$$f(w) = (|w|_1, \ldots, |w|_d),$$

where $|w|_i$ is the sum of the powers of the occurences of the letter i in w. Then, the *incidence matrix* of an endomorphism σ of F_d, denoted by M_σ, is the $d \times d$ integer matrix whose i-th column is the vector $f(\sigma(i))$. Last, an endomorphism of F_d is said to be *unimodular* if its incidence matrix has determinant ± 1.

Example 1. Let σ be the endormorphism of F_3 defined by $\sigma(1) = 12$, $\sigma(2) = 13$ and $\sigma(3) = 1$. Note that σ is a substitution (often called *Rauzy* substitution). One computes, for example, $\sigma(1^{-1}2) = \sigma(1)^{-1}\sigma(2) = 2^{-1}1^{-1}13 = 2^{-1}3$, and $f(2^{-1}3) = e_3 - e_2$. This substitution is unimodular since its incidence matrix (below) has determinant 1:

$$M_\sigma = \begin{pmatrix} 1 & 1 & 1 \\ 1 & 0 & 0 \\ 0 & 1 & 0 \end{pmatrix}.$$

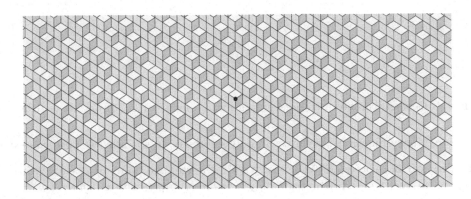

Fig. 2. Geometrical interpretation of the stepped plane $\mathcal{P}_{(24,9,10),0}$ (highlighted origin). This can be seen either as faces of unit cubes, or as a lozenge tiling of the plane.

We are now in a position to define *dual maps*:

Definition 3. *The* dual map *of a unimodular endomorphism σ of F_d, denoted by $E_1^*(\sigma)$, maps any function $\mathcal{F} \in \mathfrak{F}_d$ to the function $E_1^*(\sigma)(\mathcal{F})$ defined by:*

$$E_1^*(\sigma)(\mathcal{F}) : (\boldsymbol{x}, i) \mapsto \sum_{j | \sigma(i) = p \cdot j \cdot s} \mathcal{F}(M_\sigma \boldsymbol{x} + \boldsymbol{f}(p), j) - \sum_{j | \sigma(i) = p \cdot j^{-1} \cdot s} \mathcal{F}(M_\sigma \boldsymbol{x} + \boldsymbol{f}(p) - \boldsymbol{e}_j, j).$$

Note that the value of $E_1^*(\sigma)(\mathcal{F})$ in $(\boldsymbol{x}, i)$ is finite since it depends only on the values of $\mathcal{F}$ over a finite subset of $\mathbb{Z}^d \times \{1, \ldots, d\}$. This yields that $E_1^*(\sigma)$ is an endomorphism of $\mathfrak{F}_d$.

Example 2. The dual map of the substitution σ introduced in Ex. 1 satisfies:

$$E_1^*(\sigma) \ : \ \begin{cases} (\boldsymbol{0}, 1^*) \mapsto (\boldsymbol{0}, 1^*) + (\boldsymbol{0}, 2^*) + (\boldsymbol{0}, 3^*), \\ (\boldsymbol{0}, 2^*) \mapsto (-\boldsymbol{e}_3, 1^*), \\ (\boldsymbol{0}, 3^*) \mapsto (-\boldsymbol{e}_3, 2^*). \end{cases}$$

The image of any function of $\mathfrak{F}_d$, that is, of a weighted sum of faces, can then be easily computed by linearity. Fig. 3 illustrates this.

The following theorem, proved in [3], connects dual maps and stepped planes:

Theorem 1 ([3]). *Let σ be a unimodular endomorphism of F_d. Let $\boldsymbol{\alpha} \in \mathbb{R}_+^d \backslash \{\boldsymbol{0}\}$ and $\rho \in \mathbb{R}$. If $M_\sigma^\top \boldsymbol{\alpha} \in \mathbb{R}_+^d$, then the image of the stepped plane $\mathcal{P}_{\boldsymbol{\alpha}, \rho}$ by $E_1^*(\sigma)$ is the stepped plane $\mathcal{P}_{M_\sigma^\top \boldsymbol{\alpha}, \rho}$. Otherwise, this image is not a binary function[1].*

Note that, although the image by $E_1^*(\sigma)$ of a stepped plane is a stepped plane, the image of each face of this stepped plane is a weighted sum of faces (in particular, not necessarily binary). Note also that if σ is a substitution, then $M_\sigma^\top \boldsymbol{\alpha} \in \mathbb{R}_+^d$ holds for any $\boldsymbol{\alpha} \in \mathbb{R}_+^d \backslash \{\boldsymbol{0}\}$: the image of a stepped plane by the dual map of a substitution is thus always a stepped plane.

---

[1] See Def. 1.

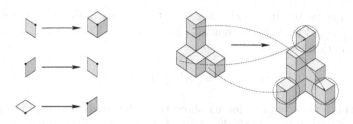

Fig. 3. Action of the dual map of Ex. 2 on faces (left) and on a binary function (right). Let us stress that the image of a binary function is not necessarily binary (unlike here).

2   Brun Expansions of Real Vectors

We here recall the Brun algorithm (see *e.g.* [10]) and use dual maps to connect it with normal vectors of stepped planes (recall Def. 2).

Definition 4. *The* Brun map T *is the map from* $[0,1]^d \setminus \{\mathbf{0}\}$ *to* $[0,1]^d$ *defined on* $\boldsymbol{\alpha} = (\alpha_1, \ldots, \alpha_d)$ *by:*

$$T(\alpha_1, \ldots, \alpha_d) = \left(\frac{\alpha_1}{\alpha_i}, \ldots, \frac{\alpha_{i-1}}{\alpha_i}, \frac{1}{\alpha_i} - \left\lfloor \frac{1}{\alpha_i} \right\rfloor, \frac{\alpha_{i+1}}{\alpha_i}, \ldots, \frac{\alpha_d}{\alpha_i} \right),$$

where $i = \min\{j \mid \alpha_j = \|\boldsymbol{\alpha}\|_\infty\}$. *Then, the* Brun expansion *of a vector* $\boldsymbol{\alpha} \in [0,1]^d$ *is the sequence* $(a_n, i_n)_{n \geq 0}$ *of* $\mathbb{N}^* \times \{1, \ldots, d\}$ *defined, while* $T^n(\boldsymbol{\alpha}) \neq \mathbf{0}$, *by:*

$$a_n = \lfloor \|T^n(\boldsymbol{\alpha})\|_\infty^{-1} \rfloor \quad and \quad i_n = \min\{j \mid \langle T^n(\boldsymbol{\alpha}) | e_j \rangle = \|T^n(\boldsymbol{\alpha})\|_\infty\}.$$

Let us stress that, in the $d = 1$ case, the Brun map T is nothing but the classic Gauss map, and if $(a_n, i_n)_{n \geq 0}$ is the Brun expansion of $\alpha \in [0,1]$, then $(a_n)_n$ is the continued fraction expansion of α, while, for all n, $i_n = 1$.

Example 3. The Brun expansion of $(3/8, 5/12)$ is $(2,2), (1,1), (2,2), (4,1), (1,2)$.

Let us mention that, as in the case of continued fractions, it turns out that a vector has a finite Brun expansion if and only if it has only rational entries. Let us now give a matrix viewpoint of the Brun map T. For $(a, i) \in \mathbb{N} \times \{1, \ldots, d\}$, one introduces the following $(d+1) \times (d+1)$ symmetric matrix:

$$B_{a,i} = \begin{pmatrix} a & & 1 & \\ & I_{i-1} & & \\ 1 & & 0 & \\ & & & I_{d-i} \end{pmatrix}, \tag{1}$$

where I_p stands for the $p \times p$ identity matrix. Then, consider a vector $\boldsymbol{\alpha} = (\alpha_1, \ldots, \alpha_d) \in [0,1]^d \setminus \{\mathbf{0}\}$. A simple computation shows that, with $i = \min\{j \mid \alpha_j = \|\boldsymbol{\alpha}\|_\infty\}$ and $a = \lfloor \alpha_i^{-1} \rfloor$, one has:

$$(1, \boldsymbol{\alpha}) = \|\boldsymbol{\alpha}\|_\infty B_{a,i}(1, T(\boldsymbol{\alpha})), \tag{2}$$

where, for any vector $\boldsymbol{u}$, $(1, \boldsymbol{u})$ stands for the vector obtained by adding to $\boldsymbol{u}$ a first entry equal to 1. Note that $B_{a,i}$ is invertible. Thus, one can rewrite the previous equation as follows:

$$(1, T(\boldsymbol{\alpha})) = ||\boldsymbol{\alpha}||_\infty^{-1} B_{a,i}^{-1}(1, \boldsymbol{\alpha}). \tag{3}$$

To conclude this section, let us show that this matrix viewpoint allows to connect Brun expansions with the stepped planes and dual maps introduced in the previous section. Let us introduce *Brun substitutions*:

Definition 5. *Let $a \in \mathbb{N}^*$ and $i \in \{1, \ldots, d\}$. The* Brun substitution $\beta_{a,i}$ *is the endomorphism of F_{d+1} defined by:*

$$\beta_{a,i}(1) = 1^a \cdot (i + 1), \quad \beta_{a,i}(i + 1) = 1, \quad \forall j \notin \{1, i + 1\}, \ \beta_{a,i}(j) = j.$$

One checks that $\beta_{a,i}$ is unimodular and has $B_{a,i}$ for incidence matrix.[2] Note also that $\beta_{a,i}$ is invertible, since one computes:

$$\beta_{a,i}^{-1}(1) = (i + 1), \quad \beta_{a,i}^{-1}(i + 1) = (i + 1)^{-a} \cdot 1, \quad \forall j \notin \{1, i + 1\}, \ \beta_{a,i}^{-1}(j) = j.$$

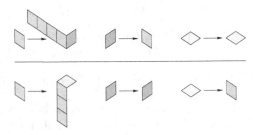

Fig. 4. Action on faces of the dual maps $E_1^*(\beta_{4,1})$ (top) and $E_1^*(\beta_{3,2})$ (bottom)

One then can consider dual maps of Brun substitutions (see Fig. 4), and one deduces from Th. 1 that Eq. (2) and (3) respectively yield:

$$E_1^*(\beta_{a,i})(\mathcal{P}_{||\boldsymbol{\alpha}||_\infty(1,T(\boldsymbol{\alpha})),\rho}) = \mathcal{P}_{(1,\boldsymbol{\alpha}),\rho}, \tag{4}$$

$$\mathcal{P}_{||\boldsymbol{\alpha}||_\infty(1,T(\boldsymbol{\alpha})),\rho} = E_1^*(\beta_{a,i}^{-1})(\mathcal{P}_{(1,\boldsymbol{\alpha}),\rho}). \tag{5}$$

3   Generation of Stepped Planes

We here show how dual maps and Brun expansions can be used to easily generate arbitrarily big patches of a stepped plane (that is, binary functions less or equal to it), provided that its normal vector has rational entries. Indeed, one proves:

---

[2] Let us recall that $B_{a,i}$ is the symmetric matrix defined Eq. (1).

Theorem 2. *Let $\alpha \in [0,1]^d \cap \mathbb{Q}^d$ with the finite Brun expansion $(a_n, i_n)_{0 \leq n \leq N}$ and $\rho \in \mathbb{R}$. Let $\rho' = \rho/\|B_{a_0,i_0} \times \ldots \times B_{a_N,i_N} e_1\|_\infty$ and $\mathcal{D}_{(1,\alpha),\rho}$ be the binary function defined by:*

$$\mathcal{D}_{(1,\alpha),\rho} = E_1^*(\beta_{a_0,i_0}) \circ \ldots \circ E_1^*(\beta_{a_N,i_N})(\lfloor \rho' \rfloor e_1, 1^*),$$

and $L_{(1,\alpha),\rho}$ be the lattice of rank d of $\mathbb{Z}^{d+1}$ defined by:

$$L_{(1,\alpha),\rho} = B_{a_0,i_0}^{-1} \ldots B_{a_N,i_N}^{-1} \sum_{k=2}^{d+1} \mathbb{Z} e_k.$$

Then, the geometrical interpretation of the stepped plane $\mathcal{P}_{(1,\alpha),\rho}$ is the union of all the translations along $L_{(1,\alpha),\rho}$ of the geometrical interpretation of $\mathcal{D}_{(1,\alpha),\rho}$.

Example 4. Fig. 5 shows the generation of the binary function $\mathcal{D}_{(1,3/8,5/12),0}$ by the dual maps of the Brun substitutions associated with the Brun expansion of the vector $(3/8, 5/12)$ (recall Ex. 3). One also computes:

$$L_{(1,3/8,5/12),0} = \mathbb{Z}(e_1 + 4e_2 - 6e_3) + \mathbb{Z}(2e_1 - 2e_2 - 3e_3).$$

Thus, according to Th. 2, the geometrical interpretation of the rational stepped plane $\mathcal{P}_{(1,3/8,5/12),0} = \mathcal{P}_{(24,9,10),0}$ (see Fig. 2) is the union of all the translations along $L_{(1,3/8,5/12),0}$ of the geometrical interpretation of $\mathcal{D}_{(1,3/8,5/12),0}$.

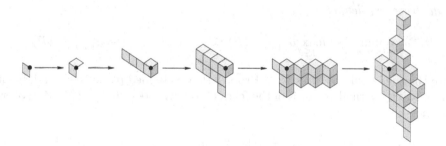

Fig. 5. Generation of $\mathcal{D}_{(1,3/8,5/12),0}$ by applications of the dual maps $E_1^*(\beta_{1,2})$, $E_1^*(\beta_{4,1})$, $E_1^*(\beta_{2,2})$, $E_1^*(\beta_{1,1})$ and $E_1^*(\beta_{2,2})$ (from left ro right – highlighted origin). According to Th. 2, the stepped plane $\mathcal{P}_{(1,3/8,5/12),0}$ can be generated by translating $\mathcal{D}_{(1,3/8,5/12),0}$.

Note that, in terms of functions, one has $\mathcal{D}_{(1,\alpha),\rho} \leq \mathcal{P}_{(1,\alpha),\rho}$. This means that the geometrical interpretation of $\mathcal{D}_{(1,\alpha),\rho}$ is included in the one of $\mathcal{P}_{(1,\alpha),\rho}$. To conclude this section, let us mention that one can show that $\mathcal{D}_{(1,\alpha),\rho}$ has minimal size: such a piece of $\mathcal{P}_{(1,\alpha),\rho}$ is called a *fundamental domain* of $\mathcal{P}_{(1,\alpha),\rho}$.

4   Brun Expansions of Stepped Planes

We here show how Brun expansions of normal vectors of stepped planes can be directly computed on stepped planes relying on the notion of *run*:

Definition 6. *An (i,j)-run of a binary function $\mathcal{B}$ is a maximal sequence of contiguous faces of type i, aligned with the direction $\mathbf{e}_j$, whose geometric interpretation is included in the one of $\mathcal{B}$.*

For example, the stepped plane depicted on Fig. 2 has $(1,2)$-runs and $(1,3)$-runs of size 2 or 3, and $(3,2)$-runs of size 1 or 2 (see also Fig. 6 in the general case of a binary function). The infimum and the supremum of the sizes (Recall Def. 1) of the (i,j)-runs of a binary function $\mathcal{B}$ are respectively denoted by $a_{i,j}^-(\mathcal{B})$ and $a_{i,j}^+(\mathcal{B})$. The following proposition shows that runs contain information about the normal vector of a stepped plane:

Proposition 1. *Let $\boldsymbol{\alpha} = (\alpha_1, \ldots, \alpha_d) \in \mathbb{R}_+^d \setminus \{0\}$ and $\rho \in \mathbb{R}$. Then, for $\alpha_j \neq 0$:*

$$a_{i,j}^-(\mathcal{P}_{\boldsymbol{\alpha},\rho}) = \max(\lfloor \alpha_i/\alpha_j \rfloor, 1) \quad and \quad a_{i,j}^+(\mathcal{P}_{\boldsymbol{\alpha},\rho}) = \max(\lceil \alpha_i/\alpha_j \rceil, 1).$$

In particular, let us show that runs contain enough information to compute Brun expansions of normal vectors of so-called *expandable* stepped planes:

Definition 7. *A stepped plane $\mathcal{P} \in \mathfrak{P}_{d+1}$ is said to be* expandable *if one has:*

$$\max_{1 \leq i \leq d} a_{i+1,1}^+(\mathcal{P}) = 1 \quad and \quad \min_{1 \leq i \leq d} a_{1,i+1}^-(\mathcal{P}) < \infty.$$

In this case, we define:

$$i(\mathcal{P}) = \min_{1 \leq i \leq d} \{ i \mid \max_{1 \leq j \leq d} a_{j+1,i+1}^+(\mathcal{P}) \leq 1 \} \quad and \quad a(\mathcal{P}) = a_{1,i(\mathcal{P})+1}^-(\mathcal{P}).$$

Note that one easily deduces from Prop. 1 that a stepped plane is expandable if and only if its normal vector is of the form $(1, \boldsymbol{\alpha})$, with $\boldsymbol{\alpha} \in [0,1]^d \setminus \{0\}$. Moreover, one then has:

$$i(\mathcal{P}_{(1,\boldsymbol{\alpha}),\rho}) = \min\{i \mid \alpha_i = ||\boldsymbol{\alpha}||_\infty\} \quad and \quad a(\mathcal{P}_{(1,\boldsymbol{\alpha}),\rho}) = \lfloor ||\boldsymbol{\alpha}||_\infty^{-1} \rfloor. \tag{6}$$

This leads to the following definition:

Definition 8. *Let $\tilde{T}$ be the map defined over expandable stepped planes by:*

$$\tilde{T}(\mathcal{P}) = E_1^*(\beta_{a(\mathcal{P}),i(\mathcal{P})}^{-1})(\mathcal{P}).$$

In particular, $\tilde{T}$ has values in $\mathfrak{P}_{d+1}$. More precisely, Eq. (4) yields:

$$\tilde{T}(\mathcal{P}_{(1,\boldsymbol{\alpha}),\rho}) = \mathcal{P}_{(1,T(\boldsymbol{\alpha})),\rho}. \tag{7}$$

Thus, the Brun expansion of a vector $\boldsymbol{\alpha}$ can be computed on a stepped plane $\mathcal{P}$ of normal vector $(1, \boldsymbol{\alpha})$, since it is nothing but the sequence $(a(\tilde{T}^n(\mathcal{P})), i(\tilde{T}^n(\mathcal{P})))_n$. By abuse, this Brun expansion is called *Brun expansion of the stepped plane $\mathcal{P}$.*

5   Brun Expansions of Binary Functions

Here, we show that runs allow to define Brun expansions not only of stepped planes but also of binary functions, although the latter do not have normal vectors. We first need to refine Def. 6 (see Fig. 6):

Definition 9. *Let $\mathcal{R}$ be an (i, j)-run of a binary function $\mathcal{B}$. Thus, there is a vector $\boldsymbol{x} \in \mathbb{Z}^d$ and an interval I of $\mathbb{Z}$ (not necessarily finite) such that:*

$$\mathcal{R} = \sum_{k \in I} (\boldsymbol{x} + k\boldsymbol{e}_j, i^*).$$

This run is right-closed *if I has a right endpoint b such that $\mathcal{B}(\boldsymbol{x} + b\boldsymbol{e}_j, j^*) = 1$, and* left-closed *if I has a left endpoint a such that $\mathcal{B}(\boldsymbol{x} + (a-1)\boldsymbol{e}_j + \boldsymbol{e}_i, j^*) = 1$. The terms* closed, open, right-open *and* left-open *are then defined as for intervals.*

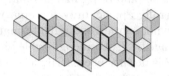

Fig. 6. This binary function has every type of $(1, 3)$-runs: left-closed, right-closed, closed and open (framed runs, from left to right). It is moreover recognizable, with $(a, i) = (2, 2)$ (see definition below).

Then, as we previously defined Brun expansions of *expandable* stepped planes, we will here restrict to *recognizable* binary functions:

Definition 10. *A binary function $\mathcal{B} \in \mathfrak{B}_{d+1}$ is recognizable if it satisfies the two following conditions. First, it shall exist $i \in \{1, \ldots, d\}$ such that:*

$$a^+_{1,i+1}(\mathcal{B}) \geq 2 \quad and \quad \min_{1 \leq j \leq d} a^+_{i+1,j+1}(\mathcal{B}) \geq 2.$$

Let $i(\mathcal{B})$ denotes the smallest such i. Second, $\mathcal{B}$ shall have closed $(1, i(\mathcal{B}) + 1)$-runs, with the smallest one having size $a^+_{1,i(\mathcal{B})+1}(\mathcal{B}) - 1$. Let $a(\mathcal{B})$ denotes this size.

Let us explain this definition. Assume that $\mathcal{B} \leq \mathcal{P}_{(1,\boldsymbol{\alpha}),\rho}$, for $\boldsymbol{\alpha} \in \mathbb{R}^d_+$ and $\rho \in \mathbb{R}$. Then, it is not hard to deduce from Prop. 1 that the first recognizability condition ensures that the $i(\mathcal{B})$-th entry of $\boldsymbol{\alpha}$ is smaller than 1 and greater than all the other entries, while the second recognizability condition ensures that $\mathcal{P}_{(1,\boldsymbol{\alpha}),\rho}$ has $(1, i(\mathcal{B}))$-runs of two different sizes, with the smallest size being equal to $a(\mathcal{B})$. In other words, recognizability ensures $\boldsymbol{\alpha} \in [0, 1]^d$, $i(\mathcal{B}) = i(\mathcal{P}_{(1,\boldsymbol{\alpha}),\rho})$ and $a(\mathcal{B}) = a(\mathcal{P}_{(1,\boldsymbol{\alpha}),\rho})$. Thus, the formula defining $\tilde{T}$ over stepped planes (Def. 8) can still be used to define $\tilde{T}$ over recognizable binary functions. This leads to define the Brun expansion of a recognizable binary function $\mathcal{B}$ as the sequence $(a(\tilde{T}^n(\mathcal{B})), i(\tilde{T}^n(\mathcal{B})))_n$, for n such that $\tilde{T}^n(\mathcal{B})$ is a recognizable binary function.

6 Recognition of Stepped Planes

We are here interested in the following recognition problem: given a binary function $\mathcal{B} \in \mathfrak{B}_{d+1}$ whose size $|\mathcal{B}|$ is finite, decide whether the following convex polytope of $\mathbb{R}^{d+1}$ is empty or not:

$$P(\mathcal{B}) = \{(\alpha, \rho) \in [0,1]^d \backslash \{\mathbf{0}\} \times \mathbb{R} \mid \mathcal{B} \leq \mathcal{P}_{(1,\alpha),\rho}\}.$$

The idea is that if the map $\tilde{T}$ previously defined would satisfy, for any $\mathcal{B} \in \mathfrak{B}_{d+1}$:

$$0 \leq \mathcal{B} \leq \mathcal{P} \iff 0 \leq \tilde{T}(\mathcal{B}) \leq \tilde{T}(\mathcal{P}), \tag{8}$$

then, $P(\mathcal{B})$ would be not empty if and only if computing the sequence $(\tilde{T}^n(\mathcal{B}))_{n \geq 0}$ would lead to a binary function of the form $\sum_{x \in X}(x, 1^*)$, with the vectors of X having all the same first entries (such a binary function is easily recognizable). But Eq. (8) does not always hold. The first problem is that $\tilde{T}$ is defined only over expandable stepped planes and recognizable binary functions. However, this problem turns out to generally appear only for binary functions whose size is small, because their runs do not contain enough information. The second problem seems more tedious: the image by $\tilde{T}$ of a recognizable binary function less than or equal to a stepped plane $\mathcal{P}$ is neither necessarily less than or equal to $\tilde{T}(\mathcal{P})$, nor even always a binary function. Let us first consider this problem. We introduce three rules acting over binary functions (see Fig. 7, and also Fig. 8, left):

Definition 11. *Let $a \in \mathbb{N}^*$ and $i \in \{1, \ldots, d\}$. The rule $\phi_{a,i}$ left-extends any right-closed and left-open $(1, i+1)$-run into a run of size a; the rule $\psi_{a,i}$ right-closes any right-open $(1, i+1)$-run of size greater than a; the rule χ_i removes any left-closed and right-open $(1, i+1)$-run.*

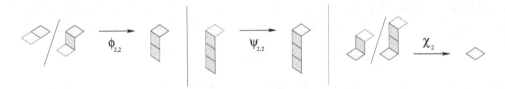

Fig. 7. The rules $\phi_{2,2}$, $\psi_{2,2}$ and χ_2 (dashed edges represent missing faces)

The following proposition then shows that one can replace any recognizable binary function $\mathcal{B}$ by a recognizable binary function $\tilde{\mathcal{B}}$ which satisfies Eq. (8) under an additional hypothesis:

Proposition 2. *Let $\mathcal{B} \in \mathfrak{B}_{d+1}$ be a recognizable binary function and $\tilde{\mathcal{B}}$ be the binary function obtained by successively applying $\phi_{a(\mathcal{B}),i(\mathcal{B})}$, $\psi_{a(\mathcal{B}),i(\mathcal{B})}$ and $\chi_{i(\mathcal{B})}$. Then, for any stepped plane $\mathcal{P} \in \mathfrak{P}_{d+1}$, one has $\mathcal{B} \leq \mathcal{P}$ if and only if $\tilde{\mathcal{B}} \leq \mathcal{P}$. Moreover, if $\tilde{\mathcal{B}}$ does not have open $(1, i(\mathcal{B})+1)$-runs, then one has:*

$$0 \leq \tilde{\mathcal{B}} \leq \mathcal{P} \iff 0 \leq \tilde{T}(\tilde{\mathcal{B}}) \leq \tilde{T}(\mathcal{P}).$$

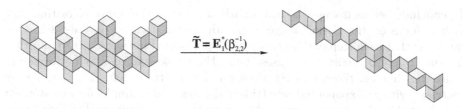

$$\tilde{T} = E_1^*(\beta_{2,2}^{-1})$$

Fig. 8. The recognizable binary function $\mathcal{B}$ of Fig. 6 is transformed by applying the rules of Fig. 7 into a binary function $\tilde{\mathcal{B}}$ (left) such that $0 \leq \mathcal{B} \leq \mathcal{P} \Leftrightarrow 0 \leq \tilde{\mathcal{B}} \leq \mathcal{P}$. Here, since $\tilde{\mathcal{B}}$ does not have open $(1,3)$-run, its image by $\tilde{T}$ (right) is such that, for any stepped plane $\mathcal{P}$, one has: $0 \leq \tilde{\mathcal{B}} \leq \mathcal{P} \Leftrightarrow 0 \leq \tilde{T}(\tilde{\mathcal{B}}) \leq \mathcal{P}$.

Thus, we still have the problem that a binary function $\mathcal{B}$ is not always recognizable (Def. 10), while the second problem is now that $\tilde{\mathcal{B}}$ can have problematic open runs. However, it is expected that, in practice, both not recognizable binary functions and problematic open runs have rather small sizes. Hence, it is worth considering a hybrid algorithm. Given a recognizable binary function $\mathcal{B}$, we compute $\tilde{\mathcal{B}}$, remove problematic open runs and apply the map $\tilde{T}$. We iterate this up to obtain an unrecognizable binary function. Then, we use an already existing algorithm (*e.g.* a preimage algorithm, see [4]) to recognize this binary function and, finally, to refine the recognition by considering the previously removed open runs. More precisely, consider the following algorithm, where XReco is an algorithm which computes the set $P(\mathcal{B})$ and $B'_{a,i}$ is the $(d+2) \times (d+2)$ block matrix whose first block is $B_{a,i}$ and the second the 1×1 identity matrix:

HybridBrunReco($\mathcal{B}$)

---
1. $n \leftarrow 0$;
2. $\mathcal{B}_0 \leftarrow \mathcal{B}$;
3. **while** $\mathcal{B}_n$ is recognizable **do**
4. $(a_n, i_n) \leftarrow (a(\mathcal{B}_n), i(\mathcal{B}_n))$;
5.   compute $\tilde{\mathcal{B}}_n$;
6. $L_n \leftarrow$ open runs of $\tilde{\mathcal{B}}_n$;
7. $\mathcal{B}_{n+1} \leftarrow E_1^*(\beta_{a_n,i_n}^{-1})(\tilde{\mathcal{B}}_n - L_n)$;
8. $n \leftarrow n+1$;
9. **end while**;
10. $P_n \leftarrow$ XReco($\mathcal{B}_n$);
11. **for** $k = n-1$ **downto** k=0 **do**
12. $P_k \leftarrow B'_{a_k,i_k} P_{k+1}$;
13. $P_k \leftarrow P_k \cap$ XReco(L_k);
14. **end for**;
15. **return** P_0;

One shows:

Theorem 3. *The algorithm* HybridBrunReco *with a binary function $\mathcal{B}$ of finite size as input returns the set $P(\mathcal{B})$ in finite time.*

To conclude, let us discuss the computational cost of the above algorithm. Let us first focus on the "Brun" stage of the algorithm, that is, on lines 3–9. One can show that each step of this stage can be performed in time $\mathcal{O}(|\mathcal{B}_n|)$ and that the size $|\mathcal{B}_n|$ of $\mathcal{B}_n$ strictly decreases. Thus, the whole stage can be performed in quadratic time (in the size of $\mathcal{B}$). However, let us stress that $(|\mathcal{B}_n|)_n$ generally decreases with an exponential rate (this is the case, for example, for any stepped plane), so that this stage is expected, in practice, to be performed in near linear time. Let us now consider the "correction" stage of the algorithm, that is, lines 10–14. Note that the sum of sizes of inputs of XReco is less than $|\mathcal{B}|$. Thus, assuming that XReco works in time no more than quadratic (this holds; for example, for a preimage algorithm, see [4]), the bound given for the first stage still holds. We also need to compute intersections of convex polytopes. The complexity of such operations is not trivial in higher dimensions, but let us stress that the intersection of k convex polytopes of $\mathbb{R}^3$ can be computed in time $\mathcal{O}(m \ln k)$, where m stands for the total size of these polytopes (see [5]). Moreover, let us recall that the first unrecognizable $\mathcal{B}_n$ as well as the sum of sizes of the L_k's are expected to be much smaller than $\mathcal{B}$. In conclusion, theoretical time complexity bounds are probably much bigger than the practical efficiency of this algorithm, so that further expriments shall be performed in order to better understand the computational cost of this hybrid algorithm.

References

1. Andres, E.: Le plan discret. In: Proc. of Discrete Geometry for Computer Imagery DGCI 1993, pp. 45–61 (1993)
2. Arnoux, P., Ito, S.: Pisot substitutions and Rauzy fractals. Bull. Bel. Math. Soc. Simon Stevin 8, 181–207 (2001)
3. Berthé, V., Fernique, Th.: Brun expansions of stepped surfaces (preprint, 2008)
4. Brimkov, V., Cœ, D.: Computational aspects of Digital plane and hyperplane recognition. In: Reulke, R., Eckardt, U., Flach, B., Knauer, U., Polthier, K. (eds.) IWCIA 2006. LNCS, vol. 4040, pp. 543–562. Springer, Heidelberg (2006)
5. Chazelle, B.: An optimal algorithm for intersecting three-dimensional convex polyhedra. SIAM J. Comput. 21, 671–696 (1992)
6. Ei, H.: Some properties of invertible substitutions of rank d and higher dimensional substitutions. Osaka Journal of Mathematics 40, 543–562 (2003)
7. Françon, J.: Discrete combinatorial surfaces. Graphical Models & Image Processing 57, 20–26 (1995)
8. Klette, R., Rosenfeld, A.: Digital straightness–a review. Elec. Notes in Theoret. Comput. Sci, vol. 46 (2001)
9. Réveillès, J.-P.: Calcul en nombres entiers et algorithmique, Ph. D Thesis, Univ. Louis Pasteur, Strasbourg (1991)
10. Schweiger, F.: Multi-dimensional continued fractions. Oxford Science Publications, Oxford Univ. Press, Oxford (2000)
11. Troesch, A.: Interprétation géométrique de l'algorithme d'Euclide et reconnaissance de segments. Theor. Comput. Sci. 115, 291–320 (1993)
12. Vuillon, L.: Combinatoire des motifs d'une suite sturmienne bidimensionelle. Theoret. Comput. Sci. 209, 261–285 (1998)

About the Frequencies of Some Patterns in Digital Planes Application to Area Estimators

Alain Daurat, Mohamed Tajine, and Mahdi Zouaoui

LSIIT CNRS UMR 7005, Université Louis Pasteur (Strasbourg 1),
Pôle API, Boulevard Sébastien Brant, 67400 Illkirch-Graffenstaden, France
{daurat,tajine,zouaoui}@dpt-info.u-strasbg.fr

Abstract. In this paper we prove that the function giving the frequency of a class of patterns of digital planes with respect to the slopes of the plane is continuous and piecewise affine, moreover the regions of affinity are precised. It allows to prove some combinatorial properties of a class of patterns called (m, n)-cubes. This study has also some consequences on local estimators of area: we prove that the local estimators restricted to regions of plane never converge to the exact area when the resolution tends to zero for almost all slope of plane. Actually all the results of this paper can be generalized for the regions of hyperplanes for any dimension $d \geq 3$.

The proofs of some results used in this article are contained in the extended version of this paper [1].

Keywords: Digital Plane, Pattern, (m, n)-cube, Area Estimator, Local Estimator, Multigrid Convergence.

1 Introduction

Digital Planes are very classical objects of Discrete Geometry. Their combinatorics have been studied in a lot of papers (for example [2, 3, 4, 5, 6, 7, 8, 9, 10, 11, 12, 13]), for a recent review on the subject, see [14]. In this paper we are interested in a class of patterns called (m, n)-cubes which are intuitively the pieces of digital planes of size $m \times n$. These objects have been studied for quite a long time, for example it is well-known that the number of (m, n)-cubes appearing in a digital plane is always less than mn ([5, 6, 7, 9, 11]). These (m, n)-cubes can be used in different domains of image analysis for example for normal vector estimation, area estimation ([15, 16], see also second section of this paper), form reconstruction. The originality of this paper is the study of not only the presence of a (m, n)-cube in a digital plane, but also of the frequency of this (m, n)-cube in all the digital planes.

The main result of the first part of this paper enunciates that the function giving the frequency of (m, n)-cubes of digital planes with respect to the slopes of the plane is continuous and piecewise affine. Moreover we will see that the study of the frequency allows to prove some combinatorial properties on the (m, n)-cubes.

D. Coeurjolly et al. (Eds.): DGCI 2008, LNCS 4992, pp. 45–56, 2008.

In a second part of the paper, we use our study about frequencies to prove some results about local estimators of area. A local estimator of area simply consists to decompose a surface into little pieces (in fact similar to (m, n)-cubes) and to sum some weights which correspond to the pieces. The study of the frequency of the (m, n)-cubes allows to prove that even for planar regions these estimators are not correct in the sense that, if the discrete regions are obtained from a continuous plane, then the estimated area does not converge to the exact area for almost all slopes of plane when the resolution of the discretization tends to zero. It is in fact a generalization to 3D of [17]. Actually we can prove with the same technics that all the results of this paper are true for the regions of hyperplanes for any dimension $d \geq 3$.

The proofs of some results used in this article are contained in the extended version of this paper [1].

2 Preliminaries

Let $a, b \in \mathbb{N}$ and $a \leq b$. The discrete interval $\{a, a + 1, \ldots, b - 1, b\}$ is denoted $[\![a, b]\!]$. For $x \in \mathbb{R}$, $\lfloor x \rfloor$ (resp. $\langle x \rangle$) denotes the integral part (resp. the fractional part) of x. So, $x = \lfloor x \rfloor + \langle x \rangle$ with $\lfloor x \rfloor \in \mathbb{Z}$, $\lfloor x \rfloor \leq x < \lfloor x \rfloor + 1$ and $0 \leq \langle x \rangle < 1$. For any set E, card(E) denotes the cardinality of E.

We refer in all the following to a subset of $\mathbb{R}^3$ of the form $R = \{(x, y, \alpha x + \beta y + \gamma) \mid a \leq x \leq b$ and $c \leq y \leq d\}$ such that $\alpha, \beta \in [0, 1]$ and $a, b, c, d \in \mathbb{R}$ as a rectangular planar region. It corresponds to a subset of plane whose projection on the XY-plane is a rectangle with sides parallel to X, Y-axes.

In this paper all the topological notions are considered relatively to the Euclidean usual topology. If E is subset of a topological space, $\overline{E}$ denotes its topological closure (the smallest closed set containing E). The measure notions are considered relatively to the Lebesgue measure on the Euclidean space, for example 'negligible set' (set with zero measure) and 'almost everywhere' are considered relatively to the Lebesgue measure on the Euclidean space.

3 Frequencies of the (m, n)-cubes

In this paper we consider naive digital planes $P_{\alpha, \beta, \gamma} = \{(x, y, \lfloor \alpha x + \beta y + \gamma \rfloor) \mid (x, y) \in \mathbb{Z}^2\}$ with $\alpha, \beta \in [0, 1]$ and $\gamma \in \mathbb{R}$. So a naive plane is functional in its x, y coordinates: $z = p_{\alpha, \beta, \gamma}(x, y) = \lfloor \alpha x + \beta y + \gamma \rfloor$ for all $(x, y, z) \in P_{\alpha, \beta, \gamma}$. Moreover we fix two positive integers m and n and we define $\mathcal{F}_{m,n} = [\![0, m - 1]\!] \times [\![0, n - 1]\!]$.

Definition 1. *A (m, n)-pattern is a function $w : \mathcal{F}_{m,n} \to \mathbb{Z}$. We note $\mathcal{M}_{m,n}$ the set of all (m, n)-patterns and the size of a (m, n)-pattern is $m \times n$.*

We can also see a (m, n)-pattern as a set of voxels which projection in the XY-plane is $\mathcal{F}_{m,n}$ and which has at most one point in each line parallel to the third coordinate direction.

In all the following, a pattern of size less than $m \times n$ corresponds to a (m', n')-pattern where $m' \leq m$, $n' \leq n$ and $(m, n) \neq (m', n')$.

A (m, n)-cube is a (m, n)-pattern which can be extracted from a naive digital plane, more precisely:

Definition 2. *The (m, n)-cube at position (i, j) of the digital plane $P_{\alpha, \beta, \gamma}$ is the (m, n)-pattern w defined by $w(i', j') = p_{\alpha, \beta, \gamma}(i + i', j + j') - p_{\alpha, \beta, \gamma}(i, j)$ for any $(i', j') \in \mathcal{F}_{m,n}$. It is denoted $w_{i,j}(\alpha, \beta, \gamma)$.*

So a (m, n)-cube is simply a piece of a digital plane which projection in the XY-plane is a translation of $\mathcal{F}_{m,n}$. Fig.1 corresponds to a $(3, 3)$-cube in a digital plane.

Note that for all $i, j \in \mathbb{Z}$ and $\alpha, \beta, \gamma \in \mathbb{R}$, $w_{i,j}(\alpha, \beta, \gamma) = w_{0,0}(\alpha, \beta, \alpha i + \beta j + \gamma)$.

Fig. 1. A $(3, 3)$-cube in a digital plane

Let $C_{i,j}^{\alpha, \beta} = 1 - \langle \alpha i + \beta j \rangle$ for $(i, j) \in \mathcal{F}_{m,n}$, and $\sigma^{\alpha, \beta}$ be a bijection from $[\![1, mn]\!]$ to $\mathcal{F}_{m,n}$ such that the sequence $(B_i^{\alpha, \beta})_{0 \leq i \leq mn}$ defined by $B_i^{\alpha, \beta} = C_{\sigma^{\alpha, \beta}(i)}^{\alpha, \beta}$ for $1 \leq i \leq mn$ and $B_0^{\alpha, \beta} = 0$, is increasing.

We recall some known results (see for example [6]).

Proposition 1. *For all $\alpha, \beta, \gamma \in \mathbb{R}$ we have:*

1. *The (k, l)-th point of the (m, n)-cube at position (i, j) of the digital plane $P_{\alpha, \beta, \gamma}$ can be computed by the formula:*

$$w_{i,j}(\alpha, \beta, \gamma)(k, l) = \begin{cases} \lfloor \alpha k + \beta l \rfloor & \text{if } \langle \alpha i + \beta j + \gamma \rangle < C_{k,l}^{\alpha, \beta} \\ \lfloor \alpha k + \beta l \rfloor + 1 & \text{otherwise} \end{cases}$$

2. *The (m, n)-cube $w_{i,j}(\alpha, \beta, \gamma)$ only depends on the interval $[B_h^{\alpha, \beta}, B_{h+1}^{\alpha, \beta}[$ containing $\langle \alpha i + \beta j + \gamma \rangle$.*

3. *For all $h \in [\![1, mn - 1]\!]$, if $[B_h^{\alpha, \beta}, B_{h+1}^{\alpha, \beta}[$ is not empty ($B_h^{\alpha, \beta} < B_{h+1}^{\alpha, \beta}$), then there exist i, j such that $\langle \alpha i + \beta j + \gamma \rangle \in [B_h^{\alpha, \beta}, B_{h+1}^{\alpha, \beta}[$ and thus the number of (m, n)-cubes in the digital plane $P_{\alpha, \beta, \gamma}$ is equal to $\mathrm{card}(\{C_{k,l}^{\alpha, \beta} \mid (k, l) \in \mathcal{F}_{m,n}\})$. We have, in particular, $\mathrm{card}(\{C_{k,l}^{\alpha, \beta} \mid (k, l) \in \mathcal{F}_{m,n}\}) \leq mn$.*

So, we have $w_{0,0}(\alpha, \beta, \gamma) = w_{0,0}(\alpha, \beta, \langle\gamma\rangle)$ and thus $w_{i,j}(\alpha, \beta, \gamma) = w_{0,0}(\alpha, \beta, \langle\alpha i + \beta j + \gamma\rangle)$ for all $\alpha, \beta, \gamma \in \mathbb{R}$ and $(i, j) \in \mathbb{Z}^2$.

By Proposition 1, the set of (m, n)-cubes of the digital plane $P_{\alpha,\beta,\gamma}$ depends only on α, β and it is denoted $\mathcal{C}_{m,n,\alpha,\beta}$. In all the following, $\mathcal{U}_{m,n}$ denotes the set of all the (m, n)-cubes. So, $\mathcal{U}_{m,n} = \bigcup_{(\alpha,\beta)\in[0,1]^2} \mathcal{C}_{m,n,\alpha,\beta}$.

Definition 3 ([18]). *Let w be a (m, n)-cube, then the pre-image $\mathrm{PI}(w)$ of w is the set of the triple $(\alpha, \beta, \gamma) \in [0,1]^3$ such that w is the (m, n)-cube at position $(0,0)$ of the digital plane $P_{\alpha,\beta,\gamma}$.*

Remark. It is easy to see that $\mathrm{PI}(w)$ is a convex polyhedron defined by the inequalities $w(k, l) \le k\alpha + l\beta + \gamma < w(k, l) + 1$ for $(k, l) \in \mathcal{F}_{m,n}$. Moreover the set of the $\gamma' \in [0,1]$ such that $(\alpha, \beta, \gamma') \in \mathrm{PI}(w_{i,j}(\alpha, \beta, \gamma))$ is exactly the interval $[B_h^{\alpha,\beta}, B_{h+1}^{\alpha,\beta}[$ containing $\langle\alpha i + \beta j + \gamma\rangle$.

The last remark leads to the following definition:

Definition 4. *The γ-frequency of a (m, n)-cube w for the slopes (α, β) (denoted $\mathrm{freq}_{\alpha,\beta}(w)$) is the length of the interval $I^{\alpha,\beta}(w) = \{\gamma \in [0,1] \mid (\alpha, \beta, \gamma) \in \mathrm{PI}(w)\}$. (so the function $TP : \mathrm{PI}(w) \to \mathbb{R}$ such that $TP(\alpha, \beta) = \mathrm{freq}_{\alpha,\beta}(w)$ is the tomographic projection of $\mathrm{PI}(w)$ w.r.t. the third coordinate direction).*

Definition 5. *The overlapping frequency of a (m, n)-cube in the digital plane $P_{\alpha,\beta,\gamma}$ is*

$$\lim_{N \to +\infty} \frac{\mathrm{card}(\{(i,j) \in [\![-N, N]\!]^2 \mid w_{i,j}(\alpha, \beta, \gamma) = w\})}{(2N + 1)^2}$$

if the limit exists. It is denoted $\mathrm{overfreq}_{\alpha,\beta,\gamma}(w)$.

So, $\mathrm{overfreq}_{\alpha,\beta,\gamma}(w) = \lim_{N \to +\infty} \frac{\mathrm{card}(\{(i,j)\in[\![-N,N]\!]^2 \mid \langle\alpha i+\beta j+\gamma\rangle \in I^{\alpha,\beta}(w)\})}{(2N+1)^2}$

We have the following properties:

Proposition 2 ([1]). *For any $\alpha, \beta \in [0,1]$ and $\gamma \in \mathbb{R}$ we have:*

1. *$w \in \mathcal{C}_{m,n,\alpha,\beta}$ if and only if $\mathrm{freq}_{\alpha,\beta}(w) > 0$.*
2. *$\mathrm{overfreq}_{\alpha,\beta,\gamma}(w) = \mathrm{freq}_{\alpha,\beta}(w)$*

(A related result is stated in [19]).

Definition 6. *A function $f : \mathbb{R}^2 \to \mathbb{R}$ is called a piecewise affine function if there exists a finite collection $(C_i)_{i\in I}$ of open convex subsets of $\mathbb{R}^2$ and affine functions $f_i : \mathbb{R}^2 \to \mathbb{R}$ for $i \in I$, such that :*

- *$C_i \cap C_{i'} = \emptyset$ for $i, i' \in I$ and $i \ne i'$,*
- *$\bigcup_{i\in I} \overline{C_i} = \mathbb{R}^2$ and*
- *the restriction of f to C_i is f_i for all $i \in I$ (for all $i \in I, f(x) = f_i(x)$ for all $x \in C_i$).*

Property 1 ([1]). Let $f, g : \mathbb{R}^2 \to \mathbb{R}$ be two piecewise affine functions. Then $-f, f + g, f - g, \max(f, g)$ and $\min(f, g)$ are also piecewise affine functions.

Theorem 1. *For any (m,n)-cube w, the function $(\alpha, \beta) \mapsto \mathrm{freq}_{\alpha,\beta}(w)$ is a continuous function which is piecewise affine.*

Proof. $PI(w) = \{(\alpha, \beta, \gamma) \in \mathbb{R}^3 \mid w(k,l) \le \alpha k + \beta l + \gamma < w(k,l) + 1 \text{ for all } (k,l) \in \mathcal{F}_{m,n}\}$.

Then $I^{\alpha,\beta}(w) = [\max_{(k,l) \in \mathcal{F}_{m,n}}(w(k,l) - \alpha k - \beta l), \min_{(k,l) \in \mathcal{F}_{m,n}}(w(k,l) + 1 - \alpha k - \beta l)[$.

So, $\mathrm{freq}_{\alpha,\beta}(w) = \max(0, \min_{(k,l) \in \mathcal{F}_{m,n}}(w(k,l) + 1 - \alpha k - \beta l) - \max_{(k,l) \in \mathcal{F}_{m,n}}(w(k,l) - \alpha k - \beta l))$. Affine functions, max and min are continuous functions. Then $(\alpha, \beta) \mapsto \mathrm{freq}_{\alpha,\beta}(w)$ is a continuous function which is piecewise affine because it is composition of continuous functions and by Property 1 it is piecewise affine function. $\qquad\square$

Proposition 3. *Let $(\alpha_1, \beta_1), (\alpha_2, \beta_2), (\alpha_3, \beta_3)$ be points of $[0,1]^2$ and T be the convex hull of these three points. Let $(\alpha_0, \beta_0) \in T$ and consider $\lambda_1, \lambda_2, \lambda_3 \ge 0$ such that $(\alpha_0, \beta_0) = \sum_{i=1}^3 \lambda_i(\alpha_i, \beta_i)$ and $\sum_{i=1}^3 \lambda_i = 1$ ($\lambda_1, \lambda_2, \lambda_3$ are barycentric coordinates of (α_0, β_0) relatively to $(\alpha_1, \beta_1), \ldots (\alpha_3, \beta_3)$). Suppose moreover that the function $(\alpha, \beta) \mapsto \mathrm{freq}_{\alpha,\beta}(w)$ is affine on T for any (m,n)-cube w, then*

$$\mathcal{C}_{m,n,\alpha_0,\beta_0} = \bigcup_{1 \le i \le 3 \text{ and } \lambda_i \ne 0} \mathcal{C}_{m,n,\alpha_i,\beta_i}$$

Proof. By affinity of $(\alpha, \beta) \mapsto \mathrm{freq}_{\alpha,\beta}(w)$ on T we have:

$$\mathrm{freq}_{\alpha_0,\beta_0}(w) = \sum_{i=1}^3 \lambda_i \mathrm{freq}_{\alpha_i,\beta_i}(w)$$

If $w \notin \mathcal{C}_{m,n,\alpha_0,\beta_0}$ then by Proposition 2, $\mathrm{freq}_{\alpha_0,\beta_0}(w) = 0$ and so for any i, $\mathrm{freq}_{\alpha_i,\beta_i}(w) = 0$ or $\lambda_i = 0$ because $\lambda_1, \lambda_2, \lambda_3 \ge 0$ which implies that for any i, if $\lambda_i \ne 0$ then $w \notin \mathcal{C}_{m,n,\alpha_i,\beta_i}$. Conversely as $\lambda_1, \lambda_2, \lambda_3 \ge 0$ and $\sum_{i=1}^3 \lambda_i = 1$, if $w \in \mathcal{C}_{m,n,\alpha_0,\beta_0}$, then by Proposition 2, $\mathrm{freq}_{\alpha_0,\beta_0}(w) > 0$ and thus, there must exist a $i \in \{1, 2, 3\}$ such that $\lambda_i \ne 0$ and $\mathrm{freq}_{\alpha_i,\beta_i}(w) > 0$ $\qquad\square$

We will now precise the domains where the function $(\alpha, \beta) \mapsto \mathrm{freq}_{\alpha,\beta}(w)$ is affine: Let $D_{u,v,w}$ be the line $\{(\alpha, \beta) \in \mathbb{R}^2 \mid \alpha u + \beta v + w = 0\}$ and

$$E_{m,n} = \bigcup_{(u,v,w) \in [\![-m+1,m-1]\!] \times [\![-n+1,n-1]\!] \times \mathbb{Z}} D_{u,v,w} \cap [0,1]^2.$$

$E_{m,n}$ involves only straight lines $D_{u,v,w}$ such that $D_{u,v,w} \cap [0,1]^2 \ne \emptyset$ and so we must only consider the straight lines $D_{u,v,w}$ such that $|w| \le |u| + |v|$ and thus $E_{m,n}$ involves only a finite number of straight lines.

$E_{m,n}$ is called Hyper Farey fan in [2] and Farey's diagram in [9].

Fig.2 corresponds to Farey's diagram for $m = 4$ and $n = 3$.

Theorem 2 ([1]). *The function $(\alpha, \beta) \mapsto \mathrm{freq}_{\alpha,\beta}(w)$ is affine on the closure of any connected component of $[0,1]^2 \setminus E_{m,n}$ for all $w \in \mathcal{U}_{m,n}$. Moreover for any $(\alpha, \beta), (\alpha', \beta') \in [0,1]^2 \setminus E_{m,n}$: $\mathcal{C}_{m,n,\alpha,\beta} = \mathcal{C}_{m,n,\alpha',\beta'}$ if and only if (α, β) and (α', β') are in the same connected component of $[0,1]^2 \setminus E_{m,n}$.*

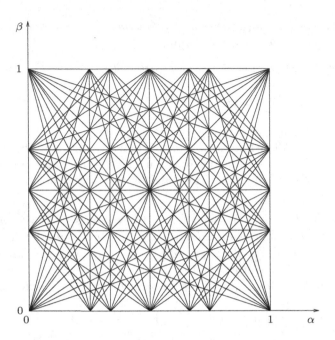

Fig. 2. Farey's diagram for $m = 4$ and $n = 3$

Corollary 1. *Let $\mathcal{O}$ be a connected component of $[0,1]^2 \setminus E_{m,n}$. Then $\mathcal{O}$ is a convex polygon and if p_1, p_2, p_3 are distinct vertexes of the polygon $\mathcal{O}$ then*

1. *for any point $p \in \mathcal{O}$, $\mathcal{C}_{m,n,p} = \mathcal{C}_{m,n,p_1} \cup \mathcal{C}_{m,n,p_2} \cup \mathcal{C}_{m,n,p_3}$ and*
2. *for any point p in the interior of the segment of vertexes p_1, p_2, $\mathcal{C}_{m,n,p} = \mathcal{C}_{m,n,p_1} \cup \mathcal{C}_{m,n,p_2}$.*

Proof. The function $(\alpha, \beta) \mapsto \mathcal{C}_{m,n,\alpha,\beta}$ is constant on $\mathcal{O}$. By Theorem 2, for all $w \in \mathcal{U}_{m,n}$, the function $(\alpha, \beta) \mapsto \text{freq}_{\alpha,\beta}(w)$ is affine on $\overline{\mathcal{O}}$ and we conclude by using Proposition 3 □

Corollary 2. *The number of (m,n)-cubes is in $O((m+n)^2 m^3 n^3)$.*

Proof. Any line of equation $ux + vy + w = 0$ with $|w| > |u| + |v|$ does not intersect the square $[0,1]^2$, so $E_{m,n}$ is composed of at most $(2m+1)(2n+1)(2(m+n+1)+1) = f(m,n)$ lines. Thanks to Theorem 2 and Corollary 3 all the (m,n)-cubes appear in the vertices of the connected components of $[0,1]^2 \cap E_{m,n}$. Each vertex is the intersection of two lines of $E_{m,n}$ so there are at most $f(m,n)^2$ such vertices. Each vertex corresponds to at most mn (m,n)-cubes, so in total there are at most $((2m+1)(2n+1)(2(m+n+1)+1))^2 mn$ (m,n)-cubes, which proves the claim. □

Corollary 2 gives an upper bound for the number of (m,n)-cubes. In the following, we will give a lower bound for this number.

Definition 7. *Let $m \in \mathbb{N} \setminus \{0\}$.*

1. *Let $\alpha, \gamma \in [0,1]$. The set $S(m, \alpha, \gamma) = \{(x, \lfloor \alpha x + \gamma \rfloor) \mid x \in [\![0, m-1]\!]\}$ is called a digital segment of size m.*
2. *$\mathcal{S}_m = \{S(m, \alpha, \gamma) \mid \alpha, \gamma \in [0,1]\}$ is the set of all digital segments of size m.*

Property 2 ([20, 21])

1. *$\mathrm{card}(\mathcal{S}_m) = 1 + \sum_{i=1}^{m}(m - i + 1)\varphi(i)$ where φ is the Euler's totient function ($\varphi(i) = \mathrm{card}(\{j \mid 1 \leq j < i \text{ and } i \text{ and } j \text{ are co-prime}\})$).*
2. *$\mathrm{card}(\mathcal{S}_m) = \frac{m^3}{\pi^2} + O(m^2 \log(m))$.*

Property 3. Let $m, n \in \mathbb{N}^*$ and $\alpha, \beta, \gamma_1, \gamma_2 \in [0,1]$ and consider the two digital segments $S(m, \alpha, \gamma_1), S(n, \beta, \gamma_2)$. Then, there exists $w \in \mathcal{U}_{m,n}$ such that $S(m, \alpha, \gamma_1) = \{(i, w(i, 0)) \mid i \in [\![0, m-1]\!]\}$ and $S(m, \beta, \gamma_2) = \{(j, w(0, j)) \mid j \in [\![0, m-1]\!]\}$.

Proof. By [21, 6], there exist $i, j \in \mathbb{Z}$ such that $S(m, \alpha, \gamma_1) = \{(x - i, \lfloor \alpha x \rfloor - \lfloor \alpha i \rfloor) \mid x \in [\![i, m+i-1]\!]\}$ and $S(n, \beta, \gamma_2) = \{(x-j, \lfloor \beta y \rfloor - \lfloor \beta j \rfloor) \mid y \in [\![j, n+j-1]\!]\}$. Then $w = w_{i,j}(\alpha, \beta, 0)$ verifies the conditions of the property. □

Corollary 3. $\mathrm{card}(\mathcal{U}_{m,n}) \geq \mathrm{card}(\mathcal{S}_m)\mathrm{card}(\mathcal{S}_n)$. So, $\mathrm{card}(\mathcal{U}_{m,n}) \geq \frac{1}{\pi^4}m^3 n^3 + O(m^2 n^2 \log(m) \log(n))$.

Proof. By Property 3, we have $\mathrm{card}(\mathcal{U}_{m,n}) \geq \mathrm{card}(\mathcal{S}_m)\mathrm{card}(\mathcal{S}_n)$ and the second assertion is a direct consequence of Property 2. □

Corollaries 2 and 3 imply that there exist two constant numbers k_1, k_2 such that $k_1 m^3 n^3 < \mathrm{card}(\mathcal{U}_{m,n}) < k_2(m + n)^2 m^3 n^3$.

4 Application to Local Estimators

A digital surface is the discretization of a surface of $\mathbb{R}^3$. We investigate in this section the local estimators of the area of digital surface in the digital space $r\mathbb{Z}^3$ of resolution r.

The local estimator of area is obtained by associating a weight $p(w)$ to each pattern $w \in \mathcal{M}(m, n)$ of size $m \times n$ where $\mathcal{M}(m, n)$ is the set of all $m \times n$-patterns, then any digital surface $\mathcal{S}_r$, can be obtained by concatenation of elements in $\mathcal{M}(m, n)$ with perhaps a pattern ε_i, ε'_j of size less that $m \times n$. In other words $\mathcal{S}_r$ can be viewed as a bi-dimensional word on the alphabet of patterns of size less or equal to $m \times n$. If

$$
\mathcal{S}_r =
\begin{matrix}
w_{1,1} & w_{2,1} & \cdots & w_{M,1} & \varepsilon_1 \\
w_{1,2} & w_{2,2} & \cdots & w_{M,2} & \varepsilon_2 \\
\vdots & \vdots & \cdots & \vdots & \vdots \\
w_{1,N} & w_{2,N} & \cdots & w_{M,N} & \varepsilon_N \\
\varepsilon'_1 & \varepsilon'_2 & \cdots & \varepsilon'_M & \varepsilon'_{M+1}
\end{matrix}
$$

where $w_{i,j} \in \mathcal{M}(m,n)$ for all i,j, then we define the area of $\mathcal{S}_r$ by $\mathfrak{S}_{r,m,n,p}(\mathcal{S}_r) = r^2 \sum_{i,j} p(w_{i,j})$ *(i.e. we neglect the contribution of the digital surfaces ε_i and ε'_j).*

Actually, we investigate the following problem:

Does there exist m,n and $p(.)$ such that for any surface $\mathcal{S} \in \mathbb{R}^3$ the areas $\mathfrak{S}_{r,m,n,p}(\mathcal{S}_r)$ converge to the area of $\mathcal{S}$ where r tends to 0? *(i.e. $\mathcal{S}_r$ is a discretization of $\mathcal{S}$).*

In this section, we study this problem for a particular class of surfaces: the set of rectangular planar regions. Moreover we suppose that the discretization operator δ_r restricted to these regions is in the class of the "Bresenham" discretization.

Let $a,b,c,d \in \mathbb{R}$ such that $a < b$ and $c < d$ and $0 \le \alpha, \beta \le 1$. Let $r > 0$ be the resolution of the discrete space $r\mathbb{Z}^3$. Let the rectangular planar region $R = \{(x,y,\alpha x + \beta y + \gamma) \mid a \le x \le b \text{ and } c \le y \le d\}$. So the "Bresenham" discretization of R in $r\mathbb{Z}^3$ is

$$R_r = r\{(x,y,\lfloor \alpha x + \beta y + \tfrac{\gamma}{r} \rfloor) \mid (x,y) \in [\![\lceil \tfrac{a}{r} \rceil, \lfloor \tfrac{b}{r} \rfloor]\!] \times [\![\lceil \tfrac{c}{r} \rceil, \lfloor \tfrac{d}{r} \rfloor]\!]\}.$$

We fix m,n as a positive integers. As it has been explained for surfaces, the discrete region R_r can be seen as the bi-dimensional word:

$$R_r = \begin{array}{ccccc} w_{1,1} & w_{2,1} & \cdots & w_{M_r,1} & \varepsilon_{1,r} \\ w_{1,2} & w_{2,2} & \cdots & w_{M_r,2} & \varepsilon_{2,r} \\ \vdots & \vdots & \ddots & \vdots & \vdots \\ w_{1,N_r} & w_{2,N_r} & \cdots & w_{M_r,N_r} & \varepsilon_{N_r,r} \\ \varepsilon'_{1,r} & \varepsilon'_{2,r} & \cdots & \varepsilon'_{M_r,r} & \varepsilon'_{M_r+1,r} \end{array} \qquad (1)$$

where $M_r = \lfloor \frac{\lfloor \frac{b}{r} \rfloor - \lceil \frac{a}{r} \rceil + 1}{m} \rfloor$ and $N_r = \lfloor \frac{\lfloor \frac{d}{r} \rfloor - \lceil \frac{c}{r} \rceil + 1}{n} \rfloor$ and for all i,j, $w_{i,j}$ is a (m,n)-cube and ε_i, ε'_j are patterns of size less than $m \times n$.

We construct $\mathfrak{S}_{r,m,n,p}$ as the local estimator of measure by using a weight function $p : \mathcal{U}_{m,n} \to \mathbb{R}$. Then $\mathfrak{S}_{r,m,n,p}$ is defined by:

$$\begin{aligned} \mathfrak{S}_{r,m,n,p}(R_r) &= r^2 \sum_{(i,j) \in [\![1, M_r]\!] \times [\![1, N_r]\!]} p(w_{i,j}) \\ &= r^2 \sum_{w \in \mathcal{U}_{m,n}} n(w, R_r, r) p(w) \end{aligned} \qquad (2)$$

Where $n(w, R_r, r) = \text{card}(\{(i,j) \in [\![1, M_r]\!] \times [\![1, N_r]\!] \mid w_{i,j} = w\})$ which is the number of occurrences of the pattern w in the bi-dimensional word R_r *(i.e. we neglect the contributions of the $\varepsilon_{i,r}$ and $\varepsilon'_{j,r}$ for $i \in [\![1, N_r]\!]$ and $j \in [\![1, M_r + 1]\!]$).*

The central question of this section can be formulated as the following, does there exist m,n and $p(.)$ such that, for any rectangular planar region R, the estimation $\mathfrak{S}_{r,m,n,p}(R_r)$ converges to the area of R when the resolution r tends to 0?

We will prove in this section that the response is almost everywhere no.

Actually, we will prove that for almost all rectangular planar regions R, the estimation $\mathfrak{S}_{r,m,n,p}(R_r)$ does not converge to the area of R when the resolution r tends to 0. The result of this section is an extension of the results of [17] for estimating area of rectangular planar regions of $\mathbb{R}^3$.

Put $DA_{m,n,p}(R) = \lim_{r \to 0} \mathfrak{S}_{r,m,n,p}(R_r)$.

In all the following, to simplify the notations we denote $E_r = (\llbracket \lceil \frac{a}{r} \rceil, \lfloor \frac{b}{r} \rfloor \rrbracket \times \llbracket \lceil \frac{c}{r} \rceil, \lfloor \frac{d}{r} \rfloor \rrbracket) \cap ((m\mathbb{Z} + \lceil \frac{a}{r} \rceil) \times (n\mathbb{Z} + \lceil \frac{c}{r} \rceil))$ and $S_r = (\lfloor \frac{b}{r} \rfloor - \lceil \frac{a}{r} \rceil + 1)(\lfloor \frac{d}{r} \rfloor - \lceil \frac{c}{r} \rceil + 1)$

Definition 8. *The non-overlapping frequency $F_r^{\alpha,\beta,\gamma,a,b,c,d}$ of a pattern w of size $m \times n$ in R_r is defined by:*

$$F_r^{\alpha,\beta,\gamma,a,b,c,d} = \frac{\mathrm{card}(\{(x,y) \in E_r \mid w_{x,y}(\alpha, \beta, \frac{\gamma}{r}) = w\})}{S_r}$$

Lemma 1. *Let $\alpha, \beta \in [0,1]$ such that α or β is irrational, $\gamma, a, b, c, d \in \mathbb{R}$, $w \in \mathcal{C}_{m,n,\alpha,\beta}$. Then*

$$F^{\alpha,\beta,\gamma,a,b,c,d} = \lim_{r \to 0} F_r^{\alpha,\beta,\gamma,a,b,c,d} = \frac{1}{mn} \mathrm{freq}_{\alpha,\beta}(w)$$

In particular $F^{\alpha,\beta,\gamma,a,b,c,d}$ does not depend on $\gamma, a, b, c,$ and d.

Proof.

$$F^{\alpha,\beta,\gamma,a,b,c,d} = \lim_{r \to 0} \frac{\mathrm{card}(\{(x,y) \in E_r \mid w_{x,y}(\alpha,\beta,\frac{\gamma}{r}) = w\})}{S_r}$$
$$= \lim_{r \to 0} \frac{\mathrm{card}(\{(x,y) \in E_r \mid \langle \alpha x + \beta y + \frac{1}{r}\gamma \rangle \in I^{\alpha,\beta}(w)\})}{S_r}$$

So, if we take $p = m$, $q = n$, $\gamma_r = \frac{1}{r}\gamma$ and $I = I^{\alpha,\beta}(w)$ in Theorem 4 of the Appendix A [1], then we have $F^{\alpha,\beta,\gamma,a,b,c,d} = \frac{1}{mn}\mu(I^{\alpha,\beta}(w)) = \frac{1}{mn}\mathrm{freq}_{\alpha,\beta}(w)$ because by Proposition 2 overfreq$_{\alpha,\beta,\gamma}(w) = \mu(I^{\alpha,\beta}(w))$ □

Theorem 3. *Let $\mathcal{O}$ be a connected component of $[0,1]^2 \setminus E_{m,n}$. Then there exist $u, v, t \in \mathbb{R}$ such that $DA_{m,n,p}(R) = (b-a)(d-c)(u\alpha + v\beta + t)$ for all rectangular planar regions $R = \{(x, y, \alpha x + \beta y + \gamma) \mid a \leq x \leq b \text{ and } c \leq y \leq d\}$ such that $\alpha, \beta \in \mathcal{O}$ and α or β is irrational.*

In other words, $DA_{m,n,p}(.)$ is an affine function in (α, β) for $(\alpha, \beta) \in (\mathcal{O} \setminus \mathbb{Q}^2)$.

Proof. By (1) and (2) we have:

$$\mathfrak{S}_{r,m,n,p}(R_r) = r^2 \sum_{1 \leq i \leq M_r} \sum_{1 \leq j \leq N_r} p(w_{i,j})$$
$$= r^2 \sum_{w \in \mathcal{U}_{m,n}} n(w, R_r, r)p(w)$$

where $n(w, R_r, r) = \mathrm{card}(\{(x,y) \in (((m\mathbb{Z} + \lceil \frac{a}{r} \rceil) \times (n\mathbb{Z} + \lceil \frac{c}{r} \rceil)) \cap (\llbracket \lceil \frac{a}{r} \rceil, \lfloor \frac{b}{r} \rfloor \rrbracket \times \llbracket \lceil \frac{c}{r} \rceil, \lfloor \frac{d}{r} \rfloor \rrbracket) \mid w_{x,y}(\alpha, \beta, \frac{\gamma}{r}) = w\})$ which is the number of occurrences of the pattern w in the bi-dimensional word R_r. So,

$$DA_{m,n,p}(R) = \lim_{r \to 0} r^2 \sum_{w \in \mathcal{U}_{m,n}} n(w, R_r, r)p(w)$$

$$= \lim_{r \to 0} r^2 S_r \sum_{w \in \mathcal{U}_{m,n}} \frac{n(w, R_r, r)}{S_r} p(w)$$

$$= (b-a)(d-c) \sum_{w \in \mathcal{U}_{m,n}} \frac{1}{mn} \text{freq}_{\alpha,\beta}(w)p(w) \quad \text{(By Lemma 1)}$$

So, according to Theorem 2, $DA_{m,n,p}(.)$ is an affine function in (α, β) for $(\alpha, \beta) \in (\mathcal{O} \setminus \mathbb{Q}^2)$ □

Corollary 4. *The set of $(\alpha, \beta) \in ([0, 1]^2 \setminus E_{m,n})$ such that α or β is irrational and $DA_{m,n,p}(R) = area(R)$ is a negligible (relatively to the Lebesgue measure on the Euclidean space) where for $a, b, c, d \in \mathbb{R}$, $R = \{(x, y, \alpha x + \beta y + \gamma) \mid a \le x \le b \text{ and } c \le y \le d\}$.*

Proof. We consider a connected component $\mathcal{O}$ of $[0, 1]^2 \setminus E_{m,n}$. By Theorem 3, there exist $u, v, t \in \mathbb{R}$ such that the estimated area of the rectangular planar region R is $DA_{m,n,p}(R) = (b-a)(d-c)(u\alpha + v\beta + t)$ for α or β is irrational. The exact area of R is $area(R) = (b-a)(d-c)\sqrt{1 + \alpha^2 + \beta^2}$. So we have:
$DA_{m,n,p}(R) = area(R) \iff (u\alpha + v\beta + t)^2 = 1 + \alpha^2 + \beta^2$
Which is equivalent to $(u^2 - 1)\alpha^2 + (v^2 - 1)\beta^2 + 2(uv\alpha\beta + ut\alpha + vt\beta) + t^2 - 1 = 0$
But, the last equation corresponds to an object of Lebesgue measure greater than 0 only when $u^2 - 1 = 0$, $v^2 - 1 = 0$, $t^2 - 1 = 0$, $uv = 0$, $ut = 0$ and $vt = 0$ which never happens. So, the last equation corresponds to a curve in $\mathbb{R}^2$ *(which is the intersection of conic and the region $\mathcal{O}$)* and thus, for $(\alpha, \beta) \in \mathcal{O}$, the estimated area can be equal to the exact area for only (α, β) in a set included in the intersection of a conic and the region $\mathcal{O}$ which corresponds to a negligible set.

But, $[0, 1]^2 \setminus E_{m,n}$ contains only a finite number of connected components. Thus, the set of $(\alpha, \beta) \in ([0, 1]^2 \setminus (E_{m,n} \cup \mathbb{Q}^2))$ such that the estimated area is equal to the exact area is a negligible set because it is a finite union of negligible sets. □

Corollary 5. *For any $m, n \in \mathbb{N}^*$ and any weight function $p(.)$ the set of $(\alpha, \beta) \in [0, 1]^2$ such that the rectangular planar region $R = \{(x, y, \alpha x + \beta y + \gamma) \mid a \le x \le b \text{ and } c \le y \le d\}$ (where $\gamma, a, b, c, d \in \mathbb{R}$) satisfies $area(R) = DA_{m,n,p}(R)$ is a negligible set. So, for any $m, n \in \mathbb{N}^*$ and any weight function $p(.)$, for all rectangular planar regions R with the parameters $\alpha, \beta \in [0, 1]$, we have $area(R) \ne DA_{m,n,p}(R)$ almost everywhere.*

Proof. By Corollary 5, we have, for almost all rectangular planar regions R with parameters $(\alpha, \beta) \in ([0, 1]^2 \setminus E_{m,n})$ $area(R) \ne DA_{m,n,p}(R)$. But $\mathbb{Q}^2$ is infinite countable set and $E_{m,n}$ is a finite set of straight lines. So $E_{m,n} \cup \mathbb{Q}^2$ is a negligible set. So, for all rectangular planar regions R with the parameters $\alpha, \beta \in [0, 1]$, $area(R) \ne DA_{m,n,p}(R)$ almost everywhere □

5   Conclusion

In this paper we have seen that the frequencies of the (m, n)-cubes of digital planes are given by a continuous piecewise affine function in the slopes of the digital planes. This has consequences on the combinatorics of (m, n)-cubes, in particular on the asymptotic behavior of the number of (m, n)-cubes when m and n tend to infinity.

Moreover it has also consequences on local estimators of area as it permits to prove that any local estimator of area is never multigrid-convergent: for almost all region of plane it does not converge to the true area. This result is a generalization of a result in dimension two proved in [17]. Actually we can prove with the same technics that this result is true for the equivalent notions for any finite dimension.

Acknowledgment

This work was supported by the IRMC program.

References

1. Daurat, A., Tajine, M., Zouaoui, M.: About the frequencies of some patterns in digital planes. Application to area estimators - extended version (preprint, 2007), http://hal.archives-ouvertes.fr/hal-00174960
2. Forchhammer, S.: Digital plane and grid point segments. Comput. Vis. Graph. Image Process 47, 373–384 (1989)
3. Françon, J., Schramm, J.M., Tajine, M.: Recognizing arithmetic straight lines and planes. In: Miguet, S., Ubéda, S., Montanvert, A. (eds.) DGCI 1996. LNCS, vol. 1176, pp. 141–150. Springer, Heidelberg (1996)
4. Schramm, J.M.: Coplanar tricubes. In: Ahronovitz, E. (ed.) DGCI 1997. LNCS, vol. 1347, pp. 87–98. Springer, Heidelberg (1997)
5. Reveillès, J.P.: Combinatorial pieces in digital lines and planes. In: Proc. SPIE Vision Geometry IV, vol. 2573, pp. 23–34 (1995)
6. Gérard, Y.: Contribution à la Géométrie Discrète. PhD thesis, Université d'Auvergne, Clermont-Ferrand (1999)
7. Gérard, Y.: Local Configurations of Digital Hyperplanes. In: Bertrand, G., Couprie, M., Perroton, L. (eds.) DGCI 1999. LNCS, vol. 1568, pp. 65–95. Springer, Heidelberg (1999)
8. Vittone, J., Chassery, J.M. (n,m)-Cubes and Farey Nets for Naive Planes Understanding. In: Bertrand, G., Couprie, M., Perroton, L. (eds.) DGCI 1999. LNCS, vol. 1568, pp. 76–90. Springer, Heidelberg (1999)
9. Vittone, J.: Caractérisation et reconnaissance de droites et de plans en géométrie discrète. PhD thesis, Université Joseph Fourier, Grenoble (1999)
10. Vuillon, L.: Combinatoire des motifs d'une suite sturmienne bidimensionnelle. Theoret. Comput. Sci. 209, 261–285 (1998)
11. Vuillon, L.: Local configurations in a discrete plane. Bull. Belg. Math. Soc. Simon Stevin 6(4), 625–636 (1999)

12. Brimkov, V.E., Andres, E., Barneva, R.P.: Object discretizations in higher dimensions. Pattern Recogn. Lett. 23(6), 623–636 (2002)
13. Veelaert, P.: Digital planarity of rectangular surface segments. IEEE Trans. Pattern Anal. Mach. Intell. 16(6), 647–652 (1994)
14. Brimkov, V., Coeurjolly, D., Klette, R.: Digital planarity - a review. Discrete Appl. Math. 155(4), 468–495 (2007)
15. Lindblad, J.: Surface area estimation of digitized 3d objects using weighted local configurations. Image Vis. Comput. 23(2), 111–122 (2005)
16. Kenmochi, Y., Klette, R.: Surface area estimation for digitized regular solids. In: Proc. SPIE, Vision Geometry IX, vol. 4117, pp. 100–111 (2000)
17. Tajine, M., Daurat, A.: On Local Definitions of Length of Digital Curves. In: Nyström, I., Sanniti di Baja, G., Svensson, S. (eds.) DGCI 2003. LNCS, vol. 2886, pp. 114–123. Springer, Heidelberg (2003)
18. Coeurjolly, D., Sivignon, I., Dupont, F., Feschet, F., Chassery, J.M.: On digital plane preimage structure. Discrete Appl. Math. 151(1-3), 78–92 (2005)
19. Berthé, V., Fiorio, C., Jamet, D., Philippe, F.: On some applications of generalized functionality for arithmetic discrete planes. Image Vis. Comput. 25(10), 1671–1684 (2007)
20. Berenstein, C., Lavine, D.: On the number of digital straight line segments. IEEE Trans. On Pattern Analysis and Machine Intelligence 10(6), 880–887 (1988)
21. Mignosi, F.: On the number of factors of Sturmian words. Theoret. Comput. Sci. 82(1), 71–84 (1991)

Combinatorial View of Digital Convexity[*]

S. Brlek[1], J.-O. Lachaud[2], and X. Provençal[1]

[1] Laboratoire de Combinatoire et d'Informatique Mathématique,
Université du Québec à Montréal,
CP 8888 Succ. Centre-ville, Montréal (QC) Canada H3C 3P8
brlek.srecko@uqam.ca, xavierprovencal@gmail.com
[2] Laboratoire de Mathématiques, UMR 5127 CNRS, Université de Savoie,
73376 Le Bourget du Lac, France
jacques-olivier.lachaud@univ-savoie.fr

Abstract. The notion of convexity translates non-trivially from Euclidean geometry to discrete geometry, and detecting if a discrete region of the plane is convex requires analysis. In this paper we study digital convexity from the combinatorics on words point of view, and provide a fast optimal algorithm checking digital convexity of polyominoes coded by the contour word. The result is based on the Lyndon factorization of the contour word, and the recognition of Christoffel factors that are approximations of digital lines.

Keywords: Digital Convexity, Lyndon words, Christoffel words.

1 Introduction

In Euclidean geometry, a given region R is said to be *convex* if and only if for any pair of points p_1, p_2 in R the line segment joining p_1 to p_2 is completely included in R. In discrete geometry on square grids, the notion does not translate trivially, since the only convex (in the Euclidean sense) regions are rectangles. Many attempts have been made to fill the gap, and a first definition of discrete convexity based on discretisation of continuous object came from Sklansky [1] and Minsky and Papert [2]. Later, Kim [3,4] then Kim and Rosenfeld [5] provided several equivalent characterizations of discrete convex sets, and finally Chaudhuri and Rosenfeld [6] proposed a new definition of digital convexity based this time on the notion of digital line segments (see [7] for a review of digital straightness).

Given a finite subset S of $\mathbb{Z}^2$ its *convex hull* is defined as the intersection of all Euclidean convex sets containing S. Of course all the vertices of the convex hull are points from S. Therefore, throughout this work, a polyomino P (which is the interior of a closed non-intersecting grid path of $\mathbb{Z}^2$) is called *convex* if and only if its convex hull contains no integer point outside P. Debled-Rennesson et al. [8] already provided a linear time algorithm deciding if a given polyomino is convex. Their method uses arithmetical tools to compute series of digital line

---

[*] With the support of NSERC (Canada).

D. Coeurjolly et al. (Eds.): DGCI 2008, LNCS 4992, pp. 57–68, 2008.
© Springer-Verlag Berlin Heidelberg 2008

segments of decreasing slope: optimal time is achieved with a moving digital straight line recognition algorithm [9,10]

Recently, Brlek et al. looked at discrete geometry from the combinatorics of words point of view, showing for instance how the discrete Green theorem provides a series of optimal algorithms for diverse statistics on polyominoes [11,12]. This method is extended to study minimal moment of inertia polyominoes in [13]. This approach also gave an elementary proof and a generalization [14,15] of a result of Daurat and Nivat [16] relating salient and reentrant points in discrete sets. Some geometric properties of the contour of polyominoes may be found in [17,18]. Recently it was successfully used to provide an optimal algorithm for recognizing tiles that tile the plane by translation [19,20]. It is worth noting that the study of these objects goes back to Bernouilli, Markov, Thue and Morse (see Lothaire's books [21,22,23] for an exhaustive bibliographic account) and as suggested in the recent survey of Klette and Rosenfeld [7] the bridge between discrete geometry and combinatorics on words will benefit to both areas.

Here we study the problem of deciding whether a polyomino coded by its contour word, also called Freeman chain code, is convex or not. To achieve this we use well known tools of combinatorics on words. The first is the existence of a unique Lyndon factorization, and its optimal computation by the linear algorithm of Duval [24]. The second concerns the Christoffel words, a class of finite factors of Sturmian words, that are discrete approximations of straight lines. After recalling the combinatorial background and basic properties, we propose another linear time algorithm deciding convexity of polyominoes. This new purely discrete algorithm is much simpler to implement. Some experiments revealed that it is 10 times faster than previous linear algorithms. Furthermore, one of its main interests lies in the explicit link between combinatorics on words and discrete geometry. Since our method does not rely on geometric and vector computations, it also shows that digital convexity is much more fundamental and abstract property than general convexity.

2   Preliminaries

A word w is a finite sequence of letters $w_1 w_2 \cdots w_n$ on a finite alphabet Σ, that is a function $w : [1..n] \longrightarrow \Sigma$, and $|w| = n$ is its *length*. Consistently its number of a letters, for $a \in \Sigma$, is denoted $|w|_a$. The set of words of length n is denoted Σ^n and the set of all finite words is Σ^*, the free monoid on Σ. The empty word is denoted ε and by convention $\Sigma^+ = \Sigma^* \setminus \{\varepsilon\}$. The k-th power of word w is defined by $w^k = w^{k-1} \cdot w$ with the convention that $w^0 = \varepsilon$. A word is said *primitive* when it is not the power of a nonempty word. A *period* of a word w is a number p such that $w_i = w_{i+p}$, for all $1 \le i \le |w| - p$.

Given a total order $<$ on Σ, the *lexicographic ordering* extends this order to words on Σ by using the following rule :

$w < w'$ if either   (i) $w' \in w\Sigma^+$,
                    (ii) $w = uav$ and $w' = ubv'$ with $a < b, a, b \in \Sigma, u \in \Sigma^*$.

Two words w, w' on the alphabet Σ are said to be *conjugate*, written $w \equiv w'$, if there exist u, v such that $w = uv$ and $w' = vu$. The conjugacy class of a word is defined as the set of all its conjugates and is equivalent to the set of all circular permutations of its letters.

Let w be a finite word over the alphabet $\{0, 1\}$. We denote by $\vec{w}$ the vector $(|w|_0, |w|_1)$. For any word w, the partial function $\phi_w : \mathbb{N} \longrightarrow \mathbb{Z} \times \mathbb{Z}$ associates to any integer j, $0 \le j \le |w|$, the vector $\phi_w(j) = \overrightarrow{w_1 w_2 \cdots w_j}$. In other words, this map draws the word as a 4-connected path in the plane starting from the origin, going *right* for a letter 0 and *up* for a letter 1. This extends naturally to

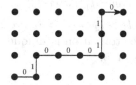

Fig. 1. Path encoded by the word $w = 01000110$

more general paths by using the four letter alphabet $\Sigma = \{0, 1, \bar{0}, \bar{1}\}$, associating to the letter $\bar{0}$ a *left* step and to $\bar{1}$ a *down* step. This notation allows to code the border of any polyomino by a 4-letter word known as the Freeman chain code.

The lexicographic order $<$ on points of $\mathbb{R}^2$ or $\mathbb{Z}^2$ is such that $(x, y) < (x', y')$ when either $x < x'$ or $x = x'$ and $y < y'$. The *convex hull* of a finite set S of points in $\mathbb{R}^2$ is the smallest convex set containing these points and is denoted by $\mathrm{Conv}(S)$. S being finite, it is clearly a polygon in the plane whose vertices are elements of S. The *upper convex hull* of S, denoted by $\mathrm{Conv}^+(S)$, is the clockwise oriented sequence of consecutive edges of $\mathrm{Conv}(S)$ starting from the lowest vertex and ending on the highest vertex. The *lower convex hull* of S, denoted by $\mathrm{Conv}^-(S)$, is the clockwise oriented sequence of consecutive edges of $\mathrm{Conv}(S)$ starting from the highest vertex and ending on the lowest vertex.

3   Combinatorics on Words

Combinatorics on words has imposed itself as a powerful tool for the study of large number of discrete, linear, non-commutative objects. Such objects appears in almost any branches of mathematics and discrete geometry is not an exception. Traditionally, discrete geometry works on characterization and recognition of discrete objects using arithmetic approach or computational geometry. However combinatorics on words provide mathematical tools and efficient algorithms to address this problem as already mentioned. Lothaire's books [21,22,23] constitute the reference for presenting a unified view on combinatorics on words and many of its applications.

3.1   Lyndon Words

Introduced as *standard lexicographic sequences* by Lyndon in 1954, Lyndon words have several characterizations (see [21]). We shall define them as words being strictly smaller than any of their circular permutations.

Definition 1. *A Lyndon word $l \in \Sigma^+$ is a word such that $l = uv$ with $u, v \in \Sigma^+$ implies that $l < vu$.*

Note that Lyndon words are always primitive. An important result about Lyndon words is that any word w admits a factorization as a sequence of decreasing Lyndon words :

$$w = l_1^{n_1} l_2^{n_2} \cdots l_k^{n_k} \tag{1}$$

where $n_1, n_2, \ldots, n_k \geq 1$ and $l_1 > l_2 > \cdots > l_k$ are Lyndon words (see Lothaire [21] Theorem 5.1.1). Such a factorization is unique and a linear time algorithm to compute it is given in Section 5.

3.2   Christoffel Words

Introduced by Christoffel [25] in 1875 and reinvestigated recently by Borel and Laubie [26] who pointed out some of their geometrical properties, Christoffel words reveal an important link between combinatorics on words and discrete geometry.

This first definition of Christoffel word, borrowed from Berstel and de Luca [27], highlights their geometrical properties and helps to understand the main result of this work stated in Proposition 2. Let $\Sigma = \{0, 1\}$. The *slope* of a word is a map

$$\rho : \Sigma^* \to \mathbb{Q} \cup \{\infty\}$$

defined by

$$\rho(\epsilon) = 1, \quad \rho(w) = |w|_1 / |w|_0, \quad \text{for } w \neq \epsilon.$$

It is assumed that $1/0 = \infty$. It corresponds to the slope of the straight line joining the first and the last point of the path coded by w. For each k, $1 \leq k \leq |w|$, we define the set

$$\delta_k(w) = \{u \in \Sigma^k | \rho(u) \leq \rho(w)\},$$

of words of length k whose slope is not greater than the slope of w. The quantity

$$\mu_k(w) = \max\{\rho(u) | u \in \delta_k(w)\}$$

is used to define Christoffel words (see Figure 2).

Definition 2. *A word w is a Christoffel word if for any prefix v of w one has $\rho(v) = \mu_{|v|}(w)$.*

A direct consequence of this definition is that given a Christoffel word $u^r = v^s$ for some $r, s \geq 1$, both words u and v are also Christoffel words. From an arithmetical point of view, a Christoffel word is a connected subset of a standard line joining upper leaning points (see Reveillès [28]). The following properties of Christoffel words are taken from Borel and Laubie [26].

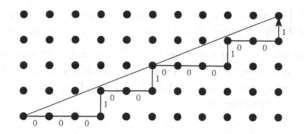

Fig. 2. The path coded by the Christoffel word $w = 00010010001001$ staying right under the straight line of slope $r = \frac{2}{5}$

Property 1. All primitive Christoffel words are Lyndon words.

Property 2. Given c_1 and c_2 two Christoffel words, $c_1 < c_2$ iff $\rho(c_1) < \rho(c_2)$.

Property 3. Given $r \in \mathbb{Q}^+ \cup \{\infty\}$, let F_r be the set of words w on the alphabet $\{0,1\}$ such that $\rho(v) \leq r$ for all non-empty prefix v of w. F_r correspond to the words coding paths, starting from the origin, that stay below the Euclidean straight line of slope r. Among these paths, those being the closest ones to the line and having their last point located on it are Christoffel words.

Originally Christoffel [25] defined these words as follows. Given $k < n$ two relatively prime numbers, a (primitive) Christoffel word $w = w_1 w_2 \ldots w_n$ is defined by :

$$w_i = \begin{cases} 0 & \text{if } r_{i-1} < r_i, \\ 1 & \text{if } r_{i-1} > r_i, \end{cases}$$

where r_i is the remainder of $(i\,k) \mod n$.

In [27] Berstel and de Luca provided an alternative characterization of primitive Christoffel words. Let CP be the set of primitive Christoffel words, PAL the set of palindromes and PER the set of words w having two periods p and q such that $|w| = p + q - 2$. The following relations hold :

$$CP = \left(\{0,1\} \cup 0 \cdot PER \cdot 1\right) \subset \left(\{0,1\} \cup 0 \cdot PAL \cdot 1\right).$$

These properties of Christoffel words are essential for deciding if a given word is Christoffel or not.

4   Digital Convexity

There are several (more or less) equivalent definitions of digital convexity, depending on whether or not one asks the digital set to be connected. We say that a finite 4-connected subset S of $\mathbb{Z}^2$ is *digitally convex* if it is the Gauss digitization of a convex subset X of the plane, i.e. $S = \text{Conv}(X) \cap \mathbb{Z}^2$.

The *border* $\text{Bd}(S)$ of S is the 4-connected path that follows clockwise the pixels of S that are 8-adjacent to some pixel not in S. This path is a word of $\{0, 1, \bar{0}, \bar{1}\}^*$, starting by convention from the lowest point and in clockwise order.

Definition 3. *A word w is said to be* digitally convex *if it is conjugate to the word coding the border of some finite 4-connected digitally convex subset of $\mathbb{Z}^2$.*

Note that implicitely, a digitally convex word is necessarily closed. Now, every closed path coding the boundary of a region is contained in a smallest rectangle such that its contour word w may be factorized as follows. Four extremal points are defined by their coordinates:

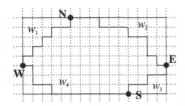

W is the lowest on the Left side;
N is the leftmost on the Top side;
E is the highest on the Right side;
S is the rightmost on the Bottom side;
So that $w \equiv w_1 w_2 w_3 w_4$.

This factorization is called the *standard decomposition*. We say that a word w_1 in $\{0, 1\}^*$ is *NW-convex* iff there are no integer points between the upper convex hull of the points $\{\phi_w(j)\}_{j=1...|w|}$ and the path w.

Define the counterclockwise $\pi/2$ circular rotation by

$$\sigma : (0, 1, \bar{0}, \bar{1}) \longmapsto (1, \bar{0}, \bar{1}, 0).$$

Then we have w_2 in $\{0, \bar{1}\}^*$ is *NE-convex* iff $\sigma(w_2)$ is NW-convex, and more

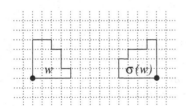

$$w = 1111001\bar{0}1\bar{1}1\bar{0}1\bar{0}000,$$
$$\sigma(w) = \bar{0}0\bar{0}\bar{0}110100101\bar{1}\bar{1}\bar{1}\bar{1}.$$

generally, in the factorization above

$$w_i \quad \text{is convex} \iff \sigma^{i-1}(w_i) \quad \text{is NW-convex.}$$

Clearly, the convexity of w requires the convexity of each w_i for $i = 1, 2, 3, 4$, and we have the following obvious property.

Proposition 1. *Let $w \equiv w_1 w_2 w_3 w_4$ be the standard decomposition of a polyomino. Then w is digitally convex iff $\sigma^{i-1}(w_i)$ is NW-convex, for all i.*

Let $Alph(w)$ be the set of letters of w. Observe that if for some i, w_i contains more than 2 letters, that is if $Alph(\sigma^{i-1}(w_i)) \not\subseteq \{0, 1\}$, then w is not digitally convex.

We are now in position to state the main result which is used in Section 5 to design an efficient algorithm for deciding if a word is convex.

Proposition 2. *A word v is NW-convex iff its unique Lyndon factorization $l_1^{n_1} l_2^{n_2} \cdots l_k^{n_k}$ is such that all l_i are primitive Christoffel words.*

In order to prove Proposition 2, we first need the following lemma.

Lemma 1. *Let $v \in \{0,1\}^*$ be a word coding an NW-convex path and let e be one of the edges of its convex hull. The factor u of v corresponding to the segment of the path determined by e is a Christoffel word.*

This is a direct consequence of Property 3 since both the starting and ending points of an edge of the convex hull of a discrete figure are necessarily part of its border. We may now proceed to the proof of Proposition 2.

Proof. Let v be a word coding an NW-convex path and let the ordered sequence of edges $(e_1, e_2, \ldots, e_k)$ be the border of its convex hull. For each i from 1 to k, let u_i be the factor of v determined by the edge e_i so that $v = u_1 u_2 \cdots u_k$. Let l_i be the unique primitive word such that $u_i = l_i^{n_i}$. By definition of NW-convexity and Lemma 1, u_i is a Christoffel word, which implies that l_i is a primitive Christoffel word. By Property 1, l_i is also a Lyndon word. Now, since $(e_1, e_2, \ldots, e_k)$ is the convex hull of w, it follows that the slope s_i of the edge e_i is greater than the slope s_{i+1} of the edge e_{i+1} leading to the following inequality :

$$\rho(l_i) = \rho(u_i) = s_i > s_{i+1} = \rho(u_{i+1}) = \rho(l_{i+1}).$$

By Property 2 we conclude that $l_i > l_{i+1}$. Thus $l_1^{n_1} l_2^{n_2} \cdots l_k^{n_k}$ is the unique factorization w as a decreasing sequence of Lyndon words.

Conversely, let $v \in \{0,1\}^+$ be such that its Lyndon factorization $l_1^{n_1} l_2^{n_2} \cdots l_k^{n_k}$ consists of primitive Christoffel words. For each i from 1 to k, let e_i be the segment joining the starting point of the path coded by $l_i^{n_i}$ to its ending point. We shall show that $(e_1, e_2, \ldots, e_k)$ is the upper convex hull of ϕ_v. Since $l_i^{n_i}$ is a Christoffel word, Property 3 ensures that no integer point is located between the path coded by $l_i^{n_i}$ and the segment e_i and, moreover, the path always stays below the segment. By hypothesis, $l_i > l_{i+1}$. Using the same argument as before we have that the slope of e_i is strictly greater than the slope of e_{i+1}.

We have just built a sequence of edges which is above the path ϕ_v, such that no integer points lies in-between, and with decreasing slopes. $(e_1, e_2, \ldots, e_k)$ is thus the upper convex hull of ϕ_v and v is NW-convex. $\qquad\square$

For example, consider the following NW-convex path $v = 1011010100010$.

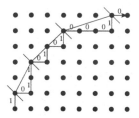

The Lyndon factorization of v is

$$v = (1)^1 \cdot (011)^1 \cdot (01)^2 \cdot (0001)^1 \cdot (0)^1,$$

where 0, 011, 01, 0001 and 0 are all Christoffel words.

5 Algorithm to Check Word Convexity

We give now a linear time algorithm checking digital NW-convexity for paths encoded on $\{0,1\}$. This is achieved in two steps: first we compute the prefix $l_1^{n_1}$ of the word w using the Fredricksen and Maiorana algorithm [29] (rediscovered by Duval [24]), and then Algorithm 2 below checks that the Lyndon factor $l_1 \in CP$. Iterating this process on all Lyndon factors of w provides the answer whether all l_i are primitive Christoffel words.

Given a word $w \in \Sigma^*$ whose Lyndon factorization is $w = l_1^{n_1} l_2^{n_2} \ldots l_k^{n_k}$, the following algorithm, taken from Lothaire's book [23], computes the pair (l_1, n_1).

Algorithm 1 (FirstLyndonFactor)
Input $w \in \Sigma^n$; **Output** (l_1, n_1)
1 : $(i,j) \leftarrow (1,2)$
2 :   **while** $j \leq n$ **and** $w_i \leq w_j$ **do**
3 :   **If** $w_i < w_j$ **then**
4 : $i \leftarrow 1$
5 :   **else**
6 : $i \leftarrow i+1$
7 :   **end if**
8 : $j \leftarrow j+1$
9 :   **end while**
10 :   **return** $(w_1 w_2 \cdots w_{j-i}, \lfloor (j-1)/(j-i) \rfloor)$

Clearly this algorithm is linear in $n_1 |l_1|$, and hence the Lyndon factorization of w is computed in linear time with respect to $|w|$. On the other hand, given a primitive word $w \in \{0,1\}^*$, checking whether it is a Christoffel word is also achieved in linear time using the definition from [25]: first, compute $k = |w|_1$ and $n = |w|$; then compute successively $r_1, r_2, \ldots, r_{\lceil n/2 \rceil}$ where $r_i = (i\,k) \bmod n$ and verify that w_i satisfies the definition. Note that since $CP \setminus \{0,1\} \subset 0PAL1$ the second half of w is checked at the same time by verifying that $w_i = w_{n-i+1}$ for $2 \leq i \leq \lceil n/2 \rceil$. This yields the following algorithm.

Algorithm 2 (IsChristoffelPrimitive)
Input $w \in \Sigma^n$
1 : $k \leftarrow |w|_1;\, i \leftarrow 1;\, r \leftarrow k$;
2 : $rejected := \mathbf{not}(w_1 = 0 \text{ and } w_n = 1)$
3 :   **while not**($rejected$) **and** $i < \lceil n/2 \rceil$ **do**
4 : $i \leftarrow i+1$; $r' \leftarrow r+k \bmod n$
5 :   **If** $r < r'$ **then**
6 : $rejected \leftarrow \mathbf{not}(0 = w_i \text{ and } 0 = w_{n-i-1})$
7 :   **else**
8 : $rejected \leftarrow \mathbf{not}(1 = w_i \text{ and } 1 = w_{n-i-1})$
9 :   **end if**
10 : $r \leftarrow r'$
11 :   **end while**
12 :   **return not**($rejected$)

Combining these two algorithms provides this following algorithm that checks NW-convexity of a given word $w \in \Sigma^*$.

Algorithm 3 (IsNW-Convex)
Input $w \in \Sigma^n$
  1 : *index* $\leftarrow 1$; *rejected* $\leftarrow$ *false*
  2 : **while not**(*rejected*) **and** *index* $\leq n$ **do**
  3 : $(l_1, n_1) \leftarrow$ **FirstLyndonFactor**($w_{index} w_{index+1} \cdots w_n$)
  4 :   *rejected* $\leftarrow$ **not**(**IsChristoffelPrimitive**(l_1))
  5 :   *index* $\leftarrow$ *index* $+ n_1 |l_1|$
  6 : **end while**
  7 : **return not**(*rejected*)

Equation (1) ensures that $\sum_i |l_i| \leq |w|$ so that this algorithm is linear in the length of the word w.

5.1 The Final Algorithm

According to Proposition 1, we have to check convexity for each term in the standard decomposition $w \equiv w_1 w_2 w_3 w_4$. Instead of applying the morphism σ to each w_i, which requires a linear pre-processing, it suffices to implement a more general version of Algorithm 1 and Algorithm 2, with the alphabet and its order relation as a parameter. For that purpose, ordered alphabets are denoted as lists $Alphabet = [\alpha, \beta]$ with $\alpha < \beta$.

The resulting algorithm is the following where we assume that w is the contour of a non-empty polyomino.

Algorithm 4 (IsConvex)
Input $w \in \Sigma^n$
  0 : *Compute the standard decomposition* $w \equiv w_1 w_2 w_3 w_4$;
  1 : *rejected* $\leftarrow$ *false*; $i \leftarrow 1$; *Alphabet* $\leftarrow [0, 1]$;
  2 : **while not**(*rejected*) **and** $i \leq 4$ **do**
  3 : $u \leftarrow w_i$; $k \leftarrow |u|$;
  4 :   **if** Alph(u) $\subseteq$ *Alphabet* **then**
  5 :   *index* $\leftarrow 1$;
  6 :   **while not**(*rejected*) **and** *index* $\leq k$ **do**
  7 : $(l_1, n_1) \leftarrow$ **FirstLyndonFactor**($[u_{index} u_{index+1} \cdots u_k]$, *Alphabet*);
  8 :   *rejected* $\leftarrow$ **not**(**IsChristoffelPrimitive**(l_1), *Alphabet*);
  9 :   *index* $\leftarrow$ *index* $+ n_1 |l_1|$;
 10 :   **end while**
 11 :   **else**
 12 :   *rejected* $\leftarrow$ *true*;
 13 :   **end if**
 14 : $i \leftarrow i + 1$; *Alphabet* $\leftarrow [\sigma^{i-1}(0), \sigma^{i-1}(1)]$;
 15 : **end while**
 16 : **return not**(*rejected*)

Remark. For more efficiency, testing that the letters of w_i belong to $\sigma^{i-1}(\{0,1\}^*)$ (Line 4) can be embedded within the algorithm **FirstLyndonFactor** or in the computation of the standard decomposition (Line 0) and returning an exception.

6   Concluding Remarks

The implementation of our algorithm was compared to an implementation of that of Debled-Rennesson et al. [8]. The results (see figure below) showed that our technique was 10 times faster than the technique of maximal segments.

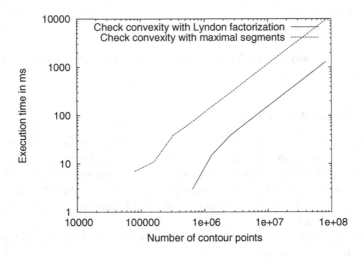

This speedup is partially due to the fact that computing maximal segments provides more geometrical informations while testing convexity is simpler. Nevertheless, our algorithm is much simpler conceptually and suggests that the notion of digital convexity might be a more fundamental concept than what is usually perceived. The fact that the combinatorial approach delivers such an elegant algorithm begs for a systematic study of the link between combinatorics on words and discrete geometry. In particular, there exist another characterization of Christoffel words that involve their palindromic structure.

Among the many problems that can be addressed with this new approach we mention the computation of the convex hull. It is also possible to improve algorithm **IsConvex** by merging some computations in one pass instead of calling independent routines. The resulting algorithm is more tricky, but providing a still faster implementation, and its description will appear in third author's PhD dissertation [30].

Acknowledgements. We are grateful to Christophe Reutenauer for his beautiful lectures on Sturmian words during the School on Combinatorics on words held in Montreal in march 2007. The fact that the Lyndon factorization of the discretization of a concave function contains only Christoffel words was first observed by C. Reutenauer in 2002. Our fruitful discussions led to a better understanding of Christoffel words and inspired the present work.

References

1. Sklansky, J.: Recognition of convex blobs. Pattern Recognition 2(1), 3–10 (1970)
2. Minsky, M., Papert, S.: Perceptrons, 2nd edn. MIT Press, Cambridge (1988)
3. Kim, C.: On the cellular convexity of complexes. Pattern Analysis and Machine Intelligence 3(6), 617–625 (1981)
4. Kim, C.: Digital convexity, straightness, and convex polygons. Pattern Analysis and Machine Intelligence 4(6), 618–626 (1982)
5. Kim, C., Rosenfeld, A.: Digital straight lines and convexity of digital regions. Pattern Analysis and Machine Intelligence 4(2), 149–153 (1982)
6. Chaudhuri, B., Rosenfeld, A.: On the computation of the digital convex hull and circular hull of a digital region. Pattern Recognition 31(12), 2007–2016 (1998)
7. Klette, R., Rosenfeld, A.: Digital straightness—a review. Discrete Appl. Math. 139(1-3), 197–230 (2004)
8. Debled-Rennesson, I., Rémy, J.-L., Rouyer-Degli, J.: Detection of the discrete convexity of polyominoes. Discrete Appl. Math. 125(1), 115–133 (2003)
9. Feschet, F., Tougne, L.: Optimal Time Computation of the Tangent of a Discrete Curve: Application to the Curvature. In: Bertrand, G., Couprie, M., Perroton, L. (eds.) DGCI 1999. LNCS, vol. 1568, pp. 31–40. Springer, Heidelberg (1999)
10. Lachaud, J.O., Vialard, A., de Vieilleville, F.: Fast, accurate and convergent tangent estimation on digital contours. Image and Vision Computing 25, 1572–1587 (2007)
11. Brlek, S., Labelle, G., Lacasse, A.: Incremental Algorithms Based on Discrete Green Theorem. In: Nyström, I., Sanniti di Baja, G., Svensson, S. (eds.) DGCI 2003. LNCS, vol. 2886, pp. 277–287. Springer, Heidelberg (2003)
12. Brlek, S., Labelle, G., Lacasse, A.: Algorithms for polyominoes based on the discrete Green theorem. Discrete Applied Math 147(3), 187–205 (2005)
13. Brlek, S., Labelle, G., Lacasse, A.: On minimal moment of inertia polyominoes. In: Coeurjolly, D., et al. (eds.) DGCI 2008. LNCS, vol. 4992, Springer, Heidelberg (2008)
14. Brlek, S., Labelle, G., Lacasse, A.: A Note on a Result of Daurat and Nivat. In: De Felice, C., Restivo, A. (eds.) DLT 2005. LNCS, vol. 3572, pp. 189–198. Springer, Heidelberg (2005)
15. Brlek, S., Labelle, G., Lacasse, A.: Properties of the contour path of discrete sets. Int. J. Found. Comput. Sci. 17(3), 543–556 (2006)
16. Daurat, A., Nivat, M.: Salient and reentrant points of discrete sets. In: del Lungo, A., di Gesu, V., Kuba, A. (eds.) Proc. IWCIA 2003, Int. Workshop on Combinatorial Image Analysis, Palermo, Italia, May 14–16. Electronic Notes in Discrete Mathematics, Elsevier Science, Amsterdam (2003)
17. Brlek, S., Labelle, G., Lacasse, A.: Shuffle operations on lattice paths. In: Rigo, M. (ed.) Proc. CANT 2006, Int. School and Conf. on Combinatorics, Automata and Number theory, Liège, Belgium, May 8–19, pp. 8–19. University of Liège (2006)
18. Brlek, S., Labelle, G., Lacasse, A.: Shuffle operations on discrete paths. In: Theoret. Comput. Sci. (in press, 2007)
19. Brlek, S., Provençal, X.: An Optimal Algorithm for Detecting Pseudo-squares. In: Kuba, A., Nyúl, L.G., Palágyi, K. (eds.) DGCI 2006. LNCS, vol. 4245, pp. 403–412. Springer, Heidelberg (2006)
20. Brlek, S., Fédou, J.-M., Provençal, X.: On the tiling by translation problem. Discrete Applied Math. (to appear, 2008)

21. Lothaire, M.: Combinatorics on words. In: Cambridge Mathematical Library, Cambridge University Press, Cambridge (1997)
22. Lothaire, M.: Algebraic combinatorics on words. Encyclopedia of Mathematics and its Applications, vol. 90. Cambridge University Press, Cambridge (2002)
23. Lothaire, M.: Applied combinatorics on words. Encyclopedia of Mathematics and its Applications, vol. 105. Cambridge University Press, Cambridge (2005)
24. Duval, J.P.: Factorizing words over an ordered alphabet. J. Algorithms 4(4), 363–381 (1983)
25. Christoffel, E.B.: Observatio arithmetica. Annali di Mathematica 6, 145–152 (1875)
26. Borel, J.P., Laubie, F.: Quelques mots sur la droite projective réelle. J. Théor. Nombres Bordeaux 5(1), 23–51 (1993)
27. Berstel, J., de Luca, A.: Sturmian words, Lyndon words and trees. Theoret. Comput. Sci. 178(1-2), 171–203 (1997)
28. Reveillès, J.P.: Géométrie discrète, calcul en nombres entiers et algorithmique. PhD thesis, Université Louis Pasteur, Strasbourg (December 1991)
29. Fredricksen, H., Maiorana, J.: Necklaces of beads in k colors and k-ary de Bruijn sequences. Discrete Math. 23(3), 207–210 (1978)
30. Provençal, X.: Combinatoire des mots, pavages et géométrie discrète. PhD thesis, Université du Québec à Montréal, Montréal (2008)

Decomposition and Construction of Neighbourhood Operations Using Linear Algebra

Atsushi Imiya, Yusuke Kameda, and Naoya Ohnishi

Institute of Media and Information Technology, Chiba University, Japan
School of Science and Technology, Chiba University, Japan

Abstract. In this paper, we introduce a method to express a local linear operated in the neighbourhood of each point in the discrete space as a matrix transform. To derive matrix expressions, we develop a decomposition and construction method of the neighbourhood operations using algebraic properties of the noncommutative matrix ring. This expression of the transforms in image analysis clarifies analytical properties, such as the norm of the transforms. We show that the symmetry kernels for the neighbourhood operations have the symmetry matrix expressions.

1 Introduction

Linear vs Nonlinear, Local vs Global, Shift-Invariant vs Shift-Variant are fundamental characteristics for classifying the mathematical properties of the operations for image processing. Geometric vs Algebraic and Combinatorial vs Analytical are fundamental methodologies for analysing and designing algorithms and operations in discrete geometry and digital image processing.

In this paper, using an algebraic method for the description of the neighbourhood operations, we introduce the norm of the local operations as a classification criterion for shift-invariant local linear operations for image processing [3]. To define the norm of the operations, we introduce a decomposition and construction method for the neighbourhood operations of digital image processing using algebraic properties of the noncommutative matrix ring. Using this decomposition and construction method, we develop a method to express the neighbourhood operations [1,2] in matrix forms [4,6,7]. The matrix expressions of the neighbourhood operations yield the operator norm [13] of the operations. This norm of the neighbourhood operations allows us to deal with the neighbourhood operations analytically, that is, we can define the spectra of the neighbourhood operations.

In signal processing and analysis, it is well known that a shift-invariant linear operation is expressed as a convolution kernel. Furthermore, a linear transform in a finite dimensional space is expressed as a matrix [4,13,12]. It is also possible to express a shift-invariant operation as a band-diagonal matrix [3,4,5]. However, this expression is not usually used in signal processing and analysis. In numerical computation of the partial differential equations, approximations of the partial differentiations in discrete operations are one of the central issues

D. Coeurjolly et al. (Eds.): DGCI 2008, LNCS 4992, pp. 69–80, 2008.

[10,11,13]. The discrete approximations of the partial differentiations are called the neighbourhood operation in digital signal and image processing. In mathematical morphology, a neighbourhood operation is expressed as a small binary matrix. This small matrix, which is called the structure element in binary mathematical morphology [2], expresses the local distribution of boolean value zero and one in a small region.

For the analysis and expression of digital image transformations from the viewpoint of functional analysis, we introduce a method to describe the neighbourhood operations in the matrix form. Two kinds of expressions for the neighbourhood operations, the tensor expression [15], which is called the matrix expression in mathematical morphology [2,3], and the convolution operation, which expresses the operation as a convolution between input and kernel, are well established methods in digital signal and image processing. The kernel of the convolution operation is expressed as a small matrix [3] in the context of digital image processing. We address the third expression of the operations in digital image processing. The matrix expression of the linear operation derives the mathematical definition of the norm of the linear operation [4,10]. In digital image analysis, geometrical properties of the results of the operation are the most important concerns, since the transformations are operated to extract geometrical features such as the boundary curves, medial axis, and corners. The norm of the linear operation allows us to analyse spectral properties of the operations. The spectrum of operations determine some analytical properties of the transformation in the space of functions.

2   Mathematical Preliminaries

We assume that the sampled image $f(i, j, k)$ exists in the $M \times M \times M$ grid region, that is, we express f_{ijk} as the value of $f(i, j, k)$ at the point $(i, j, k)^\top$ in three-dimensional discrete space $\mathbf{Z}^3$. We use the following notations to express shift-invariant local operation as three-dimensional discrete operations.

For the sequence $\boldsymbol{w} = (w_{(-1)}, w_{(o)}, w_{(+1)})$,

$$\boldsymbol{w} f_{ijk} = w_{(-1)} f_{i-1\,jk} + w_{(0)} f_{ijk} + w_{(+1)} f_{i+1\,jk}.$$

For the 3×3 matrix $\boldsymbol{W} = (\boldsymbol{w}_{(-1)}, \boldsymbol{w}_{(0)}, \boldsymbol{w}_{(+1)})$,

$$\boldsymbol{W} f_{ijk} = \boldsymbol{w}_{(-1)} f_{i\,j-1\,k} + \boldsymbol{w}_{(0)} f_{ijk} + \boldsymbol{w}_{(+1)} f_{i\,j+1\,k}.$$

For the $3 \times 3 \times 3$ tensor $\mathbf{W} = (\boldsymbol{W}_{(-1)}, \boldsymbol{W}_{(0)}, \boldsymbol{W}_{(+1)})$,

$$\mathbf{W} f_{ijk} = \boldsymbol{W}_{(-1)} f_{i\,j\,k-1} + \boldsymbol{W}_{(0)} f_{ijk} + \boldsymbol{W}_{(+1)} f_{i\,j\,k+1}.$$

where $\boldsymbol{W}_\pm$ and $\boldsymbol{W}_0$ are 3×3 matrices in the $3 \times 3 \times 3$ tensor $\mathbf{W}$. For the three-dimensional vector function $\boldsymbol{x}_{ijk} = \boldsymbol{x}(i, j, k) = (x_{ijk}, y_{ijk}, z_{ijk})^\top$ defined in $\mathbf{Z}^3$,

$$\mathbf{W}\boldsymbol{x} = (\mathbf{W} x_{ijk}, \mathbf{W} y_{ijk}, \mathbf{W} z_{ijk})^\top.$$

For the $3 \times 3 \times 3$ tensor $\mathbf{W}$, the $5 \times 5 \times 5$ tensor $\mathbf{W}^2$ is defined as

$$\mathbf{W}^2 = ((w_{ijk}^{(2)})) \quad w_{ijk}^{(2)} = \sum_{pqr} w_{i-p\,j-q,\,j-r} w_{pqr}.$$

where $((w_{ijk}))$ expresses the tensor whose ijk element is w_{ijk}. These vector, matrix, and tensor are called convolution kernels for one-, two-, and three- dimensional digital functions in digital signal and image processing. For these vectors, matrices, and tensors, we define symmetry.

Definition 1. *If $w_{-1} = w_{+1}$, $\boldsymbol{w}_{-1} = \boldsymbol{w}_{+1}$, and $\boldsymbol{W}_{-1} = \boldsymbol{W}_{+1}$, we call that vectors, matrices, and tensors are kernel symmetry.*

In this paper, we derive a method to describe the kernel symmetry local operations as matrices and define the norm of operations using matrix expressions.

Hereafter, we use $\mathbf{T}$ and $\boldsymbol{T}$ to express the tensor and the matrix, respectively, of the linear transform T to arrays in n-dimensional discrete space $\mathbf{Z}^n$ for $n \geq 3$.

We describe the fundamental numerical differentiations in tensor forms. In numerical differentiation [11,13], since

$$\frac{\partial}{\partial x} f(x, y, z) \cong \Delta_i f_{ijk} = f_{i+\frac{1}{2}\,j,k} - f_{i-\frac{1}{2}\,jk},$$

we have

$$\frac{\partial^2}{\partial x^2} f \cong \Delta_i^2 f_{ijk} = f_{i+1\,jk} - 2f_{ijk} + f_{i-1\,jk}.$$

If linear approximation is used to derive $f_{i+\frac{1}{2}\,jk}$ from f_{ijk} and $f_{i+1\,jk}$, we have

$$\Delta_i f_{ijk} = f_{i+1\,j,k} - f_{i-1\,jk}.$$

Furthermore, the gradient $\nabla f = (f_x, f_y, f_z)^\top$ and the Laplacian $\nabla^2 f$ for the numerical computation of partial differential equations are approximated as

$$\begin{pmatrix} f_x \\ f_y \\ f_z \end{pmatrix} \cong \begin{pmatrix} \Delta_i f_{ijk} \\ \Delta_j f_{ijk} \\ \Delta_k f_{ijk} \end{pmatrix} = \begin{pmatrix} (f_{i+1\,jk} - f_{ijk}) + \{-(f_{i-1\,jk} - f_{ijk})\} \\ (f_{ijk+1} - f_{ijk}) + \{-(f_{ijk-1} - f_{ijk})\} \\ (f_{ij+1\,k} - f_{ijk}) + \{-(f_{ij-1\,k} - f_{ijk})\} \end{pmatrix}$$

$$= \begin{pmatrix} f_{i+1\,jk} - f_{i-1\,jk} \\ f_{ij+1\,k} - f_{ij-1\,k} \\ f_{ijk+1} - f_{ijk-1} \end{pmatrix}$$

and

$$\nabla^2 f \cong \Delta_i^2 f_{ijk} + \Delta_j^2 f_{ijk} + \Delta_k^2 f_{ijk}$$
$$= (f_{i+1\,jk} - 2f_{ijk} + f_{i-1\,jk}) + (f_{ij+1\,k} - 2f_{ijk} + f_{ij-1\,k})$$
$$+ (f_{ijk+1} - 2f_{ijk} + f_{ijk-1})$$
$$= (f_{i+1\,jk} + f_{i-1\,jk} + f_{ij+1\,k} + f_{ij-1\,k} + f_{ijk+1} + f_{ijk-1}) - 6f_{ijk},$$

respectively. Therefore, the Laplacian for the numerical computation of the partial differential equation is six times the 6-connected Laplacian from the viewpoint of digital image analysis.

The tensors [15] $\mathbf{A}_6$, $\mathbf{A}_{12}$, and $\mathbf{A}_8$ for the computation of averages in the 6-, 18-, and 26- neighbourhoods in $\mathbf{Z}^3$ for volumetric image analysis are

$$\mathbf{A}_6 = \frac{1}{6} \left(\begin{pmatrix} 0\,0\,0 \\ 0\,1\,0 \\ 0\,0\,0 \end{pmatrix} \begin{pmatrix} 0\,1\,0 \\ 1\,0\,1 \\ 0\,1\,0 \end{pmatrix} \begin{pmatrix} 0\,0\,0 \\ 0\,1\,0 \\ 0\,0\,0 \end{pmatrix} \right)$$

$$\mathbf{A}_{12} = \frac{1}{12} \left(\begin{pmatrix} 0\,1\,0 \\ 1\,0\,1 \\ 0\,1\,0 \end{pmatrix} \begin{pmatrix} 1\,0\,1 \\ 0\,0\,0 \\ 1\,0\,1 \end{pmatrix} \begin{pmatrix} 0\,1\,0 \\ 1\,0\,1 \\ 0\,1\,0 \end{pmatrix} \right),$$

$$\mathbf{A}_8 = \frac{1}{8} \left(\begin{pmatrix} 1\,0\,1 \\ 0\,0\,0 \\ 1\,0\,1 \end{pmatrix} \begin{pmatrix} 0\,0\,0 \\ 0\,0\,0 \\ 0\,0\,0 \end{pmatrix} \begin{pmatrix} 1\,0\,1 \\ 0\,0\,0 \\ 1\,0\,1 \end{pmatrix} \right).$$

These tensors have the following properties,

$$\mathbf{A}_{18} = \frac{6}{18}\mathbf{A}_6 + \frac{12}{18}\mathbf{A}_{12}, \quad \mathbf{A}_{26} = \frac{6}{26}\mathbf{A}_6 + \frac{12}{26}\mathbf{A}_{12} + \frac{8}{26}\mathbf{A}_8.$$

Therefore, the tensors $\mathbf{L}_6$, $\mathbf{L}_{12}$, and $\mathbf{L}_8$ for the computation of Laplacians in 6-, 18-, and 26- neighbourhoods are

$$\mathbf{L}_6 = \mathbf{A}_6 - \mathbf{I}, \ \mathbf{L}_{18} = \mathbf{A}_{18} - \mathbf{I}, \ \mathbf{L}_{26} = \mathbf{A}_{26} - \mathbf{I}.$$

In the next sections, we express these operations as matrices and compute the spectral radii of these operations.

Setting $f_k(x_1, x_2, \cdots, x_n)$ to be the kth-order fundamental symmetry forms of $\{x_i\}_{i=1}^n$, that is,

$$f_k = \sum_{\text{all}} \{k\text{th-order mononimals of } \{x_i\}_{i=1}^n\},$$

for example,

$$f_1(x_1, x_2, \cdots, x_n) = x_1 + x_2 + \cdots + x_n,$$
$$f_2(x_1, x_2, \cdots, x_n) = x_1 x_2 + x_2 x_3 + \cdots + x_{n-1} x_n,$$
$$f_3(x_1, x_2, \cdots x_n) = x_1 x_2 x_3 + x_2 x_3 x_4 + \cdots + x_{n-2} x_{n-1} x_n,$$
$$\vdots$$
$$f_n(x_1, x_2, \cdots, x_n) = x_1 x_2 \cdots x_n,$$

the kth-order fundamental symmetry forms $\{f_k\}_{i=1}^n$ satisfies the relation

$$\Pi_{i=1}^n (1 + x_i) = 1 + \sum_{k=1}^n f_k(x_1, x_2, \cdots, x_n).$$

3 Operations in Matrix Form

3.1 Differential Operation in Matrix Form

From the matrix

$$D = \begin{pmatrix} -2 & 1 & 0 & 0 & \cdots & 0 & 0 \\ 1 & -2 & 1 & 0 & \cdots & 0 & 0 \\ 0 & 1 & 0 & 1 & \cdots & 0 & 0 \\ \vdots & \vdots & \vdots & \vdots & \ddots & \vdots & \vdots \\ 0 & 0 & 0 & \cdots & 0 & 1 & -2 \end{pmatrix}, \tag{1}$$

which computes the second-order derivative [10,11,13,15], we define the matrix B,

$$B = \frac{1}{2} \begin{pmatrix} 0 & 1 & 0 & 0 & \cdots & 0 & 0 \\ 1 & 0 & 1 & 0 & \cdots & 0 & 0 \\ 0 & 1 & 0 & 1 & \cdots & 0 & 0 \\ \vdots & \vdots & \vdots & \vdots & \ddots & \vdots & \vdots \\ 0 & 0 & 0 & \cdots & 0 & 1 & 0 \end{pmatrix} = \frac{1}{2}(D + 2I). \tag{2}$$

The matrix B computes the average of vectors. Setting $\rho(B)$ to be the spectrum of the matrix B, we have the next theorem.

Theorem 1. *The matrix B satisfies the relation $\rho(B) < 1$. For the proof, see the appendix.*

3.2 Matrix Ring over the Kronecker Product and Matrix Addition

Setting $A \otimes B$ to be the Kronecker product [14,15] of a pair of matrices, for the matrix A, we define the symbol $A^{\otimes n}$ as

$$A^{\otimes n} = A \otimes A \otimes \cdots \otimes A. \tag{3}$$

Over the collection of symmetry matrices, the Kronecker product $\otimes$ and the matrix sum $+$ define the noncommutative ring [8,9] $M(S, \otimes, +)$. Since the nth-order fundamental symmetry forms $\{f_k(A)\}_{k=1}^{n}$ of the symmetry matrix A on $M(S, \otimes, +)$ is defined as

$$f_k(A) = \sum_{\text{all}} \{\text{the Kronecker products of } k \ A \text{ and } (n - k) \ I\}, \tag{4}$$

the symmetry forms are generated by

$$(I + A)^{\otimes n} = I + \sum_{k=1}^{n} f_k(A). \tag{5}$$

Since, for $A \otimes A$, the eigenvalue is $\lambda_i \lambda_j$, we have the following relation.

Lemma 1. *for the $n \times n$ regular symmetry matrix A, setting $\{\lambda_k\}_{k=1}^m$ to be the eigenvalue of A,*

$$f_k(A)u = f_k(\lambda_1, \lambda_2, \cdots, \lambda_n)u. \tag{6}$$

Since the number of terms of $f_k(A)$ is $\binom{n}{k}$, Lemma 1 implies the next theorem.

Theorem 2. *The spectral radii of $f_k(A)$ satisfies the relation*

$$\rho(f_k(A)) \leq \binom{n}{k}\rho(A). \tag{7}$$

3.3   Averages in the Neighbourhood

Using $\{f_k(A)\}_{k=1}^n$, we define the operations for the computation of the average in the neighbourhood and the Laplacian in the various neighbourhoods in $\mathbf{Z}^n$.

Setting

$$N_{n(k)}(x) = \{x' \in \mathbf{Z}^n, |x - x'| = k\} \tag{8}$$

we have the relation

$$n(k) = |N_{n(k)}| = \binom{n}{k}2^k, \tag{9}$$

since the number of terms in $f_k(B)$ is $\binom{n}{k}$.

Therefore, setting

$$N_{n(k)} = f_k(B), \tag{10}$$

the matrix

$$A_{n(k)} = \binom{n}{k}^{-1}N_{n(k)} \tag{11}$$

computes the average of the points x' in $N_{n(k)}(x)$. Since $\rho(B) < 1$, eq. (7) implies the next lemma.

Lemma 2. *Since $\rho(f_k(N_{n(k)})) \leq \binom{n}{k}$, we have the relation $\rho(f_k(A_{n(k)})) \leq 1$.*

From the algebraic property of the fundamental symmetry forms, the matrices $\{N_{n(k)}\}_{k=1}^n$ are generated from $(I + B)^{\otimes n}$ as

$$(I + B)^{\otimes n} = I + \sum_{k=1}^n N_{n(k)}. \tag{12}$$

We define the matrix of the Laplacian with x and $x' \in N_{d(k)}$ as the relation

$$L_{d(k)} = f_k(D). \tag{13}$$

In particular, $L_{n(1)}$ is the discrete Laplacian with $2n$ connectivity in $\mathbf{Z}^n$, that is,

$$L_{n(1)} = D \otimes I \otimes \cdots \otimes I + I \otimes D \otimes \cdots \otimes I + \cdots + I \otimes I \otimes \cdots \otimes D. \tag{14}$$

Similarly to eq. (12), we have the relation

$$(I + D)^{\otimes n} = I + \sum_{k=1}^n L_{d(k)}. \tag{15}$$

3.4 Construction of Laplacian Matrices

We define the rth Laplacian in $\mathbf{Z}^n$ in the matrix form as

$$L_{n(k)} = A_{n(k)} - I \tag{16}$$

As the generalisation of the Horn-Schunck Laplacian [16] in $\mathbf{Z}^n$, we have

$$L_{(m)} = \sum_{i=k}^{m} \alpha_k A_{n(k)} - I, \ \ 1 \le m \le n \tag{17}$$

for

$$\alpha_1 : \alpha_2 : \cdots : \alpha_n = n : (n-1) : \cdots : 1.$$

3.5 $(m^n - 1)$ Neighbourhood Operation

We define the matrix $\mathbf{B}^{(k)} = ((b_{ij}))$ as

$$b_{ij} = \begin{cases} \frac{1}{2}\delta_{|i-j|0}, \ i \ne j, \\ 1 \qquad \text{otherwise.} \end{cases} \tag{18}$$

Then, we have the relations $\mathbf{B}^{(1)} = \mathbf{B}$ and $\mathbf{B}^{(0)} = I$.

Setting

$$N_+^m = \left\{ x = (i(1), i(2), \cdots, i(n))^\top \,|\, 0 \le i(1), i(2), \cdots i(n) \le \frac{m+1}{2} \right\}, \tag{19}$$

we define the matrix

$$\mathbf{B}^{(i(1)i(2)\cdots i(n))} = \mathbf{B}^{(i(1))} \otimes \mathbf{B}^{(i(2))} \otimes \cdots \otimes \mathbf{B}^{(i(n))}. \tag{20}$$

Matrix $\mathbf{B}^{(i(1)i(2)\cdots i(n))}$ computes the average of the points $p = (|i(1))|, |i(2)|, \cdots, |i(n)|)^\top$ in $\mathbf{Z}^n$. Therefore, the matrix

$$N_n = \sum_{(i(1)i(2)\cdots i(n)) \in N_+^n} \gamma_{(i(1)i(2)\cdots i(n))} \mathbf{B}^{(i(1)i(2)\cdots i(n))} - \gamma_{00\cdots 0} I \tag{21}$$

for

$$\sum_{(i(1)i(2)\cdots i(n)) \in N_+^n} \gamma_{(i(1)i(2)\cdots i(n))} = 1, \ 0 < \gamma_{(i(1)i(2)\cdots i(n))} < 1 \tag{22}$$

computes the weighted average of f_p, $p \in \mathbf{Z}^n$ for the points in the $(m^n - 1)$ neighbourhood of each point. For $\rho(\mathbf{B}^{(i(1)i(2)\cdots i(n))}) \le 1$, we have the property that $\rho(\mathbf{N}_n) \le 1$.

4 Operations in $\mathbf{Z}^3$

In this section, we construct the Laplacian and average operation in $\mathbf{Z}^3$ using the results in the previous section.

6-Neighbourhood Operations. Using the matrix B, we can construct the operation to compute the average of digital functions in the 6-neighbourhood.

Theorem 3. *The matrix*

$$A_6 = \frac{1}{3}(B \otimes I \otimes I + I \otimes B \otimes I + I \otimes I \otimes B) \tag{23}$$

computes the average of f_{ijk} in the 6-neighbourhood of the point $(i, j, k)^\top$.

The matrix of the 6-neighbourhood Laplacian is

$$L_6 = \frac{1}{6}(D \otimes I \otimes I + I \otimes D \otimes I + I \otimes I \otimes D). \tag{24}$$

18-Neighbourhood Operations. The simple Laplacian in $\mathbf{Z}^2$ is

$$L_8 = L_4 + \frac{1}{4}(D \otimes D). \tag{25}$$

Equation (25) shows that the 8-neighbourhood Laplacian on $\mathbf{Z}^2$ involves the fourth-order differentiation for the points $(i - 1, j - 1)^\top$, $(i + 1, j + 1)^\top$, $(i + 1, j - 1)^\top$, and $(i + 1, j + 1)^\top$. Therefore, from the viewpoint of the numerical computation, it is desired to use the Laplacian in the form

$$L_8 = \alpha_1 L_4 + \alpha_2 \frac{1}{4}(D \otimes D). \tag{26}$$

such that $\alpha_1 \gg \alpha_2 > 0$.

Equation (17) derives the next relation for $n = 2$,

$$L_{8HS} = \frac{2}{3}L_4 + \frac{1}{3}\left\{\frac{1}{4}(D \otimes D)\right\}. \tag{27}$$

This matrix is expressed in the 3×3 tensor as

$$\mathbf{L}_{8HS} = \frac{2}{3}\begin{pmatrix} 0 & 1 & 0 \\ 1 & -4 & 1 \\ 0 & 1 & 0 \end{pmatrix} + \frac{1}{3}\begin{pmatrix} 1 & 0 & 1 \\ 0 & -4 & 0 \\ 1 & 0 & 1 \end{pmatrix} = \frac{1}{12}\begin{pmatrix} 1 & 2 & 1 \\ 2 & -12 & 2 \\ 1 & 2 & 1 \end{pmatrix}. \tag{28}$$

In this section, we clarify the three-dimensional analogue of the operation of eq. (27).

For the matrix

$$A_{12} = \frac{1}{3}(B \otimes B \otimes I + B \otimes B \otimes I + B \otimes I \otimes B), \tag{29}$$

we have the next theorem.

Theorem 4. *The matrix B_{18}, such that*

$$B_{18} = \frac{6}{18}B_6 + \frac{12}{18}B_{12}, \tag{30}$$

computes the average in the 18-neighbourhood of each point.

From this theorem, the Laplacian in the 18-neighbourhood is given as

$$
\begin{aligned}
\boldsymbol{L}_{18} &= \boldsymbol{B}_{18} - \boldsymbol{I} \\
&= \frac{6}{18}\boldsymbol{B}_6 + \frac{12}{18}\boldsymbol{B}_{12} \\
&= \frac{6}{18}(\boldsymbol{B}_6 - \boldsymbol{I}) + \frac{12}{18}(\boldsymbol{B}_{12} - \boldsymbol{I}) \\
&= \frac{1}{3}\boldsymbol{L}_6 + \frac{1}{3}\{(\boldsymbol{B} \otimes \boldsymbol{B} \otimes \boldsymbol{I} + \boldsymbol{I} \otimes \boldsymbol{B} \otimes \boldsymbol{B} + \boldsymbol{B} \otimes \boldsymbol{I} \otimes \boldsymbol{B} - \boldsymbol{I}\} \\
&= \frac{1}{3}\boldsymbol{L}_6 + \frac{2}{3}\{\boldsymbol{D} \otimes \boldsymbol{D} \otimes \boldsymbol{I} + \boldsymbol{I} \otimes \boldsymbol{D} \otimes \boldsymbol{D} + \boldsymbol{D} \otimes \boldsymbol{I} \otimes \boldsymbol{D}\}.
\end{aligned}
\tag{31}
$$

The second term of the right-hand side of eq. (31) is the fourth-order numerical differentiation, since $\boldsymbol{D} \otimes \boldsymbol{D}$ corresponds to the fourth time derivatives. This algebraic property implies the next assertion.

Assertion 1. *The 18-neighbourhood Laplacian* $\mathbf{L}$ *is a sum of the 6-neighbourhood Laplacian and a fourth-order symmetry differentiation.*

The three-dimensional version of the Horn-Schunck Laplacian is expressed in the $3 \times 3 \times 3$ tensor form as

$$
\mathbf{L}_{HS} = \left(\frac{1}{36}\begin{pmatrix} 0 & 1 & 0 \\ 1 & 4 & 1 \\ 0 & 1 & 0 \end{pmatrix} \quad \frac{1}{36}\begin{pmatrix} 1 & 4 & 1 \\ 4 & -36 & 4 \\ 1 & 4 & 1 \end{pmatrix} \quad \frac{1}{36}\begin{pmatrix} 0 & 1 & 0 \\ 1 & 4 & 1 \\ 0 & 1 & 0 \end{pmatrix} \right),
\tag{32}
$$

where the elements express the weights for the computation of the Laplacian of the centre of the tensor, that is, $\mathbf{L}$ is the weighted 18-neighbourhood operation. This operation is decomposed into two tensors as

$$
\begin{aligned}
\mathbf{L}_{HS} &= \frac{2}{3}\left(\begin{pmatrix} 0 & 0 & 0 \\ 0 & 1 & 0 \\ 0 & 0 & 0 \end{pmatrix} \quad \begin{pmatrix} 0 & 1 & 0 \\ 1 & -6 & 1 \\ 0 & 1 & 0 \end{pmatrix} \quad \begin{pmatrix} 0 & 0 & 0 \\ 0 & 1 & 0 \\ 0 & 0 & 0 \end{pmatrix} \right) \\
&\quad + \frac{1}{3}\left(\begin{pmatrix} 0 & 1 & 0 \\ 1 & 0 & 1 \\ 0 & 1 & 0 \end{pmatrix} \quad \begin{pmatrix} 1 & 0 & 1 \\ 0 & -12 & 0 \\ 1 & 0 & 1 \end{pmatrix} \quad \begin{pmatrix} 0 & 1 & 0 \\ 1 & 0 & 1 \\ 0 & 1 & 0 \end{pmatrix} \right).
\end{aligned}
\tag{33}
$$

The first term and the second term of the right-hand side of eq.(33) are the 6-neighbourhood Laplacian and the 12-neighbourhood Laplacian, respectively, in three-dimensional discrete space. Therefore, the three-dimensional Horn-Schunck Laplacian is in the form

$$
\begin{aligned}
L_{HS} &= \frac{2}{3}L_6 + \frac{1}{3}L_{18} \\
&= \frac{2}{3}(\boldsymbol{D} \otimes \boldsymbol{I} \otimes \boldsymbol{I} + \boldsymbol{I} \otimes \boldsymbol{D} \otimes \boldsymbol{I} + \boldsymbol{I} \otimes \boldsymbol{I} \otimes \boldsymbol{D}) \\
&\quad + \frac{1}{3}(\boldsymbol{D} \otimes \boldsymbol{D} \otimes \boldsymbol{I} + \boldsymbol{D} \otimes \boldsymbol{I} \otimes \boldsymbol{D} + \boldsymbol{I} \otimes \boldsymbol{D} \otimes \boldsymbol{D}).
\end{aligned}
\tag{34}
$$

26-Neighbourhood Operations. In digital image analysis, the operations in the 26-neighbourhood are used, since the operations in the 26-neighbourhood derive the smoother results than the 6- and 18- neighbourhood operations. In this section, we examine the norm properties of the operations in the 26-neighbourhood. For the matrices A_6, A_{12}, and

$$A_8 = B \otimes B \otimes B, \tag{35}$$

we have the next theorem.

Theorem 5. *The matrix B_{26}, such that*

$$B_{26} = \frac{6}{26}A_6 + \frac{12}{26}A_{12} + \frac{8}{26}A_8 \tag{36}$$

computes the averages in the 26-neighbourhood of each point.

From this theorem, we can have the Laplacian in the 18-neighbourhood as

$$
\begin{aligned}
L_{26} &= B_{26} - I \\
&= \frac{6}{26}L_6 + \frac{12}{26}L_{12} + \frac{8}{26}L_8
\end{aligned}
\tag{37}
$$

As the 26-neighbourhood Horn-Schunck Laplacian is given as

$$L_{26HS} = \frac{3}{6}L_6 + \frac{2}{6}L_{12} + \frac{1}{6}L_8 = \frac{1}{2}L_6 + \frac{1}{3}L_{12} + \frac{1}{6}L_8. \tag{38}$$

4.1   Discrete Gaussian Kernel

By embedding signal sequences as

$$f = (f_1, f_2, \cdots, f_n)^\top \rightarrow (0, 0, \cdots, f_1, f_2, \cdots, f_n, 0, \cdots, 0)^\top$$

with an appropriate number of zeros, a discrete version of the heat equation

$$\frac{\partial}{\partial t}f(x, t) = \frac{1}{2}\frac{\partial^2}{\partial x^2}f(x, t) \tag{39}$$

is given as

$$f^{(t+1)} - f^{(t)} = \frac{1}{4}Df^{(t)}, \ f^{(0)} = f. \tag{40}$$

From this equation, we have the relation

$$f^{(t+1)} = Gf^{(t)} = \left(I + \frac{1}{4}D\right)f^{(t)} = \frac{1}{2}(I + B)f^{(t)} = G^t f \tag{41}$$

where

$$G = \frac{1}{4}\begin{pmatrix} 2 & 1 & 0 & \cdots & 0 \\ 1 & 2 & 1 & \cdots & 0 \\ \vdots & 1 & 2 & \cdots & 0 \\ \vdots & \vdots & \vdots & \ddots & \vdots \\ 0 & \cdots & \cdots & 1 & 2 \end{pmatrix}. \tag{42}$$

The nD kernel is expressed as

$$G_n^t = (G^{\otimes n})^t = (G \otimes G \otimes \cdots \otimes G)^t = (G^t \otimes G^t \otimes \cdots \otimes G^t) \qquad (43)$$

Since the 1D Gaussian convolution sequence is $g = \left(\frac{1}{4}, \frac{1}{2}, \frac{1}{4}\right)^\top$, the 3D kernel in the tensor form is G_3^t for $G = (G_{-1}, G_0, G_1)$ where

$$G_{-1} = \frac{1}{4}\begin{pmatrix} \frac{1}{16} & \frac{1}{8} & \frac{1}{16} \\ \frac{1}{8} & \frac{1}{4} & \frac{1}{8} \\ \frac{1}{16} & \frac{1}{8} & \frac{1}{16} \end{pmatrix}, \quad G_0 = \frac{1}{2}\begin{pmatrix} \frac{1}{16} & \frac{1}{8} & \frac{1}{16} \\ \frac{1}{8} & \frac{1}{4} & \frac{1}{8} \\ \frac{1}{16} & \frac{1}{8} & \frac{1}{16} \end{pmatrix}, \quad G_{+1} = \frac{1}{4}\begin{pmatrix} \frac{1}{16} & \frac{1}{8} & \frac{1}{16} \\ \frac{1}{8} & \frac{1}{4} & \frac{1}{8} \\ \frac{1}{16} & \frac{1}{8} & \frac{1}{16} \end{pmatrix}. \qquad (44)$$

5 Conjectures for the Neighbourhood in $\mathbf{Z}^n$

In four-dimensional discrete space $\mathbf{Z}^4$, elemental operations, which is derived as the fundamental symmetry forms of B on $M(S, \otimes, +)$, for the computation of the averages in the neighbourhoods are

$$A_8 = \frac{1}{4}(B \otimes I \otimes I \otimes I + I \otimes B \otimes I \otimes I$$
$$+ I \otimes I \otimes B \otimes I + I \otimes I \otimes I \otimes B), \qquad (45)$$
$$A_{24} = \frac{1}{6}(B \otimes B \otimes I \otimes I + I \otimes B \otimes B \otimes I + I \otimes I \otimes B \otimes B$$
$$+ B \otimes I \otimes bmI \otimes B + B \otimes I \otimes B \otimes I + I \otimes B \otimes I \otimes B), \qquad (46)$$
$$A_{32} = \frac{1}{4}(B \otimes B \otimes B \otimes I + I \otimes B \otimes B \otimes B$$
$$+ B \otimes I \otimes B \otimes B + I \otimes B \otimes I \otimes B), \qquad (47)$$
$$A_{16} = B \otimes B \otimes B \otimes B. \qquad (48)$$

In three-dimensional discrete space, A_{12} and $\frac{6}{14}A_6 + \frac{8}{14}A_8$ correspond to the averages in the FCC neighbourhood and the BCC neighbourhood, respectively. From these properties of the operations in the noncubic neighbourhoods, we have the following open problem and conjecture.

Open Problem 1. *Using the fundamental symmetry forms of B on $M(S, \otimes, +)$, characterise the noncubic grid systems in $\mathbf{Z}^n$.*

In $\mathbf{Z}^4$, A_{12} and $\langle A_6, A_8 \rangle$ define the neighbourhood point configurations for FCC and BCC grids [2], respectively. From these geometric properties of the noncubic grids embedded in $\mathbf{Z}^4$, we have the next conjecture for the noncubic grids.

Conjecture 1. *The pairs $\langle A_{n(k)}, A_{n(n-k)} \rangle$ for $k = 1, 2, \cdots, (n-1)$ derive the n-dimensional versions of FCC and BCC grids.*

6 Conclusions

We introduced a decomposition and construction method for the neighbourhood operations of the digital image processing. First, we have derived a method to express the symmetry neighbourhood operations as matrices. The matrix expressions of the neighbourhood operations derived the operator norm of the operations. This norm of the neighbourhood operations allowed us to deal with the neighbourhood operations analytically, that is, we defined the spectra of the neighbourhood operations.

References

1. Klette, R., Rosenfeld, A.: Digital Geometry: Geometric Method for Digital Picture Analysis. Morgan Kaufmann, San Francisco (2004)
2. Serra, J.: Mathematical Morphology. Academic Press, London (1982)
3. Bracewell, R.N.: Two Dimensional Imaging. Prentice-Hall, Englewood Cliffs (1995)
4. Bellman, R.: Introduction to Matrix Analysis, 2nd edn. SIAM, Philadelphia (1987)
5. Strang, G., Nguyen, T.: Wavelets and Filter Banks. Wellesley-Cambridge (1997)
6. Fuhrmann, P.A.: Polynomial Approach to Linear Algebra. Springer, Heidelberg (1996)
7. Mignotte, M., Stefanescu, D.: Polynomials: An Algorithmic Approach. Springer, Heidelberg (1999)
8. Clausen, M., Baum, J.: Fast Fourier Transform. Bibliographishes Institut & F.A, Brockhaus (1993)
9. Terras, A.: Fourier Analysis on Finite Groups and Applications, Cambridge (1999)
10. Demmel, J.W.: Applied Numerical Linear Algebra. SIAM, Philadelphia (1997)
11. Grossmann, C., Roos, H.-G.: Numerik partieller Differentialgleichungen, Teubner (1994)
12. Grossmann, C., Terno, J.: Numerik der Optimierung, Teubner (1997)
13. Varga, R.S.: Matrix Iteration Analysis, 2nd edn. Springer, Heidelberg (2000)
14. Graham, A.: Kronecker Products and matrix Calculus with Applications. Ellis Horwood (1981)
15. Ruíz-Tolosa, J.R., Castillo, E.: From Vectors to Tensors. Springer, Heidelberg (2005)
16. Horn, B.K.P., Schunck, B.G.: Determining optical flow. Artificial Intelligence 17, 185–204 (1981)

Appendix

Spectrums of Matrices

For the matrix D, setting $DU = \Lambda U$, where U is an orthogonal matrix, and $\Lambda = Diag\,(\lambda_M, \lambda_{M-1}, \cdots, \lambda_1)$, the eigenvalues are $\lambda_k = -2\left(1 - \cos\frac{\pi}{M+1}k,\right)$. Since $B = \frac{1}{2}(D+2I)$, we have the eigenvalues of B as $\mu_k = \cos\frac{\pi}{M+1}k$. Therefore, $\rho(B) < 1$. Using the same treatment on the fundamental symmetrical polynomial of μ_i, we can have the relation $\rho(A_{n(r)}) < 1$. Furthermore, since $\rho(B^{n(r)}) \leq \max_i \sum_j |b_{ij}^{n(r)}| = 1$, we have $\rho(B^{(i)}) \leq 1$.

Digitally Continuous Multivalued Functions

Carmen Escribano, Antonio Giraldo,* and María Asunción Sastre

Departamento de Matemática Aplicada, Facultad de Informática
Universidad Politécnica, Campus de Montegancedo
Boadilla del Monte, 28660 Madrid, Spain

Abstract. We introduce in this paper a notion of continuity in digital spaces which extends the usual notion of digital continuity. Our approach uses multivalued maps. We show how the multivalued approach provides a better framework to define topological notions, like retractions, in a far more realistic way than by using just single-valued digitally continuous functions. In particular, we characterize the deletion of simple points, one of the most important processing operations in digital topology, as a particular kind of retraction.

Keywords: Digital space, continuous function, simple point, retraction.

Introduction

The notion of continuous function is the fundamental concept in the study of topological spaces. Therefore it should play also an important role in Digital Topology.

There have been some attempts to define a reasonable notion of continuous function in digital spaces. The first one goes back to A. Rosenfeld [13] in 1986. He defined continuous function in a similar way as it is done for continuous maps in $\mathbb{R}^n$. It turned out that continuous functions agreed with functions taking 4-adjacent points into 4-adjacent points. He proved, amongst other results, that a function between digital spaces is continuous if and only it takes connected sets into connected sets.

More results related with this type of continuity were proved by L. Boxer in [1] and, more recently in [2,3,4]. In these papers, he introduces such notions as homeomorphism, retracts and homotopies for digitally continuous functions, applying these notions to define a digital fundamental group, digital homotopies and to compute the fundamental group of sphere-like digital images. However, as he recognizes in [3], there are some limitations with the homotopy equivalences he get. For example, while all simple closed curves are homeomorphic and hence homotopically equivalent with respect to the Euclidean topology, in the digital case two simple closed curves can be homotopically equivalent only if they have the same cardinality.

A different approach was suggested by V. Kovalevsky in [11], using multivalued maps. This seems reasonable, since an expansion as $f(x) = 2x$ must take 1

---

* Partially supported by DGES.

D. Coeurjolly et al. (Eds.): DGCI 2008, LNCS 4992, pp. 81–92, 2008.
© Springer-Verlag Berlin Heidelberg 2008

pixel to 2 pixels if the image of an interval has still to be connected. He calls a multivalued function continuous if the pre-image of an open set is open. He considers, however, that another important class of multivalued functions is what he calls "connectivity preserving mappings". By its proper definition, the image of a point by a connectivity preserving mapping is a connected set. This is not required for merely continuous functions. He finally asserts that the substitutes for continuous functions in finite spaces are the simple connectivity preserving maps, where a connectivity preserving map f is simple if for any x such that $f(x)$ has more than 1 element then $f^{-1}f(x) = \{x\}$. However, in this case it would be possible to map the center of a 3×3 square to the boundary of it leaving the points of the boundary fixed, obtaining in this way a "continuous" retraction from the square to its boundary, something impossible in the continuous realm. The number of admissible function is reduced considering restrictions, as in its notion of n-isomorphism, upon the diameter of the image of a point, although, if we want $f(x) = nx$ to be continuous, these restrictions should be dependent of the function. Nevertheless, this kind of restrictions might be useful, since allowing bigger images we gradually increase the class of allowable functions.

In [14], A. Rosenfeld and A. Nakamura introduce the notion of a local deformation of digital curves that, as they mention, "can be regarded as digitally continuous in the sense that it takes neighboring pixels into neighboring pixels, but it is not exactly the same as a digitally continuous mappings". This is mainly due to the fact that one point in a curve can be related to several points in the other.

The multivalued approach to continuity in digital spaces has also been used by R. Tsaur and M. Smyth in [15], where a notion of continuous multifunction for discrete spaces is introduced: A multifunction is continuous if and only if it is "strong" in the sense of taking neighbors into neighbors with respect to Hausdorff metric. They use this approach to prove some results concerning the existence of fixed points for multifunctions. However, although this approach allows more flexibility in the digitization of continuous functions defined in continuous spaces, it is still a bit restrictive, as shown by the fact that the multivalued function used by them to illustrate the convenience of using multivalued functions is not a strong continuous multifunction.

In this paper, we present a theory of continuity in digital spaces which extends the one introduced by Rosenfeld. In particular, most of the results in [2,3,4] are still valid in our context.

In section 1 we revise the basic notions on digital topology required throughout the paper. In particular we recall the different adjacency relations used to model digital spaces. In section 2 we introduce the notion of subdivision of a topological space. This notion is next used to define continuity for multivalued functions and to prove some basic properties concerning the behavior of digitally continuous multivalued functions under restriction and composition. In section 3 we show that the deletion of simple points can be completely characterized in terms of digitally continuous multivalued functions. In particular, a point is simple if and only if certain multivalued function is continuous. Section 4 is devoted to the definition and properties of multivalued retractions. We show that the behavior of multivalued retractions in the digital plane is completely analogous to the behavior of retractions in the continuous (real) plane, in contrast with what happens with the existing notions of digital continuity. In the last section we translate to the digital space the well known Hahn-Mazurkiewicz theorem which characterizes locally connected continua as continuous images of the interval.

For information on Digital Topology we recommend the survey [9] and the books by Kong and Rosenfeld [10], and by Klette and Rosenfeld [8].

We are grateful to the referees for their suggestions and remarks which have helped to improve the final version of this paper.

1   Digital Spaces

We consider $\mathbb{Z}^n$ as model for digital spaces.

Two points in the digital line $\mathbb{Z}$ are adjacent if they are different but their coordinates differ in at most a unit. Two points in the digital plane $\mathbb{Z}^2$ are 8-adjacent if they are different and their coordinates differ in at most a unit. They are said 4-adjacent if they are 8-adjacent and differ in at most a coordinate. Two points of the digital 3-space $\mathbb{Z}^3$ are 26-adjacent if they are different and their coordinates differ in at most a unit. They are said 18-adjacent if they are 26-adjacent and differ in at most two coordinates, and they are said 6-adjacent if they are 26-adjacent and differ in at most a coordinate. In an analogous way, adjacency relations are defined in $\mathbb{Z}^n$ for $n \geq 4$, for example, in $\mathbb{Z}^4$ there exist 4 different adjacency relations: 80-adjacency, 64-adjacency, 32-adjacency and 8-adjacency.

Given $p \in \mathbb{Z}^2$ we define $\mathcal{N}(p)$ as the set of points 8-adjacent to p, i.e. $\mathcal{N}(p) = \{p_1, p_2, \ldots, p_8\}$. This is also denoted as $\mathcal{N}_8(p)$. Analogously, $\mathcal{N}_4(p)$ is the set of points 4-adjacent to p (with the above notation $\mathcal{N}_4(p) = \{p_2, p_4, p_6, p_8\}$).

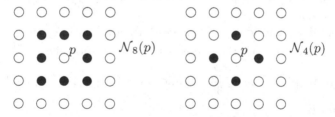

Given $p \in \mathbb{Z}^3$, $\mathcal{N}(p)$ will be the set of points 26-adjacent to p. In this case there will exist three kinds of neighborhood: $\mathcal{N}_{26}(p)$, $\mathcal{N}_{18}(p)$ and $\mathcal{N}_6(p)$.

A k-path P in $\mathbb{Z}^n$ (where k corresponds to any of the possible adjacency relations in $\mathbb{Z}^n$) is a sequence $P = \{p_0, p_1, p_2, \ldots, p_r\}$ of points such that p_i is k-adjacent to p_{i+1}, for every $i \in \{0, 1, 2, \ldots, r-1\}$. It is said then that P is a k-path from p_0 to p_r. If $p_0 = p_r$ then it is called a closed path.

A k-path $\{p_0, p_1, p_2, \ldots, p_r\}$ is called a k-arc if the only k-adjacent points are consecutive points and, possibly, the end points (i.e., if $0 \leq i < i+1 < j < r$ or $0 < i < i+1 < j \leq r$ then p_i is not k-adjacent to p_j). Every k-path P contains a k-arc with the same end points.

A set $S \subset \mathbb{Z}^n$ is k-connected if for every pair of points of S there exists a k-path contained in S joining them. A k-connected component of S is a k-connected maximal set.

2   Continuous Multivalued Functions

Definition 1. Let $f : X \subset \mathbb{Z}^m \longrightarrow \mathbb{Z}^n$ be a function between digital spaces with adjacency relations k and k'. According to [13], f is (k, k')-continuous if f sends k-adjacent points to k'-adjacent points. When $m = n$ and $k = k'$, f is said to be just k-continuous.

We will say that $f : X \subset \mathbb{Z}^m \longrightarrow \mathbb{Z}^n$ is continuous if it is (k, k')-continuous for some k and k'.

In the following definition we introduce the concept of subdivision of $\mathbb{Z}^n$.

Definition 2. The first subdivision of $\mathbb{Z}^n$ is formed by the set

$$\mathbb{Z}_1^n = \left\{ \left(\frac{z_1}{3}, \frac{z_2}{3}, \ldots, \frac{z_n}{3} \right) \mid (z_1, z_2, \ldots, z_n) \in \mathbb{Z}^n \right\}$$

and the $3^n : 1$ map $i : \mathbb{Z}_1^n \hookrightarrow \mathbb{Z}^n$ given by $i \left(\frac{z_1}{3}, \frac{z_2}{3}, \ldots, \frac{z_n}{3} \right) = (z_1', z_2', \ldots, z_n')$ where $(z_1', z_2', \ldots, z_n')$ is the point in $\mathbb{Z}^n$ closer to $\left(\frac{z_1}{3}, \frac{z_2}{3}, \ldots, \frac{z_n}{3} \right)$.

The r-th subdivision of $\mathbb{Z}^n$ is formed by the set

$$\mathbb{Z}_r^n = \left\{ \left(\frac{z_1}{3^r}, \frac{z_2}{3^r}, \ldots, \frac{z_n}{3^r} \right) \mid (z_1, z_2, \ldots, z_n) \in \mathbb{Z}^n \right\}$$

and the $3^{nr} : 1$ map $i_r : \mathbb{Z}_r^n \hookrightarrow \mathbb{Z}^n$ given by $i_r \left(\frac{z_1}{3^r}, \frac{z_2}{3^r}, \ldots, \frac{z_n}{3^r} \right) = (z_1', z_2', \ldots, z_n')$ where $(z_1', z_2', \ldots, z_n')$ is the point in $\mathbb{Z}^n$ closer to $\left(\frac{z_1}{3^r}, \frac{z_2}{3^r}, \ldots, \frac{z_n}{3^r} \right)$. Observe that $i_r = i \circ i \circ \cdots \circ i$.

Moreover, if we consider in $\mathbb{Z}^n$ a k-adjacency relation, we can consider in $\mathbb{Z}_r^n$, in an immediate way, the same adjacency relation, i.e., $\left(\frac{z_1}{3^r}, \frac{z_2}{3^r}, \ldots, \frac{z_n}{3^r} \right)$ is k-adjacent to $\left(\frac{z_1'}{3^r}, \frac{z_2'}{3^r}, \ldots, \frac{z_n'}{3^r} \right)$ if and only if $(z_1, z_2, \ldots, z_n)$ is k-adjacent to $(z_1', z_2', \ldots, z_n')$.

Proposition 1. *i_r is k-continuous as a function between digital spaces.*

Definition 3. Given $X \subset \mathbb{Z}^n$, the r-th subdivision of X is the set $X_r = i_r^{-1}(X)$.

Intuitively, if we consider X made of pixels (resp. voxels), the r-th subdivision of X consists in replacing each pixel with 9^r pixels (resp. 27^r voxels) and the map i_r is the inclusion.

Remark 1. Given $X, Y \subset \mathbb{Z}^n$, any function $f : X_r \longrightarrow Y$ induces in an immediate way a multivalued function $F : X \longrightarrow Y$ where $F(x) = \bigcup_{x' \in i_r^{-1}(x)} f(x')$.

Definition 4. Consider $X, Y \subset \mathbb{Z}^n$. A multivalued function $F : X \longrightarrow Y$ is said to be a (k, k')-continuous multivalued function if it is induced by a (k, k')-continuous (single-valued) function from X_r to Y for some $r \in \mathbb{N}$.

Remark 2. Let $F : X \longrightarrow Y$ $(X, Y \subset \mathbb{Z}^n)$ be a (k, k')-continuous multivalued function. Then

i) $F(x)$ is k'-connected, for every $x \in X$,
ii) if x and y are k-adjacent points of X, then $F(x)$ and $F(y)$ are k'-adjacent subsets of Y.
iii) F takes k-connected sets to k'-connected sets.

Note that (iii) implies, and is implied by, (i) and (ii).

On the other hand, not all multivalued functions satisfying (i), (ii) and (iii) are induced by continuous single-valued functions. For example, the following multivalued function $F : \mathcal{N}(p) \cup \{p\} \longrightarrow \mathcal{N}(p)$

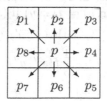

given by
$$F(x) = \begin{cases} \mathcal{N}(p) & \text{if } x = p \\ \{x\} & \text{if } x \in \mathcal{N}(p) \end{cases}$$

Remark 3. It is immediate that any digitally continuous function in the sense of Rosenfeld is also continuous as a multivalued function. In particular, any single-valued constant map is continuous as a multivalued map.

On the other hand, "strong" continuous multivalued functions in [15] satisfy (ii) above, although the image of a point does not need to be connected. Hence strong multivalued continuity does not imply our notion of continuity. There is even strong continuous multivalued function with the images of all points connected, like the example in the previous remark, which are not continuous as defined in this paper.

Conversely, the function in Example 1 in Section 4 is a continuous multivalued function (as defined here) which is not a strong continuous multivalued function.

The following result is easy to prove.

Proposition 2. *If $F : X \longrightarrow Y$ $(X, Y \subset \mathbb{Z}^n)$ is a (k, k')-continuous multivalued function and $X' \subset X$ then $F|_{X'} : X' \longrightarrow Y$ is a (k, k')-continuous multivalued function.*

If $F : X \longrightarrow Y$ and $G : Y \longrightarrow Z$ are continuous multivalued functions, then we can consider the composition $GF : X \longrightarrow Z$. However, it is not straightforward to prove that GF is continuous. We need first the following result.

Lemma 1. *Let $f : X \longrightarrow Y$ ($X \subset \mathbb{Z}^m$, $Y \subset \mathbb{Z}^n$) be a (k, k')-continuous function. Then, for every $r \in \mathbb{N}$, f induces a (k, k')-continuous function $f_r : X_r \longrightarrow Y_r$ such that $f_r(i_r^{-1}(x)) = i_r^{-1}(f(x))$ for every $x \in X$. This is equivalent to the following diagram*

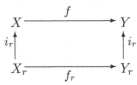

being commutative.

To be precise, we should have denoted by $(i_r)_X$ and $(i_r)_Y$ the maps $(i_r)_X : X_r \longrightarrow X$ and $(i_r)_Y : Y_r \longrightarrow Y$, respectively. In order not to complicate the notation we have denoted both maps by i_r.

Proof. It is simply an interpolation. For example, when $m = n = 2$ and $r = 1$, then we have just to define $f_1(x, y) = f(x, y)$ if $(x, y) \in \mathbb{Z}^2$, while if $(x, y) \notin \mathbb{Z}^2$ then $(x, y) = \frac{2}{3}(x', y') + \frac{1}{3}(x'', y'')$ with $(x', y'), (x'', y'') \in \mathbb{Z}^2$ adjacent points, and we define $f_1(x, y) = \frac{2}{3}f(x', y') + \frac{1}{3}f(x'', y'')$. $\square$

Theorem 1. *If $F : X \longrightarrow Y$ is a (k, k')-continuous multivalued function and $G : Y \longrightarrow Z$ is a (k', k'')-continuous multivalued functions, then $GF : X \longrightarrow Z$ is a (k, k'')-continuous multivalued function.*

Proof. Suppose that F is induced by $f : X_r \longrightarrow Y$ and G is induced by $g : Y_s \longrightarrow Z$ (observe that gf is not in general well defined).

Consider X_{r+s} the $(r + s)$-subdivision of X which is also the s-subdivision of X_r. Then f induces, according to the above lemma, a (k, k')-continuous function $f_s : X_{r+s} \longrightarrow Y_s$. Then $gf_s : X_{r+s} \longrightarrow Z$ is a (k, k'')-continuous function which induces GF. $\square$

3 Continuous Multivalued Functions and Simple Points

It may seem that the family of continuous multivalued functions could be too wide, therefore not having good properties. In this section we show that this is not the case. We show, in particular, that the existence of a k-continuous multivalued function from a set X to $X \setminus \{p\}$ which leaves invariant $X \setminus \{p\}$ is closely related to p being a k-simple point of X.

Let $X \subset \mathbb{Z}^2$ and $p \in X$. The point p is called k-simple ($k = 4, 8$) in X (see [9]) if the number of k-connected components of $\mathcal{N}(p) \cap X$ which are k-adjacent to p is equal to 1 and $\mathcal{N}_{\bar{k}}(p) \cap X^c \neq \emptyset$ (this last condition is equivalent to p being a boundary point of X). Here $\bar{k} = 4$ if $k = 8$ and $\bar{k} = 8$ if $k = 4$.

Theorem 2. *Let $X \subset \mathbb{Z}^2$ and $p \in X$. Suppose there exists a k-continuous multivalued function $F : X \longrightarrow X \setminus \{p\}$ such that $F(x) = \{x\}$ if $x \neq p$ and $F(p) \subset \mathcal{N}(p)$. Then p is a k-simple point.*

The converse is true if and only if p is not 8-interior to X (note that if p is 8-simple it can not be 8-interior, however there are 4-simple points which are 8-interior).

Proof. Suppose that F is induced by $f_r : X_r \longrightarrow X$. Then $f_r(x) = i_r(x)$ for every $x \in X_r$ such that $i_r(x) \neq p$.

Suppose that p is not k-simple. We have two possibilities: p is a boundary point with at least two different k-connected components of $\mathcal{N}(p) \cap X$ which are k-adjacent to p, or p is an interior point.

In the first case, let A and B be any two such components. Consider any $x_r \in i_r^{-1}(p)$ k-adjacent to $i_r^{-1}(A)$. Then $x = f_r(x_r)$ must be k-adjacent to A (since $F(A) = A$), and since A is a k-connected component of $\mathcal{N}(p) \cap X$, then $f_r(x_r) \in A$. On the other hand, there exists also $y_r \in i_r^{-1}(p)$ k-adjacent to $i_r^{-1}(B)$ and, hence, $f_r(y_r) = y \in B$. Consider $\{z_0 = x_r, z_1, z_2 \ldots, z_{m-1}, z_m = y_r\} \subset i_r^{-1}(p)$ such that z_i is k-adjacent to z_{i-1} for every $i = 1, 2, \ldots, m$. Then $\{f_r(z_0) = x, f_r(z_1), f_r(z_2), \ldots, f_r(z_{m-1}), f_r(z_m) = y\} \subset \mathcal{N}(p) \cap X$ is a k-path in $\mathcal{N}(p) \cap X$ from x to y. Contradiction.

Suppose now that p is an interior point and that there exists a k-continuous multivalued function $F : X \longrightarrow X \setminus \{p\}$ such that $F(p_i) = \{p_i\}$ for every $i = 1, 2, \ldots, 8$. Consider a subdivision X_r of X and a k-continuous map $f_r : X_r \longrightarrow X$ which induces F.

We divide $i_r^{-1}(p)$ into concentric paths (the first path would be its boundary, the next path would be the boundary of the interior, and so on). Then, by continuity, for the outer path, there are points in it whose images are p_2, p_4, p_6 and p_8. Therefore $F(p) \subset \{p_2, p_4, p_6, p_8\}$. If we consider now the next concentric paths of $i_r^{-1}(p)$, by the continuity of f_r, in any of them there are points whose images are p_2, p_4, p_6 and p_8. In particular, this will hold for the innermost path, making impossible to define f_r for the point in the center in a consistent way.

To prove the converse statement, consider first the following situation, with $p \in X$ a k-simple point such that $\mathcal{N}(p) \cap X = \{p_1, p_3, p_4, p_5, p_6, p_7, p_8\}$.

p_1		p_3
p_8	p	p_4
p_7	p_6	p_5

We are going to construct F as in the statement of the theorem. To do that we consider the second subdivision X_2 of X and we are going to construct a single-valued function $f : X_2 \longrightarrow X$ which induces the desired F. Since $F(x) = \{x\}$ for every $x \in \mathcal{N}(p) \cap X$, then $f(x') = x$ for every $x' \in i_2^{-1}(x)$. In order to define $f(p')$ for $p' \in i_2^{-1}(p)$ we divide $i_2^{-1}(p)$ in groups and define f as shown by the arrows in the following figure.

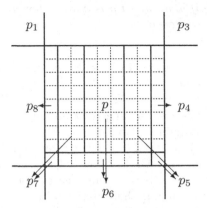

Then $f : X_2 \longrightarrow X$ is 4-continuous and 8-continuous. Therefore, if we define $F(p) = \{p_4, p_5, p_6, p_7, p_8\}$, F is a 8-continuous and 4-continuous multivalued function.

If p is any other simple point in the hypothesis of the theorem then there exists a point in $\mathcal{N}_4(p)$ which is not in X but one (or both) of the (clockwise) next points in $\mathcal{N}(p)$ are in X. We may suppose, making, if necessary, a rotation of X (by 90, 180 or 270 degrees) that p_2 is that point.

The same groups in $i_2^{-1}(p)$ can be used to define $F(p)$ in any of these cases. To show it, we label these groups as follows

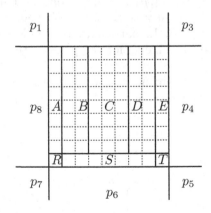

There are two possibilities: either $\mathcal{N}(p) \cap X = \{p_3\}$, in which case we define $F(p) = \{p_3\}$, or $p_4 \in X$. In the second case we have again two possibilities: either $p_6 \notin X$, in which case $\mathcal{N}(p) \cap X \subset \{p_3, p_4, p_5\}$ and we define $F(p) = \{p_4\}$, or $\{p_4, p_6\} \subset X$. If $\{p_4, p_6\} \subset X$ but $p_8 \notin X$, then $\mathcal{N}(p) \cap X \subset \{p_3, p_4, p_5, p_6, p_7\}$. If $p_5 \notin X$ we define $F(E \cup T) = p_4$ and $F(A \cup B \cup C \cup D \cup R \cup S) = p_6$, while if $p_5 \in X$ we define $F(E) = p_4$, $F(D \cup T) = p_5$ and $F(A \cup B \cup C \cup R \cup S) = p_6$.

Finally, if $\{p_4, p_6, p_8\} \subset X$, we consider two cases. If $\{p_5, p_7\} \subset X$, we define $F(E) = p_4$, $F(D \cup T) = p_5$, $F(C \cup S) = p_6$, $F(B \cup R) = p_7$ and $F(A) = p_8$. On the other hand, if $\{p_5, p_7\} \not\subset X$, then p can only be 8-simple (not 4-simple), and we can define $F(D \cup E \cup T) = p_4$, $F(C \cup S) = p_6$, $F(A \cup B \cup R) = p_8$.

We see finally that if p is 4-simple and 8-interior, then there is not a 4-continuous multivalued function $F : X \longrightarrow X \setminus \{p\}$ such that $F(x) = \{x\}$ if $x \neq p$ and $F(p) \subset \mathcal{N}(p)$. To see this, observe first that $\mathcal{N}(p) \cap X$ must be as follows (or a rotation of it)

p_1	p_2	
p_8	p	p_4
p_7	p_6	p_5

Suppose there exists a 4-continuous multivalued function $F : X \longrightarrow X \setminus \{p\}$ such that $F(x) = \{x\}$ if $x \neq p$ and $F(p) \subset \mathcal{N}(p) \cap X$. Suppose that F is induced by $f_r : X_r \longrightarrow X$. Then $f_r(x) = p_2$ for every $x \in X_r$ such that $i_r(x) = p_2$ and $f_r(x) = p_4$ for every $x \in X_r$ such that $i_r(x) = p_4$. Then if we consider the upper rightmost point $x \in i_r^{-1}(p)$, it is not possible to define $f_r(x)$ in such a way that f_r is 4-continuous.   □

Remark 4. F is not unique. For example, for $\mathcal{N}(p) \cap X$ as follows,

p_1		p_3
p_8	p	p_4
p_7	p_6	p_5

if we define $F(p) = \{p_1, p_3, p_4, p_5, p_6, p_7, p_8\}$, then F is still a 8-continuous and 4-continuous multivalued function. The same is true if we define $F(p) = \{p_3, p_4, p_5, p_6, p_7, p_8\}$. However, if we define $F(p) = \{p_5, p_6, p_7, p_8\}$, then F is nor a 8-continuous neither a 4-continuous multivalued function, since (ii) in Remark 2 does not hold in this case (p_3 and p are adjacent points of X, but $F(p_3)$ and $F(p)$ are not adjacent subsets of X).

Remark 5. It is easy to see that for $\mathcal{N}(p) \cap X$ as in Remark 4, although p is 4-simple and 8-simple, any single-valued function $f : X \longrightarrow X \setminus \{p\}$, such that $f(x) = x$ if $x \neq p$, can not be 4-continuous neither 8-continuous, hence Theorema 2 does not hold for Rosenfeld's digitally continuous functions.

Remark 6. There exist in the literature results characterizing simple points in terms of properties of certain inclusion maps, in the spirit of our theorem. For example, in [12] (see [5,6] for further and more recent results) simple surfels (the equivalent for a digital surface X of simple points in the digital plane) are characterized as points x such that the morphism $i_* : \Pi_1^n(X \setminus x) \longrightarrow \Pi_1^n(X)$ is an isomorphism.

4 Multivalued Digital Retractions

In the previous section we have characterized the deletion of a k-simple point in $\mathbb{Z}^2$ as a k-continuous multivalued function $f : X \longrightarrow X \setminus \{p\}$ such that $f(x) = x$ if $x \neq p$. This is a particular case of a wider class of functions known as multivalued retractions.

Definition 5. Consider $Y \subset X \subset \mathbb{Z}^n$. A multivalued k-retraction from X to Y is a k-continuous multivalued function $F : X \longrightarrow Y$ such that $f(y) = \{y\}$ for every $y \in Y$.

Remark 7. If we look at the notion of simple connectivity preserving mapping [11], it is clear that if a connectivity preserving mapping from X to a subset Y has to leave Y fixed, then if it is simple it must be single-valued. Therefore, in that case, retractions would agree with those in [1].

The next two results are the digital versions of two well known facts about $\mathbb{R}^2$, namely, that the boundary of a disk is not a retract of the whole disk, while an annulus (or of a punctured disk) can be retracted to its outer boundary.

Proposition 3. The boundary ∂X of a square X is not a k-retract of X $(k = 4, 8)$.

Proof. Let p_N, p_E, p_S, p_W be points in each of the four sides of ∂X, different from the corner points. Suppose that there exists a k-continuous multivalued function $F : X \longrightarrow \partial X$ such that $F(p) = \{p\}$ for every $p \in \partial X$. Consider a subdivision X_r of X and a k-continuous map $f_r : X_r \longrightarrow X$ which induces F.

Divide $i_r^{-1}(X)$ into concentric paths. Then, by continuity, for the outer path in $X \setminus \partial X$, there are points in it whose images by f_r are, respectively p_N, p_E, p_S and p_W. If we successively consider the different concentric paths of $i_r^{-1}(p)$, by the continuity of f_r, in any of them there are points whose images are p_N, p_E, p_S and p_W. In particular, this will hold for the innermost path, making it impossible to define f_r for the point at the center in a consistent way. □

Example 1. Let X be a squared annulus and consider $F : X \longrightarrow X$ defined as follows

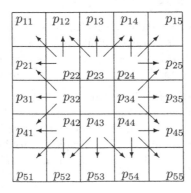

i.e. F is the identity on the outer boundary of the annulus, and for every p in the inner boundary, $F(p)$ is equal to three points in the outer boundary.

Then F is a multivalued k-retraction for $k = 4$ and $k = 8$.

Proof. F is induced by the following map $f_1 : X_1 \longrightarrow X$:

p_{11}	p_{12}	p_{13}	p_{14}	p_{15}
p_{21}				p_{25}
p_{31}				p_{35}
p_{41}				p_{45}
p_{51}	p_{52}	p_{53}	p_{54}	p_{55}

where the rest of the images are completed symmetrically. □

This result improves [1, Theorem 4.4]. There, it was proved that the boundary of a filled square is not a retract of the whole square, but the same arguments in the proof of that theorem shows that neither is the boundary of a squared annulus a digital retract (according to Boxer's definition) of it.

5   Digital Version of Hahn-Mazurkiewicz Theorem

By the well-known Hahn-Mazurkiewicz theorem (see, for example, [7]), any locally connected continuum (compact and connected set) is the continuum image of the interval $[0, 1]$ and, conversely, any continuum image of the interval $[0, 1]$ is a locally connected continuum. We end the paper with a digital version of this result.

Theorem 3 (Digital version of Hahn-Mazurkiewicz theorem). *Let $X \subset \mathbb{Z}^n$. Consider $I = \{0, 1\}$ the digital unit interval. Then, there exists a surjective $(2, k)$-continuous multivalued function $F : I \longrightarrow X$ if and only X is a bounded k-connected set (where k correspond to any of the possible adjacency relations in $\mathbb{Z}^n$).*

Proof. Suppose there exists a surjective $(2, k)$-continuous multivalued function $F : I \longrightarrow X$. Since any subdivision of I is a finite 2-connected set, then $X = F(I)$ must be finite and hence bounded and also k-connected (by Remark 2).

Suppose, on the other hand, that X is a bounded k-connected set. Then, since it is also finite, there exists a k-path $P = \{p_0, p_1, p_2, \ldots, p_n\}$ which goes through all the points in X (the path may go more than once through each point). Consider I_r subdivision of I such that $2 \cdot 3^r \geq n + 1$. Then there exists a k-continuous function $f : I_r \longrightarrow P$ which induces a k-continuous multivalued function $F : I \longrightarrow X$. □

Remark 8. In the general version of Hahn-Mazurkiewicz theorem, X is required to be compact and locally connected. These conditions do not appear in the digital version because all bounded digital sets satisfy these two properties.

Remark 9. The digital version of Hahn-Mazurkiewicz theorem can be stated in a more surprising way as follows: X is a bounded k-connected set if and only $F : \{p\} \longrightarrow X$ given by $F(p) = X$ is a k-continuous multivalued function. This is a consequence of the existence of a k-continuous multivalued function $F : \{p\} \longrightarrow I$ defined in an obvious way.

As a consequence, since a single-valued constant map is continuous (as a multivalued function), then, for every digital sets X and Y (Y k-connected and bounded), the map $F : X \longrightarrow Y$, given by $F(p) = Y$ for every $p \in X$, is a surjective k-continuous multivalued function.

References

1. Boxer, L.: Digitally continuous functions. Pattern Recognition Letters 15, 833–839 (1994)
2. Boxer, L.: A Classical Construction for the Digital Fundamental Group. Journal of Mathematical Imaging and Vision 10, 51–62 (1999)
3. Boxer, L.: Properties of Digital Homotopy. Journal of Mathematical Imaging and Vision 22, 19–26 (2005)
4. Boxer, L.: Homotopy properties of Sphere-Like Digital Images. Journal of Mathematical Imaging and Vision 24, 167–175 (2006)
5. Burguet, J., Malgouyres, R.: Strong thinning and polyhedric approximation of the surface of a voxel object. Discrete Applied Mathematics 125, 93–114 (2003)
6. Fourey, S., Malgouyres, R.: Intersection number and topology preservation within digital surfaces. Theoretical Computer Science 283, 109–150 (2002)
7. Hocking, J.G., Young, G.S.: Topology. Addison-Wesley, Reading (1961)
8. Klette, R., Rosenfeld, A.: Digital Geometry. Elsevier, Amsterdam (2004)
9. Kong, T.Y., Rosenfeld, A.: Digital Topology: Introduction and survey. Computer Vision, Graphics and Image Processing 48, 357–393 (1989)
10. Kong, T.Y., Rosenfeld, A. (eds.): Topological algorithms for digital image processing. Elsevier, Amsterdam (1996)
11. Kovalevsky, V.: A new concept for digital geometry. In: Ying-Lie, O., et al. (eds.) Shape in Picture. Proc. of the NATO Advanced Research Workshop, Driebergen, The Netherlands (1992), Computer and Systems Sciences, vol. 126. Springer-Verlag (1994)
12. Malgouyres, R., Lenoir, A.: Topology preservation within digital surfaces. Comput. Graphics Image Process. Mach. Graphics Vision 7, 417–426 (1998)
13. Rosenfeld, A.: Continuous functions in digital pictures. Pattern Recognition Letters 4, 177–184 (1986)
14. Rosenfeld, A., Nakamurab, A.: Local deformations of digital curves. Pattern Recognition Letters 18, 613–620 (1997)
15. Tsaur, R., Smyth, M.B.: Continuous multifunctions in discrete spaces with applications to fixed point theory. In: Bertrand, G., Imiya, A., Klette, R. (eds.) Digital and Image Geometry, Dagstuhl Seminar 2000. LNCS, vol. 2243, pp. 75–88. Springer, Heidelberg (2002)

Continued Fractions and Digital Lines with Irrational Slopes

Hanna Uscka-Wehlou

Uppsala University, Department of Mathematics
Box 480, SE-751 06 Uppsala, Sweden
hania@wehlou.com
http://wehlou.com/hania/index.htm

Abstract. This paper expands on previous work on relationships between digital lines and continued fractions (CF). The main result is a parsimonious description of the construction of the digital line based only on the elements of the CF representing its slope and containing only simple integer computations. The description reflects the hierarchy of digitization runs, which raises the possibility of dividing digital lines into equivalence classes depending on the CF expansions of their slopes. Our work is confined to irrational slopes since, to our knowledge, there exists no such description for these, in contrast to rational slopes which have been extensively examined. The description is exact and does not use approximations by rationals. Examples of lines with irrational slopes and with very simple digitization patterns are presented. These include both slopes with periodic and non-periodic CF expansions, i.e. both quadratic surds and other irrationals.

Keywords: digital geometry, theory of digital lines, irrational slope, continued fraction, quadratic surd.

1 Introduction

1.1 About This Paper

The aim of the present paper is to solve the following problem: given the continued fraction (CF) expansion of the slope $a \in \,]0, 1[\setminus \mathbb{Q}$ of a straight line, how is the digitization of this line constructed? The description uses only the elements of the CF and is exact, i.e. does not use the commonly applied approximations by rationals. The method is based on simple integer computations that can be easily applied to computer programming.

This description forms the Main Result (Theorem 4; *description by CFs*). The theoretical basis for this article is Uscka-Wehlou (2007) [15]. The main result there is recalled in Sect. 2 of the present paper (Theorem 1; *description by the digitization parameters*). It gives a description of digitization runs on all digitization levels for lines $y = ax$ where $a \in \,]0, 1[\setminus \mathbb{Q}$, which is based on digitization parameters defined in Def. 1 and the function Reg_a defined in Def. 2.

D. Coeurjolly et al. (Eds.): DGCI 2008, LNCS 4992, pp. 93–104, 2008.

Although Theorem 1 looks similar to Theorem 4, the former involves computations on irrational numbers, which is not the case in the latter.

The idea of the new description was to replace the heavy computations involved in the description by digitization parameters by simple computations on integers. In order to do that, the digitization parameters and the function Reg_a for each $a \in\ ]0,1[\ \setminus \mathbb{Q}$ were expressed by the elements of the CF expansion of a. The key role in this transform is played by the *index jump function* (Def. 3).

The main work leading to the successful translation of Theorem 1 into the CF description (Theorem 4) has been done in Theorems 2 and 3. The first one expresses the digitization parameters in terms of CFs and the second one does the same with the function Reg_a. These results allowed us to replace the computationally challenging conditions and formulae for run lengths by equivalent conditions and formulae based on the elements of the CF expansion of a.

In general, it is hard to perform arithmetical operations on CFs; see Khinchin 1997 (p. 20 in [7]). However, Def. 1 and Theorem 1 involve only the operations which form an exception to this rule. These operations are: finding the inverse to a CF, finding the integer (fractional) part of the inverse to a CF, and subtracting a CF from 1. The formula for the last operation is described in Lemmas 1 and 2, the others are clearly easy to perform. This made it possible to find the simple description formulated in Theorem 4.

The computations on irrationals did not disappear during the translation of Theorem 1 into the CF version (Theorem 4). They were moved into the process of finding the CF expansion of the slope. For some slopes we are able to compute the CF expansions exactly, using mathematical methods; some examples will be shown in Sect. 4, for both algebraic and transcendental numbers. For other slopes we can use algorithms for finding CF expansions.

The present CF description of digital lines is similar to the formula of Markov and Venkov ([16], p. 67), but since their method was not meant for descriptions of digital lines, it does not reflect the hierarchical structure of digitization runs on all levels, which our method does. This permits the grouping of digital lines into classes according to properties defined by the elements of the CF expansions of the slopes (to be presented in a forthcoming paper by the author).

The new method presented here is computationally simple, involving only easy computations with integers, excepting the algorithm for determining the CF expansion of the slope. The method applies to irrational slopes and gives the exact results instead of approximations by rationals. To the author's knowledge, there are no previous descriptions of digital lines with irrational slopes fulfilling all the criteria just mentioned, and reflecting the hierarchy of runs.

1.2   Earlier Developments

The use of CFs in modelling digital lines was discussed by R. Brons [3] as early as in 1974. Already then it was clear that the patterns generated in the digitization process of straight lines were related to the CF expansions of the slopes. However, the algorithm provided by Brons is only valid for rational slopes.

Some other researchers describing the construction of digital lines with rational slopes in terms of CFs were J. P. Reveillès (1991) [11], K. Voss (1993: 153–157) [18] - the *splitting formula*, I. Debled (1995: 59–66) [5] - description by the Stern-Brocot tree, P. D. Stephenson (1998) [13] - an algorithmic solution, F. de Vieilleville and J.-O. Lachaud (2006) [17] - a combinatoric approach. See also the review of R. Klette and A. Rosenfeld from 2004 [8].

Irrational numbers have been less central in research on digital line construction, possibly because irrational slopes must appear not to have direct applications for computer graphics. Nevertheless, irrational numbers may play an important role in our understanding of computer graphics theory which has a basis in digital geometry.

A CF description of digital lines was presented by L. Dorst and R. P. W. Duin (1984) [6]. Although their solution can be applied to irrational slopes, it is formulated as an algorithm. Since it is not a mathematical theorem, it will not result in descriptions of digital lines as mathematical objects, or help research on their abstract properties.

L. D. Wu formulated in 1982 a theorem describing digital straightness. Proofs of this theorem based on CFs were published in 1991 independently by A. M. Bruckstein and K. Voss; see Klette and Rosenfeld (2004: 208–209) [8]. A. M. Bruckstein described digital straightness in [4] by a number of transformations preserving it. Some of these transformations were defined by CFs.

Some work on the subject has also been done outside digital geometry and computer graphics, however, the solutions obtained in other fields do not reflect the hierarchical structure of digitization runs, which is an important feature of digital lines as mathematical objects.

For example, as far back as in 1772, astronomer Johan III Bernoulli applied the CF expansion of a to the solution of the problem of describing the sequence $(\lfloor na \rfloor)_{n \in \mathbb{N}^+}$ for an irrational a. The problem is clearly equivalent to finding the digitization of $y = ax$. Bernoulli failed to provide any proofs. Venkov catalogued the entire history of the problem and its solution (including the solution by Markov from 1882) in [16] (pp. 65–71).

Stolarsky (1976) described in [14] applications of CFs to *Beatty sequences*.

Last but not least, we have to mention the research on *Sturmian words*, because this is very closely related to the research on digital lines with irrational slopes; see chapter 2 in [9] (by J. Berstel and P. Séébold).

2   Earlier Results by Uscka-Wehlou (2007) [15]

This section presents results obtained by the author in [15]. In the current paper, as in [15], we will discuss the digitization of lines with equations $y = ax$ where $a \in\ ]0, 1[\setminus \mathbb{Q}$. Also here, like in [15], the standard Rosenfeld digitization (R-digitization) is replaced by the R'-digitization. The R'-digitization of the line with equation $y = ax$ is equal to the R-digitization of $y = ax + \frac{1}{2}$. For the definitions of both digitizations see [15]. The R'-digitization of $y = ax$ was obtained there using the following *digitization parameters*.

Definition 1. *For* $y = ax$, *where* $a \in \,]0,1[\,\backslash\, \mathbb{Q}$, *the* digitization parameters *are:*

$$\sigma_1 = \text{frac}\left(\tfrac{1}{a}\right) , \qquad (1)$$

and for all natural numbers $k > 1$

$$\sigma_k = \text{frac}\left(1/\sigma_{k-1}^{\wedge}\right), \quad \text{where} \quad \sigma_{k-1}^{\wedge} = \min(\sigma_{k-1}, 1 - \sigma_{k-1}) \in \,]0,\tfrac{1}{2}[\,\backslash\, \mathbb{Q} . \quad (2)$$

For $j \in \mathbb{N}^+$, σ_j *and* $\sigma_j^{\wedge}$ *are the* digitization parameters *and* modified digitization parameters *of the digitization level* j, *respectively.*

For each $a \in \,]0,1[\,\backslash\, \mathbb{Q}$, an auxiliary function Reg_a was introduced. This function gives for each $k \geq 2$ the number of all the digitization levels i, where $1 \leq i \leq k-1$, with digitization parameters fulfilling the condition $\sigma_i < \tfrac{1}{2}$.

Definition 2. *For a given line with equation* $y = ax$, *where* $a \in \,]0,1[\,\backslash\, \mathbb{Q}$, *we define a function* $\text{Reg}_a : \mathbb{N}^+ \longrightarrow \mathbb{N}$ *as follows:*

$$\text{Reg}_a(k) = \begin{cases} 0 & \text{if } k = 1 \\ \sum_{i=1}^{k-1} \chi_{]0,1/2[}(\sigma_i) & \text{if } k \in \mathbb{N}^+ \backslash \{1\} , \end{cases} \qquad (3)$$

where $\chi_{]0,1/2[}$ *is the characteristic function of the interval* $]0,\tfrac{1}{2}[$.

The digitization runs of level k for $k \in \mathbb{N}^+$ were defined recursively as sets of runs of level $k-1$ (if we define integer numbers as runs of level 0). We used to call $\text{run}_k(j)$ for $k, j \in \mathbb{N}^+$ a *run of digitization level* k. We used notation run_k or in plural runs_k, meaning $\text{run}_k(j)$ *for some* $j \in \mathbb{N}^+$, or, respectively, $\{\text{run}_k(i);\ i \in I\}$ *where* $I \in \mathcal{P}(\mathbb{N}^+)$. We also defined the *length of a digitization run* as its cardinality.

Function Reg_a defined in Def. 2 was very important in the description of the form of runs. It helped to recognize which kind of runs was the most frequent (also called *main*) on each level and which kind of runs was first, i.e., beginning in $(1,1)$. We showed that for a given straight line l with equation $y = ax$, where $a \in \,]0,1[\backslash\mathbb{Q}$, the R'-digitization of the positive half line of l is the following subset of $\mathbb{Z}^2$:

$$D_{R'}(l) = \bigcup_{j \in \mathbb{N}^+} \{\text{run}_1(j) \times \{j\}\} . \qquad (4)$$

This was a part of the main result achieved in [15].

The main theorem of [15] was a formalization of the well-known conditions the digitization runs fulfill. On each level k for $k \geq 1$ we have short runs S_k and long runs L_k, which are composed of the runs of level $k - 1$. Only one type of the runs (short or long) on each level can appear in sequences, the second type always occurs alone.

Notation. In the present paper we will use the notation $S_k^m L_k$, $L_k S_k^m$, $L_k^m S_k$ and $S_k L_k^m$, where $m = \lfloor 1/\sigma_k^{\wedge} \rfloor - 1$ or $m = \lfloor 1/\sigma_k^{\wedge} \rfloor$, when describing the form of digitization runs_{k+1}. For example, $S_k^m L_k$ means that the run_{k+1} we are talking about consists of m short runs_k (abbreviated S_k) and one long run_k (abbreviated

L_k) in this order, so it is a run_{k+1} with the most frequent element short. The length of such a run_{k+1}, being its cardinality, i.e., the number of runs_k contained in it, is then equal to $m + 1$. We will also use the notation $\|S_{k+1}\|$ and $\|L_{k+1}\|$ for the length of the short resp. long runs_{k+1}.

We will use the following reformulation of the main result from [15].

Theorem 1 (Main Result in [15]; description by the digitization parameters). *For a straight line with equation $y = ax$, where $a \in \ ]0,1[\setminus \mathbb{Q}$, we have $\|S_1\| = \lfloor \frac{1}{a} \rfloor$, $\|L_1\| = \lfloor \frac{1}{a} \rfloor + 1$, and the forms of runs_{k+1} (form_run_{k+1}) for $k \in \mathbb{N}^+$ are as follows:*

$$\text{form\_run}_{k+1} = \begin{cases} S_k^m L_k & \text{if } \text{Reg}_a(k+1) = \text{Reg}_a(k) + 1 \text{ and } \text{Reg}_a(k) \text{ is even} \\ S_k L_k^m & \text{if } \text{Reg}_a(k+1) = \text{Reg}_a(k) \quad\ \text{ and } \text{Reg}_a(k) \text{ is even} \\ L_k S_k^m & \text{if } \text{Reg}_a(k+1) = \text{Reg}_a(k) + 1 \text{ and } \text{Reg}_a(k) \text{ is odd} \\ L_k^m S_k & \text{if } \text{Reg}_a(k+1) = \text{Reg}_a(k) \quad\ \text{ and } \text{Reg}_a(k) \text{ is odd} , \end{cases}$$

where $m = \lfloor \frac{1}{\sigma_k^\wedge} \rfloor - 1$ if the run_{k+1} is short and $m = \lfloor \frac{1}{\sigma_k^\wedge} \rfloor$ if the run_{k+1} is long. The function Reg_a is defined in Def. 2, and σ_k for $k \in \mathbb{N}^+$ in Def. 1.

Theorem 1 shows exactly how to find the R'-digitization of the positive half line $y = ax$ for $a \in \ ]0,1[\setminus \mathbb{Q}$. We get the digitization by calculating the digitization parameters and proceeding step by step, recursively. The knowledge about the kind of the first run on each level allows us go as far as we want in the digitization. The only problem was in the heavy computation of the σ-parameters, but this will be solved now, in Sect. 3.

3 Main Result

Before presenting the description of the digitization, we provide a brief introduction on CFs. For more details see e.g. [7].

Let a be an irrational number. The following algorithm gives the regular (or simple) CF for a, which we denote by $[a_0; a_1, a_2, a_3, \ldots]$. We define a sequence of integers (a_n) and a sequence of real numbers (α_n) by:

$$\alpha_0 = a; \quad a_n = \lfloor \alpha_n \rfloor \quad \text{and} \quad \alpha_n = a_n + \frac{1}{\alpha_{n+1}} \quad \text{for } n \geq 0 .$$

Then $a_n \geq 1$ and $\alpha_n > 1$ for $n \geq 1$. The natural numbers $a_0, a_1, a_2, a_3, \ldots$ are called the *elements* of the CF. They are also called the *terms* of the CF, see p. 20 in [1]; or *partial quotients*, see p. 40 in [16]. We use the word *elements*, following Khinchin (p. 1 in [7]).

If a is irrational, so is each α_n, and the sequences (a_n) and (α_n) are infinite. A CF expansion exists for all $a \in \mathbb{R}$ and is unique if we impose an additional condition that the last element (if a is rational) cannot be 1; see [7], p. 16.

The following two lemmas concern subtracting CFs from 1.

Lemma 1. *Let $b_i \in \mathbb{N}^+$ for all $i \in \mathbb{N}^+$ and $b_1 \geq 2$. Then*

$$1 - [0; b_1, b_2, b_3, \ldots] = [0; 1, b_1 - 1, b_2, b_3, \ldots] . \tag{5}$$

Proof. Let $b = [0; b_1, b_2, b_3, \ldots]$ and $b_1 \geq 2$. Then $\frac{1}{b} = [b_1; b_2, b_3, \ldots]$ and we get
$1 - b = 1/(1 + 1/(\frac{1}{b} - 1)) = 1/(1 + 1/([b_1; b_2, b_3, \ldots] - 1)) = [0; 1, b_1 - 1, b_2, \ldots]$. $\square$

Lemma 2. *If $a_i \in \mathbb{N}^+$ for all $i > 1$, then $1 - [0; 1, a_2, a_3, \ldots] = [0; a_2 + 1, a_3, \ldots]$.*

Proof. Put $b_1 - 1 = a_2, b_2 = a_3, b_3 = a_4, \ldots, b_i = a_{i+1}, \ldots$ in Lemma 1. $\square$

Because clearly $[0; 1, a_2, a_3, \ldots] > \frac{1}{2}$ for all sequences $(a_2, a_3, \ldots)$ of positive integers, Lemma 2 illustrates the modification operation for the σ-parameters according to Def. 1. This leads us to define the following *index jump function*, which will allow us describe the digitization in terms of CFs.

Definition 3. *For each $a \in {]}0, 1[\setminus \mathbb{Q}$, the index jump function $i_a : \mathbb{N}^+ \to \mathbb{N}^+$ is defined by $i_a(1) = 1$, $i_a(2) = 2$ and $i_a(k+1) = i_a(k) + 1 + \delta_1(a_{i_a(k)})$ for $k \geq 2$, where $\delta_1(x) = \begin{cases} 1, x = 1 \\ 0, x \neq 1 \end{cases}$ and $a_1, a_2, \ldots \in \mathbb{N}^+$ are the CF elements of a.*

The following theorem translates Def. 1 into the language of CFs. It is a very important step on the way of translating our earlier results into a simple CF description. This will also serve as a springboard for future research on the connection between the sequences of the consecutive digitization parameters for given $a \in {]}0, 1[\setminus \mathbb{Q}$ and the iterations of the Gauss Map $G(a) = \text{frac}\left(\frac{1}{a}\right)$.

Theorem 2. *Let $a \in {]}0, 1[\setminus \mathbb{Q}$ and $a = [0; a_1, a_2, \ldots]$. For the digital straight line with equation $y = ax$, the digitization parameters as defined in Def. 1 are*

$$\sigma_k = [0; a_{i_a(k+1)}, a_{i_a(k+1)+1}, \ldots] \quad for \ \ k \geq 1 , \tag{6}$$

where i_a is the index jump function defined in Def. 3.

Proof. The proof will be by induction. For $k = 1$, the statement is $\sigma_1 = [0; a_2, a_3, \ldots]$, because $i_a(2) = 2$. From Def. 1 and because $a = [0; a_1, a_2, \ldots]$, we have $\sigma_1 = \text{frac}\left(\frac{1}{a}\right) = [0; a_2, a_3, \ldots]$, so the induction hypothesis for $k = 1$ is true. Let us now suppose that $\sigma_k = [0; a_{i_a(k+1)}, a_{i_a(k+1)+1}, \ldots]$ for some $k \geq 1$. We will show that this implies that $\sigma_{k+1} = [0; a_{i_a(k+2)}, a_{i_a(k+2)+1}, \ldots]$. From Def. 3 we have $i_a(k+2) = i_a(k+1) + 1 + \delta_1(a_{i_a(k+1)})$. According to Def. 1, $\sigma_{k+1} = \text{frac}(1/\sigma_k^\wedge)$. We get two cases:

- $a_{i_a(k+1)} \neq 1$ (thus $\delta_1(a_{i_a(k+1)}) = 0$). This means that $\sigma_k < \frac{1}{2}$, so $\sigma_k^\wedge = \sigma_k$. We get the statement, because
  $\sigma_{k+1} = \text{frac}(1/\sigma_k) = [0; a_{i_a(k+1)+1}, \ldots] = [0; a_{i_a(k+2)}, \ldots]$.
- $a_{i_a(k+1)} = 1$ ($\delta_1(a_{i_a(k+1)}) = 1$). This means that $\sigma_k > \frac{1}{2}$, so $\sigma_k^\wedge = 1 - \sigma_k$. Lemma 2 and Def. 3 give us the statement, because
  $\sigma_{k+1} = \text{frac}(1/(1 - \sigma_k)) = [0; a_{i_a(k+1)+2}, \ldots] = [0; a_{i_a(k+2)}, \ldots]$.

This completes the proof. $\square$

In order to get a CF description of the digitization, we will express the function Reg_a (determining the form of the digitization runs on all the levels) using the function i_a defined in Def. 3. The translation of Def. 2 into the following CF version results in a very simple relationship between the complicated Reg_a and the simple i_a. It is a very important step in translating Theorem 1 into a CF version.

Theorem 3. *For a given $a \in \,]0,1[\,\backslash\,\mathbb{Q}$, there is the following connection between the corresponding functions Reg_a and i_a. For each $k \in \mathbb{N}^+$*

$$\text{Reg}_a(k) = 2k - i_a(k+1) \ . \tag{7}$$

Proof. For $k = 1$ a direct check gives the equality. Let us assume that $\text{Reg}_a(k) = 2k - i_a(k+1)$ for some $k \geq 1$. We will show that this implies $\text{Reg}_a(k+1) = 2k + 2 - i_a(k+2)$, which will, by induction, prove our statement.

It follows from Def. 2, that for $k \geq 1$

$$\text{Reg}_a(k+1) = \text{Reg}_a(k) + \chi_{]0,1/2[}(\sigma_k) \ . \tag{8}$$

Moreover, according to Def. 3, for $k \geq 1$

$$i_a(k+2) = i_a(k+1) + 1 + \delta_1(a_{i_a(k+1)}) \ . \tag{9}$$

Putting (8) and (9) in the induction hypothesis for $k + 1$, we see that we have to show the following:

$$\text{Reg}_a(k) + \chi_{]0,1/2[}(\sigma_k) = 2k + 2 - (i_a(k+1) + 1 + \delta_1(a_{i_a(k+1)})) \ . \tag{10}$$

Due to the induction hypothesis for k, it is enough to show that for all $k \geq 1$

$$\chi_{]0,1/2[}(\sigma_k) = 1 - \delta_1(a_{i_a(k+1)}) \ . \tag{11}$$

To prove this, we use Theorem 2, which says that $\sigma_k = [0; a_{i_a(k+1)}, \ldots]$:

$$\chi_{]0,1/2[}(\sigma_k) = 1 \quad \Leftrightarrow \quad [0; a_{i_a(k+1)}, a_{i_a(k+1)+1}, \ldots] < \frac{1}{2}$$

$$\Leftrightarrow \quad a_{i_a(k+1)} \neq 1 \quad \Leftrightarrow \quad 1 - \delta_1(a_{i_a(k+1)}) = 1 \ .$$

This completes the proof. □

Now we are ready to formulate our main theorem. The theorem is more parsimonious from a computational standpoint than Theorem 1, because the function i_a is very simple and contains only computations with integers. This is an important advantage for efficient computer program development. The entire description uses only one function: the index jump function.

Theorem 4 (Main Result; description by CFs). *Let $a \in \,]0,1[\,\backslash\,\mathbb{Q}$ and $a = [0; a_1, a_2, \ldots]$. For the digital straight line with equation $y = ax$, we have $\|S_1\| = a_1$, $\|L_1\| = a_1 + 1$, and the forms of runs_{k+1} (form_run$_{k+1}$) for $k \in \mathbb{N}^+$ are as follows:*

$$\text{form\_run}_{k+1} = \begin{cases} S_k^m L_k & \text{if} \ \ a_{i_a(k+1)} \neq 1 \ \ \text{and} \ \ i_a(k+1) \ \text{is even} \\ S_k L_k^m & \text{if} \ \ a_{i_a(k+1)} = 1 \ \ \text{and} \ \ i_a(k+1) \ \text{is even} \\ L_k S_k^m & \text{if} \ \ a_{i_a(k+1)} \neq 1 \ \ \text{and} \ \ i_a(k+1) \ \text{is odd} \\ L_k^m S_k & \text{if} \ \ a_{i_a(k+1)} = 1 \ \ \text{and} \ \ i_a(k+1) \ \text{is odd} \ , \end{cases} \tag{12}$$

where $m = b_{k+1} - 1$ if the run_{k+1} is short and $m = b_{k+1}$ if the run_{k+1} is long. The function i_a is defined in Def. 3 and $b_{k+1} = a_{i_a(k+1)} + \delta_1(a_{i_a(k+1)})a_{i_a(k+1)+1}$.

Figure 1 illustrates the connection between the hierarchy of runs (the first five levels), the index jump function and the digitization parameters for $y = ax$, where $a = [0; 1, 2, 1, 1, 3, 1, 1, a_8, a_9, \ldots]$ for some $a_8, a_9, \ldots \in \mathbb{N}^+$.

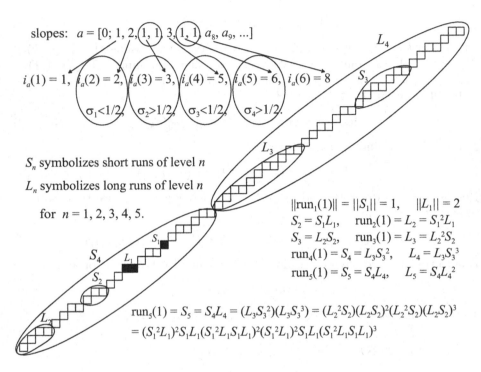

Fig. 1. The index jump function, digitization parameters and hierarchy of runs

Proof. Theorem 4 follows from Theorems 1, 2 and 3. From Theorem 1 we know that the length of the short runs_{k+1} is $\left\lfloor \frac{1}{\sigma_k^\wedge} \right\rfloor$. According to Theorem 2, $\sigma_k = [0; a_{i_a(k+1)}, a_{i_a(k+1)+1}, \ldots]$. We have to consider two possible cases:

- $a_{i_a(k+1)} > 1$. This means that $\sigma_k < \frac{1}{2}$ and $\sigma_k^\wedge = \sigma_k$, so the length of the short runs on the level $k+1$ is $a_{i_a(k+1)}$. Because $\delta_1(a_{i_a(k+1)}) = 0$, we get the statement about the run lengths.
- $a_{i_a(k+1)} = 1$. This means that $\sigma_k > \frac{1}{2}$ and (from Def. 1) $\sigma_k^\wedge = 1 - \sigma_k = [0; 1 + a_{i_a(k+1)+1}, a_{i_a(k+1)+2}, \ldots]$, so the length of the short runs on the level $k+1$ is $1 + a_{i_a(k+1)+1} = a_{i_a(k+1)} + \delta_1(a_{i_a(k+1)}) \cdot a_{i_a(k+1)+1}$.

Theorem 3 gives the statement concerning the form of runs on all levels. It says that $\text{Reg}_a(k) = 2k - i_a(k+1)$ and $\text{Reg}_a(k+1) = 2k + 2 - i_a(k+2)$, so the condition $\text{Reg}_a(k+1) = \text{Reg}_a(k)$ is equivalent to $i_a(k+2) = i_a(k+1) + 2$, thus, according to Def. 3, to $\delta_1(a_{i_a(k+1)}) = 1$, so $a_{i_a(k+1)} = 1$. In the same way we show that the condition $\text{Reg}_a(k+1) = \text{Reg}_a(k)+1$ is equivalent to $a_{i_a(k+1)} \neq 1$. Moreover, because $\text{Reg}_a(k) = 2k - i_a(k+1)$, the parity of $\text{Reg}_a(k)$ and $i_a(k+1)$ is the same for all k, so we can replace "$\text{Reg}_a(k)$ is even" from Theorem 1 by "$i_a(k+1)$ is even" in the CF description.   □

4   Some Applications of the Main Result

4.1   Slopes with Periodic CF Expansions

Period Length 1. We are looking for numbers $a \in \,]0,1[\, \setminus \mathbb{Q}$ having periodic CF expansion with period length 1, i.e., $a = [0; n, n, \ldots] = [0; \overline{n}]$. This means that a is a root of equation $a = \frac{1}{n+a}$, thus of $a^2 + na - 1 = 0$, which gives $a = \frac{1}{2}(\sqrt{n^2 + 4} - n)$, because $a \in \,]0,1[$. We have two groups of lines:

- when $n = 1$, we get a one-element group, containing the line $y = \frac{1}{2}(\sqrt{5}-1)x$ with the Golden Section as slope. Here we have $i_a(1) = 1$ and for $k \geq 2$ there is $i_a(k) = 2k - 2$, which is always even. Moreover, $b_1 = 1$ and for $k \geq 1$ we have $b_{k+1} = a_{2k} + a_{2k+1} = 2$. According to Theorem 4, we get the following digitization: $\|S_1\| = 1$, $\|L_1\| = 2$; for $k \in \mathbb{N}^+$ $S_{k+1} = S_k L_k$, $L_{k+1} = S_k L_k^2$.
- when $n \geq 2$, we have $i_a(k) = k$ for each $k \in \mathbb{N}^+$ and $b_{k+1} = a_{k+1} = n$ for all $k \in \mathbb{N}$. This means, according to Theorem 4, that for all the lines $y = \frac{1}{2}(\sqrt{n^2 + 4} - n)x$ where $n \in \mathbb{N}^+ \setminus \{1\}$, we get the following description of the digitization: $\|S_1\| = n$, $\|L_1\| = n+1$; for $k \in \mathbb{N}^+$: $S_{2k} = S_{2k-1}^{n-1} L_{2k-1}$, $L_{2k} = S_{2k-1}^n L_{2k-1}$, $S_{2k+1} = L_{2k} S_{2k}^{n-1}$, $L_{2k+1} = L_{2k} S_{2k}^n$.

Period Length 2. Now we are looking for numbers $a \in \,]0,1[\setminus \mathbb{Q}$ having periodic CF expansion with period length 2, i.e., $a = [0; \overline{n,m}]$. This means that a is a root of equation $a = [0; n, m + a]$, thus of $na^2 + mna - m = 0$, which gives $a = \frac{1}{2n}(\sqrt{m^2 n^2 + 4mn} - mn)$, because $a \in \,]0,1[$. If $m = n$, see the description for period length 1. If $m \neq n$, we get three possible classes of lines:

- when $m, n \geq 2$, we have $i_a(k) = k$ for each $k \in \mathbb{N}^+$ and $b_{k+1} = a_{k+1}$ for all $k \in \mathbb{N}$. This means, from Theorem 4, that for all the lines $y = ax$, where $a = \frac{1}{2n}(\sqrt{m^2 n^2 + 4mn} - mn)$ for some $n, m \in \mathbb{N}^+ \setminus \{1\}$, we get the following description of the digitization: $\|S_1\| = n$, $\|L_1\| = n+1$; for $k \in \mathbb{N}^+$: $S_{2k} = S_{2k-1}^{m-1} L_{2k-1}$, $L_{2k} = S_{2k-1}^m L_{2k-1}$, $S_{2k+1} = L_{2k} S_{2k}^{n-1}$, $L_{2k+1} = L_{2k} S_{2k}^n$.
- when $m = 1$ and $n \geq 2$, we have $i_a(1) = 1$ and $b_1 = n$. For $k \in \mathbb{N}^+$ there is $i_a(k+1) = 2k$ and $b_{k+1} = a_{2k} + a_{2k+1} = n + 1$. The digitization is thus: $\|S_1\| = n$, $\|L_1\| = n+1$; for $k \in \mathbb{N}^+$: $S_{k+1} = S_k L_k^n$, $L_{k+1} = S_k L_k^{n+1}$.
- when $m \geq 2$ and $n = 1$, we have $i_a(1) = 1$, $i_a(2) = 2$ and $i_a(k+1) = 2k-1$ for $k \geq 2$, which means that $i_a(k)$ is odd for all $k \neq 2$. Moreover, $b_1 = 1$, $b_2 = m$ and $b_{k+1} = a_{2k-1} + a_{2k} = 1 + m$ for $k \geq 2$. The digitization is thus as follows: $\|S_1\| = 1$, $\|L_1\| = 2$, $S_2 = S_1^{m-1} L_1$, $L_2 = S_1^m L_1$, and for $k \geq 2$ we have $S_{k+1} = L_k^m S_k$ and $L_{k+1} = L_k^{m+1} S_k$.

Generally - Quadratic Surds. Let us recall that an *algebraic number of degree* n is a root of an algebraic equation $a_0 x^n + a_1 x^{n-1} + \cdots + a_{n-1} x + a_n = 0$ of degree n with integer coefficients, but is not a root of any algebraic equation of lower degree with integer coefficients. Algebraic numbers of the second degree are called *quadratic irrationals* or *quadratic surds*. The following theorem is a merge of Lagrange's theorem from 1770 with Euler's theorem from 1737 (see [1], pp. 66–71). *Quadratic surds, and only they, are represented by periodic CFs*, meaning purely or mixed periodic ([1], p. 66). It follows from this theorem that all the lines with quadratic surds from the interval $]0, 1[$ as slopes have simple digitization patterns, which can be described by general formulae for all of the digitization levels. Moreover, in [10] on p. 88, we find the following theorem.

If d is a positive, non-square integer, then $\sqrt{d} = [x_0; \overline{x_1, x_2, \ldots, x_2, x_1, 2x_0}]$, where each partial quotient is a positive integer.

The CFs of pure quadratic irrationals all have the same structure, involving *palindromes*. Sequence **A003285** in [12] shows for each $n \in \mathbb{N}^+$ the length of the period of CF for $\sqrt{n}$ (0 if n is a square). Also in [10], on p. 89, we find some patterns in the CF expansions of quadratic surds, for example $\sqrt{k^2 + 1} = [k; \overline{2k}]$, $\sqrt{k^2 + 2} = [k; \overline{k, 2k}]$, $\sqrt{k^2 + m} = [k; \overline{2k/m, 2k}]$. These patterns make it very easy to construct the digitization of the lines with slopes $\sqrt{k^2 + 1} - k$, $\sqrt{k^2 + 2} - k$, or, generally, $\sqrt{k^2 + m} - k$, using Theorem 4 from the present paper. See [10], pp. 83–91, for both theory and examples on this subject.

4.2 Slopes with Non-periodic CF Expansions

Quadratic irrationals are not the only numbers showing simple patterns in their CF expansion. There also exist transcendental numbers with simple patterns. CF sequences for some transcendental number have *periodic forms*. Some examples can be found, among others, on p. 97 in [2]. The examples were given by Euler in 1737, but the first of them was, according to Brezinski, already given by R. Cotes in the *Philosophical Transactions* in 1714.

$$e - 2 = [0; 1, 2, 1, 1, 4, 1, 1, 6, 1, \ldots, 1, 2k, 1, \ldots] = [0; \overline{1, 2k, 1}]_{k=1}^{\infty}, \qquad (13)$$

$$\frac{e + 1}{e - 1} - 2 = [0; \overline{2 + 4k}]_{k=1}^{\infty}, \qquad \frac{e - 1}{2} = [0; 1, \overline{2 + 4k}]_{k=1}^{\infty}. \qquad (14)$$

On p. 124 in [10] we find the following. For $n \geq 2$

$$\sqrt[n]{e} - 1 = [0; \overline{(2k - 1)n - 1, 1, 1}]_{k=1}^{\infty}. \qquad (15)$$

On p. 110 in [2] we find the following formula, obtained by Euler in 1737 and Lagrange in 1776, but each using different methods:

$$\frac{e^2 - 1}{e^2 + 1} = [0; 1, 3, 5, \ldots, 2k - 1, \ldots] = [0; \overline{2k - 1}]_{k=1}^{\infty}. \qquad (16)$$

This means that we are able to describe exactly, i.e., not by using approximations by rationals, the construction of the digital lines $y = ax$, where a is equal to

$e - 2$, $\frac{e+1}{e-1} - 2$, $\frac{e-1}{2}$, $\sqrt{e} - 1$, $\sqrt[3]{e} - 1$ or $\frac{e^2-1}{e^2+1}$, using Theorem 4 from the present paper. Because of the repeating pattern in the CF expansions of the slopes, we are able to obtain general formulae for all of the digitization levels.

For example, if the slope a is equal to one of the following $\frac{e+1}{e-1} - 2$, $\frac{e-1}{2}$, $\frac{e^2-1}{e^2+1}$, then the digitization patterns can be described for all of them in the following way. For all $k \in \mathbb{N}^+$ there is $i_a(k) = k$, thus $b_k = a_k$, because there are no elements $a_k = 1$ for $k \geq 2$ in the CF expansions. This gives the following digitization pattern for these lines: $\|S_1\| = a_1$, $\|L_1\| = a_1 + 1$ and for $k \in \mathbb{N}^+$

$$(S_{k+1}, L_{k+1}) = \begin{cases} (S_k^{a_{k+1}-1}L_k, \ S_k^{a_{k+1}}L_k) & \text{if } k \text{ is odd} \\ (L_kS_k^{a_{k+1}-1}, \ L_kS_k^{a_{k+1}}) & \text{if } k \text{ is even .} \end{cases} \quad (17)$$

The only difference in the digitization patterns for the three slopes are different run lengths, defined by the elements a_k of the CF expansions (14) and (16).

Formula (13) gives the digitization of the line $y = ax$ with $a = e - 2$. Here $i_a(2k) = 3k - 1$ $(a_{i_a(2k)} = 2k \neq 1)$ and $i_a(2k+1) = 3k$ $(a_{i_a(2k+1)} = 1)$ for $k \in \mathbb{N}^+$, so we get $b_1 = 1$, $b_{2k} = 2k$ and $b_{2k+1} = 2$ for $k \in \mathbb{N}^+$, and the digitization pattern is as follows: $\|S_1\| = 1$, $\|L_1\| = 2$ and for $k \in \mathbb{N}^+$

$$(S_{k+1}, L_{k+1}) = \begin{cases} (S_k^kL_k, \ S_k^{k+1}L_k) & \text{if } k \equiv 1 \ (\text{mod } 4) \\ (S_kL_k, \ S_kL_k^2) & \text{if } k \equiv 0 \ (\text{mod } 4) \\ (L_kS_k^k, \ L_kS_k^{k+1}) & \text{if } k \equiv 3 \ (\text{mod } 4) \\ (L_kS_k, \ L_k^2S_k) & \text{if } k \equiv 2 \ (\text{mod } 4) . \end{cases} \quad (18)$$

For example, $S_5 = S_4L_4 = (L_3S_3^3)(L_3S_3^4) = (L_2^2S_2)(L_2S_2)^3(L_2^2S_2)(L_2S_2)^4 = (S_1^2L_1)^2(S_1L_1)[(S_1^2L_1)(S_1L_1)]^3(S_1^2L_1)^2(S_1L_1)[(S_1^2L_1)(S_1L_1)]^4$, where $\|S_1\| = 1$ and $\|L_1\| = 2$.

5   Conclusion

We have presented a computationally simple description of the digitization of straight lines $y = ax$ with slopes $a \in \ ]0, 1[\ \backslash \ \mathbb{Q}$, based on the CF expansions of the slopes. The description reflects the hierarchical structure of digitization runs. Moreover, it is exact, avoiding approximations by rationals.

The theoretical part of the paper was based on [15] and the examples were based on the literature concerning CFs ([2] and [10]). The examples show how to use the theory in finding digitization patterns. This description can also be useful in theoretical research on digital lines with irrational slopes. The present study may serve as a springboard for future research on some classes of digital lines, defined by the CF expansions of their slopes.

The new method gives a special treatment to the CF elements equal to 1, which makes it very powerful for some slopes with 1's in the CF expansion. To our knowledge, there exist no other methods of describing digital lines with irrational slopes by CFs which give a special treatment to CF elements equal to 1, which makes our method original.

A comparison between our new method and some other CF methods will be presented in a forthcoming paper by the author. We will show for example how to construct a slope $a \in \,]0, 1[\, \setminus \, \mathbb{Q}$ so that for each $n \geq 2$ the difference between the length of the digital straight line segment (its cardinality as a subset of $\mathbb{Z}^2$) of $y = ax$ produced in the nth step of our method and the length of the digital straight line segment of $y = ax$ produced in the nth step of the method by Venkov (described in [16] on p. 67) is as large as we decide in advance.

Acknowledgments. I am grateful to Christer Kiselman for comments on earlier versions of the manuscript.

References

1. Beskin, N.M.: Fascinating Fractions. Mir Publishers, Moscow (1986) (Revised from the 1980 Russian edition)
2. Brezinski, C.: History of Continued Fractions and Padé Approximants. Springer, Heidelberg (1991) (Printed in USA)
3. Brons, R.: Linguistic Methods for the Description of a Straight Line on a Grid. Comput. Graphics Image Processing 3, 48–62 (1974)
4. Bruckstein, A.M.: Self-Similarity Properties of Digitized Straight Lines. Contemp. Math. 119, 1–20 (1991)
5. Debled, I.: Étude et reconnaissance des droites et plans discrets. Ph.D. Thesis, Strasbourg: Université Louis Pasteur, pp. 209 (1995)
6. Dorst, L., Duin, R.P.W.: Spirograph Theory: A Framework for Calculations on Digitized Straight Lines. IEEE Transactions on Pattern Analysis and Machine Intelligence PAMI 6(5), 632–639 (1984)
7. Khinchin, A.Ya.: Continued Fractions, 3rd edn. Dover Publications (1997)
8. Klette, R., Rosenfeld, A.: Digital straightness – a review. Discrete Appl. Math. 139(1–3), 197–230 (2004)
9. Lothaire, M.: Algebraic Combinatorics on Words. Cambridge Univ. Press, Cambridge (2002)
10. Perron, O.: Die Lehre von den Kettenbrüchen. Band I: Elementare Kettenbrüche, 3rd edn (1954)
11. Réveillès, J.-P.: Géométrie discrète, calculus en nombres entiers et algorithmique, 251 pages, Strasbourg: Université Louis Pasteur, Thèse d'État (1991)
12. Sloane, N.J.A.: The On-Line Encyclopedia of Integer Sequences, http://www.research.att.com/~njas/sequences/A003285
13. Stephenson, P.D.: The Structure of the Digitised Line: With Applications to Line Drawing and Ray Tracing in Computer Graphics. North Queensland, Australia, James Cook University. Ph.D. Thesis (1998)
14. Stolarsky, K.B.: Beatty sequences, continued fractions, and certain shift operators. Canad. Math. Bull. 19, 473–482 (1976)
15. Uscka-Wehlou, H.: Digital lines with irrational slopes. Theoret. Comput. Sci. 377, 157–169 (2007)
16. Venkov, B.A.: Elementary Number Theory. Translated and edited by Helen Alderson. Wolters-Noordhoff, Groningen (1970)
17. de Vieilleville, F., Lachaud, J.-O.: Revisiting Digital Straight Segment Recognition. In: Kuba, A., Nyúl, L.G., Palágyi, K. (eds.) DGCI 2006. LNCS, vol. 4245, pp. 355–366. Springer, Heidelberg (2006)
18. Voss, K.: Discrete Images, Objects, and Functions in $\mathbb{Z}^n$. Springer-Verlag (1993)

New Characterizations of Simple Points, Minimal Non-simple Sets and P-Simple Points in 2D, 3D and 4D Discrete Spaces

Michel Couprie and Gilles Bertrand

Université Paris-Est, LABINFO-IGM, UMR CNRS 8049, A2SI-ESIEE, France
{m.couprie,g.bertrand}@esiee.fr

Abstract. In this article, we present new results on simple points, minimal non-simple sets (MNS) and P-simple points. In particular, we propose new characterizations which hold in dimensions 2, 3 and 4, and which lead to efficient algorithms for detecting such points or sets. This work is settled in the framework of cubical complexes, and some of the main results are based on the properties of critical kernels.

Introduction

Topology-preserving operators, like homotopic skeletonization, are used in many applications of image analysis to transform an object while leaving unchanged its topological characteristics. In discrete grids ($\mathbb{Z}^2$, $\mathbb{Z}^3$, $\mathbb{Z}^4$), such a transformation can be defined thanks to the notion of simple point [20]: intuitively, a point of an object is called simple if it can be deleted from this object without altering topology.

The most "natural" way to thin an object consists in removing some of its border points in parallel, in a symmetrical manner. However, parallel deletion of simple points does not guarantee topology preservation in general. In fact, such a guarantee is not obvious to obtain, even for the 2D case (see [10]). C. Ronse introduced the minimal non-simple sets [28] to study the conditions under which points may be removed simultaneously while preserving topology of 2D objects. This leads to verification methods for the topological soundness of parallel thinning algorithms. Such methods have been proposed for 2D algorithms by C. Ronse [28] and R. Hall [15], they have been developed for the 3D case by T.Y. Kong [21,16] and C.M. Ma [25], as well as for the 4D case by C-J. Gau and T.Y. Kong [13,19]. For the 3D case, G. Bertrand [1] introduced the notion of P-simple point as a verification method but also as a methodology to design parallel thinning algorithms [2,9,23,24].

Introduced recently by G. Bertrand, critical kernels [3,4] constitute a general framework settled in the category of abstract complexes for the study of parallel thinning in any dimension. It allows easy design of parallel thinning algorithms which produce new types of skeletons, with specific geometrical properties, while guaranteeing their topological soundness [7,5,6]. A new definition of a simple point is proposed in [3,4], based on the collapse operation which is a classical

D. Coeurjolly et al. (Eds.): DGCI 2008, LNCS 4992, pp. 105–116, 2008.

tool in algebraic topology and which guarantees topology preservation. Then, the notions of an *essential face* and of a *core* of a face allow to define the *critical kernel* $\mathcal{K}$ of a complex X. The most fundamental result proved in [3,4] is that, if a subset Y of X contains $\mathcal{K}$, then X collapses onto Y, hence X and Y "have the same topology".

In this article, we present new results on simple points, minimal non-simple sets (MNS), P-simple points and critical kernels. Let us summarize the main ones among these results.

First of all, we state some *confluence properties* of the collapse operation (Th. 5, Th. 6), which play a fundamental role in the proof of forthcoming theorems. These properties do not hold in general due to the existence of "topological monsters" such as Bing's house ([8], see also [27]); we show that they are indeed true in some discrete spaces which are not large enough to contain such counterexamples.

Based on these confluence properties, we derive a new characterization of 2D, 3D and 4D simple points (Th. 7) which leads to a simple, greedy linear time algorithm for simplicity checking.

Then, we show the equivalence (up to 4D) between the notion of MNS and the notion of crucial clique, derived from the framework of critical kernels. This equivalence (Th. 21) leads to the first characterization of MNS which can be verified using a polynomial method.

Finally, we show the equivalence between the notion of P-simple point and the notion of weakly crucial point, also derived from the framework of critical kernels. This equivalence (Th. 23) leads to the first local characterization of P-simple points in 4D.

This paper is self-contained, however the proofs cannot be included due to space limitation. They can be found in [12,11], together with some illustrations and developments.

1   Cubical Complexes

Abstract complexes have been promoted in particular by V. Kovalevsky [22] in order to provide a sound topological basis for image analysis. For instance, in this framework, we retrieve the main notions and results of digital topology, such as the notion of simple point.

Intuitively, a cubical complex may be thought of as a set of elements having various dimensions (*e.g.* cubes, squares, edges, vertices) glued together according to certain rules. In this section, we recall briefly some basic definitions on complexes, see also [7,5,6] for more details. We consider here n-dimensional complexes, with $0 \leq n \leq 4$.

Let S be a set. If T is a subset of S, we write $T \subseteq S$. We denote by $|S|$ the number of elements of S.

Let $\mathbb{Z}$ be the set of integers. We consider the families of sets $\mathbb{F}_0^1$, $\mathbb{F}_1^1$, such that $\mathbb{F}_0^1 = \{\{a\} \mid a \in \mathbb{Z}\}$, $\mathbb{F}_1^1 = \{\{a, a+1\} \mid a \in \mathbb{Z}\}$. A subset f of $\mathbb{Z}^n$, $n \geq 2$, which is the Cartesian product of exactly m elements of $\mathbb{F}_1^1$ and $(n-m)$ elements of

$\mathbb{F}_0^1$ is called a *face* or an *m-face* of $\mathbb{Z}^n$, m is the *dimension of f*, we write $\dim(f) = m$.

Observe that any non-empty intersection of faces is a face. For example, the intersection of two 2-faces A and B may be either a 2-face (if $A = B$), a 1-face, a 0-face, or the empty set.

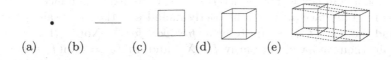

(a)   (b)   (c)   (d)   (e)

Fig. 1. Graphical representations of: (a) a 0-face, (b) a 1-face, (c) a 2-face, (d) a 3-face, (e) a 4-face

We denote by $\mathbb{F}^n$ the set composed of all m-faces of $\mathbb{Z}^n$, with $0 \leq m \leq n$. An m-face of $\mathbb{Z}^n$ is called a *point* if $m = 0$, a *(unit) interval* if $m = 1$, a *(unit) square* if $m = 2$, a *(unit) cube* if $m = 3$, a *(unit) hypercube* if $m = 4$ (see Fig. 1).

Let f be a face in $\mathbb{F}^n$. We set $\hat{f} = \{g \in \mathbb{F}^n \mid g \subseteq f\}$ and $\hat{f}^* = \hat{f} \setminus \{f\}$.
Any $g \in \hat{f}$ is a *face of f*, and any $g \in \hat{f}^*$ is a *proper face of f*.
If X is a finite set of faces in $\mathbb{F}^n$, we write $X^- = \cup\{\hat{f} \mid f \in X\}$, X^- is the *closure of X*.

A set X of faces in $\mathbb{F}^n$ is a *cell* or an *m-cell* if there exists an m-face $f \in X$, such that $X = \hat{f}$. The *boundary of a cell $\hat{f}$* is the set $\hat{f}^*$.

A finite set X of faces in $\mathbb{F}^n$ is a *complex (in $\mathbb{F}^n$)* if $X = X^-$. Any subset Y of a complex X which is also a complex is a *subcomplex of X*. If Y is a subcomplex of X, we write $Y \preceq X$. If X is a complex in $\mathbb{F}^n$, we also write $X \preceq \mathbb{F}^n$. In Fig. 2 and Fig. 3, some complexes are represented. Notice that any cell is a complex.

Let $X \subseteq \mathbb{F}^n$ be a non-empty set of faces. A sequence $(f_i)_{i=0}^{\ell}$ of faces of X is a *path in X (from f_0 to f_ℓ)* if either f_i is a face of f_{i+1} or f_{i+1} is a face of f_i, for all $i \in [0, \ell - 1]$. We say that X is *connected* if, for any two faces f, g in X, there is a path from f to g in X; otherwise we say that X is *disconnected*. We say that Y is a *connected component of X* if $Y \subseteq X$, Y is connected and if Y is maximal for these two properties (*i.e.*, we have $Z = Y$ whenever $Y \subseteq Z \subseteq X$ and Z is connected).

Let $X \subseteq \mathbb{F}^n$. A face $f \in X$ is *a facet of X* if there is no $g \in X$ such that $f \in \hat{g}^*$. We denote by X^+ the set composed of all facets of X.
If X is a complex, observe that in general, X^+ is not a complex, and that $[X^+]^- = X$.

Let $X \preceq \mathbb{F}^n$, $X \neq \emptyset$, the number $\dim(X) = \max\{\dim(f) \mid f \in X^+\}$ is the *dimension of X*. We say that X is an *m-complex* if $\dim(X) = m$.
We say that X is *pure* if, for each $f \in X^+$, we have $\dim(f) = \dim(X)$.
In Fig. 2, the complexes (a) and (f) are pure, while (b,c,d,e) are not.

2   Collapse and Simple Sets

Intuitively a subcomplex of a complex X is simple if its removal from X "does not change the topology of X". In this section we recall a definition of a simple subcomplex based on the operation of collapse [14], which is a discrete analogue of a continuous deformation (a homotopy).

Let X be a complex in $\mathbb{F}^n$ and let $f \in X$. If there exists one face $g \in \hat{f}^*$ such that f is the only face of X which strictly includes g, then g is said to be *free for X* and the pair (f, g) is said to be a *free pair for X*. Notice that, if (f, g) is a free pair, then we have necessarily $f \in X^+$ and $\dim(g) = \dim(f) - 1$.

Let X be a complex, and let (f, g) be a free pair for X. The complex $X \setminus \{f, g\}$ is an *elementary collapse of X*.

Let X, Y be two complexes. We say that X *collapses onto Y* if $Y = X$ or if there exists a *collapse sequence from X to Y*, i.e., a sequence of complexes $\langle X_0, ..., X_\ell \rangle$ such that $X_0 = X$, $X_\ell = Y$, and X_i is an elementary collapse of X_{i-1}, $i = 1, ..., \ell$. If X collapses onto Y and Y is a complex made of a single point, we say that *collapses onto a point*.

Fig. 2 illustrates a collapse sequence. Observe that, if X is a cell of any dimension, then X collapses onto a point. It may easily be seen that the collapse operation preserves the number of connected components.

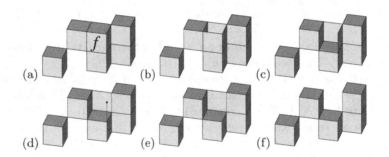

Fig. 2. (a): a pure 3-complex $X \preceq \mathbb{F}^3$, and a 3-face $f \in X^+$. (f): a complex Y which is the detachment of $\hat{f}$ from X. (a-f): a collapse sequence from X to Y.

Let X, Y be two complexes. Let Z such that $X \cap Y \preceq Z \preceq Y$, and let $f, g \in Z \setminus X$. It may be seen that the pair (f, g) is a free pair for $X \cup Z$ if and only if (f, g) is a free pair for Z. Thus, by induction, we have the following property.

Proposition 1 ([3,4]). *Let $X, Y \preceq \mathbb{F}^n$. The complex $X \cup Y$ collapses onto X if and only if Y collapses onto $X \cap Y$.*

The operation of detachment allows to remove a subset from a complex, while guaranteeing that the result is still a complex.

Definition 2 ([3,4]). *Let $Y \subseteq X \preceq \mathbb{F}^n$. We set $X \oslash Y = (X^+ \setminus Y^+)^-$. The set $X \oslash Y$ is a complex which is the detachment of Y from X.*

In the following, we will be interested in the case where Y is a single cell. For example in Fig. 2, we see a complex X (a) containing a 3-cell $\hat{f}$, and $X \oslash \hat{f}$ is depicted in (f).

Let us now recall here a definition of simplicity based on the collapse operation, which can be seen as a discrete counterpart of the one given by T.Y. Kong [17].

Definition 3 ([3,4]). *Let $Y \subseteq X$; we say that Y is* simple *for X if X collapses onto $X \oslash Y$.*

The collapse sequence displayed in Fig. 2 (a-f) shows that the cell $\hat{f}$ is simple for the complex depicted in (a).

The notion of attachment, as introduced by T.Y. Kong [16,17], leads to a local characterization of simple sets, which follows easily from Prop. 1.

Let $Y \preceq X \preceq \mathbb{F}^n$. The *attachment* of Y for X is the complex defined by $Att(Y,X) = Y \cap (X \oslash Y)$.

Proposition 4 ([3,4]). *Let $Y \preceq X \preceq \mathbb{F}^n$. The complex Y is simple for X if and only if Y collapses onto $Att(Y,X)$.*

Let us introduce informally the *Schlegel diagrams* as a graphical representation for visualizing the attachment of a cell. In Fig. 3a, the boundary of a 3-cell $\hat{f}$ and its Schlegel diagram are depicted. The interest of this representation lies in the fact that a structure like $\hat{f}^*$ lying in the 3D space may be represented in the 2D plane. Notice that one 2-face of the boundary, here the square $efhg$, is not represented by a closed polygon in the schlegel diagram, but we may consider that it is represented by the outside space.

As an illustration of Prop. 4, Fig. 3b shows (both directly and by its Schlegel diagram) the attachment of $\hat{f}$ for the complex X of Fig. 2a, and we can easily verify that $\hat{f}$ collapses onto $Att(\hat{f},X)$.

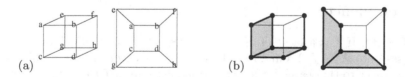

(a)   (b)

Fig. 3. (a): The boundary of a 3-cell and its Schlegel diagram. (b): The attachment of $\hat{f}$ for X (see Fig. 2a).

Representing 4D objects is not easy. To start with, let us consider Fig. 4a where a representation of the 3D complex X of Fig. 2a is given under the form of two horizontal cross-sections, each black dot representing a 3-cell.

In a similar way, we may represent a 4D object by its "3D sections", as the object Y in Fig. 4b. Such an object may be thought of as a "time series of 3D objects". In Fig. 4b, each black dot represents a 4-cell of the whole 4D complex Y.

Schlegel diagrams are particularly useful for representing the attachment of a 4D cell $\hat{f}$, whenever this attachment if not equal to $\hat{f}^*$. Fig. 5a shows the

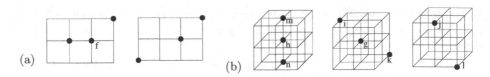

Fig. 4. (a): An alternative representation of the 3D complex X of Fig. 2a. (b): A similar representation of a 4D complex Y.

Schlegel diagram of the boundary of a 4-cell (see Fig. 1e), where one of the 3-faces is represented by the outside space. Fig. 5b shows the Schlegel diagram of the attachment of the 4-cell g in Y (see Fig. 4b). For example, the 3-cell H represented in the center of the diagram is the intersection between the 4-cell g and the 4-cell h. Also, the 2-cell I (resp. the 1-cell J, the 1-cell K, the 0-cell L) is $g \cap i$ (resp. $g \cap j$, $g \cap k$, $g \cap l$). The two 2-cells which are the intersections of g with, respectively, m and n, are both included in the 3-cell H. Observe that the cell g is not simple (its attachment is not connected).

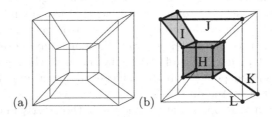

Fig. 5. (a): The Schlegel diagram of the boundary of a 4-cell. (b): The Schlegel diagram of the attachment of the 4-cell g of Fig. 4b, which is not simple.

3   Confluences

Let $X \preceq \mathbb{F}^n$. If f is a facet of X, then by Def. 3, $\hat{f}$ is simple if and only if X collapses onto $X \oslash \hat{f}$. From Prop. 4, we see that checking the simplicity of a cell f reduces to the search for a collapse sequence from $\hat{f}$ to $Att(\hat{f}, X)$. We will show in Sec. 4 that the huge number (especially in 4D) of possible such collapse sequences need not be exhaustively explored, thanks to the confluence properties (Th. 5 and Th. 6) introduced in this section.

Consider three complexes A, B, C. If A collapses onto C and A collapses onto B, then we know that A, B and C "have the same topology". If furthermore we have $C \preceq B \preceq A$, it is tempting to conjecture that B collapses onto C.

In the two-dimensional discrete plane $\mathbb{F}^2$, the above conjecture is true, we call it a confluence property. But quite surprisingly it does not hold in $\mathbb{F}^3$ (more generally in $\mathbb{F}^n, n \geq 3$), and this fact constitutes indeed one of the principal difficulties when dealing with certain topological properties, such as the Poincaré conjecture for example. A classical counter-example to this assertion is Bing's house ([8], see also [27]).

In the boundary of an n-face with $n \leq 4$, there is "not enough room" to build such counter-examples, and thus some kinds of confluence properties hold.

Theorem 5 (Confluence 1). *Let f be a d-face with $d \in \{2, 3, 4\}$, let $A, B \preceq \hat{f}^*$ such that $B \preceq A$, and A collapses onto a point. Then, B collapses onto a point if and only if A collapses onto B.*

The second confluence theorem may be easily derived from Th. 5 and the fact that $\hat{f}$ collapses onto a point.

Theorem 6 (Confluence 2). *Let f be a d-face with $d \in \{2, 3, 4\}$, and let $C, D \preceq \hat{f}^*$ such that $D \preceq C$, and $\hat{f}$ collapses onto D. Then, $\hat{f}$ collapses onto C if and only if C collapses onto D.*

4   New Characterization of Simple Cells

In the image processing literature, a (binary) digital image is often considered as a set of pixels in 2D or voxels in 3D. A pixel is an elementary square and a voxel is an elementary cube, thus an easy correspondence can be made between this classical view and the framework of cubical complexes.

If $X \preceq \mathbb{F}^n$ and if X is a pure n-complex, then we write $X \sqsubseteq \mathbb{F}^n$. In other words, $X \sqsubseteq \mathbb{F}^n$ means that X^+ is a set composed of n-faces (*e.g.*, pixels in 2D or voxels in 3D). If $X, Y \sqsubseteq \mathbb{F}^n$ and $Y \preceq X$, then we write $Y \sqsubseteq X$.

Notice that, if $X \sqsubseteq \mathbb{F}^n$ and if $\hat{f}$ is an n-cell of X, then $X \oslash \hat{f} \sqsubseteq \mathbb{F}^n$. There is indeed an equivalence between the operation on complexes which consists of removing (by detachment) a simple n-cell, and the removal of an 8-simple (resp. 26-simple, 80-simple) point in the framework of 2D (resp. 3D, 4D) digital topology (see [16,17,7,6]).

From Prop. 4 and Th. 6, we have the following characterization of a simple cell, which does only depend on the status of the faces which are in the cell.

Theorem 7. *Let $X \sqsubseteq \mathbb{F}^d$, with $d \in \{2, 3, 4\}$. Let f be a facet of X, and let $A = Att(\hat{f}, X)$. The two following statements hold:*
i) The cell $\hat{f}$ is simple for X if and only if $\hat{f}$ collapses onto A.
ii) Suppose that $\hat{f}$ is simple for X. For any Z such that $A \preceq Z \preceq \hat{f}$, if $\hat{f}$ collapses onto Z then Z collapses onto A.

Now, thanks to Th. 7, if we want to check whether a cell $\hat{f}$ is simple or not, it is sufficient to apply the following greedy algorithm:
Set $Z = \hat{f}$;
Select any free pair (f, g) in $Z \setminus A$, and set Z to $Z \setminus \{f, g\}$;
Continue until either $Z = A$ (answer yes) or no such pair is found (answer no).

If this algorithm returns "yes", then obviously $\hat{f}$ collapses onto A and by i), $\hat{f}$ is simple. In the other case, we have found a subcomplex Z of A such that $\hat{f}$ collapses onto Z and Z does not collapse onto A, thus by the negation of ii), $\hat{f}$ is not simple.

This algorithm may be implemented to run in linear time with respect to the number of elements in the attachment of a cell.

5   Critical Kernels

Let us briefly recall the framework introduced by one of the authors (in [3,4]) for thinning, in parallel, discrete objects with the warranty that we do not alter the topology of these objects. We focus here on the two-, three- and four-dimensional cases, but in fact some of the results in this section are valid for complexes of arbitrary dimension. This framework is based solely on three notions: the notion of an essential face which allows us to define the core of a face, and the notion of a critical face.

Definition 8. *Let $X \preceq \mathbb{F}^n$ and let $f \in X$. We say that f is an essential face for X if f is precisely the intersection of all facets of X which contain f, i.e., if $f = \cap\{g \in X^+ \mid f \subseteq g\}$. We denote by $Ess(X)$ the set composed of all essential faces of X. If f is an essential face for X, we say that $\hat{f}$ is an essential cell for X. If $Y \preceq X$ and $Ess(Y) \subseteq Ess(X)$, then we write $Y \lhd X$.*

Observe that a facet of X is necessarily an essential face for X, i.e., $X^+ \subseteq Ess(X)$. Observe also that, if X and Y are both pure n-complexes, we have that $Y \lhd X$ whenever Y is a subcomplex of X.

Definition 9. *Let $X \preceq \mathbb{F}^n$ and let $f \in Ess(X)$. The core of $\hat{f}$ for X is the complex $Core(\hat{f}, X) = \cup\{\hat{g} \mid g \in Ess(X) \cap \hat{f}^*\}$.*

Proposition 10 ([3]). *Let $X \preceq \mathbb{F}^n$, and let $f \in Ess(X)$. Let $K = \{g \in X \mid f \subseteq g\}$, and let $Y = X \oslash K$. We have: $Core(\hat{f}, X) = Att(\hat{f}, Y \cup \hat{f}) = \hat{f} \cap Y$.*

Corollary 11 ([3,4]). *Let $X \preceq \mathbb{F}^n$, and let $f \in X^+$. We have: $Core(\hat{f}, X) = Att(\hat{f}, X)$.*

Definition 12. *Let $X \preceq \mathbb{F}^n$ and let $f \in X$. We say that f and $\hat{f}$ are regular for X if $f \in Ess(X)$ and if $\hat{f}$ collapses onto $Core(\hat{f}, X)$. We say that f and $\hat{f}$ are critical for X if $f \in Ess(X)$ and if f is not regular for X. If $X \preceq \mathbb{F}^n$, we set $Critic(X) = \cup\{\hat{f} \mid f$ is critical for $X\}$, we say that $Critic(X)$ is the critical kernel of X. A face f in X is a maximal critical face, or an M-critical face (for X), if f is a facet of $Critic(X)$.*

In other words, f is an M-critical face if it is critical and not included in any other critical face.

Proposition 13 ([3,4]). *Let $X \preceq \mathbb{F}^n$, and let $f \in Ess(X)$. Let $Y = \cup\{\hat{g} \mid g \in X^+$ and $f \subseteq g\}$ and $Z = [X \oslash Y] \cup \hat{f}$. The face f is regular for X if and only if $\hat{f}$ is simple for Z.*

The following theorem is the most fundamental result concerning critical kernels. We use it for the proofs of our main properties in dimension 4 or less, but notice that the theorem holds whatever the dimension.

Theorem 14 ([3,4]). *Let $n \in \mathbb{N}$, let $X \preceq \mathbb{F}^n$.*
i) The complex X collapses onto its critical kernel.
ii) If $Y \trianglelefteq X$ contains the critical kernel of X, then X collapses onto Y.
iii) If $Y \trianglelefteq X$ contains the critical kernel of X, then any Z such that $Y \preceq Z \trianglelefteq X$ collapses onto Y.

If X is a pure n-complex (*e.g.*, a set of 3-cells, or voxels, in $\mathbb{F}^3$), the critical kernel of X is not necessarily a pure n-complex. The notion of crucial clique, introduced in [7], allows us to recover a pure n-subcomplex Y of an arbitrary pure n-complex X, under the constraint that X collapses onto Y.

Definition 15 ([7]). *Let $X \sqsubseteq \mathbb{F}^n$, and let f be an M-critical face for X. The set K of all the facets of X which contain f is called a* crucial clique *(for X). More precisely, K is the* crucial clique induced by f.

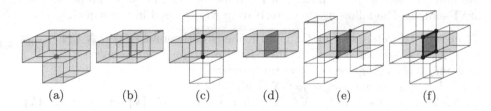

(a)   (b)   (c)   (d)   (e)   (f)

Fig. 6. Crucial cliques in $\mathbb{F}^3$ (represented in light gray): (a) induced by an M-critical 0-face; (b,c) induced by an M-critical 1-face; (d,e,f) induced by an M-critical 2-face. The considered M-critical faces are in dark gray, the core of these M-critical faces (when non-empty) is represented in black.

Some 3D crucial cliques are illustrated in Fig. 6. By Th. 14 and the above definition, if a subcomplex $Y \sqsubseteq X \sqsubseteq \mathbb{F}^n$ contains all the critical facets of X, and at least one facet of each crucial clique for X, then X collapses onto Y.

Now, let us state two properties of crucial cliques which are essential for the proof of one of our main results (Th. 21).

Proposition 16. *Let $X \sqsubseteq \mathbb{F}^d$, with $d \in \{2, 3, 4\}$, let f be an M-critical face of X, let K be the crucial clique induced by f, and let k be any facet of K. Let K' be such that $K' \subseteq K \setminus \{k\}$ and $K' \neq K \setminus \{k\}$.*
Then, k is a simple face of the complex $[X \oslash K']$.

Proposition 17. *Let $X \sqsubseteq \mathbb{F}^d$, with $d \in \{2, 3, 4\}$, let f be an M-critical face of X, let K be the crucial clique induced by f, and let k be any facet of K.*
Then, k is not a simple face of the complex $[X \oslash K] \cup \hat{k}$.

6   Minimal Non-simple Sets

C. Ronse introduced in [28] the minimal non-simple sets (MNS) to propose some conditions under which simple points can be removed in parallel while preserving topology. This leads to verification methods for the topological soundness of 2D thinning algorithms [28,15], 3D thinning algorithms [21,16,25], the 4D case has even been considered in [13,18,19].

The main result of this section (Th. 21) states the equivalence between MNS and crucial cliques in dimensions 2, 3 and 4. This equivalence leads to the first characterization of MNS which can be verified using a polynomial method. In contrast, the very definition of a MNS (see below), as well as the characterization of Th. 18, involves the examination of all subsets of a given candidate set, $e.g.$, a subset of a $2 \times 2 \times 2 \times 2$ block in 4D.

Let $X \sqsubseteq \mathbb{F}^d$, with $d \in \{2, 3, 4\}$. A sequence $\langle k_0, \ldots, k_\ell \rangle$ of facets of X is said to be a $simple$ $sequence$ for X if k_0 is simple for X, and if, for any $i \in \{1, \ldots, \ell\}$, k_i is simple for $X \oslash \{k_j \mid 0 \le j < i\}$. Let K be a set of facets of X. The set K is said to be F-$simple$ (where "F" stands for facet) for X if K is empty, or if the elements of K can be ordered as a simple sequence for X. The set K is $minimal$ non-$simple$ for X if it is not F-simple for X and if all its proper subsets are F-simple. The following characterization will be used in the sequel.

Theorem 18 (adapted from Gau and Kong [13], theorem 3). Let $X \sqsubseteq \mathbb{F}^d$, $with$ $d \in \{2, 3, 4\}$, and let $K \subseteq X^+$. $Then$ K is a $minimal$ non-$simple$ set for X if and $only$ if the two $following$ $conditions$ $hold:$
$i)$ $Each$ k of K is non-$simple$ for $[X \oslash K] \cup \hat{k}$.
$ii)$ $Each$ k of K is $simple$ for $[X \oslash K']$ $whenever$ $K' \subseteq K \backslash \{k\}$ and $K' \ne K \backslash \{k\}$.

For example, it may be seen that the sets displayed in Fig. 6 in light gray are indeed minimal non-simple sets.

Th. 19 is a key property[1] which is used to prove Prop. 20 and Th. 23.

Theorem 19. Let f be a d-$face$ $with$ $d \in \{2, 3, 4\}$, let ℓ be an $integer$ $strictly$ $greater$ $than$ 1, let $X_1, \ldots, X_\ell$ be ℓ $subcomplexes$ of $\hat{f}$. The two $following$ $asser$-$tions$ are $equivalent:$
$i)$ For all $L \subseteq \{1, \ldots, \ell\}$ $such$ $that$ $L \ne \emptyset$, $\cup_{i \in L} X_i$ $collapses$ $onto$ a $point.$
$ii)$ For all $L \subseteq \{1, \ldots, \ell\}$ $such$ $that$ $L \ne \emptyset$, $\cap_{i \in L} X_i$ $collapses$ $onto$ a $point.$

Let us now establish the link between MNS and crucial cliques.

Proposition 20. Let $X \sqsubseteq \mathbb{F}^d$, $with$ $d \in \{2, 3, 4\}$, let K be a $minimal$ non-$simple$ set for X, and let f be the $intersection$ of all the $elements$ of K. $Then,$ f is an M-$critical$ $face$ for X and K is the $induced$ $crucial$ $clique.$

If K is a crucial clique for X, then from Th. 18, Prop. 16 and Prop. 17, K is a minimal non-simple set for X. Conversely, if K is a minimal non-simple set for X, then by Prop. 20, K is a crucial clique. Thus, we have the following theorem.

Theorem 21. Let $X \sqsubseteq \mathbb{F}^d$, $with$ $d \in \{2, 3, 4\}$, and let $K \subseteq X^+$. $Then$ K is a $minimal$ non-$simple$ set for X if and $only$ if it is a $crucial$ $clique$ for X.

---

[1] Notice that a similar property holds in $\mathbb{R}^3$, in the framework of algebraic topology, if we replace the notion of collapsibility onto a point by the one of contractibility [18,26].

7   P-Simple Points

In the preceding section, we saw that critical kernels which are settled in the framework of abstract complexes allow to derive the notion of a minimal non-simple set proposed in the context of digital topology. Also in the framework of digital topology, one of the authors introduced the notion of P-simple point [2], and proved for the 3D case a local characterization which leads to a linear algorithm for testing P-simplicity. In [7], we stated the equivalence between the notion of 2D P-simple points and a notion derived from the one of crucial clique. Here, we extend this equivalence result up to 4D.

Let $X \sqsubseteq \mathbb{F}^n$, and let $C \subseteq X^+$. A facet $k \in C$ is said to be *P-simple for* $\langle X, C \rangle$ if k is simple for all complexes $X \oslash T$, such that $T \subseteq C \setminus \{k\}$.

Definition 22. *Let* $X \sqsubseteq \mathbb{F}^n$, *and let* C *be a set of facets of* X, *we set* $D = X^+ \setminus C$. *Let* $k \in C$, *we say that* k *is* weakly crucial *for* $\langle X, D \rangle$ *if* k *contains a face* f *which is critical for* X, *and such that all the facets of* X *containing* f *are in* C.

Theorem 23. *Let* $X \sqsubseteq \mathbb{F}^d$, *with* $d \in \{2, 3, 4\}$, *let* C *be a set of facets of* X, *let* $D = X \oslash C$. *Let* $k \in C$, *the facet* k *is P-simple for* $\langle X, C \rangle$ *if and only if* k *is not weakly crucial for* $\langle X, D \rangle$.

8   Conclusion

We provided in this article a new characterization of simple points, in dimensions up to 4D, leading to an efficient simplicity testing algorithm. Moreover, we demonstrated that the main concepts previously introduced in order to study topology-preserving parallel thinning in the framework of digital topology, namely P-simple points and minimal non-simple sets, may be not only retrieved in the framework of critical kernels, but also better understood and enriched. Critical kernels thus appear to constitute a unifying framework which encompasses previous works on parallel thinning.

References

1. Bertrand, G.: On P-simple points. Comptes Rendus de l'Académie des Sciences, Série Math. I(321), 1077–1084 (1995)
2. Bertrand, G.: Sufficient conditions for 3D parallel thinning algorithms. In: SPIE Vision Geometry IV, vol. 2573, pp. 52–60 (1995)
3. Bertrand, G.: On critical kernels. Technical Report IGM,2005-05 (2005), http://www.esiee.fr/~coupriem/ck
4. Bertrand, G.: On critical kernels. Comptes Rendus de l'Académie des Sciences, Série Math. I(345), 363–367 (2007)
5. Bertrand, G., Couprie, M.: New 2D Parallel Thinning Algorithms Based on Critical Kernels. In: Reulke, R., Eckardt, U., Flach, B., Knauer, U., Polthier, K. (eds.) IWCIA 2006. LNCS, vol. 4040, pp. 45–59. Springer, Heidelberg (2006)

6. Bertrand, G., Couprie, M.: A new 3D parallel thinning scheme based on critical kernels. In: Kuba, A., Nyúl, L.G., Palágyi, K. (eds.) DGCI 2006. LNCS, vol. 4245, pp. 580–591. Springer, Heidelberg (2006)
7. Bertrand, G., Couprie, M.: Two-dimensional parallel thinning algorithms based on critical kernels. Technical Report IGM,2006-02 (2006), http://www.esiee.fr/~coupriem/ck
8. Bing, R.H.: Some aspects of the topology of 3-manifolds related to the Poincaré conjecture. Lectures on modern mathematics II, 93–128 (1964)
9. Burguet, J., Malgouyres, R.: Strong thinning and polyhedric approximation of the surface of a voxel object. Discrete Applied Mathematics 125, 93–114 (2003)
10. Couprie, M.: Note on fifteen 2d parallel thinning algorithms. Technical Report IGM,2006-01 (2006), http://www.esiee.fr/~coupriem/ck
11. Couprie, M., Bertrand, G.: New characterizations, in the framework of critical kernels, of 2D, 3D and 4D minimal non-simple sets and P-simple points. Technical Report IGM2007-08 (2007), www.esiee.fr/~coupriem/ck
12. Couprie, M., Bertrand, G.: New characterizations of simple points in 2D, 3D and 4D discrete spaces. Technical Report IGM2007-07 (2007), www.esiee.fr/~coupriem/ck
13. Gau, C.-J., Kong, T.Y.: Minimal non-simple sets in 4D binary images. Graphical Models 65, 112–130 (2003)
14. Giblin, P.: Graphs, surfaces and homology. Chapman and Hall, Boca Raton (1981)
15. Hall, R.W.: Tests for connectivity preservation for parallel reduction operators. Topology and its Applications 46(3), 199–217 (1992)
16. Kong, T.Y.: On topology preservation in 2-D and 3-D thinning. International Journal on Pattern Recognition and Artificial Intelligence 9, 813–844 (1995)
17. Kong, T.Y.: Topology-preserving deletion of 1's from 2-, 3- and 4-dimensional binary images. In: Ahronovitz, E. (ed.) DGCI 1997. LNCS, vol. 1347, pp. 3–18. Springer, Heidelberg (1997)
18. Kong, T.Y.: Minimal non-simple and minimal non-cosimple sets in binary images on cell complexes. In: Kuba, A., Nyúl, L.G., Palágyi, K. (eds.) DGCI 2006. LNCS, vol. 4245, pp. 169–188. Springer, Heidelberg (2006)
19. Kong, T.Y., Gau, C.-J.: Minimal non-simple sets in 4-dimensional binary images with (8-80)-adjacency. In: procs. IWCIA, pp. 318–333 (2004)
20. Kong, T.Y., Rosenfeld, A.: Digital topology: introduction and survey. Computer Vision, Graphics and Image Processing 48, 357–393 (1989)
21. Kong, T.Y.: On the problem of determining whether a parallel reduction operator for n-dimensional binary images always preserves topology. In: procs. SPIE Vision Geometry II, vol. 2060, pp. 69–77 (1993)
22. Kovalevsky, V.A.: Finite topology as applied to image analysis. Computer Vision, Graphics and Image Processing 46, 141–161 (1989)
23. Lohou, C., Bertrand, G.: A 3D 12-subiteration thinning algorithm based on P-simple points. Discrete Applied Mathematics 139, 171–195 (2004)
24. Lohou, C., Bertrand, G.: A 3D 6-subiteration curve thinning algorithm based on P-simple points. Discrete Applied Mathematics 151, 198–228 (2005)
25. Ma, C.M.: On topology preservation in 3d thinning. Computer Vision, Graphics and Image Processing 59(3), 328–339 (1994)
26. Maunder, C.R.F.: Algebraic topology. Dover, london (1996)
27. Passat, N., Couprie, M., Bertrand, G.: Minimal simple pairs in the 3-d cubic grid. Technical Report IGM2007-04 (2007)
28. Ronse, C.: Minimal test patterns for connectivity preservation in parallel thinning algorithms for binary digital images. Discrete Applied Mathematics 21(1), 67–79 (1988)

Cancellation of Critical Points in 2D and 3D Morse and Morse-Smale Complexes

Lidija Čomić[1] and Leila De Floriani[2]

[1] FTN, University of Novi Sad (Serbia)
comic@uns.ns.ac.yu
[2] University of Genova (Italy) and University of Maryland (USA)
deflo@disi.unige.it

Abstract. Morse theory studies the relationship between the topology of a manifold M and the critical points of a scalar function f defined on M. The Morse-Smale complex associated with f induces a subdivision of M into regions of uniform gradient flow, and represents the topology of M in a compact way. Function f can be simplified by cancelling its critical points in pairs, thus simplifying the topological representation of M, provided by the Morse-Smale complex. Here, we investigate the effect of the cancellation of critical points of f in Morse-Smale complexes in two and three dimensions by showing how the change of connectivity of a Morse-Smale complex induced by a cancellation can be interpreted and understood in a more intuitive and straightforward way as a change of connectivity in the corresponding ascending and descending Morse complexes. We consider a discrete counterpart of the Morse-Smale complex, called a quasi-Morse complex, and we present a compact graph-based representation of such complex and of its associated discrete Morse complexes, showing also how such representation is affected by a cancellation.

1 Introduction

The Morse-Smale complex is a widely used topological representation which describes the subdivision of a manifold M into meaningful parts, characterized by uniform flow of the gradient of a scalar function f, defined on M. In applications, large data sets are usually interpolated by a continuous function, and then topological features are extracted, which represent the initial data in a compact way. Morse-Smale complexes can be applied for segmenting the graph of a scalar field for terrain modeling in 2D, and recently some algorithms have been developed for segmenting three-dimensional scalar fields through Morse-Smale complexes [11,5]. For the review of work in this area, see [4] and the references therein.

The Morse-Smale complex of a scalar function f, defined on a manifold M, is the intersection of two complexes, namely the ascending and the descending Morse complexes. Intuitively, the descending cell of a critical point p describes the flow of the gradient of f towards p, and (dually) the ascending cell of p describes the flow away from p. Each cell of a Morse-Smale complex is obtained as the intersection of two cells, the descending cell of a critical point p, and the

D. Coeurjolly et al. (Eds.): DGCI 2008, LNCS 4992, pp. 117–128, 2008.

ascending cell of a critical point q, and it describes the flow of the gradient vector field of f from q towards p.

One of the major issues when computing a representation of a scalar field as a Morse, or a Morse-Smale complex is the over-segmentation, due to the presence of noise in the data sets. To this aim, *simplification algorithms* have been developed, in order to eliminate less significant features from the Morse-Smale complex. The simplification is achieved by applying an operation called *cancellation* of critical points. A cancellation transforms one Morse-Smale complex into another, with fewer number of vertices, while maintaining the consistency of the underlying complex, and it enables the creation of a hierarchical representation. In 2D, the cancellation of a pair of critical points in a Morse-Smale complex consists of collapsing an extremum p and a saddle s into another extremum q [18,9,1,17]. Cancellation in 2D Morse complexes is considered in [3], where it is shown that cancellations reduce to well-known Euler operators [13]. Similar operations have been considered in the framework of 2D combinatorial maps [2].

Here, we investigate the effect of the cancellation in Morse-Smale complexes in two and three dimensions, and we show how the change of connectivity in a Morse-Smale complex, induced by a cancellation of pairs of critical points of f, can be interpreted and understood in a more intuitive and straightforward way as a change of connectivity in the corresponding ascending and descending Morse complexes. We then consider a discrete counterpart of the Morse-Smale complex, called a quasi-Morse complex, and we present a compact graph-based representation of such complex and of its associated discrete Morse complexes, showing also how such representation is affected by a cancellation. This representation, together with cancellation operations (and their inverse operations, called anticancellations) is a suitable basis for a multi-resolution representation of the topology of manifold M, induced by f.

The remainder of the paper is organized as follows. In Section 2, we recall some basic notions on cell complexes. In section 3, we review Morse theory, Morse and Morse-Smale complexes. In Section 4, we describe how a Morse-Smale complex is affected by cancellation of critical points of f, and we investigate how cancellation affects the structure of a Morse complex. In Section 5, we discuss the notion of a quasi-Morse complex, we introduce the incidence graph as a representation for encoding both the descending and ascending Morse complexes and the Morse-Smale complex, and we define modifications of incidence graph, induced by a cancellation. Concluding remarks are drawn in Section 6.

2   Background Notions

We briefly review some basic notions on cell complexes. For more details on algebraic topology, see [14].

Intuitively, a Euclidean cell complex is a collection of basic elements, called cells, which cover a domain in the Euclidean space $\mathbb{R}^n$ [13]. A k-*dimensional cell* (k-cell) γ, $1 \le k \le n$, is a subset of $\mathbb{R}^n$ homeomorphic to an open k-dimensional ball $B^k = \{x \in \mathbb{R}^k : ||x|| < 1\}$, with non-null (relative) boundary with respect to the topology induced by the usual topology of $\mathbb{R}^n$. A 0-cell is a point in $\mathbb{R}^n$. The boundary of a 0-cell is empty. k is called the *order* or *dimension* of a k-cell γ. The boundary and closure of γ are denoted $bd(\gamma)$ and $cl(\gamma)$, respectively.

A *Euclidean cell complex* is a finite set of cells Γ in $\mathbb{R}^n$ of dimension at most d, $0 \le d \le n$, such that (i) the cells are pairwise disjoint, (ii) for each cell $\gamma \in \Gamma$, the boundary $bd(\gamma)$ is the union of cells of Γ, (iii) if $\gamma, \gamma_1 \in \Gamma$, such that $cl(\gamma) \cap cl(\gamma_1) \ne \emptyset$, then $cl(\gamma) \cap cl(\gamma_1)$ is the disjoint union of cells of Γ. The maximum d of dimensions of cells γ over all cells of a complex Γ is called the *dimension* or the *order* of the complex. A subset Λ of Γ is called a *subcomplex* of Γ if and only if Λ is a cell complex. The k-skeleton of Γ is the subcomplex of Γ, which consists of all cells of Γ of dimension less than or equal to k. The *domain* (or *carrier*) $\Delta\Gamma$ of a Euclidean cell complex Γ is the subset of $\mathbb{R}^n$ spanned by the cells of Γ. We shall consider two and three dimensional cell complexes Γ, such that $\Delta\Gamma$ is homeomorphic to a smooth compact two or three dimensional manifold M without boundary. Recall that a *d-manifold M* without boundary is a (separable Hausdoff) topological space in which each point p has a neighborhood which is homeomorphic to $\mathbb{R}^d$.

An h-cell γ' which belongs to the boundary $bd(\gamma)$ of a cell γ is called an *h-face* of γ. If $\gamma' \ne \gamma$, then γ' is called a *proper face* of γ, and γ and γ' are said to be *incident*. Two distinct k-cells γ and γ' are *adjacent* if (i) for $0 < k \le d$, there exists some $(k-1)$-cell of Γ which is a face of both γ and γ', and (ii) for $0 \le k < d$, there exists some $(k+1)$-cell of Γ which contains both γ and γ' as a face.

The *space dual* Γ^* of a d-complex Γ is a cell complex such that there is a one-to-one mapping from Γ onto Γ^* such that (i) the image of a k-cell is a $(d-k)$-cell, (ii) cells γ and γ' are adjacent in Γ if and only if their images are adjacent in Γ^*. The space dual Γ^* of a complex Γ is unique up to adjacency relation between cells.

3 Morse Theory

Morse theory studies relationships between the topology of a manifold, and (the critical points of) a function defined on a manifold. We review here the basic notions of Morse theory in the case of d-manifolds. For more details, see [16,15].

3.1 Morse Functions

Let f be a C^2-differentiable real-valued function defined over a domain $D \subseteq \mathbb{R}^d$. A point $p \in \mathbb{R}^d$ is a *critical point* of f if the gradient ∇f of f vanishes on p, i.e., if $\nabla f(p) = 0$. Function f is said to be a *Morse function* if all its critical points are non-degenerate (the Hessian matrix $Hess_p f$ of the second derivatives of f at p is non-singular). Since all these properties are local, a Morse function f can be defined on a d-dimensional manifold M. In some local coordinate system, there is a neighborhood of each critical point p of f, in which

$$f(x_1, x_2, \ldots, x_d) = f(p) - x_1 - \ldots - x_i + x_{i+1} + \ldots + x_d.$$

The number of minus signs in the above equality, i.e., the number of negative eigenvalues of $Hess_p f$, is called the *index* of a critical point p. The corresponding eigenvectors show the directions in which f is decreasing. A critical point p is a *minimum* or a *maximum* if it has index 0 or d, respectively. Otherwise, if index of p is i, $0 < i < d$, p is an *i-saddle*.

An *integral line* $c(t)$ of a function f is a maximal path which is everywhere tangent to ∇f. Thus, $c(t)$ follows the direction in which the function has the maximum increasing growth. Each integral line connects two distinct critical points of f, called its *origin* and *destination*. In 2D, each saddle s is incident to two ascending and two descending integral lines, which connect s to two (not necessarily distinct) maxima and two (not necessarily distinct) minima, respectively. These ascending and descending integral lines alternate cyclically around s. In 3D, each 1-saddle (2-saddle) s is incident to two descending (ascending) integral lines, which connect s to two (not necessarily distinct) minima (maxima).

An operation called *cancellation* of critical points is defined on the set of critical points of f. It eliminates the critical points of f in pairs. If p and q are two critical points of f, such that p is of index i, q is of index $i+1$, and such that there is a unique integral line c having p as origin and q as destination, then f can be perturbed to another Morse function g defined on M, such that critical points of g coincide in position and in index with the remaining critical points of f (except p and q). A cancellation of a pair of critical points p and q modifies the gradient field along c, and in a small neighborhood of c, thus simulating the smoothing of f.

3.2   Morse and Morse-Smale Complexes

Let $f : M \to \mathbb{R}$ be a Morse function, where M is a compact d-manifold without boundary. Integral lines that converge to (originate from) a critical point p of index i form an i-cell ($(d-i)$-cell) called a *stable (unstable) manifold* (or *descending (ascending) manifold*) of p. The descending (ascending) manifolds are pairwise disjoint and they decompose M into open cells which form a complex, since the boundary of every cell is a union of lower-dimensional cells. Such complexes are called *descending* and *ascending Morse complexes*.

If p is a minimum, the descending manifold of p is p itself, while if p is a 1-saddle, the descending manifold of p is an open 1-cell bounded by two (not necessarily distinct) minima. If p is a maximum in a 2D manifold M, the descending manifold of p is an open 2-cell, bounded by a sequence of (at least one) minima and (at least one) descending 1-cells of saddles. In 3D, the boundary B of a descending three-cell of a maximum p consists of (at least one) descending 2-cells of 2-saddles, and the boundaries of these 2-cells consist of descending 1-cells of 1-saddles on B, and (at least one) minima.

A Morse function f is called a *Morse-Smale function* if and only if the descending and the ascending manifolds intersect transversally. This means that the intersection (if it exists) of the descending i-dimensional manifold of a critical point p of index i, and the ascending $(d-j)$-dimensional manifold of a critical point q of index j, $i \geq j$, is an $(i-j)$-dimensional manifold. Cells that are obtained as the intersection of descending and ascending manifolds of a Morse-Smale function f decompose M into a *Morse-Smale complex*.

In 2D, each 2-cell of a Morse-Smale complex is related to a maximum p and a minimum q, as it is obtained as the intersection of the descending 2-manifold of p and the ascending 2-manifold of q. Such 2-cell is quadrangular, with vertices q, s_1, p, s_2 (of index 0,1,2,1), in this order along the boundary. These 2-cells are obtained as union of two triangles, namely q, s_1, p and q, s_2, p, where s_1 and s_2 are the (not necessarily distinct) saddles, which are in the boundary of both the descending 2-manifold of p and the ascending 2-manifold of q. A maximum p and a minimum q may determine more than one 2-cell of a Morse-Smale complex.

In 3D, each 3-cell of a Morse-Smale complex is related to a maximum p and a minimum q. Such 3-cell is obtained as union of tetrahedra of the form q, s_1, s_2, p, where s_1 and s_2 are 1-saddles and 2-saddles, which are in the boundary of both the descending 3-manifold of p and the ascending 3-manifold of q.

If f is a Morse-Smale function, then the ascending and the descending complex of f are dual to each other. Each critical point p of index i of a Morse-Smale function f corresponds to a vertex (which we denote by p) of a Morse-Smale complex of f, and it also corresponds to the descending i-dimensional manifold of p, i.e., to an i-cell of the descending Morse complex (denoted by p), and to the ascending $(d - i)$-dimensional manifold of p. We illustrate the correspondence between Morse and Morse-Smale complexes in Figure 1.

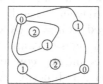 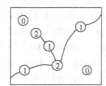 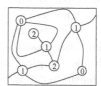

Fig. 1. A Morse-Smale function f, defined on a 2-dimensional domain. Numbers indicate the index of the critical points. (left) Descending Morse complex. (middle) Ascending Morse complex. (right) Morse-Smale complex.

4   Cancellations in Morse-Smale and Morse Complexes

Although Morse and Morse-Smale complexes describe the topology of a manifold M in a more compact way than the full dataset, it is often desirable to reduce the number of critical points of f, thus obtaining a simplified representation of the topology of M. This can be achieved by applying the operation of cancellation of (pairs of) critical points. A pair of critical points p and q of function f can be cancelled if p and q have consecutive indices i and $i + 1$, respectively, and there is a unique integral line connecting them. In the Morse-Smale complex of f, vertices p and q are adjacent and connected through a single edge, while in the Morse complex, cell p is in the boundary of cell q, and p appears only once in the boundary of q. Specifically, a vertex p and an edge q may be cancelled if q is not a loop with endpoint p. An edge p and a face q may be cancelled if in a cyclic order of edges in the boundary of q, edge p appears only once. In 2D, this means that the unique other face r incident to p is different from q. Finally, (in 3D) a face p and a 3-cell (volume) q may be cancelled if the unique other 3-cell r incident to p is different from q.

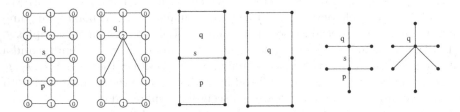

Fig. 2. Cancellation of a maximum p and a saddle s in a 2D (left) Morse-Smale complex, (middle) descending Morse complex (edge removal), (right) ascending complex, and dually, cancellation of a minimum p and a saddle s in the descending complex (edge contraction). Numbers indicate the index of the corresponding critical point.

In 2D, a saddle and an extremum are cancelled. Cancellation of a maximum p and a saddle s in a Morse-Smale complex is illustrated in Figure 2 (left). It can be viewed as merging p and s into the other maximum q adjacent to saddle s, while at the same time all four edges incident in s are deleted. If $q = p$, cancellation is forbidden. Note that we are only interested in the changes of connectivity in a Morse-Smale complex, while we do not take into account any considerations about the feasibility of a cancellation, which are induced by function values. In the example above, $f(q)$ has to be greater than the value of f at any of the saddles which become connected with q, if the cancellation of p and s is to be allowed. In a descending Morse complex, cancellation corresponds to merging the 2-cells p and q into q, i.e., to the removal of edge s, as illustrated in Figure 2 (middle). In a Morse-Smale complex, each saddle which was previously connected to p is connected to q after cancellation. Correspondingly, in the descending Morse complex, each edge which was previously on the boundary of 2-cell p is on the boundary of 2-cell q after cancellation. Similarly, the cancellation of a saddle and a minimum corresponds to an edge contraction, which is the dual operation of edge removal, as illustrated in Figure 2. The reverse is true for the ascending complex. The order in which the pairs of points are canceled can be determined based on the notion of persistence [9,1].

Simplification of 3D Morse-Smale complexes has been investigated only recently [10]. In 3D there are three possible types of cancellations: minimum and 1-saddle, 1-saddle and 2-saddle, and 2-saddle and maximum. While the two cancellations involving a minimum or a maximum are similar to the ones performed in the 2D case, the saddle-saddle cancellation does not have an analog in 2D.

A cancellation of a minimum and a 1-saddle, or of a maximum and a 2-saddle, in a Morse-Smale complex, can be interpreted as merging the two canceled points, an extremum point p and a saddle s into the unique extremum point q which is the other endpoint of the (descending or ascending) 1-manifold of saddle s. As in the 2D case, if $q = p$, cancellation is forbidden. Edges and faces which were incident in s are deleted from the Morse-Smale complex, and q will replace p in all edges and faces which had p as one of their vertices. In a 3D descending Morse complex, cancellation of a 1-saddle and a minimum corresponds to an edge contraction, while cancellation of a maximum and a 2-saddle corresponds to the removal of a face. The modifications of (a part of) a Morse-Smale and of the descending and ascending Morse complexes, induced by a cancellation of a maximum p and a 2-saddle s, are illustrated in Figure 3.

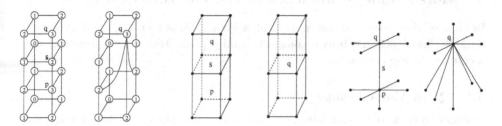

Fig. 3. Cancellation of a maximum p and a 2-saddle s in 3D in a (left) Morse-Smale complex, (middle) descending Morse complex (face removal), (right) ascending Morse complex, and dually, cancellation of a minimum p and a 1-saddle s in the descending complex (edge contraction). Numbers indicate the index of the corresponding critical point.

A cancellation of a 1-saddle p and a 2-saddle q makes the extension from 2D to 3D non-trivial. When p and q are cancelled, the descending and ascending manifolds of p and q are deleted from the Morse-Smale complex, and the remaining points on the boundary of these 2-manifolds are reconnected. Each 2-saddle, which was adjacent to p, becomes adjacent to each 1-saddle, which was adjacent to q, as illustrated in Figure 4 (left). In this way, the number of cells in the Morse-Smale complex increases, although the number of critical points (vertices of the Morse-Smale complex) decreases by two. The cancellation of a 2-saddle q and a 1-saddle p in a descending Morse complex is illustrated in Figure 4 (right). Edge p and face q are deleted, and the boundary of each face incident in p is extended to include the boundary of face q (with the exception of edge p). It can be viewed as each face incident in p is extended to include a copy of face q, keeping an infinitesimal separation between these copies. Each edge which was on the boundary of face q is, after cancellation, on the boundary of all other faces incident to p. Thus, all the vertices (minima) on the boundary of face q will be, after cancellation, on the boundary of all faces incident to edge p, and on the boundary of all volumes (maxima) incident to p. All such minima-maxima pairs induce new cells in the Morse-Smale complex, although the number of cells in the Morse complexes decreases by two.

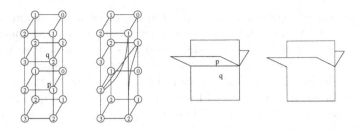

Fig. 4. Cancellation of a 1-saddle p and a 2-saddle q in 3D in a (left) Morse-Smale complex, (right) descending Morse complex. Numbers indicate the index of the corresponding critical point.

5   Morse-Smale Complexes in the Discrete Case

In this Section, we recall the notion of a quasi-Morse complex, and we introduce the incidence graph as a compact, dual representation of descending and ascending Morse complexes.

5.1   Quasi-Morse Complexes

In order to capture the combinatorial structure of a Morse-Smale complex of a manifold M, without making reference to a function f, a notion of a *quasi-Morse complex* in 2D and 3D is introduced in [9] and [8], respectively.

In 2D, a cell complex Γ is a *quasi-Morse complex* if it is a subdivision of M, in which the set of vertices (0-cells) of Γ is partitioned into three sets U, V and W and the set of edges (1-cells) of Γ is partitioned into two sets A and B with the following properties: (i) there is no edge which connects two vertices from $U \cup W$, or two vertices from V, (ii) edges in A have endpoints in $U \cup V$ and edges in B have endpoints in $V \cup W$, (iii) each vertex $p \in V$ belongs to four edges, which alternate between A and B in a cyclic order around p, (iv) all 2-cells of Γ are quadrangles, with vertices from U, V, W, V, in this order along the boundary.

In 3D, a cell complex Γ is a *quasi-Morse complex* if it is a subdivision of M, in which the set of vertices of Γ is partitioned into four sets U, V, X and Y, the set of edges of Γ is partitioned into three sets R, S, and T and the set of 2-cells of Γ is partitioned into two sets P and Q with the following properties: (i) edges from R, S, and T connect vertices from U and V, V and X, and X and Y, respectively, and 2-cells from P and Q are quadrangles with nodes from U, V, X, V, and V, X, Y, X, in that order, respectively, along the boundary, (ii) there are no vertices within 1-, 2-, and 3-cells of Γ, (iii) each edge in S is on the boundary of four quadrangles, which in a cyclic order alternate between P and Q. In 3D, a quasi-Morse complex is more restrictive than a Morse-Smale complex. In the Morse complexes corresponding to a quasi-Morse complex, faces bounded by only one vertex, and edges which do not bound a face, are not allowed, since they would imply the existence of faces in a quasi-Morse complex, which are not quadrangles.

Thus, in a quasi-Morse complex, the cancellation of an edge e and a face f is forbidden if (i) e is the only edge in the boundary of f, and there is a face different from f, bounded only by e, or (ii) e is bounding only f, and there is an edge different from e, which is bounding only f. Otherwise, a cancellation could result in a face bounded by only one vertex, and an edge not bounding a face, respectively. Cancellation of a minimum (vertex) p and a 1-saddle (edge) q is allowed if there is no 2-saddle (face) r such that q is the only edge in the boundary of r. Dually, cancellation of a maximum (volume) p and a 2-saddle (face) q is allowed if there is no 1-saddle (edge) r such that r bounds no other face but q. These situations are illustrated in Figure 5.

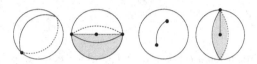

Fig. 5. (left) Forbiden face-edge cancellation. (right) Forbiden vertex-edge and volume-face cancellation.

5.2 Dual Representation of Morse Complexes

We encode the descending Morse complex Γ_d, obtained from a quasi-Morse complex of a Morse-Smale function, using an incidence graph $IG = (C, A)$.

The *Incidence Graph (IG)* [7] is an incidence-based explicit data structure for cell complexes in arbitrary dimensions. The topological information captured by the IG is the set of incidence relations among cells that differ by exactly one dimension. Formally, the IG encodes all the cells of any given d-dimensional cell complex Γ, and for each i-cell γ, its immediate boundary, and immediate co-boundary relations, namely all $(i - 1)$-cells in the boundary of γ, and all $(i + 1)$-cells in the co-boundary of γ. An incidence graph corresponds to the *Hasse diagram* describing the complex [12]. If we turn the incidence graph IG of a descending complex Γ_d 'upside-down', we obtain the incidence graph of the ascending complex Γ_a, which is dual to Γ_d. Thus, the incidence graph encodes simultaneously both the ascending and the descending Morse complexes. In Figure 6 (left), a cell complex describing the subdivision of the extended plane, its dual complex, and the corresponding incidence graph are illustrated.

An incidence graph IG encodes directly the 1-skeleton of a quasi-Morse complex, since the nodes of the IG correspond to the critical points of f, and there is an arc in the IG connecting two nodes p and q if and only if p and q are connected by an edge in the quasi-Morse complex. Other cells of a Morse-Smale complex are encoded implicitly. In 2D, all 2-cells of a Morse-Smale complex, associated with a maximum p and a minimum q, are obtained as the union of two triangles p, s_1, q and p, s_2, q, where s_1 and s_2 are two (not necessarily distinct) saddles which are connected to both p and q in IG. In 3D, all 3-cells of a Morse-Smale complex, associated with a maximum p and a minimum q, are obtained as the union of tetrahedra of the form q, s_1, s_2, p, where s_1 are 1-saddles which are connected to q, s_2 are 2-saddles which are connected to p, and s_1 and s_2 are connected to each other in IG. 2-cells in a 3D Morse-Smale complex are associated either with a maximum p and a 1-saddle s_1, or with a minimum q and a 2-saddle s_2, and are obtained in a similar way as the 2-cells in a 2D Morse-Smale complex. Note that an incidence graph IG cannot encode a general Morse complex, since in IG there is no possibility to encode, for example, a face bounded by just one vertex.

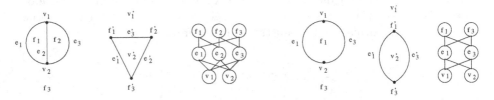

Fig. 6. (left) A 2-dimensional cell complex Γ_d, which consists of vertices v_1 and v_2, edges e_1, e_2 and e_3, and faces f_1 f_2 and f_3, the dual complex Γ_a, and incidence graph IG of Γ_d and Γ_a. (right) Γ_d, Γ_a, and IG, after the cancellation of edge e_2 and face f_2.

Cancellation of critical points in a quasi-Morse complex can be formalized in terms of the incidence graph. Let $IG = (C, A)$ be the incidence graph of the quasi-Morse complex associated with f, and let p and q be two critical points of f of consecutive index i and $i + 1$, respectively, such that there is a unique integral line of f joining p and q. Then p and q are nodes of the IG, which belong to two consecutive layers of the IG, such that p and q are connected by a unique arc of IG. Let us denote by I^+ the set of arcs incident to either p or q. We will call this set the *influence set* of the cancellation $u_c(p, q)$ of p and q. For each arc $(p, s) \in I^+$, where s is an $(i + 1)$-cell and $s \neq q$, we call s a *relevant $(i + 1)$-cell*, and similarly for each arc $(q, r) \in I^+$, where r is an i-cell and $r \neq p$, we call r a *relevant i-cell* of update $u_c(p, q)$. Cancellation of p and q eliminates p and q from IG, together with arcs from I^+, and introduces new set of arcs, denoted I^- connecting all relevant i-cells r, to all relevant $(i + 1)$-cells s. Thus, after the cancellation of p and q, the incidence graph $IG = (C, A)$ of f is transformed into another incidence graph $IG' = (C', A')$, such that $C' = C\backslash\{p, q\}$, $A' = (A\backslash I^+) \cup I^-$. Cancellation of edge e_2 and face f_2 in cell complex (and the corresponding incidence graph) from Figure 6 (left) is illustrated in Figure 6 (right).

Another condition needs to be imposed since IG' needs to have the structure of an incidence graph, as explained above. We formulate these conditions in terms of the incidence graph. Cancellation of a 1-cell p and a 2-cell q should be forbidden if (i) there are no relevant 1-cells, and there is a 2-cell s which is connected to no 1-cell other than p, or (ii) there are no relevant 2-cells, and there is a 1-cell r which is connected to no 2-cell other than q. Cancellation of a 0-cell p and a 1-cell q should be forbidden if there is a 2-cell s, which is connected to no 1-cell other than q, while cancellation of a 2-cell p and a 3-cell q should be forbidden if there is a 1-cell r which is connected to no 2-cell other than p.

6 Concluding Remarks

We have recalled the notion of cancellation of critical points of a Morse function f, and have shown how the changes of connectivity in a Morse-Smale complex of f after a cancellation are easier to understand if interpreted as changes of connectivity in a Morse complex of f.

We have considered the quasi-Morse complex as the combinatorial structure of the Morse-Smale complex in the discrete case and we have discussed the incidence graph as a representation for quasi-Morse complexes. This representation encodes both the ascending and the descending complexes associated with a quasi-Morse complex. The incidence graph for a 3D scalar field can be constructed by applying the algorithm described in [5], which computes the two dual Morse complexes through an efficient discrete approach that does not involve any floating-point computation.

We are currently working on a definition of a multi-resolution topological model for 3D scalar fields, based on the cancellation operators discussed and on the notion of dependency of the modifications of a cell complex. This will lead to a morphology-based description of a 3D scalar field at different levels of abstraction, based on the framework on multi-resolution modeling introduced in [6]. Further developments of this work are concerned with extending the formalization proposed here to the n-dimensional case.

Acknowledgements

This work has been partially supported by the European Network of Excellence AIM@SHAPE under contract number 506766, by the National Science Foundation under grant CCF-0541032, by the MIUR-FIRB project SHALOM under contract number RBIN04HWR8 and by the MIUR-PRIN project on "Multiresolution modeling of scalar fields and digital shapes".

References

1. Bremer, P.-T., Edelsbrunner, H., Hamann, B., Pascucci, V.: A Topological Hierarchy for Functions on Triangulated Surfaces. Transactions on Visualization and Computer Graphics 10(4), 385–396 (2004)
2. Brun, L., Kropatsch, W.G.: Dual Contraction of Combinatorial Maps. Technical Report PRIP-TR-54, Institute for Computer Aided Automation 183/2, Pattern Recognition and Image Processing Group, TU Wien, Austria (1999)
3. Čomić, L.: On Morse-Smale Complexes and Dual Subdivisions. In: 4th Serbian-Hungarian Joint Symposium on Intelligent Systems (SISY 2006), Subotica, Serbia, September 29-30, pp. 265–274 (2006)
4. Čomić, L., De Floriani, L., Papaleo, L.: Morse-Smale Decompositions for Modeling Terrain Knowledge. In: Cohn, A.G., Mark, D.M. (eds.) COSIT 2005. LNCS, vol. 3693, pp. 426–444. Springer, Heidelberg (2005)

5. Danovaro, E., De Floriani, L., Mesmoudi, M.M.: Topological Analysis and Characterization of Discrete Scalar Fields. In: Asano, T., Klette, R., Ronse, C. (eds.) Geometry, Morphology, and Computational Imaging. LNCS, vol. 2616, pp. 386–402. Springer, Heidelberg (2003)
6. De Floriani, L., Magillo, P., Puppo, E.: Multiresolution Representation of Shapes Based on Cell Complexes (invited paper). In: Bertrand, G., Couprie, M., Perroton, L. (eds.) DGCI 1999. LNCS, vol. 1568, pp. 3–18. Springer, Heidelberg (1999)
7. Edelsbrunner, H.: Algorithms in Combinatorial Geometry. Springer, Heidelberg (1987)
8. Edelsbrunner, H., Harer, J., Natarajan, V., Pascucci, V.: Morse-Smale Complexes for Piecewise Linear 3-Manifolds. In: Proceedings 19th ACM Symposium on Computational Geometry, pp. 361–370 (2003)
9. Edelsbrunner, H., Harer, J., Zomorodian, A.: Hierarchical Morse Complexes for Piecewise Linear 2-Manifolds. In: Proceedings 17th ACM Symposium on Computational Geometry, pp. 70–79 (2001)
10. Gyulassy, A., Natarajan, V., Pascucci, V., Bremer, P.-T., Hamann, B.: Topology-based Simplification for Feature Extraction from 3D Scalar Fields. In: Proceedings IEEE Visualization 2005, pp. 275–280. ACM Press, New York (2005)
11. Gyulassy, A., Natarajan, V., Pascucci, V., Hamann, B.: Efficient Computation of Morse-Smale Complexes for Three-Dimensional Scalar Functions. In: Proceedings IEEE Visualization 2007, Sacramento, California, October 28 - November 1, ACM Press, New York (2007)
12. Lewiner, T.: Geometric Discrete Morse Complexes. PhD thesis, Department of Mathematics, PUC-Rio, Advised by Hélio Lopes and Geovan Tavares (2005)
13. Mantyla, M.: An Introduction to Solid Modeling. Computer Science Press (1987)
14. Massey, W.S.: A Basic Course in Algebraic Topology. In: Graduate Texts in Mathematics, vol. 127. Springer, Heidelberg (1991)
15. Matsumoto, Y.: An Introduction to Morse Theory. Translations of Mathematical Monographs, vol. 208. American Mathematical Society (2002)
16. Milnor, J.: Morse Theory. Princeton University Press, New Jersey (1963)
17. Takahashi, S., Ikeda, T., Kunii, T.L., Ueda, M.: Algorithms for Extracting Correct Critical Points and Constructing Topological Graphs from Discrete Geographic Elevation Data. Computer Graphics Forum 14, 181–192 (1995)
18. Wolf, G.W.: Topographic Surfaces and Surface Networks. In: Rana, S. (ed.) Topological Data Structures for Surfaces, pp. 15–29. John Wiley, Chichester (2004)

Characterizing and Detecting Toric Loops in n-Dimensional Discrete Toric Spaces

John Chaussard, Gilles Bertrand, and Michel Couprie

Université Paris-Est, LABINFO-IGM, CNRS UMR8049
ESIEE, 2 boulevard Blaise Pascal, Cité DESCARTES
BP 99 93162 Noisy le Grand CEDEX, France
chaussaj@esiee.fr, bertrang@esiee.fr, coupriem@esiee.fr

Abstract. Toric spaces being non-simply connected, it is possible to find in such spaces some loops which are not homotopic to a point: we call them *toric loops*. Some applications, such as the study of the relationship between the geometrical characteristics of a material and its physical properties, rely on three-dimensional discrete toric spaces and require detecting objects having a toric loop.

In this work, we study objects embedded in discrete toric spaces, and propose a new definition of loops and equivalence of loops. Moreover, we introduce a characteristic of loops that we call *wrapping vector*: relying on this notion, we propose a linear time algorithm which detects whether an object has a toric loop or not.

1   Introduction

Topology is used in various domains of image processing in order to perform geometric analysis of objects. In porous material analysis, different topological transformations, such as skeletonisation, are used to study the relationships between the geometrical characteristics of a material and its physical properties.

When simulating a fluid flow through a porous material, the whole material can be approximated by the tessellation of the space made up by copies of one of its samples, under the condition that the volume of the sample exceeds the so-called Representative Elementary Volume (REV) of the material [1]. When the whole Euclidean space is tiled this way, one can remark that the result of the fluid flow simulation is itself the tessellation of the local flow obtained inside any copy of the sample (see Fig. 1-**a**). When considering the flow obtained inside the sample, it appears that the flow leaving the sample by one side comes back by the opposite side (see Fig. 1-**b**). Thus, it is possible to perform the fluid flow simulation only on the sample, under the condition that its opposite sides are joined: with this construction, the sample is embedded inside a toric space [2] [3]. In order to perform geometric analysis of fluid flow through porous materials, we therefore need topological tools adapted to toric spaces.

Considering the sample inside a toric space leads to new difficulties. In a real fluid flow, grains of a material (pieces of the material which are not connected with the borders of the sample) do not have any effect on the final results, as

D. Coeurjolly et al. (Eds.): DGCI 2008, LNCS 4992, pp. 129–140, 2008.

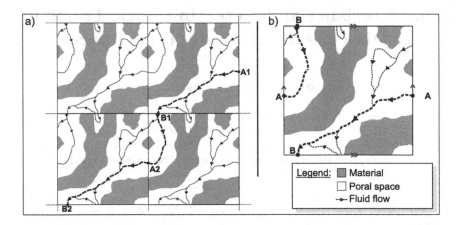

Fig. 1. Simulating a fluid flow - When simulating a fluid flow, a porous material (in gray) can be approximated by the tessellation of one of its samples (see **a**). When the results of the simulation are obtained (the dotted lines), one can see that the fluid flow through the mosaic is the tessellation of the fluid flow simulation results obtained in one sample. For example, one can look at the bold dotted line in **a**): the flow going from $A1$ to $B1$ is the same than the flow going from $A2$ to $B2$. It is therefore possible to perform the fluid flow simulation through only one sample and, in order to obtain the same results than in **a**), connect the opposite sides of the sample (see **b**): the sample is embedded inside a toric space.

these grains eventually either evacuate the object with the flow or get blocked and connect with the rest of the material. Thus, before performing a fluid flow simulation on a sample, it is necessary to remove its grains (typically, in a finite subset S of $\mathbb{Z}^n$, a grain is a connected component which does not 'touch' the borders of S). However, characterizing a grain inside a toric space, which does not have any border, is more difficult than in $\mathbb{Z}^n$. On the contrary of the discrete space $\mathbb{Z}^n$, n-dimensional discrete toric spaces are not simply connected spaces [3]: some loops, called *toric loops*, are not homotopic to a point (this can be easily seen when considering a 2D torus). In a toric space, a connected component may be considered as a grain if it contains no toric loop. Indeed, when considering a sample embedded inside a toric space, and a tessellation of the Euclidean space made up by copies of this sample, one can remark that the connected components of the sample which do not contain toric loops produce grains in the tessellation, while the connected components containing toric loops cannot be considered as grains in the tiling (see Fig. 2).

In this work, we give a new definition of loops and homotopy class, adapted to n-dimensional discrete toric spaces. Relying on these notions, we also introduce *wrapping vectors*, a new characteristic of loops in toric spaces which is the same for all homotopic loops. Thanks to wrapping vectors, we give a linear time algorithm which allows us to decide whether an n-dimensional object contains a toric loop or not.

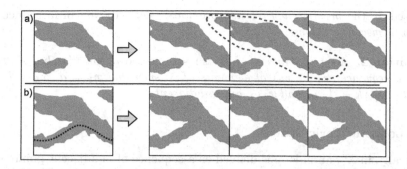

Fig. 2. Grains in toric spaces - The image in **a)** contains no grain based on the 'border criterion'; when the Euclidean space is tessellated with copies of the image, grains appear (the circled connected component is an example of grain). In **b)**, the connected component has toric loops (e.g. the dotted line) and when the Euclidean space is tessellated with copies of the image, no grain appear.

2   Basic Notions

2.1   Discrete Toric Spaces

A n-dimensional torus is classically defined as the direct product of n circles (see [2]). In the following, we give a discrete definition of toric space, based on modular arithmetic (see [4]).

Given d a positive integer. We set $\mathbb{Z}_d = \{0, ..., d-1\}$. We denote by $\oplus_d$ the operation such that for all $a, b \in \mathbb{Z}$, $(a \oplus_d b)$ is the element of $\mathbb{Z}_d$ congruent to $(a + b)$ modulo d. We point out that $(\mathbb{Z}_d, \oplus_d)$ forms a cyclic group of order d.

Let n be a positive integer, $\boldsymbol{d} = (d_1, ..., d_n) \in \mathbb{N}^n$, and $\mathbb{T}^n = \mathbb{Z}_{d_1} \times ... \times \mathbb{Z}_{d_n}$, we denote by $\oplus_{\boldsymbol{d}}$ the operation such that for all $\boldsymbol{a} = (a_1, ..., a_n) \in \mathbb{Z}^n$ and $\boldsymbol{b} = (b_1, ..., b_n) \in \mathbb{Z}^n$, $\boldsymbol{a} \oplus_{\boldsymbol{d}} \boldsymbol{b} = (a_1 \oplus_{d_1} b_1, ..., a_n \oplus_{d_n} b_n)$. The group $(\mathbb{T}^n, \oplus_{\boldsymbol{d}})$ is the direct product of the n groups $(\mathbb{Z}_{d_i}, \oplus_{d_i})_{(1 \leq i \leq n)}$, and is an *n-dimensional discrete toric space* [2].

The scalar d_i is *the size of the i-th dimension* of $\mathbb{T}^n$, and $\boldsymbol{d}$ is the *size (vector)* of $\mathbb{T}^n$. For simplicity, the operation $\oplus_{\boldsymbol{d}}$ will be also denoted by $\oplus$.

2.2   Neighbourhoods in Toric Spaces

As in $\mathbb{Z}^n$, various adjacency relations may be defined in a toric space.

Definition 1. *A vector $\boldsymbol{s} = (s_1, ..., s_n)$ of $\mathbb{Z}^n$ is an m-step $(0 < m \leq n)$ if, for all $i \in [1; n]$, $s_i \in \{-1, 0, 1\}$ and $\sum_{i=1}^{n} |s_i| \leq m$.*

Two points $\boldsymbol{a}, \boldsymbol{b} \in \mathbb{T}^n$ are m-adjacent if there exists an m-step $\boldsymbol{s}$ such that $\boldsymbol{a} \oplus \boldsymbol{s} = \boldsymbol{b}$.

In 2D, the 1- and 2-adjacency relations respectively correspond to the 4- and 8-neighbourhood adapted to bidimensional toric spaces. In 3D, the 1-, 2- and 3-adjacency relations can be respectively seen as the 6-, 18- and 26-neighbourhood adapted to three-dimensional toric spaces.

Based on the *m-adjacency relation* previously defined, we can introduce the notion of *m-connectedness*.

Definition 2. *A set of points X of $\mathbb{T}^n$ is m-connected if, for all $a, b \in X$, there exists a sequence $(x_1, ..., x_k)$ of elements of X such that $x_1 = a$, $x_k = b$ and for all $i \in [1; k-1]$, x_i and x_{i+1} are m-adjacent.*

2.3 Loops in Toric Spaces

Classically, in $\mathbb{Z}^n$, an m-loop is defined as a sequence of m-adjacent points such that the first point and the last point of the sequence are equal [5]. However, this definition does not suit discrete toric spaces: in small discrete toric spaces, two different loops can be written as the same sequence of points, as shown in the following example.

Example 3. Let us consider the bidimensional toric space $\mathbb{T}^2 = \mathbb{Z}_3 \times \mathbb{Z}_2$, and the 2-adjacency relation on $\mathbb{T}^2$. Let us also consider $x_1 = (1, 0)$ and $x_2 = (1, 1)$ in $\mathbb{T}^2$.

There are two ways of interpreting the sequence of points $\mathcal{L} = (x_1, x_2, x_1)$ as a loop of $\mathbb{T}^2$: either $\mathcal{L}$ is the loop passing by x_1 and x_2 and doing a 'u-turn' to come back to x_1, or $\mathcal{L}$ is the loop passing by x_1 and x_2, and 'wrapping around' the toric space in order to reach x_1 without any 'u-turn', as shown on Fig. 3.

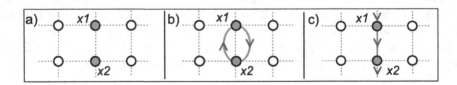

Fig. 3. Loops in toric spaces - In the toric space $\mathbb{Z}_3 \times \mathbb{Z}_2$ (see **a**), the sequence of points (x_1, x_2, x_1) can be interpreted in two different ways: **b)** and **c)**

Thus, when considering discrete toric spaces, loops cannot be considered as sequences of points since it can lead to such ambiguities. This is why we propose the following definition.

Definition 4. *Given $p \in \mathbb{T}^n$, an m-loop (of base point p) is a pair $\mathcal{L} = (p, V)$, where $V = (v_1, ..., v_k)$ is a sequence of m-steps such that $(p \oplus v_1 \oplus ... \oplus v_k) = p$.*
We call i-th point of $\mathcal{L}$, with $1 \leq i \leq k$, the point $(p \oplus v_1 \oplus ... \oplus v_{i-1})$.
The loop $(p, ())$ is called the trivial loop of base point p.

The ambiguity pinpointed in Ex. 3 is removed with this definition of loops: let v be the vector $(0, 1)$, the loop passing by x_1 and x_2 and making a u-turn is $(x_1, (v, -v))$ (see Fig. 3-b), while the loop wrapping around the toric space is $(x_1, (v, v))$ (see Fig. 3-c).

3   Loop Homotopy in Toric Spaces

3.1   Homotopic Loops

In this section, we define an equivalence relation between loops, corresponding to an homotopy, inside a discrete toric space. An equivalence relation between loops inside $\mathbb{Z}^2$ and $\mathbb{Z}^3$ has been defined in [5], however, it cannot be adapted to discrete toric spaces (see [6]). Observe that the following definition does not constrain the loops to lie in a subset of the space, on the contrary of the definition given in [5].

Definition 5. *Let* $\mathcal{K} = (p, U)$ *and* $\mathcal{L} = (p, V)$ *be two m-loops of base point* $p \in \mathbb{T}^n$, *with* $U = (u_1, ..., u_k)$ *and* $V = (v_1, ..., v_l)$. *The two m-loops* $\mathcal{K}$ *and* $\mathcal{L}$ *are directly homotopic if one of the three following conditions is satisfied:*

1. *There exists* $j \in [1; l]$ *such that* $v_j = 0$ *and* $U = (v_1, ..., v_{j-1}, v_{j+1}, ..., v_l)$.
2. *There exists* $j \in [1; k]$ *such that* $u_j = 0$ *and* $V = (u_1, ..., u_{j-1}, u_{j+1}, ..., u_k)$.
3. *There exists* $j \in [1; k-1]$ *such that*
   - $V = (u_1, ..., u_{j-1}, v_j, v_{j+1}, u_{j+2}, ..., u_k)$, *and*
   - $u_j + u_{j+1} = v_j + v_{j+1}$, *and*
   - $(u_j - v_j)$ *is an n-step.*

Remark 6. The last condition ($(u_j - v_j)$ is an n-step) is not necessary for proving the results presented in this paper. However, it is needed when comparing the loop homotopy and the loop equivalence (see [5]), as done in [6].

Definition 7. *Two m-loops* $\mathcal{K}$ *and* $\mathcal{L}$ *of base point* $p \in \mathbb{T}^n$ *are homotopic if there exists a sequence of m-loops* $(\mathcal{C}_1, ..., \mathcal{C}_k)$ *such that* $\mathcal{C}_1 = \mathcal{K}$, $\mathcal{C}_k = \mathcal{L}$ *and for all* $j \in [1; k-1], \mathcal{C}_j$ *and* $\mathcal{C}_{j+1}$ *are directly homotopic.*

Example 8. In the toric space $\mathbb{Z}_4 \times \mathbb{Z}_2$, let us consider the point $p = (0, 0)$, the 1-steps $v_1 = (1, 0)$ and $v_2 = (0, 1)$, and the 1-loops $\mathcal{L}_a$, $\mathcal{L}_b$, $\mathcal{L}_c$ and $\mathcal{L}_d$ (see Fig. 4). The loops $\mathcal{L}_a$ and $\mathcal{L}_b$ are homotopic, the loops $\mathcal{L}_b$ and $\mathcal{L}_c$ are directly homotopic, and the loops $\mathcal{L}_c$ and $\mathcal{L}_d$ are also directly homotopic.

On the other hand, it may be seen that the 1-loops depicted on Fig. 3-b and on Fig. 3-c are not directly homotopic.

3.2   Fundamental Group

Initially defined in the continuous space by Henri Poincaré in 1895 [7], the fundamental group is an essential concept of topology, based on the homotopy relation, which has been transposed in different discrete frameworks (see e.g. [5], [8], [9]).

Given two m-loops $\mathcal{K} = (p, (u_1, ..., u_k))$ and $\mathcal{L} = (p, (v_1, ..., v_l))$ of same base point $p \in \mathbb{T}^n$, the *product of* $\mathcal{K}$ *and* $\mathcal{L}$ is the m-loop $\mathcal{K}.\mathcal{L} = (p, (u_1, ..., u_k, v_1, ..., v_l))$. We set $\mathcal{K}^{-1} = (p, (-u_k, ..., -u_1))$.

The symbol $\prod$ will be used for the iteration of the product operation on loops. Given a positive integer w, we set $\mathcal{K}^w = \prod_{i=1}^{w} \mathcal{K}$ and $\mathcal{K}^{-w} = \prod_{i=1}^{w} \mathcal{K}^{-1}$. We also define $\mathcal{K}^0 = (p, ())$.

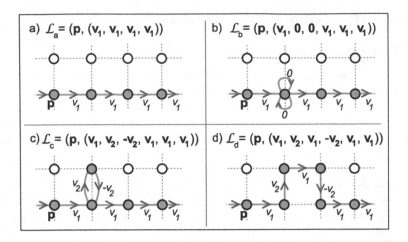

Fig. 4. Homotopic Loops - The 1-loops $\mathcal{L}_a, \mathcal{L}_b, \mathcal{L}_c$ and $\mathcal{L}_d$ are homotopic

The homotopy of m-loops is an equivalence relation and the equivalence class, called *homotopy class*, of an m-loop $\mathcal{L}$ is denoted by $[\mathcal{L}]$. The product operation can be extended to the homotopy classes of m-loops of same base point: the product of $[\mathcal{K}]$ and $[\mathcal{L}]$ is $[\mathcal{K}].[\mathcal{L}] = [\mathcal{K}.\mathcal{L}]$. We now define the fundamental group of $\mathbb{T}^n$.

Definition 9. *Given an m-adjacency relation on $\mathbb{T}^n$ and a point $p \in \mathbb{T}^n$, the m-fundamental group of $\mathbb{T}^n$ with base point p is the group formed by the homotopy classes of all m-loops of base point $p \in \mathbb{T}^n$ under the product operation.*

The identity element of this group is the homotopy class of the trivial loop.

4   Toric Loops in Subsets of $\mathbb{T}^n$

The toric loops, informally evoked in the introduction, can now be formalised using the definitions given in the previous sections.

Definition 10. *In $\mathbb{T}^n$, we say that an m-loop is a* toric m-loop *if it does not belong to the homotopy class of a trivial loop. A connected subset of $\mathbb{T}^n$ is* wrapped *in $\mathbb{T}^n$ if it contains a toric m-loop.*

Remark 11. The notion of grain introduced informally in Sec. 1 may now be defined: a connected component of $\mathbb{T}^n$ is a *grain* if it is not wrapped in $\mathbb{T}^n$.

4.1   Algorithm for Detecting Wrapped Subsets of $\mathbb{T}^n$

In order to know whether a connected subset of $\mathbb{T}^n$ is wrapped or not, it is not necessary to build all the m-loops which can be found in the subset: the Wrapped Subset Descriptor (WSD) algorithm (see Alg. 1) answers this question in linear time (more precisely, in $O(N.M)$, where N is the number of points of $\mathbb{T}^n$, and M is the number of distinct m-steps), as stated by the following proposition.

Algorithm 1. WSD(n,m,$\mathbb{T}^n$,d,X)

Data: An n-dimensional toric space $\mathbb{T}^n$ of dimension vector d and a non-empty m-connected subset X of $\mathbb{T}^n$.

Result: A boolean telling whether X has a toric m-loop or not.

1 Let p $\in$ X; Coord(p) = 0^n; S = {p };
2 **foreach** x $\in$ X **do** HasCoord(x) = false; HasCoord(p) = true;
3 **while** *there exists* x $\in$ S **do**
4 $\quad$ S = S $\setminus$ {x};
5 $\quad$ **foreach** *non-null n-dimensional m-step* v **do**
6 $\quad\quad$ y = x $\oplus_d$ v;
7 $\quad\quad$ **if** y $\in$ X *and* HasCoord(y) = true **then**
8 $\quad\quad\quad$ **if** Coord(y) $\neq$ Coord(x) + v **then**
9 $\quad\quad\quad\quad$ **return** true;
10 $\quad\quad$ **else if** y $\in$ X *and* HasCoord(y) = false **then**
11 $\quad\quad\quad$ Coord(y) = Coord(x) + v; S = S $\cup$ {y}; HasCoord(y) = true;

12 **return** false;

Proposition 12. *Let $\mathbb{T}^n$ be an n-dimensional toric space of size vector **d**. A non-empty m-connected subset X of $\mathbb{T}^n$ is wrapped in $\mathbb{T}^n$ if and only if WSD(n,m, $\mathbb{T}^n$,**d**,X) is true.*

Before proving Prop.12 (see Sec. 5.4), new definitions and theorems must be given: in particular, Th. 25 establishes an important result on homotopic loops in toric spaces. Before, let us study an example of execution of Alg. 1.

Example 13. Let us consider a subset X of points of $\mathbb{Z}_4 \times \mathbb{Z}_4$ (see Fig. 5-a) and the 2-adjacency relation. In Fig. 5-a, one element **p** of X is given the coordinates of the origin (see l. 1 of Alg. 1); then we set **x** = **p**. In Fig. 5-b, every neighbour **y** of **x** which is in X is given coordinates depending on its position relative to **x** (l. 11) and is added to the set S (l. 11).

Then, in Fig. 5-c, one element of S is chosen as **x** (l. 3). Every neighbour **y** of **x** is scanned: if **y** is in X and has already coordinates (l. 7), it is compared with **x**: as the coordinates of **x** and **y** are compatible in $\mathbb{Z}^2$ (the test achieved l. 8 returns false), the algorithm continues. Else (l. 10), **y** is given coordinates depending on its position relative to **x** (l. 11) and added to S (see Fig. 5-d).

Finally, in Fig. 5-e, an element of S is chosen as **x**. The algorithm tests one of the neighbours **y** of **x** (the left neighbour) which has already coordinates (l. 7). The coordinates of **y** and **x** are incompatible in $\mathbb{Z}^2$ (the points $(-1,1)$ and $(2,1)$ are not neighbours in $\mathbb{Z}^2$), the algorithm returns true (l. 9): according to Prop. 12, the subset X is wrapped in $\mathbb{T}^n$.

To summarize, Alg. 1 'tries to embed' the subset X of $\mathbb{T}^n$ in $\mathbb{Z}^n$: if some incompatible coordinates are detected by the test achieved on l. 8 of Alg. 1, then the object has a feature (a toric loop) which is incompatible with $\mathbb{Z}^n$. A toric 2-loop lying in X is depicted in Fig. 5-f.

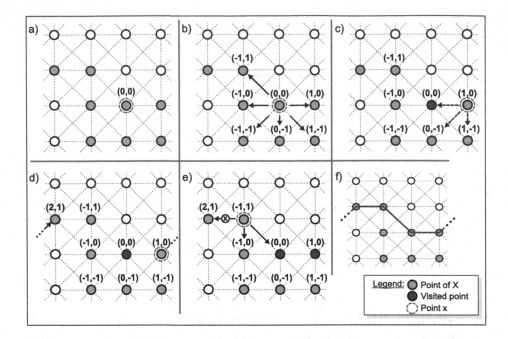

Fig. 5. Example of execution of WSD - see Ex. 13 for a detailed description

5   Wrapping Vector and Homotopy Classes in $\mathbb{T}^n$

Deciding if two loops $\mathcal{L}_1$ and $\mathcal{L}_2$ belong to the same homotopy class can be difficult if one attempts to do this by building a sequence of directly homotopic loops which 'link' $\mathcal{L}_1$ and $\mathcal{L}_2$. However, this problem may be solved using the *wrapping vector*, a characteristic which can be easily computed on each loop.

5.1   Wrapping Vector of a Loop

The *wrapping vector* of a loop is the sum of all the elements of the m-step sequence associated to the loop.

Definition 14. *Let $\mathcal{L} = (p, V)$ be an m-loop, with $V = (v_1, ..., v_k)$. Then the wrapping vector of $\mathcal{L}$ is $v_{\mathcal{L}} = \sum_{i=1}^{k} v_i$.*

Remark 15. In Def. 14, the symbol $\sum$ stands for the iteration of the classical addition operation on $\mathbb{Z}^n$, not of the operation $\oplus$ defined in Sec. 2.1.

The notion of 'basic loops' will be used in the proof of Prop. 18 and in Def. 22.

Definition 16. *Let $\mathbb{T}^n$ be an n-dimensional toric space of size vector $d = (d_1, ..., d_n)$. We denote, for each $i \in [1; n]$, by b_i the 1-step whose i-th coordinate is equal to 1, and by B_i the 1-step sequence composed of exactly d_i 1-steps b_i.*

Given $p \in \mathbb{T}^n$, for all $i \in [1; n]$, we define the i-th basic loop of base point p as the 1-loop (p, B_i).

Remark 17. For all $i \in [1; n]$, the wrapping vector of the i-th basic loop of base point p is equal to $(d_i.b_i)$.

The next property establishes that the wrapping vector of any m-loop can only take specific values in $\mathbb{Z}^n$. The proof can be found in Sec. 7.

Proposition 18. *Let $\mathbb{T}^n$ be an n-dimensional toric space of size vector $d = (d_1, ..., d_n)$. A vector $w = (w_1, ..., w_n)$ of $\mathbb{Z}^n$ is the wrapping vector of an m-loop of $\mathbb{T}^n$ if and only if, for all $i \in [1; n]$, w_i is a multiple of d_i.*

Definition 19. *Given $\mathbb{T}^n$ of size vector $d = (d_1, ..., d_n)$, let $\mathcal{L}$ be an m-loop of wrapping vector $w = (w_1, ..., w_n)$. The normalized wrapping vector of $\mathcal{L}$ is $w^* = (w_1/d_1, ..., w_n/d_n)$.*

Example 20. The normalized wrapping vector gives information on how a loop 'wraps around' each dimension of a toric space. For example, let $\mathbb{T}^3 = \mathbb{Z}_2 \times \mathbb{Z}_5 \times \mathbb{Z}_7$: a loop with normalized wrapping vector $(2,1,0)$ wraps two times in the first dimension, one time in the second, and does not wrap in the third dimension.

5.2   Equivalence between Homotopy Classes and Wrapping Vector

It can be seen that two directly homotopic m-loops have the same wrapping vector, as their associated m-step sequences have the same sum. Therefore, we have the following property.

Proposition 21. *Two homotopic m-loops of $\mathbb{T}^n$ have the same wrapping vector.*

The following definition is necessary in order to understand Prop. 24 and its demonstration, leading to the main theorem of this article.

Definition 22. *Let p be an element of $\mathbb{T}^n$, and $w^* = (w_1^*, ..., w_n^*) \in \mathbb{Z}^n$.*
*   The canonical loop of base point p and normalized wrapping vector w^* is the 1-loop $\prod_{i=1}^n (p, B_i)^{w_i^*}$, where (p, B_i) is the i-th basic loop of base point p.*

Example 23. Consider $(\mathbb{T}^3, \oplus)$, with $\mathbb{T}^3 = \mathbb{Z}_3 \times \mathbb{Z}_1 \times \mathbb{Z}_2$, $w = (3, 1, -4)$ and $p = (0, 0, 0)$. The canonical loop of base point p and wrapping vector w is the 1-loop (p, V) with:

$$V = \left(\begin{pmatrix} 1 \\ 0 \\ 0 \end{pmatrix}, \begin{pmatrix} 1 \\ 0 \\ 0 \end{pmatrix}, \begin{pmatrix} 1 \\ 0 \\ 0 \end{pmatrix}, \begin{pmatrix} 0 \\ 1 \\ 0 \end{pmatrix}, \begin{pmatrix} 0 \\ 0 \\ -1 \end{pmatrix}, \begin{pmatrix} 0 \\ 0 \\ -1 \end{pmatrix}, \begin{pmatrix} 0 \\ 0 \\ -1 \end{pmatrix}, \begin{pmatrix} 0 \\ 0 \\ -1 \end{pmatrix} \right)$$

Proposition 24. *Any m-loop of base point $p \in \mathbb{T}^n$ and wrapping vector $w \in \mathbb{Z}^n$ is homotopic to the canonical loop of base point p and wrapping vector w.*

The proof of the previous proposition can be found in Sec. 7. We can now state the main theorem of this article, which is a direct consequence of Prop. 21 and Prop. 24.

Theorem 25. *Two m-loops of $\mathbb{T}^n$ of same base point are homotopic if and only if their wrapping vectors are equal.*

Remark 26. According to Th. 25, the homotopy class of the trivial loop $(p, ())$ is the set of all m-loops of base point p that have a null wrapping vector.

5.3   Wrapping Vector and Fundamental Group

Given a point $p \in \mathbb{T}^n$, we set $\Omega = \{w^* \in \mathbb{Z}^n / \text{ there exists an m-loop in } \mathbb{T}^n \text{ of } \text{base point } p \text{ and of normalized wrapping vector } w^*\}$. From Prop. 18, it is plain that $\Omega = \mathbb{Z}^n$. Therefore, $(\Omega, +)$ is precisely $(\mathbb{Z}^n, +)$

Theorem 25 states that there exists a bijection between the set of the homotopy classes of all m-loops of base point p and Ω. The product (see Sec. 3.2) of two m-loops $\mathcal{K}$ and $\mathcal{L}$ of same base point p and of respective wrapping vectors w_k and w_l is the loop $(\mathcal{K}.\mathcal{L})$ of base point p. The wrapping vector of $(\mathcal{K}.\mathcal{L})$ is $(w_k + w_l)$, therefore we can state that there exists an isomorphism between the fundamental group of $\mathbb{T}^n$ and $(\Omega, +)$. Consequently, we retrieve in our discrete framework a well-known property of the fundamental group of toric spaces [2].

Proposition 27. *The fundamental group of $\mathbb{T}^n$ is isomorphic to $(\mathbb{Z}^n, +)$.*

5.4   Proof of Alg. 1

Proof. (of Prop. 12) For all $y \in X$ such that $y \neq p$, there exists a point x such that the test performed on l. 10 of Alg. 1 is true: we call x *the label predecessor of y*.

• If the algorithm returns false, then the test performed l. 8 of Alg. 1 was never true. Let $\mathcal{L} = (p, V)$ be an m-loop contained in X, with $V = (v_1, ..., v_k)$, and let us denote by x_i the i-th point of $\mathcal{L}$. As the test performed l. 8 was always false, we have the following:

$$\begin{cases} \text{for all } i \in [1; k-1], v_i = Coord(x_{i+1}) - Coord(x_i) \\ v_k = Coord(x_1) - Coord(x_k) \end{cases}$$

The wrapping vector of $\mathcal{L}$ is

$$w = \sum_{i=1}^{k-1}(Coord(x_{i+1}) - Coord(x_i)) + Coord(x_1) - Coord(x_k) = 0$$

Thus, if the algorithm returns false, each m-loop of X has a null wrapping vector and, according to Th. 25, belongs to the homotopy class of a trivial loop: there is no toric m-loop in X which is therefore not wrapped in $\mathbb{T}^n$.

• If the algorithm returns true, then, there exists $x, y \in X$ and an m-step a, such that $x \oplus a = y$ and $Coord(y) - Coord(x) \neq a$.

It is therefore possible to find two sequences γ_x and γ_y of m-adjacent points in X, with $\gamma_x = (p = x_1, x_2, ..., x_q = x)$ and $\gamma_y = (y = y_t, ..., y_2, y_1 = p)$, such that, for all $i \in [1; q-1]$, x_i is the label predecessor of x_{i+1}, and for all $i \in [1; t-1]$, y_i is the label predecessor of y_{i+1}. Therefore, we can set

$$\begin{cases} \text{. for all } i \in [1; q-1], u_i = Coord(x_{i+1}) - Coord(x_i) \\ \text{is an m-step such that } x_i \oplus u_i = x_{i+1} \\ \text{. for all } i \in [1; t-1], v_i = Coord(y_i) - Coord(y_{i+1}) \\ \text{is an m-step such that } y_{i+1} \oplus v_i = y_i \end{cases}$$

Let $\mathcal{N}_{x,y,a} = (p, V)$ be the m-loop such that $V = (u_1, ..., u_{q-1}, a, v_{t-1}, ..., v_1)$. The m-loop $\mathcal{N}_{x,y,a}$ is lying in X and its wrapping vector w is equal to:

$$w = \sum_{i=1}^{q-1} u_i + a + \sum_{i=1}^{t-1} v_i = a - (Coord(y) - Coord(x)) \neq 0$$

Thus, when the algorithm returns true, it is possible to find, inside X, an m-loop with a non-null wrapping vector: by Th. 25, there is a toric m-loop in X which is therefore wrapped in $\mathbb{T}^n$. $\qquad\qquad\qquad\qquad\qquad\qquad\qquad\qquad\square$

6   Conclusion

In this article, we give a formal definition of loops and homotopy inside discrete toric spaces in order to define various notions such as loop homotopy and the fundamental group. We then propose a linear time algorithm for detecting toric loops in a subset X of $\mathbb{T}^n$: the proof of the algorithm relies on the notions previously given, such as the wrapping vector which, according to Th. 25, completely characterizes toric loops.

In Sec. 1, we have seen that detecting toric loops is important in order to filter grains from a material's sample and perform a fluid flow simulation on the sample. The WSD algorithm proposed in this article detects which subsets of a sample, embedded inside a toric space, will create grains and should be removed. Future works will include analysis of the relationship between other topological characteristics of materials and their physical properties.

7   Annex

Proof. (of Prop. 18) First, let $\mathcal{L} = (\boldsymbol{p}, V)$ be an m-loop of wrapping vector $\boldsymbol{w} = (w_1, ..., w_n)$, with $\boldsymbol{p} = (p_1, ..., p_n)$. As $\mathcal{L}$ is a loop, for all $i \in [1; n]$, $p_i \oplus_{d_i} w_i = p_i$. Hence, for all $i \in [1; n]$, $w_i \equiv 0 (mod\ d_i)$.

Let $\boldsymbol{w} = (w_1, ..., w_n)$ be a vector of $\mathbb{Z}^n$ such that for all $i \in [1; n]$, w_i is a multiple of d_i. If we denote by $(\boldsymbol{p}, B_i)$ the i-th basic loop of base point $\boldsymbol{p}$, we see that $(\prod_{i=1}^{n} (\boldsymbol{p}, B_i)^{w_i/d_i})$ is an m-loop whose wrapping vector is equal to $\boldsymbol{w}$. $\qquad\square$

Proof. (of Prop. 24) Let $\boldsymbol{a}$ and $\boldsymbol{b}$ be two non-null 1-steps. Let i (resp. j) be the index of the non-null coordinate of $\boldsymbol{a}$ (resp $\boldsymbol{b}$). We say that $\boldsymbol{a}$ *is index-smaller than* $\boldsymbol{b}$ if $i < j$.

Let $\mathcal{L} = (\boldsymbol{p}, V)$ be an m-loop of normalized wrapping vector $\boldsymbol{w}^* \in \mathbb{Z}^n$.

- **1** - The m-loop $\mathcal{L}$ is homotopic to a 1-loop $\mathcal{L}_1 = (\boldsymbol{p}, V_1)$ (see Lem. 28).
- **2** - By Def. 5 and 7, the 1-loop $\mathcal{L}_1$ is homotopic to a 1-loop $\mathcal{L}_2 = (\boldsymbol{p}, V_2)$, where V_2 contains no null vector.
- **3** - Let $\mathcal{L}_3 = (\boldsymbol{p}, V_3)$ be such that V_3 is obtained by iteratively permuting all pairs of consecutive 1-steps $(\boldsymbol{v}_j, \boldsymbol{v}_{j+1})$ in V_2 such that $\boldsymbol{v}_{j+1}$ is index-smaller than $\boldsymbol{v}_j$. Thanks to Lem. 29, $\mathcal{L}_3$ is homotopic to $\mathcal{L}_2$.
- **4** - Consider $\mathcal{L}_4 = (\boldsymbol{p}, V_4)$, where V_4 is obtained by iteratively replacing all pairs of consecutive 1-steps $(\boldsymbol{v}_j, \boldsymbol{v}_{j+1})$ in V_3 such that $\boldsymbol{v}_{j+1} = (-\boldsymbol{v}_j)$ by two null vectors, and then removing these two null vectors. The loop $\mathcal{L}_4$ is homotopic to $\mathcal{L}_3$.

The 1-loop $\mathcal{L}_4$ is homotopic to $\mathcal{L}$, it has therefore the same normalized wrapping vector $\boldsymbol{w}^* = (w_1^*, ..., w_n^*)$ (see Prop. 21). By construction, each pair of consecutive

1-steps (v_j, v_{j+1}) of V_4 is such that v_j and v_{j+1} are non-null and either $v_j = v_{j+1}$ or v_j is index-smaller than v_{j+1}.

Let $d = (d_1, ..., d_n)$ be the size vector of $\mathbb{T}^n$. As the normalized wrapping vector of $\mathcal{L}_4$ is equal to w^*, we deduce that the $(d_1.|w_1^*|)$ first elements of V_4 are equal to $(w_1^*/|w_1^*|.b_1)$ (see Def. 16). Moreover, the $(d_2.|w_2^*|)$ next elements are equal to $(w_2^*/|w_2^*|.b_2)$, etc. Therefore, we have $\mathcal{L}_4 = (\prod_{i=1}^{n}(p, B_i)^{w_i^*})$. $\qquad\square$

Lemma 28. *Any m-loop $\mathcal{L} = (p, V)$ is homotopic to a 1-loop.*

Proof. Let us write $V = (v_1, ..., v_k)$ and let $j \in [1; n]$ be such that v_j is not a 1-step. The m-loop $\mathcal{L}$ is directly homotopic to $\mathcal{L}_1 = (p, V_1)$, with $V_1 = (v_1, ..., v_{j-1}, v_j, 0, v_{j+1}, ..., v_k)$. As v_j is not a 1-step, there exists an $(m-1)$-step v_j' and a 1-step v_{j1} such that $v_j = (v_{j1} + v_j')$. The m-loop $\mathcal{L}_1$ is directly homotopic to $\mathcal{L}_2 = (p, V_2)$, with $V_2 = (v_1, ..., v_{j-1}, v_{j1}, v_j', v_{j+1}, ..., v_k)$. By iteration, it is shown that $\mathcal{L}$ is homotopic to a 1-loop. $\qquad\square$

Lemma 29. *Let $\mathcal{L}_A = (p, V_A)$ and $\mathcal{L}_B = (p, V_B)$ be two m-loops such that $V_A = (v_1, ..., v_{j-1}, v_{j1}, v_{j2}, v_{j+1}, ..., v_k)$ and $V_B = (v_1, ..., v_{j-1}, v_{j2}, v_{j1}, v_{j+1}, ..., v_k)$ where v_{j1} and v_{j2} are 1-steps. Then, $\mathcal{L}_A$ and $\mathcal{L}_B$ are homotopic.*

Proof. As v_{j1} and v_{j2} are 1-steps, they have at most one non-null coordinate. If $(v_{j1} - v_{j2})$ is an n-step, the two loops are directly homotopic. If $(v_{j1} - v_{j2})$ is not an n-step, then necessarily $v_{j1} = (-v_{j2})$. Therefore, $\mathcal{L}_A$ is directly homotopic to $\mathcal{L}_C = (p, V_C)$, with $V_C = (v_1, ..., v_{j-1}, 0, 0, v_{j+1}, ..., v_k)$. Furthermore, $\mathcal{L}_C$ is also directly homotopic to $\mathcal{L}_B$. $\qquad\square$

References

1. Bear, J.: Dynamics of Fluids in Porous Media. Elsevier, Amsterdam (1972)
2. Hatcher, A.: Algebraic topology. Cambridge University Press, Cambridge (2002)
3. Stillwell, J.: Classical Topology and Combinatorial Group Theory. Springer, Heidelberg (1980)
4. Graham, R.L., Knuth, D.E., Patashnik, O.: Concrete Mathematics: A Foundation for Computer Science. Addison-Wesley Longman Publishing Co., Inc, Boston, MA, USA (1994)
5. Kong, T.Y.: A digital fundamental group. Computers & Graphics 13(2), 159–166 (1989)
6. Chaussard, J., Bertrand, G., Couprie, M.: Characterization and detection of loops in n-dimensional discrete toric spaces. Technical Report IGM2008-02, Université de Marne-la-Vallée (2008)
7. Poincaré, H.: Analysis situs. Journal de l'Ecole Polytechnique, 2eme série, cahier 1, 1–121 (1895)
8. Malgouyres, R.: Computing the fundamental group in digital spaces. IJPRAI 15(7) (2001)
9. Bertrand, G., Couprie, M., Passat, N.: 3-D simple points and simple-equivalence (submitted, 2007)

Insertion and Expansion Operations for n-Dimensional Generalized Maps*

Mehdi Baba-ali[1], Guillaume Damiand[2], Xavier Skapin[1], and David Marcheix[3]

[1] SIC-XLIM, Université de Poitiers, UMR CNRS 6172, 86962 Futuroscope
Chasseneuil Cedex, France
babaali@sic.univ-poitiers.fr, xavier.skapin@univ-poitiers.fr
[2] LaBRI, Université Bordeaux 1, UMR CNRS 5800, 33405 Talence Cedex, France
damiand@labri.fr
[3] Laboratoire d'Informatique Scientifique et Industrielle (LISI), ENSMA, France
marcheix@ensma.fr

Abstract. Hierarchical representations, such as irregular pyramids, are the bases of several applications in the field of discrete imagery. So, n-dimensional "bottom-up" irregular pyramids can be defined as stacks of successively reduced n-dimensional generalized maps (n-G-maps) [11], each n-G-map being defined from the previous level by using removal and contraction operations defined in [8]. Our goal is to build a theoretical framework for defining and handling n-dimensional "top-down" irregular pyramids. To do so, we propose in this paper to study the definition of both insertion and expansion operations that allow to conceive these kinds of pyramids.

1 Introduction

Hierarchical representations form the bases of several applications in the field of discrete imagery. Our goal is the study of basic problems related to the definition of hierarchical structures. To achieve this goal, removal and contraction operations have been defined in [8]. In this paper, we define two others basic operations: *insertion* and *expansion*, which allow to define "top-down" pyramids.

Many works deal with regular or irregular image pyramids for multi-level analysis and treatments. The first ones are [7,12,16,17]. In irregular pyramids, each level represents a partition of the pixel set into cells, i.e. *connected subsets of pixels*. There are two ways to build an irregular pyramid: "botom-up" and "top-down"[1]. In the first case, the number of cells increases between two contiguous levels of a pyramid, while in the second case this number of cells decreases. To manipulate these models, it is neccessary to handle a (topological) representation and some basic operations, for instance dual graphs [13] and removal and contraction operations [8].

---

* Partially supported by the ANR program ANR-06-MDCA-015/VORTISS.
[1] In the following sections, we use the terms "top-down" and "bottom-up" pyramids to refer to the pyramids built by using "top-down" and "bottom-up" approaches, respectively.

D. Coeurjolly et al. (Eds.): DGCI 2008, LNCS 4992, pp. 141–152, 2008.

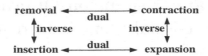

Fig. 1. Links between the basic operations which allow to handle irregular pyramids

Grasset-Simon and *al* [11] build a theoretical framework for defining and handling n-dimensional "bottom-up" irregular pyramids. To do so, they use the removal and contraction operations, defined in [8], in order to get consistent definitions of data structures for any dimension. Our goal is the same as [11]: build a theoretical framework, but for defining and handling n-dimensional "top-down" irregular pyramids. Therefore, we study the definition of insertion and expansion of i-dimensional cells within n-dimensional objects, in order to define the relations between two consecutive levels of a "top-down" pyramid. We also study the definition of these operations in order to set the links between "bottom-up" and "top-down" irregular pyramids. Indeed, insertion and expansion operations (basic operations used for the definition of "top-down" irregular pyramids) are, respectively, the inverse of removal and contraction operations (Fig. 1).

We choose to study the definitions of insertion and expansion operations for n-dimensional generalized maps, since this model enables us to unambiguously represent the topology of quasi-manifolds, which is a well-defined class of subdivisions [15]. Note that several models based on combinatorial maps [9] have been proposed for handling two-dimensional [6,10] and three-dimensional segmented or multi-level images [1,2,3]. We prefer to use generalized maps instead of combinatorial maps, since their algebraic definition is homogeneous; therefore, we can provide simpler definitions of data structures and operations with generalized maps, which are also more efficiency for the conception of softwares (note that several kernels of geometric modeling softwares are based upon data structures derived from this notion). Last, we know how to deduce combinatorial maps from generalized maps, so the results presented in this article can be extended to combinatorial maps. Precise relations between generalized, combinatorial maps and other classical data structures are presented in [14].

We recall in Section 2 the notion of generalized maps, removal and contraction operations. Then we define the insertion operation of one i-dimensional cell in Section 3, and expansion operation by duality (in Section 4). In Section 5, we show that it is possible to simultaneously insert and expand several cells of the same dimension. Last, we conclude and give some perspectives in Section 6.

2   Recalls

2.1   Generalized Maps

An n-dimensional generalized map is a set of abstract elements, called darts, and applications defined on these darts:

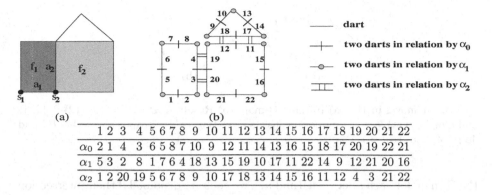

	1 2 3	4	5 6 7 8	9	10 11	12	13 14	15	16	17 18	19	20	21 22
α_0	2 1 4	3	6 5 8 7	10	9	12 11	14 13	16 15	18 17	20 19	22 21
α_1	5 3 2	8	1 7 6 4	18	13 15	19	10 17	11	22	14	9	12 21 20 16
α_2	1 2 20	19	5 6 7 8	9	10 17	18 13	14 15	16 11	12	4	3	21 22

Fig. 2. (a) A 2D subdivision. (b) The corresponding 2-G-map (involutions are explicitly given in the array). Darts are represented by numbered black segments.

Definition 1. *(Generalized map) Let $n \geq 0$. A n-dimensional generalized map (or n-G-map) is $G = (B, \alpha_0, \ldots, \alpha_n)$ where:*

1. *B is a finite set of darts;*
2. *$\forall i, 0 \leq i \leq n, \alpha_i$ is an involution[2] on B;*
3. *$\forall i, j, 0 \leq i < i + 2 \leq j \leq n, \alpha_i \alpha_j$ is an involution (condition of quasi-manifolds).*

Let G be an n-G-map, and S be the corresponding subdivison. A dart of G corresponds to an (n+1)-tuple of cells $(c_0, \ldots, c_n)$, where c_i is an i-dimensional cell that belongs to the boundary of c_{i+1} [4]. α_i associates darts corresponding with $(c_0, \ldots, c_n)$ and $(c'_0, \ldots, c'_n)$, where $c_j = c'_j$ for $j \neq i$, and $c_i \neq c'_i$ (α_i swaps the two i-cells that are incident to the same $(i - 1)$ and $(i + 1)$-cells). When two darts b_1 and b_2 are such that $b_1 \alpha_i = b_2$ ($0 \leq i \leq n$), b_1 is said i-sewn with b_2. Moreover, if $b_1 = b_2$ then b_1 is said i-free. In Fig. 2, Dart 1 corresponds to (s_1, a_1, f_1), dart $2 = 1\alpha_0$ corresponds to (s_2, a_1, f_1), $3 = 2\alpha_1$ corresponds to (s_2, a_2, f_1), and $20 = 3\alpha_2$ corresponds to (s_2, a_2, f_2). G-maps provide an implicit representation of cells:

Definition 2. *(i-cell) Let G be an n-G-map, b a dart and $i \in N = \{0, ., n\}$. The i-cell incident to b is the orbit[3] $\langle \rangle_{N - \{i\}} (b) = \langle \alpha_0, \ldots, \alpha_{i-1}, \alpha_{i+1}, \ldots, \alpha_n \rangle (b)$.*

An i-cell is the set of all darts which can be reached starting from b, by using any combination of all involutions except α_i. In Fig. 2, the 0-cell (vertex) incident to dart 2 is the orbit $\langle \alpha_1, \alpha_2 \rangle (2) = \{2, 3, 20, 21\}$, the 1-cell (edge) incident to dart 3 is $\langle \alpha_0, \alpha_2 \rangle (3) = \{3, 4, 19, 20\}$, and the 2-cell (face) incident to dart 9 is $\langle \alpha_0, \alpha_1 \rangle (9) = \{9, 10, 13, 14, 17, 18\}$. The set of i-cells is a partition of the darts of

---

[2] An involution f on S is a one mapping from S onto S such that $f = f^{-1}$.

[3] Let $\{\Pi_0, \ldots, \Pi_n\}$ be a set of permutations on B. The orbit of an element b relatively to this set of permutations is $\langle \Pi_0, \ldots, \Pi_n \rangle (b) = \{\Phi(b), \Phi \in \langle \Pi_0, \ldots, \Pi_n \rangle\}$, where $\langle \Pi_0, \ldots, \Pi_n \rangle$ denotes the group of permutations generated by $\{\Pi_0, \ldots, \Pi_n\}$.

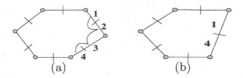

Fig. 3. 0-removal in 1D. (a) Initial 1-G-map. (b) Result. Vertex $C = \langle \alpha_1 \rangle (2) = \{2, 3\}$ and $C\alpha_0 = \{1, 4\} = B^S$. 0-removal consists in setting $1\alpha_0' = 1 (\alpha_0 \alpha_1) \alpha_0 = 4 \in B^S$ and $4\alpha_0' = 4 (\alpha_0 \alpha_1) \alpha_0 = 1 \in B^S$.

the G-map, for each i between 0 and n. Two cells are disjoined if their intersection is empty, i.e. when no dart is shared by the cells. More details about G-maps are provided in [15].

2.2   Removal and Contraction Operations

In a general way for an n-dimensional space, the removal of an i-cell consists in removing this cell and in merging its two incidents $(i + 1)$-cells: so removal can be defined for $0 \ldots (n - 1)$-cells.

Definition 3. *(i-cell removal [8]) Let $G = (B, \alpha_0, \ldots, \alpha_n)$ be an n-G-map, $i \in \{0, \ldots, n - 1\}$ and $C = \langle \rangle_{N-\{i\}} (b)$ be an i-cell, such that: $\forall b' \in C, b'\alpha_{i+1}\alpha_{i+2} = b'\alpha_{i+2}\alpha_{i+1}$[4]. Let $B^S = C\alpha_i - C$, the set of darts i-sewn to C that do not belong to C (Fig. 3). The n-G-map resulting from the removal of C is $G' = (B', \alpha_0', \ldots, \alpha_n')$ defined by:*

- $B' = B - C$;
- $\forall j \in \{0, \ldots, n\} - \{i\}, \alpha_j' = \alpha_j | B'$; [5]
- $\forall b' \in B' - B^S, b'\alpha_i' = b'\alpha_i$;
- $\forall b' \in B^S, b'\alpha_i' = b' (\alpha_i \alpha_{i+1})^k \alpha_i$ *(where k is the smallest integer such that $b' (\alpha_i \alpha_{i+1})^k \alpha_i \in B^S$).*

The last expression redefines involution α_i for each dart $b \in B^S$ in G. Indeed, the image of b by α_i' is dart $b' \in B^S$ such that b and b' both are ends of path $(\alpha_i \alpha_{i+1})^k \alpha_i$. In Fig. 3a, darts 1 and 4 are the extremities of path $(\alpha_0 \alpha_1) \alpha_0$ which is represented by a curving line.

Note that G' can contain only one n-cell, and may even be empty if G contains only one i-cell. Note also that contraction operation can be defined directly by duality (see section 4 for explanations on duality in G-map). More details about removal and contraction operations are provided in [8].

---

[4] This constraint corresponds to the fact that the local degree of i-cell C is 2 (a vertex locally incident to exactly two edges or an edge locally incident to two faces or a face locally incident to two volumes...).

[5] α_j' is equal to α_j restricted to B', i.e. $\forall b \in B', b\alpha_j' = b\alpha_j$.

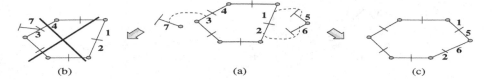

Fig. 4. 0-insertion in 1D. (a) Initial 1-G-map G corresponds to the orbit $\langle \alpha_0, \alpha_1 \rangle$ (1), 0-cell $C_1 = \langle \alpha_1 \rangle$ (7) = {7}, E_1 = {3}, F_1 = {7}, involution γ is represented by dashed lines: γ sews darts 1, 2 and 3 with darts 5, 6 and 7, respectively. 0-cell $C_2 = \langle \alpha_1 \rangle$ (5) = {5, 6}, E_2 = {1, 2}, F_2 = {5, 6}. (b) Invalid result after inserting C_1 into G (α_0 is no more an involution since $3\alpha_0 = 7$ and $4\alpha_0 = 3$). (c) Valid result after inserting C_2 into G (the precondition $b\alpha_0 = b\gamma\alpha_1\gamma$ is satisfied for all darts of E_2).

3 Insertion

The insertion of an i-cell C into an n-G-map G consists (conversely to removal operation) in adding this cell to G and splitting an $(i + 1)$-cell where C should be inserted: so insertion can be defined for $0 \ldots (n-1)$-cells. In this section, we present some useful cases of 0 and 1-insertions in 1- and 2-dimensional space before giving the general definition in nD.

3.1 Dimension 1: 0-insertion

In dimension 1, only the 0-insertion exists, which consists in adding a vertex and splitting an edge where that vertex should be inserted. Let $G = (B, \alpha_0, \alpha_1)$ be the initial 1-G-map, $C = \langle \alpha_1 \rangle$ (b) be the vertex to insert (belonging to another G-map) and E and F be two subsets of darts such that: $E \subseteq B$ and $F \subseteq C$. These subsets allow to explicit where and how cell C will be inserted in G-map G by using an additional involution[6] γ (Fig. 4). The 1-G-map resulting from 0-insertion, called $G' = (B', \alpha_0', \alpha_1')$, is obtained only by redefining α_0 for the darts of E and F as follows: $\forall b \in E \cup F, b\alpha_0' = b\gamma$, where $B' = B \cup C$ (α_0' is unchanged for the darts of $B' - (E \cup F)$, $\alpha_0' = \alpha_0$ and α_1' is unchanged for the darts of B', $\alpha_1' = \alpha_1$). Furthermore, 0-insertion can be applied only if the following preconditions are satisfied: (1) $\forall b \in C$, b is 0-free and (2) $\forall b \in E, b\alpha_0 = b\gamma\alpha_1 (\alpha_0\alpha_1)^k \gamma$ (where k is the smallest integer such that $b\gamma\alpha_1 (\alpha_0\alpha_1)^k \in F$). These constraints ensure the validity of the operation and the fact that 0-insertion is the inverse operation of 0-removal, by avoiding the following cases:

1. $b \in E$ is not 0-free and $b\gamma$ is 1-free. In such a case, we have $C = F = \{b\gamma\}$, $b' = b\alpha_0 \in B - E$ and $b' \neq b\gamma$. Thus, we have $b'\alpha_0' = b'\alpha_0 = b$ and $b'\alpha_0'\alpha_0' = b\alpha_0' = b\gamma \neq b'$. α_0' is not well-defined since it is not an involution (Fig. 4b);
2. $b \in E$ is not 0-free and it does not exist a path between darts b and $b\alpha_0$ expressed by the following composition of involutions $\gamma\alpha_1 (\alpha_0\alpha_1)^k \gamma$. Indeed, for the same reason that the previous case, α_0' is not well-defined and is not an involution;

---

[6] γ is defined on set $E \cup F$ as follows: $b \in E \Leftrightarrow b\gamma \in F$.

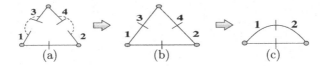

Fig. 5. Example where the precondition of the 0-insertion is not satisfied. (a) Darts of the vertex C to insert and to remove are numbered 3 and 4. Involution γ is marked with dashed line. E and F are two subsets of 0-free darts defined as: $E = \{1, 2\}$ and $F = \{3, 4\}$. (b) Resulting 1-G-map G' after sewing the darts of F with those of E (0-insertion without precondition). (c) Resulting 1-G-map after removing C from G': darts 1 and 2 are now sewed by α_0, whereas they were 0-free before insertion.

3. it exists a path[7] between two darts b and b' of F such that $b \neq b'$ and $b\gamma$ is 0-free or $b\gamma\alpha_0 \neq b'\gamma$. In such a case, we do not obtain the same 1-G-map G after successively inserting then removing a 0-cell C; hence, the 0-insertion is not the inverse operation of 0-removal (Fig. 5).

In Fig. 4c, the 0-insertion of C_2 is obtained by redefining α_0 for the darts of $E_2 = \{1, 2\}$ and $F_2 = \{5, 6\}$ such that: $1\alpha'_0 = 5$ and $2\alpha'_0 = 6$. Note that α_1 is not modified by 0-insertion.

3.2   Dimension 2

0-insertion. It consists in adding a 0-cell $C = \langle \alpha_1, \alpha_2 \rangle$ (b) into an initial 2-G-map $G = (B, \alpha_0, \alpha_1, \alpha_2)$. There are several ways to carry out this operation. Indeed, several possible combinations enable to link the darts of C with those of B (Figs. 6b and 6c). Then, it is necessary to remove this ambiguity by defining an involution that enables to characterize these links in a single way. Let γ be this involution and E and F be two subsets of darts (γ, E and F are defined as above and respect the same properties). The 2-G-map resulting from 0-insertion, called $G' = (B', \alpha'_0, \alpha'_1, \alpha'_2)$, is obtained by redefining α_0 for the darts of E and F as follows: $\forall b \in E \cup F, b\alpha'_0 = b\gamma$. In Fig. 6c, 0-insertion consists in setting $1\alpha'_0 = 5$, $2\alpha'_0 = 8$, $3\alpha'_0 = 7$ and $4\alpha'_0 = 6$. Note that this redefinition of α_0 is the same as for dimension 1 but concerns different darts, since it is a 0-cell within a 2D object (intuitively, in the general case, this operation consists in applying the 0-insertion defined for dimension 1 twice).

0-insertion can be applied only if the following preconditions are satisfied:

1. $\forall b \in E \cup F$ such that $b\alpha_2 \in E \cup F$ then $b\alpha_2\gamma = b\gamma\alpha_2$. This constraint enables to guarantee the quasi-manifold of the resulting 2-G-map: $\forall b \in E \cup F, \alpha'_0\alpha'_2$ is an involution, by using the following substitutions: $\forall b \in E \cup F, b\gamma = b\alpha'_0$ and $\forall b \in E \cup F, b\alpha_2 = b\alpha'_2$. Then, we obtain $\forall b \in E \cup F, b\alpha_2\gamma = b\gamma\alpha_2 \Leftrightarrow b\alpha'_2\alpha'_0 = b\alpha'_0\alpha'_2$. Thus, $\alpha'_0\alpha'_2$ is an involution for all darts in $E \cup F$;

---

[7] The path between two darts b and b' of F is expressed by the following composition of involutions $\alpha_1 (\alpha_0\alpha_1)^k$, where k is the smallest integer such that $b\alpha_1 (\alpha_0\alpha_1)^k = b'$.

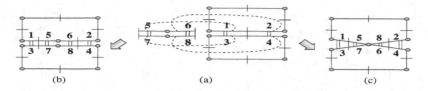

Fig. 6. 0-insertion in 2D. (a) 0-cell $C = \langle \alpha_1, \alpha_2 \rangle\,(5) = \{5, 6, 7, 8\}$, $E = \{1, 2, 3, 4\}$, $F = C$, initial 2-G-map G corresponds to the orbit $\langle \alpha_0, \alpha_1, \alpha_2 \rangle\,(1)$. Involution γ_1 is represented by dashed lines and links darts $1, 2, 3$ and 4 with $5, 6, 7$ and 8, respectively. Involution γ_2 (not shown on the figure) links darts $1, 2, 3$ and 4 with $5, 8, 7$ and 6, respectively. (b) Resulting 2-G-map after inserting C into G by using involution γ_1. (c) Resulting 2-G-map after inserting C into G by using involution γ_2.

2. $\forall b \in C, b\alpha_1\alpha_2 = b\alpha_2\alpha_1$: this constraint corresponds, in the general case, to the fact that the local degree of the vertex is equal to 2. If this constraint is not satisfied, we cannot remove C after inserting it. Indeed, $\forall b \in C, b\alpha_1\alpha_2 = b\alpha_2\alpha_1$ is also a precondition of the removal operation [8]. Thus, in order to define insertion operation as inverse of removal operation, it is necessary to check this precondition;

3. $\forall b \in E, b\alpha_0 = b\gamma\alpha_1 (\alpha_0\alpha_1)^k \gamma$ (where k is the smallest integer such that $b\gamma\alpha_1 (\alpha_0\alpha_1)^k \in F$): this constraint is the same as the one defined in the previous section.

1-insertion. It consists in adding a 1-cell $C = \langle \alpha_0, \alpha_2 \rangle\,(b)$ into an initial 2-G-map $G = (B, \alpha_0, \alpha_1, \alpha_2)$. Let E and F be two subsets of darts and γ an involution (E, F and γ are defined as in subsection 3.1). The 2-G-map G', resulting from 1-insertion, is obtained by redefining α_1 for the darts of E and F as follows: $\forall b \in E \cup F, b\alpha_1' = b\gamma$. Examples of this operation are represented in Figs. 7 and 8. 1-insertion can be applied only if the following precondition is satisfied: $\forall b \in E, b\alpha_1 = b\gamma\alpha_2 (\alpha_1\alpha_2)^k \gamma$ (where k is the smallest integer such that: $b\gamma\alpha_2 (\alpha_1\alpha_2)^k \in F$).

The first two preconditions, defined in the previous subsection, are always satisfied. Indeed, for the first case, G' is a quasi-manifold since only α_1 is redefined by 1-insertion operation, and $\alpha_0\alpha_2$ is an involution in G; for the second case, the local degree of an edge in a 2-dimensional quasi-manifold is always equal to 2.

3.3 Dimension n

The general definition of i-cell insertion for an n-dimensional G-map is a direct extension of the previous cases. Let C be the i-cell to insert, E and F be two subsets of darts. i-insertion of C in G is carried out by sewing darts of F with those of E by an involution γ. Then i-insertion consists in redefining α_i for the darts of E and F in the following way: $b\alpha_i' = b\gamma$. We obtain the general definition of the i-insertion operation:

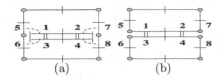
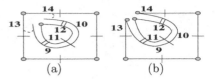

Fig. 7. 1-insertion in 2D in the general case. (a) Darts of the edge to insert are numbered $1, 2, 3$ and 4. Involution γ is represented by dashed line. (b) Result. The precondition of the 1-insertion is satisfied: $5\alpha_1 = 5\gamma\alpha_2\gamma = 6$ (here, $k = 0$).

Fig. 8. 1-insertion of a loop. (a) Initial 2-G-map with the loop to insert and involution γ. (b) Result. The precondition of the 1-insertion is satisfied: $13\alpha_1 = 13\gamma\alpha_2 (\alpha_1\alpha_2) \gamma = 14$.

Definition 4. *(i-cell insertion) Let $G = (B, \alpha_0, \alpha_1, \ldots, \alpha_n)$ be an n-G-map, $i \in \{0, \ldots, n-1\}$, $C = \langle \alpha_0, \ldots, \alpha_{i-1}, \alpha_{i+1}, \ldots, \alpha_n \rangle$ (b) be the i-cell to insert, E and F be two subsets of darts such that: $E \subseteq B$ and $F \subseteq C$ and γ be an involution defined on set $E \cup F$ such that: $b \in E \Leftrightarrow b\gamma \in F$. This operation can be applied only when:*

- *$\forall b \in F, b$ is i-free;*
- *$\forall b \in C, b\alpha_{i+1}\alpha_{i+2} = b\alpha_{i+2}\alpha_{i+1}$[8];*
- *$\forall b \in E \cup F, \forall j \ 0 \le j \le n$ such that $|i - j| \ge 2$ and $b\alpha_j \in E \cup F$ then $b\alpha_j\gamma = b\gamma\alpha_j$;*
- *$\forall b \in E, b\alpha_i = b\gamma\alpha_{i+1} (\alpha_i\alpha_{i+1})^k \gamma$ (where k is the smallest integer such that $b\gamma\alpha_{i+1} (\alpha_i\alpha_{i+1})^k \in F$).*

The n-G-map resulting from the insertion of C is $G' = (B', \alpha'_0, \ldots, \alpha'_n)$ defined by:

- *$B' = B \cup C$;*
- *$\forall j \in \{0, \ldots, n\} - \{i\}, \forall b \in B' : b\alpha'_j = b\alpha_j$;*
- *$\forall b \in B' - (E \cup F) : b\alpha'_i = b\alpha_i$;*
- *$\forall b \in E \cup F : b\alpha'_i = b\gamma$.*

The constraints of definition 4 enable to ensure that: the local degree of C is equal to 2, the resulting n-G-map is a quasi-manifold and insertion operation is the inverse of removal operation.

Note that, in the definition of G', only α_i is redefined for the darts of $E \cup F$. Indeed, i-insertion involves the sewing of darts of F with those of E by α'_i which is equal to γ. For everything else, the initial n-G-map remains unchanged. Now, we prove the validity of the operation by showing that the new structure G' is a n-G-map.

Theorem 1. *G' is an n-G-map.*

---

[8] Note that this condition does not apply for $i = n - 1$, so we can always insert any $(n - 1)$-cell.

Proof. We differentiate three cases. First for $j \neq i$, involutions α_j are not redefined but only restricted to the darts of the final G-map. Then, for $j = i$, we distinguish two cases, depending on if darts belong or not to $E \cup F$:

1. for $b_1 \in B - E$: We show that $b_2 = b_1 \alpha_i \in B - E$. Assume that $b_2 \notin B - E$, then $b_2 \in E$. In this case, $b_1 = b_2 \alpha_i \in E^9$ so $b_1 \notin B - E$: contradiction. Moreover, $b_1 \alpha_i' = b_1 \alpha_i \in B - E$ and $b_1 \alpha_i' \alpha_i' = b_1$: α_i' is well-defined and is an involution (we use the same method for the darts of $C - F$).
2. for $b_1 \in E \cup F$, We show that $b_2 = b_1 \gamma \in E \cup F$. That is true because γ is well-defined and is an involution on set $E \cup F$. Moreover, $b_1 \alpha_i' = b_1 \gamma \in E \cup F$ and $b_1 \alpha_i' \alpha_i' = b_1 \gamma \gamma = b_1$: α_i' is well-defined and is an involution.

We have now to prove that $\forall j, 0 \leq j \leq n, \forall k, j + 2 \leq k \leq n : \alpha_j' \alpha_k'$ is an involution.

- for $j \neq i$ et $k \neq i$: this is obvious since $\alpha_j' = \alpha_j$ and $\alpha_k' = \alpha_k$. As G is a G-map, $\alpha_j \alpha_k = \alpha_j' \alpha_k'$ is an involution.
- for $j = i$: we show that $\forall b_1 \in B'$, we have $b_1 \alpha_i' \alpha_k' = b_1 \alpha_k' \alpha_i'$:
  1. for $b_1 \in B - E$: $b_1 \alpha_i' = b_1 \alpha_i$ and $\alpha_k' = \alpha_k$. Since G is a G-map, $b_1 \alpha_i \alpha_k = b_1 \alpha_k \alpha_i$; since $b_1 \alpha_k = b_1 \alpha_k' \in B - E$ (indeed, assume that $b_2 = b_1 \alpha_k \in E$. In this case, $b_1 = b_2 \alpha_k \in E$: contradiction), $b_1 \alpha_k \alpha_i = b_1 \alpha_k' \alpha_i$ whence $b_1 \alpha_k \alpha_i = b_1 \alpha_k' \alpha_i'$; thus $b_1 \alpha_i' \alpha_k' = b_1 \alpha_k' \alpha_i'$.
  2. for $b_1 \in C - F$: similar to the previous case.
  3. for $b_1 \in E$: $b_1 \alpha_k \in E$ and $b_1 \alpha_k \gamma = b_1 \gamma \alpha_k$ (precondition of the insertion operation). So, $b_1 \alpha_i' = b_1 \gamma$, $b_1 \alpha_k \alpha_i' = b_1 \alpha_i' \alpha_k$. Moreover, $\forall b \in B, b \alpha_j = b \alpha_j'$; thus, $b_1 \alpha_j' \alpha_i' = b_1 \alpha_i' \alpha_j'$ (same method for the darts of F).
- for $k = i$: similar to the previous case. $\qquad\square$

4 Expansion

Informally, i-expansion consists in adding an i-cell "inside" an $(i-1)$-cell. Expansion is the dual of the insertion operation. The dual of a subdivision is a subdivision of the same space, for which an $(n-i)$-cell is associated with each initial i-cell, and incidence relations are kept the same. A nice property of G-maps is the fact that the dual G-map of $G = (B, \alpha_0, \ldots, \alpha_n)$ is $G' = (B, \alpha_n, \ldots, \alpha_0)$: we only need to reverse the involution order.

We can thus easily deduce the definition of i-expansion from the general definition of i-insertion. We only have to replace '+' with '-' for indices of involutions for preconditions and operations, i.e. $\alpha_{i+1} \alpha_{i+2} \to \alpha_{i-1} \alpha_{i-2}$ and $\alpha_i \alpha_{i+1} \to \alpha_i \alpha_{i-1}$ (see two examples of expansion in Figs. 9 and 10).

Definition 5. *(i-expansion) Let $G = (B, \alpha_0, \alpha_1, \ldots, \alpha_n)$ be an n-G-map, $i \in \{0, \ldots, n-1\}$, $C = \langle\rangle_{N-\{i\}}(b)$ be the i-cell to expand, E and F be two subsets of darts such that: $E \subseteq B$ and $F \subseteq C$, γ be an involution defined on set $E \cup F$ such that: $b \in E \Leftrightarrow b\gamma \in F$. This operation can be applied only when:*

---

[9] According to the previous preconditions, $b_2 \alpha_i = b_2 \gamma \alpha_{i+1} (\alpha_i \alpha_{i+1})^k \gamma$. Since $b_2 \gamma \alpha_{i+1} (\alpha_i \alpha_{i+1})^k \in F$ then $b_2 \alpha_i = b_2 \gamma \alpha_{i+1} (\alpha_i \alpha_{i+1})^k \gamma \in E$.

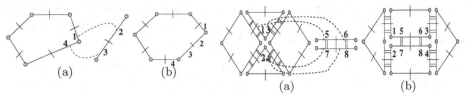

Fig. 9. 1-expansion in 1D. (a) Darts of the edge to expand are numbered 2 and 3. Involution γ are marked with dashed line. (b) Result.

Fig. 10. 1-expansion in 2D. (a) Initial 2-G-map with the edge to expand and involution γ. (b) Result.

- $\forall b \in F, b$ is i-free;
- $\forall b \in C, b\alpha_{i-1}\alpha_{i-2} = b\alpha_{i-2}\alpha_{i-1};$[10]
- $\forall b \in E \cup F, \forall j\ 0 \leq j \leq n$ such that $|i - j| \geq 2$ and $b\alpha_j \in E \cup F$ then $b\alpha_j\gamma = b\gamma\alpha_j;$
- $\forall b \in E$ et $b\alpha_i = b\gamma\alpha_{i-1}(\alpha_i\alpha_{i-1})^k\gamma$ (where k is the smallest integer such that $b\gamma\alpha_{i-1}(\alpha_i\alpha_{i-1})^k \in F$).

The n-G-map resulting from the expansion of this i-cell is $G' = (B', \alpha'_0, \ldots, \alpha'_n)$ defined by:

- $B' = B \cup C;$
- $\forall j \in \{0, \ldots, n\} - \{i\}, \forall b \in B' : b\alpha'_j = b\alpha_j;$
- $\forall b \in B' - (E \cup F) : b\alpha'_i = b\alpha_i;$
- $\forall b \in E \cup F : b\alpha'_i = b\gamma.$

Theorem 2. G' is an n-G-map.

The proof for the expansion operation is equivalent by duality (exchange $\alpha_{(i+1)}$ and $\alpha_{(i-1)}$) to the proof of theorem 1.   □

5   Generalisation

Previous definitions enable us to insert or to expand a single cell. For some applications, it could be more efficient to simultaneously apply several operations. In practice, let G be an n-G-map and G' another n-G-map[11] to insert (resp. to expand). The only difference with the definition of one insertion (resp. expansion) is that G' can contain several i-cells. Thus, the only modification in definition 4 (resp. definition 5) consists in replacing C, the i-cell to insert (resp. to expand) by G', the set of i-cells.

This allows us to apply simultaneously a set of operations and to obtain the same result as if we had successively applied the operations. For this generalisation, there is no additional precondition since i-cells are always disjoined in a n-G-map.

---

[10] Note that this condition does not apply for $i = 1$, so we can always expand any edge.
[11] By definition, a G-map is always a partition into i-cells, for $0 \leq i \leq n$.

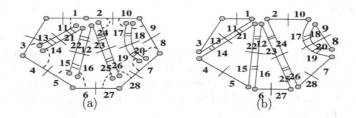

Fig. 11. An example in 2D of simultaneous insertion of 1-cells. (a) 2-G-map before operation with the 1-cells to insert and involution γ (represented by dashed lines). (b) The resulting 2-G-map. The darts belonging to inserted 1-cells are numbered $11, 12, \ldots, 26$. For instance, $1\alpha_1 = 1\gamma\alpha_2 \left(\alpha_1\alpha_2\right)^2 \gamma = 2$ since three edges $\langle\alpha_0, \alpha_2\rangle$ (11), $\langle\alpha_0, \alpha_2\rangle$ (12) and $\langle\alpha_0, \alpha_2\rangle$ (23) are inserted around the same vertex.

We now show that it is possible to simultaneously perform insertions (resp. expansions) of several i-cells for a given i ($0 \leq i \leq n$).

Generalisation 1. *We can easily prove that the previous definition of insertion (resp. expansion) stands for the insertion (resp. expansion) of a set of cells with the same dimension i.*

In order to prove this claim, we can follow the same method described in the proof of theorem 1: just consider the sewing between the darts of G and the darts of different i-cells of G'.

In Fig. 11, the 2-G-map G' to insert contains four 1-cells: $\langle\alpha_0, \alpha_2\rangle$ (11), $\langle\alpha_0, \alpha_2\rangle$ (12), $\langle\alpha_0, \alpha_2\rangle$ (23) and $\langle\alpha_0, \alpha_2\rangle$ (17). The first three edges are linked by α_1, and the fourth edge is independent. Furthermore, the simultaneous insertion of these edges consists in sewing the darts of $E = \{1, 2, \ldots, 10, 27, 28\}$ and those of $F = \{11, 12, \ldots, 20, 23, \ldots, 26\}$ (it is the same process as the one given in definition 4).

6   Conclusion and Perspectives

In this paper, we have defined insertion and expansion operations, which can be applied to one i-cell of any n-G-map, whatever their respective dimensions. Moreover, we have studied how to perform the same operations simultaneously. These definitions are homogeneous for any dimension. Since combinatorial maps can be easily deduced from orientable generalized maps, these operations can also be defined on combinatorial maps.

In order to conceive efficient algorithms, an interesting perspective is to perform different operations simultaneously: insertion and expansion of cells of different dimensions. We think that, in this case, we can apply the same preconditions as the ones used for insertion and expansion operations. Next, we will study "top-down" pyramids of n-dimensional generalized maps defined as stacks of n-G-map where each n-G-map is built from the previous level by inserting or expanding cells. This will be made for example within the Fogrimmi project whose goal is to define "top-down" pyramids to analyse biological medical images.

References

1. Bertrand, Y., Damiand, G., Fiorio, C.: Topological Encoding of 3D Segmented Images. In: Nyström, I., Sanniti di Baja, G., Borgefors, G. (eds.) DGCI 2000. LNCS, vol. 1953, pp. 311–324. Springer, Heidelberg (2000)
2. Braquelaire, J.P., Desbarats, P., Domenger, J.P.: 3d split and merge with 3 maps. In: Workshop on Graph based Representations, Ischia, Italy, May 2001, pp. 32–43, IAPR-TC15 (2001)
3. Braquelaire, J.P., Desbarats, P., Domenger, J.P., Wüthrich, C.A.: A topological structuring for aggregates of 3d discrete objects. In: Workshop on Graph based Representations, Austria, pp. 193–202, IAPR-TC15 (1999)
4. Brisson, E.: Representing geometric structures in d dimensions: topology and order. Discrete and Computational Geometry 9(1), 387–426 (1993)
5. Brun, L., Kropatsch, W.G.: Dual contraction of combinatorial maps. In: Workshop on Graph based Representations, Austria, pp. 145–154, IAPR-TC15 (1999)
6. Brun, L.: Segmentation d'images couleur à base topologique, PhD Thesis, Université de Bordeaux I (1996)
7. Burt, P., Hong, T.H., Rosenfeld, A.: Segmentation and estimation of image region properties through cooperative hierarchical computation. IEEE Transactions on Systems, Man and Cybernetics 11, 802–809 (1981)
8. Damiand, G., Dexet-Guiard, M., Lienhardt, P., Andres, E.: Removal and contraction operations to define combinatorial pyramids: application to the design of a spatial modeler. Image and Vision Computing 2(23), 259–269 (2005)
9. Edmonds, J.: A combinatorial representation for polyhedral surfaces, Notices of the American Mathematical Society 7 (1960)
10. Fiorio, C.: Approche interpixel en analyse d'images: une topologie et des algorithms de segmentation, PhD Thesis, Université Montpellier II, 24 (1995)
11. Grasset-simon, C., Damiand, G., Lienhardt, P.: nD generalized map pyramids: Definition, representations and basic operations. Pattern Recognition 39, 527–538 (2006)
12. Jolion, J., Montanvert, A.: The adaptive pyramid: a framework for 2d image analysis. Computer Vision, Graphics and Image Processing 55, 339–348 (1992)
13. Kropatsch, W.: Building irregular pyramids by dual-graph contraction. Vision, Image and Signal Processing 142, 366–374 (1995)
14. Lienhardt, P.: Topological models for boundary representation: a comparison with n-dimensional generalized maps. Computer Aided Design 23, 59–82 (1991)
15. Lienhardt, P.: N-dimensional generalized combinatorial maps and cellular quasi-manifolds. International Journal of Computational Geometry and Applications 4(3), 275–324 (1994)
16. Meer, P.: Stochastic image pyramids. Computer Vision, Graphics and Image Processing 45, 269–294 (1989)
17. Moutanvert, A., Meer, P., Rosenfeld, A.: Hierarchical image analysis using irregular tesselations. IEEE Transactions on Pattern Analysis and Machine Intelligence 13(4), 307–316 (1991)

Discrete Complex Structure on Surfel Surfaces

Christian Mercat

I3M, Université Montpellier 2 c.c. 51
F-34095 Montpellier cedex 5 France
mercat@math.univ-montp2.fr

Abstract. This paper defines a theory of conformal parametrization of digital surfaces made of surfels equipped with a normal vector. The main idea is to locally project each surfel to the tangent plane, therefore deforming its aspect-ratio. It is a generalization of the theory known for polyhedral surfaces. The main difference is that the conformal ratios that appear are no longer real in general. It yields a generalization of the standard Laplacian on weighted graphs.

1 Introduction

Conformal parametrization of surfaces is a useful technique in image processing. The key notion is to identify the tangent plane of a surface to the field of the complex numbers in a consistent way. It allows to give a meaning to the notion of angles of two crossing paths on the surface, or equivalently to the notion of *small circles* around a point. A surface with such a complex structure is called a *Riemann* surface. A conformal, holomorphic or analytic function between two Riemann surfaces is a function that preserves angles and small circles.

This notion has been tremendously successful in mathematics and engineering; an aim of this paper is to define its discrete counterpart in the context of the surfel surfaces, defining for example discrete polynomials (see Fig. 6). It is a crucial notion in *texture mapping* for example: consider a particular animal skin that is known to contain small patterns of a given shape, like round disks; if this texture is rendered in a way that stretches these patterns into ovals, the picture will be wrongly interpreted by the viewer, the *distortion* being understood as conveying an information of *tilt* of the underlying surface.

The technique has many other uses, like vector fields on surfaces, surface *remeshing*, surface recognition or surface interpolation. One of its main features is its *rigidity*: the Riemann mapping theorem tells you that a surface topologically equivalent to a disc can be conformally mapped to the unit disc, *uniquely* up to the choice of three points. In this way, surfaces which are very different can be mapped to the same space where their features can be compared. There is much less freedom than in the case of harmonic mapping for example (kernel of the Laplacian, see (4)), which depends on many arbitrary choices, which are too numerous or too sensitive in many cases. This technique is surprisingly robust to changes in the bulk of the surface and the dependency on the boundary conditions can be relaxed as well [1], putting rigidity where the data is meaningful.

D. Coeurjolly et al. (Eds.): DGCI 2008, LNCS 4992, pp. 153–164, 2008.
© Springer-Verlag Berlin Heidelberg 2008

This technique has been widely used in the polyhedral surfaces community [1, 2, 3, 4, 5, 6, 7]. **In this paper** we will describe its adaptation to the case of digital surfaces made of surfels, square boundaries of voxels in $\mathbf{Z}^3$ that constitute a simple combinatorial surface where each edgel belongs to one or two surfels. We develop in this article the theory and algorithms needed for an actual implementation on computers of these notions to the context of surfel surfaces.

The additional information that we use in order to give the digital surface a conformal structure is the data of the normal direction [8, 9, 10].

In the first section 3, we will present how this information encodes a non real discrete conformal structure. Then, in Sec. 4 we will recall elements of de Rham cohomology and apply them in Sec. 5 to define a discrete Hodge star from this conformal structure. How this leads to a theory of discrete Riemann surfaces will be explained in Sec. 6-7, first recalling the real case, then generalizing to the complex case, which is the main technical result of this paper.

2   Previous Work

Conformal maps are present in a lot of areas. They are called analytic, holomorphic, meromorphic or monodriffic. They are related to harmonic maps because they are in particular complex harmonic. In the context of discrete geometry processing, they were used mainly by the polyhedral community, for example Desbrun and al. in [1, 2] or Gu and Yau [4, 5, 6, 7]. Circle packings have been used as well to approximate conformal maps, see [3] and references therein.

3   Conformal Structure

We show here how a surfel surface equipped with normals defines a discrete conformal structure and gives a geometric interpretation to holomorphic maps.

A discrete object is a set of points in $\mathbf{Z}^3$, each center of its Voronoi cell called *voxel*. A voxel is a cube of unit side, its six faces are called *surfels*. A digital surface Σ made of surfels is a connected set of surfels. We will restrict ourselves to surfaces such that every edge in Σ belongs to at most two surfels [11]. The edges that belong to only one surfel are called *boundary edges*. Let us call (the indices stand for dimensions) $(\Diamond_0, \Diamond_1, \Diamond_2)$ the sets of vertices, edges and surfels of this cellular decomposition $\Diamond$ of the surface Σ.

Note that this cellular decomposition is *bipartite*: there exists a bicoloring of its vertices, that can be colored whether black or white, no adjacent vertices have the same color. We consider the surfels diagonals, their end points share the same color, forming *dual* black and white diagonals.

We call Γ the 1-skeleton graph, whose vertices Γ_0 are given by the black vertices and its edges Γ_1 by the black diagonals. It can be completed into a cellular decomposition of the surface by adding faces Γ_2 for each white vertex. Similarly we define its *Poincaré dual* Γ^* composed of the white vertices, white diagonals and faces associated with black vertices. We will refer to Γ as the *primal*

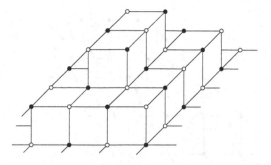

Fig. 1. A surfel surface

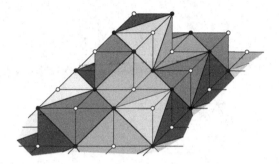

Fig. 2. The cellular decomposition Γ associated with black vertices

(black) graph and to Γ^* as its *dual* (white) graph. The k-cells of Γ are nothing else than the $2-k$ cells of Γ^*: $\Gamma_k \cong \Gamma^*_{2-k}$.

The data of a *normal direction* at each surfel is a broadly used feature of digital surfaces [8,9,10]. This normal might come from a digital scanner, or be computed from the digital surface itself by various means on which we won't elaborate. These consistent normals give an orientation to the surface.

This normal is used to project a given surfel comprising the four vertices (x, y, x', y') to the *local tangent plane*. This projection deforms the square into a parallelogram such as in Fig. 3. Its diagonals are sent to segments which are no longer orthogonal in general. We identify the tangent plane with the complex plane, up to the choice of a similitude. We call Z this local map from the cellular decomposition to the complex numbers. Each diagonal (x, x') and (y, y') is now seen as a complex number $Z(x') - Z(x)$, resp. $Z(y') - Z(y)$.

For example we can project the standard digital plane of cubes associated with $P_0 : x + y + z = 0$ onto this (constant tangent) plane P_0 and get the rhombi pattern appearing in Fig. 4.

We then associate to each diagonal $(x, x') \in \Gamma_1$ the (possibly infinite) complex ratio $i\,\rho$ of the dual diagonal by the primal diagonal, as complex numbers.

$$i\,\rho(x, x') := \frac{Z(y') - Z(y)}{Z(x') - Z(x)}.$$

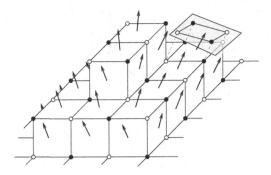

Fig. 3. A surfel projected onto the local tangent plane

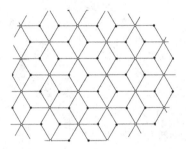

Fig. 4. The digital plane $x + y + z = 0$ projected. Note that Γ is the hexagonal lattice, Γ^* its triangular dual.

This ratio clearly does not depend on the choice of identification between the tangent plane and the field of complex numbers.

We will prefer the number ρ to the ratio $i\rho$ and still call it abusively the ratio. This number does not depend on the orientation of the edge, for $\rho(x, x') = \rho(x', x)$. The number ρ is real whenever the normal is orthogonal to (at least) one of the diagonals, that is to say, when the two diagonals are orthogonal when projected to the tangent plane. We will call this eventuality *the real case*. It is so, for example, in the standard plane case in Fig. 4 where its value is the constant $\rho_{\text{hex}} = \tan(\frac{\pi}{6}) = 1/\sqrt{3}$. The flat square grid $\mathbb{Z}^2$ is associated with the constant 1. Large or small values away from 1 appear whenever the surfels are flattened away from the square aspect ratio. The complex valuess appear when the surfel is slanted away from the orthogonal conformation.

We call this data of a graph Γ, whose edges are equipped with a complex number ρ, a *discrete conformal structure*, or a *discrete Riemann surface*.

We equip the dual edge $(y, y') \in \Gamma_1^*$ of the complex constant $\rho(e^*) = 1/\rho(e)$. In the example Fig. 4, its value is the constant $\rho_{\text{tri}} = \tan(\frac{\pi}{3}) = \sqrt{3} = 1/\rho_{\text{hex}}$.

We define a map $f : \Diamond_0 \to \mathbb{C}$ as *discrete holomorphic* with respect to ρ if and only if, it respects the ratio for each surfel:

$$\forall (x, y, x', y') \in \Diamond_2, \ f(y') - f(y) = i\,\rho(x, x')\,(f(x') - f(x)).$$

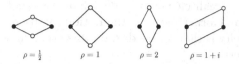

$$\rho = \tfrac{1}{2} \qquad\qquad \rho = 1 \qquad\qquad \rho = 2 \qquad\qquad \rho = 1+i$$

Fig. 5. Several surfel conformations and the associated ratio ρ

In the continuous complex analytic theory, a holomorphic function f is a complex function from the complex plane to itself, which is complex differentiable; that is to say, it is recognized by the fact that its action on a neighborhood around a point $z_0 \in \mathbf{C}$ is locally a similitude $z \mapsto a\,z + b$ where $b = f(z_0)$ and $a = f'(z_0)$, the derivative of f at z_0 with respect to the complex variable z. It sends little circles to little circles.

In the same fashion, a discrete conformal map sends each quadrilateral surfel to another quadrilateral whose diagonals have the same ratio ρ and it can be pictured as sending polygons that form the double Λ into similar polygons that still fit together. For example, the hexagonal and triangular lattices, hidden in Fig. 4, can actually be drawn on the same picture by simply joining the middles of the edgels together. The hexagons and triangles are simply both shrunk by a factor half and fit together at their vertices. A discrete conformal map is recognizable in the sense that it sends each of these regular hexagons and equilateral triangles to polygons of the same shape, touching with the same combinatorics. The discrete derivative is encoded in how much each polygon has been inflated or shrunk, and rotated.

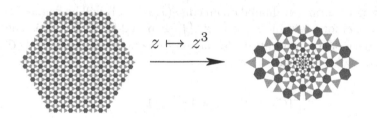

$$z \mapsto z^3$$

Fig. 6. The discrete version of the map $z \mapsto z^3$ in the hexagonal/triangular case

4   De Rham Cohomology

In the continuous theory of surfaces, the notion of complex structure relies on the existence of an operator on 1-forms such that $*^2 = -\mathrm{Id}$, called the *Hodge star*. In orthonormal local coordinates (x, y), it is defined on 1-forms by $*(f\,dx + g\,dy) = -g\,dx + f\,dy$. The discrete analogous of these local coordinates are given by these pairs of dual diagonals (x, x') and (y, y').

In order to follow further this analogy, we first have to define the spaces of discrete functions and discrete forms such as dx, dy. This is done in the theory of de-Rham cohomology [12].

In this section we recall elements of de Rham cohomology, with functions, boundary operator and forms, in which the notions of discrete analytic functions take place. The novelty is that we need to double everything to get the best out of the complex structure.

We define $\Lambda := \Gamma \sqcup \Gamma^*$, disjoint union of the two dual graphs, that we will call the *double* graph. Its vertices $\Lambda_0 = \Diamond_0$ are the same as the vertices of the surfel cellular decomposition $\Diamond$, its edges Λ_1 are the black and white diagonals, its faces Λ_2 is a set in bijection with its vertices. This doubling can look artificial at first sight but is in fact very useful in practice, allowing for nicer formulae because Λ is self-dual by construction. We could as well define complex functions as being real on the vertices Γ_0 and pure imaginary on the faces Γ_2, but it is not as practical. The case is similar to the continuous where functions and 2-forms are essentially the same set but treated differently.

The complex of *chains* $C(\Lambda) = C_0(\Lambda) \oplus C_1(\Lambda) \oplus C_2(\Lambda)$ is the vector space (over $\mathbb{R}$ the field of reals) spanned by vertices, edges and faces. It is equipped with a *boundary* operator $\partial : C_k(\Lambda) \to C_{k-1}(\Lambda)$, null on vertices and fulfilling $\partial^2 = 0$. The kernel $\ker \partial =: Z_{\bullet}(\Lambda)$ of the boundary operator are the closed chains or *cycles*. The image of the boundary operator are the *exact* chains. The *homology* of Λ is the space of cycles modulo exact chains. It encapsulates all there is to know about the topology of the surface.

The dual spaces of forms are called *cochains*, denoted with upper indices for the dimension, are functions from chains to the field of complex numbers. $C^k(\Lambda) := \mathrm{Hom}(C_k(\Lambda), \mathbb{C})$. The space $C^0(\Lambda)$ of 0-forms is the linear span of functions of the vertices, 1-forms are functions of the oriented edges, 2-forms are functions of the faces. Coupling is denoted by functional and integral notation: the value of a 1-form α evaluated on an edge $(x, x') \in \Lambda_1$ will be denoted by $\int_x^{x'} \alpha$, similarly, $f(x)$ for a 0-form on a vertex, $\iint_F \omega$ for a 2-form on a face. The dual of the boundary operator is called the *coboundary* operator $d : C^k(\Lambda) \to C^{k+1}(\Lambda)$, defined by Stokes formula:

$$\int_x^{x'} df := f(\partial(x, x')) = f(x') - f(x), \qquad \iint_F d\alpha := \oint_{\partial F} \alpha,$$

where $\oint$ denotes the circulation of a 1-form along a closed contour. A *cocycle* is a closed cochain and we note $\alpha \in Z^k(\Lambda)$. The *cohomology* of Λ is the space of cocycles modulo the exact forms.

We define an exterior wedge product, denoted $\wedge$, for 1-forms living either on edges $\Diamond_1$ or on their diagonals Λ_1, as a 2-form living on surfels $\Diamond_2$. The formula for the latter is:

$$\iint_{(x,y,x',y')} \alpha \wedge \beta := \frac{1}{2} \left(\int_{(x,x')} \alpha \int_{(y,y')} \beta - \int_{(y,y')} \alpha \int_{(x,x')} \beta \right) \tag{1}$$

The exterior derivative d is, as it should be, a derivation for the wedge product, for functions f, g and a 1-form α:

$$d(fg) = f\,dg + g\,df, \qquad\qquad d(f\alpha) = df \wedge \alpha + f\,d\alpha.$$

5 Hodge Star

After having set the spaces where our operator is going to act, we show, in this section, how the discrete conformal structure allows to define a Hodge star, that is to say, an operator verifying $*^2 = -\mathrm{Id}$ on 1-forms. It breaks forms into holomorphic and anti-holomorphic parts. Moreover, we show that this decomposition is robust to flips and that parallel planes have isomorphic decompositions.

Definition 1. *The* Hodge star *is defined on functions and 2-forms by*

$$* : C^k(\Lambda) \to C^{2-k}(\Lambda)$$

$$C^0(\Lambda) \ni f \mapsto *f : \iint_F *f := f(F^*),$$

$$C^2(\Lambda) \ni \omega \mapsto *\omega : (*\omega)(x) := \iint_{x^*} \omega\,,$$

and on 1-forms, on the dual edges $(y, y') = (x, x')^* \in \Lambda_1$, *given the (complex) discrete conformal structure* $\rho(x, x') = re^{i\theta}$, *by*

$$\begin{pmatrix} \int_x^{x'} *\alpha \\ \int_y^{y'} *\alpha \end{pmatrix} := \frac{1}{\cos\theta} \begin{pmatrix} -\sin\theta & -\frac{1}{r} \\ r & \sin\theta \end{pmatrix} \begin{pmatrix} \int_{(x,x')} \alpha \\ \int_{(y,y')} \alpha \end{pmatrix}.$$

Notice that the Hodge star is a *real* transformation. The fact that it is well defined relies on the fact that two dual diagonals are associated with inverse numbers: $\rho(y, y') = 1/\rho(x, x') = \frac{1}{r}e^{-i\theta}$ (using the former notation). It fulfills $*^2 = \mathrm{Id}$ for functions and 2-forms, and $*^2 = -\mathrm{Id}$ for 1-forms:
We have, on the surfel $(x, y, x', y') \in \Diamond_2$,

$$\frac{1}{\cos^2\theta}\begin{pmatrix} -\sin\theta & -\frac{1}{r} \\ r & \sin\theta \end{pmatrix}^2 = \frac{1}{\cos^2\theta}\begin{pmatrix} \sin^2\theta - 1 & 0 \\ 0 & \sin^2\theta - 1 \end{pmatrix} = -I_2.$$

Because of this property, we can define it on the complexified forms and functions. It breaks the complex 1-forms into two eigenspaces associated to eigenvalues $-i$ and $+i$ called respectively type $(1, 0)$ and $(0, 1)$ forms.

Definition 2. *A* holomorphic *form* $\alpha \in C^1_{\mathbb{C}}(\Lambda)$ *is a closed type* $(1, 0)$ *form, that is to say,* α *is such that* $d\alpha = 0$ *and* $*\alpha = -i\alpha$. *An* anti-holomorphic *form is a closed type* $(0, 1)$ *form.*

A function f *is* holomorphic *iff its exterior derivative* df *is as well holomorphic, that is to say, on a surfel* $(x, y, x', y') \in \Diamond_2$, $f(y') - f(y) = i\,\rho(x, x')\,(f(x') - f(x))$.

A meromorphic *form with a* pole *at* $F \in \Diamond_2$, *is a type* $(1, 0)$ *form, closed except on the surfel* F. *Its lack of closeness is called its* residue *at* F.

Holomorphic and meromorphic functions are tremendously important in mathematics, they are behind all the keys on a calculator like polynomials, inversion, cosine, tangent, exponential, logarithm... This article defines the framework in which their discrete counterparts take place.

An interesting feature of the theory is its robustness with respect to local moves. A discrete holomorphic map defined on a discrete Riemann surface is mapped by a *canonical isomorphism* to the space of discrete holomorphic maps defined on another discrete Riemann surface linked to the original one by a series of *flips*. These flips are called in the context of discrete conformal structures *electrical moves*. They come in three kinds, the third one being the flip, the others being irrelevant to our context.

Fig. 7. The electrical moves

The third move corresponds to the flip, it is called the *star-triangle transformation*. To three surfels (drawn as dotted lines in Fig. 7) arranged in a hexagon, whose diagonals form a triangle of conformal parameters ρ_1, ρ_2 and ρ_3, one associates a configuration of three other surfels whose diagonals form a three branched star with conformal parameters ρ_i' (on the opposite side of ρ_i) verifying

$$\rho_i \rho_i' = \rho_1 \rho_2 + \rho_2 \rho_3 + \rho_3 \rho_1 = \frac{\rho_1' \rho_2' \rho_3'}{\rho_1' + \rho_2' + \rho_3'}. \tag{2}$$

The value of a holomorphic function at the center of an hexagon is overdetermined with respect to the six values on the hexagon. These values have to fulfill a compatibility condition, which are the same for both hexagons, therefore a holomorphic function defined on a discrete Riemann surface can be uniquely extended to another surface differing only by a flip [13].

This means in particular that a discrete holomorphic function defined on the standard plane in Fig. 4 can be followed through all its other parallel deformations and is not sensitive to some added noise (flips deleting or inserting extra voxels) provided the normal vector is unchanged with respect to the discrete plane value: the space of holomorphic functions on these parallel or noisy planes are in one-to-one correspondence. This theoretical robustness has yet to be experimentally observed in practice because the normal vectors are not independent and a noisy plane will have noisy normal vectors as well. This will be the subject of a forthcoming article.

For real conformal structures, the formulae are simpler, we present them independently before generalizing them to the complex case.

6 The Real Case

We present in this section the formulae in the real case, for the Hodge star, the Laplacian, the scalar product and the different energies, Dirichlet and conformal.

When the complex structure is defined by a real ratio ρ, for example in the hexagonal/triangular standard discrete plane case, then, for each surfel, the Hodge star takes the simpler form

$$* : C^k(\Lambda) \to C^{2-k}(\Lambda)$$

$$C^0(\Lambda) \ni f \mapsto *f : \iint_F *f := f(F^*),$$

$$C^1(\Lambda) \ni \alpha \mapsto *\alpha : \int_e *\alpha := -\rho(e^*) \int_{e^*} \alpha, \tag{3}$$

$$C^2(\Lambda) \ni \omega \mapsto *\omega : (*\omega)(x) := \iint_{x^*} \omega.$$

The endomorphism $\Delta := -d * d * - * d * d$ is, in this real case, the usual discrete *Laplacian*: Its formula on a function at a vertex $x \in \Gamma_0$ with neighbours $x_1, \ldots, x_V \in \Gamma_0$ is the usual weighted averaged difference:

$$(\Delta(f))(x) = \sum_{k=1}^{V} \rho(x, x_k) (f(x) - f(x_k)). \tag{4}$$

The space of *harmonic forms* is defined as its kernel.

Together with the Hodge star, they give rise, in the compact case, to the usual weighted scalar product on 1-forms:

$$(\alpha, \beta) := \iint_{\Diamond_2} \alpha \wedge *\overline{\beta} = (*\alpha, *\beta) = \overline{(\beta, \alpha)} = \tfrac{1}{2} \sum_{e \in \Lambda_1} \rho(e) \int_e \alpha \int_e \overline{\beta} \tag{5}$$

The ℓ^2 norm of the 1-form df, called the *Dirichlet energy* of the function f, is the average of the usual Dirichlet energies on each independant graph

$$E_D(f) := \|df\|^2 = (df, df) = \frac{1}{2} \sum_{(x, x') \in \Lambda_1} \rho(x, x') |f(x') - f(x)|^2 \tag{6}$$

$$= \frac{E_D(f|_\Gamma) + E_D(f|_{\Gamma^*})}{2}.$$

The conformal energy of a map measures its conformality defect, relating these two harmonic functions. A conformal map fulfills the Cauchy-Riemann equation

$$* df = -i \, df. \tag{7}$$

Therefore a quadratic energy whose null functions are the holomorphic ones is

$$E_C(f) := \tfrac{1}{2} \|df - i * df\|^2. \tag{8}$$

It is related to the Dirichlet energy through the same formula as in the continuous case:

$$
\begin{aligned}
E_C(f) &= \tfrac{1}{2}\left(df - i*df,\ df - i*df\right) \\
&= \tfrac{1}{2}\|df\|^2 + \tfrac{1}{2}\|-i*df\|^2 + \mathrm{Re}(df, -i*df) \\
&= \|df\|^2 + \mathrm{Im}\iint_{\Diamond_2} df \wedge \overline{df} \\
&= E_D(f) - 2\mathcal{A}(f)
\end{aligned}
\tag{9}
$$

where the area of the image of the application f in the complex plane has the same formula

$$
\mathcal{A}(f) = \frac{i}{2}\iint_{\Diamond_2} df \wedge \overline{df}
\tag{10}
$$

as in the continuous case. For a face $(x, y, x', y') \in \Diamond_2$, the algebraic area of the oriented quadrilateral $\big(f(x), f(x'), f(y), f(y')\big)$ is given by

$$
\begin{aligned}
\iint_{(x,y,x',y')} df \wedge \overline{df} &= i\,\mathrm{Im}\left((f(x') - f(x))\overline{(f(y') - f(y))}\right) \\
&= -2i\mathcal{A}\big(f(x), f(x'), f(y), f(y')\big).
\end{aligned}
$$

When a holomorphic reference map $z : \Lambda_0 \to \mathbb{C}$ is chosen, a holomorphic (resp. anti-holomorphic) 1-form df is, locally on each pair of dual diagonals, proportional to dz, resp. $d\bar{z}$, so that the decomposition of the exterior derivative into holomorphic and anti-holomorphic parts yields $df \wedge \overline{df} = \left(|\partial f|^2 + |\bar\partial f|^2\right) dz \wedge d\bar{z}$ where the derivatives naturally live on faces.

All these concepts turn into an actual machinery that can be implemented on computer and a full featured theory mimicking the theory of Riemann surfaces to a surprisingly far extent [14]. In particular in the rhombic case, the notion of polynomials and exponentials, differential equations, logarithm (called the Green function) follow through. A striking difference is the fact that constants are of dimension two, the constants on the graph Γ being independent from the constants on the dual graph Γ^*.

7   Non Real Conformal Structure

In this section we give formulae for the scalar product, the Dirichlet energy and the Laplacian in the complex case. Nothing more is needed in order to implement computation of global conformal parametrization of surfel surfaces.

In the complex case, the dual graphs, which are independent in the real case, are no longer independent and are mixed together.

The first construction that needs an adaptation is the scalar product. The formula (5) still defines a positive definite scalar product which is preserved by $*$, even though the last equality must be replaced by a mixed sum over the two dual edges:

$$
(\alpha,\ \beta) = \tfrac{1}{2}\sum_{e \in \Lambda_1} \frac{\int_e \alpha}{\mathrm{Re}\left(\rho(e)\right)}\left(|\rho(e)|^2 \int_e \bar\beta + \mathrm{Im}\left(\rho(e)\right)\int_{e^*} \bar\beta\right).
\tag{11}
$$

The Dirichlet energy mixes the two dual graphs as well:

$$E_D(f) := \|df\|^2 = \frac{1}{2} \sum_{e \in \Lambda_1} \frac{|f(x') - f(x)|^2}{\mathrm{Re}\,(\rho(e))} \left(|\rho(e)|^2 + \mathrm{Im}\,(\rho(e)) \frac{\overline{f(y') - f(y)}}{f(x') - f(x)} \right)$$

(12)

and the Laplacian no longer splits on the two independent dual graphs: Given $x_0 \in \Lambda_0$, with dual face $x_0^* = (y_1, y_2, \ldots, y_V) \in \Lambda_2$ and neighbours $x_1, x_2, \ldots, x_V$ $\in \Lambda_0$, with dual edges $(x_0, x_k)^* = (y_k, y_{k+1}) \in \Lambda_1$, and $y_{V+1} = y_1$, we have

$$\Delta(f)(x_0) = \sum_{k=1}^{V} \frac{1}{\mathrm{Re}\,(\rho(e))} \left(|\rho(e)|^2 (f(x_k) - f(x)) + \mathrm{Im}\,(\rho(e)) (f(y_{k+1}) - f(y_k)) \right).$$

(13)

This Laplacian is still real and involves not only the neighbors along the diagonals of surfels on a weighted graph, which is the star used for the usual Laplacian, but all the vertices of the surfels that contains the vertex.

8   Conclusions and Acknowledgments

This paper defines the theory of analytic functions on digital surfaces made of surfels. It required more than a mere adaptation of the theory known in the polyhedral community since the conformal parameters are in general not real.

In forthcoming papers, this theory, already put to use in the context of polyhedral surfaces, will be implemented in the context of surfel surfaces: The first applications that come to mind are the recognition of digital surfaces, simple ones and of higher topology through the computation of their discrete period matrices [15], the analysis of vector fields on digital surfaces allowed by the Hodge theorem that decomposes vector fields into rotational and divergence free parts, the creation of vector fields with given circulation properties, in general the correct discrete treatment of partial differential equations on a digital surface which are solved by analytic functions in the continuous case, like incompressible fluid dynamics for example.

The author would like to thank the referees for useful comments, the forthcoming Géométrie Discrète ANR project, in particular Valérie Berthé (LIRMM) and Rémy Malgouyres (LLAIC) for discussions that brought the main idea that made this paper possible.

References

1. Desbrun, M., Meyer, M., Alliez, P.: Intrinsic parameterizations of surface meshes. Computer Graphics Forum 21, 209–218 (2002)
2. Desbrun, M., Kanso, E., Tong, Y.: Discrete differential forms for computational modeling. In: SIGGRAPH 2006: ACM SIGGRAPH 2006 Courses, pp. 39–54. ACM Press, New York (2006)

3. Kharevych, L., Springborn, B., Schröder, P.: Discrete conformal mappings via circle patterns. ACM Trans. Graph. 25(2), 412–438 (2006)
4. Gu, X., Yau, S.: Computing conformal structures of surfaces. Communications in Information and Systems 2(2), 121–146 (2002)
5. Gu, X., Yau, S.: Surface classification using conformal structures. International Conference on Computer Vision (2003)
6. Gu, X., Yau, S.-T.: Global conformal surface parameterization. In: SGP 2003: Proceedings of the 2003 Eurographics/ACM SIGGRAPH symposium on Geometry processing. Eurographics Association, Aire-la-Ville, Switzerland, pp. 127–137 (2003)
7. Jin, M., Wang, Y., Yau, S.-T., Gu, X.: Optimal global conformal surface parameterization. In: VIS 2004: Proceedings of the conference on Visualization 2004, pp. 267–274. IEEE Computer Society Press, Washington, DC, USA (2004)
8. Lenoir, A., Malgouyres, R., Revenu, M.: Fast computation of the normal vector field of the surface of a 3-D discrete object. In: Miguet, S., Ubéda, S., Montanvert, A. (eds.) DGCI 1996. LNCS, vol. 1176, pp. 101–112. Springer, Heidelberg (1996)
9. Lenoir, A.: Fast estimation of mean curvature on the surface of a 3d discrete object. In: Ahronovitz, E. (ed.) DGCI 1997. LNCS, vol. 1347, pp. 175–186. Springer, Heidelberg (1997)
10. Malgouyres, R.: A discrete radiosity method. In: Braquelaire, A., Lachaud, J.-O., Vialard, A. (eds.) DGCI 2002. LNCS, vol. 2301, pp. 428–438. Springer, Heidelberg (2002)
11. Malgouyres, R., Burguet, J.: Strong Thinning and Polyhedrization of the Surface of a Voxel Object. Discrete Appl. Math. 125(1) (2003). In: Nyström, I., Sanniti di Baja, G., Borgefors, G. (eds.) DGCI 2000. LNCS, vol. 1953, pp. 93–114. Springer, Heidelberg (2000)
12. Farkas, H.M., Kra, I.: Riemann surfaces, 2nd edn. Springer-Verlag, New York (1992)
13. Bobenko, A.I., Mercat, C., Suris, Y.B.: Linear and nonlinear theories of discrete analytic functions. Integrable structure and isomonodromic Green's function. J. Reine Angew. Math. 583, 117–161 (2005)
14. Mercat, C.: Discrete Riemann surfaces and the Ising model. Comm. Math. Phys. 218(1), 177–216 (2001)
15. Mercat, C.: Discrete riemann surfaces. In: Papadopoulos, A. (ed.) Handbook of Teichmüller Theory, vol. I. IRMA Lect. Math. Theor. Phys., Eur. Math. Soc., Zürich, vol. 11, pp. 541–575 (2007)

Minimal Simple Pairs in the Cubic Grid

Nicolas Passat[1], Michel Couprie[2], and Gilles Bertrand[2]

[1] LSIIT, UMR 7005 CNRS/ULP, Strasbourg 1 University, France
[2] Université Paris-Est, LABINFO-IGM, UMR CNRS 8049, A2SI-ESIEE, France
passat@dpt-info.u-strasbg.fr, {m.couprie,g.bertrand}@esiee.fr

Abstract. Preserving topological properties of objects during thinning procedures is an important issue in the field of image analysis. This paper constitutes an introduction to the study of non-trivial simple sets in the framework of cubical 3-D complexes. A simple set has the property that the homotopy type of the object in which it lies is not changed when the set is removed. The main contribution of this paper is a characterisation of the non-trivial simple sets composed of exactly two voxels, such sets being called minimal simple pairs.

Keywords: Cubical complexes, topology preservation, collapse, thinning, 3-D space.

1 Introduction

Topological properties are fundamental in many applications of image analysis. Topology-preserving operators, like homotopic skeletonisation, are used to transform an object while leaving unchanged its topological characteristics. In discrete grids ($\mathbb{Z}^2$ or $\mathbb{Z}^3$), such a transformation can be defined and efficiently implemented thanks to the notion of simple point [16]: intuitively, a point of an object is called simple if it can be deleted from this object without altering its topology.

A typical topology-preserving transformation based on simple points deletion, that we call *guided homotopic thinning* [9,8], may be described as follows. The input data consists of a set X of points in the grid (called object), and a subset K of X (called constraint set). Let $X_0 = X$. At each iteration i, choose a simple point x_i in X_i but not in K according to some criterion (*e.g.*, a priority function) and set $X_{i+1} = X_i \setminus \{x_i\}$. Continue until reaching a step n such that no simple point for X_n remains in $X_n \setminus K$. We call the result of this process a *homotopic skeleton of X constrained by K*. Notice that, since several points may have the same priority, there may exist several homotopic skeletons for a given pair X, K.

The most common example of priority function for the choice of x_i is a distance map which associates, to each point of X, its distance from the boundary of X. In this case, the points which are closest to the boundary are chosen first, resulting in a skeleton which is "centered" in the original object. In some particular applications, the priority function may be obtained through a greyscale image, for example when the goal is to segment objects in this image while respecting topological constraints (see *e.g.* [10,22]). In the latter case, the order in

D. Coeurjolly et al. (Eds.): DGCI 2008, LNCS 4992, pp. 165–176, 2008.

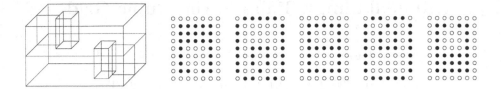

Fig. 1. Left: The Bing's house with two rooms. Right: A discrete version of the Bing's house, decomposed into its five planar slices for visualisation. The 26-adjacency relation is used for object points.

which points are considered does not rely on geometrical properties, and may be affected by noise.

In such a transformation, the result is expected to fulfil a property of minimality, as suggested by the term "skeleton". This is indeed the case for the procedure described above, since the result X_n is minimal in the sense that it contains no simple point outside of K. However, we could formulate a stronger minimality requirement, which seems natural for this kind of transformation: informally, the result X_n should not strictly include any set Y which is "topologically equivalent" to X, and which contains K. We say that a homotopic skeleton of X constrained by K is *globally minimal* if it fulfils this condition.

Now, a fundamental question arises: is any homotopic skeleton globally minimal? Let us illustrate this problem in dimensions 2 and 3. In $\mathbb{Z}^2$, consider a full rectangle X of any size, and the constraint set $K = \emptyset$. Obviously, this object X is topologically equivalent to a single point, thus only homotopic skeletons which are singletons are globally minimal. A. Rosenfeld proved in [21] that any homotopic skeleton of X is indeed reduced to a single point.

But quite surprisingly, in dimension 3, this property does not hold: if X is *e.g.* a full $10 \times 10 \times 10$ cube, we may find a homotopic skeleton of X (with empty constraint set) which is not reduced to a single point. This fact constitutes one of the main difficulties when dealing with certain topological properties, such as the Poincaré conjecture. A classical counter-example is the Bing's house with two rooms [6], illustrated in Fig. 1 (left). One can enter the lower room of the house by the chimney passing through the upper room, and vice-versa. A discrete version B of the Bing's house is displayed in Fig. 1 (right). It can be seen that the Bing's house can be carved from a full cube by iterative removal of simple points. It can also be seen that B contains no simple point: deleting any point from B would create a "tunnel".

It could be argued that objects like Bing's houses are unlikely to appear while processing real (noisy) images, because of their complex shape and their size. However, we found that there exists a large class of objects presenting similar properties, some of them being quite small (less than 50 voxels). Let us call a *lump relative to K* any object X which has no simple point outside of K, and which strictly includes a subset Y including K and topologically equivalent to X (*i.e.*, a homotopic skeleton which is not globally minimal). This notion of lump is formalised and discussed in Appendix A. One of the authors detected the existence of lumps while processing MRI images of the brain [19]. A simpler way to find lumps consists of applying a guided homotopic thinning procedure to an $N \times N \times N$ cube, using different randomly generated priority functions, until no simple point remains. The following table summarises the outcome of such an experiment, with different values of N and for $10,000$ skeletons generated using different random priority functions. We denote by p the proportion of the cases where the result is not a singleton set.

N	10	20	30	40
p	0.0001	0.0249	0.1739	0.5061

Motivated by these practical considerations, two questions arise: is it possible to detect when a thinning procedure gets stuck on a lump, and then, is it possible to find a way towards a globally minimal homotopic skeleton? For performing the latter task, a solution consists of identifying a subset of X which can be removed without changing topology; we call such a subset a *simple set*. Certain classes of simple sets have been studied in the literature dedicated to parallel homotopic thinning algorithms [20,1,12]. In these studies, the considered simple sets are composed exclusively of simple points. In our case, the situation is radically different since a lump relative to K does not contain any simple point outside of K. Then, our problem may be formulated as follows: does there exist a characterisation of certain simple sets composed of non-simple points?

We are indeed interested essentially by simple sets which are minimal, in the sense that they do not strictly include any other simple set, since it is sufficient to detect such sets in order to carry on thinning. Also, we hope that minimal simple sets have a specific structure which could make them easier to analyse.

This paper is dedicated to the study of the simplest ones among such simple sets, called simple pairs, which are those composed of two non-simple points. Our experiments showed us that these minimal simple sets are the ones which are most likely to appear in practical applications, hence the interest in understanding their structure. After proving some properties of simple pairs, we give a characterisation of these sets which allows to detect and remove them when performing homotopic thinning. This paper is self-contained, however the proofs cannot be included due to space limitation. They can be found in [18].

We shall develop this work in the framework of abstract complexes. Abstract complexes have been promoted in particular by V. Kovalevsky [17] in order to provide a sound topological basis for image analysis. In this framework, we retrieve the main notions and results of digital topology, such as the notion of simple point.

2   Cubical Complexes

Intuitively, a cubical complex may be thought of as a set of elements having various dimensions (*e.g.* cubes, squares, edges, vertices) glued together according

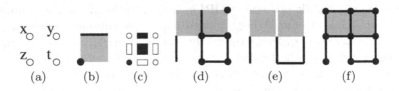

Fig. 2. (a) Four points x, y, z, t of $\mathbb{F}^2$ such that $\{x, y, z, t\}$ is a 2-face. (b,c) Two representations of the set of faces $\{\{x, y, z, t\}, \{x, y\}, \{z\}\}$. (d) A set F of faces in $\mathbb{F}^2$: we see that F is not a complex. (e) The set F^+, composed by the facets of F. (f) The set F^-, *i.e.* the closure of F, which is a complex.

to certain rules. In this section, we recall briefly some basic definitions on complexes, see also [5,3,4] for more details. For some illustrations of the notions defined hereafter, the reader may refer to Fig. 2.

Let $\mathbb{Z}$ be the set of integers. We consider the families of sets $\mathbb{F}_0^1, \mathbb{F}_1^1$, such that $\mathbb{F}_0^1 = \{\{a\} \mid a \in \mathbb{Z}\}$, $\mathbb{F}_1^1 = \{\{a, a+1\} \mid a \in \mathbb{Z}\}$. A subset f of $\mathbb{Z}^n$ $(n \geq 1)$ which is the Cartesian product of exactly m elements of $\mathbb{F}_1^1$ and $(n - m)$ elements of $\mathbb{F}_0^1$ is called a *face* or an *m-face* of $\mathbb{Z}^n$, m is the *dimension* of f, and we write $\dim(f) = m$.

We denote by $\mathbb{F}^n$ the set composed of all m-faces of $\mathbb{Z}^n$ $(m = 0$ to $n)$. An m-face of $\mathbb{Z}^n$ is called a *point* if $m = 0$, a *(unit) interval* if $m = 1$, a *(unit) square* if $m = 2$, a *(unit) cube* if $m = 3$. In the sequel, we will focus on $\mathbb{F}^3$.

Let f be a face in $\mathbb{F}^3$. We set $\hat{f} = \{g \in \mathbb{F}^3 \mid g \subseteq f\}$, and $\hat{f}^* = \hat{f} \setminus \{f\}$. Any $g \in \hat{f}$ is a *face* of f, and any $g \in \hat{f}^*$ is a *proper face* of f. If F is a finite set of faces of $\mathbb{F}^3$, we write $F^- = \bigcup\{\hat{f} \mid f \in F\}$, F^- is the *closure* of F.

A set F of faces of $\mathbb{F}^3$ is a *cell* or an *m-cell* if there exists an m-face $f \in F$ such that $F = \hat{f}$. The *boundary* of a cell $\hat{f}$ is the set $\hat{f}^*$.

A finite set F of faces of $\mathbb{F}^3$ is a *complex* (in $\mathbb{F}^3$) if for any $f \in F$, we have $\hat{f} \subseteq F$, *i.e.*, if $F = F^-$. Any subset G of a complex F which is also a complex is a *subcomplex* of F. If G is a subcomplex of F, we write $G \preceq F$. If F is a complex in $\mathbb{F}^3$, we also write $F \preceq \mathbb{F}^3$.

A face $f \in F$ is a *facet* of F if there is no $g \in F$ such that $f \in \hat{g}^*$. We denote by F^+ the set composed of all facets of F. Observe that $(F^+)^- = F^-$ and thus, that $(F^+)^- = F$ whenever F is a complex.

The *dimension* of a non-empty complex $F \in \mathbb{F}^3$ is defined by $\dim(F) = \max\{\dim(f) \mid f \in F^+\}$. We say that F is an *m-complex* if $\dim(F) = m$.

Two distinct faces f and g of $\mathbb{F}^3$ are *adjacent* if $f \cap g \neq \emptyset$. Let $F \preceq \mathbb{F}^3$ be a non-empty complex. A sequence $(f_i)_{i=0}^{\ell}$ of faces of F is a *path in F (from f_0 to f_ℓ)* if f_i and f_{i+1} are adjacent, for all $i \in [0, \ell - 1]$. We say that F *is connected* if, for any two faces f, g in F, there is a path from f to g in F. We say that $G \neq \emptyset$ is a *connected component of F* if $G \preceq F$, G is connected and if G is maximal for these two properties (*i.e.*, we have $H = G$ whenever $G \preceq H \preceq F$ and H is connected). We denote by $C[F]$ the set of all the connected components of F. We set $C[\emptyset] = \emptyset$.

3   Topology Preserving Operations

Collapse

The collapse, a well-known operation in algebraic topology [13], leads to a notion of homotopy equivalence in discrete spaces, which is the so-called simple homotopy equivalence [7]. To put it briefly, the collapse operation preserves topology.

Let F be a complex in $\mathbb{F}^3$ and let $f \in F^+$. If there exists a face $g \in \hat{f}^*$ such that f is the only face of F which includes g, then we say that the pair (f, g) is a *free pair* for F. If (f, g) is a free pair for F, the complex $F \setminus \{f, g\}$ is an *elementary collapse* of F.

Let F, G be two complexes. We say that F *collapses onto* G if there exists a *collapse sequence from F to G*, i.e., a sequence of complexes $\langle F_0, \ldots, F_\ell \rangle$ such that $F_0 = F$, $F_\ell = G$, and F_i is an elementary collapse of F_{i-1}, $i = 1, \ldots, \ell$. Steps of elementary collapse of a 3-D complex are illustrated in Fig. 3.

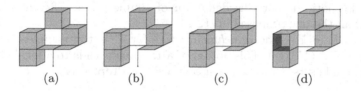

(a)   (b)   (c)   (d)

Fig. 3. (a) A complex $F \preceq \mathbb{F}^3$. (b), (c), (d) Three steps of elementary collapse of F

Let F, G be two complexes. Let H such that $F \cap G \preceq H \preceq G$, and let $f, g \in H \setminus F$. The pair (f, g) is a free pair for $F \cup H$ if and only if (f, g) is a free pair for H. Thus, by induction, we have the following proposition.

Proposition 1 ([2]). *Let $F, G \preceq \mathbb{F}^3$. The complex $F \cup G$ collapses onto F if and only if G collapses onto $F \cap G$.*

Topological Invariants

Let F be a complex in $\mathbb{F}^3$, and let us denote by n_i the number of i-faces of F, $i = 0, \ldots, 3$. The *Euler characteristic* of F, written $\chi(F)$, is defined by $\chi(F) = n_0 - n_1 + n_2 - n_3$. The Euler characteristic is a well-known topological invariant, in particular, it is easy to see that the collapse operation preserves it. This invariant will play an essential role in the proofs of this paper.

Let $F, G \preceq \mathbb{F}^3$. A fundamental and well-known property of the Euler characteristic, analog to the so-called inclusion-exclusion principle in set theory, is the following: $\chi(F \cup G) = \chi(F) + \chi(G) - \chi(F \cap G)$.

The Euler-Poincaré formula shows a deep link between the Euler characteristic and the Betti numbers, which are topological invariants defined from the homology groups of a complex. Intuitively[1], the Betti numbers b_0, b_1, b_2 correspond respectively to the number of connected components, tunnels and cavities of F. The Euler-Poincaré formula, in the case of a complex F in $\mathbb{F}^3$, states that $\chi(F) = b_0 - b_1 + b_2$. Betti numbers are also preserved by collapse.

Simplicity

Intuitively, a part of a complex F is called simple if it can be "removed" from F while preserving topology. We recall here a definition of simplicity (see [2]) based on the collapse operation, which can be seen as a discrete counterpart of the one given by T.Y. Kong [15].

Definition 2. *Let $G \preceq F \preceq \mathbb{F}^3$. We set $F \oslash G = (F^+ \setminus G^+)^-$. The set $F \oslash G$ is a complex which is the* detachment *of G from F.*
We say that G is simple *for F if F collapses onto $F \oslash G$. Such a subcomplex G is called a* simple subcomplex *of F or a* simple set *for F.*

It has to be noticed that this definition of simple set is different (and more general) than the one proposed in [12,14].

Let $G \preceq F \preceq \mathbb{F}^3$. The *attachment* of G for F is the complex defined by $Att(G, F) = G \cap (F \oslash G)$. This notion of attachment leads to a local characterisation of simple sets: Prop. 3 is a special case of Prop. 1 as $(F \oslash G) \cup G = F$.

Proposition 3. *Let $G \preceq F \preceq \mathbb{F}^3$. The complex G is simple for F if and only if G collapses onto $Att(G, F)$.*

4   Minimal Simple Pairs in $\mathbb{F}^3$

In the image processing literature, a digital image is often considered as a set of pixels in 2-D or voxels in 3-D. A voxel is an elementary cube, thus an easy correspondence can be made between this classical view and the framework of cubical complexes. In the sequel of the paper, we call *voxel* any 3-cell. If a complex $F \preceq \mathbb{F}^3$ is a union of voxels, we write $F \sqsubseteq \mathbb{F}^3$. If $F, G \sqsubseteq \mathbb{F}^3$ and $G \preceq F$, then we write $G \sqsubseteq F$. From now on, we consider only complexes which are unions of voxels.

Notice that, if $F \sqsubseteq \mathbb{F}^3$ and if $\hat{f}$ is a voxel of F, then $F \oslash \hat{f} \sqsubseteq \mathbb{F}^3$. There is indeed an equivalence between the operation on complexes that consists of removing (by detachment) a simple voxel, and the removal of a 26-simple voxel in the framework of digital topology (see [14,4]).

As discussed in the introduction, the minimal simple sets which are most likely to appear in thinning processes are those which are composed of only two voxels. In this paper, we will concentrate on this particular - but very frequent - case, and provide a definition, some properties and a characterisation of these sets.

---

[1] An introduction to homology theory can be found *e.g.* in [13].

Definition 4. *Let* $G \sqsubseteq F, G \neq \emptyset$. *The subcomplex* G *is a* minimal simple set *(for* F*) if* G *is a simple set for* F *and* G *is minimal with respect to the relation* $\sqsubseteq$ *(i.e.* $H = G$ *whenever* $H \sqsubseteq G$ *and* H *is a non-empty simple set for* F*).*

Let P *be a minimal simple set for* F *which is composed of two voxels. Then, we call* P *a* minimal simple pair, *or MSP (for* F*).*

Observe that, if a voxel is a simple cell for F, then it is also a (minimal) simple set for F. Thus, any minimal simple set which contains strictly more than one voxel cannot contain any simple voxel. In particular, if P is a simple set which contains only two voxels, then P is a MSP if and only if it does not contain any simple voxel.

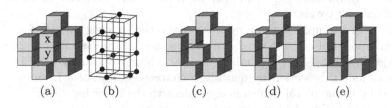

(a)   (b)   (c)   (d)   (e)

Fig. 4. Example of a MSP (voxels x and y). (a), (b): Two representations of the same complex F. (c), (d), (e): Effect of removing either x, y or both (see text).

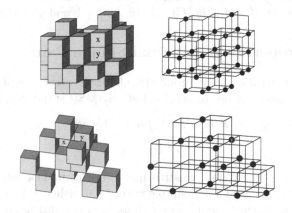

Fig. 5. Left: two complexes composed of non-simple voxels. Right: another representation of these complexes. The subset $\{x, y\}$ is a MSP for both of them (the removal of $\{x, y\}$ will not alter their topology).

Before beginning the study of MSPs (next section), let us show an example of such a configuration. Consider the complex F depicted in Fig. 4a. Another representation of this object is shown in Fig. 4b, where each cube (voxel) is represented by a black dot. It can easily be seen that the complex F is connected and has no cavity and no tunnel; furthermore it can be reduced to a single voxel by iterative deletion of simple voxels. Let us now concentrate on the set formed by the two voxels x and y.

In Fig. 4c, we can see that removing x from F creates a tunnel. Thus x is not a simple voxel. The same can be said about y (see Fig. 4d). But if both x and y are removed (see Fig. 4e), then we see that we obtain a complex G which has no tunnel. It is easily verified that the union of the cells x and y is in fact a simple subcomplex of F, so that it is a MSP for F.

Of course, the complex F of Fig. 4a is not a lump since it contains simple voxels (on its border). In Fig. 5 (1st row), we show that the same configuration can appear in a complex which has no simple voxel but is however topologically equivalent to a single voxel. This lump can be homotopically reduced by deletion of the simple pair $\{x, y\}$. The obtained result could then be further reduced to a singleton set by iterative simple voxel removal. Notice that this complex (generated by randomised homotopic thinning from a 5 voxel-width cube) is made of only 32 voxels.

There exist examples containing less points: the smallest one we found until now is composed of only 14 voxels and has some tunnels (see Fig. 5, 2nd row). We conjecture that 14 is the smallest size for a lump containing a MSP.

We conclude this section by quoting a characterisation of 3-D simple voxels proposed by Kong in [15], which is equivalent to the following theorem for subcomplexes of $\mathbb{F}^3$; this characterisation will be used in the next section. Remind that $|C[X]|$ denotes the number of connected components of X.

Theorem 5 (Adapted from Kong [15]) *Let $F \sqsubseteq \mathbb{F}^3$. Let $g \in F^+$. Then $\hat{g}$ is a simple voxel for F if and only if $|C[Att(\hat{g}, F)]| = 1$ and $\chi(Att(\hat{g}, F)) = 1$.*

5   Some Properties of Minimal Simple Pairs

We are now ready to state some properties about the structure of MSPs: first of all, a simple set need not be connected, but any MSP is indeed connected.

Proposition 6. *Let $P \sqsubseteq F$ be a MSP for F. Then:*

$$|C[P]| = 1.$$

As discussed before, the voxels constituting a MSP cannot be simple voxels. Intuitively, the attachment of a non-simple voxel $\hat{f}$ can either: i) be empty (isolated voxel), ii) be equal to the boundary of $\hat{f}$ (interior voxel), iii) be disconnected, iv) have at least one tunnel. Notice that iii) and iv) are not exclusive, the attachment of a non-simple voxel can both be disconnected and contain tunnels.

We will see that some of these cases cannot appear in a MSP. First, we prove that i) and iii) cannot hold for such a voxel, *i.e.*, the attachment of a voxel in a MSP is non-empty and connected.

Proposition 7. *Let $P \sqsubseteq F$ be a MSP for F. Then:*

$$\forall g \in P^+, |C[Att(\hat{g}, F)]| = 1.$$

Then, with the next proposition, we show that ii) cannot hold, hence, the attachment to F of any voxel g in a MSP has no cavity.

Fig. 6. Attachments of configurations of Fig. 4. From left to right: attachment of $\{x, y\}$, attachment of x, attachment of y.

Proposition 8. *Let $P \sqsubseteq F$ be a MSP for F. Then:*

$$\forall g \in P^+, Att(\hat{g}, F) \neq \hat{g}^*.$$

Recall that, according to the Euler-Poincaré formula, $\chi(Att(\hat{g}, F)) = b_0 - b_1 + b_2$, where b_0 (resp. b_2) is the number of connected components (resp. cavities) of $Att(\hat{g}, F)$. From the two previous propositions, we have $b_0 = 1$ and $b_2 = 0$. The Betti number b_1, which represents the number of tunnels, is positive. Thus, we have $\chi(Att(\hat{g}, F)) = 1 - b_1 \leq 1$. But from Theorem 5 and Prop. 7 we must have $\chi(Att(\hat{g}, F)) \neq 1$, otherwise g would be a simple voxel. This proves the following proposition, which (with Prop. 7 and Prop. 8) implies that the attachment to F of any voxel in a MSP has at least one tunnel.

Proposition 9. *Let $P \sqsubseteq F$ be a MSP for F. Then:*

$$\forall g \in P^+, \chi(Att(\hat{g}, F)) \leq 0.$$

From Prop. 6, we know that a MSP is necessarily connected. The following proposition tells us more about the intersection of the two voxels which compose any MSP.

Proposition 10. *Let $P \sqsubseteq F$ be a MSP for F, and let g_1, g_2 be the two voxels of P. Then, $g_1 \cap g_2$ is a 2-face.*

This proposition is indeed an easy consequence of the following lemma: it may be seen that Lemma 11 implies that the intersection of $Att(P, F)$ with $g_1 \cap g_2$ has at least three connected components. This is possible only when $\dim(g_1 \cap g_2) = 2$.

Lemma 11. *Let $P \sqsubseteq F$ be a MSP for F, and let g_1, g_2 be the two voxels of P. Then, $\chi(Att(P, F) \cap \hat{g}_1 \cap \hat{g}_2) \geq 3$.*

To illustrate the above properties, let us consider the attachment of the pair $P = \{x, y\}$ of Fig. 4a, which is displayed in Fig. 6 (left), and the attachments of x and y displayed in Fig. 6 (middle and right, respectively). We can see in particular that the intersection of $Att(P, F)$ with $x \cap y$ is indeed composed of three connected components (the 0-cells u, v and w), as implied by Lemma 11.

The two following propositions are necessary conditions for a MSP, which are similar to the conditions of Theorem 5 which characterise simple voxels.

From Prop. 3, P collapses onto $Att(P, F)$ whenever P is a MSP. We have $\chi(P) = 1$, and from Prop. 6, $|C[P]| = 1$. Since collapse preserves the number of connected components and the Euler characteristic, we have the following.

Proposition 12. *Let $P \sqsubseteq F$ be a MSP for F. Then:*

$$|C[Att(P, F)]| = 1.$$

Proposition 13. *Let $P \sqsubseteq F$ be a MSP for F. Then:*

$$\chi(Att(P, F)) = 1.$$

Finally, we give a characterisation of MSPs, which summarises and extends the properties shown before.

Proposition 14. *Let $P \sqsubseteq F$ be a pair. Then P is a MSP for F if and only if all the following conditions hold:*

$$\text{the intersection of the two voxels of } P \text{ is a 2-face,} \tag{1}$$
$$\forall g \in P^+, |C[Att(\hat{g}, F)]| = 1, \tag{2}$$
$$\forall g \in P^+, \chi(Att(\hat{g}, F)) \leq 0, \tag{3}$$
$$|C[Att(P, F)]| = 1, \tag{4}$$
$$\chi(Att(P, F)) = 1. \tag{5}$$

Remark 15. *Conditions (1), (3), (4), and (5) are sufficient to characterise a MSP, since condition (2) may be deduced from (1), (3), (4). Moreover, if P is a pair of non-simple voxels, then P is a MSP for F if and only if conditions (4) and (5) both hold. We retrieve a characterisation similar to Theorem 5.*

6   Conclusion

The notion of simple voxel (or simple point), which is commonly considered for topology-preserving thinning, is sometimes not sufficient to obtain reduced objects being globally minimal. The detection of MSPs (and more generally of minimal simple complexes) can then enable to improve the thinning procedures by "breaking" specific objects such as the ones studied here.

For example, let us consider again the experiment described in the introduction. Among 10,000 objects obtained by applying a homotopic thinning procedure guided by a random priority function to a $20 \times 20 \times 20$ full cube, we found 249 lumps. In 212 of these 249 cases, further thinning was made possible by the detection of a MSP. In 203 of these 212 cases, it has been possible to continue the thinning process until obtaining a single voxel.

It has to be noticed that the search of MSPs in a complex $F \sqsubseteq \mathbb{F}^3$ does not present an algorithmic complexity higher than the search of simple voxels (both being linear with respect to the number of facets of the processed complex). Consequently, it is possible to create new thinning procedures based on the detachment of both simple voxels and pairs and whose runtimes have the same order of growth as the runtimes of thinning procedures that are based only on simple voxels. Such new algorithms would be able to produce skeletons having less points than standard ones.

References

1. Bertrand, G.: On P-simple points. Comptes Rendus de l'Académie des Sciences, Série Math. I(321), 1077–1084 (1995)
2. Bertrand, G.: On critical kernels. Comptes Rendus de l'Académie des Sciences, Série Math. I(345), 363–367 (2007)
3. Bertrand, G., Couprie, M.: New 2D Parallel Thinning Algorithms Based on Critical Kernels. In: Reulke, R., Eckardt, U., Flach, B., Knauer, U., Polthier, K. (eds.) IWCIA 2006. LNCS, vol. 4040, pp. 45–59. Springer, Heidelberg (2006)
4. Bertrand, G., Couprie, M.: A new 3D parallel thinning scheme based on critical kernels. In: Kuba, A., Nyúl, L.G., Palágyi, K. (eds.) DGCI 2006. LNCS, vol. 4245, pp. 580–591. Springer, Heidelberg (2006)
5. Bertrand, G., Couprie, M.: Two-dimensional thinning algorithms based on critical kernels. Journal of Mathematical Imaging and Vision (to appear, 2008)
6. Bing, R.H.: Some aspects of the topology of 3-manifolds related to the Poincaré conjecture. Lectures on Modern Mathematics II, 93–128 (1964)
7. Cohen, M.M.: A course in simple-homotopy theory. Springer, Heidelberg (1973)
8. Couprie, M., Coeurjolly, D., Zrour, R.: Discrete bisector function and Euclidean skeleton in 2D and 3D. Image and Vision Computing 25(10), 1543–1556 (2007)
9. Davies, E.R., Plummer, A.P.N.: Thinning algorithms: a critique and a new methodology. Pattern Recognition 14(1–6), 53–63 (1981)
10. Dokládal, P., Lohou, C., Perroton, L., Bertrand, G.: Liver Blood Vessels Extraction by a 3-D Topological Approach. In: Taylor, C., Colchester, A. (eds.) MICCAI 1999. LNCS, vol. 1679, pp. 98–105. Springer, Heidelberg (1999)
11. Fourey, S., Malgouyres, R.: A concise characterization of 3D simple points. Discrete Applied Mathematics 125(1), 59–80 (2003)
12. Gau, C.-J., Kong, T.Y.: Minimal non-simple sets in 4D binary pictures. Graphical Models 65(1–3), 112–130 (2003)
13. Giblin, P.: Graphs, surfaces and homology. Chapman and Hall, Boca Raton (1981)
14. Kong, T.Y.: On topology preservation in 2-D and 3-D thinning. International Journal on Pattern Recognition and Artificial Intelligence 9(5), 813–844 (1995)
15. Kong, T.Y.: Topology-preserving deletion of 1's from 2-, 3- and 4-dimensional binary images. In: Ahronovitz, E. (ed.) DGCI 1997. LNCS, vol. 1347, pp. 3–18. Springer, Heidelberg (1997)
16. Kong, T.Y., Rosenfeld, A.: Digital topology: introduction and survey. Computer Vision, Graphics and Image Processing 48(3), 357–393 (1989)
17. Kovalevsky, V.A.: Finite topology as applied to image analysis. Computer Vision, Graphics and Image Processing 46(2), 141–161 (1989)
18. Passat, N., Couprie, M., Bertrand, G.: Minimal simple pairs in the 3-D cubic grid. Technical Report IGM2007-04, Université de Marne-la-Vallée (2007), http://igm.univ-mlv.fr/LabInfo/rapportsInternes/2007/04.pdf
19. Passat, N., Ronse, C., Baruthio, J., Armspach, J.-P., Bosc, M., Foucher, J.: Using multimodal MR data for segmentation and topology recovery of the cerebral superficial venous tree. In: Bebis, G., Boyle, R., Koracin, D., Parvin, B. (eds.) ISVC 2005. LNCS, vol. 3804, pp. 60–67. Springer, Heidelberg (2005)

20. Ronse, C.: Minimal test patterns for connectivity preservation in parallel thinning algorithms for binary digital images. Discrete Applied Mathematics 21(1), 67–79 (1988)
21. Rosenfeld, A.: Connectivity in digital pictures. Journal of the Association for Computer Machinery 17(1), 146–160 (1970)
22. Ségonne, F.: Segmentation of Medical Images under Topological Constraints. PhD thesis, MIT (2005)

A   Appendix: Simple Equivalence and MSPs

We define here the notion of lump, informally introduced in Section 1.

Definition 16. *Let* $F, G \sqsubseteq \mathbb{F}^3$. *We say that* F *and* G *are* simple-equivalent *if there exists a sequence of complexes* $\langle F_0, \ldots, F_\ell \rangle$ *such that* $F_0 = F$, $F_\ell = G$, *and for any* $i \in \{1, \ldots, \ell\}$, *we have either*
i) $F_i = F_{i-1} \oslash x_i$, *where* x_i *is a voxel which is simple for* F_{i-1} *; or*
ii) $F_{i-1} = F_i \oslash x_i$, *where* x_i *is a voxel which is simple for* F_i.

Definition 17. *Let* $G \sqsubseteq F \sqsubseteq \mathbb{F}^3$, *such that* F *and* G *are simple-equivalent. If* $F \neq G$ *and* F *does not contain any simple voxel outside of* G, *then we say that* F *is a* lump *relative to* G, *or simply a* lump.

For example, the Bing's house of Fig. 1 is a lump (relative to any one of its voxels), which is composed of 135 voxels. Another example of lump, much simpler, if given in Fig. 7 (left) (see in [18], Appendix C, some steps of a sequence which shows that Fig. 7 (left) and Fig. 7 (right) are simple-equivalent).

Remark 18. *The preceding example invites us to consider a notion based on simple-equivalence, which is more general than the one of simple set. A subcomplex* $G \sqsubseteq F$ *is called* SE-simple *for* F *if* F *and* $F \oslash G$ *are simple-equivalent. For example, the voxel* x *in the complex* F *of Fig. 7 (left) is SE-simple for* F, *although it is not a simple voxel for* F *(this kind of configuration has been analysed in [11]). Of course, any simple set is SE-simple, and the preceding example proves that the converse is not true in general. However, it is not possible to characterise locally, in the manner of Prop. 3, a voxel or a set which is SE-simple. This is why we use Def. 2 as the definition of a simple set.*

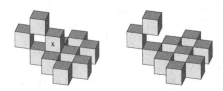

Fig. 7. On the left, the smallest lump found so far. It contains no simple voxel, and is simple-equivalent to the complex on the right, made of 10 voxels. Both objects have three tunnels.

Determining Whether a Simplicial 3-Complex Collapses to a 1-Complex Is NP-Complete

Rémy Malgouyres[1] and Angel R. Francés[2]

[1] Univ. Clermont 1, Laboratoire d'Algorithmique et Image (LAIC, EA2146),
IUT dpartement informatique Campus des Cézeaux,
B.P. 86, 63172 Aubière cedex, France
remy.malgouyres@laic.u-clermont1.fr
http://laic.u-clermont1.fr/~mr/
[2] Dpto. Informática e Ingeniería de Sistemas, Facultad de Ciencias, Universidad de
Zaragoza. C/. Pedro Cerbuna, 12. E-50009 Zaragoza, Spain
afrances@posta.unizar.es

Abstract. We show that determining whether or not a simplicial 2−complex collapses to a point is deterministic polynomial time decidable. We do this by solving the problem of constructively deciding whether a simplicial 2−complex collapses to a 1−complex. We show that this proof cannot be extended to the 3D case, by proving that deciding whether a simplicial 3−complex collapses to a 1−complex is an NP−complete problem.

Keywords: Simplicial Topology, Collapsing, Computational Complexity, NP−completeness.

Introduction

In the framework of digital topology, several authors have considered discrete deformation retraction and collapsing, in particular for characterizing simple points ([K97],[KR01],[B99]). All of these authors emphasize the importance of finding efficient algorithms for deciding whether an object can be shrunk on another object (see also [F00]). In this paper, we investigate the case when "object" is a *simplicial complex*, and "shrunk" means *collapsed*.

In [EG96], a generalized collapsing problem is proved NP−complete for simplicial 2−complexes (see Theorem 1 below). However, this problem is rather artificial and does not, contrary to the collapsing problems considered here, arise from topologists' questions on topology preservation.

This paper is intended to be readable both by topologists and by specialists in computational complexity. For this purpose, we find two (relatively long) sections recalling basic notions concerning collapsing of simplicial complexes and NP−completeness.

Then we investigate the 2D case, showing that deciding whether or not a simplicial 2−complex collapses to a 1−complex is polynomial. It follows that deciding whether a 2−complex collapses to a point is also polynomial.

D. Coeurjolly et al. (Eds.): DGCI 2008, LNCS 4992, pp. 177–188, 2008.

Finally, we prove that the corresponding 3D problem of deciding whether or not a simplicial 3−complex collapses to a 1−complex is NP−complete.

1   Basic Notions of Simplicial Topology

1.1   Simplicial Complexes

A *(finite abstract) simplicial complex* $\mathcal{C}$ is a couple $\mathcal{C} = (V, S)$, where $V = \{v_1, \ldots, v_p\}$ is a finite set, and S is a set of subsets of V, containing all singletons, and such that any subset of an element of S is also an element of S. An element of V is called a *vertex* of $\mathcal{C}$, and an element of S is called a *cell* of $\mathcal{C}$ or a *simplex* of $\mathcal{C}$. A cell s of $\mathcal{C}$ with cardinality $d + 1$ is called a $d−cell$ of $\mathcal{C}$, and the number d is called the *dimension* of s. The *dimension* of $\mathcal{C}$ is the maximal dimension of its cells. For $k \in \mathbb{N}$, we call a *simplicial k−complex* any simplicial complex with dimension less than or equal to k. In the sequel, all considered simplicial complexes are finite.

Let $\mathcal{C} = (V, S)$ be a simplicial complex. If $s' \subset s$, with $s \in S$, we say that s' is a *face* of s. If in addition we have $s \neq s'$, then s' is called a *proper face* of s. Finally, if s' is a proper face of s, s' is a $(d − 1)−$cell, and s is a $d−$cell, we call s' a *maximal proper face* of s.

1.2   Collapsing

Let $\mathcal{C} = (V, S)$ be a simplicial complex and let s' be a maximal proper face of a cell s in $\mathcal{C}$. We say that s' is a *free face* of s in $\mathcal{C}$ (or merely a free face for short) if s' is a proper face of no cell in $\mathcal{C}$ except s.

If s' is a free face of s in $\mathcal{C}$, we can define a new simplicial complex $\mathcal{C}' = (V', S')$, called an *elementary collapse* of $\mathcal{C}$, by considering the set V' of vertices of V which belong to some cell of $\mathcal{C}$ which is different from s and s', and the set $S' = S - \{s, s'\}$. We say also that there is an elementary collapse of $\mathcal{C}$ on $\mathcal{C}'$, and that the collapse is across s from s'. Though we do not need these facts, we mention that it is known (and not difficult) that if $\mathcal{C}'$ is an elementary collapse of $\mathcal{C}$, then $\mathcal{C}'$ is a strong deformation retract of $\mathcal{C}$, so that $\mathcal{C}$ and $\mathcal{C}'$ have the same homotopy type.

Now, given $\mathcal{C} = (V, S)$ a simplicial complex and $\mathcal{C}' = (V', S')$ a subcomplex of $\mathcal{C}$ (i.e. $V' \subset V$ and $S' \subset S$), we say that $\mathcal{C}$ collapses on $\mathcal{C}'$ if there is a finite sequence $(\mathcal{C}_0, \ldots, \mathcal{C}_n)$ of simplicial subcomplexes of $\mathcal{C}$, such that $\mathcal{C} = \mathcal{C}_0$, $\mathcal{C}' = \mathcal{C}_n$, and for $i = 1, \ldots, n$ the complex $\mathcal{C}_i$ is an elementary collapse of the complex $\mathcal{C}_{i-1}$. If in addition the complex $\mathcal{C}'$ is 1−dimensional, we say that $\mathcal{C}$ collapses to a 1−complex. If $\mathcal{C}'$ is reduced to a single vertex ($\mathcal{C}' = (\{v\}, \{\{v\}\})$), we say that $\mathcal{C}$ collapses to a point or, simply, that $\mathcal{C}$ is *collapsible*.

Next we give some examples of both collapsible and non−collapsible complexes that will be used in Section 3. All of these complexes are variations around the well−known Bing's house introduced in [B64].

Example 1. Let us consider the object depicted in Figure 1(a), called here the Bing's house with two thin walls. This is a 2D object, which can be described as follows. The object is a "building" with two rooms, the upper room and the lower room. We can enter in the upper room by a tunnel through the lower room. Similarly, we can enter in the lower room by a tunnel through the upper room. Finally, two thin walls (i.e. 2D walls) are added between the tunnel through each room and an exterior wall of the building, so that both rooms are simply connected. In order to realize the Bing's house with a simplicial complex $\mathcal{B} = (V, S)$, we can break all the rectangular pieces of walls into triangles (i.e. we can triangulate the 2D walls). Then we take for V the set of all vertices of the obtained triangles, and as simplexes (elements of S), the singletons of V as $0-cells$, pairs of extremities of edges as $1-cells$, and the sets consisting of the three vertices of triangles as $2-cells$. The obtained simplicial $2-complex$ $\mathcal{B}$ is not collapsible since no cell of $\mathcal{B}$ has a free face, but it has the homotopy type of a point. For this, first thicken all its walls (this is the inverse operation of a collapse) and observe that the resulting 3D object is a $3-ball$, which collapses to a point. The next example shows that it is sufficient to thicken only one appropriate wall of $\mathcal{B}$ to check this result.

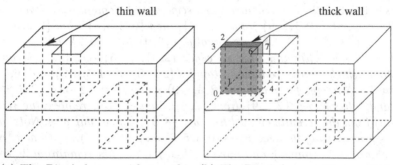

(a) The Bing's house with two thin walls.   (b) The Bing's house with one thick wall and one thin wall.

Fig. 1. Examples of simplicial complexes

Example 2. Now let us consider the object depicted in Figure 1(b). It is called the Bing's house with one thick wall. It is similar to the Bing's house with two thin walls, except that one of the walls is thickened to get a 3D paralelepipedic wall W, as represented in Figure 1(b). In order to get a simplicial complex to represent the Bing's house with one thick wall, we proceed as in Example 1 for the 2D parts, but for the 3D wall, we consider a particular triangulation $\mathcal{W}$ consisting of twelve $3-cells$, which is as follows. Consider the vertices numbers of the cube representing the thick wall of Figure 1(b). This cube is also represented in Figure 2(a), where it has been subdivided into four triangular prisms by adding a new edge $\{8, 9\}$ with extremities at the barycentres of the rectangles $(0, 3, 5, 6)$ and $(1, 2, 4, 7)$, respectively, and joining these barycentres to the vertices of the corresponding rectangles. Then, each of these prisms is in turn subdivided into three

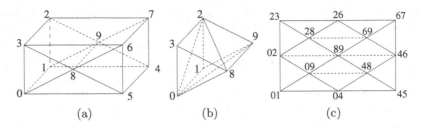

Fig. 2. Decomposition into simplexes of a cube

$3-cells$ as follows. First, divide each rectangular face of the prism with a diagonal containing the vertex with the lowest label, and then consider all tetrahedra thus obtained. For example, the prism with vertices $0, 1, 2, 3, 8, 9$ is decomposed into the $3-cells$ $\{0, 2, 3, 8\}$, $\{0, 2, 8, 9\}$ and $\{0, 1, 2, 9\}$ (see Figure 2(b)). Thus the Bing's house with one thick wall is realized as a simplicial $3-complex$ $\mathcal{B}_2$ which is collapsible.

Indeed, we can first remove each prism from the free 2D faces on the rectangles $(0, 1, 2, 3)$, $(2, 3, 6, 7)$, $(4, 5, 6, 7)$ and $(0, 1, 4, 5)$. For example, the prism with vertices $0, 1, 2, 3, 8, 9$ can be removed by a sequence of elementary collapses across the following cells: $\{0, 1, 2, 9\}$ from $\{0, 1, 2\}$, then $\{0, 2, 3, 8\}$ from $\{0, 2, 3\}$, afterwards $\{0, 2, 8, 9\}$ from $\{0, 2, 8\}$, and finally $\{0, 2, 9\}$ from $\{0, 2\}$, and similarly for the other prisms. After this process, edges $\{2, 3\}$, $\{4, 5\}$ and $\{6, 7\}$ become free faces, from which we can remove rectangles $(2, 3, 8, 9)$, $(4, 5, 8, 9)$ and $(6, 7, 8, 9)$, respectively. Namely, for rectangle $(2, 3, 8, 9)$ collapse the $2-cell$ $\{2, 3, 8\}$ from $\{2, 3\}$ and, afterwards, $\{2, 8, 9\}$ from $\{2, 8\}$. Now we can remove the rectangle $(0, 1, 8, 9)$ from the recently created free face $\{8, 9\}$ in a similar way. Note that the 2D faces representing the rectangles $(0, 3, 5, 6)$ and $(1, 2, 4, 7)$, up to now, remain in the obtained $2-complex$ and have no free faces. This remaining $2-complex$ $\mathcal{B}_2^{\{0,1\}}$ has $\{0, 1\}$ as a free face, which enables us to begin to remove the bottom part of the Bing's house, and subsequently, all the object until we get a single vertex.

Remark 1. Notice that despite the $2-cells$ lying in the rectangles $(0, 3, 5, 6)$ and $(1, 2, 4, 7)$ are free faces of $\mathcal{B}_2$, none of them can be used in order to collapse the Bing's house with one thick wall to a point.

To check this, let us consider the intersection of the thick wall W with a plane P parallel to rectangles $(0, 3, 5, 6)$ and $(1, 2, 4, 7)$. If $W = (V, S_1 \cup S_2)$ is the triangulation of W described in Example 2, where S_1 is the set of cells of W lying in the rectangles $(0, 3, 5, 6)$ and $(1, 2, 4, 7)$ and S_2 contains the remaining cells of W, the intersection of P with the cells in S_2 induces a subdivision of $W \cap P$ that can be extended (without introducing new vertices) to get a collapsible simplicial $2-complex$ $\mathcal{R}$. This complex is shown in Figure 2(c), where the label of each vertex has been made up of the vertices numbers of the corresponding edge in W, and the solid lines represent the intersection of a triangle in the set S_2 with the plane P.

Now, let $\mathcal{B}_2 = \mathcal{C}_0, \mathcal{C}_1, \ldots, \mathcal{C}_p$ be a sequence of subcomplexes of the Bing's house $\mathcal{B}_2$ such that $\mathcal{C}_{i-1}$ elementary collapses on $\mathcal{C}_i$, for $1 \leq i \leq p$, and $\mathcal{C}_p$ consists of a single point. Moreover, let i_0 be the lowest index such that the corresponding collapse is across a cell not contained in the thick wall $\mathcal{W}$. Notice that $i_0 > 1$ necessarily and $\mathcal{C}_{i_0-1}$ collapses on $\mathcal{C}_{i_0}$ from either the edge $\{0,1\}$ or from an edge in the border of the rectangle $(0,3,5,6)$ or $(1,2,4,7)$. Finally, for $0 \leq i < i_0$, let us consider the 2$-$complex $\mathcal{R}_i$ obtained by intersecting the cells of $\mathcal{W}_i = \mathcal{W} \cap \mathcal{C}_i$ with the plane P as above.

Then, for $0 < i < i_0$, if $\mathcal{C}_{i-1}$ collapses on $\mathcal{C}_i$ (or, equivalently, $\mathcal{W}_{i-1}$ collapses on $\mathcal{W}_i$) across a cell s from the free face s', it is not difficult to check that $s' \in S_2$ if and only if $\mathcal{R}_i$ is also an elementary collapse of $\mathcal{R}_{i-1}$, and the collapse is across $s \cap P$ from $s' \cap P$. For example, the elementary collapse of $\mathcal{B}_2$ across $\{0,2,3,8\}$ from $\{0,2,3\}$ corresponds with the collapse of $\mathcal{R} = \mathcal{R}_0$ across $\{02,23,28\}$ from $\{02,23\}$. Hence, any collapsing sequence of $\mathcal{R}$ to the point $\{01\}$ (i.e., to the complex $(\{01\}, \{\{01\}\})$) corresponds with a collapsing sequence of $\mathcal{B}_2$ on the complex $\mathcal{B}_2^{\{0,1\}}$, in which the edge $\{0,1\}$ is a free face, and conversely. On the other hand, if $\mathcal{C}_{i-1}$ collapses on $\mathcal{C}_i$ from a face $s' \in S_1$, then $\mathcal{R}_i$ is obtained by removing from $\mathcal{R}_{i-1}$ the cell $s \cap P$ without deleting any of its faces. For example, the collapse across $\{0,2,3,8\}$ from $\{0,3,8\}$ corresponds with the deletion of the 2$-$cell $\{02,23,28\}$ in $\mathcal{R}$. In this case $\mathcal{R}_i$ is not longer collapsible since it has a hole, and hence $\mathcal{C}_i$ cannot collapse on $\mathcal{B}_2^{\{0,1\}}$ just across elementary collapses from cells in the set S_2. Moreover, it can be proved, by induction on the number of collapses, that any elementary collapse of $\mathcal{C}_i$ from a face in S_1 will not destroy the hole in $\mathcal{R}_i$ (on the contrary, it will create a new hole or, at most, it will merge two holes) and, in addition, that the edges of the rectangles $(0,3,5,6)$ and $(1,2,4,7)$ never become free. This is a contradiction with our hypothesis, since then the edge $\{0,1\}$ cannot become free either.

This shows, in particular, that in order to collapse the Bing's house with one thick wall $\mathcal{B}_2$ to a point, the edge $\{0,1\}$ necessarily becomes a free face at some step of the collapsing sequence.

Example 3. We finish this section introducing a 3$-$complex $\mathcal{B}_3$ which does not collapse to a 1$-$complex. It is composed of two Bing's houses, $\mathcal{B}_2$ and $\mathcal{B}_2'$, each with one thick wall triangulated as in Example 2 and whose central edges $\{8,9\}$ and $\{8',9'\}$ have been identified (we use $0', 1', \ldots, 9'$ for labeling the vertices of the thick wall of $\mathcal{B}_2'$ accordingly to Figure 2(a)). In more intuitive words, we "paste" the two central edges together, by identifying vertices 8 and 9 with $8'$ and $9'$, respectively.

In order to collapse $\mathcal{B}_3$ to a 1$-$complex, we should be able to start removing the 2D bottom part of either $\mathcal{B}_2$ or $\mathcal{B}_2'$, and for this either the edge $\{0,1\}$ or $\{0',1'\}$ must become a free face after collapsing the corresponding thick wall $\mathcal{W}$ or $\mathcal{W}'$. Let us assume, without lose of generality, that $\{0,1\}$ becomes a free face. According to Remark 1, $\mathcal{W}$ must collapse from faces which does not lie in rectangles $(0,3,5,6)$ and $(1,2,4,7)$. However, the edge $\{8,9\} = \{8',9'\}$ should become a free face at some step, and this is only possible if $\mathcal{W}'$ collapses from its faces on either the rectangle $(0',3',5',6')$ or $(1',2',4',7')$; that is, if $\mathcal{B}_2'$ collapses

on a 2−complex consisting of a Bing's house with two thin walls and the 1−cell
$\{8', 9'\}$, in which the edge $\{0', 1'\}$ is not a free face. This enables us to remove
all the cells in B_2, after which there are not more free faces. Hence, B_3 cannot
collapse to a 1−complex, however it collapses on a Bing's house with two thin
walls instead.

2   The 2D Case

First we recall an NP−completeness result concerning simplicial 2−complexes
from [EG96]. Given a simplicial 2−complex C, a 2−cell of C is called *internal* if
it has no free faces. We denote by $er(C)$ the minimum number of internal 2−cells
which need to be removed from C so that the resulting complex collapses to a
1−complex. For instance, C collapses to a 1−complex iff $er(C) = 0$, and for a 2D
hollow cube (or sphere) C, we have $er(C) = 1$.

GENERALIZED 2D 1-COLLAPSING:
    INSTANCE: a finite simplicial 2-complex C and a non-negative integer k
    QUESTION: is er(C) equal to k ?

Theorem 1 ([EG96]). *GENERALIZED 2D 1-COLLAPSING is NP−complete.*

Now, let us consider the following problem for a fixed $k \in \mathbb{N}$:

k-GENERALIZED 2D 1-COLLAPSING:
    INSTANCE: a finite simplicial 2-complex C
    QUESTION: is er(C) equal to k ?

Theorem 2. *For any fixed k, the problem $k−GENERALIZED$ 2D 1-COLLAPSING
is polynomial.*

Lemma 1. *Let C be a simplicial 2−complex which is collapsible to a 1−complex
and let c be a 2−cell of C with a free face f. Then the complex C_1 obtained by
elementary collapsing c from f also collapses to a 1−complex.*

Lemma 1 follows from the fact that the 2−cells of C_1 can be collapsed in the
same order as they are collapsed when reducing C to a 1−complex, since the
removal of c and f does not affect the free character of 1−cells. Note that the
1−complexes resulting from collapsing of C and C_1 might be different.

Proof of Theorem 2: First we observe that the case $k = 0$ follows from the
polynomial character of the algorithm consisting in searching an arbitrary 2−cell
with a free face, collapsing this cell, and recursively treating the resulting complex C_1. From Lemma 1, either this algorithm constructs a collapsing sequence
of C to a 1−complex, or C cannot be collapsed to a 1−complex.

Now, for the case $k \neq 0$, since k is fixed, we can try all the possibilities of
removal of k 2−cells, and try to collapse the resulting complex using the case
$k = 0$. The resulting algorithm, for a fixed k, is polynomial (in spite of an
exponent k in the complexity). $\qquad\square$

Now consider the following problem:

```
2D POINT COLLAPSING:
  INSTANCE: a finite simplicial 2-complex C
  QUESTION: does C collapse to a point ?
```

Corollary 1. *The 2D POINT COLLAPSING problem is polynomial.*

Proof. We have a constructive algorithm to decide whether or not an input simplicial 2−complex collapses to a 1−complex. If C does not collapse to a 1−complex, then, *a forciori*, it does not collapse to a point. Otherwise, we can construct a 1−complex K (i.e. a graph) such that C collapses to K. Then, if K is not a tree, then C is not simply connected and does not collapse to a point. If K is a tree, then K (and therefore C) collapses to a point. This gives a polynomial procedure to solve 2D POINT COLLAPSING.

This proof shows, in particular, that the order in which we collapse the cells of a given collapsible 2−complex is not important to reduce it to a single point. In the 3D case, however, we could get blocked on a non-collapsible complex if we choose a wrong free face at some step of a possible collapsing sequence. For instance, let C be a triangulated solid cube such that a Bing's house with two thin walls $\mathcal{B}$ (see Example 1) is a subcomplex of C. Despite C is collapsible, it can be reduced to the non-collapsible 2−complex $\mathcal{B}$ by collapsing the 3−cells which are filling both rooms from the free faces on the tunnels (see Remark 1 for another example). Anyway, we may wonder whether Theorem 2 can be generalized to the 3D complexes case. Section 3 shows that this is impossible (unless $P = NP$), since the problem of deciding whether a simplicial 3−complex collapses to a 1−complex is proved to be NP−complete.

3   The 3D Collapsing Problem

In the sequel, we shall consider the following problem:

```
3D 1-COLLAPSING:
  INSTANCE: a finite simplicial 3-complex C
  QUESTION: does C collapse to a 1-complex ?
```

Concerning the encoding of the instance, which is a simplicial 3−complex $\mathcal{C}$, we suggest that the number of vertices can be written in binary, followed by, for each simplex s of $\mathcal{C}$, the binary expansions of all the numbers of vertices of s. Note that we could have written the number of vertices in unary and listed only the maximal simplices (i.e. those which are proper face of no simplices), but, for 3D complexes, this makes no significant difference concerning the size of the input.

Now we can state our main result:

Theorem 3. *The 3D 1 − COLLAPSING problem is NP−complete.*

First note that 3D 1 − COLLAPSING is easily seen to be in NP since, given a simplicial 3−complex $\mathcal{C} = (V, S)$, we can guess a sequence $\sigma = ((s_1, f_1), \ldots, (s_p, f_p))$ of couples of cells, and check in polynomial time that σ is a collapsing sequence to a 1−complex, that is, the cell f_i is a maximal proper face of s_i, for $i = 1, \ldots, p$, and f_i is a free face in the complex obtained by removing the cells $s_1, f_1, \ldots, s_{i-1}, f_{i-1}$, and the cells which are not of the form s_i or f_i have dimension less than or equal to 1.

The proof of Theorem 3 therefore reduces to proving that 3D 1 − COLLAPSING is NP−hard. This is made by polynomially reducing the 3 − SAT problem to 3D 1 − COLLAPSING. So, in the sequel of this section, U is a finite set of n boolean variables, and $C = \{c_1, \ldots, c_m\}$ is a finite collection of m 3−clauses. These represent an instance I of the 3 − SAT problem, and we are going to construct an instance of a simplicial 3−complex $\mathcal{C}(I)$ which is collapsible to a 1−complex iff I is satisfiable.

The complex $\mathcal{C}(I)$ is made up of several other complexes, representing literals, the clauses and the conjunction of clauses of I, which are connected by identifying some of their edges. The main idea is to represent the literals u and $\overline{u}$, for each variable $u \in U$, by two complexes $\mathcal{C}(u)$ and $\mathcal{C}(\overline{u})$, respectively, which are related in such a way that only one of them may collapse on a 1−complex at a first stage of a possible collapsing sequence of $\mathcal{C}(I)$. If the complex $\mathcal{C}(l)$, representing the literal l, could not be reduced to a 1−complex at this stage, we will obtain the necessary feedback to collapse it through a particular edge $f(l)$, which is identified with an edge in the complex $\mathcal{C}_{\mathrm{and}}$ representing the conjunction of clauses. In a second stage, each complex $\mathcal{C}(c)$ representing a clause $c = l_1 \vee l_2 \vee l_3$ will be collapsible to a 1−complex if and only if at least one of the complexes $\mathcal{C}(l_j)$, $j \in \{1, 2, 3\}$, has collapsed in the previous stage. Then $\mathcal{C}_{\mathrm{and}}$, and hence $\mathcal{C}(I)$ through the edges $f(l)$, turns out to be collapsible if all the complexes $\mathcal{C}(c_i)$, $i \in \{1, \ldots, m\}$, have been collapsed. We next describe the complexes involved in this construction.

The complex $\mathcal{C}_{\mathrm{and}}$ is just a Bing's house with one thick wall whose 2D walls have been triangulated in such a way that, for each literal l, we get an edge $f(l)$ lying at the intersection between two outer walls of the lower room (see Figure 3(a)). As it was shown in Remark 1, the only possibility to reduce $\mathcal{C}_{\mathrm{and}}$ to a 1−complex is to collapse the thick wall until the edge $e_{\mathrm{and}} = \{0, 1\}$ becomes a free face. For this reason, we paste to this edge a simplicial representation $\mathcal{C}(c_i)$ of each clause c_i, $i \in \{1, \ldots, m\}$. This way, $\mathcal{C}_{\mathrm{and}}$ intuitively represents the "conjunction" of the complexes $\mathcal{C}(c_i)$, since it will be collapsible only if each $\mathcal{C}(c_i)$ is previously reduced to a 1−complex.

In order to describe the complexes $\mathcal{C}(c_i)$ we may assume that a boolean variable u appears at most (positively or negatively) once in a given clause. If c_i contains only one literal, then $\mathcal{C}(c_i)$ consists just of a single edge, specifically the edge e_{and} of $\mathcal{C}_{\mathrm{and}}$, which is also labeled with this literal. If the clause c_i contains two literals, then it is simply represented by a triangle, one edge of this triangle being the edge e_{and} of $\mathcal{C}_{\mathrm{and}}$, and the two other edges being labeled each by one literal appearing in the clause c_i. Finally, if c_i contains three literals (see

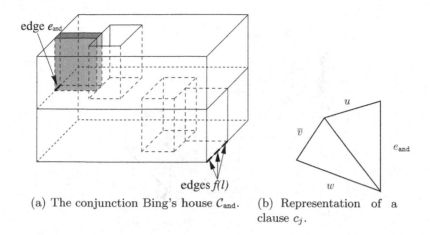

(a) The conjunction Bing's house $\mathcal{C}_{\text{and}}$.   (b) Representation of a clause c_j.

Fig. 3. How to represent 3−clauses

Figure 3(b) for an example with $c_i = u \vee \overline{v} \vee w$), the clause is represented by two adjacent triangles, being e_{and} one the four edges which are not shared by the two triangles and the three other of these edges being labeled each by one of the three literals of the clause c_i. By identifying all these edges, the union of the complexes $\mathcal{C}_{\text{and}}, \mathcal{C}(c_1), \ldots, \mathcal{C}(c_m)$ yields a simplicial complex $\mathcal{C}$ in which some edges are labeled by literals, and $\mathcal{C}$ can be collapsed to a 1−complex if for each clause we use, as a free face, an edge labeled by a literal of the clause.

Now we have to paste to $\mathcal{C}$ some representation of literals. This is the most tricky point. For this purpose we shall consider, for each variable $u \in U$, two Bing's houses with two thick walls such as the one represented in Figure 4, one, denoted by $\mathcal{C}(u)$, for the literal u, and one, denoted by $\mathcal{C}(\overline{u})$, for the literal $\overline{u}$. More precisely, for a given literal l, the two thick 3D walls of the complex $\mathcal{C}(l)$, denoted by $\mathcal{W}(l)$ and $\mathcal{F}(l)$ respectively, are triangulated as the thick wall in Example 2, while the triangulation of its 2D walls provides us with m edges $e_1(l), \ldots, e_m(l)$ (we remind the reader that m is the number of clauses), these edges lying on an outer wall (or at the intersection between two outer walls as represented in Figure 4) of the lower room of $\mathcal{C}(l)$. Then, each Bing's house $\mathcal{C}(l)$ associated to a literal l is double linked to the complex $\mathcal{C}$. Firstly, for each clause c_j, $j \in \{1, \ldots, m\}$, in which the literal l appears, we identify the edge $e_j(l)$ with the edge labeled by l in the complex $\mathcal{C}(c_j)$ representing the clause c_j. Secondly, the edge $f(l)$ of the thick wall $\mathcal{F}(l)$ of $\mathcal{C}(l)$ is also identified with the edge of the same label in the complex $\mathcal{C}_{\text{and}}$ representing the conjunction of clauses (see Figures 3(a) and 4). Finally, for each variable $u \in U$, the literal u must somehow be related to the literal $\overline{u}$ by an exclusion principle. For this, we simply identify the central edges of the thick walls $\mathcal{W}(u)$ and $\mathcal{W}(\overline{u})$ as in Example 3.

Remark 2. As it was mentioned above, in order to collapse the complex $\mathcal{C}$, and thus $\mathcal{C}(I)$, to a 1−complex, we need for each clause c_j at least one free edge labeled by a literal l appearing in the clause. Since such an edge is identified with

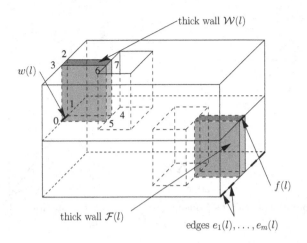

Fig. 4. The Bing's house $\mathcal{C}(l)$ with two thick walls associated to a literal l

the edge $e_j(l)$ *lying in the bottom room of the Bing's house* $\mathcal{C}(l)$, *we must first collapse these complexes. For this, we can proceed as in Example 2 to collapse the thick wall* $\mathcal{F}(l)$ *of* $\mathcal{C}(l)$. *But in the complex obtained from* $\mathcal{C}(I)$ *after this process, there are three 2−cells having the edge* $f(l)$ *as a face (two of them in* $\mathcal{C}_{\mathrm{and}}$ *and the third one in* $\mathcal{C}(l)$*). So,* $f(l)$ *is not a free face and it cannot be used to remove the remaining cells of* $\mathcal{C}(l)$. *Thus, we must follow a different collapsing sequence. Indeed, we must collapse the thick walls* $\mathcal{W}(l)$ *and* $\mathcal{W}(\bar{l})$ *simultaneously as in Example 3 in order the edge* $w(l)$ *of* $\mathcal{W}(l)$ *becomes a free face of the resulting complex (see Figure 4), from which we can remove the 2−cells of (the bottom room of)* $\mathcal{C}(l)$. *Notice that, after this process (in which the thick walls* $\mathcal{W}(l)$ *and* $\mathcal{W}(\bar{l})$ *play the role of* $\mathcal{W}$ *and* $\mathcal{W}'$ *in Example 3, respectively), the edge* $w(\bar{l})$ *of* $\mathcal{W}(\bar{l})$ *is not free in the resulting complex; so that, we must wait until the edge* $f(\bar{l})$ *becomes free (i.e., until* $\mathcal{C}_{\mathrm{and}}$ *has been collapsed) to reduce* $\mathcal{C}(\bar{l})$ *to a 1−complex.*

The following proposition clearly follows from the construction of $\mathcal{C}(I)$.

Proposition 1. *Given the data of the instance I of the $3-\mathrm{SAT}$, we can construct the complex $\mathcal{C}(I)$ in polynomial time.*

Then the NP−completeness of the **3D 1 − COLLAPSING** problem immediately follows from the following

Theorem 4. *The $3D$ simplicial complex $\mathcal{C}(I)$ collapses to a 1−complex if and only if the instance I of the $3-\mathrm{SAT}$ problem is satisfiable.*

Proof. First, assume that the instance I is satisfiable, and let $G : U \longrightarrow \{0,1\}$ be a truth assignment which satisfies the collection C of clauses of I. Let us show how to collapse $\mathcal{C}(I)$ to a 1−complex.

We extend the truth assignment G to literals by setting $G(\bar{u}) = \neg G(u)$ for $u \in U$. For each literal l such that $G(l) = 1$, we collapse the Bing's house $\mathcal{C}(l)$, as it suggested in Remark 2, so that the edges labeled by l in the complex $\mathcal{C}$ become free. Thus, since G is a satisfying truth assignment, for each clause $c_j \in C$ there is a free edge $e_j(l)$ in the resulting complex from which we can remove the (at most) two triangles of $\mathcal{C}(c_j)$. Therefore, we can collapse the thick wall of the Bing's house $\mathcal{C}_{\text{and}}$ as in Example 2 until the edge e_{and} becomes free, which allows us to remove the remaining 2−cells of $\mathcal{C}_{\text{and}}$. Finally, since the faces $f(l)$, for all literals l (in particular for those with $G(l) = 0$), are faces of no 2−cells of the remaining of $\mathcal{C}_{\text{and}}$, we can begin to collapse the thick wall $\mathcal{F}(l)$ similarly to Example 2, and then collapse all Bing's houses $\mathcal{C}(l)$ to a 1−complex.

Conversely, suppose that $\mathcal{C}$ collapses to a 1−complex K. For l literal of I, let $G(l)$ be equal to 1 if there exists and edge $e_j(l)$ in $\mathcal{C}(l)$ which is removed before the edge $f(l)$ (in particular if $f(l)$ is an edge of K) and equal to 0 otherwise. Let us prove that for any literal l we have $G(l) = 1$ implies $G(\bar{l}) = 0$, and that the truth assignment G' defined by $G'(l) = G(l)$ if $G(l) = 1$ or $G(\bar{l}) = 1$ and by $G'(u) = 1$ and $G'(\bar{u}) = 0$ for all $u \in U$ such that $G(u) = G(\bar{u}) = 0$ is a satisfying truth assignment for I.

First, assume that $G(l) = 1$. We need collapse (at least part of) the bottom room of $\mathcal{C}(l)$ in order to remove the edge $e_j(l)$. For this, according to Remark 1, we must start collapsing from either the edge $w(l)$ or $f(l)$ of $\mathcal{C}(l)$. Thus, since $e_j(l)$ is removed before $f(l)$, this process starts being the edge $w(l)$ a free face, for which necessarily the thick walls $\mathcal{W}(l)$ and $\mathcal{W}(\bar{l})$ must be collapsed simultaneously as in Example 3. As a consequence the edge $w(\bar{l})$ is not free in the remaining of $\mathcal{C}(\bar{l})$, and hence no edge $e_j(\bar{l})$ can be removed before $f(\bar{l})$. This shows that $G(\bar{l}) = 0$.

Now let us prove that G' is a satisfying truth assignment for I. In order to remove the 2−cells of the Bing's house $\mathcal{C}_{\text{and}}$ by collapsing, we have to use the edge e_{and} as a free face by Remark 1. Therefore, for each clause c_j, $j \in \{1, \ldots, m\}$, there is an edge labeled by some literal of the clause in the (at most) two triangles representing the clause c_j (see Figure 3(b)) which is used as a free face before any edge of the form $f(l)$ is removed. This shows that in each clause c_j appears a literal l such that $G(l) = 1$, so that G' is a satisfying truth assignment for I.

4   Conclusion

We proved that deciding whether a simplicial 3−complex collapses to a 1−complex is an intractable problem. This is, as far as we know, the first result of that kind concerning topology preservation. Several kindred questions remain:

- Characterizing the complexity of the problem of deciding whether a simplicial 3−complex collapses to a point;
- Studying the complexity of characterization of simple points in high dimensional digital spaces ([K97],[KR01],[B99]);
- Characterizing the complexity of the problem of deciding whether a 3D digital object can be reduced to another object by sequential deletion of simple points ([F00]).

References

[B99]   Bertrand, G.: New Notions for Discrete Topology. In: Bertrand, G., Couprie, M., Perroton, L. (eds.) DGCI 1999. LNCS, vol. 1568, Springer, Heidelberg (1999)

[B64]   Bing, R.H.: Some aspects of the topology of 3−manifolds related to the Poincaré conjecture. In: Saaty, T.L. (ed.) Lectures on Modern Mathematics, vol. II, pp. 93–128. Wiley, Chichester (1964)

[C71]   Cook, S.A.: The complexity of Theorem Proving Procedures. In: Proc. 3rd Ann. ACM Symp. on Theory of Computing, Association for Computing Machinery, New-York, pp. 151–158

[EG96]   Egecioglu, O., Gonzalez, T.F.: A Computationally Intractable Problem on Simplicial Complexes. Computational Geometry, Theory and Applications 6, 85–98 (1996)

[F00]   Fourey, S., Malgouyres, R.: A concise characterization of 3D simple points. Discrete Applied Mathematics 125(1), 59–80 (2003)

[GJ79]   Garey, M.R., Johnson, D.S.: Computers and Intractability: a guide to the theory of NP−completeness. W.H. Freeman and Company publishers, New York

[K97]   Kong, T.Y.: Topology-Preserving Deletion of 1's from 2−, 3− and 4− Dimensional Binary Images. In: Ahronovitz, E. (ed.) DGCI 1997. LNCS, vol. 1347, pp. 3–18. Springer, Heidelberg (1997)

[KR01]   Kong, T.Y., Roscoe, A.W.: Simple Points in 4−dimensional (and Higher-Dimensional) Binary Images (paper in preparation)

[VL90]   : Handbook of theoretical computer science. In: Van Leeuwen, J. (ed.) Algorithms and complexity, vol. A, pp. 67–161. Elsevier Science Publishers, Amsterdam (1990)

Medial Axis LUT Computation for Chamfer Norms Using $\mathcal{H}$-Polytopes

Nicolas Normand and Pierre Évenou

IRCCyN UMR CNRS 6597, École polytechnique de l'Université de Nantes,
Rue Christian Pauc, La Chantrerie,
44306 Nantes Cedex 3, France
Nicolas.Normand@polytech.univ-nantes.fr,
Pierre.Evenou@polytech.univ-nantes.fr

Abstract. Chamfer distances are discrete distances based on the propagation of local distances, or weights defined in a mask. The medial axis, *i.e.* the centers of the maximal disks (disks which are not contained in any other disk), is a powerful tool for shape representation and analysis. The extraction of maximal disks is performed in the general case with comparison tests involving look-up tables representing the covering relation of disks in a local neighborhood. Although look-up table values can be computed efficiently [1], the computation of the look-up table neighborhood tend to be very time-consuming. By using polytope [2] descriptions of the chamfer disks, the necessary operations to extract the look-up tables are greatly reduced.

1 Introduction

The distance transform DT_X of a binary image X is a function that maps each point x with its distance to the background *i.e.* with the radius of the largest open disk centered in x included in the image. Such a disk is said to be maximal if no other disk included in X contains it. The set of centers of maximal disks, the medial axis, is a convenient description of binary images for many applications ranging from image coding to shape recognition. Its attractive properties are reversibility and (relative) compactness.

Algorithms for computing the distance transform are known for various discrete distances [3,4,5,6]. In this paper, we will focus on chamfer (or weighted) distances. The classical medial axis extraction method is based on the removal of non maximal disks in the distance transform. It is thus mandatory to describe the covering relation of disks, or at least the transitive reduction of this relation. For simple distances this knowledge is summarized in a local maximum criterion [3]. The most general method for chamfer distances uses look-up tables for that purpose [7].

In this paper we propose a method to both compute the look-up tables and the look-up table mask based on geometric properties of the balls of chamfer norms. Basic notions, definitions and known results about chamfer disks and medial

D. Coeurjolly et al. (Eds.): DGCI 2008, LNCS 4992, pp. 189–200, 2008.

axis look-up tables are recalled in section 2. Then section 3 justifies the use of polytope formalism in our context and presents the principles of the method. In section 4, algorithms for the 2D case are given.

2 Chamfer Medial Axis

2.1 Chamfer Distances

Definition 1 (Discrete distance, metric and norm). *Consider a function* $d : \mathbb{Z}^n \times \mathbb{Z}^n \to \mathbb{N}$ *and the following properties* $\forall x, y, z \in \mathbb{Z}^n$, $\forall \lambda \in \mathbb{Z}$:

1. *positive definiteness* $d(x, y) \geq 0$ and $d(x, y) = 0 \Leftrightarrow x = y$,
2. *symmetry* $d(x, y) = d(y, x)$,
3. *triangle inequality* $d(x, z) \leq d(x, y) + d(y, z)$,
4. *translation invariance* $d(x + z, y + z) = d(x, y)$,
5. *positive homogeneity* $d(\lambda x, \lambda y) = |\lambda| \cdot d(x, y)$.

d is called a distance *if it verifies conditions 1 and 2, a* metric *with conditions 1 to 3 and a* norm *if it also satisfies conditions 4 and 5.*

Most discrete distances are built from a definition of neighborhood and connected paths (path-generated distances), the distance from x to y being equal to the length of the shortest path between the two points [8]. Distance functions differ by the way path lengths are measured: as the number of displacements in the path for simple distances like d_4 and d_8, as a weighted sum of displacements for chamfer distances [4] or by the displacements allowed at each step for neighborhood sequence distances [8,4], or even by a mixed approach of weighted neighborhood sequence paths [6].

For a given distance d, the closed ball B^c and open ball B^o of center c and radius r are the sets of points of $\mathbb{Z}^n$:

$$B^o(c, r) = \{p : d(c, p) < r\}$$
$$B^c(c, r) = \{p : d(c, p) \leq r\} \quad (1)$$

Since the codomain of d is $\mathbb{N}$: $\forall r \in \mathbb{N}, d(c, p) \leq r \Leftrightarrow d(c, p) < r + 1$. So:

$$\forall r \in \mathbb{N}, B^c(c, r) = B^o(c, r + 1) \quad (2)$$

In the following, the notation B will be used to refer to closed balls.

Definition 2 (Chamfer mask [9]). *A weighting* $M = (\vec{v}; w)$ *is a vector* $\vec{v}$ *of* $\mathbb{Z}^n$ *associated with a weight* w *(or local distance). A chamfer mask* $\mathcal{M}$ *is a central-symmetric set of weightings having positive weights and non-null displacements, and containing at least one basis of* $\mathbb{Z}^n$: $\mathcal{M} = \{M_i \in \mathbb{Z}^n \times \mathbb{N}^*\}_{1 \leq i \leq m}$

The grid $\mathbb{Z}^n$ is symmetric with respect to the hyperplanes normal to the axes and to the bisectors (G-symmetry). This divides $\mathbb{Z}^n$ in $2^n.n!$ sub-spaces (8 octants for $\mathbb{Z}^2$). Chamfer masks are usually G-symmetric so that weightings may only be given in the sub-space $0 \leq x_n \leq \ldots \leq x_1$.

Paths between two points x and y can be produced by chaining displacements. The length of a path is the sum of the weights associated with the displacements and the distance between x and y is the length of the shortest path.

Definition 3 (Chamfer distance [9]). *Let $\mathcal{M} = \{(\overrightarrow{v_i}, w_i) \in \mathbb{Z}^n \times \mathbb{N}^*\}_{1 \leq i \leq m}$ be a chamfer mask. The chamfer (or weighted) distance between two points x and y is:*

$$d(x, y) = \min \left\{ \sum \lambda_i w_i : x + \sum \lambda_i \overrightarrow{v_i} = y, \lambda_i \in \mathbb{N}, 1 \leq i \leq m \right\} . \tag{3}$$

Any chamfer masks define a metric [10]. However a chamfer mask only generates a norm when some conditions on the mask neighbors and on the corresponding weights permits a triangulation of the ball in influence cones [9,11]. When a mask defines a norm then all its balls are convex.

2.2   Geometry of the Chamfer Ball

We can deduce from (1) and (3) a recursive construction of chamfer balls:

$$B(O, r) = B(O, r - 1) \cup \bigcup_{0 \leq i \leq m} B(O + \overrightarrow{v_i}, r - w_i) . \tag{4}$$

Definition 4 (Influence cone [12]). *Let $\mathcal{M} = \{(\overrightarrow{v_i}, w_i) \in \mathbb{Z}^n \times \mathbb{N}^*\}_{1 \leq i \leq m}$ be a chamfer mask generating a norm. An influence cone is a cone from the origin spanned by a subset of the mask vectors $\{\overrightarrow{v_i}, \overrightarrow{v_j}, \overrightarrow{v_k}, \ldots\}$ in which only the weightings $M_i, M_j, M_k, \ldots$ of the mask are involved in the computation of the distance from O to any point of the cone.*

In each influence cone, the discrete gradient of the distance function is constant and equal to [9]:

$$(w_i, w_j, w_k, \ldots) \cdot (\overrightarrow{v_i} | \overrightarrow{v_j} | \overrightarrow{v_k} | \ldots)^{-1} , \tag{5}$$

where $(\overrightarrow{v_i} | \overrightarrow{v_j} | \overrightarrow{v_k} | \ldots)$ stands for the column matrix of the vectors spanning the cone. The distance $d_C(O, p)$ from the origin to any point p of this cone C is then:

$$d_C(O, p) = (w_i, w_j, w_k, \ldots) \cdot (\overrightarrow{v_i} | \overrightarrow{v_j} | \overrightarrow{v_k} | \ldots)^{-1} \cdot p . \tag{6}$$

For instance, with chamfer norm $d_{5,7,11}$, the point $(3, 1)$ is in the cone spanned by the vectors $a = (1, 0)$ and $c = (2, 1)$ and the weights involved are 5 and 11. The distance between the origin and the point $(3, 1)$ is then:

$$d_{C_{a,c}}(O, (3, 1)) = \begin{pmatrix} 5 & 11 \end{pmatrix} \cdot \begin{pmatrix} 1 & 2 \\ 0 & 1 \end{pmatrix}^{-1} \cdot \begin{pmatrix} 3 \\ 1 \end{pmatrix} = \begin{pmatrix} 5 & 1 \end{pmatrix} \cdot \begin{pmatrix} 3 \\ 1 \end{pmatrix} = 16 .$$

2.3 Chamfer Medial Axis

For simple distances d_4 and d_8, the medial axis extraction can be performed by the detection of local maxima in the distance map [13]. Chamfer distances raise a first complication even for small masks as soon as the weights are not unitary. Since all possible values of distance are not achievable, two different radii r and r' may correspond to the same set of discrete points. The radii r and r' are said to be *equivalent*. Since the distance transform labels pixels with the greatest equivalent radius, criterions based on radius difference fail to recognize equivalent disks as being covered by other disks. In the case of 3×3 2D masks or $3 \times 3 \times 3$ 3D masks, a simple relabeling of distance map values with the smallest equivalent radius is sufficient [14,15]. However this method fails for greater masks and the most general method for medial axis extraction from the distance map involves look-up tables (LUT) that represent for each radius r and displacement $\overrightarrow{v_i}$, the minimal open ball covering $B^o(O, r_1,)$ in direction $\overrightarrow{v_i}$ [7]:

$$\mathrm{Lut}_{\overrightarrow{v_i}}(r_1) = \min\{r_2 : B^o(O + \overrightarrow{v_i}, r_1) \subseteq B^o(O, r_2)\} \ .$$

Equivalently using closed balls (considering (2)):

$$\mathrm{Lut}_{\overrightarrow{v_i}}(r_1) = 1 + \min\{r_2 : B(O + \overrightarrow{v_i}, r_1 - 1) \subseteq B(O, r_2)\} \ . \tag{7}$$

Consider for instance the $d_{5,7,11}$ distance [16, Fig. 14]. $\mathrm{Lut}_{(1,0)}(10) = 12$ means that $B^o(O, 10) \subseteq B^o((2,1), 12)$ but $B^o(O, 10) \not\subseteq B^o((2,1), 11)$. Or, in terms of closed balls, $B(O, 9) \subseteq B((2,1), 11)$ but $B(O, 9) \not\subseteq B((2,1), 10)$.

Medial Axis LUT Coefficients. A general method for LUT coefficient computation was given by Rémy and Thiel [12,17,9]. The idea is that the disk covering relation can be extracted directly from values of distance to the origin. If $d(O, p) = r_1$ and $d(O, p + \overrightarrow{v_i}) = r_2$, we can deduce the following:

$$p \in B(O, r_1) = B^o(O, r_1 + 1) \ ,$$
$$p + \overrightarrow{v_i} \notin B(O, r_2 - 1) = B^o(O, r_2) \ ,$$

hence $B^o(O + \overrightarrow{v_i}, r_1 + 1) \not\subseteq B^o(O, r_2)$ and $\mathrm{Lut}_{\overrightarrow{v_i}}(r_1 + 1) > r_2$. If $\forall p, d(O, p) \leq r_1 \Rightarrow d(O, p + \overrightarrow{v_i}) \leq r_2$ then $\mathrm{Lut}_{\overrightarrow{v_i}}(r_1 + 1) = r_2 + 1$.

Finally: $\mathrm{Lut}_{\overrightarrow{v_i}}(r) = 1 + \max(d(O, p + \overrightarrow{v_i}) : d(O, p) < r)$.

This method only requires one scan of the distance function for each displacement $\overrightarrow{v_i}$. Moreover, the visited area may be restricted according to the symmetries of the chamfer mask. The order of complexity is about $O(mL^n)$ if we limit the computation of the distance function to a L^n image.

Medial Axis LUT Mask. Thiel observed that the chamfer mask is not adequate to compute the LUT and introduced a LUT Mask $\mathcal{M}_{\mathrm{Lut}}(R)$ for that purpose [12, p. 81]. $\mathcal{M}_{\mathrm{Lut}}(R)$ is the minimal test neighbourhood sufficient to detect the medial axis for shapes whose inner radius (the radius of a greatest ball) is less than or equal to R. For instance, with $d_{14,20,31,44}$: $\mathrm{Lut}_{(2,1)}(291) =$

321 and $\text{Lut}_{(2,1)}(321) = 352$ but the smallest open ball of center O covering $B^o((4,2), 291)$ is $B^o(O, 351)$. In this particular case, the point $(4,2)$ is not in the chamfer mask but should be in $\mathcal{M}_{\text{Lut}}(R)$ for R greater than 350.

A mask incompleteness produces extra points in the medial axis (undetected ball coverings). A general method for both detecting and validating $\mathcal{M}_{\text{Lut}}$ is based on the computation of the medial axis of all disks. When $\mathcal{M}_{\text{Lut}}$ is complete, the medial axis is restricted to the center of the disk, when extra points remains, they are added to $\mathcal{M}_{\text{Lut}}$. This neighborhood determination was proven to work in any dimension $n \geq 2$. However it is time consuming even when taking advantage of the mask symmetries.

3   Method Basics

3.1   General $\mathcal{H}$-Polytopes [2]

Definition 5 (Polyhedron). *A convex polyhedron is the intersection of a finite set of half-hyperplanes.*

Definition 6 (Polytope). *A polytope is the convex hull of a finite set of points.*

Theorem 1 (Weyl-Minkowski). *A subset of Euclidean space is a polytope if and only if it is a bounded convex polyhedron.*

As a result, a polytope in $\mathbb{R}^n$ can be represented either as the convex hull of its k vertices ($\mathcal{V}$-description): $P = \text{conv}(\{p_i\}_{1 \leq i \leq k})$ or by a set of l half-planes ($\mathcal{H}$-description):

$$P = \{x : Ax \leq y\} , \tag{8}$$

where A is a $l \times n$ matrix, y a vector of n values that we name $\mathcal{H}$-coefficients of P. Having two vectors $\overrightarrow{u}$ and $\overrightarrow{v}$, we denote $\overrightarrow{u} \leq \overrightarrow{v}$ if and only if $\forall i$, $\overrightarrow{u}_i \leq \overrightarrow{v}_i$.

Definition 7 (Discrete polytope). *A discret polytope is the intersection of a polytope with $\mathbb{Z}^n$.*

Minimal Representation. Many operations on $\mathbb{R}^n$ polytopes in either $\mathcal{V}$ or $\mathcal{H}$ representation often require a minimal representation. The *redundancy removal* is the removing of unnecessary vertices or inequalities in polytopes. Since our purpose is mainly to compare $\mathcal{H}$-polytopes defined with the same matrix A, no inequality removal is needed. However, for some operations, $\mathcal{H}$-representations of discrete polytopes must be minimal in terms of $\mathcal{H}$-coefficients.

Definition 8 (Minimal parameter representation). *We call minimal parameter $\mathcal{H}$-representation of a discrete polytope P, denoted $\widehat{\mathcal{H}}$-representation, a $\mathcal{H}$-representation $P = \{x : Ax \leq y\}$ such that y is minimal:*

$$P = \{x \in \mathbb{Z}^n : Ax \leq y\} \text{ and } \forall i \in [1..l], \exists x \in \mathbb{Z}^n : A_i x = y_i , \tag{9}$$

where A_i means line i in matrix A.

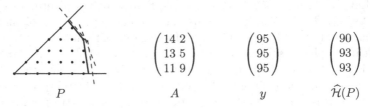

$$
A \begin{pmatrix} 14 & 2 \\ 13 & 5 \\ 11 & 9 \end{pmatrix} \quad y \begin{pmatrix} 95 \\ 95 \\ 95 \end{pmatrix} \quad \widehat{\mathcal{H}}(P) \begin{pmatrix} 90 \\ 93 \\ 93 \end{pmatrix}
$$

Fig. 1. $\mathcal{H}$-representations of a discrete G-symmetrical polytope P (restricted to the first octant). Dashed lines: $\mathcal{H}$-representation of P. Thick lines: $\widehat{\mathcal{H}}$-representation of P. In the $\widehat{\mathcal{H}}$ case, the three equalities are verified for the same point $(6,3)$. Notice that although coefficient values are minimal, this representation is still redundant: the second inequality could be removed.

The $\widehat{\mathcal{H}}$ function which gives the minimal parameters for a given polytope P is introduced for convenience: $\widehat{\mathcal{H}}(P) = \max\{Ax : x \in P\}$. $\{x : Ax \leq \widehat{\mathcal{H}}(P)\}$ is the $\widehat{\mathcal{H}}$-representation of $P = \{x : Ax \leq y\}$. Figure 1 depicts two representations of the same polytope P in $\mathbb{Z}^2$.

$\mathcal{H}$-Polytope Translation. Let $P = \{x : Ax \leq y\}$ be a $\mathcal{H}$-polytope. The translated of P by $\overrightarrow{v}$ which is also the Minkowski sum of P and $\{\overrightarrow{v}\}$ is:

$$(P)_{\overrightarrow{v}} = P \oplus \{\overrightarrow{v}\} = \{x + \overrightarrow{v} : Ax \leq y\} = \{x : Ax \leq y + A\overrightarrow{v}\} \ . \tag{10}$$

The translation of a minimal representation gives a minimal representation.

Covering Test. Let $P = \{x : Ax \leq y\}$ and $Q = \{x : Ax \leq z\}$ be two polyhedrons represented by the same matrix A but different sets of $\mathcal{H}$-coefficients y and z. P is a subset of Q if (sufficient condition):

$$y \leq z \Rightarrow P \subseteq Q \ . \tag{11}$$

Furthermore, if the $\mathcal{H}$-description of the enclosed polyhedron has minimal coefficients, the condition is also necessary:

$$y = \widehat{\mathcal{H}}(P) \leq z \Leftrightarrow P \subseteq Q \ . \tag{12}$$

3.2   Chamfer $\mathcal{H}$-Polytopes

Describing balls of chamfer norms as polygons in 2D and polyhedra in higher dimensions is not new [10]. Thiel and others have extensively studied chamfer ball geometry from this point of view [12,11,18]. Our purpose is to introduce properties specific to the $\mathcal{H}$-representation of these convex balls.

Proposition 1 (Direct distance formulation)

$$d(O, x) = \max_{1 \leq i \leq l} (d_{C_i}(x)) \tag{13}$$

where l is the number of cones, d_{C_i} is the distance function in the ith cone. Note that this formula does not require to determine in which cone lies x.

Proof. Equation (6) states that for all the influence cones C that contain p, $d_C(O, p) = d(O, p)$. For other cones, this relation does not hold, but it is still possible to compute $d_C(O, p)$. The influence cone C corresponds to a facet of the unitary real ball supported by the hyperplane $\{x \in \mathbb{R}^n : d_C(x) = 1\}$. Due to convexity, the unitary ball is included in the halfspace $\{x \in \mathbb{R}^n : d_C(x) \leq 1\}$. This applies to the vertices of the unitary ball $\frac{\vec{v_i}}{w_i}$: $d_C(\frac{\vec{v_i}}{w_i}) \leq 1$. By linearity of d_C, $d_C(\vec{v_i}) \leq w_i$ and $d_C(\sum_i \lambda_i \vec{v_i}) = \sum_i \lambda_i d_C(\vec{v_i}) \leq \sum_i \lambda_i w_i$. $d_C(\sum_i \lambda_i \vec{v_i})$ is always less than or equal to the length of the path $\sum_i \lambda_i \vec{v_i}$.

Chamfer Balls $\mathcal{H}$-Representation. The $\mathcal{H}$-representation of chamfer balls is directly derived from (13):

$$x \in B(O, r) \Leftrightarrow \max_{1 \leq i \leq l}\{d_{C_i}(x)\} \leq r \Leftrightarrow A_{\mathcal{M}} \cdot x \leq y \qquad (14)$$

where $A_{\mathcal{M}}$ is a $\mathcal{H}$-representation matrix depending only on the chamfer mask $\mathcal{M}$. The number of rows in $A_{\mathcal{M}}$ is equal to the number l of influence cones, each line of the matrix $A_{\mathcal{M}}$ is computed with (5) and y is a column vector whose values are r. For instance, the $\mathcal{H}$-representation matrix for $d_{5,7,11}$ is $A_{\mathcal{M}} = \begin{pmatrix} 5 & 1 \\ 4 & 3 \end{pmatrix}$ where (5 1) and (4 3) are the distance gradients in the two cones and $B(O, r) = \left\{ x \in \mathbb{Z}^n : \begin{pmatrix} 4 & 3 \\ 5 & 1 \end{pmatrix} \cdot x \leq \begin{pmatrix} r \\ r \end{pmatrix} \right\}$.

Proposition 2 (Furthest point). *Let $A_{\mathcal{M}}$ be the matrix defined by the chamfer mask $\mathcal{M}$ generating a norm. The furthest point from the origin in the $\widehat{\mathcal{H}}$-polytope $P = \{x : A_{\mathcal{M}} \cdot x \leq y\}$ is at a distance equal to the greatest component of y.*

Proof. By construction of $A_{\mathcal{M}}$, (13) is equivalent to $d(O, x) = \max_i\{A_{\mathcal{M}i} \cdot x\}$.

$$\max_{x \in P}\{d(O, x)\} = \max_{x \in P}\{\max_{1 \leq i \leq l}\{A_{\mathcal{M}i} \cdot x\}\} = \max\{\widehat{\mathcal{H}}_i(P)\}$$

Proposition 3 (Minimal covering ball). *The radius of the minimal ball centered in O that contains all points of a discrete $\widehat{\mathcal{H}}$-polytope P represented by the matrix $A_{\mathcal{M}}$ and the vector y is equal to the greatest component of y.*

Proof. The smallest ball that covers the polytope P must cover its furthest point from the origin.

$$\min\{r \in \mathbb{N} : P \subseteq B(O, r)\} = \max_{x \in P}\{d(O, x)\} \qquad (15)$$

Note that if P is not centered in O, the simplification due to symmetries do not hold and the full set of $\mathcal{H}$-coefficients is needed, unless we ensure that the $\mathcal{H}$-coefficents for the hyperplanes in the working sub-space are greater than $\mathcal{H}$-coefficients for the corresponding symmetric cones. This is the case when a G-symmetric polytope is translated by a vector in the sub-space.

Definition 9 (Covering function). *We call* covering function *of a set X of points of $\mathbb{Z}^n$ the function $\mathcal{C}_X$ which assigns to each point p of $\mathbb{Z}^n$, the radius of the minimal ball centered in p covering X:*

$$\mathcal{C}_X : 2^{\mathbb{Z}^n} \times \mathbb{Z}^n \to \mathbb{N}$$
$$X \ , \ p \ \to \min\{r : X \subseteq B(p, r)\} \ .$$

The covering function of the chamfer ball $B(O, r)$ at point p gives the radius of the minimal ball centered in p that contains $B(O, r)$. It is equal using central symmetry to the minimal ball centered in O covering $B(p, r)$ and therefore it is the maximal component of the $\widehat{\mathcal{H}}$-representation of $B(p, r)$:

$$\mathcal{C}_{B(O,r)}(p) = \max\{\widehat{\mathcal{H}}(B(p, r))\} = \max\{\widehat{\mathcal{H}}(B(O, r)) + A_{\mathcal{M}} \cdot p\} \ . \tag{16}$$

One can notice that the covering function of the zero radius disk is equal to the distance function, as is the distance transform of the complement of this disk:

$$\mathcal{C}_{B(O,0)}(p) = DT_{\mathbb{Z}^n \setminus \{O\}}(p) = d(p, O) = d(O, p) \ .$$

Definition 10 (Covering cone). *A* covering cone $C_{o,U}$ *in $\mathcal{C}_X$ is a cone defined by a vertex o and a subset U of the chamfer mask neighbor set with $\det(U) = \pm 1$, $C_{o,U} = \{o + \sum \lambda_i \overrightarrow{u_i}, \lambda_i \in \mathbb{N}, \overrightarrow{u_i} \in U\}$, such that:*

$$\forall p \in C_{o,U}, \forall \overrightarrow{u} \in U, B(p, \mathcal{C}_X(p)) \subsetneq B(p + \overrightarrow{u}, \mathcal{C}_X(p + \overrightarrow{u})) \ .$$

Proposition 4. *If $C_{o,U}$ is a covering cone in $\mathcal{C}_{B(O,r)}$ then for any point q in $C_{o,U} \setminus U \setminus \{o\}$ there exists p such that:*

$$B(O, r) \subsetneq B(p, \mathcal{C}_{B(O,r)}(p)) \subsetneq B(q, \mathcal{C}_{B(O,r)}(q)) \ .$$

Proof. Since $q \neq o$, there always exists another point p in $C_{o,U}$, distinct from q, such that $B(p, \mathcal{C}_{B(O,r)}(p)) \subsetneq B(q, \mathcal{C}_{B(O,r)}(q))$. $B(O, r) \subseteq B(p, \mathcal{C}_{B(O,r)}(p))$ always holds by definition of $\mathcal{C}_{B(O,r)}$. A sufficient condition for $B(O, r) \neq B(p, \mathcal{C}_{B(O,r)})$ is that a point p distinct from O exists in $C_{o,U}$ which is always the case if q is not equal to one of the generating vectors of $C_{o,U}$.

Proposition 5. *If there is a integer $j \in [1 \ldots l]$ and a point o such that $A_{\mathcal{M}j} \cdot \overrightarrow{u_i}, \forall \overrightarrow{u_i} \in U$ and $\widehat{\mathcal{H}}_j(B(o, r))$ are maximal then $C_{o,U}$ is a covering cone in $\mathcal{C}_{B(O,r)}$.*

Proof. Let j be the row number of a maximal component of $\widehat{\mathcal{H}}(B(o, r))$ and $A_{\mathcal{M}} \cdot \overrightarrow{u_i}, \forall \overrightarrow{u_i} \in U$, then j is a maximal component of any positive linear combination of these vectors. Let p be any point in $C_{o,U}$, $p = o + \sum_i \lambda_i \overrightarrow{u}_i, \lambda_i \in \mathbb{N}$ and B_p the minimal ball covering $B(O, r)$. From (16) we deduce that $\mathcal{C}_{B(O,r)}$ is affine in $C_{o,U}$:

$$\mathcal{C}_{B(O,r)}(p) = \max \left\{ \widehat{\mathcal{H}}\left(B(o, r)\right) + \sum \lambda_i A_{\mathcal{M}} \overrightarrow{u_i} \right\}$$
$$= \widehat{\mathcal{H}}_j(B(o, r)) + \sum \lambda_i A_{\mathcal{M}j} \overrightarrow{u_i} \ .$$

In the same way, the radius of the minimal ball centered in $p + \vec{u}, \vec{u} \in U$ covering $B(O, r)$ is:

$$\mathcal{C}_{B(O,r)}(p + \vec{u}) = \widehat{\mathcal{H}}_j(B(o, r)) + \sum \lambda_i A_{\mathcal{M}_j} \vec{u_i} + A_{\mathcal{M}_j} \vec{u}$$
$$= \mathcal{C}_{B(p, \mathcal{C}_{B(O,r)}(p))}(O + \vec{u}) \ .$$

In other words, $B(p, \mathcal{C}_{B(O,r)}(p))$ is a subset of $B(p + \vec{u}, \mathcal{C}_{B(O,r)}(p + \vec{u}))$.

4 LUT and $\mathcal{M}_{\mathrm{Lut}}$ Computation for 2D Chamfer Norms

LUT and $\mathcal{M}_{\mathrm{Lut}}$ computation methods for 2D chamfer norms are presented here. Both are based on the minimal $\widehat{\mathcal{H}}$-representation of the chamfer balls, from which we compute the covering function and deduce the LUT values. $\mathcal{M}$ vectors are ordered by angle so that each influence cone is defined by two successive angles $(O, \{\vec{v_i}, \vec{v_{i+1}}\})$.

4.1 $\widehat{\mathcal{H}}$-Representation of Chamfer Balls

The computation of the LUT is based on a $\widehat{\mathcal{H}}$-representation of the chamfer norm balls. All share the same matrix $A_{\mathcal{M}}$ which depends only on the chamfer mask (14). $\widehat{\mathcal{H}}$-coefficients of balls are computed iteratively from the ball of radius 0, $B(O, 0) = \{x : A_{\mathcal{M}} x = 0\}$ using (4) and (10).

$$\widehat{\mathcal{H}}(B(O, r)) = \max\{ \ \widehat{\mathcal{H}}(B(O, r - 1)),$$
$$\widehat{\mathcal{H}}(B(O, r - w_1)) + A_{\mathcal{M}} \vec{v_1}, \dots, \widehat{\mathcal{H}}(B(O, r - w_m)) + A_{\mathcal{M}} \vec{v_m}\} \ .$$

Each LUT value is obtained from the covering function (16):

$$\mathrm{Lut}_{\vec{v_i}}[r] = 1 + \mathcal{C}_{B(O, r-1)}(O + \vec{v_i}) = 1 + \max(\widehat{\mathcal{H}}(B(O, r - 1)) + A_{\mathcal{M}} \cdot \vec{v_i}) \ .$$

4.2 LUT Mask

The algorithm starts with an empty mask and balls with increasing radii are tested for direct covering relations as in [1]. However, in our case, the covering relations are seen from the perspective of the covered balls (in the covering function) whereas in [1], they are considered from the point of view of the covering balls (in the distance map). Another difference lies in the computation of covering radii which does not require the propagation of weights thanks to a direct formula (16). In order to remove useless points, all known LUT neighborhoods are checked for an indirect covering by the procedure visitPoint.

In each influence cone $(O, \{\vec{v_i}, \vec{v_{i+1}}\})$, a set of covering cones are detected to limit the search space: a 2D covering cone $((a, a), \{(1, 0), (0, 1)\})$ with a vertex chosen on the cone bissector $[O, (1, 1))$, then 1D covering cones $((b, \alpha), \{(1, 0)\})$, $((\alpha, b), \{(0, 1)\})$ for α varying from 1 to $a - 1$ (coordinates relative to $(O, \{\vec{v_i}, \vec{v_{i+1}}\})$). Fig. 2 shows the working sub-space partitioning, covering cones and visit order of points for the chamfer norm $d_{14, 20, 31, 44}$ and the inner radius 20.

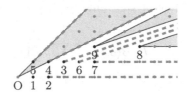

Fig. 2. 1D (dashed lines) and 2D (filled areas) covering cones and visit order of points

---

Algorithm 1. Computation of the look-up table mask

Input: chamfer mask $\mathcal{M} = \{(\vec{v_i}; w_i)\}_{1 \leq i \leq m}$, maximal radius R
$\mathcal{M}_{\mathrm{Lut}} \leftarrow \emptyset$;
for $r \leftarrow 1$ **to** R **do**   // current inner radius, increasing order
    for $i \leftarrow 1$ **to** m **do**   // visit vector axes
        visit1DCone($\mathcal{M}_{\mathrm{Lut}}$, r, O, $\vec{v_i}$);
    end
    for $i \leftarrow 1$ **to** l **do**   // visit the interior of influence cones
        $a \leftarrow$ findCoveringConeVertex(r, $O + \alpha\vec{v_i} + \alpha\vec{v_{i+1}}$, $\vec{v_i} + \vec{v_{i+1}}$);
        for $\alpha \leftarrow 1$ **to** $a - 1$ **do**   // skip the covering cone
            visitPoint($\mathcal{M}_{\mathrm{Lut}}$, r, $O + \alpha\vec{v_i} + \alpha\vec{v_{i+1}}$);
            visit1DCone($\mathcal{M}_{\mathrm{Lut}}$, r, $O + \alpha\vec{v_i} + \alpha\vec{v_{i+1}}$, $\vec{v_i}$);
            visit1DCone($\mathcal{M}_{\mathrm{Lut}}$, r, $O + \alpha\vec{v_i} + \alpha\vec{v_{i+1}}$, $\vec{v_{i+1}}$);
        end
        visitPoint($\mathcal{M}_{\mathrm{Lut}}$, r, $a \cdot M_i + a \cdot M_{i+1}$);
    end
end

Procedure visit1DCone($\mathcal{M}_{\mathrm{Lut}}$, r, p, $\vec{u}$)

Input: $\mathcal{M}_{\mathrm{Lut}}$, inner radius r, vertex and direction of the tested cone p, $\vec{u}$
for $\alpha \leftarrow 1$ **to** findCoveringConeVertex(r, p, $\vec{u}$) **do**
    visitPoint($\mathcal{M}_{\mathrm{Lut}}$, r, $\beta \cdot M_i + \alpha \cdot M_{i+1}$);
end

Procedure visitPoint($\mathcal{M}_{\mathrm{Lut}}$, r, p)

Input: $\mathcal{M}_{\mathrm{Lut}}$, inner radius r, ball center p to test
if $\forall \vec{v} \in \mathcal{M}_{\mathrm{Lut}} : B(O + \vec{v}, \mathcal{C}_{B(O,r)}(O + \vec{v})) \not\subset B(p, \mathcal{C}_{B(O,r)}(p))$ **then**
    $\mathcal{M}_{\mathrm{Lut}} \leftarrow \mathcal{M}_{\mathrm{Lut}} \cup M$
end

Function findCoveringConeVertex(r, p, $\vec{u}$)

Input: inner radius r, vertex and direction of the searched cone p, $\vec{u}$
Output: Integer
$y = \widehat{\mathcal{H}}(B(p, r)); z = A_{\mathcal{M}} \cdot \vec{u}$;
$i_0 \leftarrow \operatorname{argmax}_i \{y_i : z_i = \max\{z\}\}$ // component i_0 is maximal...
return $a = \max\left\{\left\lceil \frac{y_i - y_{i_0}}{z_{i_0} - z_i} \right\rceil\right\}$ // in vector $p + \alpha\vec{u}, \forall \alpha \geq a$

Table 1. Run times (in seconds)

L	5,7,11		14,20,31,44		62,88,139,196,224		68,96,152,215,245, 280,314,346,413	
	reference	$\mathcal{H}$	reference	$\mathcal{H}$	reference	$\mathcal{H}$	reference	$\mathcal{H}$
200	0.468524	0.001109	1.388558	0.005708	6.366711	0.053448	8.867095	0.346118
500	8.315003	0.002784	25.262300	0.017302	125.293637	0.145683	177.506492	0.975298
1000	90.670806	0.007007	268.807796	0.036611	1267.910045	0.276737	1684.583989	1.778505

4.3 Results

An implementation of these algorithms was developed in C language. It produces output in the same format as the reference algorithm [16] so that outputs can be compared character-to-character. Tests were done on various chamfer masks and different maximal radii. The results are almost always identical except for insignificant cases close to the maximal radius for which covering radii exceed the maximum. Other differences may occur in the order weightings are added to $\mathcal{M}_{\mathrm{Lut}}$ (sorted with respect to the covered radius or to the covering radius).

The run times of both reference and proposed algorithms are given in Table 1 for various sizes an distances.

5 Conclusion and Future Works

In this paper methods to compute both the chamfer LUT and chamfer LUT mask were presented. Speed gains from the reference algorithm [1] are attributable to the representation of chamfer balls as $\mathcal{H}$-polytopes. This description allows to avoid the use of weight propagation in the image domain and permits a constant time covering test by the direct computation of covering radii. Although not thoroughly tested, we think that LUT value computation is faster with our method due to the smaller size of the test space (linear with the radius of the maximal ball). For $\mathcal{M}_{\mathrm{Lut}}$ determination, results show must faster computation especially for large radii. This is due to the reduced search space eliminating covering cones and the constant time covering test.

While applications always using the same mask can use precomputed $\mathcal{M}_{\mathrm{Lut}}$ and LUT, other applications that potentially use several masks, adaptive masks, variable input image size can benefit from these algorithms. A faster computation of $\mathcal{M}_{\mathrm{Lut}}$ is also highly interesting to explore chamfer mask properties. Beyond improved run times, the $\mathcal{H}$-polytope representation helped to prove new properties of chamfer masks. And a new formula of distance which doesn't need to find in which cone lies a point was given.

Whereas the underlying theory ($\mathcal{H}$-representation of balls, translation, covering test and covering cone) does not depend on the dimension, the algorithms were given only for dimension 2. In the 2D case, vectors can be ordered by angle so two consecutive vectors define a cone. Higher dimensions require a Farey triangulation of chamfer balls.

A paper will greater details and results is being prepared for presentation in a journal. The source code for algorithms presented here is available from the IAPR technical committee on discrete geometry (TC18)[1].

References

1. Rémy, É., Thiel, É.: Medial axis for chamfer distances: computing look-up tables and neighbourhoods in 2d or 3d. Pattern Recognition Letters 23(6), 649–661 (2002)
2. Ziegler, G.M.: Lectures on Polytopes (Graduate Texts in Mathematics). Springer, Heidelberg (2001)
3. Rosenfeld, A., Pfaltz, J.L.: Sequential operations in digital picture processing. Journal of the ACM 13(4), 471–494 (1966)
4. Borgefors, G.: Distance transformations in arbitrary dimensions. Computer Vision, Graphics, and Image Processing 27(3), 321–345 (1984)
5. Coeurjolly, D., Montanvert, A.: Optimal separable algorithms to compute the reverse euclidean distance transformation and discrete medial axis in arbitrary dimension. IEEE Transactions on Pattern Analysis and Machine Intelligence 29(3), 437–448 (2007)
6. Strand, R.: Weighted distances based on neighborhood sequences. Pattern Recognition Letters 28, 2029–2036 (2007)
7. Borgefors, G.: Centres of maximal discs in the 5-7-11 distance transforms. In: Proc. 8th Scandinavian Conf. on Image Analysis, Tromsø, Norway (1993)
8. Rosenfeld, A., Pfaltz, J.L.: Distances functions on digital pictures. Pattern Recognition Letters 1(1), 33–61 (1968)
9. Thiel, É.: Géométrie des distances de chanfrein. In: mémoire d'habilitation à diriger des recherches (2001), http://www.lif-sud.univ-mrs.fr/~string~thiel/hdr/~
10. Verwer, B.J.H.: Local distances for distance transformations in two and three dimensions. Pattern Recognition Letters 12(11), 671–682 (1991)
11. Rémy, É.: Normes de chanfrein et axe médian dans le volume discret. Thèse de doctorat, Université de la Méditerranée (2001)
12. Thiel, É.: Les distances de chanfrein en analyse d'images: fondements et applications. Thèse de doctorat, Université Joseph Fourier, Grenoble 1 (1994), http://www.lif-sud.univ-mrs.fr/~thiel/these/~
13. Pfaltz, J.L., Rosenfeld, A.: Computer representation of planar regions by their skeletons. Communications of the ACM 10(2), 119–122 (1967)
14. Arcelli, C., di Baja, G.S.: Finding local maxima in a pseudo-Euclidian distance transform. Computer Vision, Graphics, and Image Processing 43(3), 361–367 (1988)
15. Svensson, S., Borgefors, G.: Digital distance transforms in 3d images using information from neighbourhoods up to 5×5×5. Computer Vision and Image Understanding 88(1), 24–53 (2002)
16. Rémy, É., Thiel, É.: Medial axis for chamfer distances: computing look-up tables and neighbourhoods in 2d or 3d. Pattern Recognition Letters 23(6), 649–661 (2002)
17. Remy, E., Thiel, E.: Computing 3D Medial Axis for Chamfer Distances. In: Nyström, I., Sanniti di Baja, G., Borgefors, G. (eds.) DGCI 2000. LNCS, vol. 1953, pp. 418–430. Springer, Heidelberg (2000)
18. Borgefors, G.: Weighted digital distance transforms in four dimensions. Discrete Applied Mathematics 125(1), 161–176 (2003)

---

[1] http://www.cb.uu.se/~tc18/

Weighted Neighbourhood Sequences in Non-Standard Three-Dimensional Grids – Metricity and Algorithms

Robin Strand[1] and Benedek Nagy[2]

[1] Centre for Image Analysis, Uppsala University,
Box 337, SE-75105 Uppsala, Sweden
[2] Department of Computer Science, Faculty of Informatics, University of Debrecen,
PO Box 12, 4010, Debrecen, Hungary
`robin@cb.uu.se, nbenedek@inf.unideb.hu`

Abstract. Recently, a distance function was defined on the face-centered cubic and body-centered cubic grids by combining weights and neighbourhood sequences. These distances share many properties with traditional path-based distance functions, such as the city-block distance, but are less rotational dependent. We present conditions for metricity and algorithms to compute the distances.

1 Introduction

When using non-standard grids such as the face-centered cubic (fcc) grid and the body-centered cubic (bcc) grid for 3D images, less samples are needed to obtain the same representation/reconstruction quality compared to the cubic grid [1]. This is one reason for the increasing interest in using these grids in, e.g., image acquisition [1], image processing [2,3,4], and visualization [5,6].

Measuring distances is of great importance in many applications. Because of its low rotational dependency, the Euclidean distance is often used as distance function. There are, however, applications where other distance functions are better suited. For example, when minimal cost-paths are computed, a distance function defined as the minimal cost-path between any two points is better suited, see, e.g., [7], where the constrained distance transform is computed using the Euclidean distance resulting in a complex algorithm. The corresponding algorithm using a path-based approach is simple, fast, and easy to generalize to higher dimensions [8,9]. Examples of path-based distances are weighted distances, where weights define the cost (distance) between neighbouring grid points [10,3,2], and distances based on neighbourhood sequences (NS-distances), where the adjacency relation is allowed to vary along the path, but all steps are of unit weights [11,4]. These path-based distance functions are generalizations of the well-known city-block and chessboard distance function defined for the square grid in [12].

Neighbourhood sequences was mentioned in [13], where vectors (steps) of various vector sets were periodically used to construct a shortest path between two

D. Coeurjolly et al. (Eds.): DGCI 2008, LNCS 4992, pp. 201–212, 2008.

points. A general definition of digital distances allowing both weights and neighbourhood sequences was presented in [14]. For the case with no weights, they presented non-metrical distances and conditions for the neighbourhood sequence to define metrical distance.

Das and his coathors [15] chosed the natural neighbourhood conditions of $\mathbb{Z}^n$ and computed several results on digital distances by neighbourhood sequences but without using any weights. Note that in the square grid one of these so-called octagonal distances is mentioned already in 1968 [11] as a relatively good and simple approximation to the Euclidean distance in a digital way.

Metrical distances are of importance both in theory and in applications. The necessary and sufficient condition for a neighbourhood sequence (without weights) to define a metric distance can be found in [15] for periodic neighbourhood sequences (NS) on $\mathbb{Z}^n$, in [16] for the non-periodic case, in [4] for the fcc and bcc grids.

In [17] NS and weights were together used in the sense of [14], but with the well-known natural neighbourhood structure of $\mathbb{Z}^2$. In [17], the basic theory for weighted distances based on neighbourhood sequences (weighted NS-distances) is presented including a formula for the distance between two points, conditions for metricity, optimal weight calculation, and an algorithm to compute the distance transform. In [18], some results for weighted NS-distances on the fcc and bcc grids were presented. In this paper, we further develop the theory for weighted NS-distances on the fcc and bcc grids by presenting sufficient conditions for metricity and algorithms that can be used to compute the distance transform and a minimal cost-path between two points.

Note that the results presented here also apply to weighted distances and NS-distances, since they are both special cases of the proposed distance function.

2   Basic Notions and Previous Results

The following definitions of the fcc and bcc grids are used:

$$\mathbb{F} = \{(x, y, z) : x, y, z \in \mathbb{Z} \text{ and } x + y + z \equiv 0 \pmod{2}\}. \tag{1}$$
$$\mathbb{B} = \{(x, y, z) : x, y, z \in \mathbb{Z} \text{ and } x \equiv y \equiv z \pmod{2}\}. \tag{2}$$

When the result is valid for both $\mathbb{F}$ and $\mathbb{B}$, the notation $\mathbb{G}$ is used. Two distinct points $\mathbf{p}_1 = (x_1, y_1, z_1), \mathbf{p}_2 = (x_2, y_2, z_2) \in \mathbb{G}$ are ρ-neighbours, $1 \leq \rho \leq 2$, if

1. $|x_1 - x_2| + |y_1 - y_2| + |z_1 - z_2| \leq 3$ and
2. $\max \{|x_1 - x_2|, |y_1 - y_2|, |z_1 - z_2|\} \leq \rho$

The points $\mathbf{p}_1, \mathbf{p}_2$ are *adjacent* if $\mathbf{p}_1$ and $\mathbf{p}_2$ are ρ-neighbours for some ρ. The ρ-neighbours which are not $(\rho - 1)$-neighbours are called *strict* ρ-neighbours. The neighbourhood relations are visualized in Figure 1 by showing the Voronoi regions, i.e. the voxels, corresponding to some adjacent grid points.

A NS B is a sequence $B = (b(i))_{i=1}^{\infty}$, where each $b(i)$ denotes a neighbourhood relation in $\mathbb{G}$. If B is periodic, i.e., if for some fixed strictly positive $l \in \mathbb{Z}_+$,

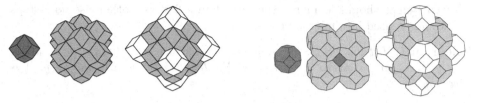

Fig. 1. The grid points corresponding to the dark and the light grey voxels are 1-neighbours. The grid points corresponding to the dark grey and white voxels are (strict) 2-neighbours. Left: fcc, right: bcc.

$b(i) = b(i + l)$ is valid for all $i \in \mathbb{Z}_+$, then we write $B = (b(1), b(2), \ldots, b(l))$. A *path*, denoted $\mathcal{P}$, in a grid is a sequence $\mathbf{p}_0, \mathbf{p}_1, \ldots, \mathbf{p}_n$ of adjacent grid points. A path between $\mathbf{p}_0$ and $\mathbf{p}_n$ is denoted $\mathcal{P}_{\mathbf{p}_0, \mathbf{p}_n}$. A path is a *B-path* of length n if, for all $i \in \{1, 2, \ldots, n\}$, $\mathbf{p}_{i-1}$ and $\mathbf{p}_i$ are $b(i)$-neighbours. The notation 1- and (strict) 2-steps will be used for a step to a 1-neighbour and step to a (strict) 2-neighbour, respectively. The number of 1-steps and strict 2-steps in a given path $\mathcal{P}$ is denoted $\mathbf{1}_{\mathcal{P}}$ and $\mathbf{2}_{\mathcal{P}}$, respectively.

Definition 1. *Given the NS B, the NS-distance $d(\mathbf{p}_0, \mathbf{p}_n; B)$ between the points $\mathbf{p}_0$ and $\mathbf{p}_n$ is the length of (one of) the shortest B-path(s) between the points.*

Let the real numbers α and β (the *weights*) and a path $\mathcal{P}$ of length n, where exactly l ($l \leq n$) adjacent grid points in the path are strict 2-neighbours, be given. The *length of the (α, β)-weighted B-path $\mathcal{P}$* is $(n - l)\alpha + l\beta$. The B-path $\mathcal{P}$ between the points $\mathbf{p}_0$ and $\mathbf{p}_n$ is a *minimal cost (α, β)-weighted B-path between the points $\mathbf{p}_0$ and $\mathbf{p}_n$* if no other (α, β)-weighted B-path between the points is shorter than the length of the (α, β)-weighted B-path $\mathcal{P}$.

Definition 2. *Given the NS B and the weights α, β, the weighted NS-distance $d_{\alpha,\beta}(\mathbf{p}_0, \mathbf{p}_n; B)$ is the length of (one of) the minimal cost (α, β)-weighted B-path(s) between the points.*

The following notation is used:

$$\mathbf{1}_B^k = |\{i : b(i) = 1, 1 \leq i \leq k\}| \text{ and}$$
$$\mathbf{2}_B^k = |\{i : b(i) = 2, 1 \leq i \leq k\}|.$$

We now recall from [18] the following two theorems giving the distance between two grid points $(0, 0, 0)$ and (x, y, z), where $x \geq y \geq z \geq 0$. We remark that by translation-invariance and symmetry, the distance between any two grid points is given by the formula below.

Remark 1. Hereafter, we will refer to "the weights α and β" as real numbers α and β such that $0 < \alpha \leq \beta \leq 2\alpha$. This is natural since

− a 2-step should be more expensive than a 1-step since strict 2-neighbours are intuitively at a larger distance than 1-neighbours (projection property) and

– two 1-steps should be more expensive than a 2-step – otherwise no 2-steps would be used in a minimal cost-path.

Theorem 1. *Let the NS B, the weights α, β and the point $(x, y, z) \in \mathbb{F}$, where $x \geq y \geq z \geq 0$, be given. The weighted NS-distance between $\mathbf{0}$ and (x, y, z) is given by*

$$d_{\alpha,\beta}\left(\mathbf{0}, (x, y, z); B\right) = \begin{cases} k \cdot \alpha & \text{if } x \leq y + z \\ (2k - x) \cdot \alpha + (x - k) \cdot \beta & \text{otherwise,} \end{cases}$$

$$\text{where } k = \min_{k} : k \geq \max\left(\frac{x + y + z}{2}, x - 2\frac{k}{B}\right).$$

Theorem 2. *Let the NS B, the weights α, β, and the point $(x, y, z) \in \mathbb{B}$, where $x \geq y \geq z \geq 0$, be given. The weighted NS-distance between $\mathbf{0}$ and (x, y, z) is given by*

$$d_{\alpha,\beta}\left(\mathbf{0}, (x, y, z); B\right) = (2k - x) \cdot \alpha + (x - k) \cdot \beta$$

$$\text{where } k = \min_{k} : k \geq \max\left(\frac{x + y}{2}, x - 2\frac{k}{B}\right).$$

We collect some previous results about paths generating NS-distances and weighted NS-distances:

Lemma 1. *Let the NS B, the weights α, β, and the point $\mathbf{p} = (x, y, z) \in \mathbb{G}$, where $x \geq y \geq z \geq 0$ be given. Then there is a path $\mathcal{P}_{\mathbf{0},\mathbf{p}}$ such that*

i *$\mathcal{P}_{\mathbf{0},\mathbf{p}}$ is a shortest B-path,*
ii *$\mathcal{P}_{\mathbf{0},\mathbf{p}}$ is a minimal cost (α, β)-weighted B-path, and*
iii *$\mathcal{P}_{\mathbf{0},\mathbf{p}}$ consists only of the steps*
$(2, 0, 0), (1, 1, 0), (1, -1, 0), (0, 1, 1), \text{ and } (1, 0, 1)$ *(fcc)*
$(2, 0, 0), (1, 1, -1), (1, 1, 1), \text{ and } (1, -1, -1)$ *(bcc)*

Proof. i and iii follow from Theorem 2 and Theorem 5 and their proofs in [4] for the fcc and bcc grids, respectively and ii follows from Lemma 4 in [18].   □

Remark 2. It follows from Lemma 1 that i and ii are valid for *any* point $\mathbf{p} \in \mathbb{G}$. Also, for any point $\mathbf{p} \in \mathbb{G}$, the paths can be limited to a few local steps as in iii by changing the sign and permuting the positions of the coordinates of the local steps.

3   Metricity of the Distance Functions

Example 1. Let $B = (2, 2, 1)$, $\alpha = 2$, and $\beta = 3$. For the points $\mathbf{p}_1 = (0, 0, 0)$, $\mathbf{p}_2 = (4, 0, 0)$, and $\mathbf{p}_3 = (6, 0, 0)$ in fcc *or* bcc, we have

$$d_{\alpha,\beta}(\mathbf{p}_1, \mathbf{p}_3; B) = 10 > 9 = 6 + 3 = d_{\alpha,\beta}(\mathbf{p}_1, \mathbf{p}_2; B) + d_{\alpha,\beta}(\mathbf{p}_2, \mathbf{p}_3; B),$$

so the triangular inequality is violated and therefore, $d_{2,3}(\cdot, \cdot; (2, 2, 1))$ is *not* a metric on neither $\mathbb{F}$ nor $\mathbb{B}$.

It is obvious that the distance functions are positive definite for positive weights. They are also symmetric since the fcc and bcc grids are point-lattices. To establish the condition of metricity, only the triangular inequality must be proved. As Example 1 shows, the triangular inequality is related to the weights and the number of occurences of 1 and 2 and their positions in the NS. We keep the weights within the interval in Remark 1 and to prove the main result about metricity, Theorem 3, we need some conditions also on the NS, namely: Given three points $\mathbf{p}_1$, $\mathbf{p}_2$, and $\mathbf{p}_3$ in $\mathbb{G}$, there are minimal cost-paths $\mathcal{P}_{\mathbf{p}_1,\mathbf{p}_2}$, $\mathcal{P}_{\mathbf{p}_2,\mathbf{p}_3}$, $\mathcal{P}_{\mathbf{p}_1,\mathbf{p}_3}$ such that

$$1_{\mathcal{P}_{\mathbf{p}_1,\mathbf{p}_3}} + 2_{\mathcal{P}_{\mathbf{p}_1,\mathbf{p}_3}} \leq 1_{\mathcal{P}_{\mathbf{p}_1,\mathbf{p}_2}} + 2_{\mathcal{P}_{\mathbf{p}_1,\mathbf{p}_2}} + 1_{\mathcal{P}_{\mathbf{p}_2,\mathbf{p}_3}} + 2_{\mathcal{P}_{\mathbf{p}_2,\mathbf{p}_3}} \text{ and} \tag{3}$$

$$1_{\mathcal{P}_{\mathbf{p}_1,\mathbf{p}_3}} + 2 \cdot 2_{\mathcal{P}_{\mathbf{p}_1,\mathbf{p}_3}} \leq 1_{\mathcal{P}_{\mathbf{p}_1,\mathbf{p}_2}} + 2 \cdot 2_{\mathcal{P}_{\mathbf{p}_1,\mathbf{p}_2}} + 1_{\mathcal{P}_{\mathbf{p}_2,\mathbf{p}_3}} + 2 \cdot 2_{\mathcal{P}_{\mathbf{p}_2,\mathbf{p}_3}}. \tag{4}$$

Eq. (3) is the triangular inequality for NS-distances and is given by the metricity of NS-distance proved in [4]. Eq. (4) can be interpreted the lengths of the paths obtained by replacing each 2-step by two 1-steps. Eq. (4) is not fulfilled for all shortest paths, as the following example shows.

Example 2. Consider the fcc grid with $B = (2)$ and $\mathbf{p}_1 = (0,0,0)$, $\mathbf{p}_3 = (2,2,0)$. The path $(0,0,0), (2,0,0), (2,2,0)$ is a shortest path, and (3) is fulfilled for any point $\mathbf{p}_2$. Eq (4) is not fulfilled with $\mathbf{p}_2 = (1,1,0)$ since $1_{\mathcal{P}_{\mathbf{p}_1,\mathbf{p}_3}} = 0$, $2_{\mathcal{P}_{\mathbf{p}_1,\mathbf{p}_3}} = 2$, $1_{\mathcal{P}_{\mathbf{p}_1,\mathbf{p}_2}} = 1$, $2_{\mathcal{P}_{\mathbf{p}_1,\mathbf{p}_2}} = 0$, $1_{\mathcal{P}_{\mathbf{p}_2,\mathbf{p}_3}} = 1$, and $2_{\mathcal{P}_{\mathbf{p}_2,\mathbf{p}_3}} = 0$

There are, however, always shortest paths (that are also minimal cost-paths) for any $\mathbf{p}_1$, $\mathbf{p}_2$, and $\mathbf{p}_3$ such that also (4) is fulfilled. This is proved in Lemma 3 and 5 below. For Example 2, (4) is fulfilled with the path $\mathcal{P}_{\mathbf{p}_1,\mathbf{p}_3}$ defined by $(0,0,0)$, $(1,1,0)$, $(2,2,0)$ and any $\mathbf{p}_2$. We will see that (3) together with $\alpha \leq \beta$ and (4) together with $\beta \leq 2\alpha$ implies metricity also for the (α, β)-weighted B-distance.

3.1   The fcc Grid

Lemma 2. *Let the weights α, β and the NS B be given. There is a minimal cost (α, β)-weighted B-path $\mathcal{P} = \mathcal{P}_{\mathbf{p}_1,\mathbf{p}_2}$ defining the (α, β)-weighted NS-distance between $\mathbf{p}_1 = (x_1, y_1, z_1) \in \mathbb{F}$ and $\mathbf{p}_2 = (x_2, y_2, z_2) \in \mathbb{F}$ such that*

$$1_{\mathcal{P}} + 2 \cdot 2_{\mathcal{P}} \geq \max\left\{|x_2 - x_1|, |y_2 - y_1|, |z_2 - z_1|\right\}. \tag{5}$$

The equality is attained when

$$\max\left\{|x_2 - x_1|, |y_2 - y_1|, |z_2 - z_1|\right\} \geq \frac{|x_2 - x_1| + |y_2 - y_1| + |z_2 - z_1|}{2}.$$

Proof. Given a point (x, y, z) such that $x \geq y \geq z \geq 0$, we use a path $\mathcal{P}'$ between $\mathbf{0}$ and (x, y, z) satisfying i–iii in Lemma 1. By Lemma 2 in [18], we have

$$1_{\mathcal{P}'} = \begin{cases} \hat{k} & \text{if } x \leq y + z \\ 2\hat{k} - x & \text{otherwise.} \end{cases} \quad \text{and} \quad 2_{\mathcal{P}'} = \begin{cases} 0 & \text{if } x \leq y + z \\ x - \hat{k} & \text{otherwise.} \end{cases}$$

where $\hat{k} = \min \left\{ k : k \geq \max \left(\dfrac{x+y+z}{2}, x - 2_B^k \right) \right\}$.

We recall that for $x \leq y + z$, $\hat{k} = \frac{x+y+z}{2}$ (see the proof of Theorem 2 in [4]). Thus when $x \leq y + z$, $1_{\mathcal{P}'} + 2 \cdot 2_{\mathcal{P}'} = \hat{k} = \frac{x+y+z}{2} \geq x = \max\{x, y, z\}$.

For the case $x \geq \frac{x+y+z}{2}$, i.e. $x \geq y + z$, $1_{\mathcal{P}'} + 2 \cdot 2_{\mathcal{P}'} = 2\hat{k} - x + 2\left(x - \hat{k} \right) = x$.

This shows that the lemma holds for $(x_1, y_1, z_1) = (0, 0, 0)$ and $(x_2, y_2, z_2) = (x, y, z)$. The lemma holds also for the general case by translation invariance and symmetry. □

Lemma 3. *Let the weights* α, β *and the NS B such that*

$$\sum_{i=1}^{N} b(i) \leq \sum_{i=j}^{j+N-1} b(i) \ \forall j, N \geq 1$$

be given. For any points $\mathbf{p}_1 = (x_1, y_1, z_1), \mathbf{p}_2 = (x_2, y_2, z_2), \mathbf{p}_3 = (x_3, y_3, z_3) \in \mathbb{F}$, *there are minimal cost* (α, β)*-weighted B-paths* $\mathcal{P}_{\mathbf{p}_1, \mathbf{p}_3}, \mathcal{P}_{\mathbf{p}_1, \mathbf{p}_2}, \mathcal{P}_{\mathbf{p}_1, \mathbf{p}_3}$ *such that*

$$1_{\mathcal{P}_{\mathbf{p}_1, \mathbf{p}_3}} + 2 \cdot 2_{\mathcal{P}_{\mathbf{p}_1, \mathbf{p}_3}} \leq 1_{\mathcal{P}_{\mathbf{p}_1, \mathbf{p}_2}} + 2 \cdot 2_{\mathcal{P}_{\mathbf{p}_1, \mathbf{p}_2}} + 1_{\mathcal{P}_{\mathbf{p}_2, \mathbf{p}_3}} + 2 \cdot 2_{\mathcal{P}_{\mathbf{p}_2, \mathbf{p}_3}}.$$

Proof. Let $\mathbf{p}_1$ and $\mathbf{p}_3$ be such that $|x_3 - x_1| \geq |y_3 - y_1| \geq |z_3 - z_1|$. We consider two cases:

— $|x_3 - x_1| > |y_3 - y_1| + |z_3 - z_1|$
  $1_{\mathcal{P}_{\mathbf{p}_1, \mathbf{p}_3}} + 2 \cdot 2_{\mathcal{P}_{\mathbf{p}_1, \mathbf{p}_3}} = |x_3 - x_1| \leq |x_2 - x_1| + |x_3 - x_2| \leq 1_{\mathcal{P}_{\mathbf{p}_1, \mathbf{p}_2}} + 2 \cdot 2_{\mathcal{P}_{\mathbf{p}_1, \mathbf{p}_2}} + 1_{\mathcal{P}_{\mathbf{p}_2, \mathbf{p}_3}} + 2 \cdot 2_{\mathcal{P}_{\mathbf{p}_2, \mathbf{p}_3}}$. Here we used (5) in Lemma 2 with equality, the triangular inequality for real numbers, and (5) in Lemma 2 respectively.

— $|x_3 - x_1| \leq |y_3 - y_1| + |z_3 - z_1|$
  $1_{\mathcal{P}_{\mathbf{p}_1, \mathbf{p}_3}} + 2 \cdot 2_{\mathcal{P}_{\mathbf{p}_1, \mathbf{p}_3}} = 1_{\mathcal{P}_{\mathbf{p}_1, \mathbf{p}_3}} + 2_{\mathcal{P}_{\mathbf{p}_1, \mathbf{p}_3}} \leq 1_{\mathcal{P}_{\mathbf{p}_1, \mathbf{p}_2}} + 2_{\mathcal{P}_{\mathbf{p}_1, \mathbf{p}_2}} + 1_{\mathcal{P}_{\mathbf{p}_2, \mathbf{p}_3}} + 2_{\mathcal{P}_{\mathbf{p}_2, \mathbf{p}_3}} \leq 1_{\mathcal{P}_{\mathbf{p}_1, \mathbf{p}_2}} + 2 \cdot 2_{\mathcal{P}_{\mathbf{p}_1, \mathbf{p}_2}} + 1_{\mathcal{P}_{\mathbf{p}_2, \mathbf{p}_3}} + 2 \cdot 2_{\mathcal{P}_{\mathbf{p}_2, \mathbf{p}_3}}$. Here we used that $2_{\mathcal{P}_{\mathbf{p}_1, \mathbf{p}_3}} = 0$ by Lemma 2 in [18], the triangular inequality for NS-distances from [4] ($d_{(1,1)}(\mathbf{p}_1, \mathbf{p}_3) = 1_{\mathcal{P}_{\mathbf{p}_1, \mathbf{p}_3}} + 2_{\mathcal{P}_{\mathbf{p}_1, \mathbf{p}_3}}$ by Lemma 1), and that $2_{\mathcal{P}_{\mathbf{p}_1, \mathbf{p}_2}}, 2_{\mathcal{P}_{\mathbf{p}_2, \mathbf{p}_3}} \geq 0$, respectively. □

3.2   The bcc Grid

Lemma 4. *Let the weights* α, β *and the NS B be given. There is a path* $\mathcal{P}$ *defining the* (α, β)*-weighted NS-distance between* $(x_1, y_1, z_1) \in \mathbb{B}$ *and* $(x_2, y_2, z_2) \in \mathbb{B}$ *such that*

$$1_{\mathcal{P}} + 2 \cdot 2_{\mathcal{P}} = \max\{|x_2 - x_1|, |y_2 - y_1|, |z_2 - z_1|\}. \tag{6}$$

Proof. Given a point (x, y, z) such that $x \geq y \geq z \geq 0$, we use a path $\mathcal{P}'$ between $\mathbf{0}$ and (x, y, z) satisfying i–iii in Lemma 1. By Lemma 4 in [18], $\mathcal{P}'$ is such that $1_{\mathcal{P}'} = 2\hat{k} - x$ and $2_{\mathcal{P}'} = x - \hat{k}$. Therefore, $1_{\mathcal{P}'} + 2 \cdot 2_{\mathcal{P}'} = x = \max\{x, y, z\}$. The general case follows by translation invariance and symmetry. □

Lemma 5. *Let the weights α, β and the NS B be given. For any three points $\mathbf{p}_1 = (x_1, y_1, z_1), \mathbf{p}_2 = (x_2, y_2, z_2), \mathbf{p}_3 = (x_3, y_3, z_3) \in \mathbb{B}$, there are minimal cost (α, β)-weighted B-paths $\mathcal{P}_{\mathbf{p}_1,\mathbf{p}_3}, \mathcal{P}_{\mathbf{p}_1,\mathbf{p}_2}, \mathcal{P}_{\mathbf{p}_1,\mathbf{p}_3}$ such that*

$$1_{\mathcal{P}_{\mathbf{p}_1,\mathbf{p}_3}} + 2 \cdot 2_{\mathcal{P}_{\mathbf{p}_1,\mathbf{p}_3}} \leq 1_{\mathcal{P}_{\mathbf{p}_1,\mathbf{p}_2}} + 2 \cdot 2_{\mathcal{P}_{\mathbf{p}_1,\mathbf{p}_2}} + 1_{\mathcal{P}_{\mathbf{p}_2,\mathbf{p}_3}} + 2 \cdot 2_{\mathcal{P}_{\mathbf{p}_2,\mathbf{p}_3}}.$$

Proof. Let $\mathbf{p}_1$ and $\mathbf{p}_3$ be such that $|x_3 - x_1| \geq |y_3 - y_1| \geq |z_3 - z_1|$. Then $1_{\mathcal{P}_{\mathbf{p}_1,\mathbf{p}_3}} + 2 \cdot 2_{\mathcal{P}_{\mathbf{p}_1,\mathbf{p}_3}} = |x_3 - x_1| \leq |x_2 - x_1| + |x_3 - x_2| \leq 1_{\mathcal{P}_{\mathbf{p}_1,\mathbf{p}_2}} + 2 \cdot 2_{\mathcal{P}_{\mathbf{p}_1,\mathbf{p}_2}} + 1_{\mathcal{P}_{\mathbf{p}_2,\mathbf{p}_3}} + 2 \cdot 2_{\mathcal{P}_{\mathbf{p}_2,\mathbf{p}_3}}$. Here we used (6) in Lemma 4 with equality, the triangular inequality for real numbers, and (6) in Lemma 4 (again), respectively. □

3.3 Metricity

Theorem 3. *If*

$$\sum_{i=1}^{N} b(i) \leq \sum_{i=j}^{j+N-1} b(i) \quad \forall j, N \geq 1 \quad and \tag{7}$$

$$0 < \alpha \leq \beta \leq 2\alpha \tag{8}$$

then $d_{\alpha,\beta}(\cdot, \cdot; B)$ is a metric on the fcc and bcc grids.

Proof. The positive definiteness and symmetry are trivial. We prove the triangular inequality. Let $\mathbf{p}_1, \mathbf{p}_2, \mathbf{p}_3 \in \mathbb{G}$ be given. We will prove $d_{(\alpha,\beta)}(\mathbf{p}_1, \mathbf{p}_3; B) \leq d_{(\alpha,\beta)}(\mathbf{p}_1, \mathbf{p}_2; B) + d_{(\alpha,\beta)}(\mathbf{p}_2, \mathbf{p}_3; B)$, i.e. that

$$1_{\mathcal{P}_{\mathbf{p}_1,\mathbf{p}_3}}\alpha + 2_{\mathcal{P}_{\mathbf{p}_1,\mathbf{p}_3}}\beta \leq 1_{\mathcal{P}_{\mathbf{p}_1,\mathbf{p}_2}}\alpha + 2_{\mathcal{P}_{\mathbf{p}_1,\mathbf{p}_2}}\beta + 1_{\mathcal{P}_{\mathbf{p}_2,\mathbf{p}_3}}\alpha + 2_{\mathcal{P}_{\mathbf{p}_2,\mathbf{p}_3}}\beta. \tag{9}$$

In [4], it is proved that when (7) is fulfilled, then the NS-distance defined by B is metric. Therefore, $d_{(1,1)}(\mathbf{p}_1, \mathbf{p}_3; B) \leq d_{(1,1)}(\mathbf{p}_1, \mathbf{p}_2; B) + d_{(1,1)}(\mathbf{p}_2, \mathbf{p}_3; B)$. Thus,

$$1_{\mathcal{P}_{\mathbf{p}_1,\mathbf{p}_3}} + 2_{\mathcal{P}_{\mathbf{p}_1,\mathbf{p}_3}} \leq 1_{\mathcal{P}_{\mathbf{p}_1,\mathbf{p}_2}} + 2_{\mathcal{P}_{\mathbf{p}_1,\mathbf{p}_2}} + 1_{\mathcal{P}_{\mathbf{p}_2,\mathbf{p}_3}} + 2_{\mathcal{P}_{\mathbf{p}_2,\mathbf{p}_3}} \tag{10}$$

for *any* shortest B-paths $\mathcal{P}_{\mathbf{p}_1,\mathbf{p}_3}, \mathcal{P}_{\mathbf{p}_1,\mathbf{p}_2}$, and $\mathcal{P}_{\mathbf{p}_2,\mathbf{p}_3}$. By Lemma 3 and 5, it follows that

$$1_{\mathcal{P}'_{\mathbf{p}_1,\mathbf{p}_3}} + 2 \cdot 2_{\mathcal{P}'_{\mathbf{p}_1,\mathbf{p}_3}} \leq 1_{\mathcal{P}'_{\mathbf{p}_1,\mathbf{p}_2}} + 2 \cdot 2_{\mathcal{P}'_{\mathbf{p}_1,\mathbf{p}_2}} + 1_{\mathcal{P}'_{\mathbf{p}_2,\mathbf{p}_3}} + 2 \cdot 2_{\mathcal{P}'_{\mathbf{p}_2,\mathbf{p}_3}}. \tag{11}$$

for *some* shortest B-paths $\mathcal{P}'_{\mathbf{p}_1,\mathbf{p}_3}, \mathcal{P}'_{\mathbf{p}_1,\mathbf{p}_2}$, and $\mathcal{P}'_{\mathbf{p}_2,\mathbf{p}_3}$.

By Lemma 1, there are paths such that $\mathcal{P}_{\mathbf{p}_1,\mathbf{p}_3} = \mathcal{P}'_{\mathbf{p}_1,\mathbf{p}_3}, \mathcal{P}_{\mathbf{p}_1,\mathbf{p}_2} = \mathcal{P}'_{\mathbf{p}_1,\mathbf{p}_2}$, and $\mathcal{P}_{\mathbf{p}_2,\mathbf{p}_3} = \mathcal{P}'_{\mathbf{p}_2,\mathbf{p}_3}$.

We consider three cases:

i $1_{\mathcal{P}_{\mathbf{p}_1,\mathbf{p}_3}} > 1_{\mathcal{P}_{\mathbf{p}_1,\mathbf{p}_2}} + 1_{\mathcal{P}_{\mathbf{p}_2,\mathbf{p}_3}}$
   By (10) and since $\frac{\alpha}{\beta} \leq 1$ by (8),

$$\frac{-2_{\mathcal{P}_{\mathbf{p}_1,\mathbf{p}_3}} + 2_{\mathcal{P}_{\mathbf{p}_1,\mathbf{p}_2}} + 2_{\mathcal{P}_{\mathbf{p}_2,\mathbf{p}_3}}}{1_{\mathcal{P}_{\mathbf{p}_1,\mathbf{p}_3}} - 1_{\mathcal{P}_{\mathbf{p}_1,\mathbf{p}_2}} - 1_{\mathcal{P}_{\mathbf{p}_2,\mathbf{p}_3}}} \geq 1 \geq \frac{\alpha}{\beta}.$$

Rewriting this gives (9).

ii $1_{\mathcal{P}_{\mathbf{P_1},\mathbf{P_3}}} = 1_{\mathcal{P}_{\mathbf{P_1},\mathbf{P_2}}} + 1_{\mathcal{P}_{\mathbf{P_2},\mathbf{P_3}}}$

By (10) and since $\beta > 0$ by (8), (9) is fulfilled.

iii $1_{\mathcal{P}_{\mathbf{P_1},\mathbf{P_3}}} < 1_{\mathcal{P}_{\mathbf{P_1},\mathbf{P_2}}} + 1_{\mathcal{P}_{\mathbf{P_2},\mathbf{P_3}}}$

By (11) and since $\frac{2\alpha}{\beta} \geq 1$ by (8),

$$\frac{-2 \cdot 2_{\mathcal{P}_{\mathbf{P_1},\mathbf{P_3}}} + 2 \cdot 2_{\mathcal{P}_{\mathbf{P_1},\mathbf{P_2}}} + 2 \cdot 2_{\mathcal{P}_{\mathbf{P_2},\mathbf{P_3}}}}{1_{\mathcal{P}_{\mathbf{P_1},\mathbf{P_3}}} - 1_{\mathcal{P}_{\mathbf{P_1},\mathbf{P_2}}} - 1_{\mathcal{P}_{\mathbf{P_2},\mathbf{P_3}}}} \leq 1 \leq \frac{2\alpha}{\beta}.$$

Rewriting this gives (9).

Thus, (9) holds for all possible cases, so $d_{(\alpha,\beta)}(\mathbf{p_1},\mathbf{p_3}; B) \leq d_{(\alpha,\beta)}(\mathbf{p_1},\mathbf{p_2}; B) + d_{(\alpha,\beta)}(\mathbf{p_2},\mathbf{p_3}; B)$.   □

Note that using the natural conditions for the weights α and β, the condition of metricity is exactly the same as for the non-weighted case (see [4]), but the proof is more sophisticated. Thus, to check that a weighted NS distance function is a metric can be done with the same efficiency as for the non-weighted case.

4   Algorithms

In [19], weights and neighbourhood sequences that are well-suited for applications are calculated. By minimizing a number of different error-functions measuring the deviation from the Euclidean distance, "optimal" parameters are obtained. Since we are working in digital grids, using the Euclidean distance leads to some unpleasant properties. For instance, the spheres (points with equal distance value from a given point) by the Euclidean distance are not connected (in any sense), opposite to the spheres defined by NS-distances. The Euclidean distance is not integer-valued, which is often preferred in picture analysis; NS-distances are integer-valued and using weights $\alpha, \beta \in \mathbb{N}$ the weighted NS-distance is also integer-valued. Opposite to the weighted distances, the Euclidean distance is not *regular* on digital grids in the sense of [20]. As mentioned in the introduction, the weighted NS-distance is better suited for the *computation* of the constrained distance transform. Here, we give algorithms for the constrained distance transform and for finding one minimal cost (α, β)-weighted B-path between any two points.

4.1   Algorithm to Find a Minimal Cost-Path between Two Points

The following algorithms construct a shortest weighted NS path from $\mathbf{0} = (0,0,0)$ to the point $\mathbf{p} = (x,y,z)$ with $x \geq y \geq z \geq 0$ in the bcc and fcc grids, respectively. The result is the point sequence $\mathbf{0} = \mathbf{p_0}, \mathbf{p_1}, \ldots, \mathbf{p_m} = \mathbf{q}$ as (one of) the (α, β)-weighted minimal cost B-paths and the distance d between $\mathbf{0}$ and $\mathbf{q}$. The obtained paths satisfy Lemma 1.

Algorithm 1. Constructing a minimal cost (α, β)-weighted B-path on $\mathbb{F}$

Input: NS B and weights α, β, and a grid point $\mathbf{q} \in \mathbb{F}$.

Output: A minimal cost (α, β)-weighted B-path $\mathcal{P}_{\mathbf{0},\mathbf{q}}$ and $d = d_{\alpha,\beta}(\mathbf{0}, \mathbf{q}; B)$.

Initialisation: Set $\mathbf{p}_0 \leftarrow \mathbf{0}, d \leftarrow 0, i \leftarrow 0$.

```
while p_i ≠ q
    i ← i + 1, p_i ← p_{i-1}.
    if (b(i) = 2 and x − x_i > y − y_i + z − z_i)
        p_i ← p_i + (2, 0, 0), d ← d + β
    else d ← d + α.
        if (the vector q − p_i contains at least 2 positive values)
            if (min (x − x_i, y − y_i) > z − z_i)
                p_i ← p_i + (1, 1, 0)
            else if (min (x − x_i, z − z_i) > y − y_i)
                p_i ← p_i + (1, 0, 1)
                else if (min (y − y_i, z − z_i) > x − x_i)
                    p_i ← p_i + (0, 1, 1)
                endif
            endif
        else if (the vector q − p_i contains a negative value)
            p_i ← p_i + (1, 0, −1)
        else
            p_i ← p_i + (1, 0, 1)
        endif
    endif
endwhile
```

Algorithm 2. Constructing a minimal cost (α, β)-weighted B-path on $\mathbb{B}$

Input: NS B and weights α, β, and a grid point $\mathbf{q} \in \mathbb{B}$.

Output: A minimal cost (α, β)-weighted B-path $\mathcal{P}_{\mathbf{0},\mathbf{q}}$ and $d = d_{\alpha,\beta}(\mathbf{0}, \mathbf{q}; B)$.

Initialisation: Set $\mathbf{p}_0 \leftarrow \mathbf{0}, d \leftarrow 0, i \leftarrow 0$.

```
while p_i ≠ q
    i ← i + 1, p_i ← p_{i-1}.
    if (b(i) = 2 and x − x_i > max (y − y_i, z − z_i) )
        p_i ← p_i + (2, 0, 0), d ← d + β
    else d ← d + α
        if (the vector q − p_i contains 3 positive values)
            p_i ← p_i + (1, 1, 1)
        else if (the vector q − p_i contains a negative value)
            p_i ← p_i + (1, −1, 1)
        else
            p_i ← p_i + (1, 1, −1)
        endif
    endif
endwhile
```

Algorithm 3. Computing cDT for weighted NS-distances

Input: B, α, β, an object $S \subset \mathbb{G}$, and source grid points $s \in S$.

Initialisation: Set $DT_{cost}(\mathbf{p}) \leftarrow 0$ for source grid points and $DT_{cost}(\mathbf{p}) \leftarrow \infty$ for remaining grid points. Set $DT_{length} = DT_{cost}$. For all source grid points $\mathbf{p}$: push $(\mathbf{p}, DT_{cost}(\mathbf{p}), DT_{length}(\mathbf{p}))$ to the list $\mathcal{L}$ of ordered triplets sorted by increasing $DT_{length}(\mathbf{p})$.

```
while L is not empty
    forall p in L with lowest DT_length(p)
        Pop (p, DT_cost(p), DT_length(p)) from L
        forall q that are b(DT_length(p) + 1)-neighbours with p:
            if DT_cost(q) ≥ DT_cost(p) + w_p-q
                DT_cost(q) ← DT_cost(p) + w_p-q
                DT_length(q) ← DT_length(p) + 1
                Push (q, DT_cost(q), DT_length(q)) to L
            endif
        endfor
    endfor
endwhile
```

Comment: The following notation is used: $w_{\mathbf{p,q}}$ is α if $\mathbf{p}, \mathbf{q}$ are 1-neighbours and β if $\mathbf{p}, \mathbf{q}$ are strict 2-neighbours. This algorithm works for any pair of non-negative weights α, β.

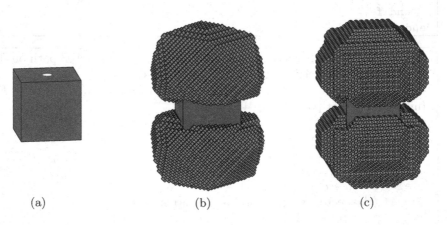

(a)	(b)	(c)

Fig. 2. The constrained distance transform on the fcc and bcc grids. Figure (a) shows the *shape* of the obstacle – a cube with a single through-hole. Inside the cube, there is a source grid point. The grid points with a constrained distance less than a threshold T is shown in (b) and (c) for the fcc and bcc grid, respectively. For the fcc grid, $T = 38$, $B = (1, 2)$, $\alpha = 2$, and $\beta = 3$ were used and for the bcc grid, $T = 110$, $B = (1, 2)$, $\alpha = 5$, and $\beta = 6$ were used.

4.2   Algorithm to Compute the Constrained Distance Transform

In Algorithm 3 below, pseudo-code for computing the cDT for weighted NS-distances is presented. The basic idea is to update grid points that are neighbours with a grid point in the wave-front. This algorithm is presented in [17] for the square grid and similar algorithms are presented in, e.g., [8,21]. When $\alpha \neq \beta$, the algorithm, in addition to the distance (the cost), must keep track of the length of the shortest path to the source grid points. The length is needed to determine which neighbourhood that is allowed in each step. Note that only the object grid points S, and not the set of obstacle grid points, are processed by the algorithm.

5   Conclusions

We have presented sufficient conditions on the weights and neighbourhood sequence for the weighted NS-distance on the fcc and bcc grids to be a metric. The conditions derived here are very natural, for the neighbourhood sequence it is the same condition as metricity for NS-distances (i.e. with unit weights) introduced in [16]. Note that the conditions for metricity on the fcc and bcc grids presented here are the same as for metricity on the square grid, see [17].

The algorithms presented in this paper give different results – Algorithm 1 and 2 compute a minimal cost-path between two points and Algorithm 3 computes the constrained distance transform. The latter have several applications, for example finding the shortest path, see [9,7] and can also be used for shape representation using centres of maximal balls, see e.g. [22].

Acknowledgements

The research is supported by *Digital geometry with applications in image processing in three dimensions (DIGAVIP)* at Center for Image Analysis, Uppsala University.

References

1. Matej, S., Lewitt, R.M.: Efficient 3D grids for image reconstruction using spherically-symmetric volume elements. IEEE Transactions on Nuclear Science 42(4), 1361–1370 (1995)
2. Strand, R., Borgefors, G.: Distance transforms for three-dimensional grids with non-cubic voxels. Computer Vision and Image Understanding 100(3), 294–311 (2005)
3. Fouard, C., Strand, R., Borgefors, G.: Weighted distance transforms generalized to modules and their computation on point lattices. Pattern Recognition 40(9), 2453–2474 (2007)
4. Strand, R., Nagy, B.: Distances based on neighbourhood sequences in non-standard three-dimensional grids. Discrete Applied Mathematics 155(4), 548–557 (2007)

5. Carr, H., Theussl, T., Möller, T.: Isosurfaces on optimal regular samples. In: Bonneau, G.-P., Hahmann, S., C.D.H. (eds.) Proceedings of the symposium on Data visualisation 2003, Eurographics Association, pp. 39–48 (2003)
6. Strand, R., Stelldinger, P.: Topology preserving marching cubes-like algorithms on the face-centered cubic grid. In: Proceedings of 14th International Conference on Image Analysis and Processing (ICIAP 2007), Modena, Italy, pp. 781–788 (2007)
7. Coeurjolly, D., Miguet, S., Tougne, L.: 2D and 3D visibility in discrete geometry: an application to discrete geodesic paths. Pattern Recognition Letters 25(5), 561–570 (2004)
8. Verwer, B.J.H., Verbeek, P.W., Dekker, S.T.: An efficient uniform cost algorithm applied to distance transforms. IEEE Transactions on Pattern Analysis and Machine Intelligence 11(4), 425–429 (1989)
9. Strand, R., Malmberg, F., Svensson, S.: Minimal cost-path for path-based distances. In: Petrou, M., Saramäki, T., Erçil, A., Loncaric, S. (eds.) ISPA 2007, pp. 379–384 (2007)
10. Borgefors, G.: Distance transformations in digital images. Computer Vision, Graphics, and Image Processing 34, 344–371 (1986)
11. Rosenfeld, A., Pfaltz, J.L.: Distance functions on digital pictures. Pattern Recognition 1, 33–61 (1968)
12. Rosenfeld, A., Pfaltz, J.L.: Sequential operations in digital picture processing. Journal of the ACM 13(4), 471–494 (1966)
13. Yamashita, M., Honda, N.: Distance functions defined by variable neighborhood sequences. Pattern Recognition 17(5), 509–513 (1984)
14. Yamashita, M., Ibaraki, T.: Distances defined by neighborhood sequences. Pattern Recognition 19(3), 237–246 (1986)
15. Das, P.P., Chakrabarti, P.P.: Distance functions in digital geometry. Information Sciences 42, 113–136 (1987)
16. Nagy, B.: Distance functions based on neighbourhood sequences. Publicationes Mathematicae Debrecen 63(3), 483–493 (2003)
17. Strand, R.: Weighted distances based on neighbourhood sequences. Pattern Recognition Letters 28(15), 2029–2036 (2007)
18. Strand, R.: Weighted distances based on neighbourhood sequences in non-standard three-dimensional grids. In: Ersbøll, B.K., Pedersen, K.S. (eds.) SCIA 2007. LNCS, vol. 4522, pp. 452–461. Springer, Heidelberg (2007)
19. Strand, R., Nagy, B.: Weighted neighbourhood sequences in non-standard three-dimensional grids – parameter optimization. In: Brimkov, V.E., Barneva, R.P., Hauptman, H. (eds.) IWCIA 2008. LNCS, vol. 4958, pp. 51–62. Springer, Heidelberg (2008)
20. Klette, R., Rosenfeld, A.: Digital geometry: Geometric methods for digital image analysis. The Morgan Kaufmann Series in Computer Graphics. Morgan Kaufmann, San Francisco (2004)
21. Piper, J., Granum, E.: Computing distance transformations in convex and non-convex domains. Pattern Recognition 20(6), 599–615 (1987)
22. Strand, R.: Shape representation with maximal path-points for path-based distances. In: Petrou, M., Saramäki, T., Erçil, A., Loncaric, S. (eds.) ISPA 2007, pp. 397–402 (2007)

Euclidean Eccentricity Transform by Discrete Arc Paving

Adrian Ion[1], Walter G. Kropatsch[1], and Eric Andres[2],*

[1] Pattern Recognition and Image Processing Group,
Faculty of Informatics, Vienna University of Technology, Austria
{ion,krw}@prip.tuwien.ac.at
[2] University of Poitiers
SIC, FRE CNRS 2731, France
andres@sic.sp2mi.univ-poitiers.fr

Abstract. The eccentricity transform associates to each point of a shape the geodesic distance to the point farthest away from it. The transform is defined in any dimension, for simply and non simply connected sets. It is robust to Salt & Pepper noise and is quasi-invariant to articulated motion. Discrete analytical concentric circles with constant thickness and increasing radius pave the 2D plane. An ordering between pixels belonging to circles with different radius is created that enables the tracking of a wavefront moving away from the circle center. This is used to efficiently compute the single source shape bounded distance transform which in turn is used to compute the eccentricity transform. Experimental results for three algorithms are given: a novel one, an existing one, and a refined version of the existing one. They show a good speed/error compromise.

Keywords: eccentricity transform, discrete analytical circles.

1 Introduction

A major task in image analysis is to extract high abstraction level information from an image that usually contains too much information and not necessarily the correct one (e.g. because of noise or occlusion). Image transforms have been widely used to move from low abstraction level input data to a higher abstraction level that forms the output data (skeleton, connected components, etc.). The purpose is to have a reduced amount of (significant) information at the higher abstraction levels. One class of such transforms that is applied to 2D shapes, associates to each point of the shape a value that characterizes in some way it's relation to the rest of the shape, e.g. the distance to some other point of the shape. Examples of such transforms include the well known distance transform [1], which associates to each point of the shape the length of the **shortest** path to the border, the Poisson equation [2], which can be used to associate to each point the **average** time to reach the border by a random path (average

---

* Partially supported by the Austrian Science Fund under grants S9103-N13, P18716-N13. The many fruitful discussions with Samuel Peltier are gratefully acknowledged.

D. Coeurjolly et al. (Eds.): DGCI 2008, LNCS 4992, pp. 213–224, 2008.

length of the random paths from the point to the boundary), and the eccentricity transform [3] which associates to each point the length of the **longest** of the shortest paths to any other point of the shape. Using the transformed images one tries to come up with an abstracted representation, like the skeleton [4] or shock graph [5] build on the distance transform, which could be used in e.g. shape classification or retrieval. Minimal path computation [6] as well as the distance transform [7] are commonly used in 2D and 3D image segmentation.

This paper is focusing on the eccentricity transform (ECC) which has its origins in the graph based eccentricity [8,9]. It has been defined in the context of digital images in [3,10]. It was applied in the context of shape matching in [11]. It has been shown that the transform is very robust with respect to noise. The eccentricity transform can be defined for discrete objects of any dimension, closed (e.g. typical 2D binary image) or open sets (surface of an ellipsoid), and for continuous objects of any dimension (e.g. 3D ellipsoid or the 2D surface of the 3D ellipsoid, etc.). For the case of discrete shapes, a naive algorithm and a more efficient one for 2D shapes without holes, have been presented in [3], with experimental results only for the 4 and 8 neighbourhoods. For simply connected shapes on the hexagonal and dodecagonal grid, an efficient algorithm was given in [12]. Regarding continuous shapes, a detailed study has been made for the case of an ellipse, and some preliminary properties regarding rectangles, and a class of elongated convex shapes, have been given [13]. An algorithm for finding the eccentric/furthest points for the vertices of a simple polygon was given in [14].

This paper presents an algorithm for efficiently computing the single source shape bounded distance transform using discrete circles. The idea is to use discrete circles to propagate the distance in a shape following an idea already proposed for discrete wave propagation simulation [15]. In addition, a novel algorithm and a refined one are given to compute the eccentricity transform. It has been shown that, so called, eccentric points play a major role in the transform. These points are shape border points. Different ideas are proposed in order to attempt to identify these eccentric points. Distance transforms, originating at these candidate eccentric points, are computed and accumulated to obtain the eccentricity transform. A comparison between these three methods is provided.

Section 2 gives a short recall of the eccentricity transform, the main properties relevant for this paper and gives the important facts about discrete circles. Sections 3 and 4 present the proposed algorithms, followed by experimental results. Section 5 concludes the paper and gives an outlook of the future work.

2   Recall

In this section basic definitions and properties of the eccentricity transform and discrete circles are given.

2.1   Recall ECC

The following definitions and properties follow [3,11].

Let the shape S be a closed set in $\mathbb{R}^2$ and ∂S be its border[1]. A (geodesic) path π is the continuous mapping from the interval $[0,1]$ to S. Let $\Pi(\mathbf{p_1}, \mathbf{p_2})$ be the set of all paths between two points $\mathbf{p_1}, \mathbf{p_2} \in S$ within the set S. The geodesic distance $d(\mathbf{p_1}, \mathbf{p_2})$ between two points $\mathbf{p_1}, \mathbf{p_2} \in S$ is defined as the length $\lambda(\pi)$ of the shortest path $\pi \in \Pi(\mathbf{p_1}, \mathbf{p_2})$ between $\mathbf{p_1}$ and $\mathbf{p_2}$

$$d(\mathbf{p_1}, \mathbf{p_2}) = \min\{\lambda(\pi(\mathbf{p_1}, \mathbf{p_2})) | \pi \in \Pi\} \text{ where } \lambda(\pi(t)) = \int_0^1 |\dot{\pi}(t)| dt \qquad (1)$$

where $\pi(t)$ is a parametrization of the path from $\mathbf{p_1} = \pi(0)$ to $\mathbf{p_2} = \pi(1)$.

The eccentricity transform of S can be defined as, $\forall \mathbf{p} \in S$

$$ECC(S, \mathbf{p}) = \max\{d(\mathbf{p}, \mathbf{q}) | \mathbf{q} \in S\} \qquad (2)$$

i.e. to each point $\mathbf{p}$ it assigns the length of the shortest geodesics to the points farthers away from it. In [3] it is shown that this transformation is quasi-invariant to articulated motion and robust against salt and pepper noise (which creates holes in the shape).

This paper considers defined by points on a square grid $\mathbb{Z}^2$. Paths need to be contained in the area of $\mathbb{R}^2$ defined by the union of the support squares for the pixels of S. The distance between any two pixels whose connecting segment is contained in S is computed using the ℓ_2-norm.

2.2 Properties of Eccentric Points

In general, an *extremal point* is a point where a function reaches an extremum (local or global). In the case of the geodesic distance d in a shape S we call a point $\mathbf{x} \in S$ *maximal* iff $d(\mathbf{x}, \mathbf{p})$ is a local maximum for a given point $\mathbf{p} \in S$. $X(S)$ denotes the set of all maximal points of shape S.

An *eccentric point* of a shape S is a point $\mathbf{e} \in S$ that is farthest away in S from at least one other point $\mathbf{p} \in S$ i.e. $\exists \mathbf{p} \in S$ s.t. $ECC(S, \mathbf{p}) = d(\mathbf{p}, \mathbf{e})$. For a shape S, $E(S) = \{\mathbf{e} \in S | \mathbf{e}$ is eccentric for at least one point of $S\}$ denotes the set of all its eccentric points. The set of eccentric points $E(S)$ is a subset of the set of all maximal points $X(S)$ i.e. $E(S) \subseteq X(S)$ (eccentric points are global maxima for d, while maximal points are local maxima for a given point $\mathbf{p}$).

Knowing $E(S)$ can speedup the computation of the $ECC(S)$. A naive algorithm for $ECC(S)$ computes all geodesic distances $d(\mathbf{p}, \mathbf{q})$ for all pairs of points in S and takes the maximum at each point $\mathbf{p} \in S$. Since the geodesic distance is commutative, e.g. $d(\mathbf{p}, \mathbf{q}) = d(\mathbf{q}, \mathbf{p})$, it is sufficient to restrict $\mathbf{q}$ in (2) to the eccentric points $E(S)$. Instead of computing the length of the geodesics from $\mathbf{p} \in S$ to all the other points $\mathbf{q} \in S$ and taking the maximum, one can compute the length of the geodesics from all $\mathbf{q} \in E(S)$ to all the points $\mathbf{p} \in S$. This reduces the number of steps from $|S|$ to $|E(S)|$. The key question is therefore how to estimate $E(S)$ without knowing $ECC(S)$.

The following properties of eccentric points are relevant for this paper and concern bounded 2D shapes.

---

[1] This definition can be generalized to any dimension, continuous and discrete objects.

Property 1. All eccentric points $E(\mathcal{S})$ of a shape $\mathcal{S}$ lie on the border of $\mathcal{S}$ i.e. $E(\mathcal{S}) \subseteq \partial \mathcal{S}$. (*Proof* due to [3]).

Property 2. Being an eccentric point is not a local property i.e. $\forall B \subset \partial \mathcal{S}$ a boundary part (a 2D open and simple curve), and a point $\mathbf{b} \in B$, we can construct the rest of the boundary $\partial \mathcal{S} \setminus B$ s.t. $\mathbf{b}$ is not an eccentric point of $\mathcal{S}$. (*Proof* due to [13]).

Property 3. Not all eccentric points $\mathbf{e} \in E(\mathcal{S})$ are local maxima in the eccentricity transform (the ellipse in [13] is an example).

E.g. some eccentric points $\mathbf{e} \in E(\mathcal{S})$ may find larger eccentricity values $ECC(\mathcal{S}, \mathbf{e})$ $< ECC(\mathcal{S}, \mathbf{q})$ in their neighborhood $|\mathbf{e} - \mathbf{q}| < \epsilon$ for $\epsilon > 0$. They typically form connected clusters containing also a local maximum in $ECC(\mathcal{S})$.

Property 4. Not all eccentric points are local maxima in the distance transform from any point of the shape i.e. there is no guarantee that all eccentric points of a shape $\mathcal{S}$ will be local maxima in the distance transform $DT(\mathcal{S}, \mathbf{p})$ for any random point $\mathbf{p} \in \mathcal{S}$ (see [13] for an example).

2.3   Recall Discrete Circles

The shape bounded distance transform algorithm presented in this paper propagates the distance computation in the shape. This idea has already been used to perform discrete wave propagation simulation [15]. The propagation is performed with discrete concentric circles where each circle corresponds to a given distance range. The propagation provides a cheap distance computation. The discrete circle definition has to verify specific properties as the center coordinates and the radius aren't necessarily integers. The most important property is that circles centered on a point must fill, preferably pave, the discrete space. We can't miss points during our propagation phase. Of course, the arc/circle generation algorithm has to be linear in the number of pixels generated or we would loose the whole point of proposing a new method with a reasonable complexity. There exists several different discrete circle definitions. The best known circle is an algorithmic definition proposed by Bresenham but this doesn't correspond to the requirements of our problem. The circle doesn't pave the space, the center coordinates and radius are only integer. The definition that fits our problem is the discrete analytical circle proposed by Andres [16]:

Definition 1. *(Discrete Analytical Circle) A discrete analytical circle $C(R, \mathbf{o})$ is defined by the double inequality:*

$$\mathbf{p} \in C(R, \mathbf{o}) \;\; iff \;\; \left(R - \frac{1}{2}\right) \le \|\mathbf{p} - \mathbf{o}\| < \left(R + \frac{1}{2}\right)$$

with $\mathbf{p} \in \mathbb{Z}^2, \mathbf{o} \in \mathbb{R}^2$, and $R \in \mathbb{R}$ the radius using euclidean distance.

This circle is a circle of arithmetical thickness 1. The circle definition is similar to the discrete analytical line definition proposed by Reveillès [17].

The fact that this circle definition is analytical, and not just algorithmic like Bresenham's circle, has many advantages. Point localisation, i.e. knowing if a point is inside, outside or on the circle, is trivial. The center coordinates and the radius don't have to be integers. The circle definition can easily be extended to any dimension (see [18]). Circles with the same center pave the space:

$$\biguplus_{R=0}^{\infty} C(R, \mathbf{o}) = \mathbb{Z}^2$$

Each integer coordinate point in space belongs to one and only one circle. When you draw Bresenham circles with the same center and increasing radii there are discrete points that don't belong to any of the concentric circles. What is less obvious is that this circle has good topological properties. In fact, as shown in [18] for the general case in dimension n for discrete analytical hyperspheres, a discrete circle of arithmetical thickness equal to 1 (this is the case with the given definition) is at least 8-connected. The discrete points of the circle can be ordered and there exists a linear complexity generation algorithm [16,18].

3   Shape Bounded Single Source Distance Transform

Given a discrete connected shape $\mathcal{S}$ and an initial point $\mathbf{o} \in \mathcal{S}$. The shape bounded single source distance transform DT assigns to every point $\mathbf{q} \in \mathcal{S}$ its geodesic distance to $\mathbf{o}$:

$$DT(\mathcal{S}, \mathbf{o}) = \{d(\mathbf{o}, \mathbf{q})\}, \quad \mathbf{q} \in \mathcal{S} \tag{3}$$

Another formulation would consider the time needed for a wavefront initiated at $\mathbf{o}$ traveling in the homogeneous medium $\mathcal{S}$ to reach each point $\mathbf{q}$ of $\mathcal{S}$. Following the previous formulation we propose to model the wavefront using discrete arcs and record the time when each point $\mathbf{q}$ is reached for the first time. The wavefront travels with speed 1 i.e. 1 distance unit corresponds to 1 time unit.

The wavefront propagating from a point $\mathbf{o}$ will have the form of a circle centered at $\mathbf{o}$. If the wavefront is blocked by obstacles, the circle is "interrupted" and disjoint arcs of the same circle continue propagating in the unblocked directions. For a start point $\mathbf{o}$, the wavefront at time t is the set of points $\mathbf{q} \in \mathcal{S}$ at distance $t - 0.5 \leqslant d(\mathbf{o}, \mathbf{q}) < t + 0.5$. The wavefront at any time t consists of a set of arcs $W(t) \subset \mathcal{S}$. Each arc $A \in W(t)$ lies on a circle centered at the point where the path from $\mathbf{q} \in A$ to $\mathbf{p}$ first touches $\partial \mathcal{S}$.

The computation starts with a circle of radius 1 centered at the source point $\mathbf{o}$, $W(1) = \{C(1, \mathbf{o})\}$. It propagates and clusters as presented above, with the addition that pixels with distance smaller than the current wavefront also block the propagation. An arc $A \in W(t)$ of the wavefront $W(t)$, not touching $\partial \mathcal{S}$ at time t, but touching at $t-1$, diffracts new circles centered at the endpoints $\mathbf{e} \in A$. The added arcs start with radius one and handicap $d(\mathbf{p}, \mathbf{e})$. The handicap of an

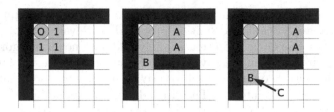

Fig. 1. The three steps during wavefront propagation (shape white, background black). Left, radius 1: circle with radius 1 and center O is bounded to an arc. Middle, radius 2: the front is splitted in two arcs A, B. Right, radius 3: arc B, touching the hole at radius 2 but not at radius 3, creates arc C with the center at the current point of B.

arc accumulates the length of the shortest path to the center of the arc such that the distance of the wavefront from the initial point o is always the sum of the handicap and the radius of the arc. No special computation is required for the initial angles of arcs with centers on the boundary pixels of S. They will be corrected by the clustering and bounding command when drawing with radius 1. See Fig. 1 and Alg. 1.

The complexity of Alg. 1 is determined by the number of pixels in S, denoted $|S|$, and the number of arcs of the wave. Arcs are drawn in $O(n)$ where n is the number of pixels of the arc. Pixels in the shadow of a hole, where different parts of the wavefront meet ('shocks'), are drawn by each part. All other pixels are drawn only once. Adding and extracting arcs to/from the wavefront (W in Alg. 1) can be done in $\log(\text{size}(W))$.

For convex shapes, the size of W is 1 all the time, so the algorithm executes in $O(|S|)$. For simply connected shapes, each pixel is drawn only once. Assuming an exaggerated upper bound $\text{size}(W) = |S|$ and each arc only draws one pixel, the complexity for simply connected shapes is below $O(|S| \log |S|)$. Each hole creates an additional direction to reach a point, e.g. no hole: 1 direction; 1 hole: 2 directions - one on each side of the hole; 2 holes: maximum 3 directions - one side, between holes, other side, etc. Note that we don't count the number of possible paths, but the number of directions from which connected wavefronts can reach a point at the shortest geodesic distance. For a non-simply connected shape with k holes, a pixel is set a maximum of k times (worst case). Thus, the complexity for non-simply connected shapes with k holes is below $O(k|S| \log |S|)$.

4   Progressive Refinement Eccentricity Transform

In [3] an algorithm for approximating the ECC of discrete shapes was presented (will be denoted by ECC06) (see Alg. 2). The algorithm is faster than the naive one (see [3]), and computes the exact values for a class of simply connected shapes. For all other shapes it gives an approximation. Based on Properties 3 and 4 we refine algorithm ECC06 by adding a third phase to ECC06. This step finds the limits of clusters of eccentric points for which at least one eccentric point

---

Algorithm 1. $DT(\mathcal{S}, \mathbf{p})$ - Compute distance transform using discrete circles.

Input: Discrete shape $\mathcal{S}$ and pixel $\mathbf{p} \in \mathcal{S}$.

1: **for all** $\mathbf{q} \in \mathcal{S}$ **do** $D(\mathbf{q}) \leftarrow \infty$ /*initialize distance matrix*/
2: $D(\mathbf{p}) \leftarrow 0$
3: $W \leftarrow \text{Arc}(\mathbf{p}, 1, [0; 2\pi], 0, \emptyset)$ /*Arc(center, radius, angles, handicap, parent)*/
4:
5: **while** $W \neq \emptyset$ **do**
6:     $A \leftarrow \arg\min\{A.r + A.h | A \in W\}$ /*select and remove arc with smallest radius+handicap*/
7:
8:     /*draw arc points with lower distance than known before, use real distances*/
9:     $D(\mathbf{m}) \leftarrow \min\{D(\mathbf{m}), A.h + d(A.\mathbf{c}, \mathbf{m}) | \mathbf{m} \in A \cap \mathcal{S}\}$
10:
11:     $P_1, \ldots, P_k \leftarrow$ actually drawn (sub)arcs/parts of A /*split and bound*/
12:     $W \leftarrow W + \text{Arc}(A.\mathbf{c}, A.r + 1, P_i.\mathbf{a}, A.h, A), \forall i = 1..k$ /*propagate*/
13:
14:     /*diffract if necessary*/
15:     **if** $A.\mathbf{p}$ touches $\partial\mathcal{S}$ on either side **then**
16:         $\mathbf{e} \leftarrow$ last point of A, on side where $A.\mathbf{p}$ was touching $\partial\mathcal{S}$
17:         $W \leftarrow W + \text{Arc}(\mathbf{e}, 1, [0; 2\pi], D(\mathbf{e}), A)$
18:     **end if**
19: **end while**

Output: Distances D.

---

has been found (this is phase 2, lines 19-28 of Alg. 3). The refined algorithm is denoted by ECC06'.

Algorithms ECC06 and ECC06' try to identify the ECC centers (smallest ECC value). Computing the $DT(\mathcal{S}, \mathbf{c})$ for a center point $\mathbf{c}$ is expected to create local maxima where eccentric points lie. For non-simply connected shapes the center can become very complex, it can contain many points and it can be disconnected. This makes identifying all center points harder, as not all eccentric points are farthest away from all center points. Missing center points can lead to missing eccentric points which leads to an approximation errors.

4.1 New Algorithm (ECC08)

Algorithm ECC08 first attempts to identify at least one point from each cluster of eccentric points. Like in ECC06', scanning along $\partial\mathcal{S}$ is then used to find the limits of each cluster.

Inspecting a shape $\mathcal{S}$ is done by repeatedly computing $DT(\mathcal{S}, \mathbf{q})$ for the highest local maximum in the current approximation of $ECC(\mathcal{S})$ and accumulating the results (using max) to obtain the next approximation. After some steps, this process can enter an infinite loop if all reachable points have been visited already. Such a configuration is called an oscillating configuration and the visited points are called oscillating points [19]. If $DT(\mathcal{S}, \mathbf{q})$ with $\mathbf{q} \in \mathcal{S}$ is considered as an approximation for $ECC(\mathcal{S})$, the error is expected to be higher around $\mathbf{q}$

Algorithm 2. $ECC06(\mathcal{S})$ - Eccentricity transform by progressive refinement.

Input: Discrete shape $\mathcal{S}$.

1: **for all** $q \in \mathcal{S}, ECC(q) \leftarrow 0$ /*initialize distance matrix*/
2: $p \leftarrow$ random point of $\mathcal{S}$ /*find a starting point*/
3:
4: /*Phase 1: find a diameter*/
5: **while** p not computed **do**
6:     $ECC \leftarrow \max\{ECC, DT(\mathcal{S}, \mathbf{p})\}$ /*accumulate & mark **p** as computed*/
7:     $\mathbf{p} \leftarrow \arg\max\{ECC(\mathbf{p})|\mathbf{p} \in \mathcal{S}\}$ /*highest current ECC (farthest away)*/
8: **end while**
9:
10: /*Phase 2: find center points and local maxima*/
11: $pECC \leftarrow 0$ /*make sure we enter the loop*/
12: **while** $pECC \neq ECC$ **do**
13:     $pECC \leftarrow ECC$
14:     $C \leftarrow \arg\min\{ECC(\mathbf{p})|\mathbf{p} \in \mathcal{S}\}$ /*find all points with minimum ECC*/
15:     **for all** $\mathbf{c} \in C$, **c** not computed **do**
16:         $D \leftarrow DT(\mathcal{S}, \mathbf{c})$ /*do a distance transform from the center*/
17:         $ECC \leftarrow \max\{ECC, D\}$ /*accumulate & mark **c** as computed*/
18:
19:         $M \leftarrow \{\mathbf{q} \in \mathcal{S}|D(\mathbf{q})$ local maximum in $\mathcal{S}$ & **q** not computed$\}$
20:         **for all** $\mathbf{m} \in M$, **m** not computed **do**
21:             $ECC \leftarrow \max\{ECC, DT(\mathcal{S}, \mathbf{m})\}$ /*accumulate & mark **m** as computed*/
22:         **end for**
23:     **end for**
24: **end while**

Output: Distances ECC.

---

and smaller around the points farther away from **q** i.e. the points with highest values in $DT(\mathcal{S}, \mathbf{q})$. Whenever an oscillating configuration is reached all points of $\partial \mathcal{S}$ which are local minima in the current ECC approximation are selected for distance computation. If the last operation does not produce any unvisited points as ECC local maxima, the search is terminated (Alg. 3).

All three algorithms try to find $E(\mathcal{S})$ and compute $DT(\mathcal{S}, \mathbf{e})$ for all **e** in the current approximation of $E(\mathcal{S})$. As $E(\mathcal{S})$ is actually known only after computing $ECC(\mathcal{S})$, all algorithms incrementally refine an initial approximation of $ECC(\mathcal{S})$ by computing $DT(\mathcal{S}, \mathbf{q})$ for candidate eccentric points **q** that are identified during the progress of the approximation. For ECC06' and ECC08, one eccentric point per cluster is sufficient to produce the correct result.

4.2   Experimental Results

We have compared the three algorithms, ECC06, ECC06', and ECC08, on 70 shapes from the MPEG7 CE-Shape1 database [20], 6 from [21], and one additional new shape (see Table 5).

Algorithm 3. $ECC08(\mathcal{S})$ - Eccentricity transform by progressive refinement.

Input: Discrete shape $\mathcal{S}$.

1: **for all** $\mathbf{q} \in \mathcal{S}, ECC(\mathbf{q}) \leftarrow 0$ /*initialize distance matrix*/
2: $ToDo \leftarrow$ random point of $\mathcal{S}$ /*find a starting point*/
3:
4: /*Phase 1: inspect shape*/
5: **while** $ToDo \neq \emptyset$ **do**
6:    $\mathbf{p} \leftarrow \arg\max\{ECC(\mathbf{p})|\mathbf{p} \in ToDo\}$ /*remove point with highest current ECC*/
7:    $ECC \leftarrow \max\{ECC, DT(\mathcal{S}, \mathbf{p})\}$ /*accumulate & mark $\mathbf{p}$ as computed*/
8:
9:    /*add not computed local maxima to ToDo*/
10:    $ToDo \leftarrow ToDo \cup \{\mathbf{q} \in \mathcal{S}|ECC(\mathbf{q})$ local maximum in $\mathcal{S}$ & $\mathbf{q}$ not computed$\}$
11:
12:    /*test if an oscillating configuration was found*/
13:    **if** $ToDo = \emptyset$ **then**
14:       $ToDo \leftarrow ToDo \cup \{\mathbf{q} \in \partial\mathcal{S}|ECC(\mathbf{q})$ local minimum in $\partial\mathcal{S}$ & $\mathbf{q}$ not computed$\}$
15:    **end if**
16: **end while**
17:
18: /*Phase 2: find limits of clusters of eccentric points */
19: $ToDo \leftarrow$ all neighbours in $\partial\mathcal{S}$ of all eccentric points in ECC
20: **while** $ToDo \neq \emptyset$ **do**
21:    $\mathbf{p} \leftarrow \arg\max\{ECC(\mathbf{p})|\mathbf{p} \in ToDo\}$ /*remove point with highest current ECC*/
22:    $ECC \leftarrow \max\{ECC, DT(\mathcal{S}, \mathbf{p})\}$ /*accumulate & mark $\mathbf{p}$ as computed*/
23:
24:    /*do we need to continue in this direction?*/
25:    **if** ECC changed previously i.e. $\mathbf{p}$ is an eccentric point **then**
26:       $ToDo \leftarrow ToDo \cup \{\mathbf{q} \in \partial\mathcal{S}|\mathbf{q}$ is a neighbour of $\mathbf{p}$ in $\partial\mathcal{S}$ & $\mathbf{q}$ not computed$\}$
27:    **end if**
28: **end while**

Output: Distances ECC.

The MPEG7 database contains 1400 shapes from 70 object classes. One shape from each class was taken (the first one) and reduced to about 36,000 pixels (aspect ratio and connectivity preserved). Table 1 summarizes the main characteristics of the 70 shapes, their sizes and the range of smallest and largest eccentricity values. The smallest eccentricity appears at the center of the shape and the largest eccentricity corresponds to its diameter.

Correct ECC values are computed by the naive algorithm RECC as the maximum of the distance transforms of all boundary points $\partial\mathcal{S}$ (according to Property 1). Table 2 compares the performance of the 3 algorithms:

max.pixel error: maximum difference between RECC and ECC per pixel / max.eccentricity for this shape;

max.error size: maximum number of pixels that differ between RECC and ECC / size of this shape;

Table 1. Characteristics of shapes from the MPEG7 database

measure	ranges from	to
sizes in pixel	683	28821
smallest eccentricity in pixel (ECC$_{min}$)	28	235
maximum eccentricity in pixel (ECC$_{max}$)	55	469

Table 2. Results of 70 images from the MPEG7 database

measure	ECC06	ECC06'	ECC08
max.pixel error	4.45 / 221.4	4.27 / 221.4	6.57 / 266.6
max.error size	4923 / 19701	2790 / 19701	2359 / 19701
#DT(ECC) / #DT(RECC)	8%	10%	15%
100% correct	44 / 70	60 / 70	56 / 70

Table 3. 'Worst' results from the MPEG7 database

nb.	name	ECC$_{min}$	ECC$_{max}$	size	ECC06	ECC06'	ECC08	ECC06	ECC06'	ECC08
	shape characteristics				max.ECC.diff.			size of ECC.diff.		
58	pocket	170.7	266.6	13815	3.750	0.000	6.568	2241	0	1318
48	hat	126.5	221.4	19701	4.454	4.274	4.274	4923	2790	2359
5	Heart	108.1	213.4	24123	2.784	0.731	0.731	2378	482	482
4	HCircle	127.0	250.2	28821	0.000	0.000	1.454	0	0	404
18	cattle	99.6	198.2	9764	1.223	1.223	1.223	2154	258	258
11	bird	116.0	230.1	14396	1.209	0.000	0.000	3963	0	0

#DT(ECC) / #DT(RECC): average number of times the distance transform(DT) is called wrt. RECC (in percent);

100% correct: the number of shapes for which the error was 0 for all pixels / the total number of shapes.

All three algorithms produce a good ECC approximation in about 8% to 15% of the time of RECC. There are only a few shapes for which the approximation was not 100% correct and the highest difference was about 7 pixels less than correct eccentricity in an image where the eccentricities varied from 170 to 266.6 pixels. Table 3 lists the 6 worst results with the three algorithms. Each shape is characterized by its number, its name, the range of eccentricities of RECC and the number of pixels (size). The next columns list the largest difference in eccentricity value and the number of pixels that were different. To judge the quality of the results we selected the example *hat* which had errors in all three algorithms (Fig. 2 shows the results by a contour line plot with the same levels). Algorithms ECC06' and ECC08 compute the correct eccentricity transform for all of the "problem" shapes showing the improvement of the discrete arc paving with respect to 4- and 8-connectivity used in [21] (see Table 4).

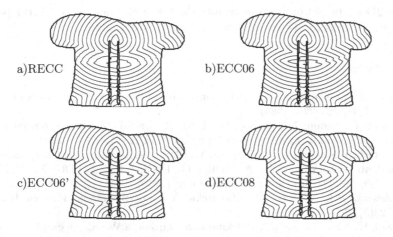

a)RECC   b)ECC06   c)ECC06'   d)ECC08

Fig. 2. Results of example shape *hat*

Table 4. Results on the 6 "problem" shapes from [21]

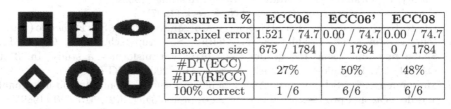

measure in %	ECC06	ECC06'	ECC08
max.pixel error	1.521 / 74.7	0.00 / 74.7	0.00 / 74.7
max.error size	675 / 1784	0 / 1784	0 / 1784
$\frac{\#DT(ECC)}{\#DT(RECC)}$	27%	50%	48%
100% correct	1 /6	6/6	6/6

Table 5. Results for image '3holes'

measure	ECC06	ECC06'	ECC08
max.pixel error	1.913 /409.2	1.100 / 409.2	12.773 / 409.2
max.error size	360 / 19919	119 / 19919	698 / 19919
$\frac{\#DT(ECC)}{\#DT(RECC)}$	7%	8%	9%

Table 5 shows the results of the three algorithms on an example 2D shape. On this example ECC06 and ECC06' produce better results than ECC08.

Overall ECC06' produces the best results with a computation speed between ECC06 and ECC08. ECC06 is the fastest in this experiment.

5 Conclusion

This paper presents a method for efficiently computing the shape bounded distance transform using discrete circles, which in turn is used by the three approximation strategies: ECC08, a novel one; ECC06, an existing one; and ECC06', a refined version of ECC06. They all approximate the eccentricity transform of a discrete 2D shape. Experimental results show the excellent speed/error performance. Extensions to 3D and higher dimensions are planned.

References

1. Rosenfeld, A.: A note on 'geometric transforms' of digital sets. Pattern Recognition Letters 1(4), 223–225 (1983)
2. Gorelick, L., Galun, M., Sharon, E., Basri, R., Brandt, A.: Shape representation and classification using the poisson equation. CVPR (2), 61–67 (2004)
3. Kropatsch, W.G., Ion, A., Haxhimusa, Y., Flanitzer, T.: The eccentricity transform (of a digital shape). In: Kuba, A., Nyúl, L.G., Palágyi, K. (eds.) DGCI 2006. LNCS, vol. 4245, pp. 437–448. Springer, Heidelberg (2006)
4. Ogniewicz, R.L., Kübler, O.: Hierarchic Voronoi Skeletons. Pattern Recognition 28(3), 343–359 (1995)
5. Siddiqi, K., Shokoufandeh, A., Dickinson, S., Zucker, S.W.: Shock graphs and shape matching. International Journal of Computer Vision 30, 1–24 (1999)
6. Paragios, N., Chen, Y., Faurgeras, O.: Handbook of Mathematical Models in Computer Vision, vol. 06, pp. 97–111. Springer, Heidelberg (2006)
7. Soille, P.: Morphological Image Analysis. Springer, Heidelberg (1994)
8. Harary, F.: Graph Theory. Addison-Wesley, Reading (1969)
9. Diestel, R.: Graph Theory. Springer, New York (1997)
10. Klette, R., Rosenfeld, A.: Digital Geometry. Morgan Kaufmann, San Francisco (2004)
11. Ion, A., Peyré, G., Haxhimusa, Y., Peltier, S., Kropatsch, W.G., Cohen, L.: Shape matching using the geodesic eccentricity transform - a study. In: 31st OAGM/AAPR, Schloss Krumbach, Austria, May 2007, OCG (2007)
12. Maisonneuve, F., Schmitt, M.: An efficient algorithm to compute the hexagonal and dodecagonal propagation function. Acta Stereologica 8(2), 515–520 (1989)
13. Ion, A., Peltier, S., Haxhimusa, Y., Kropatsch, W.G.: Decomposition for efficient eccentricity transform of convex shapes. In: Kropatsch, W.G., Kampel, M., Hanbury, A. (eds.) CAIP 2007. LNCS, vol. 4673, pp. 653–660. Springer, Heidelberg (2007)
14. Suri, S.: The all-geodesic-furthest neighbor problem for simple polygons. In: Symposium on Computational Geometry, pp. 64–75 (1987)
15. Mora, F., Ruillet, G., Andres, E., Vauzelle, R.: Pedagogic discrete visualization of electromagnetic waves. In: Eurographics poster session, EUROGRAPHICS, Granada, Spain (January 2003)
16. Andres, E.: Discrete circles, rings and spheres. Computers & Graphics 18(5), 695–706 (1994)
17. Reveillès, J.-P.: Géométrie Discrète, calcul en nombres entiers et algorithmique (in french). University Louis Pasteur of Strasbourg (1991)
18. Andres, E., Jacob, M.A.: The discrete analytical hyperspheres. IEEE Trans. Vis. Comput. Graph. 3(1), 75–86 (1997)
19. Schmitt, M.: Propagation function: Towards constant time algorithms. In: Acta Stereologica: Proceedings of the 6th European Congress for Stereology, Prague, September 7-10, vol. 13(2) (December 1993)
20. Latecki, L.J., Lakämper, R., Eckhardt, U.: Shape descriptors for non-rigid shapes with a single closed contour. In: CVPR 2000, Hilton Head, SC, USA, June 13-15, pp. 1424–1429 (2000)
21. Flanitzer, T.: The eccentricity transform (computation). Technical Report PRIP-TR-107, PRIP, TU Wien (2006)

Statistical Template Matching under Geometric Transformations

Alexander Sibiryakov

Mitsubishi Electric ITE-BV, Guildford, United Kingdom
Alexander.Sibiryakov@vil.ite.mee.com

Abstract. We present a novel template matching framework for detecting geometrically transformed objects. A template is a simplified representation of the object of interest by a set of pixel groups of any shape, and the similarity between the template and an image region is derived from the F-test statistic. The method selects a geometric transformation from a discrete set of transformations, giving the best statistical independence of such groups Efficient matching is achieved using 1D analogue of integral images - integral lines, and the number of operations required to compute the matching score is linear with template size, comparing to quadratic dependency in conventional template matching. Although the assumption that the geometric deformation can be approximated from discrete set of transforms is restrictive, we introduce an adaptive subpixel refinement stage for accurate matching of object under arbitrary parametric 2D-transformation. The parameters maximizing the matching score are found by solving an equivalent eigenvalue problem. The methods are demonstrated on synthetic and real-world examples and compared to standard template matching methods.

1 Introduction

Template matching (TM) is a standard computer vision tool for finding objects or object parts in images. It is used in many applications including remote sensing, medical imaging, and automatic inspection in industry. The detection of real-word objects is a challenging problem due to the presence of illumination and color changes, partial occlusions, noise and clutter in the background, and dynamic changes in the object itself.

A variety of template matching algorithms have been proposed, ranging from extremely fast computing simple rectangular features [1,2] to fitting rigidly or non-rigidly deformed templates to image data [3,4,13].

The exhaustive search strategy of template matching is the following: for every possible location, rotation, scale, or other geometric transformation, compare each image region to a template and select the best matching scores. This computationally expensive approach requires $O(N_l N_g N_t)$ operations, where N_l is the number of locations in the image, N_p is the number of transformation samples, and N_t is the number of pixels used in matching score computation. Many methods try to reduce the computational complexity. N_l and N_g are usually reduced by the multiresolution approach (*e.g.*,[4]) or by projecting the template and image patches onto a rotation-invariant basis [12], but while excluding rotation, these projection-based methods still need a strategy of scale selection. Often the geometric transformations are not included in the matching strategy at all, assuming that the template and the image patch differ by translation only [11].

D. Coeurjolly et al. (Eds.): DGCI 2008, LNCS 4992, pp. 225–237, 2008.

Another way to perform TM is direct fitting of the template using gradient descent or ascent optimization methods to iteratively adjust the geometric transformation until the best match is found [10]. These techniques need initial approximations that are close to the right solution.

In rapid TM methods [1,2,5,6,7], the term N_t in the computational complexity defined above is reduced by template simplification, *e.g.,* by representing the template as a combination of rectangles. Using special image preprocessing techniques (so-called integral images) and computing a simplified similarity score (the normalized contrast between "positive" and "negative" image regions defined by the template), the computat- ional speed of rapid TM is independent of the template size and depends only on the template complexity (the number of rectangles comprising the template). However, such Haar-like features are not rotation-invariant, and a few extensions [5-7] of this framework have been proposed to handle the image rotation. [5] proposed additional set diagonal rectangular templates. [6] proposed 45° twisted Haar-like features computed via 45° rotated integral images. [7] further extended this idea and used multiple sets of Haar-like features and integral images rotated by whole integer-pixel based rotations.

Rapid TM framework has a few implicit drawbacks, which are not presented in computationally expensive correlation-based TM methods:

- It is not easy to generalize two-region Haar-like features to the case of three or more pixel groups.
- Rectangle-based representation is redundant for curvilinear object shapes, *e.g.* circles. Usage of the curved templates instead of the rectangular ones should result in such cases in higher matching scores and, therefore, in better detector performance.
- Impressive results with Haar-like features were achieved by using powerful classifiers based on boosting [1]. They require training on large databases and, therefore, matching using single object template (achievable at no additional cost in correlation-based TM using grayscale template) cannot be easily performed in this framework, or it can be performed only for objects having simple shape and bimodal intensity distribution.

This paper proposes a new approach that can be placed in between rapid TM methods and standard correlation-based TM methods. The proposed approach solves all limitations listed above and can also be extended to an iterative refinement framework for precise estimation of object location and transformation. The method is based on *Statistical Template Matching* (STM), first introduced in [2]. STM framework is very similar to rapid TM framework discussed above; the main difference is that STM uses a different matching score derived from the F-test statistic, supporting multiple pixel groups. The STM method is overviewed in Section 2. Section 3 presents a new extension of STM for rotated and scaled objects based on *integral lines*. Although 1D integral lines technique may seem an obvious and particular case of 2D integral images, the area of applicability of the proposed template matching is much wider when using integral lines, than when using integral images. The integral images technique requires object shape combined of rectan- gles, whereas integral lines method requires just a combination of line segments, which is obviously more general case, because any rasterized 2D shape can be represented as a combination of segments. Section 4 presents another new extension, Adaptive Subpixel (AS) STM, suitable for accurate estimation of parametric 2D-transformation of the object. An efficient solution for a particular case of Haar-like templates is given. Section 5 demonstrates the methods in a few computer vision tasks. Section 6 concludes the paper.

2 Statistical Template Matching

The name *Statistical Template Matching* originates from the fact that only statistical characteristics of pixel groups, such as mean and variance, are used in the analysis. These pixel groups are determined by a *topological template*, which is the analogue of the Haar-like feature in a two-group case. The topological template is a set of N regions $T_0=T_1\cup\ldots\cup T_N$, representing spatial relation of object parts. Each region T_i may consist of disconnected sub-regions of arbitrary shape. *If image pixel groups, defined by template regions, statistically differ from each other, it is likely that these pixel groups belong to the object of interest.* This principle can be demonstrated by a simplified example shown in Fig.1, where the template $T_0= T_1\cup T_2\cup T_3$ is matched to image regions R_1 and R_2. In the first case, three pixel groups are similar, as they have roughly the same mean value. In the second case, the pixel groups are different (black, dark-gray and light-gray mean colours), from which we conclude that R_2 is similar to the template.

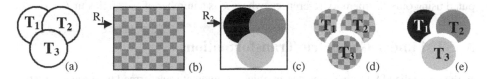

Fig. 1. Simplified example of STM: **(a)** A template consisting of three regions of circular shape; **(b)** 1st region of interest (R_1) in an image; **(c)** 2nd region in the image (R_2); **(d)** Decomposition of R_1 into three regions by the template: pixel groups are similar; **(e)** Decomposition of the R_2: pixel groups are different.

Formally, such a similarity (the matching score) between the template T_0 and an image region $R(\mathbf{x})$, centered at some pixel $\mathbf{x}=(x,y)$, is derived from the F-test statistic. Denote the number of pixels, mean and variance in the region T_i, $(i=0,\ldots,N)$ as n_i, m_i and σ_i^2 respectively. Assuming normal distribution of pixel values and equal variances, and using the standard Analysis Of Variances (ANOVA) technique, we define the *Between-group variation* V_{BG} and *Within-group variation* V_{WG}:

$$V_{BG}(T_1,\ldots,T_N) = -n_0 m_0^2 + \sum_{i=1}^{N} n_i m_i^2 , \qquad V_{WG}(T_1,\ldots,T_N) = \sum_{i=1}^{N} n_i \sigma_i^2 , \qquad (1)$$

Taking into account degrees of freedom of V_{BG} and V_{WG}, they relationship $V_{BG}+V_{WG}=n_0\sigma_0^2$ and applying equivalent transformations, the F-variable becomes

$$F = \frac{V_{BG}}{V_{WG}} \frac{n_0 - N}{N - 1} = \left(\frac{n_0 \sigma_0^2}{n_1 \sigma_1^2 + \ldots + n_N \sigma_N^2} - 1 \right) \frac{n_0 - N}{N - 1} . \qquad (2)$$

Removing constant terms in (2), we obtain the expression for matching score [2]:

$$S(\mathbf{x}) = \frac{n_0 \sigma_0^2}{n_1 \sigma_1^2 + \ldots + n_N \sigma_N^2} . \qquad (3)$$

Computed for all pixels $\mathbf{x}$, the matching scores (3) form a confidence map, in which the local maxima correspond to likely object locations. Application-dependent analysis of statistics m_i, σ_i helps to reduce the number of false alarms. When photometric properties of the object parts are given in advance, *e.g.*, some of the regions are darker or less textured than the others, additional constraints, such as (4), reject false local maxima.

$$m_i < m_j, \qquad \sigma_i < \sigma_j \qquad (4)$$

For Haar-like features ($N=2$), the matching score (3) can also be derived from the squared t-test statistic, which is the squared signal-to-noise ratio (SNR), ranging from 1 (noise), corresponding to the case when all groups are similar, to infinity (pure signal), corresponding to the case when the template strictly determines the layout of pixel groups and all pixels in a group are equal. The distribution of pixel values in image patches can be arbitrary and usually does not satisfy the above assumptions (normal distribution, equal variances); therefore, in practice, it is convenient to interpret (3) as SNR. Instead of using statistical tables for the F-variable, a reasonable SNR threshold above 1 can determine if the similarity (3) between the template and the image region is large enough.

The real-time implementation of STM [2] uses templates with regions T_i consisting of the union of rectangles. Using the integral images, the pixel variances from (3) are computed using only $8k_i$ memory references, where k_i is a number of rectangles in T_i.

3   STM under Geometric Transformations

In the generalized STM we consider an object of interest transformed by a transformation **P** with unknown parameters $\mathbf{p}=(p_1,\ldots,p_k)^T$. This is schematically shown in Fig.2. In order to match the object accurately, the template should be transformed using the same model **P**. As the parameters are unknown, all combinations ($p_1^{(j_1)},\ldots,p_k^{(j_k)}$) of their discrete values $p_i^{(j)} = p_{i\,min}+j\Delta p_i$ are used to transform the template and compute the best matching score:

$$S(\mathbf{x}) = \max_{p_1,\ldots,p_k} S(\mathbf{x};p_1,\ldots,p_k) \qquad (5)$$

By storing the indexes of the best parameter combination

$$(j_1,\ldots,j_k)^* = \arg\max_{j_1,\ldots,j_k} S(\mathbf{x};p_1^{(j_1)},\ldots,p_k^{(j_k)}) , \qquad (6)$$

it is possible to recover an approximated object pose. The number of parameter combinations and computational time grow exponentially with the number of parameters; therefore, it is essential to use a minimal number of parameters. Many approaches [4-7,12,13] use the fact that moderate affine and perspective distortions are approximated well by the similarity transform requiring only two additional parameters for rotation and scale. In our method we also apply a set of similarity transforms to the template and select for each location those rotation and scale parameters giving the best matching score (5)-(6). Although the assumption that the geometric deformations are small enough to be approximated by similarity transform is restrictive, in the next section we describe an iterative technique of recovering full parametric 2D-transformation using similarity transform as initial approximation.

The transformed template is rasterized, and each region is represented by a set of line segments (Fig.3): $T_i=\{s_{i,j}\mid s_{i,j}=(x_1,x_2y)_{i,j}\}$. Each segment is a rectangle of one-pixel height, and integral images technique can still be used to compute the variances in (3). This is not optimal way of computation, and to handle segments efficiently, we propose a one-dimensional analogue of integral images, *integral lines*, defined as follows:

$$I_1(x,y) = \sum_{a \leq x} f(a,y) \; ; \qquad I_2(x,y) = \sum_{a \leq x} f^2(a,y) \qquad (7)$$

A similar definition can be given for *integral vertical lines*, where integration is performed along the y axis. The sums required for computation of the variances in (3), can now be computed via integral lines as follows:

$$u_i \equiv \sum_{(x,y) \in T_i} f(x,y) = \sum_{(x_1,x_2,y) \in T} (I_1(x_2,y) - I_1(x_1-1,y)), \quad v_i \equiv \sum_{(x,y) \in T_i} f^2(x,y) = \sum_{(x_1,x_2,y) \in T} (I_2(x_2,y) - I_2(x_1-1,y)) \qquad (8)$$

where $I_1(-1,y) = I_2(-1,y) = 0$. Thus, the number of memory references is reduced from the number of pixels to the number of lines in the rasterized template.

For efficient implementation, we rewrite (3) in a more convenient form (9) using definitions (8):

$$S = \frac{v_0 - u_0^2 / n_0}{v_0 - \left(\sum_{i=1}^{N-1} \frac{u_i^2}{n_i} \right) - \frac{1}{n_N} \left(u_0 - \sum_{i=1}^{N-1} u_i \right)^2} . \qquad (9)$$

Thus, the algorithm does not requite multiple sums of squared pixels v_i to compute the matching score. It is sufficient to compute only the sum of squared pixels in the entire template T_0 and N sums of pixels in $T_0, T_1,..., T_{N-1}$. Moreover, for a rotationally-symmetrical template, v_0 and u_0 remain constant for each rotation angle, and only $u_1,..,u_{M-1}$ need recomputing. Excluding one region T_N from computations gives additional advantage in computation speed, as we can denote as T_N the most complex region, consisting of the largest number of lines. Line configurations change during template rotation, thus alternating the most complex region at each rotation angle.

Rapid STM [2] requires $\Sigma 8k_i$ memory references independently on template size, where k_i is a number of rectangles in the region T_i. Correlation-based TM requires N_t (the number of pixels) memory references, quadratically dependent on the template size. In the generalized STM, the number of memory references is $4k_0 + 2k_1 + ... + 2k_{N-1}$, where k_i is the number of lines in the template region T_i. The total number of lines is roughly proportional to the template height multiplied by the number of regions N; therefore, it depends linearly on template size. Thus, the computational efficiency of the proposed method lies between that of the rapid TM and correlation-based TM methods.

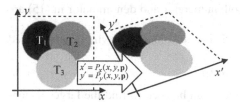

Fig. 2. Example of object transformation (perspective model

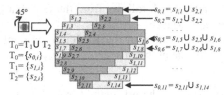

Fig. 3. Rotation of a two-region template by $45°$ and its representation by a set of lines

4   Adaptive Subpixel STM

The method proposed in this section is not restricted by rotation and scale only, and uses full transformation **P** (Fig.2) to iteratively estimate object location and transformation with high accuracy. In this paper, we use the perspective model for all simulations, but any other parametric transformation is also applicable. The goal of the iterative STM method is to compute transformation parameters **p** adaptively from image data, maximizing the matching score $S(\mathbf{x},\mathbf{p})$ at a particular object location **x**. The discrete method from Section 3 can be used to find an initial approximation of the object location $\mathbf{x_0}=(x_0,y_0)$ and initial transformation parameters $\mathbf{p_0}$. Following the standard technique of iterative image registration [10], we obtain a linear approximation of the transformed pixels (x_0',y_0') near their initial location (x_0,y_0). Such an approximation is given by

$$
\begin{aligned}
f'(x_0',y_0') &\approx f(x_0,y_0)+\frac{\partial f(x,y)}{\partial x}\Delta x'+\frac{\partial f(x,y)}{\partial y}\Delta y'\\
&= f(x_0,y_0)+\frac{\partial f(x,y)}{\partial x}\sum\frac{\partial x'}{\partial p_i}\Delta p_i+\frac{\partial f(x,y)}{\partial y}\sum\frac{\partial y'}{\partial p_i}\Delta p_i \equiv \mathbf{f}^T(x_0,y_0)\Delta\mathbf{p},
\end{aligned}
\tag{10}
$$

where $\Delta\mathbf{p} = (1,\Delta p_1,...,\Delta p_k)^T$ is a vector of parameter amendments and

$$
\mathbf{f}^T(x_0,y_0) = (f(x_0,y_0),f_{p_1}(x_0,y_0),...,f_{p_k}(x_0,y_0))
\tag{11}
$$

$$
f_{p_j} = \frac{\partial f(x,y)}{\partial x}\frac{\partial x'}{\partial p_j}+\frac{\partial f(x,y)}{\partial y}\frac{\partial y'}{\partial p_j}
\tag{12}
$$

From (8), the linearized expressions for u_i^2 and v_i have the following matrix form:

$$
v_i \approx \Delta\mathbf{p}^T\left(\sum_{(x,y)\in T_i}\mathbf{f}(x,y)\mathbf{f}^T(x,y)\right)\Delta\mathbf{p} \equiv \Delta\mathbf{p}^T\mathbf{V_i}\Delta\mathbf{p}
\tag{13}
$$

$$
\frac{u_i^2}{n_i} \approx \Delta\mathbf{p}^T\frac{1}{n_i}\left(\sum_{(x,y)\in T_i}\mathbf{f}(x,y)\right)\left(\sum_{(x,y)\in T_i}\mathbf{f}(x,y)\right)^T\Delta\mathbf{p} \equiv \Delta\mathbf{p}^T\mathbf{U_i}\Delta\mathbf{p}
\tag{14}
$$

Substituting (13) and (14) to (9), we obtain the linearized matching score in the form of the Rayleigh quotient:

$$
S = \frac{\Delta\mathbf{p}^T\mathbf{A}\Delta\mathbf{p}}{\Delta\mathbf{p}^T\mathbf{B}\Delta\mathbf{p}},
\tag{15}
$$

where $\mathbf{A}=\mathbf{V_0}-\mathbf{U_0}$, $\mathbf{B}=\mathbf{V_0}-\mathbf{U_1}-...-\mathbf{U_k}$. The matrices **A** and **B** are one-rank modifications of the same covariance matrix $\mathbf{V_0}$. They are symmetric by definition and positive-definite, which follows from the fact that both numerator and denominator in (15) are image variances.

Maximization of the Rayleigh quotient (15) is equivalent to solving a generalized eigenvalue problem

$$
\mathbf{A}\Delta\mathbf{p} = S\mathbf{B}\Delta\mathbf{p},
\tag{16}
$$

Any state-of-the-art method from linear algebra can be used to find the largest eigenvalue S (which is also the maximized matching score) and corresponding eigenvector $\Delta\mathbf{p}$ (the amendments to the image transformation parameters). Examples of such methods are power iterations and inverse iterations (see [8] for a detailed review).

When the eigenvector $\Delta\mathbf{p}$ is found, any vector of the form $\alpha\Delta\mathbf{p}$ is also a solution of (16). Selecting an optimal α that improves the convergence and prevents the solution from oscillations around the maximum is an important part of the algori- thm, and we found that Linesearch strategy provides a robust solution. A detailed review of this and other strategies can be found in [9].

The original non-linear problem can be solved by iteratively applying the linearized solution. The iterations stop when the matching score, the center of the image patch and/or parameter amendments do not change significantly. Below is the outline of the AS STM algorithm that starts at iteration $n=0$ from initial values S_0, $\mathbf{x_0}$, $\mathbf{p_0}$:

1. Resample image patch centered at $\mathbf{x_n}$ using current $\mathbf{p_n}$
2. Compute image derivatives from resampled image patch $f(x,y)$; compute partial derivatives of the transformation model $\mathbf{P}$ in (12) using current values of $\{p_i\}$.
3. Compute matrices $\mathbf{V_0}$, $\mathbf{U_1}$,...,$\mathbf{U_k}$, $\mathbf{A}$, $\mathbf{B}$ and solve the optimization problem (15) by finding maximal eigenvalue S_{max} and eigenvector $\Delta\mathbf{p_n}$ of (16)
4. Use the Linesearch strategy to find α_n maximizing $S_{max}(\mathbf{p_n}+\alpha_n \Delta\mathbf{p_n})\equiv S_{n+1}$
5. Update parameters: $\mathbf{p_{n+1}} = \mathbf{p_n}+\alpha_n \Delta\mathbf{p_n}$ and a new object location $\mathbf{x_{n+1}} = \mathbf{P}(\mathbf{x_n}, \mathbf{p_{n+1}})$.
6. If $|\alpha_n\Delta\mathbf{p_n}|<\varepsilon_1$ and/or $|S_{n+1}-S_n|<\varepsilon_2$ then stop; else go to step 1 for a next iteration $n=n+1$.

If the template consists of two regions, $T_0=T_1\cup T_2$, there is an analytic solution of the eigenproblem (16) that does not require iterations. In this case the matrices $\mathbf{A}$, $\mathbf{B}$ are $\mathbf{A}=\mathbf{V_0}-\mathbf{U_0}$, $\mathbf{B}=\mathbf{V_0}-\mathbf{U_1}-\mathbf{U_2}$. They are related by $\mathbf{A}-\mathbf{B}=a\mathbf{w}\mathbf{w}^T$, derived from the definition of $\mathbf{U_i}$ (14), where $a=n_1n_2/n_3$ and

$$\mathbf{w} = \frac{1}{n_1} \sum_{(x,y)\in T_1}\mathbf{f}(x,y) - \frac{1}{n_2} \sum_{(x,y)\in T_2}\mathbf{f}(x,y) \tag{17}$$

The vector $\mathbf{w}$ is the linearized contrast between regions T_1 and T_2. The solution of (16) is given by the following statement.

Statement: *The largest eigenvalue of the eigenproblem* (16) *and the corresponding eigenvector are:*

$$\Delta\mathbf{p_{max}} = \mathbf{B}^{-1}\mathbf{w} \tag{18}$$
$$S_{max}=a\mathbf{w}^T \Delta\mathbf{p_{max}} + 1 \tag{19}$$

Proof: Consider an equivalent eigenvalue problem $(\mathbf{A}-\mathbf{B})\mathbf{x} =a\mathbf{w}\mathbf{w}^T\mathbf{x}=\lambda\mathbf{B}\mathbf{x}$, having the same eigenvectors as (16) and eigenvalues transformed as $\lambda=S-1$. Using Cholecky decomposition $\mathbf{B}=\mathbf{L}\mathbf{L}^T$, where $\mathbf{L}$ is a bottom-triangular matrix, and introducing a vector transformation $\mathbf{y}=\mathbf{L}^T\mathbf{x}$, we obtain another equivalent eigenvalue problem $\mathbf{w_1}\mathbf{w_1}^T\mathbf{y}=\lambda\mathbf{y}$, where $\mathbf{w_1}=\mathbf{L}^{-1}\mathbf{w}$. One-rank matrix $\mathbf{w_1}\mathbf{w_1}^T$ whose size is $(k+1)\times(k+1)$ has k-dimensional eigenspace corresponding to $\lambda_1=0$. The vectors $\mathbf{y_1},...,\mathbf{y_k}$ from this eigenspace satisfy $\mathbf{w_1}^T\mathbf{y_i}=0$. The remaining eigenvector $\mathbf{y_{k+1}}$ corresponding to $\lambda_2\neq0$ can be found from the orthogonality condition $\mathbf{y_{k+1}}^T\mathbf{y_i}=0$. Therefore, $\mathbf{y_{k+1}}=\mathbf{w_1}$, from which $\mathbf{x_{k+1}}=\mathbf{B}^{-1}\mathbf{w}$. Substituting $\mathbf{x_{k+1}}$ into the eigenvalue equation, we obtain $a\mathbf{w}\mathbf{w}^T\mathbf{B}^{-1}\mathbf{w}=\lambda\mathbf{B}\mathbf{B}^{-1}\mathbf{w}$, from which $\lambda_2=a\mathbf{w}^T\mathbf{B}^{-1}\mathbf{w}$. $\mathbf{B}$ is positive-definite; therefore, $\lambda_2>0$ and the largest eigenvalue of the problem (16) is $S_{max}=\lambda_2+1=a\mathbf{w}^T\Delta\mathbf{p_{max}}+1$.

The adaptive subpixel STM method (15),(18),(19) has the form similar to geometrically-adaptive subpixel correlation coefficient between images, derived in [14]. Using the obtained results, step 3 of the ASSTM algorithm is implemented as follows: 3.1) Compute matrices V_0, U_1, U_2, B and vector w; 3.2) Solve the system $B\Delta p_n = w$ by Cholecky decomposition; and 3.3) Find the eigenvalue (19).

5 Experiments

The presented methods have been applied to a variety of object detection scenarios and tasks, including target detection in remote sensing images, hand, face and facial feature detection, and detection of camera calibration pattern and its structure.

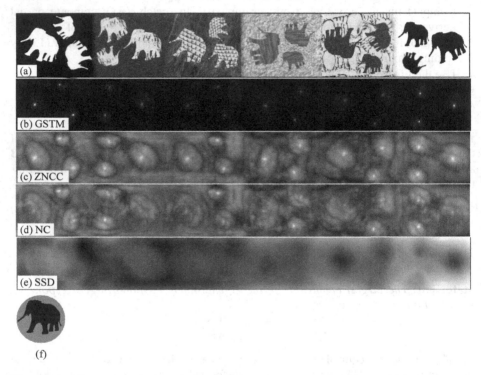

Fig. 4. Testing the TM methods in the synthetic data. (a) Test image combined from binary images and textured objects on textured backgrounds; (b) GSTM matching score map: large distinctive peaks indicate object locations. (c) Correlation coefficient map: well-recognizable peaks indicate object locations, background matching score values are also large; (d) Normalized contrast map, some object location peaks are less prominent than in ZNCC method; (e) SSD score map; SSD method is not invariant to contrast inversion, so the object locations are defined by the maxima and minima; also, the score map is very smooth (resulting in low accuracy of localization) as there is no local normalization. (f) Two-region template used in all the template matching methods.

For the first experiment, shown in Fig.4, an object was filled by different textures, arbitrarily rotated, and scaled by three factors 1, 1.25 and 1.25^2. The resulting images were placed on textured backgrounds and, together with uniform black-and-white and inverted images, were combined into a single test image (Fig.4a). The two-region template, shown in Fig.4f, was used to detect all 18 objects in the test image, using the following methods: 1) GSTM followed by AS STM; 2) TM by zero-mean normalized cross-correlation (ZNCC); 3) TM by normalized contrast (NC), similar to [1]; and 4) TM by the sum of squared differences (SSD). In all these methods, the discrete steps of $10°$ for rotation and of 1.25 for scale were used to transform the template and select the largest matching score in each pixel location. The absolute value of the correlation coefficient was used in the ZNCC method to handle inverse contrast. The following matching score was used in the NC method: $|m_1-m_2|/\sigma_0$. In the ZNCC and SSD methods, the grayscale template was obtained from Fig.4b by setting 255 in the gray region and 0 in the black region. Both the GSTM and NC methods were implemented using integral lines, and they performed ~50 times faster than ZNCC or SSD. The following observations can be made comparing the matching score maps (Fig.4b-e) and absolute numerical values of their peaks (Table1): 1) GSTM produced very large and distinctive peaks (this fact becomes more evident after taking ratio of a peak to the background). 2) The peak values can be further improved by AS STM adjusting the object transfor-mation; 3) ZNCC produced similar results, but the peak values are very close to the background matching scores; 4) NC performs worse than both GSTM and ZNCC, as many peaks become undistinguishable from the back-ground; 5) SSD does not solve the problem at all, because: a) it is not invariant to contrast inversion; b) as there is no intensity normalization, the score mapis smooth, resulting in low localization accuracy.

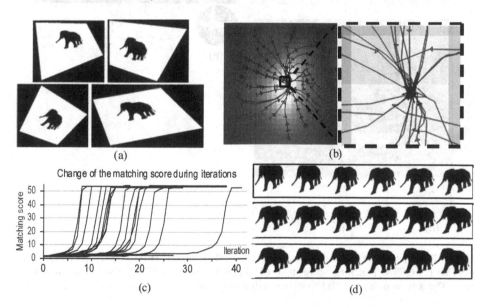

(a) (b)

Change of the matching score during iterations

(c) (d)

Fig. 5. Testing the Adaptive Subpixel STM method in synthetic data. (**a**) Examples of the perspective transformation of the test image; (**b**) Trajectories of the image patch center during iterations. Initial positions are selected in the vicinity of matching score peak; almost all trajectories converge to the same subpixel position, which is visible in the zoomed view; (**c**) Graphs of the matching score along the image patch trajectories. Starting from low values the matching score converges to the true large value, indicating perfect fit of the image data; (**d**) Example of image patch transformation along one of the trajectories. In this example, the lower-bottom image from figure (a), whose initial orientation estimated by GSTM has been compensated, iteratively transformed so that at the final 18th iteration it coincides with the template (Fig.4f).

Table 1. Numerical values of the matching score peaks corresponding to 12 objects from Fig.4a out of 18. Object counting starts from the leftmost object in Fig.4a. Average matching score, computed from the entire score map, is shown in the last column "Bakground". The last row shows improvement of the matching score after applying ASSTM at the object locations.

Method \ Object	1	2	3	4	5	6	7	8	9	10	11	12	Backgr.
ZNCC	0.98	0.95	0.97	0.91	0.91	0.93	0.81	0.83	0.80	0.87	0.88	0.87	0.34
NC	19.05	28.87	23.74	27.80	22.35	18.10	15.71	25.39	19.57	17.04	26.83	21.43	9.21
GSTM	16.44	6.02	13.41	6.09	5.40	6.29	3.07	3.75	2.90	4.70	4.39	3.92	1.21
AS STM	19.07	12.83	16.87	6.23	6.03	6.71	3.20	4.09	3.17	5.33	5.11	4.84	1.21

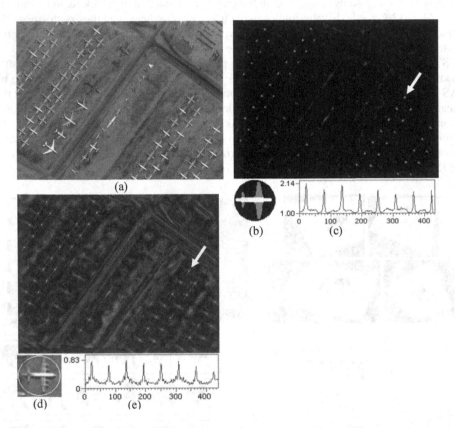

Fig. 6. Template matching in an aerial image. (a) Test image; (b) Three-region template for GSTM; (c) GSTM matching score map; (d) Template for ZNCC; (e) ZNCC matching score map. The arrows in (c) and (d) show the starting points and directions of the 1D-profiles.

In the second experiment (Fig.5), we generated 100 random perspective transformations of the test image (Fig.5a), and applied the combination GSTM+AS STM with the same template (Fig.4b) to recover the transformation parameters. GSTM was used to roughly estimate object rotation, then AS STM started 30 times from different locations within the vicinity of the matching score peak (Fig.5b). The parameters obtain- ed were compared to ground truth parameters by computing the correlation coefficient

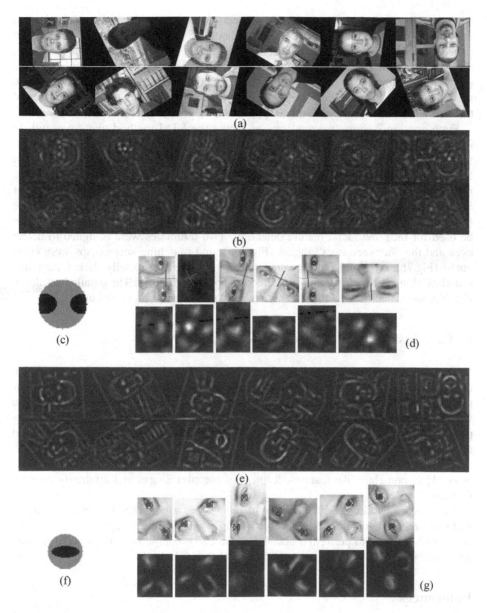

Fig. 7. STM application to facial feature detection. **(a)** 12 face images from the Caltech database (http://www.vision.caltech.edu/html-files/archive.html) rotated by random angle; **(b)** Matching score map resulted from matching with the template in (c); **(c)** Template for detecting the "between eyes" region; **(d)** Close-ups of the first 6 images from (a) and corresponding matching score maps from (b). The crosses indicate matching score local maxima; orientation of the crosses corresponds to orientation of the template giving the best matching score; **(e)** Matching score map resulted from matching with the template in (f); **(f)** Template for detecting eyes; **(g)** Close-ups of the images 7-12 from (a) and corresponding matching score maps from (e). The markers indicate matching score local maxima.

(cosine of the angle between 8-element vectors). In all tests (except for a few cases of distant start), this correlation coefficient was about 1, indicating high accuracy of estimated parameters. This was also confirmed by high matching score values (Fig.5c). Also, the accuracy of the final location of the image patch was computed. The average distance between the ground truth value and obtained object location and its standard deviation were 0.07 and 0.11 pixels respectively. This experiment confirms the high accuracy of the method and its robustness to initial approximation of object location.

Fig.6 and Fig.7 show two applications where the STM method may be useful. In Fig.6, the method is used to detect airplanes in an aerial image. It was found that the three-region template from Fig.6b gives higher matching scores than a two-region template, in which one region outlines an entire airplane shape. The matching score map clearly shows locations of the object, which can be obtained by searching for local maxima. ZNCC method (Fig.6e) applied using a template taken from the image (Fig.6d), produces similar but more noisy output. Fig.7 demonstrates how GSTM can be used for face and facial feature detection. Two templates were designed to detect eyes and the "between eyes" region (Fig.7c,f), and matching score maps were computed (Fig.7b,e). Taking into account that eye regions are usually darker than surrounding skin, we used filters (4) to remove opposite situation. The detailed results in Fig.7d,g show that the method reliably detected the required facial features.

6 Conclusions and Future Work

We have presented novel extensions of the Statistical Template Matching framework that can now be efficiently used to match geometrically transformed objects. Two main contributions of this paper are: 1) integral lines technique for computing the matching score with number of operations linearly dependent on template size; 2) a novel adaptive sub-pixel refinement method that reduces the problem of transformation parameter estimation to a standard eigenvalue problem.

The proposed methods can also be used to generalize rapid object detection framework [1] to non-Haar-like features, features of complex shape, and arbitrarily oriented features.

Future work will investigate how real image patches can be used to create the simplified multi-region templates. One possible way is to quantize the image into a few levels. We would also like to perform a detailed, quantitative comparison of the STM method with standard template matching methods on real-world data.

References

1. Viola, P., Jones, M.: Rapid object detection using a boosted cascade of simple features. IEEE CVPR, 511–518 (2001)
2. Sibiryakov, A., Bober, M.: A Method of Statistical Template Matching and Applications to Face and Facial Feature Detection, WSEAS Trans. on Information Science and Applications (September 2005)
3. Jain, A., Zhong, Y., Lakshmanan, S.: Object Matching Using Deformable Templates. IEEE TPAMI 18(3), 267–278 (1996)

4. Yoshimura, S., Kanade, T.: Fast template matching based on the normalized correlation by using multiresolution eigenimages. In: IEEE/RSJ/GI Int. Conf. on Intelligent Robots and Systems (IROS 1994), vol. 3, pp. 2086–2093 (1994)
5. Jones, M., Viola, P.: Fast Multi-view Face Detection. IEEE CVPR (June 2003)
6. Lienhart, R., Maydt, J.: An extended set of Haar-like features for rapid object detection. In: ICIP 2002, vol. 1, pp. 900–903 (2002)
7. Messom, C.H., Barczak, A.L.: Fast and Efficient Rotated Haar-like Features using Rotated Integral Images. In: Australasian Conf. on Robotics and Automation (2006)
8. Golub, G., Van Loan, C.: Matrix computations. Johns Hopkins University Press, Baltimore (1996), ISBN: 0-8018-5414-8
9. Gould, N., Leyffer, S.: An introduction to algorithms for nonlinear optimization. In: Blowey, J.F., Craig, A.W., Shardlow, T. (eds.) Frontiers in Numerical Analysis, pp. 109–197. Springer, Berlin (2003)
10. Lucas, B., Kanade, T.: An iterative image registration technique with an application to stereo vision. In: Proc. of Imaging understanding workshop, pp. 121–130 (1981)
11. Zitova, B., Flusser, J.: Image Registration Methods: a Survey. Image and Vision Computing 24, 977–1000 (2003)
12. Choi, M.-S., Kim, W.-Y.: A novel two stage template matching method for rotation and illumination invariance. Pattern recognition 35(1), 119–129 (2002)
13. Tanaka, K., Sano, M., Ohara, S., Okudaira, M.: A parametric template method and its application to robust matching. IEEE CVPR, 620–627 (2000)
14. Zheltov, S., Sibiryakov, A.: Adaptive Subpixel Cross-Correlation in a Point Correspondence Problem. In: Gruen, A., Kuhbler, O. (eds.) Optical 3D Measurement Techniques, Zurich, September 29 - October 2, pp. 86–95 (1997)

Distance Transformation on Two-Dimensional Irregular Isothetic Grids

Antoine Vacavant[1,*], David Coeurjolly[2], and Laure Tougne[1]

[1] LIRIS - UMR 5205, Université Lumière Lyon 2
5, avenue Pierre Mendès-France
69676 Bron cedex, France
[2] LIRIS - UMR 5205, Université Claude Bernard Lyon 1
43, boulevard du 11 novembre 1918
69622 Villeurbanne cedex, France
{antoine.vacavant,david.coeurjolly,laure.tougne}@liris.cnrs.fr

Abstract. In this article, we propose to investigate the extension of the E^2DT (squared Euclidean Distance Transformation) on irregular isothetic grids. We give two algorithms to handle different structurations of grids. We first describe a simple approach based on the complete Voronoi diagram of the background irregular cells. Naturally, this is a fast approach on sparse and chaotic grids. Then, we extend the separable algorithm defined on square regular grids proposed in [22], more convenient for dense grids. Those two methodologies permit to process efficiently E^2DT on every irregular isothetic grids.

1 Introduction

The definition of discrete distances is a very important concept in image analysis and shape description [20,21]. Here, we are interested in the definition of a distance and its application on Irregular Isothetic Grids (I-grids for short) [5], where the cells are rectangles defined by variable positions and sizes, and may be determined by subdivision rules. Those grids are very common, and permit to represent an image in a more compact and adapted manner. Here, we will use two classical I-grids: the Quadtree decomposition [23] and the Run-Length Encoding (or RLE) [11]. In our study, we have chosen to consider that the distance between two cells in a two dimensional (2-D) I-grid is the distance between their centers (which is the same as the regular square grid, where we compute the distance between points of $\mathbb{Z}^2$). The squared Euclidean Distance Transformation (E^2DT) of a binary image is a tool that has been largely investigated for decades, and represents a very common way to analyze the shape of graphical objects, for various applications (see [18] and references of [16] for more details). The purpose of this process is to label each (*foreground*) cell of an object with the distance to the closest cell of its complement (or *background*). Since we consider

---

* The authors would like to thank Pr. Annick Montanvert for her constructive suggestions during the preparation of this article.

D. Coeurjolly et al. (Eds.): DGCI 2008, LNCS 4992, pp. 238–249, 2008.
© Springer-Verlag Berlin Heidelberg 2008

the distance between the center of grid cells, this process can be naturally linked with the computation of the Voronoi Diagram (VD) [6,19]. Moreover, we can notice that the VD may also be extracted from the distance map [13,22].

The E^2DT of a binary image generally considers it as a regular square grid. Many studies have aimed to develop a special Distance Transformation (DT) for non-square grids, but these approaches can not be extended to every $\mathbb{I}$-grids. In fact, we want to propose a global model which could be then applied to each sort of $\mathbb{I}$-grid. The DT of Quadtree or Octree based grids [14,23,26] are dependent on the specific structure of the concerned trees. Those approaches compute the DT by propagating distance values from parent nodes to their children nodes. To handle medical images digitized on elongated grids, where the cells are longer along an axis, a lot of methodologies that perform the chamfer distance have been adapted [4,7,9,25]. To use the same technique on an $\mathbb{I}$-grid $\mathbb{I}$, we would have to extend chamfer masks and change them for each cell of $\mathbb{I}$ (non-stationary computation of the DT). In the same way, the algorithms designed on other non-standard grids [10,27], $e.g.$ Face-Centered Cubic (FCC) or Body-Centered Cubic (BCC) grids, suppose the regularity of the neighbors of a cell, and may not be adapted to every $\mathbb{I}$-grids. Indeed, we make no hypothesis about the configuration (number, position, size, $etc.$) of the neighbors of a cell in $\mathbb{I}$. In the regular square grid case, most VD-based algorithms [2,16,24] build a partial VD to compute the E^2DT and lead to an optimal linear time complexity for the E^2DT computation, in the number of cells of the grid. Since these methods are separable, $i.e.$ they perform operations independently along the two axis, they propose a natural extension to handle d-dimensional images. Here, we propose to extend the separable and linear-time algorithm presented in [22] to compute the squared Euclidean distance transform on $\mathbb{I}$-grids (or $\mathbb{I}$-DT). Thus, we propose an original and efficient approach to perform the $\mathbb{I}$-DT, which can be used for every $\mathbb{I}$-grids we have cited before.

In this article, we first introduce discrete distances on $\mathbb{I}$-grids, and we present the algorithm to compute the $\mathbb{I}$-DT by implementing the complete VD. For some sparse irregular grids, the complete VD based approach seems to be the best way to compute the $\mathbb{I}$-DT. In Section 3, we give details about our extension of [22] on $\mathbb{I}$-grids. Indeed, thanks to a data structure to represent every $\mathbb{I}$-grids, we insure that the $\mathbb{I}$-DT is then error-free, and this allows us to fix an upper bound for the complexity of this method. Then, we propose to compare the speed and the complexity of our method with respect to the direct approach based on the complete VD. We thus propose to measure the performance of the two algorithms, to study what kind of grids they efficiently handle. We finally discuss the applications and possible extensions of our approach.

2   A Simple Approach Based on the Complete Voronoi Diagram Implementation

We first define an $\mathbb{I}$-grid as a tiling of the plane with isothetic rectangles. We shortly recall that each rectangle R (also called $cell$) of $\mathbb{I}$ is defined by its center

$(x_R, y_R) \in \mathbb{R}^2$ and a size $(l_R^x, l_R^y) \in \mathbb{R}^2$. The position and the size of R may be controlled by different level of constraints [5]. Each cell of the grid is also associated with a foreground label (1) or a background label (0). Here, we denote $\mathbb{I}_F$ the set of foreground cells in the grid, and $\mathbb{I}_B$ the set of background ones. The sizes of those sets are denoted n_F and n_B respectively. The distance between two cells R and R' is the distance between their centers. If we denote p (respectively p') the center of a cell R (respectively R'), we have

$$d^2(R, R') = d^2(p, p') = (x_R - x_{R'})^2 + (y_R - y_{R'})^2 \qquad (1)$$

for the square of the Euclidean distance. For a cell $R \in \mathbb{I}$, the squared Euclidean distance transformation on $\mathbb{I}$-grids ($\mathbb{I}$-DT) is given by:

$$\mathbb{I}\text{-DT}(R) = \min_{R'} \{ d^2(R, R'); \ R' \in \mathbb{I}_B \} \qquad (2)$$

and is exactly the E^2DT if we consider a regular square grid $\mathbb{I}$. This irregular discrete distance computed between cell centers can be applied in Geographical Information Systems (GIS), where an irregular spatial structure [1,23] permits to quickly locate points and to compute the distance between them. We could also choose to extend the distance proposed by H. Samet [23] (used in the framework of a chessboard distance transformation, that refers to the underlying regular square grid) on every $\mathbb{I}$-grids . In this case, we would have to compute the distance between a foreground cell center p and the frontier between the object containing it and the background. Let $\mathcal{S}$ be the set of segments so that the intersection between a cell $R \in \mathbb{I}_F$ and a cell $R' \in \mathbb{I}_B$ adjacent with R belongs to $\mathcal{S}$. In this case, the $\mathbb{I}$-DT should be defined as follows:

$$\mathbb{I} - \text{DT}(R) = \min_{\alpha} \{ d^2(p, \alpha); \ \alpha \in \mathcal{S} \}. \qquad (3)$$

Contrary to the $\mathbb{I}$-DT given in Equation 2, this process does not take into account the representation of the background. More precisely, we do not consider the centers of the backgound cells of $\mathbb{I}$.

In this article, we propose a VD-based algorithm to compute the $\mathbb{I}$-DT defined in Equation 2. The first step of this algorithm is to compute the VD where the sites are the centers of the background cells in $\mathbb{I}_B$. Then, we perform the $\mathbb{I}$-DT of each foreground cell R by locating its center p in the VD. We search for one of the nearest Voronoi site s to finally compute the $\mathbb{I}$-DT of R (see Algorithm 1 for more details). We show in Figure 1 examples of results obtained with a small binary image digitized with a regular square grid, a Quadtree grid, and a RLE along X grid. We have chosen those two $\mathbb{I}$-grids because they are common in image analysis applications, but we could also consider Kd-tree based grids [1] for example. The Voronoi Diagram $\mathcal{V}$ is computed thanks to the CGAL library [3,15], and has an optimal time complexity $\mathcal{O}(n_B \log n_B)$ and fill $\mathcal{O}(n_B)$ space [8]. Then, the main loop of Algorithm 1 consists in locating each foreground cell P in $\mathcal{V}$ (with its center p), and in searching for its nearest Voronoi site s in $\mathcal{V}$. The location query is proved to have a $\mathcal{O}(\log n_B)$ complexity [8]. Thus, we perform this loop in $\mathcal{O}(n_F \log n_B)$ time. Indeed, searching for the nearest site

Algorithm 1. 𝕀-DT based on the complete Voronoi diagram

input : A labelled 𝕀-grid 𝕀.
output: The 𝕀-DT of 𝕀.

1 Compute the Voronoi Diagram $\mathcal{V}$ of the points $\{p;\ R \in \mathbb{I}_B\}$;
2 **foreach** *cell* $R \in \mathbb{I}_F$ **do**
3   Locate p in $\mathcal{V}$;
4   **if** *p belongs to a Voronoi vertex v of $\mathcal{V}$* **then**
5          $s :=$ Voronoi site of an adjacent cell of v;
6   **else if** *p belongs to a Voronoi edge e in $\mathcal{V}$* **then**
7          $s :=$ Voronoi site of an adjacent cell of e;
8   **else**
        // p belongs to a Voronoi cell c of V
9          $s :=$ Voronoi site of c;
10   𝕀-DT$(R) := d^2(s, p)$;
11 **foreach** *cell* $R \in \mathbb{I}_B$ **do**
12   𝕀-DT$(R) := 0$;

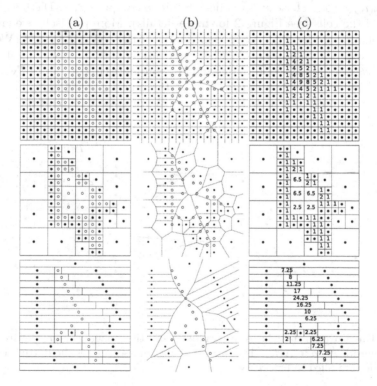

Fig. 1. Examples of results of Algorithm 1 with a small binary image of size 16 x 16. The image *cursor* is digitized with a regular square grid, a Quadtree decomposition, and a RLE along the X axis *(a)*. Background points are illustrated in black, and foreground ones in white. We present the Voronoi diagram for each case in dotted lines *(b)*, and the final distance map for the d^2 distance *(c)*.

s is processed in constant time, in the three cases: p is inside a Voronoi cell, p stands on a Voronoi edge, and p is on a Voronoi vertex. When p stands on a vertex or on an edge of $\mathcal{V}$, the choice of the Voronoi cell containing s is arbitrary. Thus, Algorithm 1 is performed in $\mathcal{O}(N \log n_B)$ time and $\mathcal{O}(N)$ space, where $N = n_B + n_F$ is the total number of cells in $\mathbb{I}$.

3   A d-Dimensional Algorithm for $\mathbb{I}$-DT

We first propose a data structure to represent any $\mathbb{I}$-grids, and to simplify the cell scanning. An *irregular matrix* $\mathbf{A}$ associated to the labelled $\mathbb{I}$-grid $\mathbb{I}$ is built by organizing aligned cells along X and Y axis. The value of a node in $\mathbf{A}$ is fixed according to two cases: (1) $\mathbf{A}(i,j)$ is the center of a cell in $\mathbb{I}$, this node is a foreground node ($\mathbf{A}(i,j) = 1$) or a background node ($\mathbf{A}(i,j) = 0$); (2) it does not correspond to a cell center in $\mathbb{I}$, and we set it as a foreground *extra node* ($\mathbf{A}(i,j) = 1$). Those extra nodes permit to compute the $\mathbb{I}$-DT between the centers of the cells (see Figure 2 for more details). More precisely, we create as many columns as the different X-coordinates of the cell centers of $\mathbb{I}$. We make the same with the columns of $\mathbf{A}$, when we consider the Y-coordinate of the cell centers. We store the X-coordinates and Y-coordinates in two tables T_X and T_Y, and we denote $n_X = |T_X|$ and $n_Y = |T_Y|$ the number of columns and rows of $\mathbf{A}$. So, the X-coordinate of $\mathbf{A}(i,j)$ is $T_X(i)$. We propose to extend the separable

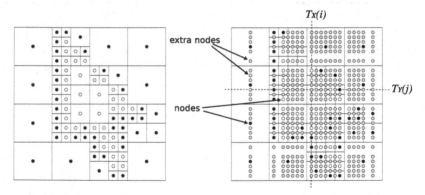

Fig. 2. The irregular matrix $\mathbf{A}$ associated with the image *cursor*, digitized on a Quadtree grid. A node $\mathbf{A}(i,j)$ is depicted as the intersection of the dotted lines, and extra nodes are filled in light grey.

algorithm described in [22] and we adapt it on the irregular matrix. The $\mathbb{I}$-DT of an $\mathbb{I}$-grid $\mathbb{I}$ given in Equation 2 is represented by the irregular matrix:

$$\mathbf{C}(i,j) = \min_{x,y} \left\{ (T_X(i) - T_X(x))^2 + (T_Y(j) - T_Y(y))^2 ; \right.$$

$$\left. x \in \{0, ..., n_X - 1\}, \ y \in \{0, ..., n_Y - 1\}, \ \mathbf{C}(x,y) = 0 \right\}. \quad (4)$$

This min operation is processed over all the elements of $\mathbf{C}$ (nodes and extra nodes). To compute this $\mathbb{I}$-DT, our algorithm may be decomposed into two steps:

1. Let $\mathbf{A}$ be the irregular matrix built from the $\mathbb{I}$-grid $\mathbb{I}$. We perform here a one-dimensional $\mathbb{I}$-DT along X axis, stored in the irregular matrix $\mathbf{B}$ such that:

$$\mathbf{B}(i,j) = \min_x \left\{ |T_X(i) - T_X(x)|; \ x \in \{0, ..., n_X - 1\}, \ \mathbf{A}(x,j) = 0 \right\}. \quad (5)$$

2. We perform then a Y axis process to build the final irregular matrix $\mathbf{C}$:

$$\mathbf{C}(i,j) = \min_y \left\{ \mathbf{B}(i,y)^2 + (T_Y(j) - T_Y(y))^2; \ y \in \{0, ..., n_Y - 1\} \right\}. \quad (6)$$

We also present in Algorithm 2 the two steps of our approach. In the first step, we can notice that the only difference with the regular square case [6,22] is the computation of the distance, lines 2 and 2. Indeed, we have to consider in those operations the distance between the point $\mathbf{B}(i,j)$ and its neighbor (*e.g.* $|T_X(i) - T_X(i - 1)|$ in line 2). The second step of our algorithm is in fact the computation of the lower envelope of a set of parabolas [6]. After Step 1, we can consider the set of parabolas $\mathcal{F}_y^i(j) = \mathbf{B}(i,y)^2 + (T_Y(j) - T_Y(y))^2$ on the column $\{\mathbf{B}(i,y)\}_{0 \le y \le n_Y}$. With Step 2, the column $\{\mathbf{C}(i,y)\}_{0 \le y \le n_Y}$ is the lower envelope of the set $\{\mathcal{F}_y^i\}_{0 \le y \le n_Y}$. The function $Sep^i(u,v)$ is the exact coordinate of the intersection point between two parabolas [6,17] (we will simply denote $\mathcal{F}_y(j) = \mathcal{F}_y^i(j)$ and $Sep(u,v) = Sep^i(u,v)$ when the parameter i is fixed):

$$Sep^i(u,v) = \left(T_Y(v)^2 - T_Y(u)^2 + \mathbf{B}(i,v)^2 - \mathbf{B}(i,u)^2\right) / \left(2(T_Y(v) - T_Y(u))\right). \quad (7)$$

Algorithm 2. $\mathbb{I}$-DT by a separable approach

```
input  : A labelled I-grid I.                      // Step 2 along Y-axis
output : The I-DT of I, stored in an irregu-    19  for i = 0 to nX - 1 do
         lar matrix C.                           20      q := 0; s[0] := 0; t[0] := 0;
                                                 21      for j = 1 to nY - 1 do
2  Build the irregular matrix A associated to I; 22          while q ≥ 0 ∧ Fs[q](t[q]) > Fj(t[q])
   // Step 1 along X-axis                                    do
3  for j = 0 to nY - 1 do                         23              q := q - 1;
4      if A(0, j) = 0 then                        24          if q < 0 then
5          B(0, j) := 0;                          25              q := 0; s[q] := j;
6      else                                       26          else
7          B(0, j) := ∞;                          27              w := Sep(s[q], j);
8      for i = 1 to nX - 1 do                     28              if w ≤ TY(nY - 1) then
9          if A(i, j) = 0 then                    29                  Find    the    node
10             B(i, j) := 0;                                          B(i, k),    k      ∈
11         else                                                      {s[q], ..., nY - 1} such
12             B(i, j) := |TX(i) - TX(i-1)| +                         that TY(k) > w;
                 B(i - 1, j);                     30                  q := q + 1; s[q] := k; t[q] := w;
13     for i = nX - 2 to 0 do
14         if B(i + 1, j) < B(i, j) then          31      for j := nY - 1 to 0 do
15             B(i, j) := |TX(i) - TX(i+1)| +     32          C(i, j) := Fs[q](j);
                 B(i + 1, j);                     33          if TY(j) = t[q] then
                                                  34              q := q - 1;
17
```

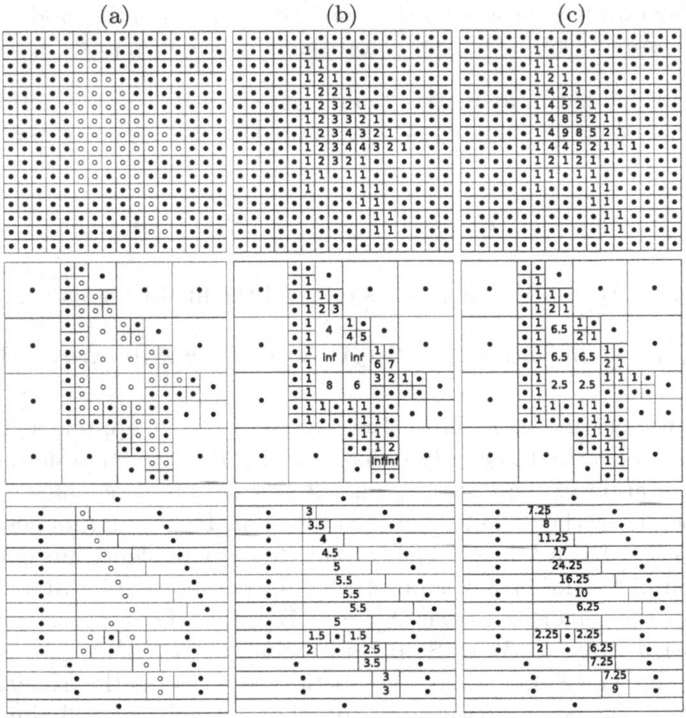

Fig. 3. We present the temporary distance map obtained with Step 1 *(b)* on the image *cursor* digitized with different $\mathbb{I}$-grids *(a)*. We can notice that the *inf* node means that no background node exists on the row containing this point. So, the last *for* loop in Algorithm 2 line 2 does not change $\mathbf{B}(i,j) = \infty$. The final distance transformation is depicted as Figure 1 *(c)*.

In comparison with the regular grid case, we can see that the operator *div* has been replaced by the floating-point operator / in this equation to compute the exact intersection point. For $\mathbb{I}$-grids computed from a cell subdivision or a cell grouping process, we still have an exact arithmetic division. In those cases, coordinates may be half-integers, and we just have to multiply grid cells coordinates by four to compute the integer intersection point with $Sep()$. The computation of w (line 2) only depends on the function $Sep(u,v)$ and then permits to find the intersection point in $\mathbf{B}$ (line 2). Here, this find command is performed with a dichotomous search through the ordered set of nodes $\{\mathbf{B}(i,k)\}_{s[q] \leq k \leq n_Y - 1}$, and has a $\mathcal{O}(\log n_Y)$ time complexity in the worst case. But in our experiments, we have observed that this is a fast operation, since we begin the search from the last intersection point (with index $s[q]$). In Figure 3, we present some results of Algorithm 2 on the small binary image *cursor* used in Figure 1. To apply the $\mathbb{I}$-DT given in Equation 3 (between a foreground cell and the foreground/background frontier), we would have to link each node and each extra node in the irregular matrix to the cell in $\mathbb{I}$ that contains it. This permits to change equations 4, 5, and 6 to take into account the size of the cells. For example, if we denote $R_{i,j}$ the cell in $\mathbb{I}$ associated with the node $\mathbf{A}(i,j)$, Equation 5 becomes:

$$\mathbf{B}(i,j) = \min_x \left\{ |T_X(i) - T_X(x)| - \frac{l^x_{R_{x,j}}}{2}; \; x \in \{0,...,n_X - 1\}, \; \mathbf{A}(x,j) = 0 \right\}.$$
(8)

We have presented here a separable algorithm on $\mathbb{I}$-grids. The first operation (build the irregular matrix) is performed in $\mathcal{O}(n_X n_Y)$ time. More precisely, we first scan all the cells of $\mathbb{I}$ to get the n_Y rows and n_X columns of $\mathbf{A}$. Then, we consider each node of $\mathbf{A}$ and assign its value by checking if it coincides with a cell center in $\mathbb{I}$. This algorithm has a global time complexity in $\mathcal{O}(n_X n_Y \log n_Y)$. It can be easily extended to higher dimensions: the Step 1 stands as an initialization step, and for each greater dimension, a mixing process, as Step 2, permits to combine results obtained in the lower dimensions. If we consider a d-dimensional labelled $\mathbb{I}$-grid, the cost of the consecutive steps is in $\mathcal{O}(n^d \log^{d-1} n)$, where the dimension of the irregular matrix $\mathbf{A}$ associated to $\mathbb{I}$ is n^d. The size of $\mathbf{A}$ clearly depends on the organization of the cells of $\mathbb{I}$; a matrix $\mathbf{A}$ built with a regular grid $\mathbb{I}$ would have the same size as $\mathbb{I}$. The more an $\mathbb{I}$-grid has an irregular structure, the more the difference between $n_X n_Y$ and N, the number of cells of $\mathbb{I}$, is important. The space required, in $\mathcal{O}(n_X n_Y)$, is principally occupied by the irregular matrix $\mathbf{A}$, $\mathbf{B}$ and $\mathbf{C}$. Furthermore, when we have implemented this algorithm, we have used only one matrix that stores initial and temporary distance values. Those two elements about the complexity of our contribution will be discussed in the conclusion.

4 Experiments

We first illustrate in Figure 4 the irregular grids we have generated with the sample binary images we have chosen: *canon* which is a big image containing a single binary object, *lena*, a more complex binary image, and finally an image generated with gaussian noise, named *noise*. In Figure 5, we depict the distance maps obtained by our two algorithms processed on those grids and the regular grids. The grey level gl of a cell corresponds to the value of the distance d with a simple modulo operator ($gl = d \mod 255$). We present in Table 2 the time of execution for each algorithm, and in Table 1 the important features for each $\mathbb{I}$-grid (*e.g.* number of background and foreground cells, size of the irregular matrix). We have performed those experiments on a mobile workstation with a 1.5 Ghz Intel Pentium M processor, and 1 Gb RAM. Algorithm 1 is a fast way to compute the $\mathbb{I}$-DT on irregular sparse grids (in our tests, the Quadtree and RLE grids) and is faster than Algorithm 2 on those grids. But, our separable approach is very competitive and fast for the regular grid and the irregular dense grids (Quadtree and RLE grids for the image *noise*). The time of execution of Algorithm 2 is slighlty the same as the original version of [22] in the regular square grid case. Indeed, the irregular matrix and the image that generates the grid have the same size ($n_X \times n_Y = N$), and the only difference between the two versions of this algorithm is the find operation in the last scan, which is fastly processed (for the image *canon*, the original algorithm takes 0.90 seconds). Thanks to Table 1, we can see that the numbers of background and foreground cells do not significally modify the behaviour of our algorithms. The structure of the considered grid seems to be the only factor that slows down them. Algorithm 1 hardly handles regular grids, and returns the I-DT in more than 1 minute for the image *canon*. On the contrary, irregular sparse grids make the irregular matrix more complex than the initial image structuration for Algorithm 2.

In conclusion, to anticipate the speed of our algorithm, we should consider the size and the density of the irregular grid. If we compare the complexity of our algorithms, it is clear that Algorithm 2 has generally a worse time complexity in $\mathcal{O}(n_X n_Y \log n_Y)$, where n_X and n_Y are the size of the irregular matrix. However, in practice, this approach is faster than Algorithm 1 which time complexity is $\mathcal{O}((n_B + n_F) \log n_B)$, where n_F and n_B are the number of foreground and background cells. This could be explained by the simplicity of our separable algorithm, more precisely two independent scans along the X and Y axis. In Algorithm 1, we have to search for each foreground point p the Voronoi cell where p stands. Even if this operation is bounded at $\mathcal{O}(\log n_B)$, the global execution time suffers from the data structure implemented.

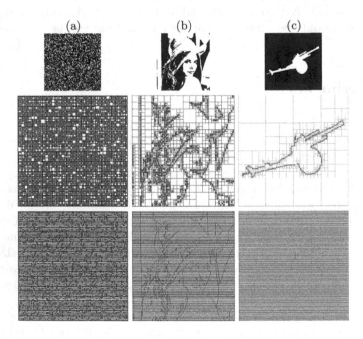

Fig. 4. We depict here the input irregular grids - *(top)*: Quadtree, *(bottom)*: RLE along X axis- we consider for each sample image *(a): noise, (b): lena, (c): canon*

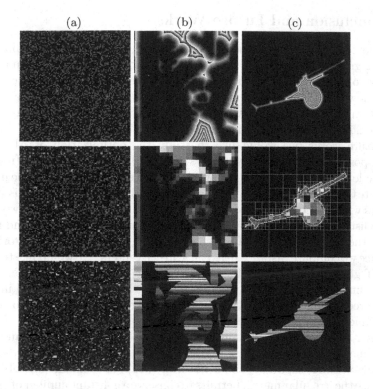

Fig. 5. Distance maps for our sample images, for the square grid *(top)*, the Quadtree grid *(middle)*, and the RLE grid along X axis *(bottom)*

Table 1. For each $\mathbb{I}$-grid, we illustrate the image size, the number of cells, the number of background and foreground cells, and the size of the associated irregular matrix

Image	Image size	$\mathbb{I}$-grid	N	n_B	n_F	$n_X \times n_Y$	$\frac{N}{n_X \times n_Y}$
		Square	10 000	7414	2586	100×100	1.000
noise	100×100	Quadtree	7 457	5 003	2 454	138×138	0.390
		RLE	3 903	1 973	1 930	199×100	0.200
		Square	46 176	18 789	27 387	208×222	1.000
lena	208×222	Quadtree	8 171	4 143	4 028	308×365	0.070
		RLE	3 462	1 695	1 767	392×222	0.040
		Square	262 144	234 302	27 842	512×512	1.000
canon	512×512	Quadtree	4 177	2 162	2 015	920×527	0.010
		RLE	1 432	972	460	577×512	0.005

Table 2. The execution time in seconds for every $\mathbb{I}$-grids

Image	Algorithm 1 (complete VD algorithm)		
	Square	Quadtree	RLE
noise	0.77	0.34	0.22
lena	2.8	**0.36**	**0.22**
canon	104.4	**0.38**	**0.27**

Image	Algorithm 2 (proposed approach)		
	Square	Quadtree	RLE
noise	**0.15**	0.21	**0.15**
lena	**0.26**	0.40	0.32
canon	**1.0**	1.2	1.3

5 Conclusion and Future Works

In this article, we have proposed two completing algorithms to compute the $\mathbb{I}$-DT on $\mathbb{I}$-grids. The execution time of our approaches mainly depends on the structure of the grid and on the size of the image treated. The sparser the grid is, the slower the $\mathbb{I}$-DT will be performed thanks to our separable approach (Algorithm 2), and inversely, Algorithm 1 (based on the Voronoi diagram of the background cells) hardly handles dense grids.

In future works, we would like to study the optimization of Algorithm 2 and propose an optimal time process in $\mathcal{O}(N)$ time complexity (instead of $\mathcal{O}(n_X n_Y \log n_Y)$). Indeed, we compute the distance in every nodes of the matrix to propagate the distance values in the rest of the matrix. Since the size of this structure increases with the complexity of the $\mathbb{I}$-grid (see Table 1, last column), we propose to use a list-based data structure to reduce the computational time and memory space of the $\mathbb{I}$-DT [2,12]. We have also depicted applications of our algorithms in GIS, since we compute distance between grid cell centers. We would like to develop the $\mathbb{I}$-DT given in Equation 3. In this case, the representation of the background does not impact the result obtained with the $\mathbb{I}$-DT. We are interested in investigating three dimensions (3-D) $\mathbb{I}$-DT, since we can easily extend our approaches to higher dimensions. To build the medial axis of binary irregular objects, we should make the $\mathbb{I}$-DT reversible and compute the REDT (Reversible Euclidean Distance Transformation) on $\mathbb{I}$-grids, as it was proposed in [6] for the regular square grid. This REDT algorithm is separable and is naturally adapted to handle 3-D objects. Finally, since the irregular matrix permits to conserve a constant number of neighbors for each node, a chamfer mask-based approach (like in [10]) may be adapted to this structure. The main problem is to change the mask and consider the configuration of each node of the matrix during the scan (non-stationary distance computation).

References

1. Bentley, J.L.: Multidimensional Binary Search Trees Used for Associative Searching. Communications of the ACM 18(9), 509–517 (1975)
2. Breu, H., Gil, J., Kirkpatrick, D., Werman, M.: Linear Time Euclidean Distance Algorithms. IEEE Transactions on Pattern Analysis and Machine Intelligence 17(5), 529–533 (1995)
3. CGAL, Computational Geometry Algorithms Library, http://www.cgal.org
4. Chehadeh, Y., Coquin, D., Bolon, P.: A Skeletonization Algorithm Using Chamfer Distance Transformation Adapted to Rectangular Grids. In: 13th International Conference on Pattern Recognition (ICPR 1996), vol. 2, pp. 131–135 (1996)
5. Coeurjolly, D.: Supercover Model and Digital Straight Line Recognition on Irregular Isothetic Grids. In: Andrès, É., Damiand, G., Lienhardt, P. (eds.) DGCI 2005. LNCS, vol. 3429, pp. 311–322. Springer, Heidelberg (2005)
6. Coeurjolly, D., Montanvert, A.: Optimal Separable Algorithms to Compute the Reverse Euclidean Distance Transformation and Discrete Medial Axis in Arbitrary Dimension. IEEE Transactions on Pattern Analysis and Machine Intelligence 29(3), 437–448 (2007)
7. Cuisenaire, O.: Distance Transformations: Fast Algorithms and Applications to Medical Image Processing. PhD Thesis, Université Catholique de Louvain, Louvain-La-Neuve, Belgium (October 1999)

8. Devillers, O.: Improved Incremental Randomized Delaunay Triangulation. In: 14th Annual ACM Symposium on Computational Geometry, 106–115 (1998)
9. Fouard, C., Malandain, G.: 3-D Chamfer Distances and Norms in Anisotropic Grids. Image and Vision Computing 23(2), 143–158 (2005)
10. Fouard, C., Strand, R., Borgefors, G.: Weighted Distance Transforms Generalized to Modules and their Computation on Point Lattices. Pattern Recognition 40(9), 2453–2474 (2007)
11. Golomb, S.W.: Run-length Encodings. IEEE Transactions on Information Theory 12(3), 399–401 (1966)
12. Guan, W., Ma, S.: A List-Processing Approach to Compute Voronoi Diagrams and the Euclidean Distance Transform. IEEE Transactions on Pattern Analysis and Machine Intelligence 20(7), 757–761 (1998)
13. Hesselink, W.H., Visser, M., Roerdink, J.B.T.M.: Euclidean Skeletons of 3D Data Sets in Linear Time by the Integer Medial Axis Transform. In: Proceedings of 7th International Symposium on Mathematical Morphology, pp. 259–268 (2005)
14. Jung, D., Gupta, K.K.: Octree-Based Hierarchical Distance Maps for Collision Detection. In: IEEE International Conference on Robotics and Automation, vol. 1, pp. 454–459 (1996)
15. Karavelas, M.I.: Voronoi diagrams in CGAL. In: 22nd European Workshop on Computational Geometry (EWCG 2006), pp. 229–232 (2006)
16. Maurer, C.R., Qi, R., Raghavan, V.: A Linear Time Algorithm for Computing Exact Euclidean Distance Transforms of Binary Images in Arbitrary Dimensions. IEEE Transactions on Pattern Analysis and Machine Intelligence 25(2), 265–270 (2003)
17. Meijster, A., Roerdink, J.B.T.M., Hesselink, W.H.: A General Algorithm for Computing Distance Transforms in Linear Time. In: Mathematical Morphology and its Applications to Image and Signal Processing, pp. 331–340 (2000)
18. Paglieroni, D.W.: Distance Transforms: Properties and Machine Vision Applications. In: CVGIP: Graphical Models and Image Processing, vol. 54, pp. 56–74 (1992)
19. Preparata, F.P., Shamos, M.I.: Computational Geometry - An Introduction. Springer, Heidelberg (1985)
20. Rosenfeld, A., Pfaltz, J.L.: Sequential Operations in Digital Picture Processing. Journal of the ACM 13(4), 471–494 (1966)
21. Rosenfeld, A., Pfalz, J.L.: Distance Functions on Digital Pictures. Pattern Recognition 1, 33–61 (1968)
22. Saito, T., Toriwaki, J.: New Algorithms for n-dimensional Euclidean Distance Transformation. Pattern Recognition 27(11), 1551–1565 (1994)
23. Samet, H.: The Design and Analysis of Spatial Data Structures. Addison-Wesley Longman Publishing Co., Inc, Amsterdam (1990)
24. Schouten, T., Broek, E.: Fast Exact Euclidean Distance (FEED) Transformation. In: 17th International Conference on Pattern Recognition (ICPR 2004), vol. 3, pp. 594–597 (2004)
25. Sintorn, I.M., Borgefors, G.: Weighted Distance Transforms for Volume Images Digitized in Elongated Voxel Grids. Pattern Recognition Letters 25(5), 571–580 (2004)
26. Vörös, J.: Low-Cost Implementation of Distance Maps for Path Planning Using Matrix Quadtrees and Octrees. Robotics and Computer-Integrated Manufacturing 17(6), 447–459 (2001)
27. Wang, X., Bertrand, G.: Some Sequential Algorithms for a Generalized Distance Transformation Based on Minkowski Operations. IEEE Transactions on Pattern Analysis and Machine Intelligence 14(11), 1114–1121 (1992)

Self-similar Discrete Rotation Configurations and Interlaced Sturmian Words

Bertrand Nouvel*

Center for Frontier Medical Engineering
Chiba University
1-33 Yayoi-cho, Inage-ku, Chiba 263-8522, Japan

Abstract. Rotation configurations for quadratic angles exhibit self-similar dynamics. Visually, it may be considered as quite evident. However, no additional details have yet been published on the exact nature of the arithmetical reasons that support this fact. In this paper, to support the existence of self-similar dynamic in 2d-configuration, we will use the constructive 1-d substitution theory in order to iteratively build quadratic rotation configurations from substitutive Sturmian words. More specifically : the self-similar rotation configurations are first shown to be an interlacing of configurations that are direct product of superposition of Sturmian words.

1   Introduction

Throughout the history of the Digital Geometry, the analysis of the characteristics of discrete lines, discrete planes, discrete spheres, intersections and distances in digital spaces had been a major issue of the field. And, this is generally true for the digital counterparts of all the fundamental objects from Euclidean Geometry Within these studies, regular discrete patterns have occurred quite often. Some of them can be shown to be quasi-periodic, however, the self-similar characteristics that these patterns may have , are generally less obvious.

For instance, the characteristics of the dragon-shape tiles that appear as equivalence class under quasi-affine transform has remained relatively difficult to explain with constructive discrete tools (See [NR95], [LKV04]). Nowadays, through some constructive theories, including continued fraction theory and generalized Sturmian sequences and their generalizations, it seems that we get closer to a wider understanding of connection in-between real dynamics and substitutions occurring in symbolic dynamical systems. Similarly, using above theories, any Sturmian word with specified slope and intercept can be constructed by the mean of S-adic systems. In this paper, we add another example of applications of these theories. We explain how to connect these results to prove the existence of substitutions that underlies the dynamics in discrete rotation configurations.

* Thanks to LAVOISIER Program of the French Ministry of Foreign Affairs and to the French Television Channel TF1, they both support financially this research.

D. Coeurjolly et al. (Eds.): DGCI 2008, LNCS 4992, pp. 250–261, 2008.

The main question that this paper deals with is to test whether knowledge that we have one dimensional self-similar dynamics is useful or not to analyze and generate 2d-self-similar configurations such as rotation configurations ?

This paper provides the details for constructing rotation configurations of quadratic angles from Sturmian words. So far, there was no English publication that really provides the evidences for the self-similar aspects of this process.

At first, we will review the necessary framework : basic definitions, rotation configurations, their properties, constructive results about Sturmian words. Then we will focus on self-similar configurations. We shall empirically notice the self-similarity of the configurations. Then we will considerate some of the properties of quadratic angles. This will provide a useful decomposition of the configurations. This will allow us to emphasize the validity of the approach and to show that the computed substitutions output the correct words.

2   Vocabulary and Notations

We use $\lfloor x \rfloor$ to denote the usual *floor function* (The biggest integer that is equal or smaller to x). The *rounding to the closest integer* point is then defined by $[x] = \lfloor x + \frac{1}{2} \rfloor$. These functions from $\mathbb{R}$ to $\mathbb{Z}$ extend to higher dimensions by independent applications on each component. Let $\{x\} = x - [x]$, $x \mapsto \{x\}$ will be considered as a *canonical projection* to the interval $[-\frac{1}{2}, \frac{1}{2}[$, which is also a representant of the torus $\mathbb{T} = \mathbb{R}/\mathbb{Z}$. However, for arithmetical issues, we may tend to prefer projection to the interval $[0, 1[$, another representant of $\mathbb{T}$. We will then use : $\lfloor x \rceil = x - \lfloor x \rfloor$.

Generally, we will also work in the complex plane $\mathbb{C}$. For any point $z \in \mathbb{C}$, its real part is denoted $\Re(z)$, and its imaginary parts is denoted $\Im(z)$. They are reals such that $z = \Re(z) + \Im(z)i$. The set of points whose real and imaginary parts are both integers form the set of Gaussian integers, $\mathbb{Z}[i]$.

Let α denote an angle in radians, i.e. an element of $\mathcal{A} = \mathbb{R}/(2\pi\mathbb{Z})$. The *Euclidean rotation* r_α is the one-to-one isometry of $\mathbb{C}$, $z \mapsto ze^{i\alpha}$. *The discretized rotation* $[r_\alpha]$ is the successive computation of the Euclidean rotation of angle α and of the discretization operator $z \mapsto [z]$.

Let Q be a finite set whose elements will be called *letters*. Q is called the *Alphabet*. Any finite sequence of elements of Q is called a *word* on Q . Bi-infinite sequence of letters are *bi-infinite words* ($\mathbb{Z} \to Q$). If all the letters of a word w_f occur consecutively in a word w then w_f is a *factor* of w.

3   Rotation Configurations (C_α)

3.1   Definition

A *configuration* is an application C that maps each Gaussian integer z of $\mathbb{Z}[i]$ to an element $C(z)$ of a finite set Q.

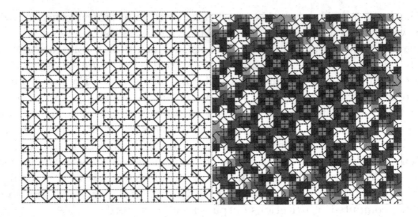

Fig. 1. The configuration $C_{0.4}$. In further drawing, projections of the elements of $2^{\mathbb{A}}$ to arbitrary color-set, in order to observe the dynamic at a broader scale.

For a specified angle α, we are interested in the *rotation configurations C_α*, which is the application that *to any Gaussian integer z associates the relative position of its 4-neighbors after rotation*. The set of states Q used in this paper is the set of subsets of the set of arrows $\mathbb{A} = \{-1, 0, 1\}[i]$, $Q = 2^{\mathbb{A}}$. Formally:

$$C_\alpha : \mathbb{Z}[i] \mapsto 2^{\mathbb{A}} : z \to \bigcup_{\delta \in \{1, -1, i, -i\}} \{[e^{i\alpha}(z + \delta)] - [e^{i\alpha}z]\}$$

The details of the construction are illustrated on Figure 2, and can be stated as follows : Figure 2(a), within the Gaussian integers of $\mathbb{Z}[i]$ displayed in gray, a point z has been evinced. Its image under rotation $e^{i\alpha}\mathbb{Z}[i]$ is highlighted. The different discretization cells are separated by thin black dotted lines. These line provide a representation of the frontier in-between the equivalence class for the discretization operator. The Figure 2(b) z, the point that is being considered, and its four 4-neighbors; there we compute their image under discretized rotation. Figure 2(c), we have displayed $C_\alpha(z)$, the code will associate to z. This is the set of arrows that link the image of z with the image of its 4-neighbors. When this code is actually computed for all $z \in \mathbb{Z}[i]$ this leads to figures such as Figure 1.

3.2   General Properties

We can refer to [No06] to ensure that :

- The rotation configurations encode efficiently the discretized rotation $[r_\alpha]$;
- The configurations are bi-periodic if and only if α is such that $cos(\alpha)$ and $\sin(\alpha)$ are rational;
- For any angle α, the resulting configuration C_α is quasi-periodic;

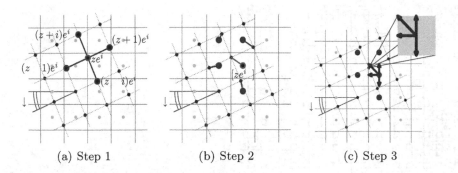

(a) Step 1   (b) Step 2   (c) Step 3

Fig. 2. Step-by-step construction of the C_α configuration as explained in main text. The resulting symbol is designed using a arrows, it directly represents the code associated to z.

- The configuration may be reinterpreted as the coding of a $\mathbb{Z}^2$-action on a labeled torus $(\mathbb{R}/\mathbb{Z})^2$. Formally :

$$\exists l : \mathbb{T}[i] \to \mathbb{A}, T_x, T_y \in \mathbb{T}[i] \to \mathbb{T}[i] | \exists T_x^{-1}, T_y^{-1} \wedge \forall \mathbf{v} \in \mathbb{Z}^2, C(\mathbf{v}) = l(T_x^{v_x}(T_y^{v_y}(\mathbf{v})))$$

For a specified angle of rotation α, the following system is precisely known and it has been described in [No06]:

- $T_x : z \mapsto z + e^{i\alpha}$ and $T_y : z \mapsto z + i\, e^{i\alpha}$;
- $l : z \mapsto Q$ is such that there exists a morphism $\phi : Q \to Q$, for any $z \in \mathbb{T}[i], l(z) = \phi(l(iz))$; moreover $l' : \mathbb{T} \to Q$ is such that there exists an application $\psi : Q^2 \to Q$ such that for any $z \in \mathbb{T}[i] l(z) = \psi(l'(\Re(z)), l'(\Im(z)))$.

As a consequence of this we retain that C is defined by

$$C(z) = \psi(l'(\Re(\{ze^{i\alpha}\})), l'(\Im(\{ze^{i\alpha}\}))) \tag{1}$$

3.3 Self-similarity

Any angle such that $\cos(\alpha)$ and $\sin(\alpha)$ belong to the same quadratic field is called a *quadratic* angle. Note that if $\cos(\alpha)$ and $\sin(\alpha)$ belong to the same quadratic field then necessarily there exists $p, q, k \in \mathbb{Z}\ p\cos(\alpha) + q\sin(\alpha) = k$.

In this paper, we aim at finding a way to explain the self-similarity of the rotation configurations for quadratic angles. Hence, we have first to conjecture that rotation configurations may be self-similar. In order to get convinced, we show two simple sample configurations for which it is quite easy to notice that the resulting configuration exhibits some self-similar patterns, namely $C_{\pi/4}$ and $C_{\pi/6}$. Please have a look at Figure 3.

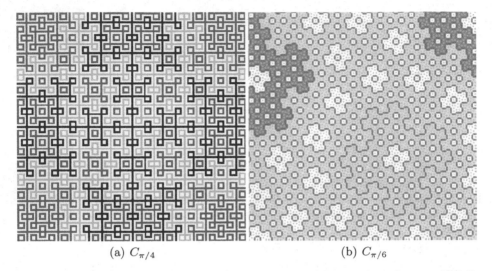

(a) $C_{\pi/4}$ (b) $C_{\pi/6}$

Fig. 3. Sample auto-similar configurations. The configurations have been artificially recolored using bucket-fill tool under "the gimp" in order to make more visible the self-similar patterns.

4   Fundamentals for Constructive Sturmian Theory

4.1   Sturmian Words and Rotation Words

According to Morse-Hedlund Theorem [MH38], for any bi-infinite word w, if there is a positive integer n such that the number of factor of size n is smaller than n then w is periodic. If we define the *complexity function* of word w as the function that to $n \in \mathbb{N}$ associates the number of factors of size n, then the contraposition of the theorem is that any aperiodic bi-infinite word is aperiodic has necessarily a complexity function $p(n)$ that is such that $p(n) > n$ for any $n \in \mathbb{N}$. The class of words such that $p(n) = n+1$ for any n, is the class of *Sturmian words*. It widely known that Sturmian words code the trajectory of straight lines of irrational slopes on the discrete grid ([Fo02]). Hence any Sturmian word $\vec{w}$ (α, β) is defined by a slope $\alpha\mathbb{R}\backslash\mathbb{Q}$ and by an intercept $\beta \in \mathbb{R}$ and an orientation, in this paper[1]; we set up : $\vec{w}$ $(\alpha, \beta) = \mathbb{1}_{[1-\alpha,1[}(\lfloor n\alpha + \beta \rfloor)$. More generally, a rotation word $\overset{\circlearrowleft}{w}$ (α, β, γ) will be defined here by : $\overset{\circlearrowleft}{w}$ $(\alpha, \beta, \gamma) = \mathbb{1}_{[1-\gamma,1[}(\lfloor n\alpha + \beta \rfloor)$. Regarding the complexity of rotation words, an extensive study is available in [Al96].

4.2   Constructive Results on Sturmian Words

Due to limited space, we discuss only the essential issues, for real introductions to Sturmian theory, please consult [Fo02], and for more constructive results : [BHZ], [Ad02].

---

[1] The other case would be : $\vec{w}'$ $(\alpha, \beta) = \mathbb{1}_{]1-\alpha,1]}(\lfloor n\alpha + \beta \rfloor)$.

According to a Theorem of Berstel-Séébold, the set of Sturmian morphisms[2] has a monoid structure. Let $\sigma_{0a}, \sigma_{0b}, \sigma_{1a}, \sigma_{1b}$ are four applications from $\{0,1\}$ to $\{0,1\}^*$ that we are going to extend as monoid morphism. It can be shown that they can serve as generators for the Berstel-Séébold monoid :

$$\begin{cases} \sigma_{0a}(0) \to 0 \\ \sigma_{0a}(1) \to 10 \end{cases} \begin{cases} \sigma_{0b}(0) \to 0 \\ \sigma_{0b}(1) \to 01 \end{cases} \begin{cases} \sigma_{1a}(0) \to 01 \\ \sigma_{1a}(1) \to 1 \end{cases} \begin{cases} \sigma_{1b}(0) \to 10 \\ \sigma_{1b}(1) \to 1 \end{cases}$$

Let $0 < \alpha < 1$ be an irrational real, according to Lagrange theorem, it has an *infinite continued fraction expansion* : $[0; a_1, a_2, \ldots, a_i, \ldots]$, which denotes that $\alpha = \frac{1}{\alpha_1 + \frac{1}{\alpha_2 + \ldots}}$. If α is quadratic then sequence of the partial quotients is ultimately periodic. The rational numbers obtained by truncating the expansion of α after a certain partial quotient are called the *convergents*.

Also the error in-between the convergent and the value of α form an important quantity; we will denote it by ϵ_i. This value is major for understanding odometers-like dynamical system (such as calendars), and also the 3-distances Theorem of V.T. Sós.

As presented in Berthé [Be02], let α be an irrational real, the α-*Ostrowski expansion* (c_i) of a real number β is :

$$\beta = \sum c_{i+1}\epsilon_i, \text{ with } \forall i \in \mathbb{N}, c_i \in \mathbb{N}, \text{, and } c_i \text{ relying in } - \text{between } 0 \text{ and } a_i$$

Theorem 1 (Valérie Berthé,Charles Holton,Luca Q. Zamboni, [BHZ]).
[3] *Any Sturmian sequence S-adic representation of the form :*

$$\vec{w}(\alpha, \beta) = (T^{c_1} \circ \sigma_{0b}^{a_1}) \circ (T^{c_2} \circ \sigma_{1b}^{a_2}) \circ (T^{c_3} \circ \sigma_{0b}^{a_3}) \circ (T^{c_4} \circ \sigma_{1b}^{a_4}) \ldots \quad (2)$$

where T is the usual shift on sequences, where $\{a_i\}_{i\in\mathbb{N}}$ correspond to partial quotients of $\alpha' = \frac{\alpha}{1+\alpha}$, and the sequence $\{c_i\}_{i\in\mathbb{N}}$ matches the α'-Ostrowski-expansion of the intercept $\beta' = \frac{\beta}{1+\alpha}$.

We should then remark that, for any Sturmian word w, one has $T^{c_i} \circ \sigma_{0b}^{a_i}(w)$ (resp. $T^{c_i} \circ \sigma_{1b}^{a_i}(w)$) coincide with $\sigma_{0a}^{c_i} \circ \sigma_{0b}^{a_i-c_i}(w)$ (resp. $\sigma_{1a}^{c_i} \circ \sigma_{1b}^{a_i-c_i}(w)$) on the first $|\sigma_{0b}^{a_i}(w)| - c_1$ letters. Thus Equation 2 may be rewritten as :

$$\vec{w}(\alpha, \beta) = (\sigma_{0b}^{c_1} \circ \sigma_{0a}^{a_1-c_1}) \circ (\sigma_{1b}^{c_2} \circ \sigma_{1a}^{a_2-c_2}) \circ (\sigma_{0b}^{c_3} \circ \sigma_{0a}^{a_3-c_3}) \circ (\sigma_{1b}^{c_4} \circ \sigma_{1a}^{a_4-c_4}) \ldots \quad (3)$$

Referring to [BHZ], we know that a Sturmian sequence of slope α of intercept x is a if and if only if α is a quadratic irrational number and $x \in \mathbb{Q}(\alpha)$.

Also, one should note an important corollary of previous remarks and theorems. Any Sturmian word whose slope is quadratic and whose intercept is in the

---

[2] Endomorphism on Sturmian sequences.
[3] The fix-up of α, and β into α' and β' through $x \mapsto (x)/1 + \alpha$ is a classical operation. It corresponds to the fact that the system we used for defining Sturmian words is different from traditional odometers.

quadratic extension of the slope is substitutive. It is the image under morphism of a fixed point of substitution. Hence it exhibits some self-similar characteristics.

Generalization are possibles to rotation words: In [Ad02], we may find an interesting constructive results on S-adic systems by the mean of generalized continued fraction like expansions. Using this algorithm, we can find the S-adic expansion associated to a rotation word.

5   Grid-Configurations and Sturmian Words

From now, we shall continuously assume that α is a quadratic angle. The interlacing vector z_i associated with a quadratic angle α is a non null Gaussian integer $z_I = p + qi$ such that $z_I e^{i\alpha}$ has an integer component. The notation does not mention the angle α, the link is however existent and implicit.

5.1   Grid Configurations

Let $z_M \in \mathbb{Z}[i]/(z_I\mathbb{Z}[i])$, in order to analyze the configurations C_α, we introduce the *grid configurations* defined by $D_{\alpha,z_M}(z) = C_\alpha(z_I z + z_M)$.

By application of Equation 1, we immediately have

$$D_{\alpha,z_M}(z) = \psi(l'(\Re(\{(zz_I + z_M)e^{i\alpha}\})), l'(\Im(\{(zz_I + z_M)e^{i\alpha}\}))) \qquad (4)$$

Since $z_I e^{i\alpha}$ has one integer coordinate, these configurations will have some useful properties in further discussions. Practically, a configuration D_{α,z_M} contains points regularly extracted from C_α. They extracted points form the set : $z_I\mathbb{Z}[i] + z_M$. Such networks are represented within different symbols on rotated grids in the Figures 6 and 5. Conversely, from D_{α,z_M}, we may redefine $C_\alpha(z) = D_{\alpha,z_M}(z_R)$; with $z_M \in D(\mathbb{Z}[i]/(z_I\mathbb{Z}[i]))$ and $z_R \in \mathbb{Z}[i]$ such that $z = z_I z_R + z_M$, where $D(\mathbb{Z}[i]/(z_I\mathbb{Z}[i]))$ denotes the canonical projection domain of $\mathbb{Z}[i]/(z_I\mathbb{Z}[i])$ on $\mathbb{Z}[i]$.

For a specified α, the number of configurations D_{α,z_M} can be seen as $N = \#(\mathbb{Z}[i]/(z_I\mathbb{Z}[i]))$; recall that $z_I = p + qi$, then $N = p^2 + q^2$, it is the number of integer point in a square[4] of side z_i.

5.2   Grid Configurations from Rotation Words

Since zz_I has an integer component then, from Equation 4, we can observe that the bi-dimensional configuration $l'(\{\Re(zz_I + z_M)\})$ is actually completely defined one by one infinite word. Either all lines or columns are constant. More precisely, we have, either for all $z \in \mathbb{Z}[i]$, $\{\Re(1z_I e^{i\alpha})\} = 0$ or for all $z \in \mathbb{Z}[i]$, $\{\Im(iz_I e^{i\alpha})\} = 0$. The same remark holds also for the imaginary part and the configuration $l'(\{\Im(zz_I + z_M)\})$.

---

[4] Square with half opened borders.

Let's assume without any loss of generality, since the same argumentation hold in the other case, that for all $z \in \mathbb{Z}[i]$ we have : $\lfloor\Re(z_I e^{i\alpha})\rceil = 0$

We can now define the word : $w_v(n) = l'(\{\Re((niz_I + z_M)e^{i\alpha})\})$. This one can be rewritten in : $w_v(n) = l'(\{(\Re(z_m) - nq)\cos(\alpha) - (\Im(z_m) + pn)(\sin(\alpha))\})$ Hence, we have

$$w_v(n) = l'(\{i_0 - n\alpha_0\})$$

with $i_0 = (\Re(z_m)\cos(\alpha) - \Im(z_m)\sin(\alpha))$ and $\alpha_0 = (q\cos(\alpha) + pn(\sin(\alpha)))$. Since l is a function to a finite set whose antecedents are forms connected sets of $\mathbb{T}$, we can rewrite l in such way :

$$l'(x) = \sum_{i=0}^{N-1} i \mathbb{1}_{[k(b(i)),k_{d(i)}[}(x)$$

Where N is the number of areas that are necessary to define l'. S, the set of cuts, is then formally defined by:

$$S = \{\overline{l^{\{-1\}}(i)} \cup \overline{l^{\{-1\}}(j)}, (i,j) \in \{0, \ldots, N-1\}^2\} = \{k(i), i \in \{0, \ldots, N-1\}\}$$

Hence k is a function from $\{0, \ldots, N-1\}$ to that associated to an integer the ith cut of the partition that underlies the l function.

We can now write

$$w_v(n) = \sum_{i=0}^{N-1} i \overset{\circlearrowleft}{w} (\alpha_0, i_0 - k(b(i)), k(d(i)) - k(b(i)))(n)$$

5.3 From Rotation Words to Sturmian Words

All the cuts are actually copies under translation by $ze^{i\alpha}$ of the discontinuity of the discretization operator located at $\frac{1}{2}$, for some z Gaussian integer z. Therefore, all the cuts belong to the quadratic field as $\cos(\alpha)$ and $\sin(\alpha)$.

The conclusion is that actually all the rotation words expressed in previous section can be expressed as product of Sturmian words where the intercepts of the word belong to the quadratic to the same quadratic field.

Proposition 1. *Let α be a quadratic angle. All the configurations C_α can be constructed by superposing, adding, multiplying substitutive Sturmian words according to the scheme that has been explained in this section. Moreover, the sequence of morphism to generate this word from $(01)^\omega$ is precisely known.*

The proof of this proposition is a corollary of the previous argumentation.

5.4 Application to $C_\pi/4$

To be more concrete, we apply our construction to $C_{\pi/4}$.

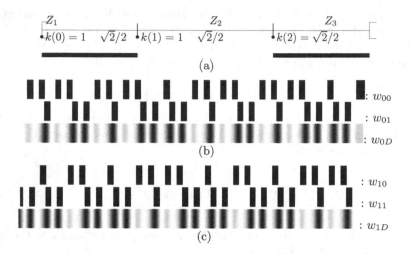

Fig. 4. Decomposition of the words that generate the configurations $D_{\pi/4,0}, D_{\pi/4,1}$. (a) : The basic partition function l' that is used for $C_{\pi/4}$ contains only 3 areas. (b) and (c) : The details of the construction of the words W_{1D} and W_{2D} from the words w_{00}, w_{01}, w_{10}, w_{11} which contains the odd letters of Sturmian words whose characteristics have already been specified.

First, we notice $\sin(\alpha) = \cos(\alpha) = \frac{\sqrt{2}}{2}$. The interlacing vector is $z_I = (1 + i)$, we have $\Re((1 + i)e^{i\alpha}) = 0$ and $z_i \bar{z}_i = 2$. Thus, two grid-configurations exist: $D_{\pi/4,0}$ and $D_{\pi/4,1}$. From previous argumentation, we have that we can define these configurations from two words : $D_{\pi/4,0} = \psi(w_{0D}(\Im(z)), w_{0D}(\Re(z)))$, (resp. $D_{\pi/4,1} = \psi(w_{1D}(\Im(z)), w_{1D}(\Re(z))))$. It is easy to notice that in this case horizontal and vertical words are the same. We thus have two construct the following words : $w_{1D}(n) = l'(\lceil n\sqrt{2} + \frac{\sqrt{2}}{2}\rceil)$ and $w_{0D}(n) = l'(\lceil n\sqrt{2}\rceil)$.

We choose to compute l', the morphism issued from the characterization of the zones, By using the characteristic function of the zones $Z_1 \cup Z_2$ and $Z_2 \cup Z_3$, (referring to the notations of Figure 4(a)), because they have the appropriate size to make Sturmian words. We obtain that : $w_{0D}(n) = \psi(w_{00}(n), w_{01}(n))$ and $w_{1D}(n) = \psi(w_{10}(n), w_{11}(n))$ where: $w_{01}(n) = \vec{w}\, (\frac{\sqrt{2}}{2}, \frac{1}{2})(2n)$, $w_{00}(n) = w_{11}(n) = \vec{w}\, (\frac{\sqrt{2}}{2}, \frac{\sqrt{2}}{2} - \frac{1}{2})(2n)$, $w_{10}(n) = \vec{w}\, (\frac{\sqrt{2}}{2}, \sqrt{2} - \frac{1}{2})(2n)$.

Proposition 2. *The following words may be used to generate $C_\pi/4$, according to the scheme explained :*
$$w_{01} = \sigma_{1b}^2(w_B), \quad w_{00} = w_{11} = \sigma_{1a} \circ \sigma_{1b}(w_B) \ , \quad w_{10} = \sigma_{1a}^2(w_B), \ with$$

$$w_B = (\sigma_{0a} \circ \sigma_{0b} \circ \sigma_{1a} \circ \sigma_{1b})^\omega(0)$$

The remaining numerical details that concludes the proof are left to the reader.

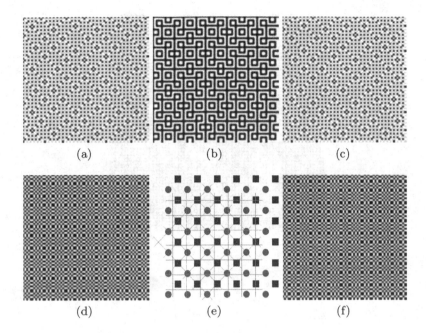

Fig. 5. In (b), the configuration $C_{\pi/4}$. $D_{\pi/4,0}$ and $D_{\pi/4,1}$ on their rotated network have been displayed as (a) and (c) (i.e. has they are inserted in (b)). In (d) and (f), we show the same grid-configurations $D_{\pi/4,0}$ and $D_{\pi/4,1}$ but normally, on an unrotated networks. The sub-figure (e) explains the way the two configurations are interleaved for $C_{\pi/4}$.

6 Perspectives

In this article, we have shown that any quadratic rotation configuration can be reinterpreted as a superposition and an interlacing Sturmian words.

In further work, the question of compatibility of the grid-configuration that are interlaced is to be addressed. Can the substitutions issued from each grid can be rearranged to form a connected rectangular substitutions on the grid ? Also, one may explore the exact limits of the generating power of substitutions as planar substitutive patterns on rotated and interlaced grids may be explored. This generalization of the studied configurations would includes the whole class of configurations that may be obtained by interweaving different grid configurations with different angles. So far, in our experiments, we have noticed that a wide diversity of patterns may appear. The methods that may be used to guess from a configuration whether it belongs to this class are also to be investigated. It may not be possible for all configurations, but it would be nice to know for which configurations this is possible The Rauzy graph, as a tool for the analysis of the topology induced by the patterns of the configurations, seems to be a promising solution, see [No06].

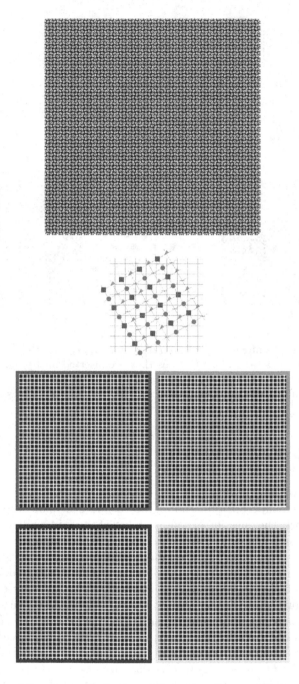

Fig. 6. The largest figure represents the configuration $C_{\frac{\pi}{6}(-)}$, with a very limited color-set. The four underlying figure represents the grid-configurations $\{D_{\frac{\pi}{6},b}\}_{b\in\{0,1,i,1+i\}}$, here $z_I = 2$. In between, the central figures recall visually the way the interlacing of the grid-configurations is done for $C_{\pi/6}$.

References

[Ad02]   Adamczewski, B.: Approche dynamique et combinatoire de la notion de discrépance. PhD thesis, Institut de Mathématiques de Luminy (2002)

[Al96]   Alessandri, P.: Codages de rotations et basse complexité. PhD thesis, Université de la Méditerranée (1996)

[Be02]   Berthé, V.: Autour du systéme de numération d'Ostrowski. Source: Bull. Belg. Math. Soc. Simon Stevin, 8(2), 209–239 (preprint, 2001) (English version: About Ostrowski Numeration System)

[BHZ]   Berthé, V., Holton, C., Zamboni, L.Q.: Initial Power of Sturmian Sequences (preprint)

[Fo02]   Fogg, P.: Substitutions in Dynamics, Arithmetics and Combinatorics. In: Berthé, V., Ferenczi, S., Mauduit, C., Siegel, A.: Lecture Notes in Mathematics, vol. 1794, 402 pages, (2002), ISBN: 3-540-44141-7

[LKV04] Lowenstein, J.H., Koupstov, K.L., Vivaldi, F.: Recursive tiling and geometry of piecewise rotations by pi/7. Nonlinearity 17, 371–395 (2004)

[MH38]   Morse, M., Hedlund, G.A.: Symbolic Dynamics. American Journal of Mathematics 60(4), 815–866 (1938)

[No06]   Nouvel, B.: Rotations Discrétes et Automates Cellulaires. PhD thesis, École Normale Supérieure de Lyon (2006)

[NR95]   Nehlig, P.W., Réveilles, J.-P.: Fractals and Quasi-Affine Transformations. Computer Graphics Forum 14(2), 147–157 (1995)

[NR05]   Nouvel, B., Rémila, É.: Configurations Induced by Discrete Rotations: Periodicity and Quasiperiodicity Properties. Discrete Applied Mathematics 127(2-3), 325–343 (2005)

Segmenting Simplified Surface Skeletons

Dennie Reniers[1] and Alexandru Telea[2]

[1] Department of Mathematics and Computer Science
Eindhoven University of Technology, Eindhoven, The Netherlands
d.reniers@tue.nl
[2] Institute for Mathematics and Computing Science
University of Groningen, Groningen, The Netherlands
a.c.telea@rug.nl

Abstract. A novel method for segmenting simplified skeletons of 3D shapes is presented. The so-called simplified Y-network is computed, defining boundaries between 2D sheets of the simplified 3D skeleton, which we take as our skeleton segments. We compute the simplified Y-network using a robust importance measure which has been proved useful for simplifying complex 3D skeleton manifolds. We present a voxel-based algorithm and show results on complex real-world objects, including ones containing large amounts of boundary noise.

1 Introduction

The skeleton, also known as medial axis, is a compact shape descriptor. The skeleton is of a lower dimensionality than the shape it describes, and makes the shape's structure, such as its topology and articulation, more explicit. Hence, it is a useful tool in a wide range of computer vision and visualization applications, such as shape analysis, recognition, shape alignment, motion planning, and collision detection. One way of analyzing the structure of the skeleton is by segmenting it in its logical parts. To give just one of many examples, Zhang *et al.* use such a skeleton segmentation as a step in their 3D model retrieval pipeline [1]. Besides the explicit capture of a shape's logical components, skeletons offer pose invariance under many types of shape deformation. However, skeletons are notoriously unstable to small boundary perturbations, which affects the segmentation robustness. What is needed is a robust segmentation of a simplified skeleton, i.e., a skeleton from which spurious, small-scale, parts are removed.

In this paper, we present a voxel-based algorithm for segmenting simplified skeletons of 3D shapes into disjoint segments. A 3D skeleton consists of a set of compact 2D manifolds, called *sheets*. A sheet boundary consists of 1D curves which are either part of the 3D skeleton's boundary, or curves where at least three sheets intersect. Given this property, the sheet intersection curves are also called *Y-curves* [2]. Our aim is to robustly segment a noisy 3D skeleton in its sheets.

We compute a robust simplified skeleton using [3], and extend this method to compute the simplified network of Y-curves by combining information from

D. Coeurjolly et al. (Eds.): DGCI 2008, LNCS 4992, pp. 262–274, 2008.
© Springer-Verlag Berlin Heidelberg 2008

two different simplification levels. Our segmentation, based on this simplified Y-network, can handle objects with large amounts of noise on the boundary, without having to prune the segmentation afterward. Besides the segmentation itself, the simplified Y-network is also a useful result of our approach, which can be used in shape analysis tasks. We demonstrate our voxel-based algorithm on several real-world examples, and show that our approach can handle objects containing large amounts of noise on the boundary.

As far as we know, this is the first segmentation method specifically designed for simplified 3D skeletons. Existing surface-segmentation approaches, such as the well-known topological segmentation for discrete surfaces of Malandain et al. [4], fail on our simplified skeletons, as we show in Sec. 5. To achieve a better segmentation, we consider not only the skeleton itself, but also its relation to the object boundary.

The outline of this paper is as follows. Section 2 introduces the definition of the skeleton and details on its structure. Section 3 briefly outlines our previous work on simplified skeletons necessary for a good understanding of the following material. Section 4 details our novel method for computing the simplified Y-network Y_τ, at simplification level τ, and the decomposition of Y_τ into its constituent Y-curves. In Section 5, we extend Y_τ using information from a less-simplified Y-network Y_υ to correctly and robustly segment the simplified skeleton at level of detail τ. Section 6 presents and discusses results. Section 7 concludes this paper.

2 Preliminaries

2.1 Skeleton Definition

Let Ω^d be a d-dimensional shape with boundary $\partial\Omega$. Let $D : \Omega \to \mathbb{R}^+$ be the distance transform, assigning to each object point the minimum Euclidean distance to the object's boundary. Let $F : \Omega \to \mathcal{P}(\partial\Omega)$ be the feature transform [5] (where $\mathcal{P}$ is the power set), assigning to each object point the set of boundary points at minimum distance:

$$F(p \in \Omega) = \left\{ q \in \partial\Omega \mid \|p - q\| = D(p) \right\}, \tag{1}$$

where $\|\cdot\|$ denotes Euclidean distance. The *skeleton* $\mathcal{S}$ of Ω can be defined in multiple equivalent ways: as the locus of centers of maximally inscribed balls, as the singularities of D, or as those object points having at least two feature points. We use the latter definition:

$$\mathcal{S}(\Omega) = \left\{ p \in \Omega \mid |F(p)| \geq 2 \right\}. \tag{2}$$

This definition can be used both in 2D and 3D. In 3D, $\mathcal{S}$ is also called the medial surface, or *surface skeleton*, to distinguish it from the curve skeleton, or centerline [6]. It has been proved that the skeleton is homotopic to the original shape in any dimension [7].

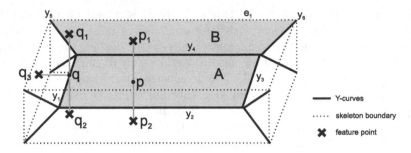

Fig. 1. Skeleton structure of a box. Point p is a sheet point having two feature points p_1, p_2, point q is a Y-curve point having three feature points q_1, q_2, q_3.

2.2 Skeleton Structure and Segment Definition

It is well known that the skeleton of a 3D object consists of manifolds, called *sheets*, that intersect in curves, called *Y-curves* [2]. A sheet's boundary consists of Y-curves and/or skeleton boundary curves. In addition to the above sheet boundary curves, a 3D skeleton can also contain isolated curves in some degenerate cases, such as a cylinder. We assume for the time being that the skeleton contains no such degeneracies. The set of Y-curves is called the *Y-network*, denoted $Y = \{y_1, ..., y_n\}$. In the following, a *skeleton segment* is equivalent to a sheet. For example, Fig. 1 shows the skeleton of a box. This contains 12 skeleton boundary curves (dotted lines), and a Y-network containing 12 Y-curves (thick lines), yielding 13 sheets, or segments. For illustration, segment A is bounded by Y-curves $y_1...y_4$, whereas B is bounded by Y-curves y_4, y_5, y_6 and skeleton boundary curve e_1.

Skeleton points can be classified by the number of feature points they have [8]. A skeleton boundary-curve point has an infinite amount of feature points, because the inscribed ball at such a point has a contiguous arc of contact points with $\partial\Omega$. A sheet point p has two feature points, $|F(p)| = 2$. Within a sheet, the associated feature points evolve smoothly over the object boundary. That is, for a point $p + \Delta p$ close to p, the feature points $F(p + \Delta p)$ neighbor the feature points $F(p)$ of p. Finally, a Y-curve point q, where three sheets intersect, has three feature points, $|F(q)| = 3$. Each sheet contributes two feature points, but each feature point is shared with one of the two other sheets. Fig. 1 illustrates such a sheet point p and Y-curve point q and their respective feature points (crosses).

3   Simplified Skeletons

Following from Eq. 2, skeletons are inherently sensitive to small boundary perturbations [9]. Boundary noise, for example due to sampling or acquisition artifacts, produces spurious skeleton parts. To handle this, some skeletonization methods define an importance measure $\rho : \mathcal{S} \rightarrow \mathbb{R}^+$ indicating the importance for each skeleton point to the shape representation. Together with a subsequent

Fig. 2. Simplified skeletons $\mathcal{S}_{\tau=10}, \mathcal{S}_{\tau=20}, \mathcal{S}_{\tau=70}$ (from left to right) of a noisy box. The intensity encodes the importance measure ρ. In the right image a voxel was selected which has 4 feature voxels, resulting in 6 paths.

pruning strategy, this delivers a simplified skeleton [10]. Well-known measures are the distance and angle between feature points [11]. The problem with these measures is that they are non-monotonic, which means that a non-trivial, often non-intuitive, pruning strategy is needed to enforce connectedness of the simplified skeleton, which is a desirable property as the skeleton should be connected (if the original object is connected). A frequent step of such a pruning strategy is an erosion step (e.g. [12]). To eliminate these complications, a monotonic importance measure can be used, so that simply thresholding the measure delivers a connected skeleton. In 2D, this is achieved by the monotonic boundary-distance measure, which takes as importance for a point the smaller distance along the boundary between its two feature points [13,14,15]. A robust, progressively simplified, connected skeleton is obtained simply by thresholding this measure with increasing values.

In [3], we extended the boundary-distance measure to 3D, and obtained the first monotonic importance measure for 3D skeletons. We assign to each point $p \in \mathcal{S}$ the length of the shortest geodesic on the surface $\partial\Omega$ between the two feature points $F(p)$. This measure is continuous on sheets, over which the feature points evolve smoothly, may contain jumps at Y-curves (cf. Fig. 2), and has a local maximum ridge that forms a 1D connected structure on $\mathcal{S}$. This last property has been shown in [16] and was used to formally define a curve skeleton.

The next section explains how we compute our measure in a discrete setting, a step which is needed prior to computing the simplified Y-network (Sec. 4).

3.1   Computing Simplified Skeletons

Most skeletonization methods work on discretized objects, using e.g. a boundary sampling [13] or a volumetric sampling on a regular voxel grid $\mathbb{Z}^3$. We take the latter, voxel-based, approach. Def. 1 and 2 for the continuum have to be adapted accordingly.

The object Ω, its boundary $\partial\Omega$, and skeleton $\mathcal{S}$ are represented as sets of voxels. We compute the feature transform of Ω by the raster-scanning approach of [17]. Voxelization introduces the problem that when placing a discretized inscribed ball at a skeleton voxel, this ball does not always touch the boundary exactly in two voxels: It might touch it only in one [5]. Thus, we *extend* the feature transform by merging the feature set of a voxel $p = (p_x, p_y, p_z)$ with the feature sets of p's 26-neighbors in the first octant:

$$\overline{F}(p \in \Omega) = \bigcup_{\Delta x, \Delta y, \Delta z \in \{0,1\}} F(p_x + \Delta x, p_y + \Delta y, p_z + \Delta z) . \qquad (3)$$

In the rest of this paper, the extended feature transform will simply be referred to as feature transform. Hence, skeletons based on $\overline{F}$ can be up to two voxels thick.

In order to compute our importance measure ρ for a voxel $p \in \Omega$, we compute the *shortest-path set* Γ, by computing between each two feature voxels a, b the shortest path $\gamma(a, b)$, as discrete equivalent to the shortest geodesic, on the boundary $\partial \Omega$:

$$\Gamma(p) = \bigcup_{a,b \in \overline{F}(p)} \gamma(a, b) . \qquad (4)$$

Shortest paths are computed as 26-connected voxel chains using Dijkstra's algorithm on the boundary graph, in which the voxels $\partial \Omega$ are the nodes, and the 26-neighborhood relations represent the edges. By caching computed paths, we avoid computing the same path twice, and thereby accelerate the method, as detailed in [3].

The importance measure ρ is now defined for each object voxel p as the maximum shortest-path length in $\Gamma(p)$:

$$\rho(p \in \Omega) = \max_{\gamma \in \Gamma(p)} \|\gamma\|, \qquad (5)$$

where $\|\gamma\|$ denotes the length of path γ, computed using the geodesic length estimator of [18]. In Figure 2 (right), the shortest-path set Γ containing 6 paths for a voxel with 4 feature voxels is shown. The resulting ρ is robust as it maximizes path length.

Finally, the simplified skeleton $\mathcal{S}_\tau$ is defined by imposing a threshold τ on ρ:

$$\mathcal{S}_\tau(\Omega) = \{p \in \Omega \mid \rho(p) > \tau\} . \qquad (6)$$

The threshold τ functions as a continuous scale-parameter controlling the simplification level. Small τ values eliminate less important skeleton parts that are due to small-scale surface features such as noise. Larger τ values can be used to retain the most salient parts of the skeleton. Thresholding ρ combines both skeleton detection *and* simplification in a single step. Experimental studies show that for $\tau \geq 5$ all non-skeleton voxels are pruned. Figure 2 illustrates the use of τ for a noisy box. Whereas $\mathcal{S}_{\tau=10}$ (left) contains some spurious sheets. $\mathcal{S}_{\tau=20}$ (middle) can be considered robust, as it contains 13 sheets like a non-noisy box. In $\mathcal{S}_{\tau=70}$ (right), only the center sheet is retained, which can be seen as a coarse scale representation of the box.

The simplified skeleton is homotopic to the original object as long as the local maximum ridge in the thresholded ρ forms a connected structure. For completeness, we note that an alternative definition of ρ (Eq. 5) can be formulated on this ridge in order to obtain a monotonic measure for *all* skeleton points [3,19], so that the simplified skeleton is homotopic for every τ. However, this is outside this paper's scope.

4   The Simplified Y-network

In this section, we show how we extend the previous work on simplified skeletons to compute a simplified Y-network, on which our skeleton segmentation is based.

4.1   Computing the Y-network

In order to find the Y-network, i.e. the sheet-intersection curves, of a simplified skeleton $\mathcal{S}_\tau$ we must check if a voxel is on a Y-curve or not. In the continuous $\mathbb{R}^3$ space, an intersection curve point has (at least) three feature points: $|F| \geq 3$ (Sec. 2.2). However, as indicated in Section 3, the feature transform $\overline{F}(p)$ for a skeleton voxel p consists of several voxels. For example, $\overline{F}(p)$ in Fig. 3(a) contains 4 feature voxels $p_1...p_4$ and $\overline{F}(q)$ contains 6 voxels: $q_1..q_6$. If we naively use the cardinality of $\overline{F}$ to detect the Y-curves, then both p and q would be selected, which is wrong for p. To solve this problem, we group each two feature points that have a small geodesic distance (shortest-path length) on the boundary. More formally, we define an equivalence relation $a \sim b$ on $\overline{F}$:

$$a \sim b \Leftrightarrow \|\gamma(a,b)\| < \tau , \tag{7}$$

where τ is the same threshold we used to simplify the skeletons (Eq. 6). This relation gives rise to a number equivalence classes. We now replace $\overline{F}$ by the *simplified feature transform* $\overline{F}_\tau$, defined as a set of class representatives following Eq. 7, i.e. a subset of $\overline{F}$ containing one point from each equivalence class. Which particular point we choose in a class is not important, as we are only interested in the cardinality of $\overline{F}_\tau$.

Using the simplified feature transform $\overline{F}_\tau$, we replace the definition of $\mathcal{S}_\tau$ in Eq. 6 with:

$$\overline{\mathcal{S}}_\tau(\Omega) = \{p \in \Omega \mid |\overline{F}_\tau(p)| \geq 2\}, \tag{8}$$

which more closely parallels Eq. 2. In $\overline{\mathcal{S}}_\tau$ all sheet points p that have a shortest path between their two feature points that is shorter than τ, or in other words, that have a lower importance than τ, are pruned.

Now, the *simplified Y-network* is straightforwardly defined as:

$$Y_\tau = \{p \in \Omega \mid |\overline{F}_\tau(p)| \geq 3\} , \tag{9}$$

which is a subset of $\overline{\mathcal{S}}_\tau$. For a Y-curve point q, where three sheets meet in q's neighborhood, this means that if one of the sheets in its neighborhood is pruned because its importance is lower than τ, the point is not considered a Y-curve point in Y_τ. It is important to note that the simplified Y-network is not simply a post-processing of the simplified skeleton. Instead, the Y-network is computed

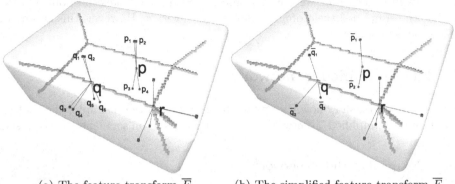

(a) The feature transform $\overline{F}$.   (b) The simplified feature transform $\overline{F}_\tau$.

Fig. 3. A (non-axis aligned) box with its detected network $Y_{\tau=10}$. Three selected points p, q, r: a sheet point p, a Y-curve point q, and a Y-curve intersection r.

directly out of the shape, using the integral quantity of geodesic distance on the object boundary. This is more stable than extracting the sheet-intersection curves from the voxelized skeleton, and also offers a natural scale parameter.

Fig. 3 shows a non-axis aligned box with its simplified network Y_τ, $\tau = 10$. Like in the continuous case, the discrete Y-network forms a connected structure. Three points p, q, r are selected: a sheet point p, a Y-curve point q, and a Y-curve intersection point r. Their feature voxels are connected to the corresponding points by line segments. Fig. 3(a) uses the non-simplified feature transform $\overline{F}$ for the selected points, while Fig. 3(b) uses the simplified feature transform $\overline{F}_\tau$. We see the merit of $\overline{F}_\tau$: p and q can be classified as a sheet and a Y-curve voxel respectively based on the cardinality $|\overline{F}_\tau|$, but not on $|\overline{F}|$. For point q for instance, $\overline{F}_\tau$ gives us exactly three feature points $\overline{q}_1, \overline{q}_2, \overline{q}_3$, i.e. the representations of the classes $\{q_1, q_2\}, \{q_3, q_4\}, \{q_5, q_6\}$.

4.2   Y-network Decomposition

Although Eq. 9 enables the detection of the Y-network voxels, it does not make explicit the structure of the Y-network as a collection of Y-curves. This section presents how to compute such a decomposition. The next section shows how to use the decomposition to compute a skeleton segmentation.

To decompose the network Y_τ into its n Y-curves $\{y_1, ..., y_n\}$, we define two points $p, q \in Y_\tau$ to be on the same Y-curve when there is no junction in the Y-network between them. For illustration, Figure 4(a) sketches (part) of a Y-network. Let p, q be two points on Y_τ, each one having three feature points, with the indexes p_1, p_2, p_3 and q_1, q_2, q_3 chosen in such a way that they are "aligned", that is, the sum of the geodesic distances $\sum_i \|\gamma(p_i, q_i)\|$ is minimal. Suppose now there is a junction j between p, q, due to a Y-curve y_a as shown (dotted) in the figure. Such a Y-curve y_a results from a sheet A with a local importance higher than τ, otherwise y_a would have been pruned and would not be present in Y_τ.

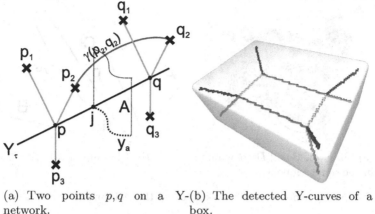

(a) Two points p, q on a Y-(b) The detected Y-curves of a
network.   box.

Fig. 4. Y-network decomposition into Y-curves

Thus, we can detect whether there is a junction j between p and q by looking
at the geodesic distance $\|\gamma\|$ between the feature points of p and q. Because the
feature points evolve smoothly within a skeletal sheet, we argue that if there
is a Y-curve y_a between p, q due to sheet A, the geodesic distance between the
feature points p_2 and q_2 must be larger than τ, because A's importance is at
least τ. Generalizing, two points $p, q \in Y_\tau$ belong to the same Y-curve y if and
only if all three geodesic distances between each pair of feature points (p_i, q_i)
are smaller than τ:

$$p, q \in y \in Y_\tau \Leftrightarrow \forall_{i \in \{1,2,3\}} \|\gamma(p_i, q_i)\| \leq \tau . \tag{10}$$

One remark is due. Voxels where several Y-curves come together have more
than three feature points (see e.g. Fig. 3(b), point r). To handle these cases,
Eq. 10 is applied for all subsets of both p and q, having 3 feature points. One
might ask why we do not just detect the Y-curve intersection points and use
these to separate the Y-curves, which is more trivial. The first reason is that our
approach is of sub-voxel precision: a single voxel may contain multiple Y-curves.
Second, our method does not use a topological analysis of the Y-network voxels
and is thus more accurate in cases where Y-curves meet under small angles.
Figure 4(b) shows the segmentation of the Y-network of a box obtained using
Eq. 10. The Y-curves are distinctively colored[1].

5   Skeleton Segmentation

Although it seems natural to use the simplified network Y_τ to segment the skeleton $\overline{S}_\tau$, both computed at the same scale τ, this can sometimes deliver unex-
pected results. Figure 5(a) shows the segmentation produced for the skeleton

---

[1] Please view the images in color.

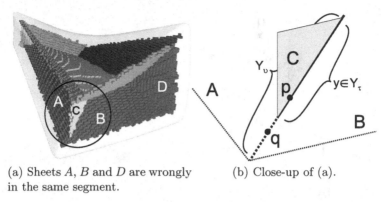

(a) Sheets A, B and D are wrongly   (b) Close-up of (a).
in the same segment.

Fig. 5. Segmented simplified skeleton of the Deformed Box using Y_τ

$\overline{S}_\tau$ ($\tau = 10$) of a twisted box. We expect 13 segments, similar to the skeleton segmentation of a non-deformed box (Sec. 2.2). Yet, for the deformed box some segments get merged inadvertently. For example, the skeleton sheets A, B and D are incorrectly merged, whereas they should be distinct segments as for a non-deformed box.

Figure 5(b) is a schematic close-up of the situation. The problem is that the Y-curve $y \in Y_\tau$ does not extend all the way to the skeleton boundary. The missing part is indicated by the dotted line. A narrow tunnel connects the areas A and B on the skeleton manifold, so that they end up in the same segment. The reason that y is too short is that sheet C is simplified for lower values of τ than A and B at the dotted line segment. The cardinality of the simplified feature transform for these points is 2, so that point q, for example, is not detected as a Y-curve point by Eq. 9. In other words, only two sheets are found to come together at q in the *simplified* skeleton, namely A and B. For the same reason, other surface segmentation approaches based on local topology, e.g. [4], would fail segmenting this *simplified* skeleton. Note that it would also fail because our skeletons are two voxels thick.

We solve the issue noticing that curve y in Fig. 5(b) would be longer for less simplified skeletons. For non-simplified skeletons, we would not have the problem at all. Hence, to segment a skeleton $\overline{S}_\tau$ we use a less simplified Y-network Y_v, $v < \tau$, which contains longer, extended, versions of the Y-curves which ended prematurely in Y_τ. However, we must be careful to only extend Y-curves from Y_τ, and not to incorporate *any* Y-curves that only occur in Y_v. Hence, we only consider those Y-curves $y \in Y_v$ for which there is at least one point $p \in y$ in Y_τ. We call this set the extension of Y_τ using Y_v, denoted as $Y_{\tau,v}$:

$$Y_{\tau,v} = \left\{ y \in Y_v \mid \exists_{p \in y} \, p \in Y_\tau \right\}. \qquad (11)$$

Finally, we compute the decomposition of $Y_{\tau,v}$ into its respective Y-curves by taking both scales τ and v into account and adapting Eq. 10 accordingly:

$$p, q \in y \in Y_{\tau,v} \Leftrightarrow \forall_{i \in \{1,2,3\}} \|\gamma(p_i, q_i)\| \leq \begin{cases} \tau & \text{if } p \in Y_\tau \wedge q \in Y_\tau \\ v & \text{otherwise.} \end{cases} \qquad (12)$$

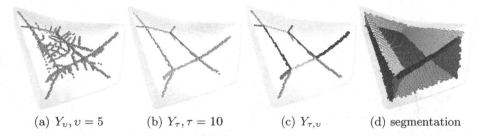

(a) $Y_v, v = 5$ (b) $Y_\tau, \tau = 10$ (c) $Y_{\tau,v}$ (d) segmentation

Fig. 6. Correctly segmenting the simplified skeleton of a deformed box

We have not yet specified the value of v. For all objects we tested, a fixed value of $v = 5$ gives very good results. As explained in Sec. 3, this value essentially robustly yields the entire skeleton of a 3D shape.

After computing $Y_{\tau,v}$ we can segment $\overline{S}_\tau$ as follows. We consider the skeleton $\overline{S}_\tau$ as a 26-connected graph, from which we remove the voxels occupied by $Y_{\tau,v}$, and then determine the connected components in the remaining skeleton graph using e.g. a flood-fill. Since $\overline{S}_\tau$ can be up to two voxels thick [3] and the Y-network is only one voxel thick, we first dilate the Y-network by 1 voxel in each 26-direction, before removing them from the skeleton graph. Hereafter we erode the dilated Y-network so that these voxels are also part of a segment. Clearly, many other alternative implementations are possible once we have a robust Y-network.

Figure 6 shows Y_v, Y_τ, $Y_{\tau,v}$ (decomposed in its Y-curves), and the segmentation based on $Y_{\tau,v}$ respectively, which is a correct segmentation of the deformed box skeleton as opposed to Fig. 5(a).

6   Results and Discussion

We have tested our method on shapes of varying complexity and amount of boundary noise. As input objects we used 3D triangle meshes, voxelized using binvox (http://www.cs.princeton.edu/~min/binvox/) in various resolutions ranging up to two million object voxels. For all images in this paper, the original object mesh representation is rendered instead of its voxel representation for nicer display.

Figure 7 shows a selection of the results. For each object, τ is empirically chosen based on the noise level of the object, and we show both the extended simplified network $Y_{\tau,v}$, decomposed into its Y-curves (top image), and the segmentation of the simplified skeleton (bottom image) using $Y_{\tau,v}$. In Figures 7(a) and 7(e), we added 5-10% noise to show the robustness of our approach.

We highlight some of the results. For the Noisybox, we see that our method correctly detects 13 segments and 12 Y-curves. For the Dinosaur and Noisydino, the four legs and feet are all assigned different segments. Our method correctly segments the skeleton of the Dodecahedron (Fig. 7(f)). This is a difficult skeleton to segment as it contains degeneracies, in the sense that each Y-curve actually separates five sheets instead of the usual three, which is why the Y-network is thicker than for the other objects.

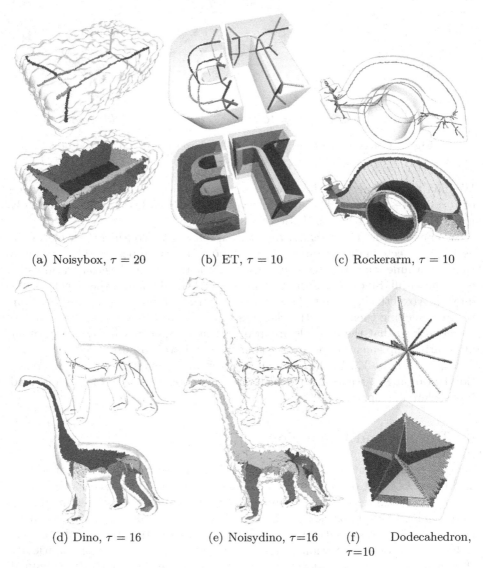

(a) Noisybox, $\tau = 20$ (b) ET, $\tau = 10$ (c) Rockerarm, $\tau = 10$

(d) Dino, $\tau = 16$ (e) Noisydino, $\tau = 16$ (f)   Dodecahedron, $\tau = 10$

Fig. 7. Results. Y-network decomposition (top images), $\overline{\mathcal{S}}_\tau$ segmentation (bottom images)

Table 1 shows timing measurements performed on a Pentium IV 3 GHz, with 1 GB of RAM for all objects in this paper. Columns "dim", "$|\Omega|$", and $|\partial\Omega|$ denote the grid resolution, and the amount of object and boundary voxels respectively. Columns "$\overline{\mathcal{S}}$ time" and "segm. time" denote the wall-clock time for computing the skeleton and segmentation respectively. The most time is spent computing shortest paths in the boundary graph, needed in both stages.

Table 1. Timing measurements

Object	dim	$|\Omega|$	$|\partial\Omega|$	$\mathcal{S}$ time (s)	segm. time (s)
Deformed Box	65x64x124	247k	24k	21s	11s
Dodecahedron	124x124x124	945k	39k	48s	17s
ET	125x78x173	1.045k	93k	304s	37s
Noisydino	99x325x365	1.128k	100k	62s	33s
Rockerarm	366x188x112	2.000k	151k	470s	62s

7 Conclusion

We have presented a voxel-based approach for robustly segmenting simplified skeletons of 3D shapes, based on the simplified feature transform and simplified Y-network. The Y-network is decomposed into its respective Y-curves. The simplified Y-network could prove to be more useful in certain shape analysis or retrieval tasks than the segmentation itself, because the Y-curves change more continuously under shape deformations than the segmentation does.

Our entire method relies upon the choice of a single parameter τ, which controls the simplification of the skeleton [3], but *not* the segmentation itself, thereby yielding a fully autonomous segmentation method. One limitation of our current implementation is that cylindrical object-parts having degenerate curve skeletons may locally result in over-segmentation. Second, the two-voxel thickness of the skeletons might yield undesirable topological changes in the skeleton for too complex objects, when compared to thin skeletons.

In future work, we would like to investigate whether our skeleton segmentation can be used to achieve a robust patch-type segmentation of the object surface. As opposed to traditional methods acting only on the surface, a skeleton-based segmentation has the benefit that it captures a sense of symmetry, and is potentially more robust, being based on the object's volume, not on its surface.

References

1. Zhang, J., Siddiqi, K., Macrini, D., Shokoufandeh, A., Dickinson, S.: Retrieving articulated 3-d models using medial surfaces and their graph spectra. In: Int. Workshop On Energy Minimization Methods in Computer Vision and Pattern Recognition, pp. 285–300 (2005)
2. Damon, J.: Global medial structure of regions in R^3. Geometry and Topology 10, 2385–2429 (2006)
3. Reniers, D., Van Wijk, J.J., Telea, A.: Computing multiscale curve and surface skeletons of genus 0 shapes using a global importance measure. IEEE TVCG 14(2), 355–368 (2008)
4. Malandain, G., Bertrand, G., Ayache, N.: Topological segmentation of discrete surfaces. International Journal of Computer Vision 10(2), 183–197 (1993)
5. Reniers, D., Telea, A.: Tolerance-based feature transforms. In: Advances in Computer Graphics and Computer Vision, CCIS, vol. 4, pp. 187–200. Springer, Heidelberg (2008)

6. Cornea, N.D., Silver, D., Min, P.: Curve-skeleton properties, applications and algorithms. IEEE Transactions on Visualization and Computer Graphics 13(3), 530–548 (2007)
7. Lieutier, A.: Any open bounded subset of $\mathbb{R}^3$ has the same homotopy type as its medial axis. In: Proc. of the 8th ACM symposium on Solid modeling and applications, pp. 65–75 (2003)
8. Giblin, P., Kimia, B.B.: A formal classification of 3d medial axis points and their local geometry. IEEE Trans. on Pattern Analysis and Machine Intelligence 26(2), 238–251 (2004)
9. Chazala, F., Lieutier, A.: The λ-medial axis. Graphical Models 67(5), 304–331 (2005)
10. Shaked, D., Bruckstein, A.M.: Pruning medial axes. Computer Vision and Image Understanding 69(2), 156–169 (1998)
11. Malandain, G., Fernández-Vidal, S.: Euclidean skeletons. Image and Vision Computing 16(5), 317–327 (1998)
12. Siddiqi, K., Bouix, S., Tannenbaum, A., Zucker, S.W.: The hamilton-jacobi skeleton. In: Proc. of the Int. Conference on Computer Vision (ICCV 1999), pp. 828–834 (1999)
13. Ogniewicz, R.L., Kübler, O.: Hierarchic voronoi skeletons. Pattern Recognition 28(3), 343–359 (1995)
14. Costa, L.F., Cesar Jr, R.M.: Shape analysis and classification, pp. 416–419. CRC Press, Boca Raton (2001)
15. Telea, A., Van Wijk, J.J.: An augmented fast marching method for computing skeletons and centerlines. In: Proc. of the Symposium on Data Visualisation (VisSym 2002), pp. 251–259 (2002)
16. Dey, T.K., Sun, J.: Defining and computing curve-skeletons with medial geodesic function. In: Proc. of Eurographics Symposium on Geometry Processing, pp. 143–152 (2006)
17. Mullikin, J.C.: The vector distance transform in two and three dimensions. CVGIP: Graphical Models and Image Processing 54(6), 526–535 (1992)
18. Kiryati, N., Székely, G.: Estimating shortest paths and minimal distances on digitized three-dimensional surfaces. Pattern Recognition 26, 1623–1637 (1993)
19. Reniers, D., Telea, A.C.: Skeleton-based hierarchical shape segmentation. In: Proc. of the IEEE Int. Conf. on Shape Modeling and Applications (SMI 2007), pp. 179–188 (2007)

Geometric Feature Estimators for Noisy Discrete Surfaces*

L. Provot and I. Debled-Rennesson

LORIA Nancy
Campus Scientifique - BP 239
54506 Vandœuvre-lès-Nancy Cedex, FRANCE
{provot,debled}@loria.fr

Abstract. We present in this paper robust geometric feature estimators on the border of a possibly noisy discrete object. We introduce the notion of patch centered at a point of this border. Thanks to a width parameter, attached to a patch, the noise on the border of the discrete object can be considered, and an extended flat neighborhood of a border point is computed. Stable geometric features are then extracted around this point. A normal vector estimator is proposed as well as a detector of convex and concave parts on the border of a discrete object.

1 Introduction

Geometric feature estimation on the surface of a discrete object is very important in digital geometry. For instance, in computer vision, rendering algorithms rely on the normal vector estimation on the surface of an object to produce a realistic view of this object. In image analysis, features such as area or curvature can be used for producing classification functions able to sort objects by shape or by size. In bioinformatics, in the protein-protein docking framework, one tries to associate two proteins according to their geometrical complementarity. Thus, it is important to locate critical areas on the surface, like holes (concave parts) and knobs (convex parts) or even big flat areas.

Several studies [1,2,3,4,5,6,7] have been done on different types of estimators but the estimated geometric features are by definition very local. Thus, the noise introduced by acquisition tools like scanners or MRI, even weak, is disturbing for the estimator, and classical algorithms of digital geometry which rely on the regularity of discrete primitives do not always yield good results.

Therefore, for the last few years, the interest around the geometry of noisy objects has grown [8,9,10,11]. New discrete primitives such as *blurred segments* and *blurred pieces of discrete planes* have been introduced and, thanks to a width parameter, enable to take into account the noise in data while controlling the approximation. Encouraging results have been obtained in 2D for the curvature estimation, relying on blurred segments [12].

---

* This work is supported by the ANR in the framework of the GEODIB project.

D. Coeurjolly et al. (Eds.): DGCI 2008, LNCS 4992, pp. 275–286, 2008.

In this paper we deal with the 3D case. In section 2, after recalling the definition of blurred pieces of discrete planes, we introduce the notion of *a width-ν patch centered at p*. In section 3, we then present some features (normal vector, area, outline) of a width-ν patch centered at p. With this material we develop, in section 4, an estimator able to distinguish between points that belong to concave, convex or flat parts of the surface of a possibly noisy discrete object. The paper ends up with some conclusions and perspectives in section 5.

2   Basic Notions and Definitions

Hereafter we denote by $\mathcal{O}_b$ a possibly noisy 6-connected discrete object. We call *surface* or *border* of $\mathcal{O}_b$ the set of points $\mathcal{B}_b$ which have a 6-neighbor that does not belong to $\mathcal{O}_b$.

Before presenting the notion of *a width-ν patch centered at p*, we recall the definition of a width-ν blurred piece of discrete plane [11], which is the underlying discrete primitive of the patch.

2.1   Blurred Pieces of Discrete Planes

One can see a blurred piece of discrete plane as an arithmetic discrete plane for which some points are missing. More formally:

Definition 1. *Let N be a norm on $\mathbb{R}^3$ and $\mathcal{E}$ a set of points in $\mathbb{Z}^3$. We say that the discrete plane $\mathcal{P}(a,b,c,\mu,\omega)^1$ is a **bounding plane of** $\mathcal{E}$ if all the points of $\mathcal{E}$ belong to $\mathcal{P}$, and we call **width of** $\mathcal{P}(a,b,c,\mu,\omega)$, the value $\frac{\omega-1}{N(a,b,c)}$.*

*A bounding plane of $\mathcal{E}$ is said **optimal** if its width is minimal.*

Definition 2. *A set $\mathcal{E}$ of points in $\mathbb{Z}^3$ is a **width-ν blurred piece of discrete plane** if and only if the width of its optimal bounding plane is less than or equal to ν.*

Two recognition algorithms of blurred pieces of discrete planes have been proposed in [11]. The first one considers the Euclidean norm and, for a set of points P in $\mathbb{Z}^3$, it solves the recognition problem by using the geometry of the convex hull of P. The second one considers the infinity norm and uses methods from linear programming to solve the recognition problem.

In the following sections, the results we present have been obtained by using the geometrical approach which uses the Euclidean norm.

2.2   Width-ν Discrete Patches

Here we are interested in the computation of different geometric features of a possibly noisy discrete surface, such as the normal vector, the area or the

---

[1] An arithmetic discrete plane $\mathcal{P}(a,b,c,\mu,\omega)$ is the set of integer points (x,y,z) verifying $\mu \le ax+by+cz < \mu+\omega$, where $(a,b,c) \in \mathbb{Z}^3$ is the normal vector of the plane. $\mu \in \mathbb{Z}$ is named the translation constant and $\omega \in \mathbb{Z}$ the arithmetical thickness.

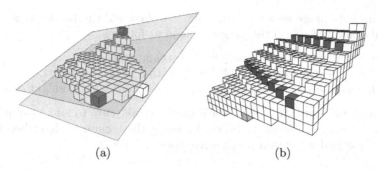

(a)   (b)

Fig. 1. (a) A width-3 blurred piece of discrete plane and (b) its optimal bounding plane $\mathcal{P}(4, 8, 19, -80, 49)$, using the Euclidean norm

curvature. Numerous methods exist in the literature to compute such features in the discrete space. One can refer to [1,13,2,4,5] for normal vectors estimation and [3,7] for curvature estimators.

The problem is that these geometric features are by definition very local. And most of the defined estimators are very noise sensitive. But if we know that we are working on a noisy surface, we would like to use estimators that take into account the irregularity of this surface. A way to estimate these features at a point p of the noisy surface is to use the information of the points lying in a neighborhood of p. The notion of patch we present hereafter takes place in this framework, considering an adaptive neighborhood around p.

Definition 3. *Let $\mathcal{B}_b$ be the border of a discrete object and p a point in $\mathcal{B}_b$. Let T be a scan process of the neighboring points of p in $\mathcal{B}_b$ and ν the greatest real value allowed. We call* **width-ν patch centered at** *p, and denote by $\Gamma_\nu(p)$, a width-ν blurred piece of discrete plane incrementally recognized from p by adding points of $\mathcal{B}_b$ following the scan process T.*

About the Incremental Recognition: We construct a *width-ν patch centered at p* using the incremental recognition algorithm of blurred pieces of discrete planes introduced in [11]. We add the points following the scan process T and as soon as the width of the blurred piece of discrete plane becomes greater than ν we stop the recognition process.

About the Scan Process T: To uniformly spread the patch in all directions the best solution would be to scan the neighborhood of p according to a geodesic distance. Nevertheless, for efficiency, we have chosen to scan this neighborhood according to a chamfer mask $\langle 3, 4, 5 \rangle$ which is a good approximation of the geodesic distance [14,15]. The aim is to have a well-balanced patch around p which looks almost circular. To implement this behaviour we use a priority queue Q. This method has originaly been proposed by Verwer *et al.* [16] and first used by Coeurjolly on digital surfaces [17]. We start by pushing p into the queue with a weight equals to zero. Then, while we do not exceed the limit width ν, we pop out of Q the point v with the lowest weight w and we add v to the blurred piece of discrete plane we are recognizing. We then add the border voxels of the 26-neighborhood of v into the priority queue as follows:

- the 6-neighbors are added in Q with a weight of $w + 3$,
- the strict 18-neighbors are added in Q with a weight of $w + 4$,
- and finally the strict 26-neighbors are added in Q with a weight of $w + 5$.

To stay homeomorphic to a topological disk, we take care to add a point only if it does not create a hole in the patch, by using the technique described in [18]. With this method we obtain patches like those in Fig. 2.

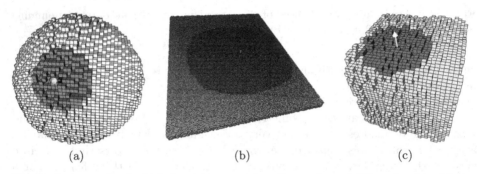

(a)   (b)   (c)

Fig. 2. An example of width-2 patches spread on the surface of different noisy objects. (a) A sphere of radius 20, (b) a $150 \times 6 \times 200$ box and (c) a cube of edge 25.

3   Patch Features

3.1   Intuition

A patch $\Gamma_\nu(p)$, as previously defined, characterizes the planarity of the surface around p (with respect to the width ν). Thus, the more the patch is spread, the lesser the surface around p is bent.

In addition, if the growth of $\Gamma_\nu(p)$ stopped, it means that the close neighboring points outside $\Gamma_\nu(p)$ would bend the patch too much if they were added. In that case the patch could no longer be regarded as flat. Therefore, it is possible to deduce a conformation of the discrete surface around p by studying the patches centered along the points of the outline of $\Gamma_\nu(p)$.

To quantify these observations we have to compute the patch area, extract the outline of $\Gamma_\nu(p)$ and study the lie of the patches in the neighborhood of $\Gamma_\nu(p)$. This is what we present in the following sections.

3.2   Width-ν Normal

With the previous intuition we can see that the normal vector of $\Gamma_\nu(p)$ is a good estimation of the normal at p. Thus, assimilating the normal vector of $\Gamma_\nu(p)$ to the normal of the surface at p, we define a normal vector estimator for each point of the surface of a possibly noisy discrete object.

Definition 4. *Let* $\mathcal{B}_b$ *be the border of a discrete object and* p *a point of* $\mathcal{B}_b$. *We call* **width-ν normal at** p *the normal vector*

$$\overrightarrow{n_\nu}(p) = \overrightarrow{n}\left(\Gamma_\nu(p)\right)$$

where $\overrightarrow{n}\left(\Gamma_\nu(p)\right)$ *is the normal vector of the patch* $\Gamma_\nu(p)$.

Other normal estimators have been proposed before. For instance, L. Papier [4] estimates the normal vector at a point p of the surface using a weighted mean of the elementary unit normals of the surfels in the neighborhood of p (*umbrella of order* n). The size of the neighborhood around p can be set with the parameter n. In [13], A. Lenoir uses the slices of the object along the three canonical planes. He first computes tangential lines through p in these slices using partial derivatives computation (implemented by convolution products). He then gives a normal vector estimation using these tangential lines. The size of the neighborhood around p can be set with a scale factor. Although these methods can deal with noise, the considered neighborhood is not adaptive, contrary to our approach which fits the shape of the surface. Some results on noisy and

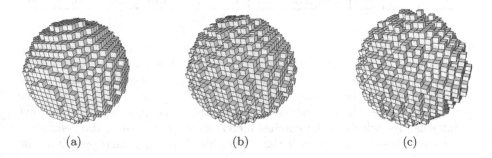

(a)   (b)   (c)

Fig. 3. A discrete sphere of radius 10 (a) without noise, (b) weakly noisy and (c) strongly noisy

non-noisy spheres (see Fig. 3) of different radius are presented in Tab. 1. To synthetize noisy spheres we start to generate a discrete sphere (i.e. the set of points $\left\{(x, y, z) \in \mathbb{Z}^3 \mid x^2 + y^2 + z^2 < r(r+1)\right\}$ where r is the radius of the sphere). We then add noise by randomly moving outward a border voxel of the sphere. To add more noise we just repeat the last step.

We have compared the estimated normal vectors to the theoretical normal vectors[2] at p. The mean angular value δ_{moy}, in degrees, the maximal angular value δ_{max} and the standard deviation σ are presented in Tab 1.

We can see that with our method the greater the radius of the spheres is, the better the estimation is, contrary to the method based on the 26-neighborhood where there is no real improvement. It is due to the fact that our method is adaptive and considers a wider neighborhood when the sphere radius increases.

---

[2] The theoretical normal vector is the vector starting from the center of the sphere to p.

Table 1. Comparison between the proposed normal vector estimator and the one using only the 26-neighborhood. Angle values δ are in degrees $°$.

Tested spheres			Without noise			Weakly noisy			Strongly noisy		
Radius			10	20	40	10	20	40	10	20	40
Width	$\nu = 1$	δ_{moy}	3.66	2.05	1.04	10.36	9.08	9.04	15.30	14.10	14.33
		δ_{max}	8.21	6.10	3.25	52.53	91.25	101.24	95.46	101.42	117.80
		σ	1.83	1.07	0.58	7.70	7.50	7.82	12.46	11.90	11.95
	$\nu = 2$	δ_{moy}	2.33	1.25	0.77	3.66	2.39	1.46	5.65	4.57	4.35
		δ_{max}	9.92	3.35	2.94	16.86	10.16	5.83	46.98	40.54	52.57
		σ	1.49	0.61	0.44	2.19	1.35	0.78	3.77	3.33	3.97
	$\nu = 3$	δ_{moy}	1.85	1.17	0.70	2.75	1.74	1.02	3.29	2.37	1.70
		δ_{max}	6.49	3.93	2.58	8.90	7.46	3.89	14.63	8.80	7.32
		σ	0.98	0.69	0.40	1.54	0.92	0.55	1.90	1.30	0.93
26-neighborhood		δ_{moy}	4.61	4.50	4.22	7.55	7.50	7.51	8.83	8.69	8.57
		δ_{max}	9.45	15.34	17.19	28.35	25.34	31.14	24.29	37.15	33.80
		σ	2.70	2.75	2.81	4.19	4.10	4.19	4.52	4.87	4.80

It behaves like this because when the sphere radius increases the surface around p tends to be flat. For a given width the patch can thus spread more around p.

Moreover, when the noise level increases the estimation of the normal vector is less accurate. The bad results obtained when $\nu = 1$ can be attributed to the tiny size of patches which do not always cover the 26-neighborhood of the point p. But when the width of the patches increases, say $\nu = 2$ or 3, the results are better than those obtained with the 26-neighborhood approach, and it is the case for noisy spheres as well as for non-noisy ones.

3.3   Width-ν Patch Area

Here we use the notations from [6]. Let $\mathcal{S}$ be an Euclidean surface and $\{\overrightarrow{n}\}$ its normal vector field. We can compute the area of $\mathcal{S}$ in the continuous space with the formula:

$$\mathcal{A}(\mathcal{S}) = \int_{\mathcal{S}} \overrightarrow{n}(s) \, ds$$

Now, if we consider a digitization $D(\mathcal{S})$ of $\mathcal{S}$, we can replace the integral over $\mathcal{S}$ by a finite sum over the surfels s of $D(\mathcal{S})$, the vector $\overrightarrow{n}(s)$ by an estimation of the normal vector at s and ds by a dot product with the unit orthogonal vector of s pointing outward. The idea is to compute the contribution of each surfel to the global area of $D(\mathcal{S})$ by projecting the surfel according to the direction of the normal vector. The discrete version of the previous equation is:

$$E_{\mathcal{A}}(D(\mathcal{S})) = \sum_{s \in D(\mathcal{S})} \overrightarrow{n}^{*}(s) . \overrightarrow{n}_{el}(s)$$

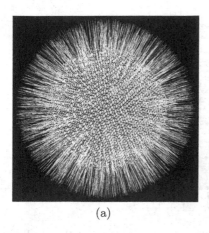 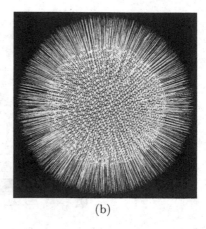

(a)   (b)

Fig. 4. Normal vector estimation on a weakly noisy sphere of radius 20 (a) using only the 26-neighborhood and (b) using the patch based approach with $\nu = 3$

where $\overrightarrow{n}^*(s)$ is an estimation of the normal vector associated to the surface element s in $D(\mathcal{S})$ and $\overrightarrow{n}_{el}(s)$ is the elementary normal vector of s.

This method was first proposed in 3D by A. Lenoir in [13]. Then, in [6], Coeurjolly et al. proved that this estimator is convergent if and only if the normal vector estimation is convergent.

In our case we can use the normal estimator proposed in section 3.2 with the previous formula to obtain an estimation of the area for the surface of a width-ν patch:

$$E_{\mathcal{A}}(\Gamma_\nu(p)) = \sum_{s \in \mathcal{S}_{\Gamma_\nu(p)}} \overrightarrow{n_\nu}(p).\overrightarrow{n}_{el}(s) = \overrightarrow{n_\nu}(p).\sum_{s \in \mathcal{S}_{\Gamma_\nu(p)}} \overrightarrow{n}_{el}(s)$$

where $\mathcal{S}_{\Gamma_\nu(p)}$ denotes the set of surface surfels of $\Gamma_\nu(p)$.

3.4 Patch Outline

To study the surrounding patches of $\Gamma_\nu(p)$, we have to clarify the notion of *surrounding*. We have chosen to study the conformation of the patches which are centered on points belonging to the outline of $\Gamma_\nu(p)$.

Let $\mathcal{B}_b$ be the border of a discrete object $\mathcal{O}_b$. We denote by $\mathcal{S}_b$ the set of surfels of $\mathcal{B}_b$ which are incident to a point that does not belong to $\mathcal{O}_b$, and $\mathcal{S}_{\Gamma_\nu(p)}$ the subset of $\mathcal{S}_b$ that belongs to $\Gamma_\nu(p)$. A point q belongs to the outline of $\Gamma_\nu(p)$ if the voxel representation of q has a surfel $s \in \mathcal{S}_{\Gamma_\nu(p)}$ and if there exists a surfel $s' \in \mathcal{S}_b \setminus \mathcal{S}_{\Gamma_\nu(p)}$ such that s and s' are adjacent by edge. An example of the outline of a patch is shown in Fig. 5.

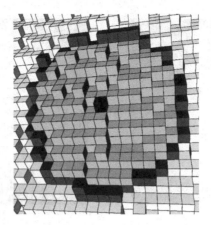

Fig. 5. The outline (in dark grey) of a width-2 grey patch centered at the black point

4   Shape

4.1   Concave and Convex Parts

With the different notions defined in the previous sections, we present an estimator that permits to characterize the shape (concave, convex or flat) around a border point of a discrete object.

The idea is to start by spreading a patch $\Gamma_\nu(p)$ and to compute its outline C. For each point $q_{i,1 \leq i \leq |C|}$ in C, we then develop the neighboring patches $\Gamma_\nu(q_i)$. From these patches we obtain a vector field $\{\overrightarrow{n_\nu}(q_i)\}$ on the outline C. This vector field is shown in Fig. 6 for a convex and a concave area.

Observation: In the concave parts the vector field $\{\overrightarrow{n_\nu}(q_i)\}$ points towards the normal vector $\overrightarrow{n_\nu}(p)$ of the central patch, but in the convex parts the vector field has the reverse tendency. It is thus possible to distinguish concave and convex parts with the value of the oriented angle between the normal vector $\overrightarrow{n_\nu}(p)$ and the vectors $\{\overrightarrow{n_\nu}(q_i)\}$. Our estimator is based on this observation. Let C be the set of points that belong to the outline of $\Gamma_\nu(p)$. Our shape estimator of the surface around a point p is then given by the formula :

$$\mathcal{F}_\nu(p) = \frac{1}{|C|} \sum_{\forall q \in C} (\widehat{\overrightarrow{n_\nu}(p), \overrightarrow{n_\nu}(q)}) \cdot \frac{E_\mathcal{A}(\Gamma_\nu(q))}{E_\mathcal{A}(\Gamma_\nu(p))}$$

where $(\widehat{\overrightarrow{n_\nu}(p), \overrightarrow{n_\nu}(q)})$ is the oriented angle value between the two normal vectors. So, the estimator $\mathcal{F}_\nu(p)$ is a weighted mean of the angle values between $\overrightarrow{n_\nu}(p)$ and the $\overrightarrow{n_\nu}(q_i)_{1 \leq i \leq |C|}$.

Angle Orientation: We determine the sign of the angle value $(\widehat{\overrightarrow{n_\nu}(p), \overrightarrow{n_\nu}(q)})$ as shown in Fig 7. To implement this process, we compute the cross products

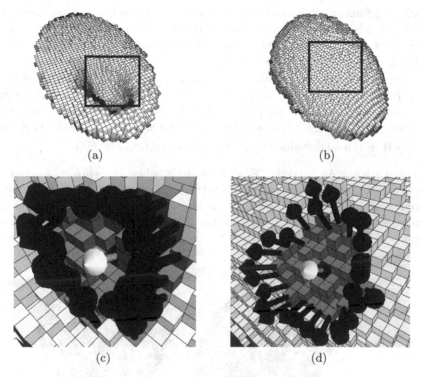

(a)   (b)

(c)   (d)

Fig. 6. Vectors layout on the outline of a patch (a, c) in a convex area and (b, d) in a concave area

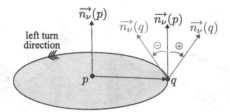

Fig. 7. Angle orientation

$\overrightarrow{n_\nu}(p) \wedge \overrightarrow{pq}$ and $\overrightarrow{n_\nu}(p) \wedge \overrightarrow{n_\nu}(q)$. If the two resulting vectors more or less point towards the same direction (i.e. if the scalar product $(\overrightarrow{n_\nu}(p) \wedge \overrightarrow{pq}) \cdot (\overrightarrow{n_\nu}(p) \wedge \overrightarrow{n_\nu}(q))$ is positif) then the angle value is counted positively, otherwise the angle value is counted negatively. The computation of the previous cross products are feasible only if the vectors are not collinear. And the scalar product also does not have to be nil. Usually, these conditions are always verified. But sometimes, especially when the patch is very small, some degenerate cases can occur. Because they do not bring a useful shape information we can easily get rid of these cases by ignoring them and going on with the next point of the outline.

Weighting Factor: It is clear that a wide patch $\Gamma_\nu(p)$ means that the surface around p tends to be flat (according to the width ν). Thus, the wider $\Gamma_\nu(p)$ is, the lesser the influence of the surrounding patches is to consider. Hence, the weighting factor $\frac{1}{E_{\mathcal{A}}(\Gamma_\nu(p))}$ is used in $\mathcal{F}_\nu(p)$.

On the other hand, if a neighboring patch $\Gamma_\nu(q_i)$ is wider than another neighboring patch $\Gamma_\nu(q_j)$, then it means that the surface around q_i is a little more stable than the surface around q_j. Thus we choose to lend more influence to information given by bigger patches. Therefore, we associate the weighting factor $E_{\mathcal{A}}(\Gamma_\nu(q))$ to the angle value $(\overrightarrow{n_\nu(p)}, \widehat{\overrightarrow{n_\nu(q)}})$ in the formula $\mathcal{F}_\nu(p)$.

Interpretation: As a result, $\mathcal{F}(p)$ is positive when the surface around p is rather convex and $\mathcal{F}(p)$ is negative when the surface around p is rather concave. An increasing value of $|\mathcal{F}(p)|$ means that the surface around p is more strongly concave or convex. If $\Gamma_\nu(p)$ is big, a value $\mathcal{F}(p)$ close to 0 means that the area around p is almost flat (according to the width ν we chose). If $\Gamma_\nu(p)$ is small,

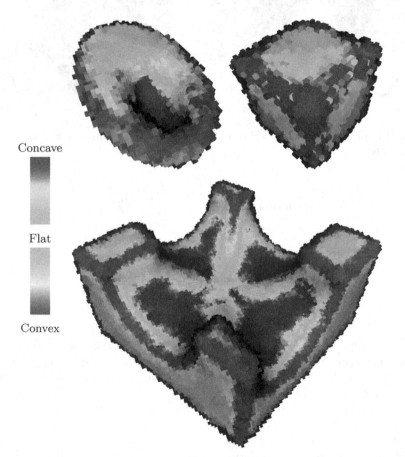

Fig. 8. Concave, convex and flat areas (according to the color scale on the left) on different objects using width-3 patches

then the border around p is strongly distorted, but in a way we can neither qualify concave, nor qualify convex (a saddle point for instance).

Some results obtained with this estimator are given in Fig. 8. To render the object we have associated to each face of a voxel v the normal $\overrightarrow{n_v}(v)$.

5 Conclusion

In this paper we have introduced a new notion related to a point on a discrete surface : the *width-ν patch centered at p*. From this patch we have designed a robust normal vector estimator at the surface of a possibly noisy discrete object. Moreover, based on the computation of the patch area, we have proposed an estimator that characterizes the shape (concave, convex or flat) of the surface around the point p.

Although the approaches presented in this paper give plausible results, more formal studies on the notion of noise have to be led. A work on the convergence of the estimators could then be envisaged.

In addition more comparative studies with other methods have to be done to confirm the reliability of these estimators. In a theoretical point of view it would be interesting to see the relation of our shape estimator with the classical notion of surface curvature.

References

1. Yagel, R., Cohen, D., Kaufman, A.E.: Normal estimation in 3d discrete space. The Visual Computer 8(5&6), 278–291 (1992)
2. Thürmer, G., Wüthrich, C.A.: Normal computation for discrete surfaces in 3d space. Computer Graphics Forum 16(3), 15–26 (1997)
3. Lenoir, A.: Fast estimation of mean curvature on the surface of a 3d discrete object. In: Ahronovitz, E. (ed.) DGCI 1997. LNCS, vol. 1347, pp. 175–186. Springer, Heidelberg (1997)
4. Papier, L., Françon, J.: Évaluation de la normale au bord d'un objet discret 3d. Revue internationale de CFAO et d'Infographie 13(2), 205–226 (1998)
5. Tellier, P., Debled-Rennesson, I.: 3d discrete normal vectors. In: Bertrand, G., Couprie, M., Perroton, L. (eds.) DGCI 1999. LNCS, vol. 1568, pp. 447–457. Springer, Heidelberg (1999)
6. Coeurjolly, D., Flin, F., Teytaud, O., Tougne, L.: Multigrid Convergence and Surface Area Estimation. In: Asano, T., Klette, R., Ronse, C. (eds.) Geometry, Morphology, and Computational Imaging. LNCS, vol. 2616, pp. 101–119. Springer, Heidelberg (2003)
7. Hermann, S., Klette, R.: Multigrid analysis of curvature estimators. In: Proceedings of Image and Vision Computing New Zealand, pp. 108–112 (2003)
8. Veelaert, P.: Uncertain geometry in computer vision. In: Andrès, É., Damiand, G., Lienhardt, P. (eds.) DGCI 2005. LNCS, vol. 3429, pp. 359–370. Springer, Heidelberg (2005)
9. Debled-Rennesson, I., Rémy, J.L., Rouyer-Degli, J.: Linear segmentation of discrete curves into fuzzy segments. Discrete Applied Math. 151, 122–137 (2005)

10. Debled-Rennesson, I., Feschet, F., Rouyer-Degli, J.: Optimal blurred segments decomposition of noisy shapes in linear time. Computers & Graphics 30(1), 30–36 (2006)
11. Provot, L., Buzer, L., Debled-Rennesson, I.: Recognition of blurred pieces of discrete planes. In: Kuba, A., Nyúl, L.G., Palágyi, K. (eds.) DGCI 2006. LNCS, vol. 4245, pp. 65–76. Springer, Heidelberg (2006)
12. Nguyen, T.P., Debled-Rennesson, I.: Curvature estimation in noisy curves. In: Kropatsch, W.G., Kampel, M., Hanbury, A. (eds.) CAIP 2007. LNCS, vol. 4673, pp. 474–481. Springer, Heidelberg (2007)
13. Lenoir, A., Malgouyres, R., Revenu, M.: Fast computation of the normal vector field of the surface of a 3-d discrete object. In: Miguet, S., Ubéda, S., Montanvert, A. (eds.) DGCI 1996. LNCS, vol. 1176, pp. 101–112. Springer, Heidelberg (1996)
14. Thiel, E., Montanvert, A.: Chamfer masks: discrete distance functions, geometrical properties and optimization. In: 11th ICPR, The Hague, The Netherlands, vol. 3, pp. 244–247 (1992)
15. Borgefors, G.: On digital distance transforms in three dimensions. Computer Vision and Image Understanding 64(3), 368–376 (1996)
16. Verwer, B., Verbeek, P., Dekker, S.: An efficient uniform cost algorithm applied to distance transforms. IEEE Transactions on Pattern Analysis and Machine Intelligence 11(4), 425–429 (1989)
17. Coeurjolly, D.: Visibility in discrete geometry: An application to discrete geodesic paths. In: Braquelaire, A., Lachaud, J.-O., Vialard, A. (eds.) DGCI 2002. LNCS, vol. 2301, pp. 326–337. Springer, Heidelberg (2002)
18. Provot, L., Debled-Rennesson, I.: Segmentation of noisy discrete surfaces. In: 12th International Workshop on Combinatorial Image Analysis, Buffalo, NY, USA, vol. 4958, pp. 160–171 (2008)

Normals and Curvature Estimation for Digital Surfaces Based on Convolutions

Sébastien Fourey[1] and Rémy Malgouyres[2]

[1] GREYC, UMR6072 – ENSICAEN, 6 bd maréchal Juin 14050 Caen CEDEX, France
Sebastien.Fourey@greyc.ensicaen.fr
[2] LAIC, EA2146 – Université Clermont 1, BP 86, Aubière CEDEX France
Remy.Malgouyres@laic.u-clermont1.fr

Abstract. In this paper, we present a method that we call *on-surface convolution* which extends the classical notion of a 2D digital filter to the case of digital surfaces (following the *cuberille* model). We also define an averaging mask with local support which, when applied with the iterated convolution operator, behaves like an averaging with large support. The interesting property of the latter averaging is the way the resulting weights are distributed: they tend to decrease following a "continuous" geodesic distance within the surface. We eventually use the iterated averaging followed by convolutions with differentiation masks to estimate partial derivatives and then normal vectors over a surface. We provide an heuristics based on [14] for an optimal mask size and show results.

1   Introduction

Estimation of geometrical properties and quantities of objects known through their digitizations is an important goal of discrete geometry. One of the classical problems is simply to measure the length of a curve (or a perimeter) in the digital plane [6,4]. One may also quote the estimation of tangents or normals to a curve [11], normal vectors over a surface [12], or area of a digital surface [3,9,20].

In 2D, a whole set of methods rely on the *digital straight segments* recognition algorithm [5] used to find maximal line segments in a curve, which may in turn be used to estimated the curve's length or its tangent vectors [11]. These methods have been extended to the 3D case with digital plane recognition [17]. Directional tangent estimation based on segments recognition was used in [19] to compute normal vectors on a digital surface and in [10] for the nD case. All of these methods are sensitive to noise.

In the case of digital surfaces, another method was introduced by Papier and Françon ([16,15]) to estimate the normal vector field. It is based on a weighted averaging of the canonical normals in a neighborhood of each surfel. Their method generalizes to large neighborhoods the approach proposed by Chen *et al.* in [2] and is very close to the one we propose here, although it differs for at least two points: First, the size of the neighborhood taken into account, called the "order of the umbrellas" in their paper, is a parameter of the method and no hints to

D. Coeurjolly et al. (Eds.): DGCI 2008, LNCS 4992, pp. 287–298, 2008.

find an optimal size is provided. Here, we rely on a new theoretical result in the 1D functional case to set this parameter. Tests on spheres and tori show that the chosen size provides experimental *convergence* (see Section 4.1). Second, umbrellas in Papier's method grow following a breadth-first traversal of the surfels v-adjacency graph, whereas our method may be seen as the result of an averaging process using masks which grow in a more *geodesic* and isotropic way (see Section 4.2).

The normal estimation method introduced here is based on the notion of *on-surface convolution* (Section 3) which extends to digital surfaces the classical 2D filters used in image processing. Using an averaging mask defined locally, we apply an iterated convolution operation on the centers of the surfels. Then, we use two orthogonal differentiation operators on the resulting centers to estimate partial derivatives, and by a cross product we obtain normal vectors. As previously mentioned, we follow the criterion given in [14] to set the optimal number of iterations depending on the digitization step (voxel size).

In the last section, we consider the problem of curvature estimation, which requires second order derivatives estimation, and show an encouraging experiment.

2   Digital Objects and Discrete Surfaces

2.1   Digital Objects

In this paper, we simply call a *digital object* a subset of $\mathbb{Z}^3$, the classical three dimensional grid. Such an object is seen as a set of unit cubes called *object voxels* centered on points with integer coordinates. *Background voxels* are voxels that do not belong to the object.

2.2   Digital Surface

A digital object can be visualized slice per slice, i.e., as a series of 2D binary images. More often, what is rendered is a 2D view of its external surface (see [1,2]). In this case, the basic rendering primitives are the *surfels*. Surfels are unit squares that are shared by two 6-adjacent voxels: one belonging to the object and one belonging to the background. There are exactly six types of surfels according to the direction of their normal vectors. Thus, a surfel can be uniquely defined by the data of its center's coordinates and its orientation. In the sequel, a surfel is a pair (p, n) where $p \in \mathbb{R}^3$ (the center) and $n \in \{(\pm 1, 0, 0), (0, \pm 1, 0), (0, 0, \pm 1)\}$ (the normal vector). Eventually, a *digital surface* is a set of surfels which is the set of all the surfels of a digital object.

We will use in the sequel the two functions σ and ν which associate to a surfel $s = (p, n)$, respectively, its centre $\sigma(s) = p$ and its normal vector $\nu(s) = n$.

2.3   Adjacency Relations between Surfels an Surfels Neighborhoods

We can define two adjacency relations between surfels: the e-adjacency and the v-adjacency relations.

As depicted in Figure 1, a surfel x might share one of its four edges with at most three other surfels. In the cases depicted in Figure 1(a)(b)(c), only one surfel y shares the bold edge with x. We say that this surfel y is e-*adjacent* to x (y is also called an e-*neighbor* of x). In the case of Figure 1(d), we choose[1] to define the surfel y as the e-neighbor of x associated with the bold edge. Thus, we have defined with examples the e-adjacency relation ("e" for *edge*) so that a surfel has exactly four e-neighbors: one per edge.

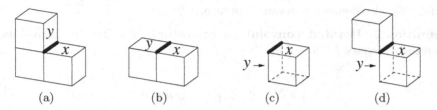

(a)   (b)   (c)   (d)

Fig. 1. The e-neighbor y of a surfel x given one of its edges (bold line)

Next, we define a *loop* in a digital surface Σ as an e-connected component of the set of the surfels of Σ which share a given vertex w. For example, if Σ is the surface of the object depicted in Figure 2(a) (which is made of three voxels), then the vertex w defines two loops: one that contains the six gray surfels, and another one in the back with three surfels. Eventually, we say that two surfels are v-*adjacent* (v for *vertex*) if they belong to a common loop of Σ.

These two adjacency relations allow us to define the e-*neighborhood* (resp. v-*neighborhood*) of a surfel x denoted by $N_e(x)$ (resp. $N_v(x)$) as the set of surfels that are v-adjacent (resp. e-adjacent) to x. Figure 2(b) shows an example of a v-neighborhood.

(a)   (b)

Fig. 2. (a) A loop of surfels (in gray), (b) A v-neighborhood

3   On-Surface Convolution

The work presented in the next sections illustrates the use of *on-surface convolution*, which we introduce here. In the sequel of the paper, Σ is a digital surface and S is a vector space over $\mathbb{R}$. We define the space of *digital surface filters over* Σ as the set of functions from $\Sigma \times \Sigma$ to $\mathbb{R}$.

---

[1] By this choice, the interior of a surface component is 6-connected [7].

Definition 1 (Convolution operator). *For $f : \Sigma \mapsto S$ and $F : \Sigma \times \Sigma \mapsto \mathbb{R}$, we define the operator Ψ as follows:*

$$\begin{aligned}
\Psi_{f,F} : \Sigma &\longrightarrow S \\
x &\longmapsto \sum_{y \in \Sigma} F(x,y) \cdot f(y)
\end{aligned} \tag{1}$$

Intuitively, Ψ acts like a convolution of the values of f on the surface with a convolution kernel whose values should depend on the relative positions of two surfels. We also define the iterated operator $\Psi^{(n)}$.

Definition 2 (Iterated convolution operator). *The* iterated convolution operator *is defined for $n \in \mathbb{Z}$ by:*

$$\begin{cases}
\Psi_{f,F}^{(0)} = f \\
\Psi_{f,F}^{(n)} = \Psi_{\Psi_{f,F}^{(n-1)},F} & \text{if } n > 0.
\end{cases} \tag{2}$$

Next, we define an averaging and two derivative filters which we will use in Section 4.1 to estimate the normal field on a digital surface.

3.1   The Averaging Filter

We define here a local averaging mask $W_{\text{avg}} : \Sigma \times \Sigma \mapsto \mathbb{R}$. This mask should be seen as a wrapping of the 2D classical mask (Figure 3(a)) which follows the *local* shape of the digital surface. We define this wrapping in such a way that the weights remain balanced, the way they are in the 2D pattern, but considering the local orientation within a v-neighborhood.

Let x and y be two surfels of Σ such that $y \in N_v(x)$. If y is e-adjacent to x, then y has exactly two e-neighbors in $N_v(x)$, say s and t. We define $\delta_x(y)$ as the number of surfels in $\{s, t\}$ which are e-adjacent to x. If y is v-adjacent but not e-adjacent to x then there is a single loop L of Σ that contains both x and y. In this case, we define $\delta_x(y) = \text{card}(L) - 3$.

The number $\delta_x(y)$ is used to take into account the number of surfels in a loop containing x which are not e-adjacent or equal to x. Within a loop, all these surfels will end up with a total contribution of $\frac{1}{16}$. If there are no such surfels, the weight $\frac{1}{16}$ is split among the two n-neighbors of x in the loop.

Now, let x be a surfel of Σ. For any surfel $y \in \Sigma$ we define the *weight* $W_{\text{avg}}(x, y)$ as follows:

$$W_{\text{avg}}(x,y) = \begin{cases}
\frac{1}{4} & \text{if } y = x, \\
\frac{1}{8} + \frac{\delta_x(y)}{32} & \text{if } y \in N_e(x), \\
\frac{1}{16 \cdot \delta_x(y)} & \text{if } y \in N_v(x) \setminus N_e(x), \\
0 & \text{if } y \notin N_v(x).
\end{cases} \tag{3}$$

One can check that for all $x \in \Sigma$, we have $\sum_{y \in \Sigma} W_{\text{avg}}(x,y) = 1$.

See Figure 3(b) for an example of a v-neighborhood and the associated values of W_{avg}.

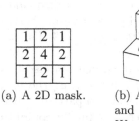

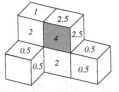

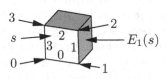

(a) A 2D mask.   (b) A surfel x (in gray) and the values of $16 \cdot W_{\mathrm{avg}}(x,y)$ for the surfels y of $N_v(x) \cup \{x\}$.   (c) Ordering of vertices and edges for a given surfel s.

Fig. 3. Illustrations of the masks definition

3.2   The First Order Derivative Filters

We introduce here two directional derivative masks which may be used with the convolution operator to obtain two orthogonal differentiation operators.

For each surfel s of Σ we define a numbering of the surfel vertices and edges as illustrated by Figure 3(c), following the coherent orientation around the outward normal. We denote by $E_i(s)$ the e-neighbor of s that shares with s its ith edge. Then, we define the derivative masks $D_{\mathrm{u}}(x,y)$ and $D_{\mathrm{v}}(x,y)$ for $x,y \in \Sigma$ as follows:

$$D_{\mathrm{u}}(x,y) = \begin{cases} \frac{1}{2} & \text{if } y = N_0(x), \\ -\frac{1}{2} & \text{if } y = N_2(x), \\ 0 & \text{otherwise.} \end{cases} \qquad D_{\mathrm{v}}(x,y) = \begin{cases} \frac{1}{2} & \text{if } y = N_1(x), \\ -\frac{1}{2} & \text{if } y = N_3(x), \\ 0 & \text{otherwise.} \end{cases} \quad (4)$$

Given the derivative masks D_{u} and D_{v}, we may define two derivative operators $\Psi_{f,D_{\mathrm{u}}}$ and $\Psi_{f,D_{\mathrm{v}}}$ which act on a function f defined on Σ.

4   Normal Estimation

In this section, we address the problem of estimating the normal vectors on the surface of a digital object. This estimation is achieved using iterated convolutions of the surfel centers with the averaging mask W_{avg}, followed by a step of differentiation using the derivative filters D_{u} and D_{v} defined in Section 3.2.

4.1   Surface Normal Estimation

For different purposes, such as shading methods for visualization or for area estimation, it is interesting to compute normal vectors on the surface of a digital object. In so doing we expect that the computed normals will be as closed as possible to the normals on the surface of a continuous object from which the digital object could have been obtained by a digitization process. In other words, given a continuous object, we expect that the estimated normals can get as close as required to the real normals as soon as the digitization step is chosen sufficiently small. Roughly speaking, this property is what is called *multigrid convergence* in the literature. Several surface normal estimation methods have been proposed, among which we may cite [13,19,10] is past DGCI's, but no method has been proved to be convergent in that sense.

Here, we show that the normal vectors of a digital surface can also be estimated by first averaging the positions of the surfel centers using the iterated convolution operator, then computing approximations of two partial derivatives to obtain vectors in the tangent space to the surface, and finally computing a normal vector by a simple cross product of the tangent vectors.

More formally, using the iterated convolution operator Ψ, the averaging mask W_{avg}, and the derivative masks D_u and D_v we define a function $\Gamma^{(n)} : \Sigma \longrightarrow \mathbb{R}^3$ for $n \in \mathbb{Z}$ such that $\Gamma^{(n)}(s)$ is the *estimated normal vector* of Σ at the center of s (after n on-surface convolutions). We define the function $\Gamma^{(n)}$ for $n \in \mathbb{Z}$ by:

$$\Gamma^{(n)}(s) = \frac{\Delta_u^{(n)}(s) \wedge \Delta_v^{(n)}(s)}{\|\Delta_u^{(n)}(s) \wedge \Delta_v^{(n)}(s)\|} \quad \text{with } \Delta_u^{(n)} = \Psi_{\Psi_{\sigma,W_{\mathrm{avg}}}^{(n)}, D_u} \quad (\text{resp. for } \Delta_v^{(n)}) \quad (5)$$

As the initial averaging process is iterated, the size of the neighborhood taken into account grows accordingly and the precision of the estimate increases as we get closer to the optimal number of iterations. (This number is discussed in Section 4.3.)

From the definition of the operator $\Psi^{(n)}$, we see that $\Gamma^{(n)}(s)$ is the result of a computation which involves all the surfels of Σ whose distance to s is at most n in the v-adjacency graph of Σ. In fact, the size of the neighborhood taken into account when computing $\Gamma^{(n)}(s)$ grows with n but the weights tend to follow a geodesic distance which does not coincide with the distance in the v-adjacency graph. This point, which we claim is a good point, is discussed in the next section. (The way the weights are distributed when the number of iterations increases is illustrated by Figures 5(b) and 4(b).) We will present in Section 4.3 the results of some experiments.

4.2   Comparison with Papier's Averaging Process

In [16,15], Papier *et al.* define averaging weights on possibly large neighborhoods obtained using a breadth-first visiting algorithm of the surfels v-adjacency graph. Their approach generalizes the one of [2] who used only the only e-neighborhood to estimate the normals by averaging the elementary normals (among the six possible ones). Both methods are based on the averaging of the canonical normals $\nu(s)$ of the surfels to estimate the exact normals. This point slightly differs from our method since we are not averaging the normal vectors but simply the surfel centers.

Furthermore, when considering a large neighborhood in an averaging process, one should expect that the boundary of the neighborhood is equidistant to its center, according to the geodesic distance within the continuous surface which has been digitized. The weights should also decrease according to this geodesic distance. As depicted in Figure 5(a) on a digitized plane with normal vector

$(1, 1, 1)$, as well as in Figure 4(c) on a digitized paraboloid, neighborhoods obtained by a breadth-first traversal of the surfels graph do not share the former property. Therefore, we think that the neighborhoods used by Papier *et al.* are not optimal.

In our case, the neighborhood taken into account by the averaging process grows after iterations of convolutions with a local mask, designed to adapt itself to the local geometry of the surface (Section 3.1). Although the actual size of the masks resulting of iterated convolutions also follow the v-adjacency graph, we observe that the weights in these masks tends to share the above mentioned properties (isotropy and decreasing according to a "continuous" geodesic distance). In order to illustrate how the averaging mask grows, we use a diffusion process: with $S = \mathbb{R}$ we choose a surfel $s_0 \in \Sigma$ and define the function $\delta_{s_0} : \Sigma \longrightarrow \mathbb{R}$ such that $\delta_{s_0}(s_0) = 1$ and $\delta_{s_0}(s) = 0$ for $s \in \Sigma \setminus \{s_0\}$. Then, we compute $\Psi^{(n)}_{\delta_{s_0}, W_{\mathrm{avg}}}$ for a given n. A result of this diffusion process that we call an *impulse response of the averaging filter* $\Psi^{(n)}_{f, W_{\mathrm{avg}}}$ is depicted in Figure 5(b) for the same plane as mentioned previously. Another example is given on the paraboloid depicted in Figure 4(b) with the surfel s_0 at the saddle point, where one can get convinced that the impulse response of our iterated convolution mask behaves as if following a geodesic distance function on the surface.

4.3 Experiments

We have evaluated the precision of the estimation that can be achieved with our method. For this purpose, we have used digitized spheres and tori with several radii and measured the average angular error between the estimated and the exact normal vectors for all surfels. In these experiments, we use a number n of convolution iterations inspired by the result of [14] (Theorem 1) for continuous functions from $\mathbb{R}$ to $\mathbb{R}$ known through their digitizations. Following the latter result, if h is the width of the pixels used for the digitization process, then a convergence at rate $h^{\frac{2}{3}}$ for the estimation of the first derivative of the function may be obtained by using a convolution mask with a width $w = \lfloor h^{-\frac{4}{3}} \rfloor$. Given this width, we deduce the number n of convolution iterations required when using the mask W_{avg} (see Section 3.1): $n = \frac{w}{2}$ if $w \in 2\mathbb{Z}$ and $n = \frac{w-1}{2}$ if $w + 1 \in 2\mathbb{Z}$.

The results of our experiments are presented in Figure 6. For Figure 6(a), we have used digital spheres with radii from 10 to 100 (i.e. h goes from $\frac{1}{20}$ to $\frac{1}{200}$). It appears clearly that the method achieves a better estimation when the size of the sphere increases (i.e. the digitization step decreases). This tends to show that our estimator is multigrid convergent.

As a comparison, tests conducted by Papier in [15] do not clearly show an improvement of the precision when the radius of the sphere increases. Furthermore, they only used umbrellas of order 1 to 5. It is however clear that the size of the mask should be set according to the resolution, as we do here. With the approach mentioned in the introduction, the best result in [10] is obtained on a sphere with radius 100 and an average error of $1.51°$ (std. dev. 2.34), when the average error of our method is $0.58°$ (std. dev. 0.27). On a sphere with radius 50, they obtain $2.19°$ (std. dev. 3.46) when we have $0.85°$ (std. dev. 0.33). The

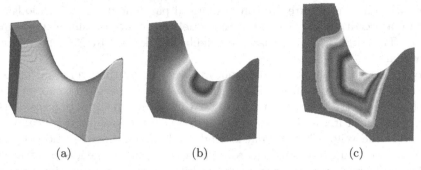

(a)   (b)   (c)

Fig. 4. (a) View of a paraboloid shaded using normals estimated with our method. (b) Response of the iterated averaging filter over the paraboloid compared with a breadth-first traversal (c). (The volume size is $150 \times 150 \times 150$ and has 104926 surfels.)

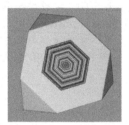
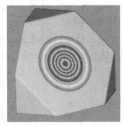

(a) Breadth-first traversal.   (b) Iterated convolution response.

Fig. 5. Breadth-first traversal of the surfels graph and convolution over a digitized plane $x + y + z = 0$

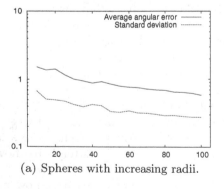
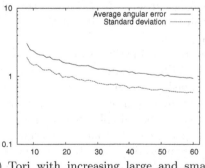

(a) Spheres with increasing radii.   (b) Tori with increasing large and small radii. (Large radius on the horizontal axis.)

Fig. 6. Average angular error, in degree and using a logarithmic scale, of the estimated normals on spheres and tori. The error, on each surfel, is computed as the angle between the estimated normal vector and the direction from the center of sphere to the center of the surfel (resp. from the skeleton of the torus).

earlier paper by Tellier and Debled-Rennesson [19] reports the best average error of 2.84° (std. dev. 2.24) for a sphere with radius 25, when our method obtains 1.16° (std. dev. 0.47).

Furthermore, spheres are not general enough to put to the test a surface normal estimator. This at least because a sphere has a constant and positive Gaussian curvature. Therefore, we have tested our method on several tori with increasing radii, rotated along the three axes. Tori are nice for this test because they have both positive and negative Gaussian curvatures. The results are depicted in Figure 6(b), where each torus had a small radius of half its large one. Again, the average error and its standard deviation decrease with an increasing resolution.

5 Curvature Estimation

In this section, we present an attempt to estimate second order quantities after a similar averaging process as the one used to estimate normals. Although the theoretical work done in [14] shows that higher orders estimates may be computed in a similar way for 1D digitized functions, it seems from our first experiments that the precision is not so good when using on-surface convolution.

5.1 Second Derivatives Operators

Because the masks D_u and D_v defined in Section 3.2 do not generally preserve the orientations of the differentiations when applied to one surfel and one of its neighbors, they should not be applied iteratively to compute second derivatives. Hence, we need to define new differentiation operators.

First, given a surfel s and its ith edge we denote by $EE_i(s)$ the e-neighbor of $E_i(s)$ which is not included in a loop containing s. Intuitively, $EE_i(s)$ is the next surfel one encounters on the surface after $E_i(s)$ when traveling in the direction from s to $E_i(s)$. Furthermore, given the jth vertex of s we denote by $EV_i^j(s)$ the e-neighbor of $N_i(s)$ that shares with s its vertex j. (See Figure 3(c) for the vertices and edges ordering.) Eventually, for $n \in \mathbb{Z}$ and $s \in \Sigma$ we define (remember that $\sigma(s)$ is the center of s)

$$\Delta_{uu}^{(n)}(s) = \frac{\left(\tilde{\sigma}^{(n)}(EE_0(s)) - \tilde{\sigma}^{(n)}(s)\right) - \left(\tilde{\sigma}^{(n)}(s) - \tilde{\sigma}^{(n)}(EE_2(s))\right)}{4}$$

$$\Delta_{vv}^{(n)}(s) = \frac{\left(\tilde{\sigma}^{(n)}(EE_1(s)) - \tilde{\sigma}^{(n)}(s)\right) - \left(\tilde{\sigma}^{(n)}(s) - \tilde{\sigma}^{(n)}(EE_3(s))\right)}{4}$$

$$\Delta_{uv}^{(n)}(s) = \frac{\left(\tilde{\sigma}^{(n)}(EV_0^0(s)) - \tilde{\sigma}^{(n)}(EV_0^1(s))\right) - \left(\tilde{\sigma}^{(n)}(EV_2^3(s)) - \tilde{\sigma}^{(n)}(EV_2^2(s))\right)}{4}$$

where

$$\tilde{\sigma}^{(n)} = \Psi_{\sigma,W_{\text{avg}}}^{(n)} \tag{6}$$

5.2   Curvature Estimation

Given a *regular parametric surface* $S : D \subset \mathbb{R}^2 \longrightarrow \mathbb{R}^3$, $(u, v) \longmapsto S(u, v) = (x(u, v), y(u, v), z(u, v))$ where D is an open and connected subset of $\mathbb{R}^2$, the *Gaussian* (K) and *mean* (H) *curvatures* are defined as follows [18]:

$$K = \frac{e.g - f^2}{E.G - F^2} \qquad\qquad H = \frac{e.G - 2.f.F + g.E}{2(E.G - F^2)} \tag{7}$$

with

$$E = \frac{\partial S}{\partial u}\frac{\partial S}{\partial u}, \; F = \frac{\partial S}{\partial u}\frac{\partial S}{\partial v}, \; G = \frac{\partial S}{\partial v}\frac{\partial S}{\partial v}, \; e = N \cdot \frac{\partial^2 S}{\partial u^2}, \; f = N \cdot \frac{\partial^2 S}{\partial u \partial v}, \; g = N \cdot \frac{\partial^2 S}{\partial v^2}$$

where N is the unit normal field over S.

For a given number n of averaging iterations, we approximate $\frac{\partial S}{\partial u}$ as $\Delta_u^{(n)}$, $\frac{\partial S}{\partial v}$ as $\Delta_v^{(n)}$, $\frac{\partial^2 S}{\partial u^2}$ as $\Delta_{uu}^{(n)}$, $\frac{\partial^2 S}{\partial v^2}$ as $\Delta_{vv}^{(n)}$, $\frac{\partial^2 S}{\partial v^2}$ as $\Delta_{uv}^{(n)}$, and N as $\Gamma^{(n)}$ (Eq. 5). Eventually, we obtain an estimation of the Gaussian or mean curvature for each surfel of a digital surface using equation 7.

5.3   Experiment

We have conducted an experiment with a large digitized torus which show that the Gaussian curvature may be only roughly estimated. The torus had a 80 voxels large radius, and a 40 voxels small radius. Figure 7 shows the estimated and exact Gaussian curvatures for all the surfels sorted according to their increasing exact Gaussian curvatures. The size of the averaging mask to be used was set according to the result of [14] about higher order derivatives (Theorem 3). Actually, we divided the estimation process in two steps. First, an estimated Gaussian curvature $\tilde{g}_1(s)$ for each surfel $s \in \Sigma$ was computed after $\frac{n}{2}$ averaging

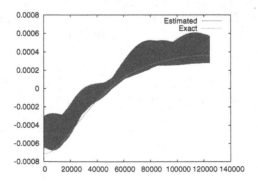

Fig. 7. Estimated and exact Gaussian curvatures on a torus (large radius is 80, small radius is 40). Surfels are numbered on the horizontal axis according to their increasing exact Gaussian curvatures. The estimated values appear as a thick area because they are plotted with lines and vary very quickly.

iterations (i.e., using $\Gamma^{\left(\frac{n}{2}\right)}$ and the operators $\Delta_*^{\left(\frac{n}{2}\right)}$ as decribed before) and we used $\tilde{g} = \Psi_{g_1, W_{\text{avg}}}^{\left(\frac{n}{2}\right)}$ as the actual estimate. Figure 7 shows that the curvature is only approximated.

6 Conclusion

We have defined an averaging operator based on a convolution over the surface of a digital object, as well as directional derivative operators. When combined, these operators may be used to estimate the normal vectors on the surface of a digitized object. The implementation of this method is straightforward (compared with methods which involve, say, DSS recognition). Furthermore, since it relies on an averaging that may be seen as low-pass filtering, this method is less noise sensitive, just as the planar case ([14]).

We also tackled the problem of curvature estimation. This estimator should be investigated in the future since, as far as we know, there is no multigrid convergent second derivative estimator known for digital surfaces.

The complexity of the convolution approach should also be compared with existing geometric estimators.

References

1. Artzy, E., Frieder, G., Herman, G.T.: The theory, design, implementation and evaluation of a three dimensional surface detection algorithm. Computer graphics and image processing 15(1), 1–23 (1981)
2. Chen, L.-S., Herman, G.T., Reynolds, R.A., Udupa, J.K.: Surface shading in the cuberille environment. IEEE Journal of computer graphics and Application 5(12), 33–43 (1985)
3. Coeurjolly, D., Flin, F., Teytaud, O., Tougne, L.: Multigrid convergence and surface area estimation. In: Asano, T., Klette, R., Ronse, C. (eds.) Geometry, Morphology, and Computational Imaging. LNCS, vol. 2616, pp. 119–124. Springer, Heidelberg (2003)
4. Coeurjolly, D., Klette, R.: A comparative evaluation of length estimators of digital curves. IEEE Transactions on Pattern Analysis and Machine Intelligence 26(2), 252–257 (2004)
5. Debled-Rennesson, I., Reveillès, J.-P.: A linear algorithm for segmentation of digital curves. International Journal of Pattern Recognition and Artificial Intelligence 9(4), 635–662 (1995)
6. Dorst, L., Smeulders, A.W.M.: Length estimators for digitized contours. Computer Vision, Graphics, and Image Processing 40(3), 311–333 (1987)
7. Herman, G.T.: Discrete multidimensional Jordan surfaces. CVGIP: Graphical Models and Image Processing 54(6), 507–515 (1992)
8. Herman, G.T., Liu, H.K.: Three-dimensional display of human organs from computed tomograms. Computer Graphics and Image Processing 9, 1–29 (1979)
9. Klette, R., Sun, H.J.: A global surface area estimation algorithm for digital regular solids. Technical report, Computer science department of the University of Auckland (September 2000)

10. Lachaud, J.-O., Vialard, A.: Geometric measures on arbitrary dimensional digital surfaces. In: Nyström, I., Sanniti di Baja, G., Svensson, S. (eds.) DGCI 2003. LNCS, vol. 2886, pp. 434–443. Springer, Heidelberg (2003)
11. Lachaud, J.-O., Vialard, A., de Vieilleville, F.: Fast, accurate and convergent tangent estimation on digital contours. Image and Vision Computing 25(10), 1572–1587 (2007)
12. Lenoir, A.: Des outils pour les surfaces discrètes. PhD thesis, Université de Caen (in French) (1999)
13. Lenoir, A., Malgouyres, R., Revenu, M.: Fast computation of the normal vector field of the surface of a 3-d discrete object. In: Miguet, S., Ubéda, S., Montanvert, A. (eds.) DGCI 1996. LNCS, vol. 1176, pp. 101–112. Springer, Heidelberg (1996)
14. Malgouyres, R., Brunet, F., Fourey, S.: Binomial convolutions and derivatives estimation from noisy discretizations. In: Coeurjolly, D., et al. (eds.) DGCI 2008. LNCS, vol. 4992, pp. 369–377. Springer, Heidelberg (2008)
15. Papier, L.: Polyédrisation et visualisation d'objets discrets tridimensionnels. PhD thesis, Université Louis Pasteur, Strasbourg, France (in French) (1999)
16. Papier, L., Françon, J.: Évalutation de la normale au bord d'un objet discret 3D. Revue de CFAO et d'informatique graphique 13, 205–226 (1998)
17. Sivignon, I., Dupont, F., Chassery, J.-M.: Discrete surfaces segmentation into discrete planes. In: Klette, R., Žunić, J. (eds.) IWCIA 2004. LNCS, vol. 3322, pp. 458–473. Springer, Heidelberg (2004)
18. Stoker, J.J.: Differential geometry. Wiley, Chichester (1989)
19. Tellier, P., Debled-Rennesson, I.: 3d discrete normal vectors. In: Bertrand, G., Couprie, M., Perroton, L. (eds.) DGCI 1999. LNCS, vol. 1568, pp. 447–458. Springer, Heidelberg (1999)
20. Windreich, G., Kiryati, N., Lohmann, G.: Voxel-based surface area estimation: from theory to practice. Pattern Recognition 36(11), 2531–2541 (2003)

On Minimal Moment of Inertia Polyominoes*

Srečko Brlek, Gilbert Labelle, and Annie Lacasse

Laboratoire de Combinatoire et d'Informatique Mathématique,
Université du Québec à Montréal,
CP 8888, Succ. Centre-ville, Montréal (QC) Canada H3C3P8
{brlek.srecko,labelle.gilbert}@uqam.ca, annie.lacasse@gmail.com

Abstract. We analyze the moment of inertia I(S), relative to the center
of gravity, of finite plane lattice sets S. We classify these sets according to
their roundness: a set S is rounder than a set T if I(S) < I(T). We show
that roundest sets of a given size are strongly convex in the discrete sense.
Moreover, we introduce the notion of quasi-discs and show that roundest
sets are quasi-discs. We use weakly unimodal partitions and an inequality for the radius to make a table of roundest discrete sets up to size 40.
Surprisingly, it turns out that the radius of the smallest disc containing
a roundest discrete set S is not necessarily the radius of S as a quasi-disc.

Keywords: Discrete sets, moment of inertia, polyominoes, lattice paths.

1 Introduction

In this paper we consider plane sets up to translations. By a *discrete set* we
mean a finite set of lattice points or a finite union of lattice closed unit squares
(*pixels*) (Figure 1 (a)).

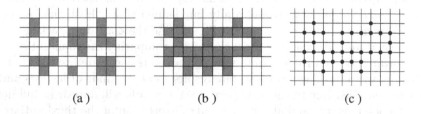

Fig. 1. (a) Discrete set (b) a typical polyomino and (c) its corresponding animal

In particular, the word *polyomino* means a finite union of pixels in the plane
whose boundary consists of a disjoint union of simple closed polygonal paths
using 4-connectedness (Figure 1(b)). These sets are well-known combinatorial
objects in discrete geometry. The dual of a polyomino, usually called *animal*,

---

* With the support of NSERC (Canada).

D. Coeurjolly et al. (Eds.): DGCI 2008, LNCS 4992, pp. 299–309, 2008.

consists of the set of centers of its pixels. Using a $(\frac{1}{2}, \frac{1}{2})$ shift, we may always assume that an animal is a subset of the discrete plane $\mathbb{Z} \times \mathbb{Z}$. Moreover, a polyomino is called *v-convex* (resp. *h-convex*) if all its columns (resp. rows) are connected (Figure 2(a),(b)). A polyomino is *hv-convex* (Figure 2(c)) if all its columns and rows are connected and *strongly-convex* (Figure 2(d)) if given any two points u and v in its corresponding animal, the lattice points w in the segment $[u, v]$ are all in the animal. This notion coincides with the MP-convexity of Minsky and Papert [1] since animals are connected.

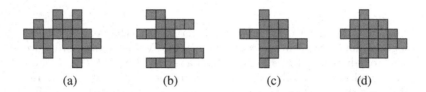

(a)   (b)   (c)   (d)

Fig. 2. A polyomino (a) v-convex (b) h-convex (c) hv-convex (d) strongly-convex

Our goal is to study the roundest discrete sets S of N pixels (or N points) having minimal moment of inertia $I(S)$ relatively to the center of gravity, a problem raised in a previous paper [2] in the context of designing incremental algorithms based on discrete Green theorem. This notion of roundness is distinct from the one given in [3] where they consider minimizing the *site perimeter* of lattice sets, that is the number of points with *Manhattan* distance 1 from the sets. For a given N, minimizing $I(S)$ is equivalent to minimizing $I(A)$, where A is the associated animal. To simplify notations and computations, the plane is identified with the complex plane $\mathbb{C}$ and $\mathbb{Z} \times \mathbb{Z}$ is identified with $\mathbb{Z} + i\mathbb{Z}$.

Section 2 recalls some basics about moment of inertia of discrete sets. Section 3 is devoted to properties of roundest discrete sets. More precisely, we establish a useful lemma concerning the moment of inertia of a union of discrete sets. We also introduce *discrete quasi-discs* and apply our lemma to establish that roundest discrete sets are strongly convex quasi-discs. Using the `combinat` package of the Maple formal tool [4] a method is developed for computing the roundest discrete sets according to size (Table 1) and some parameters associated to them (Table 2). Finally, we show how to extend our results to other kinds of lattices and to higher dimensions. Due to the lack of space, the proofs will appear in full detail in an extended version and also in the PhD dissertation of the third author [5].

2   Continuous and Discrete Moments of Inertia

We recall the definitions of the basic geometric parameters:

Definition 1. *Let S be a measurable subset of $\mathbb{C}$ such that*

$$\int\int_S |z|^2 dx\,dy < \infty. \tag{1}$$

The center of gravity g and the moment of inertia I(S), relative to the center of gravity are defined by the following equations:

$$g = g(S) = \frac{1}{\text{Area}(S)} \int \int_S z\, dx\, dy$$

and

$$I(S) = \int \int_S |z - g|^2 dx\, dy = \int \int_S |z|^2 dx\, dy - \frac{1}{\text{Area}(S)} \left| \int \int_S z\, dx\, dy \right|^2,$$

where $\text{Area}(S) = \int \int_S dx\, dy$.

In particular, if $S = P_1 \cup P_2 \cup \cdots \cup P_N$ is a union of N distinct pixels, the condition $\int \int_{P_1 \cup P_2 \cup \cdots \cup P_N} |z|^2 dx\, dy < \infty$ is obviously satisfied and $g(S)$ and $I(S)$ are well-defined. Moreover, for any single pixel P we have $I(P) = \frac{1}{6}$ and its center of gravity corresponds to its geometrical center.

Definition 2. *Let* $T = \{a_1, a_2, \cdots, a_N\} \subseteq \mathbb{C}$ *be a set of N distinct points in the complex plane where the point* a_k *has a mass* m_k *for* $k = 1, \cdots, N$. *The center of gravity g and the moment of inertia I(T), relative to the center of gravity are defined by*

$$g = g(T) = \frac{1}{m_1 + \cdots + m_N} \sum_{k=1}^{N} m_k a_k,$$

and

$$I(T) = \sum_{k=1}^{N} m_k |a_k - g|^2 = \sum_{k=1}^{N} m_k |a_k|^2 - \frac{1}{m_1 + \cdots + m_N} \left| \sum_{k=1}^{N} m_k a_k \right|^2$$

$$= \frac{1}{m_1 + \cdots + m_N} \sum_{k<l} m_k m_l |a_k - a_l|^2.$$

3   Properties of Roundest Discrete Sets

Any measurable (or finite) subset $S \subseteq \mathbb{C}$ satisfying (1), can be represented by its center of gravity g with mass $\int \int_S dx\, dy$ (or mass $|S|$). More generally, any family $S_1, S_2, \cdots, S_N$ of measurable (or finite) subsets $S_k \subseteq \mathbb{C}$ can be represented by the family $(g_1, g_2, \cdots, g_N)$ of their corresponding centers of gravity g_k having mass $m_k = \int \int_{S_k} dx\, dy$ (or mass $m_k = |S_k|$). From now on, we assume that every measurable subset of $\mathbb{C}$, satisfies (1).

The following useful lemma is a consequence of the classical parallel axis theorem [6] stating that, for any point p and any measurable (or finite) set S, the moment of inertia of S relative to p, denoted $I_p(S)$ and defined by

$$I_p(S) = \int \int_S |z - p|^2 dx\, dy \quad \left(\text{or} \sum_{k=1}^{N} m_k |a_k - p|^2 \right),$$

satisfies
$$I_p(S) = I(S) + m|p - g|^2, \tag{2}$$
where $g = g(S)$ and m is the mass of S.

Lemma 1. *Let $S_1, S_2, \cdots, S_N$ be disjoint measurable subsets $\subseteq \mathbb{C}$. Then*

$$I(S_1 \cup \cdots \cup S_N) = \sum_{k=1}^{N} I(S_k) + I(\{g_1, \cdots, g_N\})$$

$$= \sum_{k=1}^{N} I(S_k) + \sum_{k=1}^{N} m_k |g_k|^2 - \frac{1}{m}\left|\sum_{k=1}^{N} m_k g_k\right|^2$$

$$= \sum_{k=1}^{N} I(S_k) + \frac{1}{m} \sum_{k<l} m_k m_l |g_k - g_l|^2,$$

where g_k is the center of gravity of S_k with mass $m_k = \int\int_{S_k} dx\, dy$ (or mass $m_k = |S_k|$) and $m = m_1 + \cdots + m_N$.

3.1   Roundest Discrete Sets Are Strongly Convex

In order to analyse convexity properties of sets of minimal moment of inertia we need the following lemma.

Lemma 2. *Let L be a fixed line in the complex plane $\mathbb{C}$ and $c \notin L$ an arbitrary point with mass p. Let d be the point on L such that $[c, d]$ is the perpendicular segment to L. Let a and b be two distinct points in L having mass m, n respectively. We consider the following two cases:*

Case 1. *If a and b are both of the same side of d on L and*

$$0 \le |a - d| < |b - d|,$$

then the moment of inertia, $I(\{a, b, c\})$, strictly decreases as b moves towards a along the line L (Figure 3 (i)).

Case 2. *If a and b are on different sides of d on L, then $I(\{a, b, c\})$ strictly decreases as a or b moves towards d along the line L (Figure 3 (ii)).*

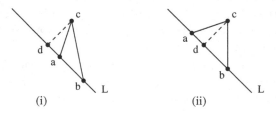

(i)   (ii)

Fig. 3. (i) Case 1 and (ii) Case 2 of Lemma 2

Theorem 1. *Let $N \geq 1$ and $S \subseteq \mathbb{C}$ be an arbitrary union of N distinct lattice pixels (or N lattice points) each having unit mass. If S is of minimal moment of inertia, then S is a strongly convex polyomino.*

Proof. *(Sketch)* It is sufficient to consider the case where S is a set of $N \geq 2$ lattice points. Then, by Definition 2, $\mathrm{I}(S) = q/N$, where $q \in \mathbb{N}^*$. Suppose that $\mathrm{I}(S)$ is minimal and that S is not strongly convex. Then, $\exists u, v \in S$ such that the segment $[u, v]$ contains a point (hole) $w \in \mathbb{Z} + i\mathbb{Z}$ such that $w \notin S$. Let

$$
\begin{aligned}
A &= \{z \in S \ : \ \exists t \in \mathbb{Q}, z = u + t(v - u), t \leq 0\}, \\
B &= \{z \in S \ : \ \exists t \in \mathbb{Q}, z = u + t(v - u), t \geq 1\}, \\
C &= S \setminus (A \cup B),
\end{aligned}
$$

where $a = g(A), b = g(B), c = g(C)$. Then by Lemmas 1 and 2, we can translate A (or B) towards w to obtain a set A' (or B') such that the set $S' = A' \cup B \cup C$ (or $S' = A \cup B' \cup C$) has smaller moment of inertia $\mathrm{I}(S') < \mathrm{I}(S)$ and such that $|S'| = |S| = N$ and $w \in S'$. If the set S' still contains a hole we repeat the previous construction to sets $S', S^{(2)}, \cdots$ such that $\mathrm{I}(S) > \mathrm{I}(S') > \mathrm{I}(S^{(2)}) > \mathrm{I}(S^{(3)}) > \cdots$. At each step, filling such a hole, decreases the moment of inertia by at least $\frac{1}{N}$. After a finite number of steps, this process must terminate and the resulting set $S^{(k)}$ must be a strongly convex set which is also an animal since it is, in particular, hv-convex. ∎

3.2 Roundest Discrete Sets Are Discrete Quasi-Discs

Much more can be said. We now show that roundest polyominoes are nearly discs in the following sense:

Definition 3. *Let $c \in \mathbb{C}$, $S \subseteq \mathbb{Z} + i\mathbb{Z}$ be a finite set of lattice points, and $r = \max_{s \in S} |s - c|$. Then S is called a discrete*

(i) *disc centered at c of radius r if*

$$
S = \{z : \ |z - c| \leq r\} \cap (\mathbb{Z} + i\mathbb{Z}),
$$

(ii) *quasi-disc centered at c of radius r if*

$$
\{z : \ |z - c| < r\} \cap (\mathbb{Z} + i\mathbb{Z}) \subseteq S \subseteq \{z : \ |z - c| \leq r\} \cap (\mathbb{Z} + i\mathbb{Z}).
$$

A disc and a quasi-disc of radius $r = 5$ are shown in Figure 4 (a) and (b) respectively. Note that every lattice point on the circumference must belong to a disc while at least only one is necessary in the case of quasi-disc. In both cases, every lattice point lying within the circumference must belong to the disc and quasi-disc.

Theorem 2. *Let S be a polyomino having N pixels with minimal moment of inertia, that is a roundest polyomino. Let A be its associated animal and $g = g(A)$ be its center of gravity. Then A is a quasi-disc centered at g with radius $r = \max_{a \in A} |g - a| = |g - a_0|$, with $a_0 \in A$.*

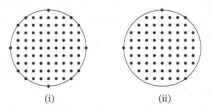

Fig. 4. (i) A discrete disc   (ii) a discrete quasi-disc

Figure 5 (a) illustrates Theorem 2 with $N = 5$. By contraposition, the 7×7 lattice set A is not minimal since the disc $C_{a_0} = \{z \in \mathbb{C} : |z - g| \leq |g - a_0|\}$ contains lattice points not in A (Figure 5 (b)). Note that the converse is false since, for $N = 3$, the quasi-disc of Figure 5 (c) is not minimal (with I = 2). The minimal one for $N = 3$ (with I $= \frac{4}{3} < 2$) is shown in Figure 5(d).

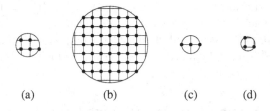

Fig. 5. Illustration of Theorem 2 and of the falsity of its converse

To pursue our study of roundest discrete sets we need a finer analysis. In particular, given N, the following result gives an upper bound for the radius r of the disc C_{a_0}.

Lemma 3. *Let A be a roundest animal having N points. The radius $r = |a_0 - g| = \max_{a \in A} |a - g|$ of the disc C_{a_0} centered at $g = g(A)$ satisfies*

$$r \leq \frac{1}{\sqrt{2}} + \sqrt{\frac{N}{\pi}}.$$

3.3   Generation of Roundest Discrete Sets According to Size

In order to generate all the roundest animals of a given size N, we classify them according to their vertical projections. Let A be a roundest animal of size N with projections $(n_1, n_2, \cdots, n_s)$ with $N = n_1 + n_2 + \cdots + n_s$. Then, by convexity property, the sequence $n_1, n_2, \cdots, n_s$ must satisfy

$$0 < n_1 \leq \cdots \leq n_k < n_{k+1} = \cdots = n_{l-1} > n_l \geq \cdots \geq n_{s-1} \geq n_s > 0.$$

Such sequences are called *weakly unimodal partitions* of N (or *stack* or *planar partitions* of N), see Stanley [7], Section 2.5, p. 76. Surprisingly, it turns out that any such sequence of projections corresponds to 0, 1 or 2 roundest animal of size N. More precisely, we have the following result.

Lemma 4. *Let $(n_1, n_2, \cdots, n_s)$ be a weakly unimodal sequence with $N = n_1 + \cdots + n_s$. Then among all animals having these vertical projections,*

(i) *there is a unique animal A, with minimal moment of inertia, if $n_1, n_2, \cdots, n_s$ have the same parity;*
(ii) *otherwise, there are exactly two animals A, A', having these projections, with minimal moment of inertia.*

These strongly convex animals A and A' are maximally symmetrically disposed along the x-axis (Figure 6). Moreover, the moment of inertia of A (and A') is given by the formula

$$
I(A) = \frac{1}{12} \sum_{k=1}^{s} n_k^3 - \frac{1}{12} N + \sum_{k=1}^{s} k^2 n_k - \frac{1}{N} \left(\sum_{k=1}^{s} k n_k \right)^2 \\
+ \frac{1}{4N} \left(\sum_{n_k \text{even}} n_k \right) \left(\sum_{n_k \text{odd}} n_k \right). \tag{3}
$$

Using the computer algebra software Maple [4], we now generate all roundest animals with size $N \leq 40$. Our strategy is the following: We first encode the weakly unimodal sequences by

$$(\lambda, b, h, \mu)$$

where λ, μ are integer partitions and $b, h \in \mathbb{N}^*$. The sequences $(n_1, \cdots, n_s)$ are given by

$$\lambda_1 \leq \lambda_2 \leq \cdots \leq \lambda_k < \underbrace{h = \cdots = h}_{b} > \mu_l \geq \mu_{l-1} \geq \cdots \geq \mu_1,$$

with $|\lambda| + bh + |\mu| = N$ (Figure 6).

Then, using the `combinat` package, we generate all (λ, b, h, μ) such that the associated animal A minimizes the moment of inertia $I(A)$ given by formula

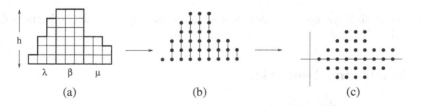

(a)   (b)   (c)

Fig. 6. (a) (λ, b, μ),   (b) $(n_1, \cdots, n_s)$,   (c) animal S to test

(3). Taking Lemma 3 into account, we restrict the generation of the 4-tuples (λ, b, h, μ) to those satisfying the further conditions

$$s \leq 2r + 1 \quad \text{and} \quad h \leq 2r + 1,$$

that is

$$\max(s, h) \leq \lfloor \sqrt{2} + 2\sqrt{\frac{N}{\pi}} + 1 \rfloor.$$

In appendix, Table 1 gives, for each $N \leq 40$, a set of representatives, up to dihedral symmetry, of the roundest animals of size N. Various parameters associated to these roundest animals are given in Table 2. The first five columns give the size N, the vertical projections, the moment of inertia, the center of gravity and the radius of the circle C_{a_0} of the roundest animals, up to dihedral symmetry.

Remark. Let $C_{\min} = \{z : |z - c| \leq r_{\min}\}$, be the smallest closed disc containing a given roundest animal A of size N. The value of c and $r_{\min}$ are given in the last two columns of Table 2. One may think that A is a quasi-disc centered at c having radius $r_{\min}$. In other words, we can replace the disc C_{a_0} of Theorem 2 by $C_{\min}$. It turns out that this is false in general. As an example, let A be the roundest animal of size $N = 17$ having projections $[4, 5, 5, 3]$ and $r_{\min} = \sqrt{5}$ be the radius of the smallest closed disc containing A shown below.

Then A *is not* a quasi-disc of radius $r_{\min} = \sqrt{5}$: indeed, the animal A together with the discs C_{a_0} and $C_{\min}$ shows that the lattice point + (cross) belongs to the open disc $C^\circ_{\min}$ but is not an element of A. This means that A is not a quasi-disc corresponding to $C_{\min}$.

Conjecture. *There exists an infinite family of roundest animals A which are not a quasi-disc of radius $r_{\min}$, where $r_{\min}$ is the radius of the smallest closed disc containing A.*

It turns out that the only occurence of such animal, up to $N = 40$, is the animal above with $N = 17$.

4   Concluding Remarks

While the recognition of digital straight lines is a well understood problem [8], both from the Euclidean and combinatorial approaches, the recognition of circles is still a challenging problem in discrete geometry. The border of discs or quasi-discs introduced in section 3.2, appears as a good candidate for a circle. This fact begs for a thorough study of their properties, perhaps useful for circle recognition.

The above results can be extended to other families of lattices. For instance, in the context of regular triangular lattices, a discrete set S is a union of N distinct closed equilateral triangles and the set A of the centers of these triangles satisfies the following formula,

$$I(S) = I(A) + N \cdot I(T),$$

where T is the unit triangle of the lattice. The lattice set $\mathbb{Z} + i\mathbb{Z}$ must be replaced by the set $\mathbb{T} \subseteq \mathbb{C}$ of the centers of all lattice triangles. The associated notions of strong convexity, (discrete) disc and quasi-disc are easily defined using $\mathbb{T}$. Theorems 1 and 2 still hold. The constant $\frac{1}{\sqrt{2}}$ of Lemma 3 must be replaced by a suitable constant α (according to the lattice). The computation of the roundest discrete sets can be established using an adaptation of the strategy described in Section 3.4. Moreover, extensions to higher dimensional lattices are also possible.

Acknowledgement. The authors are grateful to the anonymous referees for the useful and accurate comments provided.

References

1. Minsky, M., Papert, S.: Perceptrons: An Introduction to Computational Geometry. MIT Press, Cambridge (1969)
2. Brlek, S., Labelle, G., Lacasse, A.: The discrete green theorem and some applications in discrete geometry. Theoret. Comput. Sci. 346(2), 200–225 (2005)
3. Altshuler, Y., Yanovsky, V., Vainsencher, D., Wagner, I.A., Bruckstein, A.M.: On minimal perimeter polyminoes. In: Kuba, A., Nyúl, L.G., Palágyi, K. (eds.) DGCI 2006. LNCS, vol. 4245, pp. 17–28. Springer, Heidelberg (2006)
4. Heck, A.: Introduction to Maple, 3rd edn. Springer, New York (2003)
5. Lacasse, A.: Contributions à l' analyse de figures discrètes en dimension quelconque. PhD thesis, Université du Québec à Montréal, Montréal (2008)
6. Feynman, R.P., Leighton, R.B., Sands, M.: The Feynman lectures on physics. Mainly mechanics, radiation, and heat, vol. 1. Addison-Wesley Publishing, Reading (1963)
7. Stanley, R.P.: Enumerative combinatorics. Cambridge Studies in Advanced Mathematics, vol. 1, vol. 49. Cambridge University Press, Cambridge (1997), with a foreword by Gian-Carlo Rota, Corrected reprint of the 1986 original
8. Klette, R., Rosenfeld, A.: Digital straightness—a review. Discrete Appl. Math. 139(1-3), 197–230 (2004)

Appendix

Table 1. The roundest animals of size $N \leq 40$ (up to dihedral symmetry)

Table 2. Parameters associated to the roundest animals ($N \leq 40$), up to symmetry

N	vertical projections	$I(A)$	$g(A)$	$r = \max_{a \in A}\|a - g\|$	$c(A)$	$r_{\min}(A)$
1	$[1]$	0	$(1,0)$	0	$(0,0)$	0
2	$[1,1]$	$\frac{1}{2}$	$(\frac{3}{2},0)$	$\frac{1}{2}$	$(\frac{3}{2},0)$	$\frac{1}{2}$
3	$[1,2]$	$\frac{4}{3}$	$(\frac{5}{3},\frac{1}{3})$	$\frac{1}{3}\sqrt{5}$	$(\frac{3}{2},\frac{1}{2})$	$\frac{1}{2}\sqrt{2}$
4	$[2,2]$	2	$(\frac{3}{2},\frac{1}{2})$	$\frac{1}{2}\sqrt{2}$	$(\frac{3}{2},\frac{1}{2})$	$\frac{1}{2}\sqrt{2}$
5a	$[2,2,1]$	4	$(\frac{9}{5},\frac{2}{5})$	$\frac{2}{5}\sqrt{10}$	$(2,\frac{1}{2})$	$\frac{1}{2}\sqrt{5}$
5b	$[1,3,1]$	4	$(2,0)$	1	$(2,0)$	1
6	$[2,2,2]$	$\frac{33}{6}$	$(2,\frac{1}{2})$	$\frac{1}{2}\sqrt{5}$	$(2,\frac{1}{2})$	$\frac{1}{2}\sqrt{5}$
7	$[2,3,2]$	$\frac{52}{7}$	$(2,\frac{2}{7})$	$\frac{9}{7}$	$(2,\frac{1}{4})$	$\frac{5}{4}$
8	$[3,3,2]$	$\frac{78}{8}$	$(\frac{15}{8},\frac{1}{8})$	$\frac{1}{8}\sqrt{130}$	$(2,0)$	$\sqrt{2}$
9	$[3,3,3]$	$\frac{108}{9}$	$(2,0)$	$\sqrt{2}$	$(2,0)$	$\sqrt{2}$
10	$[3,3,3,1]$	$\frac{156}{10}$	$(\frac{11}{5},0)$	$\frac{9}{5}$	$(\frac{7}{3},0)$	$\frac{5}{3}$
11a	$[2,4,4,1]$	$\frac{212}{11}$	$(\frac{26}{11},\frac{5}{11})$	$\frac{1}{11}\sqrt{349}$	$(\frac{5}{2},\frac{1}{2})$	$\frac{1}{2}\sqrt{10}$
11b	$[3,4,3,1]$	$\frac{212}{11}$	$(\frac{24}{11},\frac{2}{11})$	$\frac{2}{11}\sqrt{101}$	$(\frac{9}{4},\frac{1}{4})$	$\frac{5}{4}\sqrt{2}$
12	$[2,4,4,2]$	$\frac{264}{12}$	$(\frac{5}{2},\frac{1}{2})$	$\frac{1}{2}\sqrt{10}$	$(\frac{5}{2},\frac{1}{2})$	$\frac{1}{2}\sqrt{10}$
13	$[3,4,4,2]$	$\frac{340}{13}$	$(\frac{31}{13},\frac{5}{13})$	$\frac{18}{13}\sqrt{2}$	$(\frac{23}{10},\frac{3}{10})$	$\frac{13}{10}\sqrt{2}$
14	$[3,4,4,3]$	$\frac{425}{14}$	$(\frac{5}{2},\frac{2}{7})$	$\frac{3}{14}\sqrt{85}$	$(\frac{5}{2},\frac{1}{6})$	$\frac{1}{6}\sqrt{130}$
15	$[4,4,4,3]$	$\frac{528}{15}$	$(\frac{12}{5},\frac{2}{5})$	$\frac{1}{5}\sqrt{113}$	$(\frac{5}{2},\frac{1}{2})$	$\frac{3}{2}\sqrt{2}$
16a	$[2,4,4,4,2]$	$\frac{640}{16}$	$(3,\frac{1}{2})$	$\frac{1}{2}\sqrt{17}$	$(3,\frac{1}{2})$	$\frac{1}{2}\sqrt{17}$
16b	$[4,4,4,4]$	$\frac{640}{16}$	$(\frac{5}{2},\frac{1}{2})$	$\frac{3}{2}\sqrt{2}$	$(\frac{5}{2},\frac{1}{2})$	$\frac{3}{2}\sqrt{2}$
17a	$[4,5,5,3]$	$\frac{780}{17}$	$(\frac{41}{17},\frac{2}{17})$	$\frac{40}{17}$	$(2,0)$	$\sqrt{5}$
17b	$[2,4,5,4,2]$	$\frac{780}{17}$	$(3,\frac{6}{17})$	$\frac{40}{17}$	$(3,\frac{1}{6})$	$\frac{13}{6}$
18	$[3,4,5,4,2]$	$\frac{925}{18}$	$(\frac{26}{9},\frac{5}{18})$	$\frac{1}{18}\sqrt{1685}$	$(3,0)$	$\sqrt{5}$
19	$[3,5,5,4,2]$	$\frac{1084}{19}$	$(\frac{54}{19},\frac{3}{19})$	$\frac{1}{19}\sqrt{1937}$	$(3,0)$	$\sqrt{5}$
20	$[3,5,5,5,2]$	$\frac{1255}{20}$	$(\frac{29}{10},\frac{1}{20})$	$\frac{1}{20}\sqrt{2165}$	$(3,0)$	$\sqrt{5}$
21	$[3,5,5,5,3]$	$\frac{1428}{21}$	$(3,0)$	$\sqrt{5}$	$(3,0)$	$\sqrt{5}$
22	$[3,5,5,5,4]$	$\frac{1664}{22}$	$(\frac{34}{11},\frac{1}{11})$	$\frac{21}{11}\sqrt{2}$	$(\frac{45}{14},\frac{3}{14})$	$\frac{25}{14}\sqrt{2}$
23	$[5,5,5,5,3]$	$\frac{1916}{23}$	$(\frac{65}{23},0)$	$\frac{2}{23}\sqrt{970}$	$(\frac{21}{8},0)$	$\frac{5}{8}\sqrt{17}$
24	$[1,5,5,5,5,3]$	$\frac{2183}{24}$	$(\frac{89}{24},0)$	$\frac{65}{24}$	$(\frac{18}{5},0)$	$\frac{13}{5}$
25	$[3,5,5,5,5,2]$	$\frac{2474}{25}$	$(\frac{17}{5},\frac{1}{25})$	$\frac{1}{25}\sqrt{4801}$	$(\frac{7}{2},0)$	$\frac{1}{2}\sqrt{29}$
26	$[3,5,5,5,5,3]$	$\frac{2769}{26}$	$(\frac{7}{2},0)$	$\frac{1}{2}\sqrt{29}$	$(\frac{7}{2},0)$	$\frac{1}{2}\sqrt{29}$
27	$[1,4,6,6,6,4]$	$\frac{3116}{27}$	$(\frac{35}{9},\frac{13}{27})$	$\frac{13}{27}\sqrt{37}$	$(\frac{15}{4},\frac{1}{2})$	$\frac{5}{4}\sqrt{5}$
28	$[4,6,6,6,4,2]$	$\frac{3464}{28}$	$(\frac{45}{14},\frac{1}{2})$	$\frac{1}{14}\sqrt{1570}$	$(\frac{13}{4},\frac{1}{2})$	$\frac{5}{4}\sqrt{5}$
29	$[2,5,6,6,6,4]$	$\frac{3852}{29}$	$(\frac{108}{29},\frac{12}{29})$	$\frac{10}{29}\sqrt{74}$	$(\frac{7}{2},\frac{1}{2})$	$\frac{1}{2}\sqrt{34}$
30	$[4,6,6,6,5,3]$	$\frac{4258}{30}$	$(\frac{101}{30},\frac{11}{30})$	$\frac{1}{30}\sqrt{7922}$	$(\frac{7}{2},\frac{1}{2})$	$\frac{1}{2}\sqrt{34}$
31	$[3,6,6,6,6,4]$	$\frac{4688}{31}$	$(\frac{111}{31},\frac{14}{31})$	$\frac{1}{31}\sqrt{8642}$	$(\frac{7}{2},\frac{1}{2})$	$\frac{1}{2}\sqrt{34}$
32	$[4,6,6,6,6,4]$	$\frac{5120}{32}$	$(\frac{7}{2},\frac{1}{2})$	$\frac{1}{2}\sqrt{34}$	$(\frac{7}{2},\frac{1}{2})$	$\frac{1}{2}\sqrt{34}$
33a	$[4,6,6,6,6,5]$	$\frac{5680}{33}$	$(\frac{118}{33},\frac{14}{33})$	$\frac{80}{33}\sqrt{2}$	$(\frac{67}{18},\frac{5}{18})$	$\frac{41}{18}\sqrt{2}$
33b	$[4,6,6,6,6,4,1]$	$\frac{5680}{33}$	$(\frac{119}{33},\frac{16}{33})$	$\frac{80}{33}\sqrt{2}$	$(\frac{23}{6},\frac{1}{2})$	$\frac{1}{6}\sqrt{370}$
34	$[4,6,7,7,6,4]$	$\frac{6241}{34}$	$(\frac{7}{2},\frac{5}{17})$	$\frac{1}{34}\sqrt{12833}$	$(\frac{7}{2},\frac{1}{6})$	$\frac{1}{6}\sqrt{370}$
35	$[5,6,7,7,6,4]$	$\frac{6816}{35}$	$(\frac{24}{7},\frac{8}{35})$	$\frac{1}{35}\sqrt{13309}$	$(\frac{129}{38},\frac{7}{38})$	$\frac{1}{38}\sqrt{15170}$
36	$[3,5,6,7,7,5,3]$	$\frac{7406}{36}$	$(\frac{145}{36},\frac{1}{12})$	$\frac{1}{36}\sqrt{13546}$	$(4,0)$	$\sqrt{10}$
37	$[3,5,7,7,7,5,3]$	$\frac{7992}{37}$	$(4,0)$	$\sqrt{10}$	$(4,0)$	$\sqrt{10}$
38	$[4,5,7,7,7,5,3]$	$\frac{8689}{38}$	$(\frac{149}{38},\frac{1}{19})$	$\frac{37}{38}\sqrt{13}$	$(\frac{23}{6},\frac{1}{6})$	$\frac{1}{6}\sqrt{410}$
39	$[3,5,7,7,7,6,4]$	$\frac{9388}{39}$	$(\frac{161}{39},\frac{5}{39})$	$\frac{1}{39}\sqrt{17873}$	$(\frac{25}{6},\frac{1}{6})$	$\frac{1}{6}\sqrt{410}$
40	$[3,6,7,7,7,6,4]$	$\frac{10127}{40}$	$(\frac{163}{40},\frac{1}{5})$	$\frac{1}{40}\sqrt{19433}$	$(\frac{235}{58},\frac{15}{58})$	$\frac{1}{58}\sqrt{39442}$

Gift-Wrapping Based Preimage Computation Algorithm

Yan Gerard[1], Fabien Feschet[1], and David Coeurjolly[2]

[1] Univ. Clermont 1, LAIC, Campus des Cézeaux,63172 Aubière, France
{gerard,feschet}@laic.u-clermont1.fr
[2] LIRIS, Univ. Lyon 1, Bât Nautibus, 69622 Villeurbanne Cedex, France
david.coeurjolly@liris.cnrs.fr

Abstract. The aim of the paper is to define an algorithm for computing preimages - roughly the sets of naive digital planes containing a finite subset S of $\mathbb{Z}^3$. The method is based on theoretical results: the preimage is a polytope that vertices can be decomposed in three subsets, the upper vertices, the lower vertices and the intermediary ones (equatorial). We provide a geometrical understanding (as facets on S or $S \ominus S$) of each kind of vertices. These properties are used to compute the preimage by gift-wrapping some regions of the convex hull of S or of $S \ominus S \cup \{(0,0,1)\}$.

1 Introduction

Digital straightness is an important concept in computer vision. In dimension two, for nearly half a century many digital straight line characterizations have been proposed with interactions with many fields such as arithmetic or theory of words (refer to [1] for a survey on digital straight line). A convenient framework is to consider the set of Euclidean straight lines, so-called *preimage* whose digitizations contain the input set of pixels. Based on a parametrization of digital straight lines, the preimage can be simply defined and correspond to a convex polygon in a given parameter space [2,3,4]. An important result is that such a preimage has got an important arithmetical structure that limits to four the number of vertices. This result is useful for a better understanding of this simple digital object and thus to design efficient digital straight line recognition algorithms.

In dimension 3, the same approach leads to define the notion of digital plane (see [5] or a survey) and of preimage [6,7,8]. The preimage becomes however more complex than a single polytope with 4 vertices. Questions about its arithmetical structure are open: even if it is a convex polyhedron in the digital plane parameter space, we have some difficulties to clearly understand its arithmetical structure and to bound the number of its vertices [9,10].

In this paper, we focus on a geometrical interpretation of preimage vertices and facets in order to design a fast preimage computation algorithm. The proposed algorithm is based on the computation of a surface of the convex hull of S and of the chords set $S \ominus S \cup \{(0,0,1)\}$ (as already used in [11] for recognition).

D. Coeurjolly et al. (Eds.): DGCI 2008, LNCS 4992, pp. 310–321, 2008.

The first sections are dedicated to the preliminaries and to the analyse of the preimage geometry. Then Section 4 details the proposed algorithm. Finally, Section 5 presents some experiments.

2 Preliminaries

2.1 Digital Planes and Digital Plane Recognition Problems

We start by defining naive digital planes according to J.P Reveilles definition[12]. We recall that for any vector $x \in \mathbb{R}^d$, its uniform norm is $||x||_\infty = \max\{|x_i|/1 \leq i \leq d\}$.

Definition 1. *A naive digital plane is a subset of $\mathbb{Z}^3$ characterized by a double inequality $h \leq ax + by + cz < h + ||(a, b, c)||_\infty$ where the normal vector $(a, b, c) \in \mathbb{R}^3$ is different from $(0, 0, 0)$ and where $h \in \mathbb{R}$.*

For a survey on digital plane characterization and alternative definitions, see [5].

According to the value of $||(a, b, c)||_\infty$, there exist three classes of naive digital planes obtained by rotation of the coordinates. In the following, we focus our attention on the special case where $||(a, b, c)||_\infty = |c|$. We introduce therefore the notion of z-slice which is related to naive digital planes with $|a| \leq |c|$ and $|b| \leq |c|$ while x-slices and y-slices are the equivalent objects corresponding to naive digital planes verifying $|a| = ||(a, b, c)||_\infty$ and $|b| = ||(a, b, c)||_\infty$ respectively.

Definition 2. *A z-slice is a subset of $\mathbb{Z}^3$ characterized by a double inequality $h \leq ax + by + cz < h + |c|$ where c is in $\mathbb{R}^\star$ and where a, b and h are real numbers.*

If $|c|$ is greater than $|a|$ and $|b|$, the z-slice of double-inequality $h \leq ax + by + cz < h + |c|$ is a naive digital plane. Otherwise (if $|c| < |a|$ or $|c| < |b|$), the z-slice of double-inequality $h \leq ax + by + cz < h + |c|$ is a subset of the naive digital plane $h \leq ax + by + cz < h + ||(a, b, c)||_\infty$. Thus in any case, a z-slice is contained in a naive digital plane and conversely, a naive digital plane is either an x-slice, either a y-slice or a z-slice. It means obviously that naive digital planes recognition as well as generalized preimage computation can be decomposed into x-slices, y-slices and z-slices recognition. Since these three problems only differ by a rotation of the coordinates, we focus our attention on the problem of recognition of z-slices.

2.2 Digital Plane Recognition Problems

We now state the problems that we shall address in this framework:

Problem 1. Input: a finite subset $S \subset \mathbb{Z}^3$

- P-EXI: Does there exist a z-slice containing S?
- P-ONE: Provide a z-slice containing S
- P-ALL: Provide a description of all z-slices containing S

As z-slices are described by double-inequalities $h \leq ax + by + cz < h + |c|$, the question is to find a, b, c and h. This inequation is almost linear. The only non-linear term is the absolute value $|c|$. Let us reduce the problem to the case $c = 1$.

We first notice that the set of possible $(a, b, c, h) \in \mathbb{R}^4$ is a positive cone since for any solution (a, b, c, h) of inequalities and any positive real $\lambda \in \mathbb{R}^{+\star}$, the point $(\lambda a, \lambda b, \lambda c, \lambda h)$ is also a solution. This comes from the homogeneity of the inequalities. Hence, instead of computing the whole cone of solutions, it is more convenient to reduce its computation to a section. As c is assumed different from 0, we can take the sections by the hyperplanes $c = 1$ and $c = -1$. It is clear that the whole set of solutions is characterized by its two sections.

We also notice that the solutions for a direction (a, b, c) and the solutions for direction $(-a, -b, -c)$ are closely related. By denoting respectively h_{min} and h_{max} the minimum and maximum of the finite set of values $\{ax_i + by_i + cz_i / (x_i, y_i, z_i) \in S\}$ (S is finite), the set S belongs to the z-slice $h \leq ax + by + cz < h + |c|$ iff $h \in]h_{max} - |c|, h_{min}]$ while it belongs to the z-slice $h \leq -ax - by - cz < h + |c|$ with a symmetric normal $(-a, -b, -c)$ iff $h \in] - h_{min} - |c|, -h_{max}]$. It means that one interval can be easily obtained from the other. Thus we can reduce the computations of possible parameters h to vectors (a, b, c) with a positive c (and then with $c = 1$).

2.3   The Preimage Polyotope

According to previous remarks, problems P-Exi, P-One and P-All can all be reduced to the case $c = 1$. Thus these problems can be reduced to a system of linear inequalities $h \leq ax_i + by_i + z_i < h + 1$ for $(x_i, y_i, z_i) \in S$. Classical Linear Programming algorithms such for instance simplex method can be used for solving the problems P-Exi and P-One but we can notice that usually, they do not provide the whole set of solutions. Thus the problem P-All does not enter in their framework of application.

Definition 3. *The z-preimage of a finite set S is the 3-dimensional set of values $(a, b, h) \in \mathbb{R}^3$ verifying for any point $(x_i, y_i, z_i) \in S$ the double-inequality $h \leq ax_i + by_i + z_i < h + 1$.*

We can as well define x-preimage and y-preimage. Usually, z-preimage will simply be called *preimage*. It is a convex set since it is obtained by intersection of finitely many open and closed half-spaces. If we assume that the preimage is not empty, we notice that its interior is neither empty with the consequence that its vertices and faces are also the vertices and faces of its closure. Hence, under this assumption, we can work with the polyhedron defined by non-strict inequalities $h \leq ax_i + by_i + z_i \leq h + 1$ without changing the vertices and faces of the preimage.

If we consider the 3D space of solutions (a, b, h), we can notice that the half-spaces $h \leq ax_i + by_i + z_i$ are directed downwards while the open half-spaces $ax_i + by_i + z_i \leq h + 1$ are directed upwards. It leads to introduce the upper bound of the preimage of S according to direction h and its lower bound. We denote them respectively $preimage^+(S)$ and $preimage^-(S)$. They meet in a curve that we denote $preimage^0(S)$ and call the *equator* of the preimage (Fig 1). Due to orientation considerations, the faces of $preimage^-(S)$ are among the inequalities $ax_i + by_i + z_i \leq h + 1$ while the faces of $preimage^+(S)$ are of the form $h \leq ax_i + by_i + z_i$.

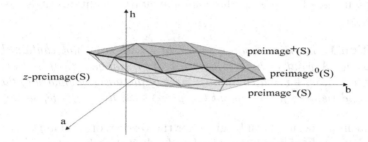

Fig. 1. We decompose the boundary of $preimage(S)$ in its upper bound $preimage^+(S)$, its lower bound $preimage^-(S)$ and its equator $preimage^0(S)$ (the points that vertical projection are on the boundary of the vertical projection of $preimage(S)$ (according to h))

In the following, we assume that the preimage is not empty and that the points of S do not belong to a vertical plane. We can eliminate this degenerated case since it can be replaced in a 2D framework. With these conditions, the preimage of S is a polytope (and the equator is closed) which can be described either by providing its vertices or a minimal set of linear constraints.

3   Geometry of the Digital Plane Preimage

Our algorithm of preimage computation requires to put in relation the preimage elements (vertices and faces) with the geometry of S. This structural analysis was initiated in [9] in which a similar result of Proposition 2 is given. This result is reformulated and improved by new properties using a uniform formalism.

3.1   The Vertices of $preimage^+(S)$ and $preimage^-(S)$

Let us consider a vertex (a', b', h') of the upper boundary $preimage^+(S)$ of the preimage of S. The vertex (a', b', h') satisfies the inequalities $h' \leq a'x + b'y + z$ (and even $h' \leq a'x + b'y + z \leq h' + 1$) for any point $(x, y, z) \in S$ and at least three independent inequalities are equalities. It follows that there exist three affinely independent points (x_i, y_i, z_i), (x_j, y_j, z_j) and (x_k, y_k, z_k) in S such that we have exactly $h' = a'x_i + b'y_i + z_i$, $h' = a'x_j + b'y_j + z_j$ and $h' = a'x_k + b'y_k + z_k$. It means that $h' = a'x + b'y + z$ is the affine plane of a facet of the lower boundary of the convex hull of S. Conversely, let us take a facet $h' = a'x + b'y + z$ of the lower boundary of the convex hull of S verifying the complementary condition that for any $(x, y, z) \in S$: $h' \leq a'x + b'y + z \leq h' + 1$. There exist three affinely independent points (x_i, y_i, z_i), (x_j, y_j, z_j) and (x_k, y_k, z_k) on this facet. They verify $h' = a'x_i + b'y_i + z_i$, $h' = a'x_j + b'y_j + z_j$ and $h' = a'x_k + b'y_k + z_k$. All the points (a, b, h) of the preimage verify $h \leq ax_i + by_i + z_i$, $h \leq ax_j + by_j + z_j$ and $h \leq ax_k + by_k + z_k$. In the parameter space (a, b, h), these three conditions define a cone of vertex (a', b', h') containing the preimage. As (a', b', h') is at least in preimage of S (due to the complementary condition), it is a vertex of the preimage.

Proposition 1. *Let S be a finite subset of $\mathbb{Z}^3$ which is not contained by any vertical plane. A point (a, b, h) is a vertex of $preimage^+(S)$ if and only if the affine plane $ax + by + z = h$ contains a facet of the lower boundary of the convex hull of S and we have for any point $(x_i, y_i, z_i) \in S$: $h \leq ax_i + by_i + z_i \leq h + 1$.*

An equivalent proposition can be given for the vertices of $preimage^-(S)$: a point (a', b', h') is a vertex of $preimage^-(S)$ if and only if the affine plane $a'x + b'y + z = h' + 1$ contains a facet of the upper boundary of the convex hull of S and we have for any point $(x_i, y_i, z_i) \in S$: $h' \leq a'x_i + b'y_i + z_i \leq h' + 1$. It means that the vertices of $preimage^+(S)$ and $preimage^-(S)$ are both derived from a subset of facets of the convex hull of S. These two subsets of facets (if non empty) can be introduced as the *caps* of S.

Definition 4. *The upper (resp. lower) cap of S is the set of points (x_i, y_i, z_i) of the upper (resp. lower) boundary of the convex hull of S for which there exist $(a, b, h) \in \mathbb{R}^3$ with $h + 1 = ax_i + by_i + z_i$ (resp. $h = ax_i + by_i + z_i$) and $h \leq ax + by + z \leq h + 1$ for all points (x, y, z) of S (see Fig 2).*

Two cases are possible: the upper (lower) cap contains facets and according to Proposition 1, these facets are one to one with the vertices of $preimage^-(S)$ ($preimage^+(S)$) -the correspondence is described in Fig 2- or there is no facet in the upper (lower) cap (it is reduced to points or edges) and $preimage^-(S)$ ($preimage^+(S)$) has no vertex (the vertices are equatorial). Authors of [9] notice that the caps usually contain exactly one facet with the consequence that there is one non-equatorial vertex in $preimage^-(S)$ and another one in $preimage^+(S)$. They prove that this case occurs necessarily under some assumptions on the input sets (large enough to contain "leaning points").

3.2   The Vertices of $preimage^0(S)$: Chords and Visibility Cone

The situation on the equator of the preimage is different because a vertex (a, b, h) of the equator is either the intersection of two faces $h \leq ax_i + by_i + z_i$ of $preimage^+(S)$ and one face $ax_j + by_j + z_j \leq h + 1$ of $preimage^-(S)$, either the intersection of one face $h \leq ax_i + by_i + z_i$ of $preimage^+(S)$ and two faces $ax_j + by_j + z_j \leq h + 1$ of $preimage^-(S)$. We thus introduce a new tool to describe them differently: the notion of *chords set*.

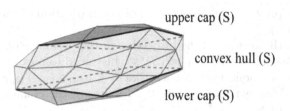

upper cap (S)

convex hull (S)

lower cap (S)

Fig. 2. The upper cap of the convex hull is made of the points of upper convex hull of S contained by a plane $h + 1 = ax + by + z$ that translation by $(0, 0, -1)$ is under S. The lower cap contains the points of the lower convex hull of S contained by a plane $h = ax + by + z$ that translation by $(0, 0, +1)$ is above S. The support planes $h = ax + by + z$ of facets of the upper cap provide the vertices $(a, b, h - 1)$ of $preimage^-(S)$ while the facets $h = ax + by + z$ of the lower cap provide the vertices (a, b, h) of $preimage^+(S)$.

Definition 5. *The set of the differences $\{x' - x / x \in S, x' \in S\}$ is called the chords set of S and we denote it $S \ominus S$.*

The interest of the chords set comes from next lemma.

Lemma 1. *Given (a, b), there exist a real h such that the finite set S belongs to the z-slice $h \leq ax + by + z \leq h + 1$ if and only if there exists a real $h' \leq 1$ such that the plane $ax + by + z = h'$ separates $(0, 0, 1)$ from the chords set of S (equality is permitted).*

Proof. We denote again $[h_{min}, h_{max}]$ the range of the values $ax_i + by_i + z_i$ for (x_i, y_i, z_i) in S. The first proposition means exactly that $h_{max} - h_{min} \leq 1$ while the second one means that any pair of indices i and j, the difference between $ax_i + by_i + z_i$ and $ax_j + by_j + z_j$ is ≤ 1. By taking the index i providing h_{min} and the index j providing the value h_{max}, the equivalence is easy to obtain. $\square$

Lemma 1 means more generally that the chords set of S contains all the information necessary to compute the projection of the preimage of S on the plane of coordinates a and b: a point (a, b) is in the projection of the preimage of S if and only if there exist a plane of normal direction $(a, b, 1)$ separating the chords set $S \ominus S$ from $(0, 0, 1)$. This last proposition makes the link between the projection of the preimage (according to direction h) and the set of planes separating $S \ominus S$ from a point. Thus it can be useful to recall that the set of directions of the planes separating a set S' from a point P can be given by the cone of visibility of S' from P. More precisely, the directions (a, b, c) of the planes separating S' from P are convex combination of the normal directions of the faces of the visibility cone. In the present framework lemma 1 means that the directions $(a, b, 1)$ of the preimage of S (or namely the projection of the preimage on the plane (a, b)) are convex combinations of the directions $(a, b, 1)$ of the faces of the cone of visibility of $S \ominus S$ from $(0, 0, 1)$ (see Fig 3). It means that the cone of visibility of $S \ominus S$ from $(0, 0, 1)$ provides directly the vertices of the projection of the preimage on the plane (a, b): a face of the cone of visibility with equation

$ax + by + z = 1$ provides the vertex (a, b) of the projection of the preimage of S on the plane (a, b).

To pass from the vertices of the projection of the preimage on the plane (a, b) to the ones of the equator $preimage^0(S)$, we just need to pick up the point (a, b) by computing the unique value h such that $h \leq ax_i + by_i + z_i \leq h + 1$ for any $(x_i, y_i, z_i) \in S$. Thus we obtain easily the vertices of the equator $preimage^0(S)$ of the preimage of S.

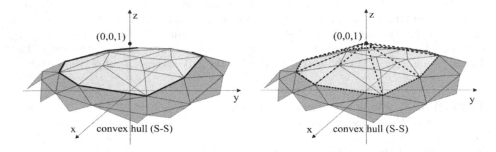

Fig. 3. The cone of visibility of the convex hull of $S \ominus S$ from $(0, 0, 1)$. The normal direction $(a, b, 1)$ of a face of the cone provides a vertex (a, b) of the projection of the preimage on plane (a, b). By picking it with height h (the unique value verifying $h \leq ax_i + by_i + z_i \leq h + 1$ for any point (x_i, y_i, z_i) of S), we obtain the vertices of the equator $preimage^0(S)$ of the preimage of S.

3.3   The Faces of the Preimage

After the geometrical description of the vertices of the preimage of S, next step is the characterization of its faces. It is obvious that the faces of $preimage^+(S)$ are among the inequalities $h \leq ax_i + by_i + z_i$ with $(x_i, y_i, z_i) \in S$ while the faces of $preimage^-(S)$ are among the inequalities $ax_j + by_j + z_j \leq h + 1$ with $(x_j, y_j, z_j) \in S$. The question is to know which points of S define the faces of $preimage^+(S)$ and which ones define the faces of $preimage^-(S)$.

Proposition 2. *The points (x_i, y_i, z_i) and (x_j, y_j, z_j) providing the upper faces $h \leq ax_i + by_i + z_i$ and lower faces $ax_j + by_j + z_j \leq h + 1$ of the polyhedron preimage of S are respectively some vertices of the lower and upper caps of S.*

Proposition 2 does not mean that all vertices of the caps provide some faces of the preimage. Let (a', b', h') be the center of an upper face $h \leq ax_i + by_i + z_i$ of the preimage, the point (x_i, y_i, z_i) is unique in S to verify $h' = a'x_i + b'y_i + z_i$ ($h' < a'x_i + b'_i + z_i \leq h' + 1$ for all other points of S). Hence, (x_i, y_i, z_i) is in the lower cap but it means more: a point (x_i, y_i, z_i) without any (a', b', h') satisfying $h' = a'x_i + b'y_i + z_i$ and $h' < a'x_i + b'_i + z_i \leq h' + 1$ for all other points of S can not provide a face of the preimage. This necessary condition is even sufficient but we do not know if there exist points in the lower cap which do not satisfy it!

4   The Recognition Algorithm

4.1   Sketch of the Algorithm

The task of the paper is to provide an efficient algorithm to solve problem P-ALL. We recall that problems P-EXI and P-ONE are already solved efficiently (quasi-linear time) by linear programming or other algorithms derived from Computational Geometry [11]. Problem P-ALL consists in computing the whole polyhedron of solutions. The algorithm is based on the computation of its vertices :

- a function *upperVertices* computes the vertices of $preimage^+(S)$ (according to proposition 1, they correspond to the faces of the lower cap of S);

- a second function *lowerVertices* computes the vertices of $preimage^-(S)$ by using their correspondence with the vertices to the upper cap of S;

- a third function *equatorVertices* is in charge of the computation of the vertices of the equator by using the cone of visibility of $S \ominus S$ from $(0, 0, 1)$. Instead of working with the real cone, we compute the facets of the convex hull of $(S \ominus S) \cup \{(0, 0, 1)\}$ containing vertex $(0, 0, 1)$.

The common point of each function is to explore a set of facets, the facets of the upper and lower caps for functions *upperVertices* or *lowerVertices* and the facets of the convex hull of $(S \ominus S) \cup \{(0, 0, 1)\}$ containing $(0, 0, 1)$ for *equatorVertices*. This exploration can be done according to a gift-wrapping principle [13,14]: starting from a facet F of the convex hull, we compute the neighboring facet F' according to a given edge e of F (thus F and F' share e).

4.2   Initialization

We start the computation with any algorithm solving P-ONE. If there exists no solution, the preimage is empty and we stop. Otherwise, the algorithm provides a double inequality $h \leq ax + by + z < h + 1$ satisfied by all points (x, y, z) of S. Normal direction $(a, b, 1)$ provides three initial points:

- we compute the point $(x_i, y_i, z_i) \in S$ realizing the minimum of $ax + by + z$ in S. We have $ax + by + z = h_i$. It is straightforward that all points (x, y, z) of S verify $h_i \leq ax + by + z \leq h_i + 1$. This proves that (x_i, y_i, z_i) is in the lower cap;

- we compute the point $(x_j, y_j, z_j) \in S$ realizing the maximum of $ax + by + z$ in S. We have $ax + by + z = h_j$. As previously, (x_j, y_j, z_j) is in the upper cap because all points of S verify $h_j - 1 \leq ax + by + z < h_j$;

- the point $(x_j - x_i, y_j - y_i, z_j - z_i)$ is a vertex of the convex hull of $S \ominus S$. It belongs to the plane $ax + by + z = h_j - h_i$ above $S \ominus S$ and cutting the vertical axis between 0 and $(0, 0, 1)$.

4.3   From a Starting Point to a Starting Facet

We have a starting point and we explore the facets that contain it until finding a satisfying one.

- We explore the facets of the convex hull of S with vertex (x_i, y_i, z_i). We can do it by turning around (x_i, y_i, z_i) by gift-wrapping until finding a facet in the lower cap of S. It is also possible that no facet adjacent with (x_i, y_i, z_i) belongs to the lower cap: it simply means that $preimage^+(S)$ has no vertex and closes this computation.

- We explore the facets of the convex hull of S with vertex (x_j, y_j, z_j) by gift-wrapping. If we do not find any facet of the upper cap of S, it ends the computation ($preimage^-(S)$ has no vertex).

- We search a facet of the convex hull of $(S \ominus S) \cup \{(0, 0, 1)\}$ having $(0, 0, 1)$ as vertex but at this point, we have only a vertex $(x_j - x_i, y_j - y_i, z_j - z_i)$ of the convex hull of $S \ominus S$ (Fig 4). We can decompose the convex hull of $S \ominus S$ between the facets which are destroyed when we add $(0, 0, 1)$ to $S \ominus S$ (region A) and the ones which are preserved (region B). By construction, the point $(x_j - x_i, y_j - y_i, z_j - z_i)$ is in region A or at least on its boundary. The challenge is to go from this point in region A in the direction of region B until finding the boundary. In this goal, we compute a first facet containing the vertex $(x_j - x_i, y_j - y_i, z_j - z_i)$ and cross a "line" of facets until an edge of the boundary (the choice of the edge that we share is made in order to advance in a given direction so to avoid loops). We go from a facet to next one by gift-wrapping. Although the number of points in $S \ominus S$ is the square of the cardinality of S, this computation can be done $O(card(S))$ time (see next Sect. 4.4). We end this computation with an edge of the boundary of region A: it remains only to add the vertex $(0, 0, 1)$ to have a facet of the convex hull of $(S \ominus S) \cup \{(0, 0, 1)\}$

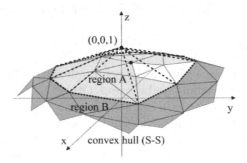

Fig. 4. Region A is made of the facets of the convex hull of $S \ominus S$ which disappear when we add the point $(0, 0, 1)$. The remaining facets are in region B. The equator of the preimage is given by the visibility cone of $S \ominus S$ from $(0, 0, 1)$. The facets of this cone are bounding the two regions. We can go from region A to region B by using a gift-wrapping process following a given direction, as drawn for instance in white.

4.4   Gift-Wrapping on the Chords Set $S \ominus S$

Gift-wrapping on a set of cardinality n takes $O(n)$ time. Given a face F of the convex hull of S and one of its edges e, we find the other facet sharing e by projecting the points of S in a plane according to direction e. To determine the new vertex to associate with e, we research the point providing the maximum angle with the projection of F. Thus the time necessary for this computation is linear in n (find the maximum of n angles through the ratio of the coordinates).

For the computation of the vertices of the equator, we use a gift-wrapping procedure working on the chords set $S \ominus S$ that cardinality can be quadratic. For thousands of points, if we still use a procedure linear in the number of points, the cost could be important. We can however improve the research of the new vertex by reducing the number of angles that should be compared. As drawn in Fig 5, the edge issued from a vertex $u - v$ in the convex hull of $S \ominus S$ is necessarily equivalent with an edge issued from u or from v. It allows to work directly on S: we compute the point of S that angle with u or v is maximal. It reduces the number of angles to compare from n^2 to $2n$ with the consequence that gift-wrapping on the chords set remains linear.

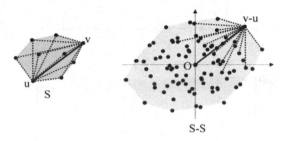

Fig. 5. If we consider the edges issued from a vertex $v - u$ of the chords set $S \ominus S$ (S is here in 2D), they are both equivalent either to an edge issued from u, either to an edge issued from v in the convex hull of S

4.5   Last Step: From a Facet to the Others

The exploration of the whole upper and lower caps from an initial facet is a standard procedure of Computational Geometry. We just have to stop the exploration in a direction as soon as next facet does not satisfy the condition to be in the considered cap. Each lower facet $h' = a'x + b'y + z$ or upper facet $a'x + b'y + z = h' + 1$ computed this way provides a vertex (a, b, h) of $preimage^+(S)$ or $preimage^-(S)$. We have however to take care that they are not on the equator by verifying that there does not exist two points (x_i, y_i, z_i) and (x_j, y_j, z_j) of S with $a'(x_j - x_i) + b'(y_j - y_i) + z_j - z_i = 1$.

   In the case of the equator, we start from a facet of the convex hull of $(S \ominus S) \cup \{(0, 0, 1)\}$ containing $(0, 0, 1)$ and we just have to turn around $(0, 0, 1)$ to obtain the whole cone of visibility of $S \ominus S$ from $(0, 0, 1)$. The faces $a'x + b'y + z = 1$ of this cone provide the values (a', b') of the vertices of the equator. It remains to compute h as the minimum of $a'x + b'y + z$ for (x, y, z) in S.

4.6   Complexity Analysis

The advantage of Gift-Wrapping algorithms is that they are output-sensitive. The complexity of computation of each new facet is linear [13]. It provides a theoretical complexity in $O(nh)$ where n is the cardinality of the set S and h the number of vertices of the preimage.

5   Experiments

In this part, we provide some experiments about the behavior of our algorithm with respect to the number n of voxels in the set S. To do this, we have used the random generator of the TC18 challenge. We have generated around 6000 tests with various sizes in the coordinates of the voxels. The number of voxels in each test was also randomly generated. We have segmented the results into classes corresponding to set size that varies from 0 to 1000, from 1000 to 2000 and so on. For each class, we have computed the minimum, the maximum and the mean value of the execution time. All those experiments have been plotted on Fig. 6.

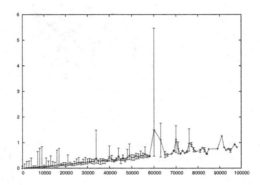

Fig. 6. Variation of the execution time (vertically in seconds) in function of the number of voxels

It is clear that both dependence on n and h are visible in the graph. For instance, the running time globally increases linearly with n but when complex instances are found the h factor leads to an increase in the running time. The maximal time obtained around 60000 voxels comes probably from a bug in our current implementation. We are currently studying this problem.

To compare to already available codes, we have used another set of around 4000 voxels not used in previous experiments. The implementation of Sivignon [15] took 9.78 seconds, the lrs code [16] took 66.76 seconds and our new implementation took 0.56 seconds on the same computer. This behavior was confirmed during all our experiments where our new method is at least 10 times faster than other preimage computation algorithms.

6   Conclusion

In this article, we have precisely studied the preimage set of digital planar sets. We have decomposed the vertices of the preimage into three sets and we have presented algorithms for compute them. We have provided an implementation of this algorithm as well as experiments showing that our new method is fast in comparison with other available algorithm. We plan to test our algorithm in the TC18 Challenge and our implementation will be shortly available on the TC18 pages (http://www.cb.uu.se/~tc18/).

References

1. Rosenfeld, A., Klette, R.: Digital straightness. In: Int. Workshop on Combinatorial Image Analysis. Electronic Notes in Theoretical Computer Science, vol. 46, Elsevier Science Publishers, Amsterdam (2001)
2. Dorst, L., Smeulders, A.N.M.: Discrete representation of straight lines. IEEE Trans. on Pattern Analysis and Machine Intelligence 6, 450–463 (1984)
3. McIlroy, M.D.: A note on discrete representation of lines. AT&T Technical Journal 64(2), 481–490 (1985)
4. Lindenbaum, M., Bruckstein, A.: On recursive, $O(n)$ partitioning of a digitized curve into digital straigth segments. IEEE Trans. on Pattern Analysis and Machine Intelligence 15(9), 949–953 (1993)
5. Brimkov, V., Coeurjolly, D., Klette, R.: Digital planarity - a review. Discrete Applied Mathematics (2006)
6. Veelaert, P.: Geometric constructions in the digital plane. Journal of Mathematical Imaging and Vision 11, 99–118 (1999)
7. Vittone, J., Chassery, J.M.: Recognition of digital naive planes and polyhedization. In: Nyström, I., Sanniti di Baja, G., Borgefors, G. (eds.) DGCI 2000. LNCS, vol. 1953, pp. 296–307. Springer, Heidelberg (2000)
8. Dexet, M., Andres, E.: A generalized preimage for the standard and supercover digital hyperplane recognition. In: Kuba, A., Nyúl, L.G., Palágyi, K. (eds.) DGCI 2006. LNCS, vol. 4245, pp. 639–650. Springer, Heidelberg (2006)
9. Coeurjolly, D., Sivignon, I., Dupont, F., Feschet, F., Chassery, J.M.: On digital plane preimage structure. Discrete Applied Mathematics 151(1–3), 78–92 (2005)
10. Coeurjolly, D., Brimkov, V.: Computational aspects of digital plane and hyperplane recognition. In: Reulke, R., Eckardt, U., Flach, B., Knauer, U., Polthier, K. (eds.) IWCIA 2006. LNCS, vol. 4040, Springer, Heidelberg (2006)
11. Gérard, Y., Debled-Rennesson, I., Zimmermann, P.: An elementary digital plane recognition algorithm. Discrete Applied Mathematics 151, 169–183 (2005)
12. Reveillès, J.P.: Géométrie discrète, calcul en nombres entiers et algorithmique. Thèse d'etat, Université Louis Pasteur, Strasbourg, France (1991)
13. Preparata, F.P., Shamos, M.I.: Computational Geometry : An Introduction. Springer, Heidelberg (1985)
14. de Berg, M., Schwarzkopf, O., van Kreveld, M., Overmars, M.: Computational Geometry: Algorithms and Applications. Springer, Heidelberg (2000)
15. Sivignon, I.: De la caractérisation des primitives à reconstruction polyhèdrique de surfaces en géométrie discrète. PhD thesis, Laboratoire des Images et des Signaux, INPG, Grenoble (2004)
16. Avis, D.: lrs: A Revised Implementation of the Reverse Search Vertex Enumeration Algorithm (1999), http://cgm.cs.mcgill.ca/~avis/C/lrs.html

Digital Planar Surface Segmentation Using Local Geometric Patterns

Yukiko Kenmochi[1], Lilian Buzer[1], Akihiro Sugimoto[1,2], and Ikuko Shimizu[3]

[1] Université Paris-Est, LABINFO-IGM, UMR CNRS 8049, A2SI-ESIEE, France
[2] National Institute of Informatics, Japan
[3] Tokyo University of Agriculture and Technology, Japan

Abstract. This paper presents a hybrid two-step method for segmenting a 3D grid-point cloud into planar surfaces by using discrete-geometry results. Digital planes contain a finite number of local geometric patterns (LGPs). Such a LGP possesses a set of normal vectors. By using LGP properties, we first reject non-linear points from a point cloud (edge-based step), and then classify non-rejected points whose LGPs have common normal vectors into a planar-surface-point set (region-based step).

1 Introduction

This paper presents a method for segmenting a 3D grid-point cloud into planar surfaces by using discrete-geometry results, providing that the point cloud is obtained from polyhedral objects.

In computer vision, conventional methods for surface segmentation are classified into three categories: region-based, edge-based, and hybrid methods. The first ones merge points having similar region properties calculated from their neighboring points such as normal vectors and curvatures [1]. As calculated properties are sensitive to noise and quantization errors, they cause over-segmentation. The second methods search edges that separate regions by using depth discontinuities [13]. As edges are not always extracted as connected curves, they cause under-segmentation. The third methods are hybrid between the two [10,12]. In particular, for the case where our interesting object is polyhedral, a hybrid method using locally planar points is proposed [10]. In that method, points not locally planar are considered to be potentially edge points.

In discrete geometry, a digital plane is defined as a set of grid points lying between two parallel planes with a small distance [7]. Local geometric patterns (LGPs) appearing on digital planes are called linear LGPs, and their number is finite. It is known that linear LGPs are related to arithmetic planes [8]. Their arithmetic properties were studied for digital plane recognition [4,11], and used to develop region-based methods for digital planar surface segmentation [9]. However, those region-based methods require an incremental plane recognition process, which causes another problem of incremental point tracking.

In this paper, to avoid plane recognition involving incremental point tracking for segmentation, we present a discrete version of the hybrid methods, consisting of an edge-based and a region-based parts, using linear LGPs. Our idea is similar

D. Coeurjolly et al. (Eds.): DGCI 2008, LNCS 4992, pp. 322–333, 2008.

to [10]. We first generate all linear LGPs in a cubic region of $(2k+1) \times (2k+1) \times (2k+1)$ grid points for an arbitrary size k, by using their arithmetic properties. We then reject a point from a grid-point set if its LPG is not linear; as rejected points define candidates of edge points (the edge-based part), the remaining non-rejected points define candidates of planar points. For the region-based part, we use the normal vectors of linear LPGs. Each linear LGP possesses a set of feasible normal vectors, called preimages [3]. We merge non-rejected points whose linear LGPs have common normal vectors to obtain digital planar surfaces. We show our experimental results demonstrating that our method is robust against not only quantization errors but also noise.

2 Digital Plane Patches and Their Preimages

2.1 Digital Planes

Let $\mathbb{R}$ be the set of real numbers. A plane $\mathbf{P}$ in the 3D Euclidean space $\mathbb{R}^3$ is defined by the following expression:

$$\mathbf{P} = \{(x, y, z) \in \mathbb{R}^3 \ : \ \alpha x + \beta y + \gamma z + \delta = 0\} \tag{1}$$

where $\alpha, \beta, \gamma, \delta \in \mathbb{R}$.

Let $\mathbb{Z}$ be the set of integers, so that $\mathbb{Z}^3$ denotes the set of grid points whose co-ordinates are integers. We now consider a digitization of $\mathbf{P}$, which is a grid-point subset in $\mathbb{Z}^3$, called a digital plane. There are various digitization techniques for $\mathbf{P}$ [7]. In this paper we adopt a grid-line digitization such that

$$\mathbf{D}(\mathbf{P}) = \{(x, y, z) \in \mathbb{Z}^3 \ : \ 0 \le \alpha x + \beta y + \gamma z + \delta < \omega\} \tag{2}$$

where $\omega = \max(|\alpha|, |\beta|, |\gamma|)$. From the definition, it is obvious that we obtain a unique digital plane $\mathbf{D}(\mathbf{P})$ from any $\mathbf{P}$.

2.2 Preimages of a Digital Plane Patch

Since a point cloud is acquired by a sensor, we can assume that it is bounded. In other words, our grid space $\mathbf{X} \subset \mathbb{Z}^3$ is bounded such that $\mathbf{X} = [X_1, X_2] \times [Y_1, Y_2] \times [Z_1, Z_2]$ where X_i, Y_i, Z_i for $i = 1, 2$ are finite integers. Then, our digital plane $\mathbf{D}(\mathbf{P})$ is also bounded such that

$$\mathbf{D_X}(\mathbf{P}) = \{(x, y, z) \in \mathbf{X} \ : \ 0 \le \alpha x + \beta y + \gamma z + \delta < \omega\}, \tag{3}$$

called a digital plane patch.

Given $\mathbf{D_X}(\mathbf{P})$, we can find a set of Euclidean planes $\mathbf{P}$ such that the digitization of each $\mathbf{P}$ in $\mathbf{X}$ is equal to $\mathbf{D_X}(\mathbf{P})$. The set of all such $\mathbf{P}$ is called the preimage of $\mathbf{D_X}(\mathbf{P})$ [3]. Note that the correspondence between $\mathbf{D_X}(\mathbf{P})$ and $\mathbf{P}$ is not one-to-one but one-to-many. Thus, the preimage of $\mathbf{D_X}(\mathbf{P})$ is represented by a set of feasible parameters $\alpha, \beta, \gamma, \delta$ such that all points of $\mathbf{D_X}(\mathbf{P})$ satisfy the inequalities of (3). It means that the preimage is given by a convex polytope in the parameter space [3]. Because all interesting parameters in this paper are translation-invariant, we focus on the three parameters α, β, γ indicating the normal vector of $\mathbf{P}$, distinguished from the intercept δ of $\mathbf{P}$. Further discussion is given in Section 4.

3   Local Geometric Patterns on Digital Planes

3.1   Local Geometric Patterns and Their Linearity

We define a local point set around a point $\boldsymbol{x}$ in $\mathbb{Z}^3$ such that

$$\mathbf{Q}_k(\boldsymbol{x}) = \{\boldsymbol{y} \in \mathbb{Z}^3 : \|\boldsymbol{x} - \boldsymbol{y}\|_\infty \leq k\} \tag{4}$$

where $k = 1, 2, \ldots$. $\mathbf{Q}_k(\boldsymbol{x})$ is a cubical grid-point set whose edge length is $2k+1$. Let us consider that each grid point in $\mathbb{Z}^3$ has a binary value such as either 1 or 0. Such a pattern of binary points in $\mathbf{Q}_k(\boldsymbol{x})$ is called local geometric patterns, abbreviated to LGP hereafter. There are $2^{(2k+1)^3-1}$ different LGP for $\mathbf{Q}_k(\boldsymbol{x})$ providing that the central point $\boldsymbol{x}$ always has a fixed value, such as 1. This indicates that $\boldsymbol{x}$ is considered to be not a background point but an object point.

In this section, we investigate, among these $2^{(2k+1)^3-1}$ LGPs, which LGPs can appear on digital planes. Note that we set binary values of points of $\mathbf{D}_{\mathbf{Q}_k(\boldsymbol{x})}(\mathbf{P})$ to be 1 and those of other points to be 0. This problem is mathematically written as follows. Let $\mathbf{F}$ be a set of points whose values are 1 in $\mathbf{Q}_k(\boldsymbol{x})$. If there is a plane $\mathbf{P}$ such that

$$\mathbf{F} = \{(p, q, r) \in \mathbf{Q}_k(\boldsymbol{x}) : 0 \leq \alpha p + \beta q + \gamma r + \delta < \omega\}, \tag{5}$$

we say that $\mathbf{F}$ forms a digital plane patch in $\mathbf{Q}_k(\boldsymbol{x})$. Therefore, our problem is solved by looking for all possible $\mathbf{F}$, namely LGP, satisfying (5). Such LGPs are called linear LGPs. Since this problem is considered to be the feasibility of the inequalities of (5) for all $(p, q, r) \in \mathbf{F}$, we check if there are feasible solutions $\alpha, \beta, \gamma, \delta$ for each different $\mathbf{F}$, namely LGP. If they exist, such an LGP appears on digital planes and becomes a linear LGP.

3.2   Linear LGP Generation by Arithmetic Planes

In this paper, however, in order to avoid computing the feasibility test for all $2^{(2k+1)^3-1}$ LGPs of $\mathbf{Q}_k(\boldsymbol{x})$, we take another approach to generate all linear LGPs. Our approach is based on arithmetic planes [7,8], and similar to [4].

From the discussion in Subsection 2.2, we know that there are many possible $\mathbf{P}$ corresponding to a given $\mathbf{D}_{\mathbf{Q}_k(\boldsymbol{x})}(\mathbf{P})$ and that the preimage is represented by a set of feasible parameters $\alpha, \beta, \gamma, \delta$. This implies that, given $\mathbf{D}_{\mathbf{Q}_k(\boldsymbol{x})}(\mathbf{P})$, we can find a corresponding $\mathbf{P}$ with only rational parameters. In addition, the denominators of those rational numbers are bounded by the size of $\mathbf{Q}_k(\boldsymbol{x})$, namely k. It is also known that a digital plane $\mathbf{D}(\mathbf{P})$ with rational slopes is equivalent to an arithmetic plane [7,8], which is defined such that

$$\mathbf{A} = \{(x, y, z) \in \mathbb{Z}^3 : 0 \leq ax + by + cz + d < w\} \tag{6}$$

where a, b and c are relatively prime integers, and d and w are integers. We call w a thickness of $\mathbf{A}$, and set $w = \max(|a|, |b|, |c|)$, similarly to ω for $\mathbf{D(P)}$. When the thickness is set as mentioned above, $\mathbf{A}$ is called a naive plane [7]. For any $\mathbf{D}_{\mathbf{Q}_k(\boldsymbol{x})}(\mathbf{P})$, it is known that we can find a corresponding $\mathbf{A}$ as its representation.

Based on this fact, we generate all linear LGPs by using naive planes $\mathbf{A}$ instead of digital planes $\mathbf{D(P)}$. Our algorithm mainly consists of the following three steps: **(I)** set parameters $a, b, c, d \in \mathbb{Z}$; **(II)** from those parameters, construct a grid point set $\mathbf{A}$ from (6); **(III)** for each point $\boldsymbol{x} \in \mathbf{A}$, observe the LGP of $\mathbf{Q}_k(\boldsymbol{x})$ for a given k. In the followings, we detail each step of **(I)** and **(II)**.

Parameter Setting for Naive Planes. As we mentioned in Subsection 2.2, we distinguish the three parameters a, b, c indicating a normal vector of $\mathbf{A}$ from the intercept d indicating a translation of $\mathbf{A}$. We are first concerned with the setting of value d. It is known that $\mathbf{A}$ always has grid points $(p, q, r) \in \mathbb{Z}^3$ satisfying the equality of (6) such that

$$ap + bq + cr + d = 0, \tag{7}$$

called leaning points [8]. Chosen a leaning point $\boldsymbol{p} \in \mathbf{A}$, even if we translate $\mathbf{A}$ with a vector $-\boldsymbol{p}$ so that the origin becomes a leaning point, it is certain that such a translation does not influence LGPs on $\mathbf{A}$. We therefore simply set

$$d = 0 \tag{8}$$

so that we consider only naive planes $\mathbf{A}$ one of whose leaning points is the origin.

Concerning the other parameters $a, b, c \in \mathbb{Z}$, we give the following constraints:

$$0 \le a \le b \le c, \; c \ne 0. \tag{9}$$

All $\mathbf{A}$ which do not satisfy (9) can be generated from the $\mathbf{A}$ which satisfy (9) by their rotations around the origin, since we set $d = 0$, and their symmetries with respect to the xy-, yz- and xz-planes. We can also bound c such that

$$c \le 8k^2 \tag{10}$$

from the size of $\mathbf{Q}_k(\boldsymbol{x})$ [2]. Therefore, once the value of k is given, we can automatically generate a set of all relatively-prime integer triplets (a, b, c), denoted by V_k, by using the Euclidean algorithm, with the constraints (9) and (10).

Constructed Part of a Naive Plane. With (9), we see that the principal projection plane of $\mathbf{A}$ is the xy-plane. Then, it is known that there are maximum $(2k + 1)^2$ different LGPs on $\mathbf{A}$ and that all different LGPs can appear in the region which is projected in the principal projection plane, *i.e.*, in the xy-plane, as a $(4k + 1) \times (4k + 1)$ squared region and whose central point is a leaning point of $\mathbf{A}$, *i.e.*, the origin [11].

Concerning to z-coordinates, thanks to the periodicity of $\mathbf{A}$ [7], we see that the maximum difference of the z-coordinates of any pair of points in the region which is projected in the xy-plane as a $(4k + 1) \times (4k + 1)$ rectangle does not exceed twice those of the x- and y-coordinates. Therefore, we can set a constructed part of $\mathbf{A}$ in a finite grid space $\mathbf{X}_k$ such that

$$\mathbf{X}_k = [-2k, 2k] \times [-2k, 2k] \times [-4k, 4k]. \tag{11}$$

Algorithm 1. Generation of all linear LGP

---

    input : a size k of $\mathbf{Q}_k(\boldsymbol{x})$
    output: a set T of linear LGPs
1 **begin**
2   initialize a set T;
3   make a set of integer normal vectors V_k with (9) and (10) ;
4   **foreach** $(a, b, c) \in V_k$ **do**
5   construct a finite grid-point set $\mathbf{A}$ in $\mathbf{X}_k$ from (6) and (11) with (8) ;
6   **foreach** $\boldsymbol{x} \in \mathbf{A} \cap \mathbf{X}_k$ **do**
7   **if** LGP *of* $\mathbf{Q}_k(\boldsymbol{x})$ *is not included in* T **then** put it in T
8   **return** T;
9 **end**

3.3 Algorithm and Results

From the above discussion, we now present Algorithm 1 for generating all linear LGPs of $\mathbf{Q}_k(\boldsymbol{x})$ for a given k. By executing Algorithm 1, we obtain 34 linear LGPs for $k = 1$, similarly 1574 for $k = 2$, 23551 for $k = 3$, and 181735 for $k = 4$, up to translations, rotations and symmetries, thanks to the constraints (9). Figure 1 shows all linear LGPs for $k = 1$.

Remark that our LGP around a point $\boldsymbol{x} = (x, y, z)$ in $\mathbf{A}$ for a given k is slightly different from the $(2k + 1, 2k + 1)$-cube [11], which is defined as a set of points $\boldsymbol{p} = (p, q, r)$ in $\mathbf{A}$ such that $|x - p| \leq 1$ and $|y - q| \leq 1$ for the case of (9). A linear LGP of $\mathbf{Q}_k(\boldsymbol{x})$ can be smaller than a $(2k + 1, 2k + 1)$-cube, so that there are less linear LGPs than $(2k + 1, 2k + 1)$-cubes; for example, there are 40 $(3, 3)$-cubes, called tricubes [4,11], against the 34 linear LGPs for $k = 1$.

3.4 Linear LGPs and Non-linear LGPs

From the above results, we see that there are a few linear LGPs relatively to non-linear ones. However, we have learned from experience that many border points of a digital object have linear LGPs, if its object surface is very smooth. Indeed, this is not difficult to understand, since any local surface patch on a smooth surface can be approximated to a planar surface when the size of the patch becomes small. In other words, even if a point has a linear LGP, we are uncertain whether the point appears on a planar surface or a non-planar surface. Contrarily, if a point has a non-linear LGP, it is certain that the point never appears on a planar surface.

4 Feasible Normal Vectors and Discrete Gaussian Spheres

In the continuous framework, each surface point has a unique normal vector, and a mapping from a surface point to its normal vector on the Gaussian sphere is called a Gaussian image. In [5], an extended Gaussian image is presented as a mapping from a point $\boldsymbol{n}$ on the Gaussian sphere to the area of the surface whose normal vector is $\boldsymbol{n}$. The extended Gaussian images for all n are useful for

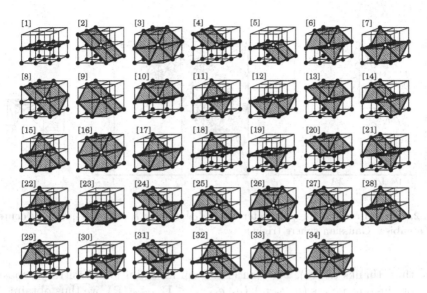

Fig. 1. The 34 linear LGPs for $k = 1$ up to translations, rotations and symmetries, with polyhedral meshes made by a discrete-marching-cube-like method for the 18-neighborhood system [6]

representing surface shapes. In this paper, we need a discrete version of extended Gaussian images, called unified discrete Gaussian images, which we use for planar surface segmentation in the next section. In the discrete framework, each linear LGP has a set of normal vectors, i.e., the preimage, represented by a set of points on the Gaussian sphere. In this section, we first investigate the preimage for every linear LGP, and study how all the preimages divide the Gaussian sphere. Such a divided Gaussian sphere is called a discrete Gaussian sphere. We finally define unified discrete Gaussian images using a discrete Gaussian sphere.

4.1 Feasible Normal Vectors of Each LGP

From the discussion in the previous section, if a surface point x of a digital object has a linear LGP of $Q_k(x)$, we can say that such x is locally linear with the size of $Q_k(x)$. In other words, the value of k indicates the absolute size of a planar surface around x. Based on this fact, we therefore calculate normal vectors at x by using (2) from a set of points on a digital plane in $Q_k(x)$.

From each linear LGP, we obtain $\mathbf{D}_{Q_k(x)}(\mathbf{P})$ in $Q_k(x)$, so that for all $(p, q, r) \in \mathbf{D}_{Q_k(x)}(\mathbf{P})$, we obtain a set of linear inequalities (2), namely,

$$0 \leq \alpha p + \beta q + \gamma r + \delta < \omega \tag{12}$$

where $\alpha, \beta, \gamma, \delta \in \mathbb{R}$. Because we assumed (9) for generating the linear LGPs in Section 3, we also consider the similar constraints such that

$$0 \leq \alpha \leq \beta \leq \gamma, \ \gamma \neq 0. \tag{13}$$

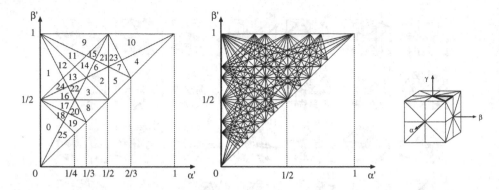

Fig. 2. Normal cells in the parameter space (α', β') for $k = 1$ (left), $k = 2$ (center), and a cubical Gaussian sphere (right)

Note that, thanks to (13), $\omega = \gamma$ in (12) from its definition. Arbitrary choosing pairs of different points $(p_i, q_i, r_i), (p_j, q_j, r_j) \in \mathbf{D}_{\mathbf{Q}_k(\boldsymbol{x})}(\mathbf{P})$, we thus obtain:

$$0 \leq \alpha p_i + \beta q_i + \gamma r_i + \delta < \gamma, \tag{14}$$

$$0 \leq \alpha p_j + \beta q_j + \gamma r_j + \delta < \gamma. \tag{15}$$

Because $\gamma \neq 0$ from (13), we divide both (14) and (15) by γ and substitute α', β' and γ' for α, β and γ such that

$$\alpha' = \frac{\alpha}{\gamma}, \ \ \beta' = \frac{\beta}{\gamma}, \ \ \gamma' = \frac{\delta}{\gamma}. \tag{16}$$

In order to eliminate γ' by Fourier-Motzkin elimination [14], we then obtain

$$-p_i \alpha' - q_i \beta' - r_i \leq \gamma' < -p_i \alpha' - q_i \beta' - r_i + 1, \tag{17}$$

$$-p_j \alpha' - q_j \beta' - r_j \leq \gamma' < -p_j \alpha' - q_j \beta' - r_j + 1, \tag{18}$$

and finally derive

$$(p_i - p_j)\alpha' + (q_i - q_j)\beta' + r_i - r_j + 1 > 0 \tag{19}$$

for any pair $(p_i, q_i, r_i), (p_j, q_j, r_j) \in \mathbf{D}_{\mathbf{Q}_k(\boldsymbol{x})}(\mathbf{P})$. The solution gives a feasible region which is a convex polygon in the space (α', β'). Remark that all calculations are done by using only integers, *i.e.*, they cause no rounding errors. Figure 2 illustrates that some of those convex polygons share the region in (α', β') and that such common regions make triangular or quadrilateral cells, called normal cells. Note that the constraints (13) make our interesting space (α', β') narrow down to a triangle bounded by linear inequalities $0 \leq \alpha' \leq \beta' \leq 1$. Table 1 shows a set of normal cells which represents a set of feasible normal vectors for each linear LGP of $\mathbf{Q}_1(\boldsymbol{x})$ depicted in Fig. 1.

We can find a figure similar to the left one of Fig. 2 in [11], which is generated by an approach based on 2D Farey series in number theory. It is clear that our

Table 1. Normal cells, illustrated in Fig. 2 (left), corresponding to each linear LGP for $k = 1$, illustrated in Fig. 1, with the constraints (13)

linear LGP for $k = 1$	corresponding normal cells
1	0 25
2	1 9 11 12
3	4 5 7 10 23
4,5	0 1 16 17 18 24
6,17	2 3 4 5 7 8
7	2 3 5 8
8,9	6 9 10 11 14 15 21 23
10,12	8 19 20 25
11	8 17 18 19 20
13,28	2 3 4 5 6 7 9 10 11 12 13 14 15 21 22 23
14	2 3 6 13 14 15 16 21 22 24
15	2 3 6 11 12 13 14 22
16	4 5 7 10 23
18,19	0 18 19 25
20,23	0 1 3 8 12 13 16 17 18 19 20 22 24 25
21,22	3 8 16 17 20 22
24,25	1 9 11 12 13 14 15 24
26,34	2 4 5 6 7 10 21 23
27	2 5 6 7 21 23
29,30	0 17 18 19 20 25
31,32	1 12 13 16 22 24
33	6 9 11 14 15 21

normal cells are related to 2D Farey series since the values of our inputs $p_i - p_j$, $q_i - q_j$ and $r_i - r_j$ of (19) are bounded by the size of $\mathbf{Q}_k(\boldsymbol{x})$. However, our result is slightly different from that in [11]. It is caused by the difference between the definition of linear LGPs and that of $(2k + 1, 2k + 1)$-cubes [11], as we already discussed at the end of Subsection 3.4. The difference can be seen in Fig. 2 (left), such that the normal cells "0" and "10" are not symmetrical with respect to a line of $\alpha' + \beta' = 1$, while the cells in [11] are symmetrical.

4.2   Discrete Gaussian Spheres

The 26 and 910 normal cells in Fig. 2 (left, center) are generated with the constraints (13). We embed these normal cells into the 3D space (α, β, γ) by using (16) with $\gamma = 1$. The triangle surrounded by thick lines in Fig. 2 (right) corresponds to the triangular region which is the union of normal cells in Fig. 2 (left, center). Once the normal cells are embedded into the space (α, β, γ), we make the congruous ones by applying to them 48 transformations of rotations and symmetries of a cube of edge length 2, centered at the origin of the 3D space. We see, in Fig. 2 (right), that there are the 48 triangles on the cube, so that the whole cube contains 1248 and 43680 normal cells for $k = 1, 2$, respectively. Such a cube is called a cubical Gaussian sphere.

We now project normal cells tiled on the cubical Gaussian sphere onto a unit sphere centered at the origin. The unit sphere separated by normal cells is called a discrete Gaussian sphere with respect to k, because the size of normal cells indicates the resolution of digitized normal vectors which are calculated from linear LGPs. Therefore, the resolution of a discrete Gaussian sphere is determined by the size of $\mathbf{Q}_k(\boldsymbol{x})$, $i.e.$, the value of k. In the remainder, we denote

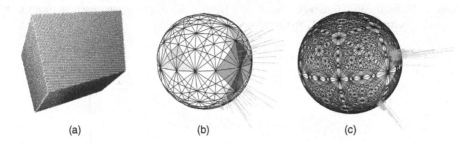

(a)   (b)   (c)

Fig. 3. A synthetic 3D image of a box (a) and its unified discrete Gaussian images for $k = 1$ (b) and $k = 2$ (c). Concerning cell colors on the discrete Gaussian sphere (b,c), the darker the blue cell, the larger the value of $u(c)$, and the red cell has the maximum value. The length of the pale blue needle for each cell c also corresponds to the value of $u(c)$. On a digitized box (a), red and blue points have linear LGPs for $k = 1$, while green points have non-linear LGPs. Note that red points correspond to the red cell.

$\mathbf{G}_k$ the set of all normal cells on the discrete Gaussian sphere with respect to k. Remark that we use only integer or rational numbers to calculate all normal cells, which are related to a cubical Gaussian sphere.

4.3   Unified Discrete Gaussian Images

By using the discrete Gaussian sphere, we give a discrete version of extended Gaussian images [5], called unified discrete Gaussian images. Let us first consider a discrete version of the Gaussian image. Let $\mathbf{V}$ be a set of object surface grid-points. For a point $\boldsymbol{x} \in \mathbf{V}$, we define a discrete Gaussian image $\mathbf{I}_k(\boldsymbol{x})$ as the set of normal cells corresponding to the linear LGP of $\boldsymbol{x}$, if its LGP is linear with respect to k; otherwise, $\mathbf{I}_k(\boldsymbol{x})$ is defined as empty. For each cell $c \in \mathbf{G}_k$, we consider a point subset of $\mathbf{V}$ such that

$$\mathbf{R}(c) = \{\boldsymbol{x} \in \mathbf{V} : c \in \mathbf{I}_k(\boldsymbol{x})\}. \tag{20}$$

We then obtain the number of points in $\mathbf{R}(c)$ for every $c \in \mathbf{G}_k$, called a unified discrete Gaussian image, such that

$$u(c) = |\mathbf{R}(c)|. \tag{21}$$

From the definition, we see that our unified discrete Gaussian image represents a distribution of normal cells of a digital object surface.

Figure 3 shows examples of unified discrete Gaussian images for a digitized box. In the figures, we see that we can extract a set of grid points which belong to a digital plane $\mathbf{D}(\mathbf{P})$ by choosing a "good" cell, for example, a red one. This is based on the following fact; if (α, β, γ) be a normal vector of $\mathbf{D}(\mathbf{P})$, (α, β, γ) is included in the common cell(s) of $\mathbf{I}_k(\boldsymbol{x})$ for all $\boldsymbol{x} \in \mathbf{D}(\mathbf{P})$.

Algorithm 2. Planar surface segmentation

> **input** : a grid-point set $\mathbf{V}$, a size k of LGP, and a minimum point-set size s
>
> **output**: an edge point set $\mathbf{E}$ and planar-surface point sets $\mathbf{S}_i$ for $i = 1, 2, 3, \ldots$

```
 1 begin
 2    initialize u(c) = 0 and R(c) = ∅ for every c ∈ Gₖ;
 3    foreach point x ∈ V do
 4       if x has a linear LGP then
 5          ⌊ increment u(c) and add x to R(c) for all c ∈ Iₖ(x);
 6       else put x in E;

 7    initialize a label such that l = 0;
 8    repeat
 9       make a priority queue Dₖ of cells c with u(c) such that u(c) ≥ s;
10       increment l and initialize Sₗ = ∅;
11       set h to be the highest priority cell in Dₖ and remove it from Dₖ;
12       while u(h) > |Sₗ| do
13          calculate the maximum connected component of R(h), set to be C;
14          if |C| > |Sₗ| then  set Sₗ = C;
15          ⌊ reset h to be the highest priority cell in Dₖ and remove it from Dₖ;
16       if |Sₗ| ≥ s then
17          forall c such that u(c) ≠ 0 and R(c) ∩ Sₗ ≠ ∅ do
18             ⌊ reset R(c) = R(c) \ Sₗ and u(c) = |R(c)|;

19    until |Sₗ| < s ;
20    return Sᵢ for i = 1, 2, …, l − 1;
21 end
```

5   Planar Surface Segmentation Using LGPs

In this section, we present an algorithm for planar surface segmentation from a given set of object surface points $\mathbf{V}$. We formulate our problem as follows: we assign each point $x \in \mathbf{V}$ into either an edge set $\mathbf{E}$ or one of surface sets $\mathbf{S}_i$ for $i = 1, 2, \ldots$ such that the points in each $\mathbf{S}_i$ have a common property. In our case, each $\mathbf{S}_i$ corresponds to a connected subset of a digital planar surface, and the points in $\mathbf{E}$ constitute the joints of digital plane patches. We assume that two surface sets have no overlap, namely, $\mathbf{S}_i \cap \mathbf{S}_j = \emptyset$ for any i, j such that $i \neq j$.

5.1   Algorithm

Our method is founded on the following hypothesis; if there is a connected point set $\mathbf{S} \subseteq \mathbf{V}$ such that $\cap_{x \in \mathbf{S}} \mathbf{I}_k(x) \neq \emptyset$, $\mathbf{S}$ may constitute a digital plane patch. Based on this, we propose an algorithm mainly consisting of the following three parts: (**A**) reject all points having non-linear LGPs for a given k and consider them as edge points; (**B**) make a unified discrete Gaussian image $u(c)$, $c \in \mathbf{G}_k$, with the point sets $\mathbf{R}(c)$; (**C**) by using $u(c)$ and $\mathbf{R}(c)$, look for connected point sets, corresponding to planar surfaces, each of whose points having a common normal cell. The parts (**A**) and (**B**) are considered to be pre-processing of the

Fig. 4. A result of planar surface detection from a 3D range image of blocks for $k = 1$ (left) and $k = 2$ (right) with $s = 100$. Four (resp. five) planar surfaces are detected for $k = 1$ (resp. $k = 2$) so that each surface is colored with a different color except for light green and blue. Light green points have non-linear LGPs and blue points have linear LGPs but cannot constitute a planar surface whose size is not less than s.

part **(C)** which is the core of the algorithm. We present Algorithm 2, in which **(A)** is achieved in Steps 2 and 2, and **(B)** is achieved from Step 2 to 2.

The core part **(C)**, which starts from Step 2, is a loop of seeking planar surfaces $\mathbf{S}_i$. Each $\mathbf{S}_i$ is a maximally connected point set, whose points have a common normal cell, as shown in Steps from 2 to 2. Once we find $\mathbf{S}_i$, we check the size of $\mathbf{S}_i$ in Step 2, and if $|\mathbf{S}_i| \geq s$, we remove all points of $\mathbf{S}_i$ from every point set $\mathbf{R}(c)$ and sometimes modify $u(c)$ in Step 2. After such modification and incrementing i, we seek a new $\mathbf{S}_i$. For finding each $\mathbf{S}_i$, we look for the maximum connected component $\mathbf{C}$ of each $\mathbf{R}(c)$ and then set $\mathbf{S}_i$ to be the maximum among all $\mathbf{C}$. In order to reduce the frequency of calculation of connected components which is a global operation, we make a priority queue D_k of cells with $u(c)$ in Step 2. We then repeat dequeue of a cell h from D_k to obtain the maximum connected component $\mathbf{C}$ of $\mathbf{R}(h)$ in Step 2. Comparing the size of $\mathbf{C}$ with the maximum among those of other cells which are already dequeued from D_k, we finally obtain the currently maximum point set $\mathbf{S}_l$ in Step 2. Note that this loop is repeated until $u(h)$ is not more than the size of $\mathbf{S}_l$ as described in Step 2. For calculating the maximum connected component of $\mathbf{R}(h)$, we apply a simple method based on a depth-first strategy by using a queue [7]. The time complexity is linear with respect to $u(h)$, namely, $O(n)$, where n is the number of grid points in $\mathbf{V}$, in the worst case. Thus, the time complexity of the whole algorithm is $O(ln)$, where l is the number of segmented regions.

5.2   Experimental Results

Figure 4 shows the results of planar surface segmentation from a 3D range image of blocks. We see that four (resp. five) planar surfaces are detected for $k = 1$ (resp. $k = 2$), even if the range image contains noise. We see that one of planes for $k = 1$ is split into two for $k = 2$ because of the different normal-cell resolutions between $k = 1$ and $k = 2$. Remark that non-linear LGPs appear around edges and ridge lines of blocks, and also appear in faces of blocks because of noise in the range image. We also see that there are more non-linear LGPs for $k = 2$ than $k = 1$. In other words, the higher k, the more sensitive the result. Linear LGPs whose corresponding normal cells are not shared by those of the neighboring LGPs mostly appear beside non-linear LGPs.

6 Conclusion

We present a discrete version of the hybrid method for planar surface segmentation from a 3D grid-point set. We first propose an algorithm to generate all linear LGPs for $k = 1, 2, \ldots$ by using the arithmetic plane properties. Based on those linear LGPs, we also introduce new notions of discrete Gaussian spheres and unified discrete Gaussian images, in order to process normal vectors in the discrete-geometry framework. By using those discrete notions, we succeed in detecting an edge-point set and planar-surface point sets. With our experiments, we show that our method is robust against not only quantization errors but also noise. As our method gives a rough segmentation result, it may be useful to obtain an initial segmentation before applying plane recognition.

For future work, the comparison with conventional methods in both computer vision and discrete geometry in terms of segmentation results and computational costs are left. In addition, we need further work on surface segmentation for non-polyhedral objects.

References

1. Besl, P.J., Jain, R.C.: Segmentation through variable-order surface fitting. IEEE Transactions on PAMI 10(2), 167–192 (1988)
2. Buzer, L.: A composite and quasi linear time method for digital plane recognition. In: Kuba, A., Nyúl, L.G., Palágyi, K. (eds.) DGCI 2006. LNCS, vol. 4245, pp. 331–342. Springer, Heidelberg (2006)
3. Coeurjolly, D., Sivignon, I., Dupont, F., Feschet, F., Chassery, J.-M.: On digital plane preimage structure. Discrete Applied Mathematics 151(1–3), 78–92 (2005)
4. Debled-Rennesson, I.: Etude et reconnaissance des droites et plans discrets. Ph.D. thesis, Université Louis Pasteur, Strasbourg (1995)
5. Horn, B.K.P.: Extended Gaussian images. Proceedings of the IEEE 72(12), 1671–1686 (1984)
6. Kenmochi, Y., Imiya, A.: Combinatorial boundary of a 3D lattice point set. Journal of Visual Communication and Image Representation 17(4), 738–766 (2006)
7. Klette, R., Rosenfeld, A.: Digital Geometry: Geometric Methods for Digital Picture Analysis. Morgan Kauffmann, San Francisco (2004)
8. Reveillès, J.-P.: Combinatorial pieces in digital lines and planes. In: Vision Geometry IV. SPIE, vol. 2573, pp. 23–34 (1995)
9. Sivignon, I., Dupont, F., Chassery, J.-M.: Discrete surfaces segmentation into discrete planes. In: Klette, R., Žunić, J. (eds.) IWCIA 2004. LNCS, vol. 3322, pp. 458–473. Springer, Heidelberg (2004)
10. Stamos, I., Allen, P.K.: 3D model construction using range and image data. In: Proceedings of IEEE Conference on CVPR, vol. 1, pp. 531–536 (2000)
11. Vittone, J., Chassery, J.-M.: (n, m)-cubes and Farey nets for naive planes understanding. In: Bertrand, G., Couprie, M., Perroton, L. (eds.) DGCI 1999. LNCS, vol. 1568, pp. 76–87. Springer, Heidelberg (1999)
12. Yokoya, N., Levine, M.D.: Range image segmentation based on differential geometry: a hybrid approach. IEEE PAMI 11(6), 643–649 (1989)
13. Zhao, D., Zhang, X.: Range-data-based object surface segmentation via edges and critical points. IEEE Transactions on IP 6(6), 826–830 (1997)
14. Ziegler, G.M.: Lectures on Polytopes. Springer, New York (1998)

Robust Estimation of Curvature along Digital Contours with Global Optimization

Bertrand Kerautret[1] and Jacques-Olivier Lachaud[2,*]

[1] LORIA, Nancy- Campus Scientifique
54506 Vandœuvre -lès-Nancy Cedex
kerautre@loria.fr
[2] LAMA, University of Savoie
73376 Le Bourget du Lac
jacques-olivier.lachaud@univ-savoie.fr

Abstract. In this paper we introduce a new curvature estimator based on global optimisation. This method called Global Min-Curvature exploits the geometric properties of digital contours by using local bounds on tangent directions defined by the maximal digital straight segments. The estimator is adapted to noisy contours by replacing maximal segments with maximal blurred digital straight segments. Experimentations on perfect and damaged digital contours are performed and in both cases, comparisons with other existing methods are presented.

1 Introduction

Estimating geometric characteristics of digital shapes is an essential step in many image analysis and pattern recognition applications. We focus here on the geometry of digital 4-connected contours. These contours arise naturally as the inter-pixel boundary of digital regions in images. We present here a new method for estimating the curvature at any point of such digital contours, i.e. we estimate the curvature field of the contour. We are interested here by the *quantitative* estimation of the curvature field and not only the detection of dominant or inflexion points, as opposed to many methods proposed in the pattern recognition community (see [1]). Note however that detecting these points is a natural byproduct of curvature computation, provided curvature estimations are stable enough. Furthermore digital contours are rarely perfect digitizations of regular shapes, as may be seen on Fig. 1. In order to be useful, curvature estimation techniques should thus be able to take into account local perturbations, provided the digital contour holds more significant information than noise.

To estimate a geometric characteristic from a digital contour, it is necessary to suppose that there is an underlying real shape, although its geometry is generally unknown. A "good" estimator aims at approaching at best the corresponding geometric characteristic of this real shape. It is however difficult if not impossible to compare objectively the respective accuracy of several estimators, since

---

* This work was partially funded by the ANR project GeoDIB. Bertrand Kerautret was partially funded by a BQR project of Nancy University.

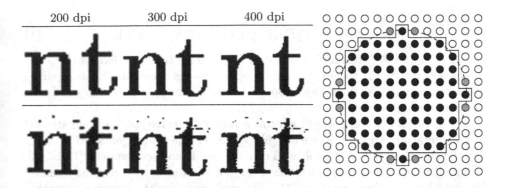

Fig. 1. Left. Perfect versus noisy digital contours. Top row: the letters "nt" written in a roman font family of size 14pt are digitized at several increasing resolution. Bottom row: the same text has been printed at 600dpi on a laser printer and then scanned at the corresponding resolution. The digital contour of both letters has been damaged in the process and presents some irregularities which are very visible on straight parts. Pepper and salt noise is also visible. Furthermore, these phenomena are visible at each resolution. **Right.** Digital object O depicted with black disks •. Its shape of reference is the disk sketched with dotted lines, whose radius is constrained by the gray disks. The perfect curvature map is thus $\kappa(t) = 1/\sqrt{5^2 + 1}$.

for a given digital contour there exists infinitely many shapes with the same digitization. This problem is even greater here, since infinitesimal perturbations in the input shape may induce huge variations in the curvature field.

Before going further, we introduce some notations and definitions (mainly taken from [2]). Let Dig_h be the Gauss digitization process of step h (i.e. the intersection of $\mathbb{Z}^2$ with the magnification of the shape by $1/h$). Let $\mathbb{F}$ be a family of shapes in $\mathbb{R}^2$ with appropriate properties [1]. Let G be some geometric feature defined for any shape X of $\mathbb{F}$. A *discrete geometric estimator* E_G *of* G is a map that associates to a digitization $\mathrm{Dig}_h(X)$ an estimation of $G(X)$. Properties of geometric estimators are classically defined for global shape features like area or perimeter (see [2]). We adapt this definition to *local geometric features* like tangent or curvature as follows (other definitions may be found [3,4]). Let s be the arc length parameterization of the topological boundary ∂X of X, and let t be s divided by the length of ∂X. Here the feature G is the curvature field, which may be then represented as a map $\kappa : t \in [0, 1] \mapsto \kappa(t) \in \mathbb{R}$. Our purpose is therefore to build a function E_κ which approaches κ at best as possible, for instance in the $L^2([0, 1])$ sense.

As far as we know, the only property of geometric estimators that can objectively be compared is the *multigrid convergence*, which indicates that the finer the digitization step, the better is the estimation. Generally speaking, the estimator E_G is *multigrid convergent* toward G for the family $\mathbb{F}$ iff, for any shape $X \in \mathbb{F}$, there exists some $h_X > 0$ for which

---

[1] We take here the family of connected compact shapes whose boundary is rectifiable and whose curvature map is in L^2. This avoids fractal-like shapes and the curvature, while not compulsory defined everywhere, is therefore square integrable.

$$\forall h, 0 < h < h_X, \|E_G(\mathrm{Dig}_h(X)) - G(X)\| \leq \tau(h), \tag{1}$$

where $\tau : \mathbb{R}^+ \rightarrow \mathbb{R}^{+*}$ has limit value 0 at $h = 0$. This function defines the *speed of convergence* of E_G toward G. This property seems appealing for comparing geometric estimators, since a good speed of convergence guarantees a good estimation at a high enough scale.

The multigrid convergence of several geometric estimators has been studied in the literature: area [5], moments [6], perimeter [7,8,9], tangents [10,11,3,12]. For the curvature field, the multigrid convergence is not yet achieved [10,13,14], although a very recent approach based on global filtering by a carefully chosen binomial kernel seems promising [15]. The multigrid convergence may nevertheless be criticized on the following points: (i) these estimators are guaranteed to be precise only at very fine resolution; (ii) this property has meaning only for perfect shape digitizations: it is no more valid when the input data has been — more or less slightly — damaged.

We propose a new objective criterion for curvature estimation which is valid also at coarse resolution (addressing criticism (i)). Furthermore, we show how to compute a numerical approximation of the optimal solution for this criterion and then how to adapt this algorithm to corrupted or noisy data (addressing criticism (ii)). More precisely, a good objective criterion should take into account not only one real shape but *all* the real shapes that have the input digital contour as digitized boundary. Of course, not all those real shapes should have the same probability to be the true shape of interest. For instance, a very classical tool in image analysis are the deformable models of Kass *et al.* [16] or the *geodesic active contours* which considers shapes with short perimeters as more likely than shapes with winding contours [17]. These methods consider shapes with smooth contours as more preferable than shapes with many points of high curvature.

Following this analogy, given a digital object O (a non-empty finite subset of $\mathbb{Z}^2$), we define the *shape of reference* $R_{O,h}$ to O at grid step h as the shape of $\mathbb{F}$ which minimizes its squared curvature along its boundary and such that $\mathrm{Dig}_h(R_{O,h}) = O$. An illustration is given in Fig. 1, right. We therefore define

Definition 1. *Given a curvature estimator E_κ, its min-curvature criterion relative to object O and step h is the positive quantity $\|E_\kappa - \kappa_{O,h}\|$, where $\kappa_{O,h}$ is the curvature map of $R_{O,h}$ and $\|\cdot\|$ is the L^2-norm.*

A good curvature estimator should therefore have a low min-curvature criterion for a large family of shapes. One may notice that a similar criterion is implicitly used for perimeter estimation by Sloboda *et al.* [8]: their perimeter estimator is defined as the perimeter of the polygon which has the same digitization as the input digital object and which minimizes its perimeter. In other terms, their shape of reference is the one that minimizes $\int_{\partial X} 1$. Their perimeter estimator has good properties at low scale but is also multigrid convergent with speed $O(h)$.

We present here a curvature estimator, called Global Min-Curvature estimator (GMC) whose min-curvature criterion is numerically zero for the family of shapes $\mathbb{F}$, i.e. it is the best possible curvature estimator for this criterion. Our method exploits the specific geometric properties of digital contours. The maximal digital straight segments of the input digital contour are used to define local bounds on the tangent directions (Section 2). These bounds are casted into the space of tangent directions. The curvature of the reference shape O is then computed by optimization in this tangent space (GMC estimator (Section 3)). In Section 4, we adapt our estimator to noisy or damaged digital contours by replacing maximal segments with maximal blurred digital straight segments. Section 5 validates our curvature estimator with several experiments on perfect or damaged digital contours, and with some comparisons with other curvature estimators.

2   Tangential Cover and Tangent Space

We assume that the input data is the inter-pixel boundary of some digital objects, that we will call later on a *digital contour*. It is given as a 4-connected closed path C in the digital plane, whose discrete points C_i are numbered consecutively. A sequence of connected points of C going increasingly from C_i to C_j is conveniently denoted by $C_{i,j}$.

Such a sequence is a *digital straight segment* iff its points are included in some standard digital straight line, i.e. $\exists (a, b, \mu) \in \mathbb{Z}^3, \forall k, i \leq k \leq j, \mu \leq ax_{C_k} - by_{C_k} < \mu + |a| + |b|$. The standard line with smallest $|a|$ and containing the sequence, defines the characteristics (a, b, μ) of the digital straight segment. In particular, the *slope* of the segment is a/b. Let us now denote by $S(i, j)$ the predicate "$C_{i,j}$ is a digital straight segment". A *maximal segment* of C is a sequence $C_{i,j}$ such that $S(i, j) \wedge \neg S(i, j+1) \wedge \neg S(i-1, j)$. The *maximal segments* are thus the inextensible digital straight segments of C. Together, they constitute the *tangential cover* of C, as illustrated on Fig. 2, left.

The tangential cover of a digital contour can be efficiently computed in linear time with respect to its number of points [18,12]. The directions of maximal segments may be used to estimate the tangent direction of the underlying shape [11]. Here we also make use of the direction of maximal segments, but to estimate locally the geometries of all possible underlying shapes. We proceed as follows.

1. Each maximal segment tells us some information on the local geometry of the underlying continuous shape (Fig. 2, right). In particular, the direction of maximal segment gives bounds on the possible tangent directions of the continuous shape. Although this can be done in several ways, we choose to estimate these bounds from the leaning points. If M and M' are the two furthest upper or furthest lower leaning points, then the minimal slope (resp. maximal slope) is chosen as the slope of the segment joining M to $M' + (0, -1)$ (resp. M to $M' + (0, 1)$). For a maximal segment of characteristics (a, b) and with only two upper leaning points and at most two lower leaning points, its *extremal slopes* are thus $\frac{a}{b} \pm \frac{1}{b}$.

2. We represent a closed C^1-curve $\mathcal{C}$ parameterized by its arc length s as a function graph which maps s to the tangent direction at $\mathcal{C}(s)$. The domain is $[0, L[$, L being the curve length and the range is $[0, 2\pi[$. Such a representation, that we call hereafter *tangent space*, defines the closed curve geometry up to a translation.

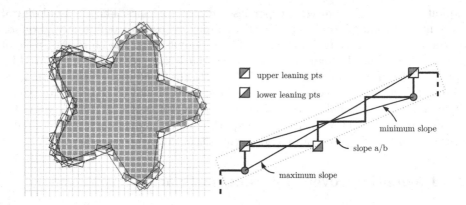

Fig. 2. Left: tangential cover of the boundary of a digitized shape, where each maximal segment is drawn as a black bounding box aligned with its slope. Right: slope of a maximal segment and estimation of maximal and minimal slopes with leaning points.

3. We fix C_0 as the starting point of the arc length parameterization. Given a digital length estimator E_L, we can estimate the arc length s_i of any point C_i as $E_L(C_0C_i)$. For each maximal segment $C_{i,j}$, we then draw in the tangent space an axis aligned box spanning abscissas s_i to s_j and whose ordinates are the inverse tangent of the extremal slopes (Fig. 3, left).

4. Although infinitely many shapes have the same digitization as the input digital object, the boundary of shapes with smooth enough tangents should have a tangent space representation that stays within the boxes defined above. On first approximation, the boundary contour is indeed locally a straight line.

3   Curvature Computation by Optimization

We extract the shape of reference to the input digital object O from the tangent space representation. Indeed, if the shape boundary $\mathcal{C}$ is smooth enough, its geometry is entirely defined by the mapping $\theta_{\mathcal{C}}$ which associates to an arc length s the direction of the tangent at point $\mathcal{C}(s)$ ($\theta_{\mathcal{C}} = \angle(0x, \mathcal{C}')$). Since the curvature is the derivative of the tangent direction, the integral $J[\mathcal{C}]$ along $\mathcal{C}$ of its squared curvature is then

$$J[\mathcal{C}] = \int_{\mathcal{C}} \kappa^2 = \int_0^L \kappa^2(s)ds = \int_0^L \left(\frac{d\theta_{\mathcal{C}}}{ds}\right)^2 ds. \tag{2}$$

The shape of reference to O is the shape in $\mathbb{F}$ of boundary $\mathcal{C}$ which minimizes $J[\mathcal{C}]$ and which is digitized as O. From the preceding section, the tangent space representation of $\mathcal{C}$ should stay within the bounds given by the maximal segments. Let us now denote (i_l) the increasing sequence of indices of the digital points that are starting or ending point of a maximal segment. To each point C_{i_l}, we

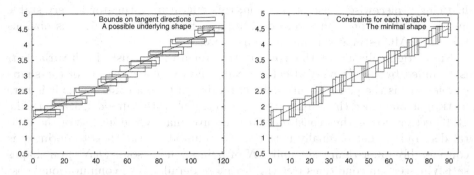

Fig. 3. The shape of interest is a digitized disk of radius 30.5. Left: bounds given by each maximal segment on the possible local tangent direction (abscissa: index of the contour point, ordinate: angle wrt axis x). A possible underlying space should have its tangent space representation staying within these boxes. Right: Each variable has a possible range given by its vertical line. The shape that minimizes its squared curvature is represented by the dashed line. The obtained curvature field is constant and corresponds to the one of a disk of radius 30.5.

associate the smallest possible tangent direction a_l as the smallest bound given by the maximal segments strictly containing C_{i_l}. The largest possible tangent direction b_l is defined symmetrically (see. Fig. 3). Finally let t_l be the tangent direction of the shape of reference at its point of arc length s_{i_l}.

The digitization constraint imposes $\forall l, a_l \leq t_l \leq b_l$. We look for a shape of reference with smooth enough boundary (which we will precise later). Its mapping θ_C is therefore continuous and interpolates points (s_{i_l}, t_l). Looking now at an arbitrary portion $[s_{i_l}, s_{i_{l+1}}]$ of the curve, standard variation calculus on Equ (2) immediately gives the necessary condition $2\frac{d}{ds}\frac{d\theta_C}{ds} = 0$, otherwise said the straight segment $\theta_C(s) = t_l + \frac{t_{l+1}-t_l}{s_{i_{l+1}}-s_{i_l}}(s - s_{i_l})$. A straight segment in the tangent space is a circular arc in the plane. It is straightforward to check that this segment stays within the bounds defined by the maximal segments.

We have therefore found the optimal boundary of the shape of reference, if the family $\mathbb{F}$ of shapes is composed of shapes with C^0 and piecewise C^1 continuity of tangent direction. It must be found in the family $C[\ldots, t_l, \ldots]$ whose tangent directions are piecewise linear functions. Curvature estimation of reference shape to O is thus reduced to

$$\text{Find } (t_l)_l, \text{ which minimizes } J[C[\ldots, t_l, \ldots]] = \sum_l \left(\frac{t_{l+1} - t_l}{s_{i_{l+1}} - s_{i_l}}\right)^2 (s_{i_{l+1}} - s_{i_l}),$$

$$\text{subject to} \quad \forall l, a_l \leq t_l \leq b_l.$$

We use classical iterative numerical techniques to solve this optimization problem, simply following $\frac{\partial J}{\partial t_l}$ for each variable t_l while staying in the given interval. Geometrically, each variable t_l is moved toward the straight segment joining $(s_{i_{l-1}}, t_{l-1})$ to $(s_{i_{l+1}}, t_{l+1})$. The *GMC estimator* E_κ^{GMC} is then simply defined as the derivative of the piecewise linear function joining points (s_{i_l}, t_l), rescaled by

h. Being a piecewise constant function, this curvature estimator is very stable and is undefined only on a zero-measure set. From Definition 1, it is obvious that the GMC estimator is (numerically) zero.

Fig. 3, right, illustrates this computation for a digital disk. Each variable t_l is bounded by the drawn vertical interval. The tangent direction of the shape of reference is the piecewise linear function drawn with a dashed line: it is here a straight line since there exists a continuous disk with same digitization. The GMC estimation is thus constant and is approximately the inverse of the digital disk radius. Let us finally note that the number of variables to optimize is considerably lower than the number N of points, and is some $O(N^{\frac{2}{3}})$ on shapes satisfying certain conditions (see [14] for more details). The computational cost and the iteration number needed for the optimisation are illustrated in Section 5.

4   Tangential Cover of Noisy Digital Contours

If we consider noisy digital contours it seems natural to adapt the concept of tangential cover by using blurred segments. A linear time recognition algorithm of blurred segments was proposed by Debled et al. [19]. This approach is based on the computation of the convex hull and on its vertical geometric width. Note that Buzer proposed a very similar approach [20]. Both of these methods assume points are added with increasing x coordinate in the recognition process.

In the same way of Roussillon et al. [21], we adapt the algorithm of Debled et al. [19] to avoid the restrictive hypothesis which assumes that points are added with increasing x coordinate (or y coordinate). The figure on the right shows an example of such a contour (light gray) with a maximal segment of width $\nu = 5$ (dark gray). Despite non increasing x (and y) coordinate, the resulting maximal blurred segment is well detected.

The tangent directions range $R = [\theta_{\min}, \theta_{\max}]$ associated to a maximal blurred segment can be determined by the slope of its bounding line and by its width ν. More precisely, if $V(v_x, v_y)$ is the vector given by the two leaning points of the upper bounding line , the interval is defined by $R = \left[\tan^{-1}(\frac{v_y + \nu}{v_x}), \tan^{-1}(\frac{v_y - \nu}{v_x})\right]$. Here the value of ν was determined from the vertical width of the blurred segment convex hull. The figure on the right illustrates the maximal blurred segments obtained on a noisy circle with initial radius 15. The maximal width ν of the blurred segment was set to 2.

To compute curvature by optimization, the maximal and minimal slopes for each point need to be evaluated. It may happen that two maximal segments including the points of interest have opposite directions. Therefore, taking the minimum (resp. maximum) of the minimal (resp. maximal) slope of these maximal segments is not always coherent. So for each point we define an oriented interval associated to the minimal and maximal tangent directions: $I_k = [\theta_{\min}^k, \theta_{\max}^k]$.

Then the strategy to define the global interval I is to merge each interval to each other. The merging process between two intervals I_1 and I_2 can be done according to the following conditions:

(1) I_2 is included in I_1: $\theta^2_{min}, \theta^2_{max} \in [\theta^1_{min}, \theta^1_{max}]$ and $\theta^2_{min} \in [\theta^1_{min}, \theta^1_{max}]$ and $\theta^2_{max} \in [\theta^1_{min}, \theta^1_{max}]$ then the resulting interval is obviously I_1 (see for example Fig. 4a).
(2) I_1 is included in I_2: same case as (1) by substituting I_1, I_2 and S_1, S_2.
(3) $\theta^2_{min} \in [\theta^1_{min}, \theta^1_{max}]$ and $\theta^2_{max} \notin [\theta^1_{min}, \theta^1_{max}]$ then $I = [\theta^1_{min}, \theta^2_{max}]$.
(4) $\theta^2_{max} \in [\theta^1_{min}, \theta^1_{max}]$ and $\theta^2_{min} \notin [\theta^1_{min}, \theta^1_{max}]$ then $I = [\theta^2_{min}, \theta^1_{max}]$ (Fig. 4b).
(5) $I_1 \cup I_2$ is not connected: we compute the distances between the two intervals $d_1 = |\theta^2_{min} - \theta^1_{max}|$ and $d_2 = |\theta^1_{min} - \theta^2_{max}|$. The resulting interval is determined according to the following conditions:
   - $d_1 < d_2$: the resulting interval is $I = [\theta^1_{min}, \theta^2_{max}]$.
   - $d_2 < d_1$: the resulting interval is $I = [\theta^2_{min}, \theta^1_{max}]$ (see Fig. 4c).
(6) In all the other cases the resulting interval is set to $I = [0, 2\pi]$. Fig. 4d shows such an example.

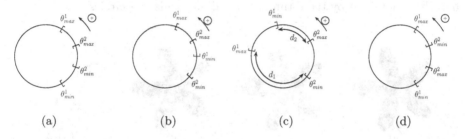

(a)   (b)   (c)   (d)

Fig. 4. Illustration of several cases which can appear in the merging process

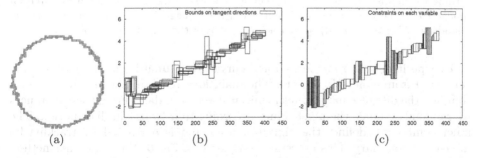

(a)   (b)   (c)

Fig. 5. Illustration of a distorted digital circle (a) with its tangent directions (b) and the constraints on each variable (c). The maximal blurred segments were obtained with width $\nu = 3$.

Note that cases (1-4) appear most of the time. But on very noisy contours, the more the width ν is small, the more cases (5) and (6) appear. In such cases, the local contour geometry is not well approached by the tangential cover. Fig. 5b shows the slope range of maximal blurred segments obtained from the noisy contour represented in (a). The width of the blurred segments was set to 3. Fig. 5c shows the resulting slope range obtained after the merging process. The total number of merging configurations (1-4) was equal to 410 while configuration (5) appears 18 times in the merging process. With width $\nu = 5$ the number of configurations (5) and (6) was equal to 0. This distribution of the different cases may define a new strategy to automatically choose the width ν that is best adapted to a given noisy contour.

5   Experimental Validation

The GMC estimator was first experimented on a circle of radius 20 with different grid steps (Fig. 7a). Even with coarse grid step, the estimator gives precise result in comparison to the method of osculating circles [2] [10] (see Fig. 7b). A second experimentation was made on the flower of Fig. 2, left, whose inner radius is 15 and outer radius is 25. In the same way as the previous experimentations, results are very precise with the different grid steps and once again results are better than those obtained by the estimator based on osculating circles.

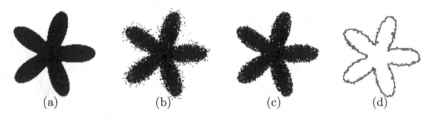

(a)   (b)   (c)   (d)

Fig. 6. Generation of noisy digital contours. Pixels from image (a) were changed according to the distance to the initial shape and according to a Gaussian distribution (image (b)). The inner component extracted from image (b) is shown in (c), and its boundary is displayed in (d). It is a noisy version of the original flower contour.

To experiment the robustness of the curvature estimator, noisy contours were generated from synthetic objects. The main idea to obtain such contours is to compute the distance map from the initial shape boundary. From these distances we randomly change the pixel values according to a Gaussian distribution. The inner component defines the digital contour and is extracted by tracking its inter-pixel boundary. This process is illustrated in Fig. 6. Note that this method of digital contour perturbation follows the local document degradation model proposed by Kanungo [22].

The curvature estimation was experimented on several noisy shapes with different width ν of blurred segments. Fig. 7e-h shows the curvature fields obtained on distorted disk and flower contours. The two contours were obtained from ideal smooth

---

² Other experimental results can be found at the following address:
http://www.loria.fr/$\sim$kerautre/dgci08/

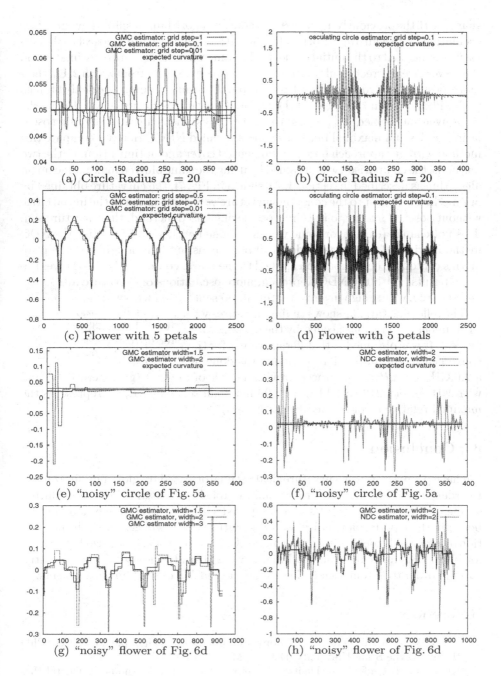

Fig. 7. Experimentations on **regular (a-d)** and **damaged (e-h)** contours. Plot (a) (resp. (c)) shows the curvature precision obtained on a quarter of circle R=20 (resp. flower) with several grid steps. Plots (b) and (d) show the results obtained by the osculating circle estimator. Curvature estimation precision on damaged shapes is shown on (e) and (g). A comparison between the GMC and NDC estimators is shown in (f,h).

shapes with the previously described degradation algorithm. Despite a few curvature oscillations with width $\nu = 1.5$, plot (e) shows that the curvature is closest to the curvature associated to the initial circle. In the same way, the curvature estimation on the flower gives correct results without concave/convex error from width 2. Furthermore, despite the strong contour degradation, the resulting curvature field obtained with width $\nu = 2$ and 3 appears stable (g), and curvature extrema are well located.

Nguyen and Debled [23] proposed a method to estimate curvature by using osculating circles and maximal blurred segments (called NDC). Here we compare this method with our approach based on tangential cover and optimization. In the two methods, the comparison was obtained by using the same blurred maximal segment algorithm as described in section 4. As shown in Fig. 7f, the curvature obtained by our method is closest to the constant curvature value associated to the initial circle without noise. Contrary to the circle osculating based method, the curvature sign does not change (except for some values obtained with $\nu = 1.5$). With the NDC method a width of $\nu = 5$ was needed to obtain a constant sign value of the curvature on this contour. The comparison applied to the flower contour (Fig. 7d-f) gives the same results. With the NDC approach, more oscillations of curvature are visible and some curvature sign inversions are still present[4], even with width $\nu = 3$.

The following tabular shows timing measures with the number of iterations needed by the optimization process. It was obtained with different values for the convergence criteria (ϵ). This measure was done on

ϵ	# iter	E_{max}	time (s)
10e-4	1648	1,38e-3	0,284
10e-6	7096	5,08e-5	0,416
10e-8	39101	5,08e-7	1,028

a 2,6 GHz *Intel Pentium* processor with a circle of radius 1 (grid step=0.01) and with 804 surfels, 212 variables and 106 maximal segments. E_{max} represents the maximal relative error defined by $(\kappa_{est}^{max} - \kappa_{est}^{min})/\kappa_{real}$.

6 Conclusion

We have proposed a new curvature field estimator based on global minimization. In order to deal with noisy contours, this approach was extended by using blurred segments. The obtained results were compared with several recent methods and are better both on perfect and noisy contours. Future works will deal with the integration of a more efficient optimization method and the use of an optimized algorithm for blurred segment recognition. Finally, we will study the use of the tangential cover in order to automatically determine the width ν of the blurred segments.

References

1. Marji, M.: On the detection of dominant points on digital planar curves. PhD thesis, Wayne State University, Detroit, Michigan (2003)
2. Klette, R., Rosenfeld, A.: Digital Geometry - Geometric Methods for Digital Picture Analysis. Morgan Kaufmann, San Francisco (2004)
3. de Vieilleville, F., Lachaud, J.O.: Convex shapes and convergence speed of discrete tangent estimators. In: Bebis, G., Boyle, R., Parvin, B., Koracin, D., Remagnino, P., Nefian, A., Meenakshisundaram, G., Pascucci, V., Zara, J., Molineros, J., Theisel, H., Malzbender, T. (eds.) ISVC 2006. LNCS, vol. 4292, Springer, Heidelberg (2006)

4. Lachaud, J.O.: Espaces non-euclidiens et analyse d'image : modèles déformables riemanniens et discrets, topologie et géométrie discrète. Habilitation diriger des recherches, Université Bordeaux 1, Talence, France (2006)

5. Huxley, M.N.: Exponential sums and lattice points. Proc. London Math. Soc. 60, 471–502 (1990)

6. Klette, R., Žunić, J.: Multigrid convergence of calculated features in image analysis. Journal of Mathematical Imaging and Vision 13, 173–191 (2000)

7. Kovalevsky, V., Fuchs, S.: Theoretical and experimental analysis of the accuracy of perimeter estimates. In: Förster, Ruwiedel (eds.) Proc. Robust Computer Vision, pp. 218–242 (1992)

8. Sloboda, F., Zaťko, B., Stoer, J.: On approximation of planar one-dimensional continua. In: Klette, R., Rosenfeld, A., Sloboda, F. (eds.) Advances in Digital and Computational Geometry, pp. 113–160 (1998)

9. Asano, T., Kawamura, Y., Klette, R., Obokata, K.: Minimum-length polygons in approximation sausages. In: Arcelli, C., Cordella, L.P., Sanniti di Baja, G. (eds.) IWVF 2001. LNCS, vol. 2059, pp. 103–112. Springer, Heidelberg (2001)

10. Coeurjolly, D.: Algorithmique et géométrie pour la caractérisation des courbes et des surfaces. PhD thesis, Université Lyon 2 (2002)

11. Lachaud, J.O., Vialard, A., de Vieilleville, F.: Analysis and comparative evaluation of discrete tangent estimators. In: Andrès, É., Damiand, G., Lienhardt, P. (eds.) DGCI 2005. LNCS, vol. 3429, pp. 140–251. Springer, Heidelberg (2005)

12. Lachaud, J.O., Vialard, A., de Vieilleville, F.: Fast, accurate and convergent tangent estimation on digital contours. IVC 25(10), 1572–1587 (2007)

13. Hermann, S., Klette, R.: A comparative study on 2d curvature estimators. Research report CITR-TR-183, CITR, The University of Auckland, New Zealand (2006)

14. de Vieilleville, F., Lachaud, J.O., Feschet, F.: Maximal digital straight segments and convergence of discrete geometric estimators. JMIV 27(2), 471–502 (2007)

15. Brunet, F.: Convolutions binomiales et dérivation de fonctions discrètes bruitées. Master's thesis, LAIC. University of Clermont-Ferrand. France (2007)

16. Kass, M., Witkin, A., Terzopoulos, D.: Snakes: Active contour models. International Journal of Computer Vision 1(4), 321–331 (1988)

17. Caselles, V., Catte, F., Coll, T., Dibos, F.: A geometric model for active contours. Numerische Mathematik 66, 1–31 (1993)

18. Feschet, F., Tougne, L.: Optimal time computation of the tangent of a discrete curve: Application to the curvature. In: Bertrand, G., Couprie, M., Perroton, L. (eds.) DGCI 1999. LNCS, vol. 1568, pp. 31–40. Springer, Heidelberg (1999)

19. Debled-Rennesson, I., Feschet, F., Rouyer-Degli, J.: Optimal blurred segments decomposition of noisy shapes in linear times. Computers and Graphics (2006)

20. Buzer, L.: An elementary algorithm for digital line recognition in the general case. In: Andrès, É., Damiand, G., Lienhardt, P. (eds.) DGCI 2005. LNCS, vol. 3429, pp. 299–310. Springer, Heidelberg (2005)

21. Roussillon, T., Tougne, L., Sivignon, I.: Computation of binary objects sides number using discrete geometry application to automatic pebbles shape analysis. In: Proc. 14th International Conference on Image Analysis and Processing (2007)

22. Kanungo, T.: Document Degradation Models and a Methodology for Degradation Model Validation. PhD thesis, University of Washington (1996)

23. Nguyen, T., Debled-Rennesson, I.: Curvature estimation in noisy curves. In: Kropatsch, W.G., Kampel, M., Hanbury, A. (eds.) CAIP 2007. LNCS, vol. 4673, pp. 474–481. Springer, Heidelberg (2007)

An Efficient and Quasi Linear Worst-Case Time Algorithm for Digital Plane Recognition

Emilie Charrier[1,2,3] and Lilian Buzer[1,2]

[1] Université Paris-Est, LABINFO-IGM, CNRS, UMR 8049
[2] ESIEE, 2, boulevard Blaise Pascal, Cité DESCARTES, BP 99
93162 Noisy le Grand CEDEX, France
[3] DGA/D4S/MRIS
charriee@esiee.fr, buzerl@esiee.fr

Abstract. This paper introduces a method for the digital naive plane recognition problem. This method is a revision of a previous one. It is the only method which guarantees an $O(n \log D)$ time complexity in the worst-case, where $(D - 1)$ represents the size of a bounding box that encloses the points, and which is very efficient in practice. The presented approach consists in determining if a set of n points in $\mathbb{Z}^3$ corresponds to a piece of digital naive hyperplane in $\lfloor 4 \log_{9/5} D \rfloor + 10$ iterations in the worst case. Each iteration performs n dot products. The method determines whether a set of 10^6 voxels corresponds to a piece of a digital plane in ten iterations in the average which is five times less than the upper bound. In addition, the approach succeeds in reducing the digital naive plane recognition problem in $\mathbb{Z}^3$ to a feasibility problem on a two-dimensional convex function. This method is especially fitted when the set of points is dense in the bounding box, i.e. when $D = O(\sqrt{n})$.

Keywords: Digital naive plane recognition, convex optimization, feasibility problem, quasi linear time complexity, chord's algorithm.

1 Introduction

Digital naive plane recognition is a deeply studied problem in digital geometry (see a review in [1]). It consists in determining whether a set of n points in $\mathbb{Z}^d$ is a piece of a digital naive plane or not. The method presented in this paper is a revision of the algorithm proposed by Buzer in [3]. Both algorithms achieve a quasi linear worst-case time complexity for the three-dimensional recognition problem but the new one is more efficient in practice and achieves an $O(n \log D)$ time complexity. The result can be extended without difficulty to the recognition of digital planes of fixed thickness.

The approaches used in recognition are usually based on linear programming [2,9,14], convex hulls and geometrical methods [4,5,12,16], combinatorial optimization [4,5,6,7,13,16] or the evenness property [15]. The methods based on linear programming can be separated into two groups. The first group [2,14] relies on the optimal result obtained by Megiddo [9]. However, even if the approaches

D. Coeurjolly et al. (Eds.): DGCI 2008, LNCS 4992, pp. 346–357, 2008.

in this group achieve an optimal linear time complexity, the resulting algorithms are too complex to be used in practice. The second group is based on efficient linear programming techniques like the simplex algorithm but their worst-case time complexity is too high in practice. Methods that partially traverse the convex hull of the chords' space of a given set of points do not override this problem. For example, the chord's algorithm [7] processes 10^6 voxels in about ten traversals of the point set. Nevertheless, this technique exhibits an $O(n^7)$ time complexity. All of the previous methods do not combine both efficiency in practice and a low worst-case time complexity. Buzer proposes in [3] an $O(n \log^2 D)$ worst-case time complexity algorithm which recognizes a digital naive plane of 10^6 voxels in $\mathbb{Z}^3$ in about 360 iterations, where $(D - 1)$ represents the size of a bounding box that encloses the points. This paper presents a revision of this algorithm that achieves an $O(n \log D)$ worst-case time complexity and that recognizes a digital naive plane of 10^6 voxels in $\mathbb{Z}^3$ in about 10 iterations. These two last approaches reduce the digital naive plane recognition in $\mathbb{Z}^d$ into a feasibility problem on a $(d-1)$-dimensional convex function. Thus, for the three-dimensional recognition problem, we only need to manage two parameters and so we can apply planar geometrical techniques to determine whether the two-dimensional solution space is empty. To solve the feasibility problem, the presented approach uses a property of the center of gravity whereas the previous method combines Megiddo oracle and one-dimensional binary search. This main change improves the time complexity and decreases the number of iterations of the algorithm.

In Sect. 2, we introduce some useful notations and definitions. In Sect. 3, we present how the recognition problem in $\mathbb{Z}^d$ can be transformed into a feasibility problem on a $(d - 1)$-dimensional convex function. Then, we focus on the three-dimensional recognition problem. In Sect. 4, we describe the solution space of the feasibility problem. Then, we sketch the algorithm and analyze its complexity in Sect. 5. Finally, in the last section, we describe some enhancements in order to improve the efficiency of our method and we present some experimental results compared to the fastest known algorithm. Note that Sect. 3 is already existing in the previous paper but we recall it for presentation convenience. The bounds in Sect. 4 are enhanced. Finally, the algorithm design is mostly new.

2   Definitions and Notations

Definition 1. *An digital plane is defined by* $P_{N,\mu,\omega} = \{p \in \mathbb{Z}^d | \mu \leq N \cdot p < \mu + \omega\}$ *where* $N = (N_1, \ldots, N_d)$ *denotes the* normal vector *in* $\mathbb{Z}^d$ *such that* $gcd(N_1, \ldots, N_d)$ *is equal to one and where* ω *denotes the* arithmetic thickness. *When* $\omega = ||N||_\infty = \max_{1 \leq i \leq d} |N_i|$ *we obtain a* naive plane *. When* $\omega = \sum_{i=1}^d |N_i|$, *we obtain a* standard plane.

In this paper, we only consider the case where the digital hyperplanes in $\mathbb{Z}^d$ are a function from $(x_1, \ldots, x_{d-1})$ to $\mathbb{Z}$. Other cases can be deduced by symmetry. Consequently, we consider that $||N||_\infty$ is equal to $|N_d|$. Moreover, we can suppose w.l.o.g. that N_d is a positive value.

Remark 1. In this paper, the expression $S = \{p^1, \ldots, p^n\}$ denotes the set of n points in $\mathbb{Z}^d$ we study. Moreover, we always use the notation $(x_1, \ldots, x_d)$ to denote the coordinates of a point x in a d-dimensional space.

We study the *digital naive plane recognition problem*. This problem consists in determining whether S corresponds to a subset of a digital naive plane. For this, we want to find a vector $N = (N_1, \ldots, N_d)$ in $\mathbb{Z}^d$ such that $gcd(N_1, \ldots, N_d) = 1$ and such that $\mu \leq N \cdot p^i < \mu + ||N||_\infty, 1 \leq i \leq n$.

3   Feasibility Problem

In this section, we show that recognizing a digital naive plane in $\mathbb{Z}^d$ is equivalent to solving a feasibility problem on a $(d-1)$-dimensional convex function.

3.1   Introduction

Proposition 1. *If there exists $\gamma \in \mathbb{R}$ and $N' \in \mathbb{R}^d$ such that $||N'||_\infty = 1$ and such that $\gamma \leq N' \cdot p^i < \gamma + 1, 1 \leq i \leq n$, then S corresponds to a piece of a digital naive plane.*

As a result of this first proposition, we can allow us to consider only normal vectors $N = (N_1, \ldots, N_d)$ such that $||N||_\infty = 1$. From the assumptions made in Sec. 2, we consider that N_d is equal to one.

Definition 2. *For $u \in \mathbb{R}^{d-1}$, the symbol N_u denotes the vector $(u_1, \ldots, u_{d-1}, 1)$. We define the function $h_S(u)$ from $\mathbb{R}^{d-1}$ to $\mathbb{R}^+$ that computes the distance, relative to the d-th axis, between the two supporting hyperplanes of normal vector N_u that enclose all the points of S (see [1]). The value $h_S(u)$ is equal to:*

$$h_S(u) = \max_{1 \leq i \leq n} (N_u \cdot p^i) - \min_{1 \leq i \leq n} (N_u \cdot p^i) \tag{1}$$

From Prop. 1, if there exists a vector u such that $h_S(u)$ is strictly less than one then S is a piece of a digital naive plane. As a result, to solve the recognition problem, we try to find a vector $u \in \mathbb{R}^{d-1}$ such that $||u||_\infty \leq 1$ and such that $h_S(u) < 1$.

Property 1. $h_S(u)$ is a convex function.

Proof. Let N_u denote a normal vector associated with the value $u \in \mathbb{R}^{d-1}$. Consider the function $g^i(u) = N_u \cdot p^i$. This function is an affine function and it is also convex. Thus, the maximum of the functions $(g^i)_{1 \leq i \leq n}$ is convex too. The function $h_S(u)$ can be rewritten as $\max_{1 \leq i \leq n}(N_u \cdot p^i) + \max_{1 \leq i \leq n}(-N_u \cdot p^i)$. By the same logic, this expression is also a convex function. Since the sum of two convex functions is convex, we deduce that h_S is convex.

In conclusion, to solve the three-dimensional recognition problem, we study the two-dimensional convex function $h_S(u) : [-1, 1]^2 \to \mathbb{R}^+$. If there exists a point u in $[-1, 1]^2$ such that the value of $h_S(u)$ is strictly less than one, we can conclude that S is a piece of a digital plane. As a result, our recognition problem is equivalent to a feasibility problem on h_S.

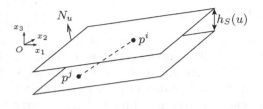

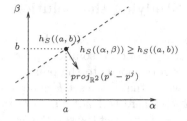

Fig. 1. Definition of h_S **Fig. 2.** A subgradient of h_S

3.2 Subgradient Computation

We first recall that a *subgradient* $g \in \mathbb{R}^d$ of a convex function $f : \mathbb{R}^d \to \mathbb{R}$ at the point x satisfies: for any $x' \in \mathbb{R}^d$, $f(x') - f(x) \geq g \cdot (x' - x)$. The subgradient indicates the steepest descent of the function f at x.

In order to solve the feasibility problem in $\mathbb{R}^{d-1}$, we have to determine how to compute a subgradient of the convex function h_S. For a given value $u \in \mathbb{R}^{d-1}$, we only have to traverse the set of points S in order to compute the value $h_S(u)$. This implies that the computation of $h_S(u)$ has a linear time complexity. Relative to the definition of h_S, we know that $h_S(u)$ is equal to $\max_{1 \leq i \leq n}(N_u \cdot p^i) - \min_{1 \leq i \leq n}(N_u \cdot p^i)$. As a result, there exist two points p^i and p^j associated with the max and the min expressions (see Fig. 1). Therefore, $h_S(u) = N_u \cdot (p^i - p^j)$. Let T denote the set of points $\{p^i, p^j\}$. As T is included in S we have: $\forall v \in \mathbb{R}^{d-1}, h_T(v) \leq h_S(v)$.

The value of $h_T(v)$ is equal to:

$$h_T(v) = |N_v \cdot p^i - N_v \cdot p^j| = |N_v \cdot (p^i - p^j)| \qquad (2)$$

and it follows:

$$\begin{aligned} h_T(v) &= |N_{v-u+u} \cdot (p^i - p^j)| \\ &= |N_u \cdot (p^i - p^j) + (v - u) \cdot proj_{\mathbb{R}^{d-1}}(p^i - p^j)| \\ &= |h_T(u) + (v - u) \cdot proj_{\mathbb{R}^{d-1}}(p^i - p^j)| \end{aligned}$$

By definition of the absolute value, we obtain:

$$\forall v \in \mathbb{R}^{d-1}, h_S(v) \geq h_T(v) \geq h_T(u) + (v - u) \cdot proj_{\mathbb{R}^{d-1}}(p^i - p^j) \qquad (3)$$

We recall that $h_S(u) = h_T(u)$ and so it follows:

$$\forall v \in \mathbb{R}^{d-1}, h_S(v) \geq h_S(u) + (v - u) \cdot proj_{\mathbb{R}^{d-1}}(p^i - p^j) \qquad (4)$$

We can now conclude that the expression $proj_{\mathbb{R}^{d-1}}(p^i - p^j)$ is a subgradient of h_S at the point u (see Fig. 2). This value corresponds to the projection of the vector $p^i p^j$ in the $\mathbb{R}^{d-1}$ space. This means that the first $d-1$ components of the vector $p^i p^j$ are sufficient to locally determine the variation of the function h_S. When we evaluate $h_S(u)$, we indirectly deduce the two points p^i and p^j. Thus, in constant time, we determine one of its subgradients.

In the next section, we study the solution space of the feasibility problem.

4   Studying the Solution Space

From now on, we focus on the three-dimensional recognition problem.

Let F denote the solution space of the feasibility problem. The set F is convex and it corresponds to the points u in $[-1,1]^2$ that satisfy $h_S(u) < 1$. We show afterwards that if the set F is not empty, then it contains a square of side length $1/(2D^3)$ where D denotes the size of a bounding box including S. This property allows us to restrict the search space to a regular grid of step $1/(2D^3)$ on $[-1,1]^2$.

We suppose that the set S lies in a bounding box of size $D - 1$ and that the origin is located at the center of the bounding box. Thus, any point p^i of S satisfies $||p^i||_\infty \leq D/2$ for $1 \leq i \leq n$ and any vector k whose endpoints correspond to two points of S satisfies:

$$||k||_\infty \leq D - 1 \tag{5}$$

Considering the convex hull of the set S, we know that there exist two supporting planes of normal vector $N = (N_1, N_2, N_3)$ such that $h_S((N_1/N_3, N_2/N_3))$ is minimal for S and such that these two planes are supported by four vertices of the convex hull (see [4]). From [12], N corresponds to the cross product of two vectors supported by the given vertices. Therefore, there exist two vectors k and k' in S such that N is equal to $k \wedge k'$. From (5), we have $||N||_\infty < 2D^2$ and in particular:

$$N_3 < 2D^2 \tag{6}$$

From Def. 1, if S corresponds to a digital naive plane then for all (x_1, x_2, x_3) in S, N satisfies $\mu \leq N_1 x_1 + N_2 x_2 + N_3 x_3 \leq \mu + N_3 - 1$. Let (N_1', N_2') denote the vector $(N_1/N_3, N_2/N_3)$, (N_1', N_2') belongs to the solution space and we have:

$$\forall (x_1, x_2, x_3) \in S, \ \mu' \leq N_1' x_1 + N_2' x_2 + x_3 \leq \mu' + 1 - 1/N_3 \tag{7}$$

We want to find an upper bound for the positive real value Δ such that the vector $(N_{\Delta 1}, N_{\Delta 2}) = (N_1' \pm \Delta, N_2' \pm \Delta)$ belongs to the solution space too. From (5) and (7), the following inequalities hold:

$$\forall (x_1, x_2, x_3) \in S, \ \mu' - D\Delta \leq N_{\Delta 1} x_1 + N_{\Delta 2} x_2 + x_3 \leq \mu' + 1 - 1/N_3 + D\Delta \tag{8}$$

This is equivalent to:

$$\forall (x_1, x_2, x_3) \in S, \ \mu' - D\Delta \leq N_{\Delta 1} x_1 + N_{\Delta 2} x_2 + x_3 \leq \mu' - D\Delta + 1 + 2D\Delta - 1/N_3 \tag{9}$$

From Prop. 1 and (9), we deduce that the expression $2D\Delta - 1/N_3$ is strictly negative and so:

$$\Delta < 1/(2DN_3) \tag{10}$$

It follows from (6) and (10) that:

$$\Delta \leq 1/(4D^3) \tag{11}$$

As a result, we can restrict the search space to a regular grid G of step $2\Delta = 1/(2D^3)$ on $[-1,1]^2$. Indeed, for any normal vector $N = (N_1, N_2, N_3)$

such that $h_S((N_1/N_3, N_2/N_3)) < 1$ and such that $N_3 < 2D^2$, there exists a square centered on $(N_1/N_3, N_2/N_3)$ in the search space corresponding to the solutions u such that $h_S(u) < 1$. If F is not empty there exists such a vector $N = (N_1, N_2, N_3)$ and its corresponding square contains a grid point. If none of the sampled values corresponds to a solution of the feasibility problem then the solution space is empty. This step size is improved relative to the one proposed in [3].

5   Algorithm Design

To solve convex optimization problems, we compute the minimum value of a convex function f. For this, we could use gradient descent methods. Using this technique, we approach the minimum of the function by moving iteratively in the direction of the steepest descent. At the i-th iteration, the gradient $\nabla(x^i)$ of f at the point x^i is computed. By definition of the gradient, there exists a value τ_i such that $f(x^i)$ is less than $f(x^i + \tau_i \nabla(x^i))$. Thus, the point x^{i+1} is equal to:

$$x^{i+1} = x^i + \tau_i \nabla(x^i) \tag{12}$$

However, we cannot easily apply such algorithms. In fact, if we want to use exact numerical computation, the value of τ_i and consequently the value of x^i must be represented by rational numbers. From (12), the size of the numerator and denominator of x^i would increase at each iteration and so this would significantly slow down the calculation. Our method evades this problem.

5.1   Sketching the Method

We briefly introduce our method. Let us call a *valid cut* on the search space a cut by an half-plane which preserves all the feasible solutions. Our algorithm consists in recursively applying valid cuts on the search space in order to reduce it. We ensure that each cut eliminates at least a constant fraction of the current search space. Thus, our algorithm solves the problem in a logarithmic number of iterations. After each cut, we transform the remaining search space into the equivalent convex polygon whose vertices are supported by the grid. When the search space is reduced to one point, our problem is solved. In the following, we give details on the main steps of our algorithm.

Valid Cuts. At each iteration, we compute the value of the function h_S at a given point u. If $h_S(u)$ is strictly less than one, then u is a feasible solution and the problem is solved. Otherwise, we cut the search space. The cut passes through u and it is orthogonal to the subgradient at this point. By definition of the subgradient, it follows that any grid point u' eliminated by the cut satisfies $h_S(u') \geq 1$. As this point does not correspond to a feasible solution, we can conclude that each cut is a valid cut.

Computing the Current Point. At each iteration, we compute the value of the function h_S at the *center of gravity* C of the search space R. We recall that the center of gravity of a polygon is equivalent to its *centroid* (also called *barycenter*). We compute its coordinates in linear time relative to the number of vertices of the polygon.

The following proposition ensures the efficiency of our algorithm (see [10]).

Proposition 2. *Let K denote a convex body in the plane, let C denote its center of gravity, then each half-plane supported by C contains between $\frac{4}{9}$ and $\frac{5}{9}$ of the area of K.*

This means that each cut passing through the center of gravity of the search space eliminates at least $\frac{4}{9}$ of its area. Thus, at most $\frac{5}{9}$ of the area of the search space is kept. From [10], we know that this ratio is optimal in the two-dimensional case. We can notice that it is very close to the ratio $\frac{1}{2}$ of the binary search.

Search Space after a Cut. We want to avoid that the numerator and the denominator of the rational numbers we use increase at each iteration. Thus, we describe the search space as a convex polygon whose vertices are supported by the grid. Let R_i and R'_i denote the search space respectively before and after the cut at the i-th iteration. Let $\overline{R_i}$ denote the largest polygon included in R_i whose vertices are included in $R_i \cap G$. By definition of $\overline{R_i}$, no grid point can lie in $R_i \backslash \overline{R_i}$. As we are looking for a solution on the grid, we set R_{i+1} at $\overline{R_i}$. For this, we use Harvey's algorithm (see [8]).

Complexity Analysis. At each iteration, as long as no solution is found, we produce a valid cut passing through the center of gravity of the current search space. Let A_i denote the area of the search space R_i. From Prop. 2 the area of R'_i is less or equal to $5/9 A_i$. We set R_{i+1} at the largest convex polygon included in R'_i whose vertices are supported by the grid. Clearly, the area of R_{i+1} is less or equal to the area of R'_i. As a result, we have $A_{i+1} \le 5/9 A_i$ and consequently we have: $A_i \le (5/9)^i A_0$.

Figure 3 shows an example of a cut on a search space R_i passing through C. We notice that the area of the search space R_{i+1} is less or equal to the $5/9$ of the area of the search space R_i.

When the area of the search space is equal to zero, it means that it is reduced to a single point or to a straight line segment. If the search space is reduced to one point, the problem is solved in linear time. Otherwise, the center of gravity is replaced by the middle of the straight line segment. Thus, the ratio of each cut becomes $1/2$ and the problem is solved in $\log_2 D$ iterations. As a result, we can conclude that our algorithm makes $O(\log A_0)$ iterations in the worst case, where A_0 denotes the area of the initial search space.

Proposition 3. *The algorithm admits an upper bounds for the number of iterations relative to A_0:*

$$\lfloor \log_{\frac{9}{5}} A_0 + \log_{\frac{9}{5}} 2 \rfloor + 3$$

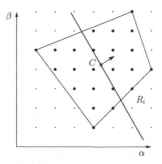
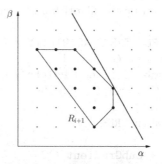

(a) The search space R_i of area 19.

(b) The reduced search space R_{i+1} of area 7.5 ($7.5 \leq \frac{5}{9} * 19$).

Fig. 3. Cut of R_i passing through the center of gravity C

Proof. While the search space R_i contains at least three grid points, we apply a cut and then the area of R_{i+1} is less or equal to $5/9 A_i$. Let I_i denote the number of interior grid points of R_i. Let B_i denote the number of boundary grid points of R_i. Thanks to Pick's theorem (see [11]) and as R_i corresponds to a polygon whose vertices are grid points, we can claim that $A_i = I_i + B_i/2 - 1$. As a result, the following inequality always holds: $I_i + B_i \leq 2A_i + 2$. Then, to determine the maximum number of iterations k used to reduce the search space to two points, we just have to solve: $2A_k + 2 < 3$ which is equivalent to $2(5/9)^k A_0 < 3$. It follows that the maximum value for k is $\log_{9/5} A_0 + \log_{9/5} 2$. As k must be an integer value, we fix it at: $\lfloor \log_{9/5} A_0 + \log_{9/5} 2 \rfloor + 1$. Finally, when only two points remain, we have to evaluate them in two iterations. Note that when the search space is reduced to one straight line segment each cut has a ratio of $1/2$ which do not increase the upper bound.

At each iteration, we generate a cut and we compute the largest convex polygon with integer vertices included in the new search space. The vertices of the convex we determine are located in an area delineated by the straight line of the cut and two straight lines supported by edges of the search space. To determine this new search space, we use Harvey's algorithm (see [8]) which runs in logarithmic time relative to the coordinates of the normal vector of these three straight lines. As the normal vector of the straight line of the cut corresponds to a subgradient, from (5) we know that its coordinates do not exceed $O(D)$. Moreover, according to the step size of the grid (see Sect. 4), the coordinates of the two other normal vectors do not exceed $O(D^3)$. As a result, this computation runs in $O(\log D)$ time. As the number of cuts does not exceed $O(\log D)$ and as each cuts adds $O(\log D)$ new vertices, the computation of the center of gravity of the search space runs in $O(\log^2 D)$ time. Moreover, from Def. 2, we know that the evaluation of the function h_S requires n dot products. In conclusion, as $O(\log^2 D)$ and $O(\log D)$ can be neglected relative to $O(n)$ (considering that $D \approx \sqrt{n}$), each cut runs in $O(n)$ time in the worst case which implies the $O(n \log D)$ time complexity of our algorithm.

We can sum up the algorithm as follows:

```
ALGORITHM FOR THE FEASIBILITY PROBLEM:
0  R_i ← [-1,1]² ∩ G    i ← 1
1  WHILE NumberOfIntegerPoints(R_i) > 1
2      C_i ← CenterOfGravity(R_i)
3      IF h_S(C_i) < 1
4          return TRUE
5      ELSE
6          sg_i ←SubGradient(C_i)
7          R'_i ←Cut(R_i, C_i, sg_i)
8          R_{i+1} ← IntegerConvexHull(R'_i)
9          i ← i + 1
10 IF NumberOfIntegerPoints(R_i) = 0
11     return FALSE
12 ELSE
13     return h_S(R_i) < 1
```

6   Improving Performance and Experimental Results

In this section, we describe some modifications already introduced in [3] that improve the performance of our method in practice. These improvements are used in the program of our algorithm.

6.1   Initial Step

Starting with a search space equal to $[-1, 1] \times [-1, 1]$ is often awkward. As we assume that the set of points is contained in a bounding box of size D, then we can claim that some points lie on the borders. Let $P_1(-D/2, y, z_1)$ and $P_2(D/2, y, z_2)$ denote two of these. They provide one constraint on the solution space. As they belong to a digital naive plane of normal $N(\alpha, \beta, 1)$, we have: $\gamma' \leq N \cdot (-D/2, y, z_1) < \gamma' + 1$ and $\gamma' \leq N \cdot (D/2, y, z_2) < \gamma' + 1$. We infer that: $|N \cdot (D, 0, z_2 - z_1)| < 1$. Thus, we can deduce the following upper and lower bound for α:

$$\frac{z_1 - z_2}{D} - \frac{1}{D} < \alpha < \frac{z_1 - z_2}{D} + \frac{1}{D} \tag{13}$$

In the same way, we can determine an upper and a lower bound for β. So, the side length of the search domain is divided by D. This stage only requires a simple traversal of the set of points in order to find the extreme ones.

6.2   Improving Valid Cuts

We previously show that the knowledge of the subgradient allows us to reject one part of the search domain relative to the point x. Nevertheless, we do not use all of the available information. In fact, when we compute the value of $h_S(x)$, we determine two points $p_i = (x_i, y_i, z_i)$ and $p_j = (x_j, y_j, z_j)$ of S such that $h_S(x) = |N_x \cdot (p_i - p_j)|$. Let T denote the set $\{p_i, p_j\}$. We have:

$$\forall x' \in \mathbb{R}^{d-1}, h_S(x') \geq h_T(x') \tag{14}$$

As a result, we can restrict the search space to the vectors $x' = (\alpha, \beta)$ which satisfy $|N_{x'} \cdot (p_i - p_j)| < 1$. By definition of the absolute value, the inequality $|N_{x'} \cdot (p_i - p_j)| < 1$ is equivalent to $N_{x'} \cdot (p_i - p_j) < 1$ and $-N_{x'} \cdot (p_i - p_j) < 1$. Thus, if the value of $h_S(x)$ is strictly less than one, the problem is solved. Otherwise, we reduce the search space by applying two valid cuts. Let $\nabla h_S(x)$ denote the subgradient of h_S at the point x, the two cuts are defined by the two following inequalities (see Fig. 4 for an example):

$$\nabla h_S(x) \cdot (\alpha, \beta) < 1 - (z_i - z_j) \tag{15}$$

$$-\nabla h_S(x) \cdot (\alpha, \beta) < 1 - (z_j - z_i) \tag{16}$$

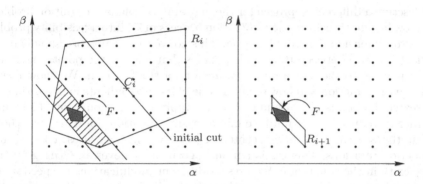

Fig. 4. The search space is reduced by a strip that contains the solution space

6.3 Experimental Results

As the fastest algorithm we know for the recognition problem is the chords' algorithm (see [7]), we compare our optimized algorithm to it. We test the two algorithms with several sets of points which correspond to a piece of a digital plane or not. As the chords' algorithm computes at each iteration n scalar products as our algorithm, we just compare their average number of iterations. Table 1 shows the experimental results. We recall that D denotes the size of a bounding box including the set of points S. To simplify the table, we call our method COBA algorithm (Convex Optimization Based Algorithm). Thanks to Prop. 3, we compute the upper bound of the number of iterations for each D whether we consider the full grid or the reduced grid introduced in Sect. 6.1. The experimental results show that our algorithm makes less iterations than the chords' algorithm in the average when the set of points is dense in the bounding box. Moreover, our algorithm makes about five times less iterations in the mean than the worst case upper bound. We have to develop our algorithm using optimal big integer operations to provide computation times in the future.

$D = \sqrt{n}$		5	10	100	300	1000
	Number of tries	100	100	100	100	20
Upper Bound	full grid $A_0 = (4D^3 + 1)^2$	25	32	55	67	79
	reduced grid $A_0 = (4D^2 + 1)^2$	19	24	40	47	55
DP	Chords' algorithm	4.80	6.03	9.85	10.75	11.85
	COBA algorithm	2.75	4.15	7.67	9.06	9.75
NDP	Chords' algorithm	3.17	2.72	2.51	2.56	2.55
	COBA algorithm	2.01	2.01	2.00	2.00	2.00

Table 1. Experimental results for digital planes (DP) and non digital planes (NDP), the experimental values correspond to the average number of iterations

7   Conclusion

We describe a different approach for the digital naive plane recognition problem. We show how to transform the recognition process in $\mathbb{Z}^3$ into a feasibility problem on a two-dimensional convex function. This convex function corresponds to the vertical distance between two parallel planes that enclose the points and whose slopes relative to the axis are the parameters of the function. We show how to evaluate this function and how to obtain one of its subgradients in linear time. As the search domain is planar and digital, we apply simple geometrical techniques in order to reduce its size until we find a feasible solution. Moreover, we choose to cut the search space at its center of gravity which ensures a logarithmic number of iterations. This version of our algorithm achieves an $O(n \log D)$ time complexity in the worst-case. We present different modifications to speed up the optimization stage. The experimental result we obtain shows that our algorithm recognizes a digital plane of 10^6 voxels in about ten iterations. These results imply that our algorithm makes less iterations in the average than the chords' algorithm when the set of points is dense. Our algorithm has been designed to be efficient in practice and especially when the set of points is dense in the bounding box. This is the only method known to the authors that achieves a $O(n \log D)$ worst-case time complexity and that is efficient in practice. We show that the recognition problem in $\mathbb{Z}^d$ can be transformed into a feasibility problem in $\mathbb{Z}^{d-1}$ for all dimensions. Nevertheless, some steps of our algorithm uses planar geometrical methods whose extension in higher dimension is not obvious. We are not able yet to extend our method in any dimension.

The authors thank the reviewers for their helpful comments.

References

1. Brimkov, V., Coeurjolly, D., Klette, R.: Digital Planarity - A review. Discrete Applied Mathematics 15(4), 468–495 (2007)
2. Buzer, L.: A linear incremental algorithm for naive and standard digital lines and planes recognition. Graphical Models 65(1-3), 61–76 (2003)

3. Buzer, L.: A composite and quasi linear time method for the digital plane recognition. In: Kuba, A., Nyúl, L.G., Palágyi, K. (eds.) DGCI 2006. LNCS, vol. 4245, pp. 331–342. Springer, Heidelberg (2006)
4. Debled-Rennesson, I.: Etude et reconnaissance des droites et plans discrets. PhD Thesis, Université Louis Pasteur, Strasbourg (1995)
5. Debled-Rennesson, I., Reveillès, J.-P.: A new approach to digital planes. In: Proc. Vision Geometry III, SPIE, vol. 2356, pp. 12–21 (1994)
6. Françon, J., Schramm, J.M., Tajine, M.: Recognizing arithmetic straight lines and planes. In: Miguet, S., Ubéda, S., Montanvert, A. (eds.) DGCI 1996. LNCS, vol. 1176, pp. 141–150. Springer, Heidelberg (1996)
7. Gerard, Y., Debled-Rennesson, I., Zimmermann, P.: An elementary digital plane recognition algorithm, 151(1-3), pp. 169–183 (2005)
8. Harvey, W.: Computing two-dimensional integer hulls. SIAM J. Compute. 28(6), 2285–2299 (1999)
9. Megiddo, N.: Linear programming in linear time when the dimension is fixed. J. ACM 31, 114–127 (1984)
10. Neumann, B.H.: On an invariant of plane regions and mass distributions. Journ. London Math. Soc. 20, 226–237 (1945)
11. Pick, G.: Geometrisches zur Zahlentheorie. Sitzenber. Lotos 19, 311–319 (1899)
12. Preparata, F.P., Shamos, M.I.: Computational Geometry: An Introduction. Springer, New York (1985)
13. Reveillès, J.-P.: Combinatorial pieces in digital lines and planes. In: Proc. Vision Geometry IV, SPIE, vol. 2573, pp. 23–34 (1995)
14. Stojmenovic, I., Tosic, R.: Digitization schemes and the recognition of digital straight lines, hyperplanes and flats in arbitrary dimensions. Vision Geometry, Contemporary Mathematics Series 119, 197–212 (1991)
15. Veelaert, P.: Digital planarity of rectangular surface segments. IEEE Trans. Pattern Analysis Machine Intelligence 16, 647–652 (1994)
16. Vittone, J., Chassery, J.-M.: Recognition of digital naive planes and polyhedrization. In: Nyström, I., Sanniti di Baja, G., Borgefors, G. (eds.) DGCI 2000. LNCS, vol. 1953, pp. 296–307. Springer, Heidelberg (2000)

Tangential Cover for Thick Digital Curves

Alexandre Faure and Fabien Feschet*

Univ. Clermont 1, LAIC, F-63172 Aubière, France,
{afaure,feschet}@laic.u-clermont1.fr

Abstract. The recognition of digital shapes is a deeply studied problem. The arithmetical framework, initiated by J.P. Reveillès in [1], provides a great theoretical basis, as well as many algorithms to deal with discrete objects. Among the many available tools, the tangential cover is a powerful one. First presented in [2], it computes the set of all maximal segments of a digital curve and allows either to obtain minimal length polygonalization, or asymptotic convergence of tangent estimations. Nevertheless, the arithmetical approach does not tolerate the introduction of irregularities, which are however inherent to the acquisition of digital shapes. In this paper, we propose a new definition for a class of so-called "thick digital curves" that applies well to a large class of discrete objects boundaries. We then propose an extension of the tangential cover to thick digital curves and provide an algorithm with a $O(n \log n)$ complexity, where n is the number of points of specific subparts of the thick digital curve.

1 Introduction

Digital straight segments are fundamental building blocks in the analysis of shapes or objects. Thus they have been thoroughly studied [3]. Usually digital straightness can be handled with efficient algorithms. Several very deep connections with the continuous case have been obtained. For instance, length [4] or tangent estimation [5] are asymptotically convergent. However, as noted in [6], digital straightness is a very rigid concept and often in practice, the recognized digital segments are short. This might be difficult then to extract significant properties of shapes without a preprocessing step which would suppress all undesirable irregularities in the shape. Those irregularities are not intrinsic, but must be suppressed because of the rigidity of the definition of digital straightness.

Several extensions of digital straightness can be found in the literature, but we here rely on the notion of blurred segment [7] (or α-thick [8] since it is the same notion) due to its geometric properties and its relations with convex hulls.

The primary goal of the present paper is to extend the notion of Tangential Cover introduced in [2,9] which has proven itself an interesting tool in the study of digital shapes. To do this, we proceed in two steps. First, we provide an algorithm for its computation using α-thick segments. The critical part of this

---

* This work was supported by the French National Agency of Research (ANR) under contract GEODIB.

D. Coeurjolly et al. (Eds.): DGCI 2008, LNCS 4992, pp. 358–369, 2008.

study is to manage two independent algorithms of Buzer [8,10], such that a low complexity is preserved, since naive connections between both algorithms lead to a quadratic algorithm, at least. Our solution has a complexity of only $O(n \log n)$. Then, in a second part, we precisely define what we call a thick digital curve. The main problem with thick curves is that the classical method must consider all points in the same fashion. Thus, interior and boundary points are treated as if they were similar. Our definition and solution allow us to avoid any confusion between interior and exterior points of thick digital curves. This leads to a solution that cannot have arbitrary cuts in the interior of the thick digital curves. Moreover, this avoids the use of any ordering of those interior points that are in fact useless for the computation. We provide an extension of our algorithm with an unchanged complexity. There, the n factor is no longer the number of points of the thick curves, but the length of the boundary of the digital curve.

This paper is organized as follows. In the first section, we recall all necessary definitions to make the paper self-content. Then, we provide a description of the algorithms of Buzer and precisely describe the necessity of a correct choice of data structures. We conclude this study by providing a good way of merging those two algorithms. Next, we provide the definition of thick digital curves and explain how to extend our algorithm to deal with such curves. We precisely describe one important problem related to the management of interior and exterior boundaries of thick digital curves. This is followed by a description of a simple to compute solution that solves the mentionned drawback. The paper ends by a conclusion and some perspectives.

2 Context and Tools

2.1 α-Thick Segments

In order to handle irregularities for discrete objects recognition and segmentation, the notion of blurred (or fuzzy) segments was introduced by I. Debled-Renesson et al [11]. It relies on the arithmetical definition of discrete lines, first proposed in [1], where a discrete line is the set S of integer points (x, y) which verify the following diophantine inequalities: $\mu \leq ax - by < \mu + \omega$, where $\frac{a}{b}$ is the line slope (usually $gcd(a, b) = 1$), μ is its lower bound and ω its thickness. Such a line is referred to as $D(a, b, \mu, \omega)$. In this scope, a line is bounded by its two real leaning lines, lower and upper, respectively determined by the equations $ax - by = \mu$ and $ax - by = \mu + \omega - 1$. Leaning points are integer points of S on those lines.

Definition 1. *[11] A set S_b of consecutive points ($|S_b| \geq 2$) of an 8-connected curve is a blurred segment with order d if there exists a discrete line $D(a, b, \mu, \omega)$, called bounding, such that all points of S_b belong to D and $\frac{\omega}{max(|a|, |b|)} \leq d$. If the abcissa interval of S_b is $[0, l - 1]$, D is named strictly bounding for S_b if D possesses at least three leaning points in $[0, l - 1]$, and S_b contains at least one lower leaning point and one upper leaning point.*

The parameter d enables the comparison between several blurred lines containing a set of points S, based on their bounding lines' thickness. The least thick of those lines (i.e. the one with the smallest order d) leads to a better-fitting approximation.

An incremental linear time algorithm of discrete curves segmentation into order d blurred segments was given in [11]. It ensured no guarantee of minimality for the built lines, as mentionned in [7]. Thus the obtained segmentation may not be optimal, in the sense that it could lead to oversegmented shapes. The study conducted in [7] suggested a way to overcome this drawback. It introduces the notion of width ν of a blurred segment. In the sequel, let us call "isothetic thickness" of a set of points the minimum value between its vertical and horizontal thicknesses. This notion is illustrated on figure 1.

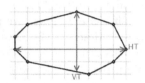

Fig. 1. The isothetic thickness of this convex hull is the minimum between its vertical (VT) and horizontal (HT) thicknesses

Definition 2. *[7] A bounding line of a digital set S_b is said optimal if its isothetic thickness is equal to the isothetic thickness of the convex hull $conv(S_b)$. S_b is a blurred segment of width ν if and only if its optimal bounding line has an isothetic distance less or equal to ν.*

This class of blurred segments with width ν has later been called by Buzer "α-thick segments". To avoid any confusion between blurred segments, fuzzy segments and fuzzy geometry, we will also refer to these segments as α-thick. Recognizing α-thick segments is the problem of computing the isothetic thickness of the convex hull of S_b. The classical method uses Melkman's algorithm [12] in order to incrementally build the successive convex hulls in linear time. The thickness parameter, α, is fixed in advance, so different values can be used for the same digital shape and so α is a user dependent parameter.

2.2 Tangential Cover for Closed Discrete Curves

The tangential cover described in this section deals with naive and standard segments, that is respectively 8- and 4-connected.

Definition 3. *A discrete curve C is an ordered set of integer points $(p_1, \ldots, p_n)$ such that the real polygonal line passing through them, in order, is a simple polygonal line.*

The curve is closed when the next point of p_n is p_1.

Definition 4. *A maximal segment T of a discrete curve C, is a subset $(p_l, \ldots, p_r)$ of C such that $\forall l \leq k \leq r$, p_k belongs to T and such that neither $T \cup \{p_{r+1}\}$ nor $T \cup \{p_{l-1}\}$ is a digital segment, for example according on the definition provided in [1].*

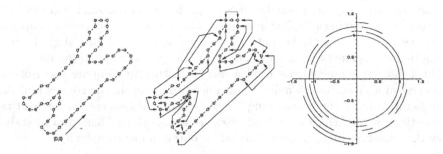

Fig. 2. Left : a chromosome digital shape. Middle : the set of its maximal segments. Right ; its tangential cover where to set apart overlapping arcs, the radius is increased or decreased accordingly.

For a given discrete curve, consecutive maximal segments overlap, and none is strictly included into another. In [2,9], this property was used to build the set $T(C)$ of all maximal segments, which is called the *tangential cover*. This construction is incremental and has a linear complexity. The incremental linear construction is obtained by using the following process: starting from any point of the 8- or 4-connected curve, points are added in $O(1)$ using the classical algorithm from [13] until this first set is no more a discrete (respectively naive or standard) segment. Then points are substracted, this time using the reverse algorithm, once again in $O(1)$ [2]. Deletions stop when a new point can be added to the right extremity of the current discrete segment, and so on. The process stops when all maximal segments have been constructed.

Each segment is represented by both its left and right ending points, p_l, p_r. Due to the properties of maximal segments, this algorithm leads to a canonical representation of closed discrete curves. In order to graphically represent this construction, $T(C)$ is mapped into the class of circular arc graphs. There, each segment $[p_l, ..., p_r]$ is associated to the arc between the angles $2l\frac{\pi}{n}$ and $2r\frac{\pi}{n}$, n being the size of the segment. Each radial line starting from the center of the graph represents one point of C, and each arc corresponds to a maximal segment containing the point.

The mapping is reversible if the slopes of the maximal segments are stored. This tool allows us to determine the polygonalization of C with the least number of vertices. This problem is known as min-DSS[3]. It is solved using the circular arcs graph rather than the original curve [14]. Solving the min-DSS problem is equivalent to finding a shortest path in the resulting circular arcs graph, using a linear algorithm described in [14].

3   Adaptation to Thin Digital Curves

3.1   Introduction

A thin digital curve is a discrete curve in the sense of definition 3. As long as we are able to compute a circular arcs graph from the maximal segments of a discrete curve, solving min-DSS remains linear in complexity. Hence it appears to be very interesting to compute the tangential cover of thin digital curves using α-thick segments. In order to adapt the tangential cover to the use of such α-thick discrete segments, we need a fast algorithm to compute the isothetic thickness of a given convex hull. We also need to maintain the convex hull that will go through the curve, searching for every maximal segment. In the next two subsections, we describe the methods used to solve both problems and then show how they must be implemented together to obtain a low complexity.

3.2   Isothetic Thickness Computation

We wish to determine whether a subset of an input polygonal chain is a maximal segment or not, in the sense of α-thick segments. The previous section shows that it is equivalent to the computation of the isothetic thickness of a convex hull. This could be performed $O(\log n)$, with n the number of vertices [8]. Let us present Buzer's approach.

It is obvious that the maximal vertical thickness in a convex hull is always located on the abscissa of a vertex. Computing the height of a vertex is done by substracting the vertex' ordinate and the ordinate of its projection onto the facing edge or vertex. In a convex hull of size n, the overall computation would cost $O(n \log n)$. Let us have a look at figure 3. Let us denote U the upper hull of the convex hull, L its lower hull, and u' (respectively l') the successive values of the slopes between vertices. Using simple mathematical and convexity properties, it is proven in [8] that the maximum height is located at the unique abscissa where u' and l' intersect. The bottom left figure represents both staircase-like graphs for u' and l'. Let MU (resp. ML) the middle point of U (resp. L). If nU (resp. nL) is the size of the upper (resp. lower) hull, then MU (resp. ML) is the point indexed by $\lfloor nU/2 \rfloor$ (resp. $\lfloor nL/2 \rfloor$). It is sufficient to know the abscissae of MU and ML, and the values of their preceding and following slopes to determine the location of the maximum with respect to MU and ML.

Here is a simple example. On figure 3, $MU.x < ML.x$ and $(MU + 1).x > ML.x$. Then we have $Slope(ML - 1, ML) > Slope(MU, MU + 1)$. In this case, due to the non-increasing (resp. non-decreasing) nature of u' (resp. l'), the two graphs cannot intersect after ML and MU. It is now possible to delete all vertices in U (resp. L) with an abscissa greater than MU's and ML's. As those were middle points, half of the convex hull's vertices have been deleted, and another iteration may be performed. All the other configurations for MU, ML and the associated slopes may be handled similarly. The algorithm stops when the intersection is reached (meaning that the maximum has been located), or when the number of remaining vertices is inferior to a desired constant value.

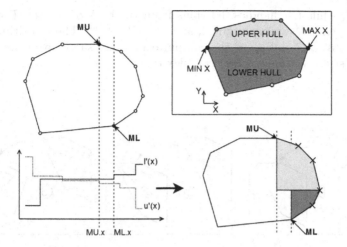

Fig. 3. An example of iteration for the logarithmic computation of the thickness of a convex hull. The upper right box represents the upper and lower hulls of a convex set.

Then, height may be computed in logarithmic time, as this operation will only be called a constant number of times. The computation of the slopes and the comparisons are performed in $O(1)$. We eliminate half of the vertices per iteration. This leads to an overall complexity of $O(\log n)$. The computation of the horizontal thickness is similar.

3.3 Convex Hull Maintenance

This adaptation of the tangential cover requires that we compute the successive convex hulls of the input curve. In other words, we need to add *and* substract points from convex hulls (i.e. managing dynamic convex hulls) in linear time. This seems difficult and known algorithms such as, Overmars and van Leeuwen [15], are at least of complexity $O(\log^2 n)$ per update. However, we do not need such a complex method, since, as shown in section 2.2, the computing of a tangential cover only requires deletions on the rear end and insertions on the front end. Thus, we are more confronted with incremental and decremental methods than to dynamic one since there is no need for insertions, deletions or updates in the middle of convex hulls.

Let us assume that the input consists in a simple polygonal chain (i.e. not self-intersecting). Recent works of Buzer [10] enable the computing of all successive convex hulls of such chains in linear time. It does so by adapting Melkman's famous algorithm [12] that computes incremental convex hulls. Buzer's algorithm performs linear time incremental *and* decremental management of convex hulls, which is exactly the tool we need. Please refer to [10] for details and complexity proofs. Let us only notice that the fast computation of additions and deletions of points in the convex hull of a simple ordered set of points C is achieved by splitting C into two consecutive subsets. C^+ is used to perform insertions in front of C, while C^- handles deletions from the back of the set. Both convex hulls of C^- and C^+ are maintained, and $C = C^- \cup C^+$. The algorithm grants a virtual access to C and keeps the computation of all successive convex hulls of a chain of size n within an $\mathcal{O}(n)$ complexity.

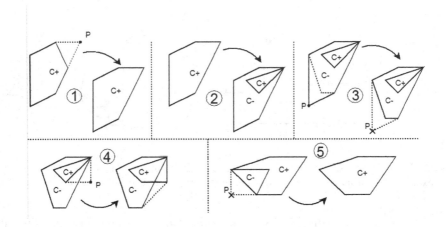

Fig. 4. Successive steps in computing an α-thick tangential cover

We now describe how to compute an α-thick tangential cover, using figure 4. C starts with any point of the input chain. At first we have $C = C^+$. Points are added to C^+ as in (1). Its isothetic thickness is computed thanks to the algorithm described in subsection 3.2. When this thickness exceeds α, the corresponding maximal segment is stored, and C^+ becomes C^-. C^+ is initialized with elements added after the last point of the convex hull, if any (2). Now $C = C^-$, as $C^+ \subset C^-$. Points are then deleted from C^- until its thickness is inferior to α (3). Then, points may be added again to C^+ (4). When C^+ is no more included in C^-, the merging $C = C^+ \cup C^-$ must be performed. It is obtained using a "smart" rotating caliper-based method that computes bridges between C^+ and C^-, without explicitly computing C (see Fig. 5). Insertions and deletions are performed in accordance with the variation of C's thickness. Obviously C^- decreases and C^+ increases. When $C^- \subset C^+$ (5), C^- is deleted, $C = C^+$ and the whole process starts again from (1) until the last maximal α-thick segment has been computed.

3.4   Putting All Together

The two methods presented above are totally independent, and were not specifically designed to work together. For a given input chain S of size n, the computation of all convex hulls costs $O(n)$, and for a given convex hull of size m, the computing of its isothetic thickness costs $O(\log m)$. We have to merge those two methods into one algorithm which computes the tangential cover of S. In

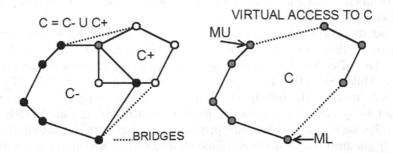

Fig. 5. C^+ and C^- are not included into each other. $C = C^+ \cup C^-$. The bridges between C^- and C^+ are computed using a rotating caliper-based method. Right, the virtual C with access to MU and ML.

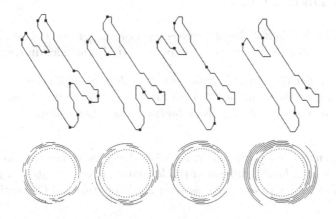

Fig. 6. The chromosome shape, with four different thickness values. $\alpha = 1, 2, 3, 4$ pixels. The dots represent a good polygonalization for each α, and on the bottom lie the associated tangential covers, represented by their circular arcs graph.

the worst case, the resulting algorithm should be in $O(n \log n)$ complexity. To achieve that, it is of the first importance that we take care of the data structures we use. The vertices of each convex hull are stored as usual in a deque, in counterclockwise order, and the first point is duplicated at the rear end of the deque. For instance, in the case (described above) where $C = C^+ \cup C^-$, C^+ and C^- not being included into each other, C is not explicitly computed but only maintained, in order to keep up with linear complexity. Meanwhile, the isothetic thickness computation algorithm requires a constant time access to the middle of C's upper and lower hulls. This is clearly the critical point for the algorithm's complexity. So, we must compute in $O(1)$ the number of points contained in the upper and lower hulls and locate the position of the bridge(s) (see Fig. 5). These values should be maintained, also in constant time, during the convex hull maintenance routine (for insertions and deletions). These operations do not generate any particular difficulties, but are crucial to keep the complexity within the $O(n \log n)$ bound.

The adaptation of the tangential cover to thin digital curves is now complete, with a total complexity of $\mathcal{O}(n \log n)$ in the worst case. It allows us to work with a class of curves not reached by algorithms mentionned in the previous section. While some other recent works use similar approaches ([16], [17]), they either only manage the incremental part [17], or use heavy algorithms which seem not as appropriate to solve our particular problem, and are slightly more costly ($\mathcal{O}(n \log n)$) [16]. Below, are presented some results of tangential covers and polygonalizations of a chromosome shape, with different thickness values, and their associated tangential covers. This particular shape is 8-connected, but our algorithm works fine with any thin digital curve.

4   Thick Digital Curves

Unformally, a thick digital closed curve is a set of integer points bounded by two distinct thin digital Jordan curves, one being strictly included in the other.

Definition 5. *Let C_{int} and C_{ext} be two thin Jordan digital curves such that $C_{int}^\circ \subset C_{ext}^\circ$ where the notation C° denotes the interior of the curve C. The set of points between C_{int} and C_{ext} (both included) is called a thick digital (closed) curve.*

Thin digital curves are particular thick digital curves where the two sets C_{int} and C_{ext} are equal. Our definition of thick digital curve is related to the notion of a simple polygon with one hole, as seen in Computational Geometry.

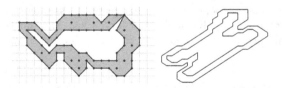

Fig. 7. Left : a thick digital closed curve. An interior Jordan thin digital curve, and an exterior Jordan thin digital curve strictly including the interior one. Right : the thickened chromosome shape.

This new definition is crucial in the sense that it avoids any questions about what lies between the two bounding thin curves. Whether it be compact blocks of points, a strongly noisy shape or totally disconnected sets of points, they still belong to the same thick digital curve. There is not even a need for any ordering of those interior points. All that is needed is an ordering of the two bounding thin curves. This can be done with a pre-processing step. Such pre-processings would frequently return two 4- or 8-connected bounding sets of points, but our algorithm is able to handle two disconnected sets of points without any problem. An example of such a curve is given on figure 7.

The tangential cover of a thick digital curve is computed in the same way as precedently, using the same methods and parameter α, the only difference being that we simultaneously try to add the next point of the interior curve and the next point of the exterior curve to the current segment. If those simultaneous insertions fail (i.e. thickness greater than α after both insertions), then we try to only insert the next point of the exterior curve. If the obtained segment's thickness is still too large, points are deleted from the back of the segment. For each bounding curve, one point at least must remain in the segment, otherwise it would not represent a true part of the thick shape. In case the segment becomes empty, the current cover stops and another one starts with the next two points. This means the input chain is disconnected according to α, and does not constitute one α-thick closed curve. Why choosing to add an exterior point in priority ? For a given shape, its exterior bounding line generally contains more points than its interior one. Let nI (resp. nO) the number of points of the interior (resp. exterior) curve. Total complexity is obviously $O((nI + nO)\log(nI + nO))$. This algorithm gives us a lightning-fast, irregularity-robust method for computing the minimal polygonalization of a thick digital curve. Nevertheless, the approach we describe could lead to some major issues, as shown on figure 8.

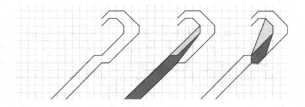

Fig. 8. A problem occuring when trying to compute maximal segments on a thick digital curve (we use the thick chromosome shape from figure 7)

In grey are drawn two consecutive maximal segments with thickness value $\alpha = 2$. Light grey represents the parts of the segments which lie outside the curve: these portions are undesirable, as the tangential cover is supposed to stick to the curve. *Segments should not lie outside the curve*, in order to give the best approximation. Moreover, this problem will carry on because the first segment englobes more points from the interior bounding curve than points from the exterior one. This induces an asymmetry that will not be fixed immediately, as shown by the next segment (right figure).

In order to solve this issue, some ideas appeared. Here is one of them: we could use a ratio to force the insertion of exterior points to a segment if the computation of previous segments led to an asymmetry. At each step, we could compute the division between the number of interior and exterior points used, and compare the result to $\frac{nI}{nO}$. If the interior border is strongly favored in a segment, then, for the next one, the algorithm will force the insertion of new

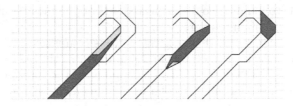

Fig. 9. Maximal segments obtained using the ratio $\frac{nI}{nO}$

points on the exterior border, until the ratio reaches $\frac{nI}{nO}$ again. Same goes if it is the exterior border that is favored. Figure 9 shows that this is an acceptable solution to this issue.

This is the same situation as figure 8. The first maximal segment uses more interior points than exterior ones. So the ratio is greater than $\frac{nI}{nO}$. The computation of the next segment is then forced to only add points from the exterior bounding curve, and the result appears on the middle: not only does the second segment stick way better to the curve, but it also cancels the asymmetry created by the first one. The recognition may then continue as usual (on the right) and leads to a better approximation of the curve.

5   Conclusion

We succeeded in extending the notion of tangential cover to thin and thick digital curves, of which we provide a definition. The latter represents a great improvement for handling irregularities and noise in digital shapes recognition. In fact, we avoid any problems generated by the lack of order within inner points of the digital shape and we do not cut arbitrarily the interior of the thick curve. By only considering the two bounding thin digital curves of a shape, we also provide a very fast algorithm. The $O((nI + nO)\log(nI + nO))$ time complexity we obtain is indeed very low regarding the size of the input shape. In further works, we plan to study the relation between triangulation of a thick digital curve and the tangential cover computation. This could lead to an even better balanced algorithm with respect to the use of both interior and exterior thin digital curves.

References

1. Réveillès, J.P.: Géométrie discrète, calcul en nombres entiers et algorithmique, Thèse d'Etat (1991)
2. Feschet, F., Tougne, L.: Optimal time computation of the tangent of a discrete curve: Application to the curvature. In: Bertrand, G., Couprie, M., Perroton, L. (eds.) DGCI 1999. LNCS, vol. 1568, pp. 31–40. Springer, Heidelberg (1999)
3. Klette, R., Rosenfeld, A.: Digital Geometry: Geometric Methods for Digital Picture Analysis. In: Computer Graphics and Geometric Modeling, Morgan Kaufman, San Francisco (2004)

4. Kovalevsky, V., Fuchs, S.: Theoretical and experimental analysis of the accuracy of perimeter estimates. In: Förster, Ruwiedel (eds.) Proc. Robust Computer Vision, pp. 218–242 (1992)
5. Lachaud, J.O., Vialard, A., de Vieilleville, F.: Fast, accurate and convergent tangent estimation on digital contours. Image and Vision Computing 25(10) (2006)
6. Bhowmick, P., Bhattacharya, B.: Fast polygonal approximation of digital curves using relaxed straightness properties. IEEE Transactions on Pattern Analysis and Machine Intelligence 29(9), 1590–1602 (2007)
7. Debled-Rennesson, I., Feschet, F., Rouyer-Degli, J.: Optimal blurred segments decomposition in linear time. In: Andrès, É., Damiand, G., Lienhardt, P. (eds.) DGCI 2005. LNCS, vol. 3429, pp. 371–382. Springer, Heidelberg (2005)
8. Buzer, L.: Digital line recognition, convex hull, thickness, a unified and logarithmic technique. In: Reulke, R., Eckardt, U., Flach, B., Knauer, U., Polthier, K. (eds.) IWCIA 2006. LNCS, vol. 4040, pp. 189–198. Springer, Heidelberg (2006)
9. Feschet, F.: Canonical representations of discrete curves. Pattern Anal. Appl. 8(1-2), 84–94 (2005)
10. Buzer, L.: Computing multiple convex hulls of a simple polygonal chain in linear time. In: 23rd European Workshop on Computational Geometry (2007)
11. Debled-Rennesson, I., Rémy, J.L., Rouyer-Degli, J.: Segmentation of discrete curves into fuzzy segments. In: 9th Intl. Workshop on Combinatorial Image Analysis. Electronic Notes in Discrete Mathematics, vol. 12, pp. 122–137. Elsevier, Amsterdam (2003)
12. Melkman, A.A.: On-line construction of the convex hull of a simple polyline. Inf. Process. Lett. 25(1), 11–12 (1987)
13. Debled-Rennesson, I., Reveillès, J.P.: A linear algorithm for segmentation of digital curves. IJPRAI 9(4), 635–662 (1995)
14. Feschet, F., Tougne, L.: On the min dss problem of closed discrete curves. Discrete Applied Mathematics 151(1-3), 138–153 (2005)
15. Overmars, M.H., van Leeuwen, J.: Maintenance of configurations in the plane. J. Comput. Syst. Sci. 23(2), 166–204 (1981)
16. Nguyen, T.P., Debled-Rennesson, I.: Curvature estimation in noisy curves. In: Kropatsch, W.G., Kampel, M., Hanbury, A. (eds.) CAIP 2007. LNCS, vol. 4673, pp. 474–481. Springer, Heidelberg (2007)
17. Roussillon, T., Tougne, L., Sivignon, I.: Computation of binary objects sides number using discrete geometry, application to automatic pebbles shape analysis. In: Cucchiara, R. (ed.) 14th International Conference on Image Analysis and Processing, pp. 763–768. IEEE Computer Society, Los Alamitos (2007)

Binomial Convolutions and Derivatives Estimation from Noisy Discretizations

Rémy Malgouyres[1], Florent Brunet[2], and Sébastien Fourey[3]

[1] Univ. Clermont 1, LAIC, IUT Dépt Informatique, BP 86, F-63172 Aubière, France
remy.malgouyres@laic.u-clermont1.fr
[2] Univ. Clermont 1, LAIC, IUT Dépt Informatique, BP 86, F-63172 Aubière, France
brunet@laic.u-clermont1.fr
[3] GREYC Image – ENSICAEN, 6 bd marchal Juin, 14050 Caen CEDEX, France
sebastien.fourey@greyc.ensicaen.fr

Abstract. We present a new method to estimate derivatives of digitized functions. Even with noisy data, this approach is convergent and can be computed by using only the arithmetic operations. Moreover, higher order derivatives can also be estimated. To deal with parametrized curves, we introduce a new notion which solves the problem of correspondence between the parametrization of a continuous curve and the pixels numbering of a discrete object.

1 Introduction

In the framework of image and signal processing, as well as shape analysis, a common problem is to estimate derivatives of functions, when only some (possibly noisy) sampling of the function is available from acquisition. This problem has been investigated through finite difference methods ([1]), scale-space ([4,8]) and discrete geometry ([6,7,5]).

We present a new approach to estimate derivatives from discretized data. As in scale-space, our method relies on simple computations of convolutions. However, our approach is oriented toward integer-only models and algorithms and is based on a discrete point of view of analysis. Unlike estimators proposed in [6], our method still works on noisy data. Moreover, we are able to estimate higher order derivatives.

Regarding the order of convergence, we have proved that our approach is as good as the one proposed in [6], that is, in $O(h^{2/3})$ for first order derivatives. Besides, this order of convergence is uniform. To the best of our knowledge, there is no uniform convergence results for estimation of derivatives from discretizations using scale-space.

The asymptotic computational complexity is worst case $O(\sqrt{-\ln(h)}/h)$ for first order derivatives which is similar to [6].

To deal with parametrized curves of $\mathbb{Z}^2$, we introduce a new notion, the *pixel length parametrization*, which solves the problem of correspondence between the pixels numbering of a discrete object and the parametrization of a continuous curve.

D. Coeurjolly et al. (Eds.): DGCI 2008, LNCS 4992, pp. 370–379, 2008.
© Springer-Verlag Berlin Heidelberg 2008

Through this paper, we also present some experimental results showing the behavior of our derivative estimator.

2   Derivatives Estimation for Real Functions

First, we consider the case of the real functions. We call *real functions* functions for which input and output sets are of dimension 1 without doing any hypothesis on the nature of these sets. We call *discrete function* a function for which the input set is such that the cardinal of all bounded subset is finite (this is the case, for example, with $\mathbb{Z}$ and $\mathbb{N}$). For simplification purposes, we consider in the sequel that a discrete function is a function from $\mathbb{Z}$ to $\mathbb{Z}$.

2.1   Principle

As said in the introduction, the principle of our derivative estimator relies on discrete convolution products. So, we are going to define what a discrete convolution product is and, then, how we are using it to construct a digital derivative estimator.

Discrete Convolution Product

Definition 1. *Let* $F : \mathbb{Z} \longrightarrow \mathbb{Z}$ *and* $K : \mathbb{Z} \longrightarrow \mathbb{Z}$ *be two discrete functions. The* discrete convolution product *of the function* F *and the kernel* K, *denoted* $F * K$, *is the discrete function defined by the following formula:*

$$F * K : \mathbb{Z} \longrightarrow \mathbb{Z}$$
$$a \longmapsto \sum_{i \in \mathbb{Z}} F(a - i)K(i) \tag{1}$$

For practical purpose, we consider that the kernel K has a finite support (i.e. K is zero out of some bounded subset I of $\mathbb{Z}$). By doing so, a computer can handle the calculation of the convolution product. Then, the computation of $F * K(a)$ can be seen as a balanced sum of values of F for abscissas close to a with coefficients given by the kernel K.

Definition 2. *Let* $F : \mathbb{Z} \longrightarrow \mathbb{Z}$ *be a discrete function and* $K : \mathbb{Z} \longrightarrow \mathbb{Z}$ *a convolution kernel. We define the operator* Ψ_K *aimed to modify the function* F *by convolution with the kernel* K:

$$\Psi_K F : \mathbb{Z} \longrightarrow \mathbb{Z}$$
$$a \longmapsto F * K(a) \tag{2}$$

We will now construct the kernel at the basis of our derivative estimator.

Finite Differences. Finite differences are a widely used method for estimating derivatives of continuous functions and solving partial differential equations.

Definition 3. *Let* $F : \mathbb{Z} \longrightarrow \mathbb{Z}$ *be a discrete function and let* $a \in \mathbb{Z}$. *The backward finite difference of* F *at the point of abscissa* a *is defined by:*

$$F(a) - F(a-1) \tag{3}$$

Considering that the derivative of F at a is the slope of the function tangent at this point, backward finite differences are a method to estimate the derivative of a discrete function.

Proposition 1. *For a discrete function* F, *the backward finite difference can be computed with* $\Psi_\delta F$ *where* δ *is the kernel defined as follow:*

$$\delta(a) = \begin{cases} 1 & \text{if } a = 0 \\ -1 & \text{if } a = 1 \\ 0 & \text{otherwise} \end{cases} \tag{4}$$

Smoothing Kernel

Definition 4. *For* $n \in \mathbb{N}$, *we define the* smoothing kernel, *denoted* H_n, *using binomial coefficients, as follow:*

$$H_n(a) = \begin{cases} \binom{n}{a+\frac{n}{2}} & \text{if } n \text{ is even and } a \in \{-\frac{n}{2}, \ldots, \frac{n}{2}\} \\ \binom{n}{a+\frac{n+1}{2}} & \text{if } n \text{ is odd and } a \in \{-\frac{n+1}{2}, \ldots, \frac{n-1}{2}\} \\ 0 & \text{otherwise} \end{cases} \tag{5}$$

H_n can be seen as the n-th line of the Pascal's triangle recentered on 0. The expression $\frac{1}{2^n}\Psi_{H_n}F(a)$ is a weighted mean of a set of values of F near the point of abscissa a giving a stronger weight for values near a and, on the contrary, a weaker weight for farther values. The division by 2^n comes from the fact that the sum of the n-th line coefficients of the Pascal's triangle is always equal to 2^n.

Derivative Kernel. Since preliminaries are set up, we can define the core kernel of our derivative estimator.

Definition 5. *For* $n \in \mathbb{N}$, *we define the* derivative kernel, *denoted* D_n, *by:*

$$D_n = \delta * H_n \tag{6}$$

So, our derivative estimator for a discrete function F is defined by:

$$\frac{1}{2^n}\Psi_{D_n}F \tag{7}$$

Composed of two parts, H_n and δ, the derivative kernel D_n has two effects. Thanks to H_n, the discrete function F is smoothed and so, the possible noise is reduced. Then, the kernel δ evaluates the derivative but on a cleansed function.

2.2 Convergence

At this point, we have defined our digital derivative estimator. Now, we want to estimate the quality of this estimator. To do so, given a continuous function and its discretization, we want to evaluate the error made by our estimator with respect to the theoretical derivative of the original continuous function.

The first thing to do is to establish a correspondence between a theoretical continuous function and its discretization. Let $\phi : \mathbb{R} \longrightarrow \mathbb{R}$ be a continuous function and let $\Gamma : \mathbb{Z} \longrightarrow \mathbb{Z}$ be a discrete function. Let h be the discretization step (i.e. the size of a pixel). Let $K > 0$ and $\alpha \in]0, 1]$ be two additional parameters. We say that Γ is a (possibly noisy) discretization of ϕ if for all $a \in \mathbb{Z}$ we have that:

$$|h\Gamma(a) - \phi(ha)| \leq Kh^\alpha \tag{8}$$

In other words, the point $(a, \Gamma(a))$ is close to the point $(a, \frac{\phi(ha)}{h})$ for all a of the domain. The deviation between these two points is bounded by the quantity Kh^α where K is a general constant and α is a parameter which aims to represent the amount of noise (considering a small discretization step, $h < 1$, the allowed noise is more important with α close to 0 than with α close to 1).

Theorem 1. *With the following hypothesis:*

- $\phi : \mathbb{R} \longrightarrow \mathbb{R}$ *is a C^3 function*
- $\phi^{(3)}$ *is bounded*
- $\alpha \in]0, 1]$, $K \in \mathbb{R}_+^*$ *and* $h \in \mathbb{R}_+^*$
- $\Gamma : \mathbb{Z} \longrightarrow \mathbb{Z}$ *is such that* $|h\Gamma(a) - \phi(ha)| \leq Kh^\alpha$
- $n = \lfloor h^{2(\alpha-3)/3} \rfloor$

there exists a function $\sigma_{\phi,\alpha} : \mathbb{R}_+ \longrightarrow \mathbb{R}_+$ *with* $\sigma_{\phi,\alpha} \in O(h^{(2/3)\alpha})$ *such that:*

$$\left| \frac{1}{2^n} \Psi_{D_n} \Gamma(a) - \phi'(ha) \right| \leq \sigma_{\phi,\alpha}(h) \tag{9}$$

In other words, we provide a derivative estimation converging at rate $h^{(2/3)\alpha}$ of functions known through their discretizations with step h and an arbitrary noise bounded by Kh^α.

We don't report here the proof of Theorem 1 for lack of space. We can just point out that our proof relies on the use of the Floater's Theorem ([2]) which gives the order of convergence of the k-th derivative of the Bernstein approximation[1] of a function and the k-th derivative of the function itself. The hypothesis made in Theorem 1 about the fact that ϕ must be C^3 with $\phi^{(3)}$ bounded come from the hypothesis required by the Floater's Theorem.

---

[1] The Bernstein approximation of a function $f : [0, 1] \longrightarrow \mathbb{R}$ is the Bézier curve with control points $\left(\frac{i}{n}, f\left(\frac{i}{n}\right) \right)_{i \in \{0,...,n\}}$.

2.3   Reducing the Complexity

Considering the maximum noise allowed by the Theorem 1 (that is, $\alpha = 0$) the computational complexity of our derivative estimator is quadratic ($O(h^{-2})$, where h is the discretization step). This complexity is not as good as the one induced by tangent estimators from discrete geometry which is in $O(h)$ for the best ([6]). However, it is possible to reduce the complexity of our estimator by considering that values at extremities of the smoothing kernel H_n are negligible.

The following theorem comes from the Hoeffding inequality ([3]).

Theorem 2. *Let $\beta \in \mathbb{N}^*$ and $n \in \mathbb{N}$. If $k = \frac{n}{2} - \sqrt{\frac{\beta n \ln(n)}{2}}$ then:*

$$\frac{1}{2^n} \sum_{j=0}^{k} \binom{n}{j} \leq \frac{1}{n^\beta} \qquad (10)$$

Theorem 2 means that the sum of the k first and the k last coefficients of the smoothing kernel are negligible with respect to the whole kernel. The parameter β enables to define what negligible mean. Typically, we chose $\beta = 2$. Using the result of Theorem 2, it is possible to reduce the size of the derivative kernel D_n recommended by Theorem 1. As a consequence, the computational complexity of our method can be improved without spoiling the quality of the derivative estimation. For example, with $\beta = 2$, the computational complexity is reduced to $O\left(\frac{\sqrt{-\ln(h)}}{h}\right)$ (with the hypothesis that the noise is maximal).

2.4   Higher Order Derivatives

One of the most interesting point about our method is that higher order derivatives can also be estimated.

Definition 6. *Let $n \in \mathbb{N}$ be the size of the kernel and $k \in \mathbb{N}^*$ the derivation order. We define the* kernel for higher order derivatives, *denoted D_n^k, as follow:*

$$D_n^k = \underbrace{\delta * \ldots * \delta}_{k \; times} * D_n \qquad (11)$$

As for first order derivatives, we have proved the convergence of the estimator for higher order derivatives.

Theorem 3. *With the following hypothesis:*

- $k \in \mathbb{N}^*$
- $\phi : \mathbb{R} \longrightarrow \mathbb{R}$ *is a C^{k+2} function*
- $\phi^{(k+2)}$ *is bounded*
- $\alpha \in]0, 1]$, $K \in \mathbb{R}_+^*$ *and $h \in \mathbb{R}_+^*$*
- $(\alpha_r)_{r \in \mathbb{N}}$ *a sequence such that $\alpha_r = (2/3)^r \alpha$*
- $\Gamma : \mathbb{Z} \longrightarrow \mathbb{Z}$ *is such that $|h\Gamma(a) - \phi(ha)| \leq Kh^\alpha$*

$$-n = \sum_{l=0}^{k-1} \left\lfloor h^{2(\alpha_l - 3)/3} \right\rfloor$$

there exists a function $\sigma_{\phi,\alpha,k} : \mathbb{R}_+ \longrightarrow \mathbb{R}_+$ with $\sigma_{\phi,\alpha,k} \in O(h^{(2/3)^k \alpha})$ such that:

$$\left| \frac{h^{-k+1}}{2^n} \Psi_{D_n^k} \Gamma(a) - \phi^{(k)}(ha) \right| \leq \sigma_{\phi,\alpha,k}(h) \qquad (12)$$

The proof of Theorem 3 is obtained by iterating the result of the Theorem 1. As a remark, we can point out that the Theorem 3 is identical to Theorem 1 if we consider the case $k = 1$.

2.5   Experimental Results

As an example, Fig. 1 shows the derivative estimation (in gray) of a noisy sine (in black). On this first example, we have used a kernel size $n = 25$.

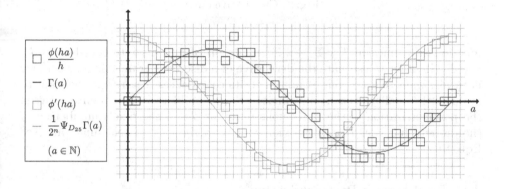

Fig. 1. Example of derivative estimation of a noisy sine wave

In Fig. 2, we start from a simple continuous sine wave. This original function is then discretized. After that, we compute the derivative from the discretized data using our estimator. Finally, the result is compared with the theoretical continuous derivative we should get. To do so, we use the cumulative error, that is, the sum of absolute differences between the theoretical values we should get and the estimated ones. These measures are repeated for several discretization steps (h) and derivative kernel widths (n). The grayed part in Fig. 2 is the place where the cumulative error happens to be minimum. This area corresponds to the optimal derivative kernel width for a given discretization step. We can see that this zone is compatible with the result of Theorem 1.

3   Derivatives Estimation for Parametrized Curves

A parametrized function is a function $\phi : \mathbb{R} \longrightarrow \mathbb{R}^2$ made up of two coordinates $\phi_x : \mathbb{R} \longrightarrow \mathbb{R}$ and $\phi_y : \mathbb{R} \longrightarrow \mathbb{R}$. In the same manner, a discrete parametrized

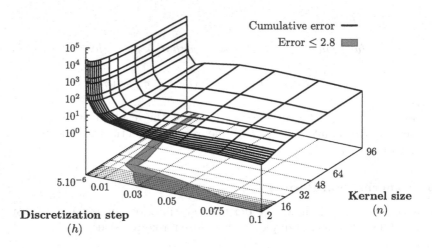

Fig. 2. Cumulative error

function is a function $\Gamma : \mathbb{Z} \longrightarrow \mathbb{Z}^2$ made up of two discrete coordinates Γ_x and Γ_y. In order to estimate the derivative of a discrete parametric function, we naturally want to apply our estimator to its coordinates. After that, we want to evaluate the quality of such an estimator. So, we build a correspondence between the parametrization of the continuous function ϕ and the pixels numbering of the discrete function Γ in a manner similar to the case of real functions (see Equ. 8).

3.1   Pixel Length Parametrization

As we have seen for the case of real functions, we need to establish a correspondence between a theoretical continuous function and its discretization in order to evaluate the quality of the derivative estimation. So, given a continuous parametric function ϕ and its discretization Γ, we would like an expression of the form:

$$\|h\Gamma(a) - \phi(ha)\| \leq \varepsilon \tag{13}$$

The problem is that Equation 13 is not necessarily satisfied if we consider ε small enough to be meaningful. This is due to the non-uniform parametrization of ϕ and the non-isotropic character of $\mathbb{Z}^2$. Therefore, we build a reparametrization $\overline{\phi}$ of the function ϕ such as:

$$\|h\Gamma(a) - \overline{\phi}(ha)\| \leq \varepsilon \tag{14}$$

Remark that the quality estimation of the derivative estimator will consider the reparametrization of the initial curve instead of the initial curve itself. As a consequence, the norm of the tangent estimation will not be meaningfull but its orientation will.

The problem of correspondence between parametrizations is solved by introducing a new notion, the pixel length parametrization. This notion is similar to the parametrization with respect to arc length but relies on the 1-norm (the taxicab norm) despite of the euclidean norm.

Definition 7. *Let* $\phi : I \subset \mathbb{R} \longrightarrow \mathbb{R}^2$ *be a parametric function. We define the length function, denoted L, by:*

$$L : I \longrightarrow \mathbb{R}_+$$
$$u \longmapsto \int_{\min(I)}^{u} \|\phi'(t)\|_1 \, dt \qquad (15)$$

The function L gives the arc length according to the 1-norm of the curve ϕ between the origin of the parametrization and the parameter u.

Definition 8. *Let $\phi : I \subset \mathbb{R} \longrightarrow \mathbb{R}^2$ a continuous parametric function. Similarly to arc length, we define the pixel length reparametrization of the function ϕ, denoted $\overline{\phi}$, by:*

$$\overline{\phi} : L(I) \longrightarrow \mathbb{R}^2$$
$$x \longmapsto \phi\big(L^{-1}(x)\big) \qquad (16)$$

We have proved that there is a correspondence between the continuous parametrization of $\overline{\phi}$ and the pixels numbering of an 8-connected discretization Γ of ϕ such that Equ. 14 is true.

Since we are able to connect a parametric continuous function and its discretization, we want to evaluate the quality of our derivative estimator. Unfortunately, there is still a problem. Indeed, the pixel length reparametrization is only piecewise C^3 and C^1 where the curve admits a vertical or horizontal tangent. This comes from the use of the 1-norm which involves absolute values. As a consequence, the hypothesis of Theorem 1 are not satisfied and it cannot be directly applied to prove the convergence of our derivative estimator for parametrized curves. At the moment, we have an another theorem for the parametric case proving the convergence of our derivative estimator which gives a convergence rate of $O(h^{2/3})$ almost everywhere on the curve and $O(h^{1/3})$ in the neighborhood of vertical and horizontal tangents. Though, from experimental results, it seems that the convergence rate is uniform $O(h^{2/3})$.

3.2 Experimental Results

Fig. 3 is an example of derivative estimation for parametrized curves. The tangents for all points of the discrete curve (black squares) are represented with black segments. Note that in this figure, the gray grid is a representation of $\mathbb{Z}^2$. From this illustration, it seems that derivative estimation is not worse for horizontal and vertical tangents than elsewhere.

Fig. 4 is a more precise experimentation intended to illustrate the behavior of our derivative estimator for vertical and oblique tangents. In this test, we start from a parametric continuous circle (centered on 0 and with radius 1). Then we

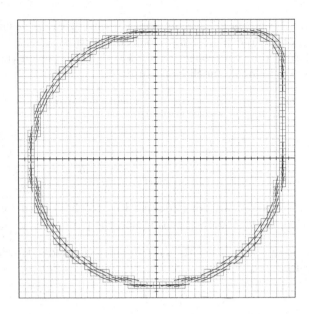

Fig. 3. Example of tangent estimation for a parametrized curve

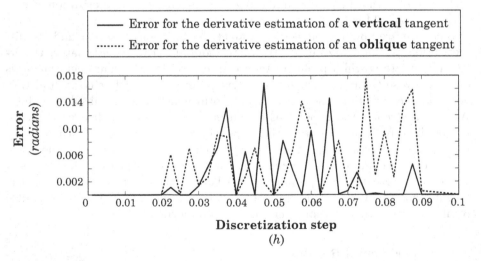

Fig. 4. Comparison of derivative estimator in the parametric case for vertical and oblique tangents

discretize this base function for several discretization steps using the pixel-length reparametrization. After that, we compute the tangent with our derivative estimator for two locations: the first one is a vertical tangent and the second one is an oblique tangent. The results are then compared with the theoretical tangents we should get (computed analytically from the continuous original function).

The error is defined as the angle between the estimated tangent and the theoretical one (in radians). From Fig. 4, we can see that our estimator is not worse for the vertical tangent than for the oblique one. Besides, we can also notice than the estimator gives better results for small discretization steps.

4   Conclusion

In this paper, we have defined a new approach based on discrete convolution products to estimate derivatives of digitized functions. We showed that under some conditions this new method has a good order of convergence even on noisy data. Given a discretization step and a noise strength, it is possible to automatically select the size of our derivative kernel to achieve the best convergence rate. As a result, the user won't have to set too many parameters.

An another benefit of this approach is that computations involved in the derivative estimation can be done using integer only arithmetic. Moreover, higher order derivatives can be estimated.

By reducing the size of the derivative kernel in such a way that the lost of quality is negligible, our computational complexity is comparable to [6].

Even if there is some technical problems for proving the exact convergence rate, experimental results showed that our derivative estimator still works for parametrized curves. Moreover, we have set up the new notion of pixel length parametrization which solves the problem of correspondence between the pixels numbering of a discrete curve and the parametrization of a continuous function.

References

1. Chen, S.H.: Finite Difference Method. High-Field Physics and Ultrafast Technology Laboratory, Taipei, Taïwan (2006)
2. Floater, M.S.: On the convergence of derivatives of Bernstein approximation. J. Approx. Theory 134(1), 130–135 (2005)
3. Hoeffding, W.: Probability inequalities for sums of bounded random variables. Journal of the American Statistical Association 58(301), 13–30 (1963)
4. Kimmel, R., Sochen, N.A., Weickert, J. (eds.): Scale-Space 2005. LNCS, vol. 3459. Springer, Heidelberg (2005)
5. Lachaud, J.-O., de Vieilleville, F.: Convex shapes and convergence speed of discrete tangent estimators. In: ISVC (2), pp. 688–697 (2006)
6. Lachaud, J.-O., Vialard, A., de Vieilleville, F.: Analysis and comparative evaluation of discrete tangent estimators. In: Andrès, É., Damiand, G., Lienhardt, P. (eds.) DGCI 2005. LNCS, vol. 3429, pp. 240–251. Springer, Heidelberg (2005)
7. Lachaud, J.-O., De Vieilleville, F.: Fast accurate and convergent tangent estimation on digital contours. Image and Vision Computing (in press, 2006)
8. Lindeberg, T.: Scale-space for discrete signals. IEEE Trans. Pattern Anal. Mach. Intell. 12(3), 234–254 (1990)

Selection of Local Thresholds for Tomogram Segmentation by Projection Distance Minimization

K. J. Batenburg and J. Sijbers

Vision Lab, Department of Physics
University of Antwerp, Belgium
joost.batenburg@ua.ac.be, jan.sijbers@ua.ac.be

Abstract. Segmentation is an important step to obtain quantitative information from tomographic data sets. To this end, global thresholding is often used in practice. However, it is usually not possible to obtain an accurate segmentation based on a single, global threshold. Instead, local thresholding schemes can be applied that use a varying threshold, depending on local characteristics of the tomogram. Selecting the best local thresholds is not a straightforward task, as local image features often do not provide sufficient information for choosing a proper threshold. Recently, the concept of *projection distance* was proposed as a new criterion for evaluating the quality of a tomogram segmentation. In this paper, we describe how Projection Distance Minimization (PDM) can be used to select local thresholds, based on the available projection data from which the tomogram was initially computed. By reprojecting the segmented image, a comparison can be made with the measured projection data. This yields a quantitative measure of the quality of the segmentation. By minimizing the difference between the computed and measured projections, optimal local thresholds can be computed.

Simulation experiments have been performed, comparing our local thresholding approach with an alternative local thresholding method and with optimal global thresholding. Our results demonstrate that the local thresholding approach yields segmentations that are significantly more accurate, in particular when the tomogram contains artifacts.

1 Introduction

Tomographic reconstructions, which are generally gray-scale images, are often segmented as to extract quantitative information, such as the shape or volume of image objects. Such segmentation is often performed by local or global thresholding [1]. However, the process of threshold selection is somewhat arbitrary and requires human interaction. Many algorithms have been proposed for selecting "optimal" thresholds with respect to various optimality measures [2]. Global thresholds are typically selected from the histogram of the image, such that the distance between the image and the segmented image is minimized. Rosenfeld and Torre employ a method that is based on analyzing the concavity points on

D. Coeurjolly et al. (Eds.): DGCI 2008, LNCS 4992, pp. 380–391, 2008.
© Springer-Verlag Berlin Heidelberg 2008

the convex hull of the image histogram. The deepest concavity points of the convex hull are potential candidates for the threshold [3]. Ridler and Calvard proposed an iterative, global thresholding method that models the gray-level distribution in an image as a mixture of two Gaussian distributions representing the background and foreground regions, respectively. Using the average of the foreground and background class means, a new threshold is established, which is iterated until convergence [4]. The clustering thresholding method of Otsu minimizes the weighted sum of intraclass variances of the foreground and background pixels to establish an optimum threshold [5]. Kapur et al. based their thresholding method on an entropy criterium. Thereby, the foreground and background classes are treated as two different sources [6]. The image is considered to be optimally thresholded when the sum of the two class entropies reaches a maximum.

Recently, Batenburg and Sijbers proposed a new approach for global threshold selection that makes use of the available tomographic projection data [7,8]. By reprojecting the segmented volume, the norm of the difference between the projections of the current segmentation and the measured projection data, called the *projection distance*, can be computed. This yields a quantitative measure of the quality of the segmentation. By minimizing the difference between the computed and measured projections (*Projection Distance Minimization*, or *PDM*), an optimal threshold can be computed. It was demonstrated in [8], that PDM leads to a significant improvement in segmentation accuracy, compared to histogram-based methods.

However, the capabilities of global threshold selection methods are limited by the maximum accuracy that can be obtained using global thresholding. If the tomogram exhibits variations in the intensity of certain image features, global thresholding can never lead to an accurate segmentation. For example, thick structures typically tend to be brighter than very thin structures in a tomogram, even if both structures consist of the same material in the original object. To account for local image variations, local thresholding methods were proposed. Abutaleb developed a local thresholding method based on the joint (2D) entropy of a pixel neighborhood [9,10]. White and Rohrer developed a nonlinear, local thresholding method in which the gray value of the pixel is compared with the average of the gray values in a small neighborhood [11]. If the pixel is significantly darker than the average, it is assigned as foreground; otherwise it is classified as background. Eikvil et al. developed a thresholding method in which a large window, with a small window positioned at its center, is moved across the image, and each pixel inside the small window is labeled on the basis of the clustering of the pixels inside the large window [12]. Blayvas et al. proposed an adaptive binarization method where the threshold is determined by interpolation of the image gray levels at points where the image gradient is high [13]. The local thresholding method of Niblack adapts the local threshold according to the local mean and standard deviation over a sliding window [14].

Most of these adaptive thresholding methods that use a varying threshold for different regions of the image, lead to better results than global thresholding in some cases. However, selection of the local thresholds becomes increasingly difficult as the size of the regions is made smaller, as the local histogram is based on only a small number of pixels. Moreover, no objective criterion for the segmentation quality is available if only the information from the reconstructed image is used for segmentation.

In this paper, we propose an extension of the projection-based threshold selection method from [8], that uses a locally varying threshold *field*, instead of a single global threshold. The same optimization criterion, PDM, is now used to find an "optimal" threshold field. The threshold field is represented on a square grid that is coarser than the pixel grid of the tomogram. The thresholds for pixels that do not coincide with grid points in the coarse grid are computed by bilinear interpolation. Computing the threshold field for which the projection distance is minimal appears to be computationally hard. We describe how a minimum of the projection distance can be computed efficiently for the case that the threshold is only allowed to vary for a single grid point in the coarse grid, while keeping the threshold values fixed for the remaining grid points. By iterating this procedure several times for all coarse grid points, a local minimum of the projection distance is reached.

Simulation experiments have been performed, comparing the result of local thresholding based on PDM with the local thresholding method of Niblack [14] and with global thresholding based on PDM [8]. Our results demonstrate that the local thresholding approach yields segmentations that are significantly more accurate, in particular when the tomogram contains artifacts due to truncation of the projection data.

2   Method

For simplicity reasons, we restrict ourselves to two-dimensional tomograms. All concepts can be generalized to a 3D setting in a straightforward manner. Similar to algebraic reconstruction methods (i.e., ART, SART, see [15]) the tomogram is represented on a rectangular grid of width w and height h. Put $n = wh$. The grey value image $v \in \mathbb{R}^n$ that we want to segment is a tomographic reconstruction of some physical object, of which projections were acquired using a tomographic scanner. Projections are measured as sets of detector values for various angles, rotating around the object. Let m denote the total number of measured detector values (for all angles) and let $p \in \mathbb{R}^m$ denote the measured data. The physical projection process in tomography can be modeled as a linear operator W that maps the image v (representing the object) to the vector p of measured data:

$$Wv = p. \tag{1}$$

For parallel projection data, the operator W is a discretized version of the well-known *Radon transform*. We represent W by an $m \times n$ matrix. From this point on, we assume that an image v has been computed that approximately satisfies Eq. (1). This image now has to be segmented using a locally varying threshold.

In this paper, we focus on the segmentation of objects that consist of a single material, so that there are only two segmentation classes, for the object and the background. We assume that the material is homogenous, i.e., a perfect reconstruction of the original object should contain only two grey levels. However, most common tomographic reconstruction algorithms yield an image that consists of a range of grey levels, instead of a binary image, even if the object in the scanner is perfectly homogeneous. This becomes particularly noticeable if a reconstruction is computed from relatively few projections or if certain parts of the projection data are missing (e.g., truncated projections, where the object is larger than the field of view of the scanner). In such cases, the reconstruction problem is severely underdetermined, and many grey level images can have the same projections. Typically, continuous reconstruction algorithms do not use the prior knowledge about the discrete grey levels, but rather compute an image that contains many grey levels.

Even if prior knowledge about the two grey levels is not used in the reconstruction algorithm, this knowledge can still be exploited by the segmentation algorithm used after reconstruction. Our segmentation approach assigns a single real-valued grey value to both segmentation classes. The projections of the segmented image are then computed and compared to the measured projection data. The difference between the computed and measured projections provides a measure for the quality of the segmentation.

Although we assume that the original object consists of a single material, we do not assume prior knowledge of the actual grey levels of the background and the interior. These grey levels are treated as variables in the segmentation problem. We denote the grey level for the background and the interior of the object by ρ_1 and ρ_2, respectively. Put $\rho = (\rho_1 \ \rho_2)^T$.

We first define a segmentation problem where the local threshold can vary independently for each of the image pixels. The set of local thresholds for all pixels is represented by a vector $\tau \in \mathbb{R}^n$. We will refer to this vector as the *threshold field*.

For any $\rho \in \mathbb{R}^2$, $t \in \mathbb{R}$, define the *threshold function* $r_{\rho,t} : \mathbb{R} \to \{\rho_1, \rho_2\}$ by

$$r_{\rho,t}(v) = \left\{ \begin{array}{cc} \rho_1 & (v \le t) \\ \rho_2 & (v > t) \end{array} \right. . \tag{2}$$

We also define the threshold function $r_{\rho,\tau}$ of an entire image $v \in \mathbb{R}^n$, which yields a vector containing the thresholded pixel values:

$$r_{\rho,\tau}(v) = (r_{\rho,\tau_1}(v_1) \ \cdots \ r_{\rho,\tau_n}(v_n))^T. \tag{3}$$

For grey levels $\rho \in \mathbb{R}^2$ and a threshold field $\tau \in \mathbb{R}^n$, define the *projection difference* $d(\rho, \tau)$ by

$$d(\rho, \tau) = ||W r_{\rho,\tau}(v) - p||_2. \tag{4}$$

The projection difference is used as the optimization criterion for finding the optimal threshold parameters. From this point, we will refer to this concept as Projection Difference Minimization (PDM).

Problem 1. Let $W \in \mathbb{R}^{m \times n}$ be a given projection matrix, let $v \in \mathbb{R}^n$ be a grey level image and let $p \in \mathbb{R}^m$ be a vector of measured projection data. Find $\tau \in \mathbb{R}^n$ and $\rho \in \mathbb{R}^2$, such that $d(\rho, \tau)$ is minimal.

In Problem 1, the threshold for each pixel is allowed to vary independently. This means that the resulting segmentation class for each pixel i (either background or interior) is independent of the grey value v_i, as the threshold τ_i can always be chosen either smaller or larger than v_i. In fact, this threshold selection problem is equivalent to a reconstruction problem from *Discrete Tomography*, where the main objective is to reconstruct a binary image from its projections [16]. Although solving this discrete tomography problem can lead to very accurate segmentation results, even if few projection are used, the problem is computationally very hard (see, e.g., [17]). In cases where it is relatively easy to acquire more projection images, continuous tomography followed by thresholding (either local or global) is often preferable.

At the other end of the granularity spectrum is the case where all entries of τ must have the same value, i.e., global thresholding. This approach was already proposed in [8]. For binary images, it was demonstrated that only the global threshold τ has to be optimized, as the optimal grey values ρ_1 and ρ_2 can be computed directly once the threshold τ has been set.

In this paper, we focus on a segmentation problem that can be considered as an "intermediate" problem, between discrete tomography and global thresholding based on PDM. Instead of allowing the threshold field τ to vary independently for each pixel, the value of the threshold is specified on a coarse grid, which is superimposed on the pixel grid of the image v. The threshold value for each pixel of v_i is then computed by bilinear interpolation from the set threshold values. In this way, the local thresholds will vary only gradually, while the threshold field can still vary significantly throughout the image. The choice for bilinear interpolation is mainly motivated by computational convenience. More sophisticated interpolation schemes (i.e., bicubic interpolation) may lead to better results. However, such schemes typically yield more variables in the resulting optimization problem.

Figure 1 illustrates how the coarse grid is superimposed onto the finer pixel grid of the image v. Note that only a small portion of the image is depicted. As an example, suppose that the thresholds are given for the four points indicated in the figure (with corresponding threshold values $t_1, \ldots, t_4$. Let the (x, y)-coordinates of these four points be given by $(0,0), (1,0), (1,1)$ and $(0,1)$, respectively. We refer to the four squares between $(0,0)$ and its surrounding coarse grid points as the *quadrants* surrounding $(0,0)$. For any pixel p with center (x_p, y_p) in the topright quadrant, the threshold t_p is defined by

$$t_p = (1 - x_p)(1 - y_p)t_1 + x_p(1 - y_p)t_2 + (1 - x_p)y_p t_3 + x_p y_p t_4. \tag{5}$$

Let k be the total number of grid points on the coarse interpolation grid. We refer to the vector of thresholds for these points by $\tau' \in \mathbb{R}^k$. The mapping $I : \mathbb{R}^k \to \mathbb{R}^n$ assigns the corresponding interpolated threshold to each pixel in the fine grid:

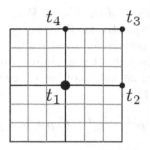

Fig. 1. A coarse grid is superimposed on the finer pixel grid of the reconstructed image. The thresholds are only specified in the coarse grid points. The threshold for each individual pixel is computed by bilinear interpolation.

$$\tau = I(\tau') \tag{6}$$

Using these definitions, we can now formulate the central problem of this paper:

Problem 2. Let $W \in \mathbb{R}^{m \times n}$ be a given projection matrix, let $v \in \mathbb{R}^n$ be a grey scale image and let $p \in \mathbb{R}^m$ be a vector of measured projection data. Find $\tau' \in \mathbb{R}^k$ and $\rho \in \mathbb{R}^2$, such that $d(\rho, I(\tau'))$ is minimal.

We will now describe how a constrained version of Problem 2 can be solved efficiently, where only one of the entries of τ' is allowed to vary, while the remaining entries are kept fixed. Again, consider the example from Figure 1. Suppose that all thresholds on the coarse grid are kept fixed, except for t_1. The only pixels for which the thresholds are affected by a change of t_1 are those in the four quadrants surrounding t_1, as shown in the figure.

For a pixel p with grey level v_p and center (x_p, y_p) in the topright quadrant, we have

$$v_p \geq t_p \iff$$
$$v_p \geq (1 - x_p)(1 - y_p)t_1 + x_p(1 - y_p)t_2 + (1 - x_p)y_p t_3 + x_p y_p t_4 \iff$$
$$\frac{v_p - x_p(1 - y_p)t_2 - (1 - x_p)y_p t_3 - x_p y_p t_4}{(1 - x_p)(1 - y_p)} \geq t_1. \tag{7}$$

The term $g(p) = \frac{v_p - x_p(1-y_p)t_2 - (1-x_p)y_p t_3 - x_p y_p t_4}{(1-x_p)(1-y_p)}$ is called the *relative grey level of p with respect to t_1.* The equation defining the relative grey level of pixels w.r.t. t_1 is different for each of the quadrants surrounding the coarse grid point. Eq. (7) shows that the problem of finding a solution of Problem 2 in case only one entry of τ' is allowed to vary, can be considered as a variant of the global thresholding problem from [8]. In this global thresholding problem, only pixels in the four quadrants surrounding the variable entry of τ' have to be considered (as the remaining pixels are unaffected by a change of this threshold) and the relative grey level of each surrounding pixel is used instead of the grey levels from v.

For any fixed $\boldsymbol{\tau}$, every entry of the vector $\boldsymbol{W}r_{\boldsymbol{\rho},\boldsymbol{\tau}}(\boldsymbol{v})$ reduces to a linear expression in ρ_1 and ρ_2. Consequently, for fixed $\boldsymbol{\tau}$, the projection difference $d(\boldsymbol{\rho},\boldsymbol{\tau})$ reduces to a second-degree polynomial in ρ_1 and ρ_2. In fact, this property holds for any segmentation of $\boldsymbol{v}$ (not just those obtained by thresholding). In [8], it is shown that for any fixed segmentation of the image $\boldsymbol{v}$, the projection difference $d(\boldsymbol{\rho})$ can be written in the form

$$d(\boldsymbol{\rho})^2 = |\boldsymbol{p}|^2 + \bar{\boldsymbol{c}}^T\boldsymbol{\rho} + \boldsymbol{\rho}^T\bar{\boldsymbol{Q}}\boldsymbol{\rho}. \tag{8}$$

(i.e., a second-degree polynomial in ρ_1 and ρ_2), where $\bar{\boldsymbol{Q}} \in \mathbb{R}^{\ell\times\ell}$ matrix and $\bar{\boldsymbol{c}} \in \mathbb{R}^2$. Explicit expressions for the matrix $\bar{\boldsymbol{Q}}$ and vector $\bar{\boldsymbol{c}}$ are also described. Once $\bar{\boldsymbol{Q}}$ and $\bar{\boldsymbol{c}}$ are known, the vector $\boldsymbol{\rho}$ for which the projection distance is minimal can be computed by minimizing the quadratic polyomial in Eq. (8). It is also shown in [8] how $\bar{\boldsymbol{Q}}$ and $\bar{\boldsymbol{c}}$ can be updated efficiently if the segmentation class for a single pixel is changed. This update step is independent of the algorithm that is used to compute the segmentation, so it can be used in our new local threshold method without much modification. Figure 2 shows the basic steps for solving the variant of Problem 2 where only one of the entries of $\boldsymbol{\tau}'$ is allowed to vary.

Computing a global minimum of $d(\boldsymbol{\rho}, I(\boldsymbol{\tau}'))$ as stated in Problem 2 appears to be computationally very hard. In fact, if the coarse grid is taken to have the same resolution as the pixel grid of the reconstructed image, solving Problem 2 is equivalent to solving a variant of the discrete tomography problem. For certain weight matrices $\boldsymbol{W}$, this problem is known to be NP-hard [17].

Instead, we propose an iterative algorithm that is guaranteed to converge to a local minimum of the projection distance. The basic steps of this algorithm are shown in Figure 3. In each iteration, a random grid point on the coarse grid is selected. The optimal threshold for this grid point is computed, while keeping the thresholds in all other coarse grid points fixed. The algorithm terminates if no improvement has been found after K iteration, where K is a constant integer. Although this algorithm may not find a global minimum of the projection distance, the projection distance of the resulting segmentation can never be worse than the projection distance found using global thresholding. The best global threshold is used to initialize the local thresholds in the coarse grid points and in each iteration the projection distance does not increase.

3   Results

Simulation experiments have been performed, starting from two phantom images: a vascular structure (referred to as 'vessel image') and a femur image, shown in Fig. 4(a) and Fig.4(b), respectively. For each experiment, simulated parallel beam projections have been computed using equally spaced projection angles. Based on the projection data, a reconstruction was computed using the SART algorithm (cfr. [15]). The resulting SART reconstructions are shown in Fig. 5(a)-5(c). Fig. 5(a) and Fig. 5(b) were generated from 180 projections, while Fig.5(c)

Let $\boldsymbol{\tau}' \in \mathbb{R}^k$ be a given vector of thresholds for the coarse grid points and assume that $\bar{Q}$ and $\bar{c}$ have already been computed (cfr. [8]);

Make a list L containing the index j of all pixels in the four quadrants surrounding the coarse grid point i, sorted in ascending order of the relative grey level $g(j)$ w.r.t. i; Denote the size of this list by $|L|$;

Let u be the largest number with $u \le |L|$ such that $g(L_u) \le \tau_i'$;

Set $u' := u$; $\bar{Q}' := \bar{Q}$; $\bar{c}' := \bar{c}$;

while $u > 1$ **do**
begin

$\quad u := u - 1$; $\tau_i' := g(L_u)$;

$\quad$ Move pixel L_u from the background class to the foreground class and update $\bar{c}$ and $\bar{Q}$ accordingly;

$\quad$ Compute the minimizer ρ of $d(\rho)$ for the current segmentation;

$\quad$ **if** a new minimum of the projection distance has been found
$\quad\quad$ save the optimal threshold τ_i';

end Set $u := u'$; $\bar{Q} := \bar{Q}'$; $\bar{c} := \bar{c}'$;

while $u < |L|$ **do**
begin

$\quad u := u + 1$; $\tau_i' := g(L_u)$;

$\quad$ Move pixel L_u from the foreground class to the background class and update $\bar{c}$ and $\bar{Q}$ accordingly;

$\quad$ Compute the minimizer ρ of $d(\rho)$ for the current segmentation;

$\quad$ **if** a new minimum of the projection distance has been found
$\quad\quad$ save the optimal threshold τ_i';

end

Fig. 2. Basic steps for finding the optimal threshold in a given coarse grid point i, while keeping the thresholds in all remaining coarse grid points fixed

shows the SART reconstruction from 180 truncated projections (i.e., the simulated beam is more narrow than the object). The truncated projections cause artifacts in the reconstruction, which are particularly difficult to segment using a single global threshold, as they result in significant grey value variations. For all local threshold experiments in this section, a spacing of 16 pixels between consecutive coarse grid points was used.

In a first phase, the results of our proposed local thresholding approach are compared with global thresholding. For global thresholding, we used the PDM algorithm from [8], which computes a global minimum of the projection distance and which was proved to yield significantly better results than conventional global thresholding techniques. The global PDM thresholding results are shown in Fig. 5(g)-5(i).

compute the optimal global threshold τ, using the algorithm described in [8];

Set $\tau_i' = \tau$ for each point $i = 1, \ldots, k$ on the coarse grid;

repeat

select a random threshold τ_i' on the coarse grid;

compute the optimal projection difference $d(\boldsymbol{\rho}, I(\boldsymbol{\tau}'))$ that can be obtained by changing τ_i', while keeping the values of the other thresholds fixed;

set the threshold τ_i' to its new optimal value (changing the local threshold field);

until (no improvement was made during the last K iterations;

Fig. 3. Basic steps of the threshold selection algorithm. The variable K in the outer loop condition refers to a constant integer.

Table 1. Comparison of the local thresholding performance of the Niblack and PDM method. The numbers represent the number of different pixels between the thresholded image and the original phantom image.

	Vessel	Femur	Femur (trunc)
Niblack	9977	3142	10694
PDM (global)	3374	1365	10743
PDM (local)	2967	1182	7584

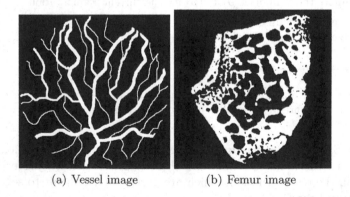

(a) Vessel image   (b) Femur image

Fig. 4. Phantom images used in our simulation experiments: (a) vessel image and (b) femur image

Next, the SART reconstructions were segmented using the local thresholding method of Niblack [14], which is commonly used as a reference for performance evaluation. The Niblack method adapts the local threshold according to the local mean and standard deviation of a sliding window. The method depends on two parameters: the width of the sliding window and the threshold weight of the

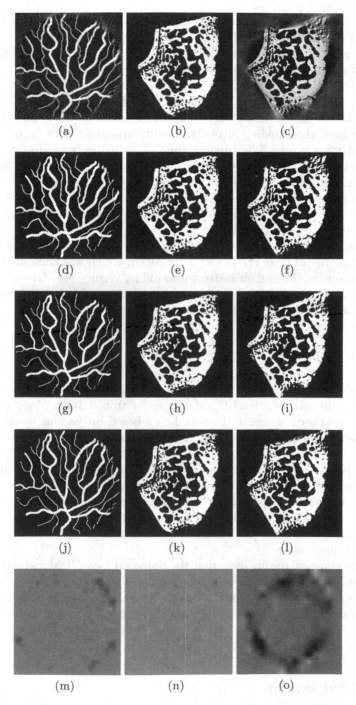

Fig. 5. (a-c) SART reconstructions from (a-b) 180 projections and (c) 180 truncated projections; (d-f) Niblack local thresholding results; (g-i) global PDM thresholding result; (j-l) local PDM thresholding result; (m-o) PDM threshold fields

standard deviation. In practice, these Niblack parameters cannot be optimized because of the lack of ground truth. In our simulation experiments, in which the ground truth was available, we selected the window width and the weight parameter such that the difference between the thresholded result and the original was minimal.

Finally, the SART reconstructions shown in Fig. 5(a)-5(c) were segmented using the local PDM thresholding scheme as proposed in this paper. In all three cases, our local thresholding algorithm results in a significant reduction of the number of pixel errors. The running time of each test was around 30s, on a Pentium IV PC running at 3GHz. The thresholding results of the local PDM method are shown in Fig. 5(j)-5(l), respectively. The corresponding threshold fields generated from the local PDM method are shown in Fig.5(m)-5(o), respectively. These fields were formed by bilinear interpolation from the threshold values τ' on the coarse grid.

To quantify the performance of each segmentation method, for each method the total number of pixel errors was determined, which was computed from the original phantom. The quantitative thresholding results are shown in Table 1. From this table, it is clear that the proposed local PDM thresholding method outperforms the other methods with respect to the pixel error.

4   Conclusion

Local grey value thresholding is a common segmentation procedure. However, finding the optimal grey level thresholds is far from trivial. Many procedures have been proposed to select the thresholds based on various image features. Unfortunately, these methods suffer from a clear objective threshold selection criterion.

In our paper, we have presented an innovative approach, called local PDM (Projection Distance Minimization), to find the optimal threshold grey levels by exploiting the available projection data. Reprojection of the segmented image and subsequent comparison with the measured projection data, yields an objective criterion for the quality of a segmentation. Our approach aims at minimizing the projection distance.

The experimental results show that the proposed local PDM method results in a small difference between the original object and the reconstruction. Simulation experiments were performed for three SART reconstructions of phantom images. In all test cases, PDM clearly leads to significantly better segmentation results than both the Niblack method for local segmentation and global thresholding based on PDM.

Acknowledgement

This work was financially supported by the F.W.O. (Fund for Scientific Research - Flanders, Belgium)

References

1. Eichler, M.J., Kim, C.H., Müller, R., Guo, X.E.: Impact of thresholding techniques on micro-CT image based computational models of trabecular bone. In: ASME Advances in Bioengineering, September 2000, vol. 48, pp. 215–216 (2000)
2. Glasbey, C.A.: An analysis of histogram-based thresholding algorithms. Graphical Models and Image Processing 55(6), 532–537 (1993)
3. Rosenfeld, A., Torre, P.: Histogram concavity analysis as an aid in threshold selection. IEEE Trans. Syst., Man, Cybern. 13, 231–235 (1983)
4. Ridler, T.W., Calvard, S.: Picture thresholding using an iterative selection method. IEEE Transactions on Systems, Man, and Cybernetics 8, 630–632 (1978)
5. Otsu, N.: A threshold selection method from gray level histograms. IEEE Trans. Syst., Man, Cybern. 9, 62–66 (1979)
6. Kapur, J., Sahoo, P., Wong, A.: A new method for gray-level picture thresholding using the entropy of the histogram. Comp Vision, Graph, and Image Proc. 29(3), 273–285 (1985)
7. Batenburg, K.J., Sijbers, J.: Automatic multiple threshold scheme for segmentation of tomograms. In: Pluim, Josien, P.W., Reinhardt, J.M. (eds.) Proceedings of SPIE Medical Imaging, San Diego, CA, USA, February 2007, vol. 6512 (2007)
8. Batenburg, K.J., Sijbers, J.: Automatic threshold selection for tomogram segmentation by reprojection of the reconstructed image. In: Kropatsch, W.G., Kampel, M., Hanbury, A. (eds.) CAIP 2007. LNCS, vol. 4673, pp. 563–570. Springer, Heidelberg (2007)
9. Abutaleb, A.S.: Automatic thresholding of grey-level pictures using two-dimensional entropies. Pattern Recognition 47, 22–32 (1989)
10. Brink, A.D.: Thresholding of digital images using two-dimensional entropies. Pattern Recognition 25, 803–808 (1992)
11. White, J.M., Rohrer, G.D.: Image thresholding for optical character recognition and other applications requiring character image extraction. IBM Journal of Research and Development 27(4), 400–411 (1983)
12. Eikvil, L., Taxt, T., Moen, K.: A fast adaptive method for binarization of document images. In: Proceedings of the International Conference on Document Analysis and Recognition, pp. 435–443 (1991)
13. Blayvas, I., Bruckstein, A., Kimmel, R.: Efficient computation of adaptive threshold surfaces for image binarization. Pattern Recognition 39(1), 89–101 (2006)
14. Niblack, W.: An introduction to image processing. Englewood Cliffs, New York (1986)
15. Kak, A.C., Slaney, M.: Principles of Computerized Tomographic Imaging. Volume Algorithms for reconstruction with non-diffracting sources, pp. 49–112. IEEE Press, New York (1988)
16. Herman, G.T., Kuba, A. (eds.): Advances in Discrete Tomography and Its Applications. Applied and Numerical Harmonic Analysis. Birkhäuser, Boston (2007)
17. Gardner, R., Gritzmann, P., Prangenberg, D.: On the computational complexity of reconstructing lattice sets from their X-rays. Discrete Math. 202, 45–71 (1999)

Reconstructing Binary Matrices with Neighborhood Constraints: An NP-hard Problem

A. Frosini[1], C. Picouleau[2], and S. Rinaldi[3]

[1] Dipartimento di Sistemi e Informatica
Università di Firenze, (Firenze, Italy)
andrea.frosini@unifi.it
[2] Laboratoire CEDRIC CNAM,
(Paris, France)
[3] Dipartimento di Scienze Matematiche ed Informatiche
Università di Siena, (Siena, Italy)

Abstract. This paper deals with the reconstruction of binary matrices having *exactly* $- 1 - 4 - adjacency$ constraints from the horizontal and vertical projections. Such a problem is shown to be non polynomial by means of a reduction which involves the classic NP-complete problem 3-color. The result is reached by bijectively mapping all the four different cells involved in 3-color into maximal configurations of 0s and 1s which show the adjacency constraint, and which can be merged into a single binary matrix.

Keywords: Discrete Tomography, polynomial time reduction, NP-complete Problem.

1   Introduction and Notations

Given a binary matrix, its horizontal and vertical projections are defined as the sum of its elements for each row and each column, respectively. The reconstruction of a binary matrix from its orthogonal projections has been studied by Ryser [7,8]. One can refer to the books of Herman and Kuba [4,5] for further information on the theory, algorithms and applications of such a kind of classical problems which fit in the wide area of the discrete tomography.

It is well-known that this basic problem, where the only constraints to verify are both projections, can be solved in polynomial time. Numerous studies deal with the same problem when additional local or global constraints are imposed. Here we consider binary matrices with a local adjacency constraint, and we show a polynomial time reduction which maps each instance of the classical NP-complete problem 3-color into an instance of their consistency problem. Then we show that such a reduction fulfills a series of requirements stated in a general framework, in [2], and allows us to set the time complexity of 3-color as lower bound to that of our problem.

Now, let us proceed by introducing basic notations and definitions: given a $m \times n$ binary matrix M we denote by $M[i,j]$ the element in position (i,j). The

D. Coeurjolly et al. (Eds.): DGCI 2008, LNCS 4992, pp. 392–400, 2008.
© Springer-Verlag Berlin Heidelberg 2008

horizontal projection of M is defined as the vector $H = (h_1, ..., h_m)$ such that for each $1 \le i \le m$, h_i counts the number of 1's on row i. Similarly, the *vertical projection* $V = (v_1, ..., v_n)$ of M, is the vector such that for each $1 \le j \le n$, v_j counts the number of 1's on column j.

In order to define the *adjacency constraint*, we rely on the definition of the neighborhood of a given internal element $M[i, j]$, $1 < i < m$, $1 < j < n$, of the binary matrix M: we speak of *4-adjacency* if its neighbors are considered to be the elements $M[i, j-1]$, $M[i, j+1]$, $M[i-1, j]$ and $M[i+1, j]$ i.e. the horizontal and vertical adjacent cells. Different kinds of adjacency can be defined as well, but they do not concern this paper.

We say that the matrix M fulfills the *exactly-1-4-adjacency* constraint if $M[i, j] = 1$ implies that there is exactly one among its 4-adjacent cells that has value 1. The class of binary matrices that fulfill exactly-1-4-adjacency is denoted by $\mathcal{N}_4^{=1}$.

The concept of adjacent constraint has a prominent role when dealing with scheduling problems (see [6]), and it has been recently studied under a tomographical perspective in [1].

Here, two classical problems related to the class $\mathcal{N}_4^{=1}$ are addressed:

Consistency$(\mathcal{N}_4^{=1}, H, V)$

Input: a couple of integer vectors H and V;

Question: does there exist an element of the class $\mathcal{N}_4^{=1}$ whose horizontal and vertical projections are the vectors H and V, respectively?

Reconstruction$(\mathcal{N}_4^{=1}, H, V)$

Input: a couple of integer vectors H and V;

Task: reconstruct an element of the class $\mathcal{N}_4^{=1}$ whose horizontal and vertical projections are the vectors H and V, respectively, if it exists, otherwise give a failure.

The condition $\sum_{i=1}^{m} h_i = \sum_{j=1}^{n} v_j$ is obviously necessary for the existence of a binary matrix respecting both projections in the two problems.

Finally, we define the class of colored matrices and their projections: given a set of colors $C = \{c_1, ..., c_k\}$, a k-colored $m \times n$ matrix A is a matrix whose elements are in $C \cup \{colorless\}$.

The projection of the ith row of A is the k dimensional vector $(h_i^{c_1}, h_i^{c_2}, ...,$ $h_i^{c_k})$, where each coordinate $h_i^{c_s}$ counts the number of elements of color $c_s \in C$ lying in the ith row. A similar definition of vertical projection can be given for a generic column j of A (see Fig. 1). The sequence of all the horizontal [resp. vertical] projections is indicated as Hk [resp. Vk].

2   Principle of the Reduction

In order to prove our main result, i.e. the NP-completeness of *Consistency* $(\mathcal{N}_4^{=1}, H, V)$, we consider the following classical problem:

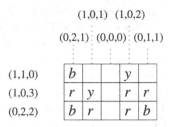

Fig. 1. A 3-color matrix with the colors $C = \{yellow, blue, red\}$, and its projections

$3 - color$:
   let $C = (yellow, blue, red)$ be a set of three colors.

Input: two 3 dimensional integer vectors

$$H3 = ((h_1^y, h_1^b, h_1^r), \ldots, (h_m^y, h_m^b, h_m^r)) \text{ and } V3 = ((v_1^y, v_1^b, v_1^r), \ldots, (v_n^y, v_n^b, v_n^r)).$$

Question: does there exists a 3-color matrix of dimension $m \times n$ whose horizontal and vertical projections are the vectors $H3$ and $V3$, respectively?

As for the class $\mathcal{N}_4^{=1}$, we can define the correspondent 3-*color reconstruction* problem.

Inside the 3-color matrix, we refer to each color in C using its initial letter, and to the colorless element using the symbol c. Furthermore, it will be useful to introduce the notations

$$h_i^c = n - h_i^y - h_i^b - h_i^r \quad \text{and} \quad v_j^c = m - v_j^y - v_j^b - v_j^r$$

to denote the number of c elements in a 3-color matrix.

Now, following [2], we define a polynomial time process, say a reduction, from an instance I of 3-color to an instance I' of *Consistency* $(\mathcal{N}_4^{=1}, H, V)$ by using the correspondence of each color c, y, b, and r of I with four different placements of 0s and 1s in a configuration of cells which satisfy the *exactly* $-1-4-adjacency$ constraints, as shown in Fig. 2.

The reduction we are going to define has to fulfill two requirements (conditions $npc1$ and $npc2$ of [2]):

i) from the instance I' we can univocally compute back the instance I;

ii) each solution of I' contains (in a set of fixed positions) only the four chosen configurations.

We achieve *ii)* by considering the placements of 0s and 1s in Fig. 2 which are the only four ones that satisfy the *exactly* $- 1 - 4 - adjacency$ constraint, and that are maximal, i.e. that have the maximum number of 1s.

Such a reduction sets the computational complexity of $3-color$ as lower bound to that of *Consistency* $(\mathcal{N}_4^{=1}, H, V)$. Since the first is known to be NP-complete (see [3]), then the same holds for the latter.

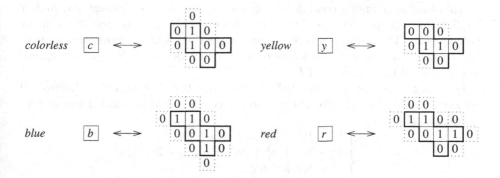

Fig. 2. The correspondence between each atom of the 3-atom problem and a configuration of elements 0 and 1 having the exactly-1-4-adjacency constraint. The values 0 in the dotted positions do not belong to the chosen configurations, but they are also fixed.

3   Going Back and Forth from Instance I to Instance I'

So, let $I = (H3, V3)$ be an instance of 3-atom, with

$$H3 = ((h_1^y, h_1^b, h_1^r), \ldots, (h_m^y, h_m^b, h_m^r)) \quad \text{and} \quad V3 = ((v_1^a, v_1^b, v_1^r), \ldots, (v_n^a, v_n^b, v_n^r)).$$

We compute an instance $I' = (H, V)$, with

$$H = (h_1, \ldots, h_{6m+1}) \quad \text{and} \quad V = (v_1, \ldots, v_{12n})$$

of $Reconstruction\,(\mathcal{N}_4^{=1}, H, V)$, as follows: for each $0 \leq i < m$, the entries of the vector H are

- $h_{6i+1} = h_{6m+1} = 8n$
- $h_{6i+2} = 2n$
- $h_{6i+3} = 4n$
- $h_{6i+4} = 2n + h_i^- + 2h_i^b + 2h_i^r$
- $h_{6i+5} = 3n + h_i^- + 2h_i^y + h_i^b + 2h_i^r$
- $h_{6i+6} = n + h_i^b$

for each $0 \leq j < n$, the entries of the vector V are

- $v_{12j+1} = v_{12j+2} = m + 1$
- $v_{12j+3} = 4m$
- $v_{12j+4} = m + 1 + v_j^b + v_j^r$
- $v_{12j+5} = m + 1 + 2v_j^- + v_j^y + v_j^b + v_j^r$
- $h_{12j+6} = 2m + v_j^y + 2v_j^b + v_j^r$
- $h_{12j+7} = m + 1 + v_j^r$
- $h_{12j+8} = 2m + 1$
- $h_{12j+9} = 2m$
- $h_{12j+10} = h_{12j+11} = 3m + 1$
- $h_{12j+12} = 0.$

As one can observe, most of the horizontal and vertical projections are independent from the vectors $H3$ and $V3$: they are introduced in order to create the exact shape of the four different configurations used in the reduction.

On the other hand, the projections which involve the entries of the vectors $H3$ and $V3$ can be easily understood by referring to the different positions of the 1s in the configurations of Fig. 2.

What we have to show is how to go back from instance I', obtained from the previous computation, to the correspondent instance I.

This last step fulfills condition i) given in the previous paragraph, and it implies that the projections of the four chosen configurations are affine linear independent. This check is crucial, since projections which are dependent may mix together, and give rise to solutions of I' which do not have a correspondent in the class of the solutions of I.

For what concerns the entry (h_i^y, h_i^b, h_i^r), with $0 \leq i < m$, of the horizontal projections $H3$ of I, its elements can be easily computed by solving the system of equations

$$\begin{cases} h_i^c + 2h_i^b + 2h_i^r = h_{6i+4} - 2n \\ h_i^c + 2h_i^y + h_i^b + 2h_i^r = h_{6i+5} - 3n \\ h_i^b = h_{6i+6} - n \\ h_i^c + h_i^y + h_i^b + h_i^r = n \end{cases}$$

The existence of an integer solution for this system of equations directly follows from the definition of the horizontal projections of the four configurations depicted in Fig. 2.

A similar system can be defined for the vector $V3$ of the vertical projections in the instance I.

4 From a Solution of I to a Solution of I'

Now we proceed in showing that the instances I and I' are equivalent, i.e. that the process of defining instance I' from I preserves the existence of its solutions. In the sequel we establish even more (but this is not unusual): there exists a one-to-one correspondence between the set of solutions of I and the set of solutions of I'.

Intuitively, a solution of I' is constructed starting from a solution of I, by means of the correspondence in Fig. 2. On the other hand, each solution of I' has some fixed positions where one can detect the configurations corresponding to a color, and only those. The horizontal and vertical projections of the instance I' defined in the previous paragraph accomplish this task.

So, let us go into details by showing, with the aid of Fig. 3, how to compute a solution M' of the instance I' from a generic solution M of I: let $1 \leq i \leq m$ and $1 \leq j \leq n$

Step 1: for each element $M[i,j]$, we insert in the matrix M the rectangular configuration of 0 and 1 depicted in Fig. 3, iii), placing its left-uppermost element in position $M[6(i-1)+1, 12(i-1)+1]'$;

Step 2: according with the color of the element $M[i,j]$, we place a configuration of 0s and 1s as in Fig. 2 inside the void elements of the corresponding rectangle;

Step 3: we copy the entries of the first row of the matrix M' in its last row of index $6m+1$.

An easy check reveals that the defined matrix M' is a solution of instance I', as desired.

5   From a Solution of I Back to a Solution of I'

On the other hand, in order to associate to each solution M' of I' a solution M of I, we need to inspect its entries and detect some of them which are fixed. Figure 3 will help the reader, showing a part of matrix M in detail: let $0 \le i < m$ and $0 \le j < n$

i) columns $12j + 12$ are completely filled with 0s; rows $6i + 1$, and row $6m + 1$ have projections $8n$, so they are determined by the exactly-1-4-adjacency constraint (see Fig. 3, $i) - ii)$);

ii) the sets of columns - $12j + 3$ whose projection is $4m$;
- $12j + 1$ and $12j + 2$ whose projection is m;
- $12j + 10$ and $12j + 11$ whose projection is $3m + 1$
are determined by the already placed entries, and the exactly-1-4-adjacency constraint(see Fig. 3, $i) - ii)$);

iii)   columns $12j + 8$ and $12j + 9$ whose projections are $2m + 1$ and $2m$, respectively, and rows $6i + 2$ and $6i + 3$ whose projections are $2n$ and $4n$, respectively, are also determined by the previous placements of entries, and again by the exactly-1-4-adjacency constraint (see Fig. 3, $iv)$).

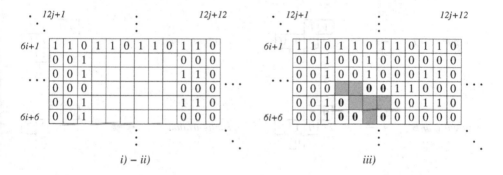

i) – ii)   iii)

Fig. 3. The reconstruction of the fixed parts of a generic solution of instance I'

Finally, the boldface elements 0 in Fig. 3, $iii)$ are set by the exactly-1-4-adjacency constraint, and finally, the remaining grey positions are those elements of M' which are not fixed, and which may eventually differ from one solution to another. We refer to (the shape of) each of these configurations by $\mathcal{S}$. The reader can check that the found $\mathcal{S}$ is the same as that associated to each color in Fig. 2.

Now the definition of the matrix M from M' is straightforward: for each $0 \le i < m$ and $0 \le j < n$

- if $M[6i + 4, 12j + 5]' = M[6i + 5, 12j + 5]' = 1$, then $M[i + 1, j + 1] = c$;
- if $M[6i + 5, 12j + 5]' = M[6i + 5, 12j + 6]' = 1$, then $M[i + 1, j + 1] = y$;
- if $M[6i + 5, 12j + 6]' = M[6i + 7, 12j + 6]' = 1$, then $M[i + 1, j + 1] = b$;
- if $M[6i + 5, 12j + 6]' = M[6i + 5, 12j + 7]' = 1$, then $M[i + 1, j + 1] = r$;

The correctness of this last step deeply relies on the assumption that no other configurations except those four coding a color are possible inside each S.

To prove it, we remind that

- each of the four colors is associated to a maximal placements of 1s in S;

- no other maximal placements exist. This fact can be easily checked by an exhaustive search.

So, let us consider the vectors of vertical projections of the four configurations S in Fig. 2:

$$v_c = (0, 2, 0, 0), \quad v_y = (0, 1, 1, 0), \quad v_b = (1, 1, 2, 0) \quad \text{and} \quad v_r = (1, 1, 1, 1),$$

and the vectors of vertical projections of the four remaining non maximal configurations of S shown in Fig.4:

$$v_1 = (1, 1, 0, 0), \quad v_2 = (0, 0, 2, 0), \quad v_3 = (0, 0, 1, 1) \quad \text{and} \quad v_0 = (0, 0, 0, 0).$$

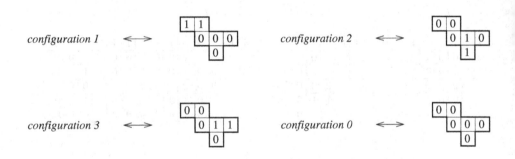

Fig. 4. The four non-maximal configurations of S

We proceed by contradiction, and we assume that there exists a solution of I which has at least one non maximal placements of elements in one of the Ss, say S'. In that case no color can be associated to S'.

The vertical projections of the $m-1$ configurations which lie above and below S', i.e. which share with it the same columns in M', together with S' itself, can be expressed both as linear combination of v_c, v_y, v_b, and v_r, and as linear combination of all and eight the vectors of projections as follows:

$$l1 : k_1 v_c + k_2 v_y + k_3 v_b + k_4 v_r \, ,$$
$$l2 : k_1' v_c + k_2' v_y + k_3' v_b + k_4' v_r + k_5' v_1 + k_6' v_2 + k_7' v_3 + k_8' v_0.$$

Since the number of the involved S in both the linear combinations has to be fixed, and since it holds that

- the sums of the elements of v_b and v_r is equal to 4, and it is greater than the sums of each of six remaining vectors;
- the sum of the elements of v_0 is zero, which is the minimum of the sums of each of the other vectors;
- the sums of the elements of each of the remaining vectors are 2,

then it follows that $k_3 + k_4 = k'_3 + k'_4$.

Since the contribution to the first entry in both the linear combinations is given only by v_b, v_r and v_1, it also holds that $k'_5 = 0$.

On the other hand, the contribution to the second entry in both the linear combinations, decreased by $k_3 + k_4$, is now restricted only to the vectors v_c and v_y, so we have $k_1 + k_2 = k'_1 + k'_2$.

As a consequence, we reach $k'_6 = k'_7 = k'_8 = 0$, against the assumption that there exists in $l2$ at least one vector of projections different from the maximal ones.

Now we can state the following

Theorem 1. *The problem Consistency* $(\mathcal{N}_4^{=1}, H, V)$ *is NP-complete.*

The proof is a direct consequence of the defined reduction, and, as a consequence, we have

Corollary 1. *The problem Reconstruction* $(\mathcal{N}_4^{=1}, H, V)$ *is NP-hard.*

The following example tries to clarify the reduction

Example 1. Let us consider the instance $I = (H3, V3)$ where

$$H3 = ((0,1,0),(1,0,1),(0,1,1)) \quad \text{and} \quad V3 = ((0,2,1),(0,1,0)).$$

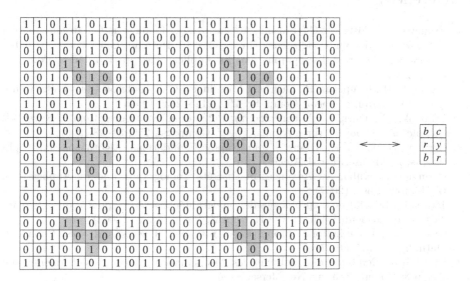

Fig. 5. The correspondence between an instance of 3-color and a matrix in $\mathcal{N}_4^{=1}$

for which the solutions are 3-color matrices of dimension 3×2.

Following what defined in paragraph 3, the correspondent instance $I' = (H, V)$ of *Consistency* $(\mathcal{N}_4^{=1}, H, V)$ is

$$H = (16, 4, 8, 7, 8, 3, 16, 4, 8, 6, 10, 2, 16, 4, 8, 8, 9, 3, 16) \text{ and}$$
$$V = (4, 4, 12, 7, 7, 11, 5, 7, 3, 10, 10, 0, 4, 4, 12, 5, 8, 8, 5, 7, 3, 10, 10, 0).$$

Figure 5 depicts a binary matrix M' which has H and V as vectors of projections, together with its associated 3-color matrix, having projections $H3$ and $V3$. Again we underline the property that there is no way of obtaining a matrix M' whose configurations of 0s and 1s inside the grey zone are not in correspondence with a color.

6   Conclusion

In this paper we have studied the computational complexity of the consistency problem for the class $\mathcal{N}_4^{=1}$. This is one the simplest examples of a class of binary matrices satisfying adjacency constraints. However, our studies in that direction are just at the beginning, and a systematical analysis of all the other cases of 4-adjacency constraints and their extension to a general framework where the adjacent constraint involve different sets of neighborhood elements are work in progress. Our attention is also attracted by the presence in $\mathcal{N}_4^{=1}$ of some subclasses which allow a fast reconstruction strategy, and which show some interesting geometrical aspects.

References

1. Brunetti, S., Costa, M.C., Frosini, A., Jarray, F., Picouleau, C.: Reconstruction of binary matrics under adjacency constraints. In: Herman, G.T., Kuba, A. (eds.) Advances in Discrete Tomography and its Applications, pp. 125–150. Birkhäuser, Basel (2007)
2. Chrobak, M., Couperus, P., Dürr, C., Woeginger, G.: A note on tiling under tomographic constraints. Theor. Comp. Sc. 290, 2125–2136 (2003)
3. Chrobak, M., Dürr, C.: Reconstructing Polyatomic Structures from X-Rays: NP Completness Proof for three Atoms. Theor. Comp. Sc. 259, 81–98 (2001)
4. Herman, G., Kuba, A.: Discrete Tomography: Foundations, Algorithms and Applications. Birkhauser, Basel (1999)
5. Herman, G., Kuba, A.: Advances in Discrete Tomography and its Applications. Birkhäuser, Basel (2007)
6. Jarray, F.: Résolution de problèmes de tomographie discrète. Applications à la planification de personnel, Ph. D. thesis, CNAM, Paris, France (2004)
7. Ryser, H.J.: Combinatorial properties of matrices of zeros and ones. Canad. J. Math. 9, 371–377 (1957)
8. Ryser, H.J.: Combinatorial mathematics, Mathematical Association of America and Quinn & Boden, Rahway, New Jersey (1963)

An Exact, Non-iterative Mojette Inversion Technique Utilising Ghosts

Shekhar Chandra[1], Imants Svalbe[1], and Jean-Pierre Guédon[2]

[1] School of Physics, Monash University, Australia
[2] IRCCyN-IVC, École polytechnique de l'Université de Nantes, France

Abstract. Mojette projections of discrete pixel arrays form good approximations to experimental parallel-beam x-ray intensity absorption profiles. They are discrete sums taken at angles defined by rational fractions. Mojette-like projections form a "half-way house" between a conventional sinogram and fully digital projection data. A new direct and exact image reconstruction technique is proposed here to invert arbitrary but sufficient sets of Mojette data. This new method does not require iterative, statistical solution methods, nor does it use the efficient but noise-sensitive "corner-based" inversion method. It instead exploits the exact invertibility of the prime-sized array Finite Radon Transform (FRT), and the fact that all Mojette projections can be mapped directly into FRT projections. The algorithm uses redundant or "calibrated" areas of an image to expand any asymmetric Mojette set into the smallest symmetric FRT set that contains all of the Mojette data without any re-binning. FRT data will be missing at all angles where Mojette data is not provided, but can be recovered exactly from the "ghost projections" that are generated by back-projecting all the known data across the calibrated regions of the reconstructed image space. Algorithms are presented to enable efficient image reconstruction from any exact Mojette projection set, with a view to extending this approach to invert real x-ray data.

1 Introduction

Tomographic reconstruction of discrete images is typically based on a sinogram of projection data obtained using finite aperture detectors oriented at some set of angles in continuous space. The mix of continuous and discrete sampling causes problems with image artefacts. To ensure the uniqueness of the reconstruction, some interpolation or smoothing is required within the Filtered Back-Projection or Fourier Inversion methods [1]. Any discrete object can, however, be reconstructed exactly from a sufficient set of discrete projections. Significant advances in discrete inversion methods have appeared in [2, 3, 4]. The discrete projections of the Mojette Transform (MT) of [5] resemble closely the form of real data projections, but the projections are restricted to angles whose tangents are rational fractions and the number of rays per projection varies with the projection angle. These MT projections are matched to the grid on which the object is reconstructed, removing the conflict between discrete and continuous sampling and

D. Coeurjolly et al. (Eds.): DGCI 2008, LNCS 4992, pp. 401–412, 2008.

provide the promise of a more faithful reconstruction process and, potentially, less dose.

A disadvantage of the very general MT approach is the lack of a direct algorithm that avoids iteration [6] or the use of noise-sensitive "corner-based" inversion approaches [5]. The discrete Finite Radon Transform (FRT) of [7] has a very simple, direct and exact inversion algorithm that is a consequence of its prime number array-size and because of its fixed length, periodically wrapped projection structure. The FRT is a particular case of the MT. Mojette projections can be mapped into FRT form. The FRT inverse could then be used to reconstruct the image, but the number of symmetric FRT projections is usually larger than the asymmetric MT set. Providing the MT set contains sufficient information to reconstruct the data exactly, we propose a method to recover the full FRT projection set and thus provide a direct inversion method for the MT via the FRT formalism. We begin with a brief overview of the MT and the FRT.

1.1   The Mojette Transform (MT)

The MT is a discrete linogram [8] transform for objects of arbitrary shape where only a minimum projection set needs to be defined [5]. The projection angles are confined to the Farey Sequence q_i/p_i, which is a set of irreducible rational fractions where $\gcd(p_i, q_i) = 1$ with $i = 1, \ldots, \mu$ and μ is the total number of projections [9]. The number of bins in a given MT projection i, $\mathcal{M}_i$, depends on the angle $\theta_i = \tan^{-1}(q_i/p_i)$ through

$$\mathcal{M}_i = |p_i|(Q - 1) + |q_i|(P - 1) + 1, \tag{1}$$

for a $P \times Q$ object. An example of a MT for a 4×4 image is given figure 1. The minimum number of projections $\mu_{\min}$ required for exact reconstruction is

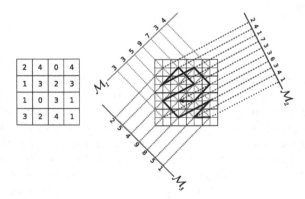

Fig. 1. An example of a Mojette Transform for a discrete image of size 4×4. The bold lines within the right-hand grid shows a possible reconstruction path using a corner-based inversion method [5].

dictated by the Katz criterion

$$N \leq 1 + \max \left(\sum_{i=1}^{\mu_{\min}} |p_i|, \sum_{i=1}^{\mu_{\min}} |q_i| \right) \qquad (2)$$

for an $N \times N$ object. Note that one can easily over-specify the projection set (i.e., $\mu > \mu_{\min}$) [10]. When the Katz criterion is not satisfied, ghosts or phantoms arise in the reconstructed image [11, 5]. These ghosts were first studied by Katz (1977, [12]) and are artefacts formed within an image that sum to zero at certain projection angles. Examples of ghosts are shown in figure 2. The goal of this paper is to enable direct the use of the FRT formalism to reconstruct exact images from arbitrary Mojette projection data.

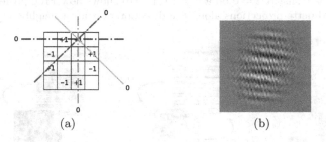

(a)   (b)

Fig. 2. Depiction of simple ghosts in an image. (a) shows image values that lead to ghosts in the four projection directions shown. (b) shows a 61×61 greyscale image of ghosts (black denotes negative and white denotes positive greyscale values). Here the image will "disappear" when viewed at 15 of the 62 FRT projection directions.

1.2   The Finite Radon Transform (FRT)

The FRT $R(t, m)$ of image $I(x, y)$ is a discrete prime-periodic transform where the image space is considered to be a torus of size $\mathfrak{p} \times \mathfrak{p}$ pixels, where $\mathfrak{p}$ is prime [7]. An example of a projection in the FRT and how these projections relate to the MT [13, 14] is given in figure 3.

The FRT projections $R(t, m)$ sum all pixels lying along the lines

$$\begin{aligned} y &\equiv mx + t \pmod{\mathfrak{p}}, & \text{for } 0 \leq m < \mathfrak{p} \\ x &\equiv t, & \text{for } m = \mathfrak{p} \end{aligned} \qquad (3)$$

where $x, y, m, t \in \mathbb{Z}$, with lines at each translate $t = 0, \ldots, \mathfrak{p} - 1$ having slope m. Each projection consists of $\mathfrak{p}$ translates which wrap around the image modulo $\mathfrak{p}$. Due to the primality of $\mathfrak{p}$, each translate of a projection sums precisely $\mathfrak{p}$ pixels, ensuring that each bin has $\mathfrak{p}$ terms within it. This in turn ensures that every pixel is sampled exactly once for each projection. When all $\mathfrak{p} + 1$ projections are taken, the FRT can then be inverted exactly as shown in figure 4(a). An algebraic example of FRT projection and inversion is shown in figure 5. When

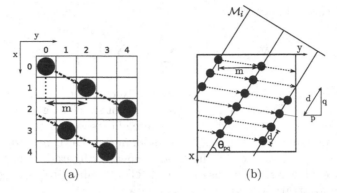

Fig. 3. (a) shows how projections are taken in $I(x, y)$ to form $R(t, m)$. Here the $m = 2$ projection has translate $t = 0$ on a $\mathfrak{p} = 5$ grid. (b) shows how FRT projections can be mapped to Mojette projections along the direction of nearest neighbour lines [13].

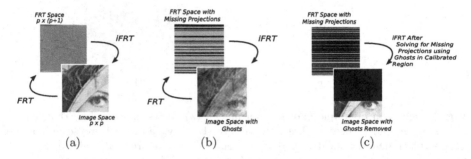

Fig. 4. (a) shows an example of an FRT space and its reconstructed image. The transform is exact and one-to-one. (b) shows the ghosts that form when projections are missing from the FRT (shown as black regions). (c) shows how a sufficient calibrated image region (black) enables exact reconstruction of a sub-region.

any FRT projections are missing, the reconstruction is only partially correct and has ghost artefacts at each pixel corresponding to a superposition of a (negative) contribution from each missing projection. The result of FRT inversion when some projections are missing is shown in figure 4(b). However, given a calibrated region of sufficient size, the missing projections can be recovered from the ghosts (see figure 4(c)). In the next two sections we discuss how ghosts are structured, how they can be used to recover missing projections and what constitutes a sufficient calibrated region.

2 Ghosts in the FRT

Consider a single missing projection or ghost, labelled as $\tilde{\mathbf{a}} = \{a_0, \ldots, a_{\mathfrak{p}-1}\}$, located at row m_a in $R(t, m)$. In the FRT inversion, each projection from $R(t, m)$

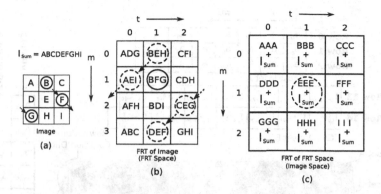

Fig. 5. Projections of the image (a) are taken using equation (3) and placed into FRT space (b) (see example pixels in closed circles). To invert the FRT, projections of the FRT space are taken in the same way as before, but with $m' = \mathfrak{p} - m$ for projections $0 < m < \mathfrak{p}$ (see example pixels in broken circles). Then every pixel in the result (c) is normalised by subtracting I_{Sum} and dividing by $\mathfrak{p}$ to recover the exact image (a).

is mapped, as a whole, onto each row in image space, translated (and wrapped periodically) by mx pixels at row x. The ghost $\tilde{a}$ also shifts periodically by $-m_a$ on each subsequent row of the image. This structure is shown in insets A and B of figure 6. The periodic shift structure of the rows can also be viewed along the column direction, as shown in inset C of figure 6. In the column data, the ghost bin indices increment at multiples of m_j. Conversion between whole rows and columns of ghost data is possible using this property. The following section presents the algorithm to exploit this periodic structure in solving for missing projections.

3   The Mojette Ghost-Recovery Algorithm (MGR)

The ghost structure represented in figure 6 forms the basis for recovering the extra FRT projections needed to invert MT data. An image containing a calibrated region (i.e., where the image values are known) has the ghosts structured as depicted in inset A of figure 6. It is possible to unscramble these ghost superpositions (or signals) exactly in this region to solve for the missing FRT projections. These signals essentially form a set of coupled linear equations that require $\mathcal{N} \times \mathfrak{p}$ calibrated row or column ghost pixels to solve for the $\mathcal{N} \times \mathfrak{p}$ missing bins of projection values. In [6] a conjugate gradient approach was used to unpack Mojette bins into consistent image data. Here the simple arithmetic nature of the FRT is exploited to provide a direct algebraic reconstruction. The idea of using redundant image regions has been used in [10] and [15] for encryption of data as projections. The MGR algorithm is demonstrated in figure 7.

Figure 7(a) shows three rows from the calibrated region where a total of three ghosts ($\tilde{a}$, $\tilde{b}$, $\tilde{c}$ and $\mathcal{N} = 3$) are superimposed. We can first eliminate ghost $\tilde{a}$ by cyclically shifting (or rotating) row 1 and row 2 by $-m_a$ and $-2m_a$ (i.e.,

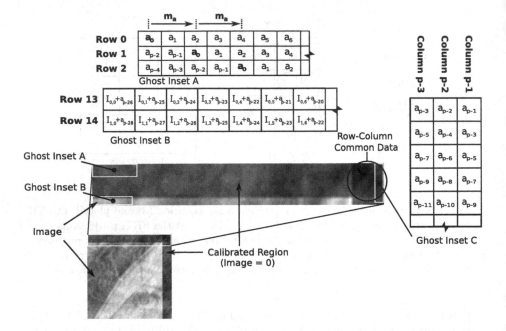

Fig. 6. Ghost structure in image space. The image (Lena) is of 100×100 pixels embedded (by padding) in a prime image space of 113×113. If we assume there is only one missing projection $\tilde{a}$ in FRT space ($\mathcal{N} = 1$) at $m_a = 2$, then inset A shows how the ghost will be structured inside the calibrated region. Inset B shows how artefacts $\tilde{a}$ affects the image data. All ghosts will be structured in the same m-dependent pattern. If multiple ghosts are present, the resulting pattern is a superposition of each ghost pattern.

to the left) respectively and differencing. The result is shown in figure 7(b) which contains a mixture of shifted ghosts $\tilde{b}$ and $\tilde{c}$ as differences. However, the differences have the same ghost order but are shifted with respect to each other. Therefore, we may remove the $\tilde{b}$ differences by aligning them by $-(m_b - m_a)$ and differencing again. The result is shown in figure 7(c), where only differences containing ghost $\tilde{c}$ remain. The differencing process leaves a systematic pattern (marked by the arrows) caused by the relative shifts with respect to m_c used to remove the other ghosts. A consecutive sum of the $\tilde{c}$ differences (where the previous result is carried on to the next) at steps $m_c - m_b$ ($\mathfrak{p}$ times) and then at steps $m_c - m_a$ ($\mathfrak{p}$ times), cancels the negative terms in the pattern, leaving only ghost $\tilde{c}$. Note that this is only made possible by the properties of prime congruence, as attempting to integrate the differences having lengths other than $\mathfrak{p}$ will result in an incomplete ghost. This sum (a discrete integration) leaves the result with a constant offset that can be removed after each integration (see figure 7(d)). The result $\tilde{c}$ can then be back substituted and the process repeated to solve for the remaining ghosts. If the redundant region is sufficient (i.e., there are at least $\mathcal{N}$ full rows or columns in the calibrated region), all

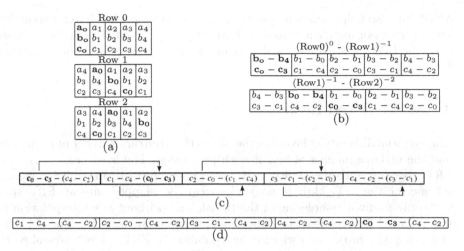

Fig. 7. An example of the MGR algorithm for $\mathfrak{p} = 5$ and $\mathcal{N} = 3$. (a) shows ghosts $\tilde{a}$, $\tilde{b}$ and $\tilde{c}$ for $m = 1, 2, 3$ superimposed in Row 0,1 and 2. The addition signs are removed for clarity. (b) shows ghost $\tilde{a}$ eliminated after alignment and pair differences of rows 0, 1 & 2. The superscripts on row labels denote direction and size of cyclic shifts. (c) shows ghost $\tilde{a}$ and $\tilde{b}$ eliminated, leaving a summation of four shifted copies of ghost $\tilde{c}$. At each step the remaining ghosts are compounded by the differencing. Integration (at steps shown by arrows) simplifies terms to leave the ghost $\tilde{c}$ with constant offset. (d) shows the first integration result at step $m_c - m_b$. Offset $-(c_4 - c_2)$ is removed and the next integration done at step $m_c - m_a$.

missing projections can be recovered exactly. In summary, the algorithm for solving missing projections is:

1. Align the first ghost (say $\tilde{a}$) by shifting the row or column data cyclically using $-m_a$ and differencing $\mathcal{N} - 1$ pairs of signals. This removes ghost $\tilde{a}$ from all those signals. Shifting by $-m_a$ also shifts all other ghosts by $-m_a$.
2. Align the next ghost (say $\tilde{b}$) by the new relative shift $-(m_b - m_a)$ and difference. This removes ghost $\tilde{b}$ from the signals.
3. Repeat until only the $\mathcal{N}^{\text{th}}$ ghost is left. Note that the solve order is completely arbitrary (i.e., in figure 7 we could have solved for $\tilde{b}$, followed by $\tilde{a}$ to get $\tilde{c}$).
4. Integrate the last ghost (removing the integration constant after each integration) to get the missing projection. Back substitute and repeat the algorithm to determine the other ghosts.

The advantageous features of this algorithm are:

1. The algorithm processes just $\mathcal{N}$ 1D cyclic blocks of data, each of length $\mathfrak{p}$.
2. The algorithm is exact and non-iterative in nature.
3. The solve order is arbitrary. When processing noisy projection data, this may be an advantage relative to "corner-based" methods.
4. The algorithm is easily and highly parallelised, because each missing projection can be solved independently.

MGR will also help to understand the nature of artefacts in CT reconstructions arising from non-uniqueness as discussed by [11], [12] and [16]. In the next section, the FRT ghosts are utilised to demonstrate a simple inversion mechanism for the Mojette transform.

4   MGR-Based Inversion

The essential difficulty of inverting the MT is the arbitrary number of projections and the variable number of bins in each projection. The fixed properties of the FRT, together with the ability to recover ghosts, provide an explicit inversion scheme for the MT. Mojette projections can be mapped into an FRT space with only a minor re-ordering of the translates t without any interpolation (as was required for reconstructing the real sinogram data in [13]). The net effect of placing MT projections without interpolation in FRT space is to embed the sought image within a larger image space. The size of the larger image space is dictated by the prime $\mathfrak{p}$ chosen so that the longest MT projection (the largest $\mathcal{M}_i$ as given by equation (1)) fits into FRT space. Merging of bins is avoided in order to preserve the information in the Mojette set under the Katz Criterion. Since MT projections tend to be longer and smaller in number than for FRT projections, the resulting FRT space largely consists of missing projections. After recovering the ghosts from the calibrated region in image space using the MGR algorithm, the Mojette projections may be inverted using the very simple inverse FRT algorithm. This final inversion of the FRT can also be done efficiently by selecting the appropriate translates from $R(t, m)$ space so only the embedded image is reconstructed (ignoring the calibrated area). Hence, the MT inversion scheme is:

1. Reorder and place MT projections into a sufficiently large FRT space.
2. Recover the missing FRT projections. This is done by creating ghosts and using the MGR algorithm to untangle each of the missing projections.
3. Invert the FRT data using only the translates in FRT space that correspond to the embedded image.

The scheme has the favourable property of not relying on the corners of the image (as required by most other exact inversion methods) and should be more noise tolerant. In the following section, initial results of the MGR inversion scheme are presented and its current limitations are discussed.

5   Results

The MGR scheme was applied to the MT inversion of a 100×100 image. The MT of the image is shown in figure 8(a) and the subsequent exact reconstruction in figure 8(b). In order to solve the system, the total number of missing projections could not be greater than the size of the calibrated region. As the FRT has $\mathfrak{p} + 1$ projections, a total of 101 MT projections had to be taken. This meant solving

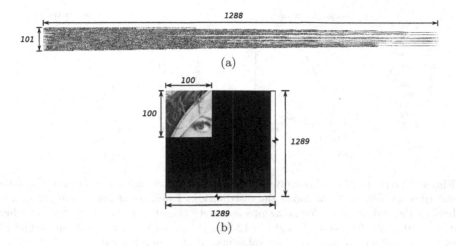

(a)

(b)

Fig. 8. Illustration of the exact and non-iterative image reconstruction using MGR inversion. (a) shows the MT for the Lena image (100×100 pixels) using 101 projections. (b) shows the exact inversion result of the MGR algorithm using rows. The larger image space is $\mathfrak{p} = 1289$ and required solving for 1188 missing projections. The reconstructed image, mostly calibrated space, has been truncated for display.

for $1289 - 101 = 1188$ missing projections, since the longest Mojette projection was $\mathcal{M}_i(\max) = 1288$ (hence $\mathfrak{p} = 1289$) corresponding to $p = -7$ and $q = 6$.

An interesting phenomenon also occurs during the differencing process. When taking differences while shifting periodically, the magnitude of the differences compound as shown in figure 7(c). When $\mathcal{N}$ is large, the pattern of resultant differences of the $\mathcal{N}^{\text{th}}$ ghost are almost purely sinusoidal. This sinusoidal effect is probably due to the differences being zero mean and periodic. The differences are always zero mean because all FRT projections must sum to the same constant I_{Sum}. The process was replicated using a computer algebra package and solved analytically. Here the ghost bin addresses were represented by variables, as in figure 7. The weight each ghost bin contributes to the ghost differencing process is shown in figure 9. This shows that the sinusoidal behaviour is independent of image data and dependent on geometry of the problem. For large $\mathcal{N}$, the same sinusoidal behaviour leads to rapid numerical growth of the difference values (reaching 10^{40} in the case of figure 8(b) for $\mathcal{N} = 1188$). An example of the rapid numerical growth is given in figure 10. This problem so far has been overcome by using arbitrary-precision integers for the differencing and integration processes in the MGR algorithm. Controlling the numerical growth is an area of future work which is discussed in the next section.

6 Further Work

The main remedy for the limitation in the number of projections required will be to fully utilise all parts of the redundant image area (i.e., a combination of

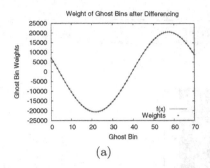

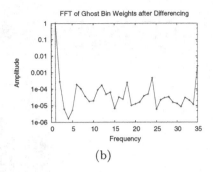

(a)   (b)

Fig. 9. The result of algebraically solving a ghost system with $\mathbf{p} = 71$ and $\mathcal{N} = 58$ as described in section 3. (a) shows the distribution of weights of the missing projection bins at the end of the differencing process. Here the fit $f(x) = -a\sin(bx + c)$ where $a = 20591.2$, $b = 0.0885 \approx \frac{2\pi}{\mathbf{p}}$ and $c = 12.212$. (b) shows Fourier transform coefficients of (a), normalised to peak absolute value and plotted on a logscale.

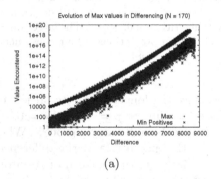

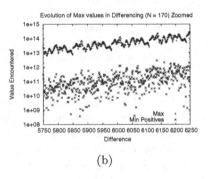

(a)   (b)

Fig. 10. The max and minimum positive values of the compounded ghosts $\mathcal{N} = 170$ during the differencing process. The max values have a fine-scale oscillatory imprint as shown in (b). Each period consists of one less point than the previous cycle, as each ghost is progressively eliminated.

a_3	a_4	$\mathbf{a_0}$	a_1	a_2
b_2	b_3	b_4	$\mathbf{b_0}$	b_1
a_4	a_2	$\mathbf{a_0}$	a_3	a_1
b_4	b_1	b_3	$\mathbf{b_0}$	b_2

(a)

a_4	a_2	$\mathbf{a_0}$	a_3	a_1
b_3	b_1	b_4	b_2	$\mathbf{b_0}$
a_4	a_2	$\mathbf{a_0}$	a_3	a_1
b_4	b_1	b_3	$\mathbf{b_0}$	b_2

(b)

$b_3 - b_4$	0	$b_4 - b_3$	$b_2 - \mathbf{b_0}$	$\mathbf{b_0} - b_2$

(c)

Fig. 11. An example of solving $\mathcal{N} = 2$ using a row and a column. Here we denote ghosts as $\tilde{\mathbf{a}}$ and $\tilde{\mathbf{b}}$ having projections $m = 2, 3$ respectively. (a) shows the initial data showing row 1 and column $\mathbf{p} - 1$. (b) shows the row put into column form and aligned so $\tilde{\mathbf{a}}$ can be removed. (c) shows $\tilde{\mathbf{a}}$ eliminated by differencing the two rows in table (b). The resulting $\tilde{\mathbf{b}}$ differences cannot be integrated because of the cancellation of like terms. This occurs regardless of the starting position of the integration. Also, the integration step required is no longer constant.

complete rows and columns of the calibrated region). Initial attempts to use both the rows and columns simultaneously have been frustrated by the interactions between row and column data (see figures 6 and 11) as $\mathfrak{p}$ independent values per row or column are needed to untangle the ghosts and integrate the result. We are also considering an alternative algorithm that uses the calibrated area to retrieve the compounded ghost values from the reconstructed image area, pixel by pixel, rather than independently solving for each ghost in its entirety.

To remedy the problem of numerical growth, the compounding of the ghost differences must be understood. Any truncation or wrapping of the differences leads to incorrect results and any re-normalisation of the values will still require high precision floating point representation. Also note from the Fourier transform of figure 9(a) (shown in figure 9(b)) that the sinusoidal signature includes very small but essential contributions from other frequencies whose origins are unclear. Future work will also address the speed relative to other Mojette inversion algorithms as well as the robustness of MGR to noise in the projection data.

7 Conclusion

An MT projection set mapped into $R(t, m)$ space has missing FRT projections or ghosts that have a known, precise structure in the reconstructed image. This structure allows the construction of algorithms to recover the missing FRT projections. The algorithms rely on a calibrated region being present with the image within a larger image space. The MGR algorithm is exact, non-iterative and easily parallelised.

The algorithm was applied to invert the MT as shown in figure 8. Known MT projections are converted to $R(t, m)$ projections. The sought image then becomes a subset of a larger image space. Since the small number of MT projections leave many missing FRT projections, portions of the larger image space are used as a calibrated area for the MGR algorithm, in order to recover these projections. Once the FRT space is filled, an efficient subset inverse FRT can be used to invert the MT.

The algorithm has two main drawbacks at the moment. The current use of either full rows or columns in the calibrated region leads to a limit on the number of MT projections able to be processed. The algorithm also leads to interesting ghost structures (see figure 9) and involves very large integer values as shown in figure 10. A further study of these ghost structures may reveal a means to control or suppress the numerical growth. The remaining area of concern is then to establish the robustness of the MGR algorithm to noise in the MT projection data. Initial results show that small levels of added noise are well tolerated, but more work is required to quantify these findings.

References

[1] Kak, A.C., Slaney, M.: Principles of Computerized Tomographic Imaging. Society of Industrial and Applied Mathematics (2001)
[2] Beylkin, G.: Discrete Radon transform. IEEE Trans. Acoust., Speech, Signal Processing 35(2), 162–172 (1987)

[3] Averbuch, A., Donoho, D., Coifman, R., Israeli, M., Walden, J.: Fast slant stack: A notion of Radon transform for data on a cartesian grid which is rapidly computable, algebraically exact, geometrically faithful and invertible. Tech. Report, Stanford University 11 (2001)

[4] Kingston, A., Svalbe, I.: Projective transforms on periodic discrete image arrays. Advances in Imaging and Electron Physics 139, 75–177 (2006)

[5] Guédon, J.P., Normand, N.: The Mojette transform: the first ten years. In: Andrès, É., Damiand, G., Lienhardt, P. (eds.) DGCI 2005. LNCS, vol. 3429, pp. 79–91. Springer, Heidelberg (2005)

[6] Servières, M., Idier, J., Normand, N., Guédon, J.P.: Conjugate gradient Mojette reconstruction. Proceedings of the SPIE - The International Society for Optical Engineering 5747(1), 2067–2074 (2005)

[7] Matúš, F., Flusser, J.: Image representation via a finite Radon transform. IEEE Transactions on Pattern Analysis and Machine Intelligence 15(10), 996–1006 (1993)

[8] Edholm, P., Herman, G.: Linograms in image reconstruction from projections: Part 1. back projecting using the Radon transform. In: Proc. Twenty-First Annual Hawaii International Conference on System Sciences, vol. 4, pp. 48–50 (1988)

[9] Hardy, G., Wright, E.: An Introduction to the Theory of Numbers, 5th edn. Clarendon Press, Oxford (1979)

[10] Normand, N., Guédon, J.P., Philippe, O., Barba, D.: Controlled redundancy for image coding and high-speed transmission. In: Proceedings of the SPIE - The International Society for Optical Engineering, vol. 2727, pp. 1070–1081 (1996)

[11] Philippe, O., Guédon, J.P.: Correlation properties of the Mojette representation for non-exact image reconstruction. In: ITG-Fachbericht Verlag, Proc. Picture Coding Symposium, vol. 143, pp. 237–242 (1997)

[12] Katz, M.: Questions of Uniqueness and Resolution in Reconstruction from Projections. Springer, Heidelberg (1977)

[13] Svalbe, I., van der Spek, D.: Reconstruction of tomographic images using analog projections and the digital Radon transform. Linear Algebra and Its Applications 339, 125–145 (2001)

[14] Servières, M., Guédon, J.P., Normand, N.: A discrete tomography approach to PET reconstruction. In: Proc. 7th International Conf. on Fully 3D Reconstruction in Radiology and Nuclear Medicine (2003)

[15] Kingston, A., Svalbe, I.: Geometric effects in redundant keys used to encrypt data transformed by finite discrete Radon projections. In: Proc. IEEE Digital Imaging Computing: techniques and applications, Cairns, Australia (2005)

[16] Boyd, J., Little, J.: Complementary data fusion for limited-angle tomography. In: Proceedings 1994 IEEE Computer Society Conference on Computer Vision and Pattern Recognition, pp. 288–294 (1994)

Approximating hv-Convex Binary Matrices and Images from Discrete Projections

Fethi Jarray[1,2,*], Marie-Christine Costa[2], and Christophe Picouleau[2]

[1] Gabes University of Sciences 6072 Gabes, Tunisia
[2] Laboratoire CEDRIC, 292 rue Saint-Martin, 75003 Paris, France
fethi_jarray@yahoo.fr, {costa,chp}@cnam.fr

Abstract. We study the problem of reconstructing hv-convex binary matrices from few projections. We solve a polynomial time case and we determine some properties of the hv-convex matrices. Since the problem is NP-complete, we provide an iterative approximation based on a longest path and a min-cost/max-flow model. The experimental results show that the reconstruction algorithm performs quite well.

Keywords: Discrete Tomography; hv-convex; Image Reconstruction.

1 Introduction

Discrete tomography deals with the reconstruction of discrete homogeneous objects regarded as binary matrices from their projections. The problem of reconstructing a $m \times n$ binary matrix from its orthogonal projections H and V is the following [12]: given $H = (h_1, \ldots, h_m)$ and $V = (v_1, \ldots, v_n)$ two nonnegative integer vectors find a binary matrix such that the number of ones in every row i (resp. column j) equals h_i (resp. v_j). Ryser [12] gives necessary and sufficient conditions for the existence of a solution. However, the problem is usually highly underdetermined and a large number of solutions may exist [13]. The reader is referred to the book of Herman and Kuba [9] for an overview on discrete tomography.

In many applications such as image processing and electron microscopy, the orthogonal projections alone are not sufficient to uniquely determine matrices or objects. Fortunately, objects that occur in practical applications usually exhibit certain properties. Hence we seek to reconstruct binary matrices under additional constraints like connectivity or convexity for instance. Woeginger [14] prove that the consistency problem for polyominoes (connected sets) is NP-complete. The consistency problem for h-convex objects (polyominoes or not) is also NP-complete [1]. The above result extends to the v-convex objects. The consistency problem for hv-convex objects is NP-complete [1,14]. Therefore, the reconstruction can be solved in polynomial time only for the hv-convex polyominoes objects [1,5].

In the present paper, we will deal with the problem of reconstructing hv-convex matrices. Since the problem is NP-complete, one way to solve it is to

---

D. Coeurjolly et al. (Eds.): DGCI 2008, LNCS 4992, pp. 413–422, 2008.

concentrate on approximative solutions. There are few papers in the literature on algorithms providing approximating results, i.e. returning binary matrices respecting the orthogonal projections but not necessary hv-convex[6,3,2]. Dahl and Flatberg [6] provide an algorithm based on a lagrangian relaxation for reconstructing a nearly hv-convex matrix respecting the prescribed projections.

In this paper, we propose a new algorithm for approximately reconstructing hv-convex matrices from their projections. Our algorithm performs a sequence of related reconstruction, each using only one projection. The algorithm uses longest path and network flows algorithms to provide a solution.

The paper is organized as follows. Notations and properties of hv-convex matrices are introduced in section 2. In section 3, we study the problem of reconstructing a hv-convex matrix respecting only one orthogonal projection (H or V). In Section 4, we propose a heuristic based on longest path and network models to approximately solve the problem with both projections. Numerical results are presented and discussed in the last section.

2   Definitions and Properties

Let x be an $m \times n$ binary matrix. The horizontal projection of x is the vector $H = (h_1, \ldots, h_m)$ such that $h_i = \sum_{j=1}^{n} x_{ij}$ is the sum of the elements lying on row i. The vertical projection of x is the vector $V = (v_1, \ldots, v_n)$ where $v_j = \sum_{i=1}^{m} x_{ij}$ is the sum of the elements on column j. Both projections H and V constitute the orthogonal projections of x. A matrix is horizontally convex (h-convex) if all the 1's of each row are adjacent. A matrix is vertically convex (v-convex) if all the 1's of each column are adjacent. A matrix is hv-convex if it is both h-convex and v-convex.

Because discrete tomography is very related to digital image processing we often refer to binary matrices as binary images and call the matrix entries pixels with values black (0) and white (1).

The related decision problem associated to the problem of reconstruction hv-convex matrix is defined as follows:

Instance: Given $H = (h_1, \ldots, h_m)$ and $V = (v_1, \ldots, v_n)$ two nonnegative integer vectors.
Question: Is there a $m \times n$ hv-convex matrix respecting the horizontal projection H and the vertical projection V?

Definition 1. *Let x and y be two $m \times n$ binary matrices. We define the resemblance $R(x, y)$ between x and y as $R(x, y) = \sum_{i=1}^{m} \sum_{j=1}^{n} x_{ij} y_{ij}$.*

$R(x, y)$ is the number of common 1's between x and y.

For any binary matrix x, denote the number of pairs of adjacent 1's (either horizontal or vertical) in x by n_x.

Proposition 1. *Let x be a binary matrix with orthogonal projections (H, V) then* *i)* $n_x = \sum_{i=1}^{m} \sum_{j=1}^{n-1} x_{ij} x_{i,j+1} + \sum_{j=1}^{n} \sum_{i=1}^{m-1} x_{ij} x_{i+1,j}$

ii) $n_x \leq 2 \sum_{j=1}^{n} v_j - m - n$

iii) x is hv-convex if and only if $n_x = 2 \sum_{j=1}^{n} v_j - m - n$.

Proof. Let x be a binary matrix respecting the projectins (H, V). The number of adjacent 1's on row i is $\sum_{j=1}^{n-1} x_{ij} x_{i,j+1} \leq h_i - 1$. Row i is h-convex if and only if the number of adjacent 1's is $h_i - 1$.

Analogously, the number of adjacent 1's on column j is $\sum_{i=1}^{m-1} x_{ij} x_{i+1,j} \leq v_j - 1$. Column j is v-convex if and only if the number of adjacent 1's is $v_j - 1$.

Thus we deduce that $n_x = \sum_{i=1}^{m} \sum_{j=1}^{n-1} x_{ij} x_{i,j+1} + \sum_{j=1}^{n} \sum_{i=1}^{m-1} x_{ij} x_{i+1,j} \leq \sum_{i=1}^{m} (h_i - 1) + \sum_{j=1}^{n} (v_j - 1) = 2 \sum_{i=1}^{m} h_i - m - n = 2 \sum_{j=1}^{n} v_j - m - n$. So $n_x \leq 2 \sum_{j=1}^{n} v_j - m - n$ is an upper bound on n_x.

A matrix is hv-convex if and only if all the rows are h-convex and all the columns are v-convex. So, a matrix x is hv-convex if and only if $n_x = 2 \sum_{j=1}^{n} v_j - m - n$.

3   Solving Polynomial Cases

We propose two algorithms to solve two important particular subproblems.

3.1   Reconstructing a hv-Convex Matrix Respecting One Projection hv-Convex(V)

In this section, we provide a polynomial time algorithm to reconstruct a hv-convex matrix respecting the vertical projection V.

We suppose that on each column j the 1's are placed on the rows from s_j to $e_j = s_j + v_j - 1$ since the matrix is hv-convex. To reconstruct a solution, it is sufficient to determine s_j or e_j for each column.

Proposition 2. *A $m \times n$ binary matrix with vertical projection V is hv-convex only if $m \geq v_1 + \sum_{j=2}^{n} \max(v_j - v_{j-1}, 0)$*

Proof. We will demonstrate the proposition by induction on n. For $n = 1$, the proposition is true because $m \geq v_1$. Suppose that the proposition is true for a $m \times (n-1)$ matrix and show that is also verified for a $m \times n$ matrix.

Let x be a $m \times n$ hv-convex binary matrix with vertical projection V. Denote by x_{n-1} the submatrix of x of size $m \times (n-1)$ composed by the first $n-1$ columns. We have $m \geq v_1 + \sum_{j=2}^{n-1} \max(v_j - v_{j-1}, 0)$ because x_{n-1} is also hv-convex.

Two cases can be distinguished according to the values of v_{n-1} and v_n.

- If $v_n \leq v_{n-1}$, the proposition is true because $v_1 + \sum_{j=2}^{n} \max(v_j - v_{j-1}, 0) = v_1 + \sum_{j=2}^{n-1} \max(v_j - v_{j-1}, 0) \leq m$.
- If $v_n > v_{n-1}$, we denote by A the set of rows of the submatrix x_{n-1} having at least a 1 from columns $1, \ldots, n-1$, B the set of rows having a 1 on column $n-1$ and C the set of rows having a 1 on column n. We have $|B| = v_{n-1}$,

$|C| = v_n$ and $|A| \geq v_1 + \sum_{j=2}^{n-1} \max(v_j - v_{j-1}, 0)$ because the submatrix x_{n-1} is hv-convex. The set of rows of x is equal to $A \cup C$. Since the matrix x is h-convex then $A \cap C \subset B$. Thus $m = |A \cap C| = |A| + |C| - A \cap C \geq |A| + |C| - |B| \geq v_1 + \sum_{j=2}^{n-1} \max(v_j - v_{j-1}, 0) + v_n - v_{n-1} = v_1 + \sum_{j=2}^{n} \max(v_j - v_{j-1}, 0)$. We conclude that the proposition is true for the matrix of size $m \times n$.

In each step j, the following greedy algorithm sets to 1 v_j cells from row s_j to row $s_j + v_j - 1$.

Greedy algorithm
$s_1 = 1$
For $j = 2$ to n **do**
    if $v_{j-1} \leq v_j$ then $s_j = s_{j-1}$
    if $v_{j-1} > v_j$ then $e_j = e_{j-1}$

The following result establishes the validity of the greedy algorithm.

Proposition 3. *The greedy algorithm solves in polynomial time the problem hv-convex(V).*

Proof. It is obvious that the reconstructed matrix is v-convex and respects the vertical projection. This matrix is also h-convex because it is v-convex, that is, $s_j \leq e_j$, $j = 1, \ldots, n$, the sequence $(s_1, \ldots, s_n)$ is monotone non decreasing and the sequence $(e_1, \ldots, e_n)$ is also monotone non decreasing. The reconstructed matrix has a minimal number of rows $(m = v_1 + \sum_{j=2}^{n} \max(v_j - v_{j-1}, 0))$.

We get simular result if the vertical projection is relaxed instead of the horizontal one. We establish the following general result:

Proposition 4. *A $m \times n$ binary matrix with orthogonal projections (H, V) is hv-convex only if $m \geq v_1 + \sum_{j=2}^{n} \max(v_j - v_{j-1}, 0)$ and $n \geq h_1 + \sum_{i=2}^{m} \max(h_i - h_{i-1}, 0)$*

3.2   Reconstructing a Nearly h-Convex Matrix Respecting the Horizontal Projection h-Convex(H,w)

The problem *h-convex(H,w)* consists of reconstructing a binary matrix x respecting the horizontal projection H, having the maximum number of adjacent 1's on the rows and being as near as possible to a given binary matrix w. We define the problem *v-convex(V,w)* analogously. We formulate the reconstruction problem as a quadratic integer program. A very natural formulation is the following:

$$P_h \quad \begin{cases} \max \sum_{i=1}^{m} \sum_{j=1}^{n} x_{ij} w_{ij} + \sum_{i=1}^{m} \sum_{j=1}^{n-1} x_{ij} x_{i,j+1} \\ \sum_{j=1}^{n} x_{ij} = h_i \quad i = 1, \ldots, m \\ x_{ij} \in \{0, 1\} \end{cases} \quad (1)$$

The objective function is the sum of the resemblance between the matrices x and w (first term) and the number of adjacent 1's on the rows of x. The constraint (1) is necessary and sufficient to guarantee the satisfaction of the horizontal projection.

We note that the rows are decoupled and the problem P_h can be decomposed into m subproblems, one per row. For a given row i, the associate subproblem to solve has the following general form:

$$Q \quad \begin{cases} \max \sum_{j=1}^{n} z_j c_j + \sum_{j=1}^{n-1} z_j z_{j+1} \\ \sum_{j=1}^{n} z_j = b \\ z_j \in \{0,1\} \end{cases} \tag{2}$$

with $x_{ij} = z_j$, $w_{ij} = c_j$ and $h_i = b$.

Dahl and Fatberg [6] show that the program Q is equivalent to a longest path problem with exactly $b+1$ vertices in the directed graph G presented in Figure 1. In this graph there is a node z_j for every column j. In addition, there is an artificial source node s and a sink node t. The arcs of G are defined as follows:

- There is an arc (s, z_j) with length $l(s, z_j) = c_j$, $j = 1, \ldots, n$.
- There is an arc (z_j, t), with length $l(z_j, t) = 0$, $j = 1, \ldots, n$.
- There is an arc (z_j, z_{j+1}), with length $l(z_j, z_{j+1}) = c_{j+1} + 1$, $j = 1, \ldots, n-1$.
- There is an arc (z_k, z_j) with length $l(z_k, z_j) = c_j$, $k = 1, \ldots, j-2$.

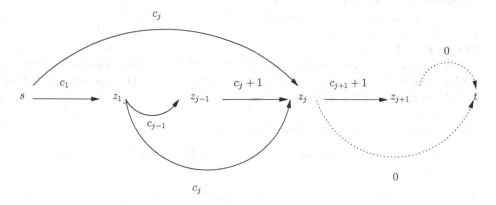

Fig. 1. The program Q and the associated longest path problem

The longest path problem with given number of vertices can be solved in polynomial time by dynamic programming [6].

We define $L(z_j, p)$ to be the length of the longest path from s to z_j using p vertices. The $L(z_j, p)$ verifies the following relations:

$$\begin{cases} L(z_j, 1) = c_j, \; j = 1, \ldots, n \\ L(z_j, p) = \max_{k<j}\{L(z_k, p-1) + l(z_k, z_j)\} \end{cases}$$

The longest path problem with exactly $b+1$ vertices from s to t is with length $L(t, b+1) = max_j\{L(z_j, b)\}$. For Q, we have a complexity of $O(bn^2)$, yielding a time complexity of $O(n^4)$ for P_h since $h_i \leq n$, $i = 1, \ldots, m$.

4 Heuristic

We provide the iterative algorithm A-hv(H,V) based on a longest path model to compute an approximate solution to the problem of reconstructing an hv-convex matrix. The aim is to maximize the resemblance between the solutions to *h-convex(H,w)* and *v-convex(V,w)*. Every iteration consists of solving *h-convex(H,w)* or *v-convex(V,w)* with an objective function depending on the previous iteration. The longest path model is used to solve the problems *h-convex(H,w)* and *v-convex(V,w)*(see Section 3.2). The length matrix w is chosen in such a way that the new reconstruction has the maximal resemblance with the previous one.

Firstly, we compute a solution y^0 to *hv-convex(V)* by using the greedy algorithm (see Section 3.1). Then, we determine x^1 solution to *h-convex(H, y^0)* by solving the associated longest path problem where the length matrix $w = y^0$, i.e. a solution that differs from y^0 in as few entries as possible. Subsequently x^1 is used as a length matrix to determine a solution to *v-convex(V, x^1)*. This procedure is repeated until an optimum is reached, i.e. the resemblance between the solutions to *h-convex(H,w)* and *v-convex(V,w)* becomes constant. As a summary we can describe the reconstruction algorithm as follows:

Algorithm A-hv(H,V)
Compute y^0 solution to $hv - convex(V)$.
$i = 0,\ R^{-1} = -1,\ R^0 = 0,$
While $R^i > R^{i-1}$ **do**
    Compute x^{i+1} solution to *h-convex(H, y^i)* maximizing the resemblance with y^i.
    Compute y^{i+1} solution to *v-convex(V, x^{i+1})* maximizing the resemblance with x^{i+1}
    $R^{i+1} = R(x^{i+1}, y^{i+1})$ and $i = i + 1$.

We will give some properties of this heuristic.

Proposition 5. *The resemblance is increasing from iteration to iteration, i.e.* $R^{i+1} \geq R^i$.

Proof. $R(x^{i+1}, y^{i+1}) \geq R(x^i, y^i)$ because $R(x^{i+1}, y^i) \geq R(x^i, y^i)$ and $R(x^{i+1}, y^{i+1}) \geq R(x^{i+1}, y^i)$.

We conclude that this heuristic is a polynomial time algorithm because the resemblance is not grater than the maximal number of adjacent 1's (n_x).

Proposition 6. *If* $R(x^i, y^i) < R(x^{i+1}, y^{i+1})$ *then* $x^{i+1} \neq x^j$ *for* $j = 1, \ldots, i$.

Proof. Suppose that $x^{i+1} = x^j$ for some j in the range $1 \leq j \leq i$. By the proposition above, $R(x^j, y^j) \leq R(x^i, y^i)$ and by the algorithm A-hv(H,V), $R(x^j, y) \leq R(x^j, y^j)$ for every solution y to *v-convex(V)*. Hence $R(x^j, y) \leq R(x^j, y^j) \leq R(x^i, y^i) < R(x^{i+1}, y^{i+1})$ for all y. In particular for $y = y^{i+1}$, we get $R(x^j, y^{i+1}) < R(x^{i+1}, y^{i+1})$, a contradiction since $x^{i+1} = x^j$.

Let x and y be the solutions provided by the previous algorithm to the problems *h-convex(H,w)* and *v-convex(V,w)*. If $x = y$ then x is an exact solution to $hv(H, V)$. Otherwise, we solve the following integer program to get a binary matrix z respecting (H, V) and nearly hv-convex.

$$
\begin{cases}
\max \sum_{i=1}^{m} \sum_{j=1}^{n} z_{ij} x_{ij} + \sum_{i=1}^{m} \sum_{j=1}^{n} z_{ij} y_{ij} \\
\sum_{j=1}^{n} z_{ij} = h_i \quad i = 1, \ldots, m \\
\sum_{i=1}^{m} z_{ij} = v_j \quad j = 1, \ldots n \\
z_{ij} \in \{0, 1\}
\end{cases}
$$

The matrix z respects both projections and resembles x and y since the objective function is $R(z, x) + R(z, y)$. This program is equivalent to a min-cost/max-flow in a complete bipartite graph [6,10].

5 Computational Results

We have implemented our algorithm in language C. The min-cost/max-flow models used by the heuristic are solved by the $CS2$ network flow library developed by Andrew Goldberg [8]. All results are obtained using a PC with 3.8 GHz processor and 512 MBs of RAM.

The main criterion to evaluate the performance of our heuristic is the ability to reconstruct hv-convex or nearly hv-convex matrix. For example, if the optimal solution has 100 adjacent 1's and our heuristic provides a solution with 95 adjacent 1's, we say that the algorithm is efficient. For each problem, the heuristic either converges to an hv-convex matrix or provides an approximate hv-convex matrix. As in [6,2], we have tested our heuristic with two sets of images.

5.1 hv-Convex Images

The first set of test cases consists of hv-convex images of various sizes. Figure 2 illustrates an approximate solution provided by the algorithm.

The results of computational experiments are summarized in Table 1. The first column contains the size of the problem. The next column gives the upper bound on the number of adjacent 1's given by Proposition 1 ($n_x \leq 2 \sum_{j=1}^{n} v_j - m - n$). Adjacent is the number of adjacent 1's on the provided solution. gap is the gap between the number of adjacent 1's on the approximated solution and the upper bound. The last column displays the total CPU time in seconds.

For the small sizes, the upper bound is tight and the algorithm gives an hv-convex matrix. For the remaining instances, the algorithm finds an approximative solution respecting both projections and having the maximum number of adjacent 1's.

The gap between the number of adjacent 1's on the approximative solution and the upper bound is about 2%. Since, the difference is negligible by regarding the size of the matrices, we conclude that the heuristic gives an approximate hv-convex matrice respecting as much as possible both projections.

Fig. 2. A 40 × 40 image instance and its reconstruction

Table 1. Reconstruction results for hv-convex images

size	bound	adjacent	gap(%)	run time
(9,23)	66	66	0	0
(40,40)	96	96	0	0.36
(50,40)	2750	2673	2.8	0.62
(50,50)	3438	3270	5	0.77
(75,75)	7064	6973	1.3	3.21
(100,100)	10322	10203	1.2	8.45

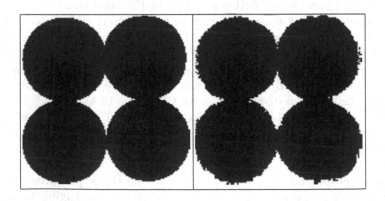

Fig. 3. A 40 × 40 image instance and its reconstruction

5.2   Random Circles

The second group of tests consists of random circles of various sizes. The matrices of this set are not necessary hv-convex (see Figure 3).

The computational results are presented in Table 2. This set of test cases exposes another advantage of the algorithm. For all the large tested problems, the algorithm provides a solution with an average number of adjacent 1's very close to the upper bound. The overall execution time of the algorithm is always under 30s.

In [6], the hv-convex problem was approximately solved by a lagrangian decomposition method. Regarding the processing time, our algorithm seems to be much faster than the previous algorithm. We note also that our heuristic is much simpler than the previous algorithm which uses a gradient procedure to solve the lagrangian problem.

Table 2. Reconstruction results for random circles

size	bound	adjacent	gap (%)	run time
(86,99)	7785	7540	3	5.13
(136,133)	18133	17591	3	22.86
(117,148)	19701	18766	4.75	29.70
(126,125)	24565	24328	1	22
(117,136)	17337	16788	3.2	22.85

6   Conclusion

In this paper, we have provided an iterative algorithm to approximate hv-convex matrices from orthogonal projections. For evaluation, we have considered two types of images: hv-convex and random circles. Given the size and the complexity of reconstructing hv-convex matrices, we believe that the computational results are very encouraging as they show that the problem can be solved to near optimality very fast.

Acknowledgments

The authors would like to thank the referees for their valuable comments and suggestions.

References

1. Barcucci, E., Del Lungo, A., Nivat, M., Pinzani, R.: Reconstructing convex polyominoes from their horizontal and vertical projections. Theoretical computer science 155(1), 321–347 (1996)

2. Batenburg, K.J.: An evolutionary algorithm for discrete tomography. Discrete applied mathematics 155(1), 36–54 (2005)
3. Boufkhad, Dubois, Nivat, M.: Reconstructing (h,v)-convex 2-dimensional patterns of objects from approximate horizontal and vertical projections. Theoretical Computer Science 290(3), 1647–1664 (2003)
4. Brualdi, R.A.: Matrices of zeros and ones with fixed row and column sum. Linear algebra and its applications 3, 159–231 (1980)
5. Chrobak, M., Dürr, C.: Reconstructing hv-convex polyominoes from orthogonal projection. Inform. Process. Lett. 69, 283–289 (1999)
6. Dahl, G., Fatberg, T.: Optimization and reconstruction of hv-convex (0,1)-matrices. Discrete applied mathematics 155(1), 93–105 (2005)
7. Gardner, R.J., Gritzmann, P., Prangenberg, D.: On the computational complexity of determining polyatomic structures by X-rays. Theoretical computer science 233, 91–106 (2000)
8. Goldberg, A.V.: An efficient implementation of a scaling minimum-cost flow algorithm. Journal of algorithms 22, 1–29 (1997)
9. Herman, G.T., Kuba, A.: Discrete Tomography: Foundations, Algorithms and Applications. Birkhäuser, Basel (1999)
10. Jarray, F.: Solving problems of discrete tomography. Applications in workforce scheduling, Ph.D. Thesis, University of CNAM, Paris (2004)
11. Picouleau, C.: Reconstruction of domino tiling from its two orthogonal projections. Theoretical computer science 255(1), 437–447 (2001)
12. Ryser, H.J.: Combinatorial Properties of Matrices of Zeros and Ones. Canad. J. Math 9, 371–377 (1957)
13. Wang, B., Zhang, F.: On the precise Number of (0,1)-Matrices in u(R,S). Discrete mathematics 187, 211–220 (1998)
14. Woeginger, G.J.: The reconstruction of polyominoes from their orthogonal projections. Information Processing Letters 77(5-6), 225–229 (2001)

Advances in Constrained Connectivity

Pierre Soille and Jacopo Grazzini

Spatial Data Infrastructures Unit
Institute for Environment and Sustainability
Joint Research Centre of the European Commission
T.P. 262, via Fermi 2749, I-21027 Ispra, Italy
{Pierre.Soille,Jacopo.Grazzini}@jrc.it

Abstract. The concept of constrained connectivity [Soille 2008, PAMI] is summarised. We then introduce a variety of measurements for characterising connected components generated by constrained connectivity relations. We also propose a weighted mean for estimating a representative value of each connected component. Finally, we define the notion of spurious connected components and investigate a variety of methods for suppressing them.

1 Introduction

A segmentation of the definition domain X of an image is usually defined as a partition of X into disjoint connected subsets $X_i, \ldots, X_n$ (called segments) such that there exists a logical predicate P returning true on each segment but false on any union of adjacent segments [1]. That is, a series of subsets X_i of the definition domain X of an image forms a segmentation of this image if and only if the following four conditions are met (i) $\cup_i(X_i) = X$, (ii) $X_i \cap X_j = \emptyset$ for all $i \neq j$, (iii) $P(X_i) = \text{true}$ for all i, and (iv) $P(X_i \cup X_j) = \text{false}$ if X_i and X_j are adjacent.

With this classical definition of image segmentation, given an arbitrary logical predicate, there may exist more than one valid segmentation. For example, the logical predicate returning true on segments containing one and only one regional minimum and false otherwise lead to many possible segmentations. The watershed transformation definition considers the additional constraint that there should exist a steepest slope path linking each pixel of the segment to its corresponding minimum for the logical predicate to return true. Still, this does not guarantee that there is a unique solution because the steepest slope path of a pixel is not necessarily unique (problem of ties).

If uniqueness of the result is required, logical predicates based on equivalence relations should be considered[1]. Indeed, it has been known for a long time that there exists a one-to-one correspondence between the partitions of a set and the equivalence relations on it, e.g., [2, p. 48]. Since connectivity relations are equivalence relations, logical predicates based on connectivity relations naturally lead

---

[1] A binary relation which is reflexive, symmetric, and transitive is called an equivalence relation.

D. Coeurjolly et al. (Eds.): DGCI 2008, LNCS 4992, pp. 423–433, 2008.

to unique segmentations. For example, the trivial connectivity relation stating that two pixels are connected if and only if they can be joined by an iso-intensity path breaks digital images into segments of uniform grey scale [3]. They are called plateaus in fuzzy digital topology [4] and flat zones in mathematical morphology [5]. In most cases, the equality of grey scale is a too strong homogeneity criterion so that it produces too many segments. Consequently, the resulting partition is *too fine*. A weaker connectivity relation consists in stating that two pixels of a grey tone image are connected if there exists a path of pixels linking these pixels and such that the grey level difference along adjacent pixels of the path (i.e., weights of the edges of the path) does not exceed a given threshold value. In this paper, we call this threshold value the local range parameter and denote it by α. Accordingly, we call the resulting connected components the α-connected components. This idea was introduced in image processing by Nagao *et al.* in the late seventies [6]. The resulting connected components are called quasi-flat zones [7] in mathematical morphology. The concept of α-connected components predates developments in image processing since it is at the very basis of the single linkage clustering method [8]. Although α-connected components often produce adequate image partitions they fail to do so when distinct image objects (with variations of intensity between adjacent pixels not exceeding α) are separated by one or more transitions going in steps having an intensity height less than or equal to α. Indeed, in this case, these objects appear in the same α-connected component so that the resulting partition is *too coarse*.

A natural solution to this problem is to limit the difference between the maximum and minimum values of each connected component by introducing a second threshold value called hereafter global range parameter and denoted by ω. This idea has been originally introduced in [9]. However, the relation at the basis of the developments of [9] is not an equivalence relation because it is not transitive and therefore does not guarantee the generation of unique connected components. This problem has been solved in [10] by introducing the notion of constrained connectivity. In the present paper, we expand on the results of [10].

The paper is organised as follows. Constrained connectivity relations originally proposed in [10] are summarised in section 2. We then propose in Sec. 3 a series of measurements that can be applied to each connected component and introduce the notion of local connectivity index leading to a weighted mean of its intensity values. Image segmentation based on constrained connectivity is studied in section 4. Particular emphasis is given on the analysis of small and usually undesirable segments. We suggest several procedures for suppressing them while preserving the hierarchical properties of partitions based on constrained connectivity. Experiments conducted on a benchmark aerial image are discussed in section 5.

2   Constrained Connectivity Relations

After a reminder about the well established notion of alpha-connectivity, this section summarises the notion of constrained connectivity recently introduced in [10].

2.1   Alpha-Connectivity

Two pixels p and q of an image f are α-connected if there exists a path going from p to q such that the range of the intensity values between two successive pixels of the path does not exceed the value of the local range parameter α [6]. By definition, a pixel is α-connected to itself. Accordingly, the α-connected component of a pixel p is defined as the set of image pixels that are α-connected to this pixel. We denote this connected component by α-CC(p):

$$\alpha\text{-CC}(p) = \Big\{p\Big\} \cup \Big\{q \mid \text{there exists a path } \mathcal{P} = (p = p_1, \dots, p_n = q), \, n > 1,$$

$$\text{such that } \mathsf{R}\{f(p_i), f(p_{i+1})\} \leq \alpha \text{ for all } 1 \leq i < n\Big\},$$

where the range function R calculates the difference between the maximum and the minimum values of a nonempty set of intensity values.

More restrictive connectivity relations detailed in Sec. 2.2 exploit the total ordering relation between the α-connected components of a pixel. Indeed, for all local range parameters α less than or equal to a given local range parameter α', the α-connected component of a pixel p is included in the α'-connected component of this pixel:

$$\alpha\text{-CC}(p) \subseteq \alpha'\text{-CC}(p) \text{ for all } \alpha \leq \alpha'. \tag{1}$$

This hierarchy is known since the fifties in the field of combinatorial optimisation, see [11] for a detailed survey till 1960. Indeed, it is at the root of the greedy algorithm of Kruskal [12] for solving the minimum spanning tree problem. In this algorithm, referred to as 'construction A' in [12], the edges of the graph are initially sorted by increasing edge weights. Then, the minimum spanning tree T is defined recursively as follows: the next edge is added to T if and only if together with T it does not form a circuit. That is, assuming the edge weights are defined by the range of the intensity values of the two nodes (pixels) they link, there is a one-to-one correspondence between (i) the α-connected components and (ii) the subtrees obtained for a distance α in Kruskal's greedy solution to the minimum spanning tree problem. This hierarchy of subtrees is itself at the very basis of the dendrogram representation of the single linkage clustering [8]. This clustering method was put forward by Sneath [13] as a convenient way of summarising taxonomic relationships in the form of taxonomic trees also called similarity trees or dendrograms.

2.2   (Alpha,Omega)-Connectivity

We define the (α, ω)-connected component of an arbitrary pixel p as the largest α_i-connected component of p such that (i) $\alpha_i \leq \alpha$ and (ii) its range is lower than or equal to ω [10]:

$$(\alpha, \omega)\text{-CC}(p) = \bigvee \Big\{\alpha_i\text{-CC}(p) \,\Big|\, \alpha_i \leq \alpha \text{ and } \mathsf{R}\Big(\alpha_i\text{-CC}(p)\Big) \leq \omega\Big\}. \tag{2}$$

The existence of a largest α_i-connected component is secured thanks to the total order relation between the α_i-connected components of a pixel (Eq. 1). Two pixels p and q are (α, ω)-connected if and only if $q \in (\alpha, \omega)$-CC(p).

Beyond range parameters, one may consider other constraints such as a connectivity index indicating the degree of cohesion of each α-connected component. This idea leads to the notion of α-strong connectivity detailed in [10]. A further generalisation to arbitrary logical predicates is presented in [14].

3   Connected Component Representation

First, a series of useful measurements that can be applied to each connected component is presented. We then indicate a method for computing a representative value for each connected component, taking into account their internal cohesion.

3.1   Measurements

Measurements performed on each connected component provide us with a set of features useful for classification purposes and subsequent processing. We propose the definition of the difference image Δ_A mapping the difference between α and the maximum value of α_i leading to the (α, ω)-connected component of p:

$$\Delta_A[\text{CC}(p)] = \alpha - \max\left\{\alpha_i \mid \alpha_i\text{-CC}(p) = (\alpha, \omega)\text{-CC}(p)\right\}.$$

Similarly, the difference image Δ_Ω measures the difference between ω and the actual range of the connected component:

$$\Delta_\Omega[\text{CC}(p)] = \omega - \text{R}\Big((\alpha, \omega)\text{-CC}(p)\Big).$$

The difference between the maximum and minimum value of α_i leading to the (α, ω)-connected component is proportional to the strength of the external isolation of the component:

$$\max\left\{\alpha_i \mid \alpha_i\text{-CC}(p) = (\alpha, \omega)\text{-CC}(p)\right\} - \min\left\{\alpha_i \mid \alpha_i\text{-CC}(p) = (\alpha, \omega)\text{-CC}(p)\right\}.$$

The connectivity index function [10] obtained for increasing threshold range values could also be used as feature vector characterising each connected component. It is used in the following section for calculating a representative value of each connected component.

3.2   Representative Value

Within the scope of image simplification, one needs to estimate a representative value for each connected segment[2]. This is also necessary when iterating the

---

[2] For the estimation of a representative value within geodesic adaptive neighbourhood instead of connected components, see [15] in this volume.

partitioning procedure. In this latter case, the estimated values at each step of the iteration influence the segments obtained in the successive steps.

A common choice for the representative value of a segment is the average of the grey levels of the pixels belonging to it. Some approaches also propose to select it as the local mode of the segment [16].

We propose here to associate with each segment a weighted mean of the intensity values of the pixels of the segment. This is achieved by calculating for each pixel of the segment the number of adjacent pixels that belong to this segment and that are within a range of α. More generally, rather than looking for adjacent pixels only, one can analyse larger neighbourhoods (of size n) to estimate a representative value $\psi_n[CC(p)]$ of any segment $CC(p)$:

$$\psi_n[CC(p)] = \left\{ \frac{\sum Cl_n(p_i)f(p_i)}{\sum Cl_n(p_i)} \mid p_i \in \alpha\text{-}CC(p) \right\}, \tag{3}$$

with the *local connectivity index* of order n of the pixel p defined as $Cl_n(p) = \text{card}(p_i \in CC(p))$ such that there exists a path $\mathcal{P}$ with length n and $R(f(p), f(p_i)) \leq \alpha$. This idea is related to the concept of (global) connectivity index function as introduced in [10] and defined at the level of a connected component.

With this definition, we have in particular $Cl_1[p] \leq N$ where N is imposed by the graph connectivity definition (4 or 8 in the square grid). This way, in the computation of the weighted mean ψ_1, large weights are assigned to the 'core' pixels of the segment, with large connectivity index, while lower weights are assigned to pixels with smaller connectivity index. Notice moreover that the border pixels of the segment, that have at least one connection with a pixel from another segment, have a connectivity index automatically forced to a value $Cl_n[p] < N$. Thus, pixels lying on the internal segment boundaries are assigned lower weights in the weighting procedure defined by Eq. 3 and they contribute less to the final grey level of the segment $(\psi_1[CC(p)])$.

4   Partition Filtering

Constrained connectivity relations partitions the image definition domain into labelled connected components. In addition, by varying the threshold values of the constraints, partition of increasing coarseness degree are obtained. This idea is applied to hierarchical image decomposition and simplification in [10]. Interestingly, any level of the hierarchy can be directly computed without requiring knowledge of the previous levels, in contrast to most alternative partition hierarchies [17].

The generated partitions deliver puzzle pieces that can be further assembled depending on application dependent rules. However, by essence, the method does not take any size criterion into account. It follows that regions as small as one pixel may survive even for large values of the constraint threshold values. We study hereafter the origin of these small regions and propose some approaches for suppressing them in cases this is required by the application at hand.

4.1   Characterisation of Small Regions

We define small regions as regions that cannot contain the elementary structuring element defined by a pixel and its adjacent neighbours (4- or 8-neighbours in the square grid). They are extracted by the following 3-step procedure:

1. perform the union of the erosion of each connected component of the labelled partition. This can be achieved by initialising the output image to 1 and then scan the input image while checking for each position that the structuring element centred at this position covers pixels with the same label value. If this is not the case, the value of the output image at the current position is set to 0. The resulting binary image is then multiplied by the input labelled partition. This image corresponds to the union of the erosion of each connected component of the labelled partition and is referred to as the *marker image* hereafter;
2. reconstruct the labelled partition from the marker image (reconstruction operation on labels, that is, the markers propagate only within the region having the same label value as the marker);
3. define *small regions* as the arithmetic difference between the initial labelled partition and the reconstructed partition as per step 2.

The resulting small segments are then categorised into two classes having different origins:

- 1 pixel thick segments containing at least one regional extremum (union of regional minima and maxima). These regions may either be due to noise or thin relevant structures.
- 1 pixel thick segments that do not contain any regional extremum. These regions are usually due to the limited resolution of the digital image leading to non ideal step edges spanning over 1 or more pixels. We call them *transition regions* (for alternative definitions of transition regions, see [18,19]). Often, transition regions are located at the boundaries between larger regions.

In both situations, these small regions cannot grow further because their growth would lead to a violation of either the local or global range constraints. We explore hereafter a number of methods for reducing or even suppressing small regions of one or both types.

4.2   Filtering Procedures

Filters can be applied either before or after the computation of the constrained connected components.

Pre-filtering. Filters reducing irrelevant local variations can help aggregating small regions into larger regions since the largest α_i-connected component satisfying the input constraints will usually be obtained for larger values of α_i. The occurrence of the first type of small regions can reduced by preprocessing the image with a filter removing isolated pixels. If necessary, more active filters such as the self-dual reconstruction of the input image from its median filter or self-dual area filters [20] can be considered.

Post-filtering. The filtering procedure consists in computing a partition given local and global range parameters. Transition regions defined as small regions not containing any regional extrema are then extracted and considered as *spurious regions*. The resulting gaps are then filled using a seeded region growing algorithm [21] (see also [22] for a version suitable for connected operators and multispectral images). This procedure ensures that all unwanted regions are suppressed but at the cost of some arbitrary decisions unless the ties are tracked (see seeded region growing algorithm enhanced in [23] to address order dependence issues).

5   Experiments

Figure 1a shows an aerial image retrieved from the miscellaneous section of image database of the University of Southern California (USC-SIPI Image Database). Figure 1b shows the partition obtained using the same value (64 grey levels) for

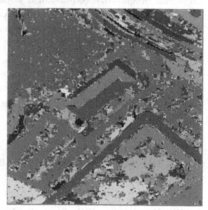

(a) 256 × 256 aerial image (54,364 iso-intensity connected components).

(b) (α, ω)-partition with $\alpha = \omega = 64$ (8,664 regions).

Fig. 1. Input aerial image and resulting partition using local and global range thresholds equal to 64. The input image corresponds to the lower left quarter of the image 5.2.09 of the miscellaneous section of the USC-SIPI image database, see http://sipi.usc.edu/database/

the local and global range parameters and considering the 4-connected graph. A simplified image can be generated by setting each region to the weighted mean as proposed in Sec. 3.2 leads to the simplified image shown in Fig. 2a. A comparison with the image obtained using the non weighted mean is shown in Fig. 2b. This image reveals that the weighted mean generate a more contrasted image even if differences can be hardly perceived through a visual comparison between mean and weighted mean representations.

Figure 3 illustrates the post-filtering procedure whereby transition regions are suppressed (1 pixel thick regions not containing any regional extremum). The

 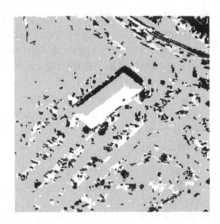

Fig. 2. Left: Edge preserving simplification of the image shown in Fig. 1a using the partition shown in Fig. 1b and weighted mean based on local connectivity index with adjacent pixels only (equation 3 with $n = 1$). Right: comparison between mean and weighted mean with grey for identical values, black for lower values with weighted mean, and white for greater values with weighted mean. In both cases, the mean was rounded to its integer part.

partition without transitions regions contains 2,897 regions contrary to the 8,664 regions of the initial partition (compare Fig. 3f with Fig. 1b by zooming on the electronic version).

6   Concluding Remarks and Perspectives

The concept of constrained connectivity offers a fruitful framework for creating image partitions and edge preserving filtering (image simplification). A non-exhaustive list of measurements characterising the generated connected components has been proposed. The use of these measurements for classification purposes will be reported in a follow-up paper together with their extension to multichannel images since the concept of constrained connectivity can be extended to these images [10]. The proposed notion of local connectivity index allows for the definition of a weighted mean for estimating a representative value of each connected component. Further improvements regard the estimation of a representative value for each connected component. For instance, rather than using the input local range threshold value α in the definition of the local connectivity index Cl_n, the actual local range threshold value α_i (Eq. 2) that varies from one connected component to another could be considered.

We have also analysed the origin of the small connected components and categorised them into two main categories. A technique for removing spurious small regions corresponding to transition regions has been proposed and allows for a drastic reduction in the number of regions of the segmented images. Other techniques based on iterative methods could also be easily designed. Finally,

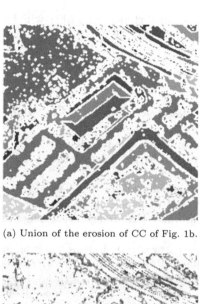

(a) Union of the erosion of CC of Fig. 1b.

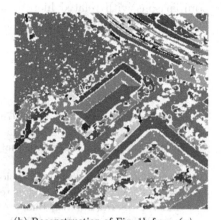

(b) Reconstruction of Fig. 1b from *(a)*.

(c) Resulting small regions.

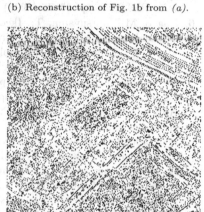

(d) Regional extrema of Fig. 1a.

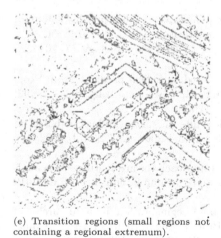

(e) Transition regions (small regions not containing a regional extremum).

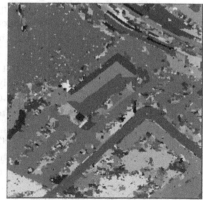

(f) Final partition with transition regions removed (2,897 regions).

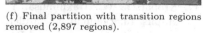

Fig. 3. Partition filtering based on the removal of transition regions (see details in text)

comparisons with related hierarchical segmentation techniques [24,17] will be addressed in an extended version of this paper.

References

1. Zucker, S.: Region growing: childhood and adolescence. Computer Graphics and Image Processing 5, 382–399 (1976)
2. Jardine, N., Sibson, R.: Mathematical Taxonomy. Wiley, London (1971)
3. Brice, C., Fennema, C.: Scene analysis using regions. Artificial Intelligence 1, 205–226 (1970)
4. Rosenfeld, A.: Fuzzy digital topology. Information and Control 40, 76–87 (1979)
5. Serra, J., Salembier, P.: Connected operators and pyramids. In: Dougherty, E., Gader, P., Serra, J. (eds.) Image Algebra and Morphological Image Processing IV, vol. SPIE-2030, pp. 65–76 (1993)
6. Nagao, M., Matsuyama, T., Ikeda, Y.: Region extraction and shape analysis in aerial photographs. Computer Graphics and Image Processing 10, 195–223 (1979)
7. Meyer, F., Maragos, P.: Morphological scale-space representation with levelings. In: Nielsen, M., Johansen, P., Fogh Olsen, O., Weickert, J. (eds.) Scale-Space 1999. LNCS, vol. 1682, pp. 187–198. Springer, Heidelberg (1999)
8. Gower, J., Ross, G.: Minimum spanning trees and single linkage cluster analysis. Applied Statistics 18, 54–64 (1969)
9. Hambrusch, S., He, X., Miller, R.: Parallel algorithms for gray-scale image component labeling on a mesh-connected computer. In: Proc. of the 4th ACM Symposium on Parallel Algorithms and Architectures, pp. 100–108 (1992)
10. Soille, P.: Constrained connectivity for hierarchical image partitioning and simplification. IEEE Transactions on Pattern Analysis and Machine Intelligence (2008), [Available as online preprint since 18 October 2007]
11. Schrijver, A.: On the history of combinatorial optimization (till 1960). In: Aardal, K., Nemhauser, G., Weismantel, R. (eds.) Handbook of Discrete Optimization, pp. 1–68. Elsevier, Amsterdam (2005)
12. Kruskal, J.: On the shortest spanning subtree of a graph and the traveling salesman problem. Proceedings of the American Mathematical Society 7, 48–50 (1956)
13. Sneath, P.: Computers in taxonomy. Journal of General Microbiology 17, 201–226 (1957)
14. Soille, P.: On genuine connectivity relations based on logical predicates. In: Proc. of 14th Int. Conf. on Image Analysis and Processing, Modena, Italy, pp. 487–492. IEEE Computer Society Press, Los Alamitos (2007)
15. Grazzini, J., Soille, P.: Adaptive morphological filtering using similarities based on geodesic time. In: Coeurjolly, D., et al. (eds.) DGCI 2008. LNCS, vol. 4992, pp. 519–528. Springer, Heidelberg (2008)
16. Comaniciu, D., Meer, P.: Mean shift: a robust approach toward feature space analysis. IEEE Transactions on Pattern Analysis and Machine Intelligence 24, 603–619 (2002)
17. Kropatsch, W., Haxhimusa, Y.: Grouping and segmentation in a hierarchy of graphs. In: Bouman, C., Miller, E. (eds.) Proc. of the 16th IS&T SPIE Annual Symposium, Computational Imaging II, vol. SPIE-5299, pp. 193–204 (2004)
18. Tanimoto, S., Pavlidis, T.: The editing of picture segmentations using local analysis of graphs. Communications of the ACM 20, 223–229 (1977)

19. Crespo, J., Schafer, R., Serra, J., Gratin, C., Meyer, F.: The flat zone approach: a general low-level region merging segmentation method. Signal Processing 62, 37–60 (1997)
20. Soille, P.: Beyond self-duality in morphological image analysis. Image and Vision Computing 23, 249–257 (2005)
21. Adams, R., Bischof, L.: Seeded region growing. IEEE Transactions on Pattern Analysis and Machine Intelligence 16, 641–647 (1994)
22. Brunner, D., Soille, P.: Iterative area filtering of multichannel images. Image and Vision Computing 25, 1352–1364 (2007)
23. Mehnert, A., Jackway, P.: An improved seeded region growing. Pattern Recognition Letters 18, 1065–1071 (1997)
24. Guigues, L., Le Men, H., Cocquerez, J.P.: The hierarchy of the cocoons of a graph and its application to image segmentation. Pattern Recognition Letters 24, 1059–1066 (2003)

On Watershed Cuts and Thinnings

Jean Cousty, Gilles Bertrand, Laurent Najman, and Michel Couprie

Université Paris-Est, LABINFO-IGM, UMR CNRS 8049, A2SI-ESIEE, France
{j.cousty,g.bertrand,l.najman,m.couprie}@esiee.fr

Abstract. We recently introduced the watershed cuts, a notion of watershed in edge-weighted graphs. In this paper, we propose a new thinning paradigm to compute them. More precisely, we introduce a new transformation, called border thinning, that lowers the values of edges that match a simple local configuration until idempotence and prove the equivalence between the cuts obtained by this transformation and the watershed cuts of a map. We discuss the possibility of parallel algorithms based on this transformation and give a sequential implementation that runs in linear time whatever the range of the input map.

Introduction

The watershed transform introduced by Beucher and Lantuéjoul [1] for image segmentation is used as a fundamental step in many powerful segmentation procedures. Many approaches [2,3,4,5] have been proposed to define and/or compute the watershed of a vertex-weighted graph corresponding to a grayscale image. The digital image is seen as a topographic surface: the gray level of a pixel becomes the elevation of a point, the basins and valleys of the topographic surface correspond to dark areas, whereas the mountains and crest lines correspond to light areas.

In a recent paper [6], we investigate the watersheds in a different framework: we consider a graph whose edges are weighted by a cost function (see, for example, [7] and [8]). A watershed of a topographic surface may be thought of as a separating line-set from which a drop of water can flow down toward several minima. Following this intuitive idea, we introduce the definition of watershed cuts in edge weighted graphs. In [6], we establish the consistency (with respect to characterizations of the catchment basins and dividing lines) of watershed cuts, prove their optimality (in terms of minimum spanning forests) and introduce an algorithm to compute them.

Our main contribution in this paper consists of a new thinning paradigm to compute watershed cuts in linear time. More precisely, we propose a new transformation, called border thinning, that lowers the values of edges which match a simple local configuration until idempotence. The minima of the transformed map constitute a minimum spanning forest relative to the minima of the original one and, hence, induce a watershed cut. Moreover, any such minimum spanning forest can be obtained by this transformation. We discuss the possibility of parallel algorithms based on this transformation and give a sequential implementation

D. Coeurjolly et al. (Eds.): DGCI 2008, LNCS 4992, pp. 434–445, 2008.

which runs in linear time. We emphasize that this algorithm, contrarily to many watershed algorithms (see, for instance, [2,3]), does not require any sorting step, nor the use of any hierarchical queue. Therefore, it runs in linear time whatever the range of the input map.

The proofs of the properties presented in this paper are given in an extended version [9].

1   Basic Notions for Edge Weighted Graphs

We present some basic definitions to handle edge-weighted graphs.

We define a *graph* as a pair $X = (V(X), E(X))$ where $V(X)$ is a finite set and $E(X)$ is composed of unordered pairs of $V(X)$, *i.e.*, $E(X)$ is a subset of $\{\{x, y\} \subseteq V(X) \mid x \neq y\}$. Each element of $V(X)$ is called a *vertex or a point (of X)*, and each element of $E(X)$ is called an *edge (of X)*. If $V(X) \neq \emptyset$, we say that X is *non-empty*.

Let X be a graph. If $u = \{x, y\}$ is an edge of X, we say that x and y are *adjacent (for X)*. Let $\pi = \langle x_0, \ldots, x_l \rangle$ be an ordered sequence of vertices of X, π is *a path from x_0 to x_l in X (or in $V(X)$)* if for any $i \in [1, l]$, x_i is adjacent to x_{i-1}. In this case, we say that x_0 and x_l are *linked for X*. If $l = 0$, then π is a trivial path in X. We say that X is connected if any two vertices of X are linked for X.

Let X and Y be two graphs. If $V(Y) \subseteq V(X)$ and $E(Y) \subseteq E(X)$, we say that Y is a subgraph of X and we write $Y \subseteq X$. We say that Y is a *connected component of X*, or simply a *component of X*, if Y is a connected subgraph of X which is maximal for this property, *i.e.*, for any connected graph Z, $Y \subseteq Z \subseteq X$ implies $Z = Y$.

Important remark: *Throughout this paper, G denotes a connected graph. In order to simplify the notations, this graph will be denoted by $G = (V, E)$ instead of $G = (V(G), E(G))$. We will also assume that $E \neq \emptyset$.*

For applications to image segmentation, we may assume that V is the set of picture elements (pixels) and E is any of the usual adjacency relations.

Let $X \subseteq G$. An edge $\{x, y\}$ of G is *adjacent to X* if $\{x, y\} \cap V(X) \neq \emptyset$ and if $\{x, y\}$ does not belong to $E(X)$; in this case and if y does not belong to $V(X)$, we say that $\{x, y\}$ is *outgoing from X* and that y *is adjacent to X*. If π is a path from x to y and y is a vertex of X, then π is a *path from x to X (in G)*.

Let $S \subseteq E$. We denote by $\overline{S}$ *the complementary set of S in E, i.e.*, $\overline{S} = E \setminus S$. *The graph induced by S* is the graph whose edge set is S and whose vertex set is made of all points that belong to an edge in S, *i.e.*, $(\{x \in V \mid \exists u \in S, x \in u\}, S)$. In the following, the graph induced by S is also denoted by S.

We denote by $\mathcal{F}$ the set of all maps from E to $\mathbb{Z}$.

Let $F \in \mathcal{F}$. If u is an edge of G, $F(u)$ is the *altitude* of u. Let $X \subseteq G$ and $k \in \mathbb{Z}$. A subgraph X of G is a *(regional) minimum of F (at altitude k)* if: *i)* X is connected ; *ii)* k is the altitude of any edge of X; and *iii)* the altitude of any edge adjacent to X is strictly greater than k.

We denote by $M(F)$ the graph whose vertex set and edge set are, respectively, the union of the vertex sets and edge sets of all minima of F.

Important remark: *Throughout this paper, F denotes an element of* $\mathcal{F}$.

For applications to image segmentation, we will assume that the altitude of u, an edge between two pixels x and y, represents the dissimilarity between x and y (*e.g.*, $F(u)$ equals the absolute difference of intensity between x and y). Thus, we suppose that the salient contours correspond to the highest edges of G.

2 Watersheds

The intuitive idea underlying the notion of a watershed comes from the field of topography: a drop of water falling on a topographic surface follows a descending path and eventually reaches a minimum. The watershed may be thought of as the separating lines of the domains of attraction of drops of water. In [6], we follow this intuitive idea to define the watersheds in an edge-weighted graph.

2.1 Extensions and Graph Cuts

We recall the notions of extension [5,6] and graph cut which play an important role for defining a watershed in an edge-weighted graph.

Intuitively, the regions of a watershed (also called catchment basins) are associated with the regional minima of the map. Each catchment basin contains a unique regional minimum, and conversely, each regional minimum is included in a unique catchment basin: the regions of the watershed "extend" the minima.

Definition 1. *Let* X *and* Y *be two non-empty subgraphs of* G.
We say that Y *is an* extension *of* X *(in* G) *if* $X \subseteq Y$ *and if any component of* Y *contains exactly one component of* X.

Let $S \subseteq E$. *We say that* S *is a (graph)* cut *for* X *if* $\overline{S}$ *is an extension of* X *and if* S *is minimal for this property, i.e., if* $T \subseteq S$ *and* $\overline{T}$ *is an extension of* X, *then we have* $T = S$.

The subgraphs in Fig. 1b and c are two extensions of the one in Fig. 1a. The set S of dashed edges in Fig. 1c is a cut for X (Fig. 1a). It can be verified that $\overline{S}$ (bold subgraph in Fig. 1c) is an extension of X and that S is minimal for this property.

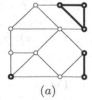 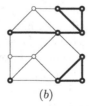 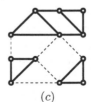 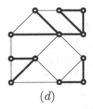

(a)   (b)   (c)   (d)

Fig. 1. A graph G. The set of vertices and edges represented in bold is: (a), a subgraph X of G; (b), an extension of X; (c), an extension Y of X which is maximal; and (d), a spanning forest relative to X. In (c), the set of dashed edges is a cut for X.

The notion of graph cut has been studied for many years in graph theory. For applications to image segmentation, a classical problem is to find a cut of minimum weight (a min-cut) for a set of terminal points. The links between these approaches and the one developed in this paper are investigated in [10].

2.2 Watersheds

We now recall the definition of a watershed cut in an edge-weighted graph. Intuitively, the catchment basins constitute an extension of the minima and they are separated by "lines" from which a drop of water can flow down towards distinct minima.

Let $\pi = \langle x_0, \ldots, x_l \rangle$ be a path in G. The path π is *descending (for F)* if, for any $i \in [1, l-1]$, $F(\{x_{i-1}, x_i\}) \geq F(\{x_i, x_{i+1}\})$.

Definition 2 (drop of water principle, Def. 3 in [6]). *Let $S \subseteq E$. We say that S is a watershed cut (or simply a watershed) of F if $\overline{S}$ is an extension of $M(F)$ and if for any $u = \{x_0, y_0\} \in S$, there exist $\pi_1 = \langle x_0, \ldots, x_n \rangle$ and $\pi_2 = \langle y_0, \ldots, y_m \rangle$ which are two descending paths in $\overline{S}$ such that:*
- x_n and y_m are vertices of two distinct minima of F; and
- $F(u) \geq F(\{x_0, x_1\})$ (resp. $F(u) \geq F(\{y_0, y_1\})$), whenever π_1 (resp. π_2) is not trivial.

In order to illustrate the previous definition, it may be seen that the set S of dashed edges in Fig. 2b is a watershed of the corresponding map F. The minima of F are depicted in bold in Fig. 2a.

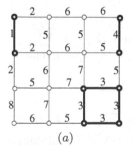
(a)

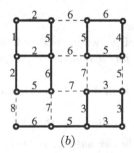
(b)

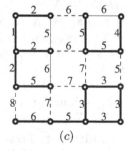
(c)

Fig. 2. A graph G and a map F. Edges and vertices in bold depict: (a), the minima of F; (b), an extension of $M(F)$ which is equal to $\overline{S}$; and (c), a MSF relative to $M(F)$. In (b) and (c), the set of dashed edges is a watershed S of F.

Let $S \subseteq E$. We remark that if S is a watershed of F, then S is necessarily a cut for $M(F)$. The converse is in general not true since a watershed of F is defined thanks to conditions that depend of the altitude of the edges whereas the definition of a cut is solely based on the structure of the graph.

A popular alternative to the drop of water principle defines a watershed exclusively by its catchment basins and does not involve any property of the divide.

In the framework of edge-weighted graph, we define a *catchment basin* as a component of the complementary of a watershed.

The following theorem (Th. 3) shows that a watershed cut can be defined equivalently by its divide line or by its catchment basins. For that purpose, we start with some definitions relative to the notion of path with steepest descent.

Important remark: *From now on, we will also denote by F the map from V to $\mathbb{Z}$ such that for any $x \in V$, $F(x)$ is the minimal altitude of an edge which contains x, i.e., $F(x) = \min\{F(u) \mid u \in E, x \in u\}$; $F(x)$ is the* altitude *of x.*

Let $\pi = \langle x_0, \dots, x_l \rangle$ be a path in G. The path π is *a path with steepest descent for F* if, for any $i \in [1, l]$, $F(\{x_{i-1}, x_i\}) = F(x_{i-1})$.

Theorem 3 (consistency, Th.1 in [6]). *Let $S \subseteq E$ be a cut for $M(F)$. The set S is a watershed of F if and only if, from each point of V to $M(F)$, there exists a path in the graph induced by $\overline{S}$ which is a path with steepest descent for F.*

Th. 3 establishes the consistency of watershed cuts: they can be equivalently defined by a steepest descent property on the catchment basins (regions) or by the drop of water principle on the cut (border) that separates them.

3 Minimum Spanning Forests and Watershed Optimality

In this section, we recall the definition of a minimum spanning forest relative to a subgraph of G. We also recall the equivalence between watershed cuts and cuts induced by minimum spanning forests relative to the minima.

Let X and Y be two non-empty subgraphs of G. We say that Y is a *forest relative to X* if:
i) Y is an extension of X; and
ii) for any extension $Z \subseteq Y$ of X, we have $Z = Y$ whenever $V(Z) = V(Y)$.
We say that Y is a *spanning forest relative to X (for G)* if Y is a forest relative to X and $V(Y) = V$.

For example, the subgraph in Fig. 1d is a spanning forest relative to the subgraph in Fig. 1a.

Let $X \subseteq G$, the *weight of X (for F)* is the value $F(X) = \sum_{u \in E(X)} F(u)$.

Definition 4. *Let X and Y be two subgraphs of G. We say that Y is a minimum spanning forest (MSF) relative to X (for F, in G) if Y is a spanning forest relative to X and if the weight of Y is less than or equal to the weight of any other spanning forest relative to X.*

For instance, the graph Y (bold edges and vertices) in Figs. 2c is a MSF relative to X (Fig. 2a).

We now have the mathematical tools to state the optimality of watershed cuts (Th. 2 in [6]).

Let X be a subgraph of G and let Y be a spanning forest relative to X. There exists a unique cut S for Y and this cut is also a cut for X. We say that this unique cut is the *cut induced by Y*. Furthermore, if Y is a MSF relative to X, we say that that S is a *MSF cut for X*.

Theorem 5 (optimality, Th. 2 in [6]). *Let* $S \subseteq E$. *The set* S *is a MSF cut for* $M(F)$ *if and only if* S *is a watershed cut of* F.

The minimum spanning tree problem is one of the most typical and well-known problems of combinatorial optimization (see [11]). In [9,6] (see also [7]), we show that the minimum spanning tree problem is equivalent to the problem of finding a MSF relative to a subgraph of G. A direct consequence is that any minimum spanning tree algorithm can be used to compute a relative MSF. Many efficient algorithms (see [11]) exist in the literature for solving the minimum spanning tree problem.

4   Optimal Thinnings

As seen in the previous section, a MSF relative to a subgraph of G can be computed by any minimum spanning tree algorithm. The best complexity for solving this problem is reached by the quasi-linear algorithm of Chazelle [12]. In this section, we introduce a new paradigm to compute MSFs relative to the minima of a map and obtain a linear time algorithm to solve this particular instance of the MSF problem. To this aim, we define a new thinning transformation that iteratively lowers the values of the edges that satisfy a simple local property. The minima of the transformed map constitute precisely a MSF relative to the minima of the original one. More remarkably, any MSF relative to the minima of a map can be obtained by this transformation. We discuss the possibility of parallel algorithms based on this transformation and give a sequential implementation (Algo. 1) which runs in linear time.

4.1   Border Thinnings and Watersheds

We introduce an edge classification based exclusively on local properties, *i.e.*, properties that depend only on the adjacent edges. This classification will be used in the definition of a lowering process (Def. 7) which allows one to extract the watersheds of a map.

Remind that, if x is a vertex of G, $F(x)$ is the minimal altitude of an edge that contains x.

Definition 6. *Let* $u = \{x, y\} \in E$.
We say that u *is* locally separating *(for* F*) if* $F(u) > \max(F(x), F(y))$.
We say that u *is* border *(for* F*) if* $F(u) = \max(F(x), F(y))$ *and* $F(u) > \min(F(x), F(y))$.
We say that u *is* inner *(for* F*) if* $F(x) = F(y) = F(u)$.

Notice that a notion similar to the one of border edge has been proposed in [13] under the name of *min-contractible edge*.

Fig. 3 illustrates the above definitions. In Fig. 4a, $\{j, n\}$, $\{a, e\}$ and $\{b, c\}$ are examples of border edges; $\{i, m\}$ and $\{k, l\}$ are inner edges and both $\{h, l\}$ and $\{g, k\}$ are locally-separating edges. Note that any edge of G corresponds exactly to one of the types presented in Def. 6.

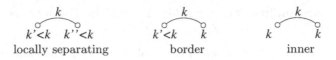

Fig. 3. Illustration of the different local configurations for edges

Let $u \in E$. The *lowering of F at u* is the map F' in $\mathcal{F}$ such that:
- $F'(u) = \min_{x \in u}\{F(x)\}$; and
- $F'(v) = F(v)$ for any edge $v \in E \setminus \{u\}$.

Definition 7 (border cut). *Let $H \in \mathcal{F}$. We say that H is a* border thinning *of F if:*
i) $H = F$; *or*
ii) there exists $J \in \mathcal{F}$ a border thinning of F such that H is the lowering of J at a border edge for J.
If there is no border edge for H, we say that H is a border kernel. *If H is a border thinning of F and if it is a border kernel, then H is a* border kernel of F. *If H is a border kernel of F, any cut for $M(H)$ is called a* border cut *for F.*

To illustrate the previous definition, we assume that F (resp. H, J) is the map of Fig. 4a (resp. b,c). The maps H and J are border thinnings of F. The map J is a border kernel of both F and H. The function depicted in Fig. 4d is another border kernel of F which is not a border kernel of H. In Fig. 4c and d, the border cuts are represented by dashed edges. Remark that the minima of the two border kernels constitute forests relative to $M(F)$, and that all edges which do not belong to the bold graphs are locally separating.

In the next subsection, we show the equivalence between border cuts and watersheds. Thus, the border thinnings constitute a new approach to compute a watershed. This approach is particularly interesting since it relies only on local conditions and produces a result that is globally optimal. It may be noticed that the optimality of the result does not depend on the order on which the values of the edges are lowered. In particular, if a set of mutually-disjoint border edges is lowered in parallel, then the resulting map is a border thinning. Consequently, this transformation suggests new strategies toward efficient parallel algorithms to compute a watershed.

4.2 Linear-Time Algorithm for Border Cut

On a sequential computer, a naive algorithm to obtain a border kernel could be the following: *i)* for all $u = \{x, y\}$ of G, taken in an arbitrary order, check the values of $F(u)$, $F(x)$ and $F(y)$ and whenever u is border, lower the value of u down to the minimum of $F(x)$ and $F(y)$; *ii)* repeat step *i)* until no border edge remains. Consider the graph G whose vertex set is $\{0, \ldots, n\}$ and whose edge set is made of all the pairs $u_i = \{i, i+1\}$ such that $i \in [0, n-1]$. Let $F(u_i) = n - i$, for all $i \in [0, n-1]$. On this graph, if the edges are processed in the order of their indices, step *i)* will be repeated exactly $|E|$ times. The cost for checking all

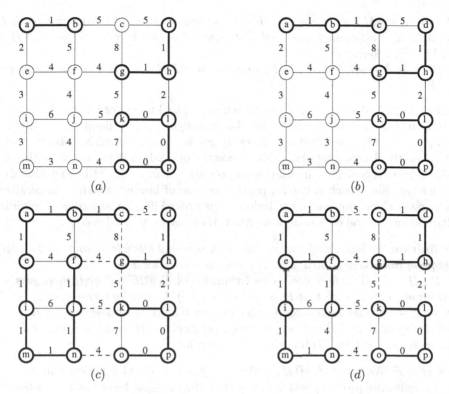

Fig. 4. A graph and some associated functions. The bold graphs superimposed are the minima of the corresponding functions; (b), a border thinning of (a); (c): a border kernel of both (a) and (b); and (d), another border kernel of (a). In (c) and (d), the border cuts are represented by dashed edges.

edges of G is $O(|E|)$. Thus, the worst case time complexity of this naive algorithm is $O(|E|^2)$. In order to reduce the complexity, we introduce a second lowering process in which any edge is lowered at most once. This process is a particular case of border thinning that also produces, when iterated until stability, a border kernel of the original map. Thanks to this second thinning strategy, we derive a linear-time algorithm to compute border kernels and, thus, watersheds.

It may be seen that an edge which is in a minimum at a given step of a border thinning sequence never becomes a border edge later. Thus, lowering first the edges adjacent to the minima seems to be a promising strategy.

Definition 8 (M-border cut). *We say that an edge u in E is* minimum-border *(for F), written M-border, if u is border for F and if exactly one of the vertices in u is a vertex of $M(F)$.*
Let $H \in \mathcal{F}$. We say that H is an M-border thinning of F if:
i) $H = F$; or
ii) there exists $J \in \mathcal{F}$ an M-border thinning of F such that H is the lowering of J at an M-border edge for J.

If there is no M-border edge for H, we say that H is an M-border kernel. *If H is an M-border thinning of F and if it is an M-border kernel, we say that H is an* M-border kernel of F.
If H is an M-border kernel of F, any cut for M(H) is called an M-border cut for F.

In Fig. 4a, the edge $\{c, d\}$ is M-border whereas $\{j, n\}$ is border but not M-border.

On an edge-weighted graph, the classical algorithm by flooding (or immersion) [3] could be described as follows: i) mark the minima with distinct labels; ii) mark the lowest edge containing exactly one labelled vertex with this label; and iii) repeat step ii) until idempotence. In fact, the M-border thinning transformation (which is itself a particular case of border thinning) generalizes immersion algorithms on edges. Indeed, it is proved [9] that any edge of minimal altitude among the edges outgoing from $M(F)$ is an M-border edge.

Definition 9 (I-cut). *If u is an edge with minimal altitude among all the edges outgoing from M(F), then u is an* immersion edge for F.

Let $H \in \mathcal{F}$. We say that H is an I-thinning *of F if H = F or if there exists a I-thinning J of F such that H is the lowering of J at an immersion edge for J. If there is no immersion edge for H, we say that H is an* I-kernel. *If H is an I-thinning of F and if it is an I-kernel, we say that H is an* I-kernel of F. *In this case, any cut for M(H) is called a* I-cut for F.

Let $H \in \mathcal{F}$. We say that $M(H)$ is the *min-graph of H*. This notion will be used in the following property which states that the min-graphs of border, M-border and I- kernels of F are MSFs relative to $M(F)$. More remarkably, any MSF relative to $M(F)$ can be obtained as the min-graph of an M-border kernel, of a flooding kernel and of an I-kernel of F.

Lemma 10. *Let $X \subseteq G$. The four following statements are equivalent:*
(i) X is the min-graph of an I-kernel of F;
(ii) X is the min-graph of an M-border kernel of F;
(iii) X is the min-graph of a border kernel of F; and
(iv) X is a MSF relative to M(F).

Since a relative MSF induces a unique graph cut, from the previous lemma, we immediately deduce that an I-kernel (resp. an M-border kernel, a border kernel) of a map defines a I-cut (resp. M-border cut, border cut). Hence, the following theorem which states the equivalence between watershed cuts, border cuts, M-border cuts and I-cuts can be easily proved.

Theorem 11. *Let $S \subseteq E$. The four following statements are equivalent:*
(i) S is an I-cut for F;
(ii) S is an M-border cut for F;
(iii) S is a border cut for F; and
(iv) S is a watershed cut for F.

Using the notions introduced in this section, we derive Algo. 1, an efficient algorithm to compute M-border kernels, hence watershed cuts. We recall that an

---

Algorithm 1. M-Border

> **Data:** (V, E, F): an edge-weighted graph;
> **Result:** F, an M-border kernel of the input map and M its minima.
> 1 $L \leftarrow \emptyset$;
> 2   Compute $M(F) = (V_M, E_M)$ and $F(x)$ for each $x \in V$;
> 3   **foreach** $u \in E$ *outgoing from* (V_M, E_M) **do** $L \leftarrow L \cup \{u\}$;
> 4   **while** *there exists* $u \in L$ **do**
> 5 | $L \leftarrow L \setminus \{u\}$;
> 6 | **if** u *is border for* F **then**
> 7 | | $x \leftarrow$ the vertex in u such that $F(x) < F(u)$;
> 8 | | $y \leftarrow$ the vertex in u such that $F(y) = F(u)$;
> 9 | | $F(u) \leftarrow F(x)$; $F(y) \leftarrow F(u)$;
> 10 | | $V_M \leftarrow V_M \cup \{y\}$; $E_M \leftarrow E_M \cup \{u\}$;
> 11 | | **foreach** $v = \{y', y\} \in E$ *such that* $y' \notin V_M$ **do** $L \leftarrow L \cup \{v\}$;

---

edge u is border for F if the altitude of one of its extremities equals the altitude of u and the altitude of the other is strictly less than the altitude of u.

In order to achieve a linear complexity, the graph G can be stored as an array of lists which maps to each point the list of all its adjacent vertices. An additional mapping can be used to access in constant time the two vertices which compose a given edge. Nevertheless, for applications to image processing, and when usual adjacency relations are used, these structures do not need to be explicit. Furthermore, to achieve a linear complexity, the minima of F must be known at each iteration. To this aim, in a first step (line 2), the minima of F are computed and represented by two Boolean arrays V_M and E_M, the size of which are respectively $|V|$ and $|E|$. This step can be performed in linear time thanks to classical algorithms. Then, in the main loop (line 4), after each lowering of F (line 9), V_M and E_M are updated (line 10). In order to access, in constant time, the edges which are M-border, the (non-already examined) edges outgoing from the minima are stored in a set L (lines 3 and 11). This set can be, for instance, implemented as a queue. Thus, we obtain the following property.

Property 12. *At the end of Algorithm 1, F is an M-border kernel of the input function F. Furthermore, Algorithm 1 terminates in linear time with respect to $|E|$.*

We emphasize that Algo. 1 does not require any sorting step nor the use of any hierarchical queue. Thus, whatever the range of the considered map, it runs in linear time with respect to the size of the input graph. Furthermore, even in the case where the range of the map is rather limited ($[0, 255]$) the proposed algorithm runs about two times faster than the watershed by flooding algorithm [3] applied on vertex-weighted graph (see the experiments in [9]).

It may be also noticed that in Algo. 1, if the set L is implemented as a hierarchical queue [3], then the resulting map is still a border kernel. Furthermore, in this case, the induced cut is "centered" (according to the distance induced

(a)   (b)   (c)

Fig. 5. Results (superimposed in black to the original image), obtained by applying a watershed on (a) an edge-weighted graph and (b, c) a vertex-weighted graph. In (a), we first assign the absolute difference of intensities to each edge and then filter the map before aplying the weatershed cut. In (b) (resp. (c)), we first assign the result of the Deriche edge detector (resp. morphological gradient) to each vertex of the graph and then filter the map before aplying a watershed. In each case, a filtering step, based on area closing [14], ensures that the watershed defines exactly 12 catchment basins (see [15] for implementation details).

by G) on plateaus. On the other hand, this version of Algo. 1 runs in linear time only if the number of distinct values of F is sufficiently small.

Conclusion and Perspectives

In this paper, we study the watersheds in edge-weighted graphs. Fig. 5 preliminary illustrates that, for some images, these watersheds produce a better delineation (see the man's helmet) than the watersheds in vertex-weighted graphs[1]. In this framework of edge weighted graphs, we introduce new thinning transformations and prove the equivalence between the produced cuts and the watershed cuts. Based on these transformations, we derive a watershed algorithm that runs in linear time whatever the range of the input map. For more details on watershed cuts, we refer to [9]. In particular in [9], we show that a watershed cut is a separation which corresponds to a separation produced by a topological watershed [4,5] defined on edge-weighted graphs. Furthermore, we study the links with shortest-path forests [8].

Future work will be focused on hierarchical segmentation schemes based on watershed cuts (including *geodesic saliency of watershed contours* [16] and *incremental MSFs*) as well as on watershed in weighted simplicial complexes, an image representation adapted to the study of topological properties. Based on border thinnings, we also expect to propose efficient parallel algorithms for watershed cuts. Furthermore, we intend to show that our watershed algorithm can be used to efficiently compute minimum spanning trees.

---

[1] We verified, on the image of Fig. 5, that similar results are obtained by different approaches [3,4] to watersheds in vertex-weighted graphs.

References

1. Beucher, S., Lantuéjoul, C.: Use of watersheds in contour detection. In: Procs. of the International Workshop on Image Processing Real-Time Edge and Motion Detection/Estimation (1979)
2. Vincent, L., Soille, P.: Watersheds in digital spaces: An efficient algorithm based on immersion simulations. IEEE Trans. PAMI 13(6), 583–598 (1991)
3. Meyer, F.: Un algorithme optimal de ligne de partage des eaux. In: Procs. of 8ème Congrès AFCET, Lyon-Villeurbanne, France, pp. 847–859 (1991)
4. Couprie, M., Bertrand, G.: Topological grayscale watershed transform. Procs. of SPIE Vision Geometry V 3168, 136–146 (1997)
5. Bertrand, G.: On topological watersheds. JMIV 22(2-3), 217–230 (2005)
6. Cousty, J., Bertrand, G., Najman, L., Couprie, M.: Watershed cuts. In: Procs. ISMM., pp. 301–312 (2007)
7. Meyer, F.: Minimum spanning forests for morphological segmentation. In: Procs. ISMM., pp. 77–84 (1994)
8. Falcão, A.X., Stolfi, J., de Alencar Lotufo, R.: The image foresting transform: theory, algorithm and applications. IEEE Trans. PAMI 26, 19–29 (2004)
9. Cousty, J., Bertrand, G., Najman, L., Couprie, M.: Watersheds, minimum spanning forest and the drop of water principle, Technical report IGM-2007-01 (Submitted, 2007), http://igm.univ-mlv.fr/LabInfo/rapportsInternes/2007/01.pdf
10. Alléne, C., Audibert, J.Y., Couprie, M., Cousty, J., Keriven, R.: Some links between min-cuts, optimal spanning forests and watersheds. In: Procs. ISMM., pp. 253–264 (2007)
11. Cormen, T.H., Leiserson, C., Rivest, R.: Introduction to algorithms, 2nd edn. MIT Press, Cambridge (2001)
12. Chazelle, B.: A minimum spanning tree algorithm with inverse-Ackermann type complexity. Journal of the ACM 47, 1028–1047 (2000)
13. Englert, R., Kropatsch, W.: Image structure from monotonic dual graph. In: Münch, M., Nagl, M. (eds.) AGTIVE 1999. LNCS, vol. 1779, pp. 297–308. Springer, Heidelberg (2000)
14. Serra, J., Vincent, L.: An overview of morphological filtering. Circuits Systems Signal Process 11(1), 48–107 (1992)
15. Najman, L., Couprie, M.: Building the component tree in quasi-linear time. IEEE Trans. Image Processing 15(11), 3531–3539 (2006)
16. Najman, L., Schmitt, M.: Geodesic saliency of watershed contours and hierarchical segmentation. IEEE Trans. PAMI 18(12), 1163–1173 (1996)

A New Fuzzy Connectivity Class Application to Structural Recognition in Images

O. Nempont[1], J. Atif[2], E. Angelini[1], and I. Bloch[1]

[1] ENST (Télécom ParisTech), Dept. TSI, CNRS UMR 5141 LTCI, Paris, France -
nempont@enst.fr[*]
[2] Unité ESPACE S140, IRD-Cayenne/UAG, Guyane Française

Abstract. Fuzzy sets theory constitutes a poweful tool, that can lead to more robustness in problems such as image segmentation and recognition. This robustness results to some extent from the partial recovery of the continuity that is lost during digitization. Here we deal with fuzzy connectivity notions. We show that usual fuzzy connectivity definitions have some drawbacks, and we propose a new definition, based on the notion of hyperconnection, that exhibits better properties, in particular in terms of continuity. We illustrate the potential use of this definition in a recognition procedure based on connected filters. A max-tree representation is also used, in order to deal efficiently with the proposed connectivity.

1 Introduction

Connectivity is a key concept in image segmentation, filtering, and pattern recognition, where objects of interest are often constrained to be connected according to some definition of connectivity. This definition depends on the selected representation of objects. The binary representation on a discrete grid remains the most widespread, and the connectivity is then generally derived from an elementary connectivity, such as 4- or 8-connectivity in 2D. The axiomatization of classes of connectivity [1,2] provides a rigorous framework to handle the concept of connectivity, which leads to the design of connected filters (e.g. [3]). These definitions were further extended to general complete lattices [2,4,5,6] and to the notion of hyperconnectivity (i.e. based on a different definition of overlapping).

In this paper we deal with connectivity of fuzzy objects. Object representation using fuzzy sets theory [7] makes it possible to model various types of imperfections, in particular related to the imprecision in images, and constitutes a powerful tool, that can lead to more robustness in problems such as image segmentation and recognition. This robustness results to some extent from the partial recovery of the continuity that is lost during the digitization process. The initial definition of fuzzy connectivity [8] provides a crisp characterization of the connectivity of a fuzzy set. Its later extension [5] leads to a characterization of the connectivity as a degree. This degree is however not continuous with respect

---

[*] This work has been partly supported by a grant from INCA.

D. Coeurjolly et al. (Eds.): DGCI 2008, LNCS 4992, pp. 446–457, 2008.

to the membership function. Therefore we propose a new definition, based on the notion of hyperconnection, that exhibits better properties, in particular in terms of continuity.

We first recall in Section 2 some previous definitions on fuzzy sets and fuzzy connectivity, and we illustrate some of their drawbacks. Section 3 is the core of the paper. We introduce a new measure of connectivity, and we show that it leads to a hyperconnection and to nice continuity properties. Hyperconnected components are defined, and an efficient representation as a max-tree is proposed. These notions allow us to build connected filters. In Section 4, we illustrate the proposed approach on an example for brain imaging.

2   Preliminaries

Fuzzy sets – Let X be a set (typically the spatial domain). A fuzzy set on X is defined as $\tilde{A} = \{(x, \mu_A(x)) | x \in X\}$, where μ_A is the membership function, which quantifies the membership degree of x to $\tilde{A}$, and takes values in $[0, 1]$. In the following, X is the digital space $\mathbb{Z}^n$, and we restrict ourselves to fuzzy sets having a bounded support. A fuzzy set is entirely characterized by the set of its α-cuts, denoted by $(\mu_A)_\alpha$: $(\mu_A)_\alpha = \{x \in X \mid \mu_A(x) \geq \alpha\}$. We denote by $\mathcal{F}$ the set of fuzzy sets defined on X. The binary relation $\leq$ on $\mathcal{F}$, defined by $\tilde{A} \leq \tilde{B} \Leftrightarrow \forall x \in X, \mu_A(x) \leq \mu_B(x)$, is a partial order, and $(\mathcal{F}, \leq)$ is a complete lattice. The supremum $\tilde{A} \vee \tilde{B}$ and infimum $\tilde{A} \wedge \tilde{B}$ are defined by their membership functions, as $\forall x \in X, \mu_{A \vee B}(x) = \max(\mu_A(x), \mu_B(x))$ and $\forall x \in X, \mu_{A \wedge B}(x) = \min(\mu_A(x), \mu_B(x))$, respectively. The smallest element is denoted by $0_{\mathcal{F}}$ and the largest element by $1_{\mathcal{F}}$. They are fuzzy sets with constant membership functions, equal to 0 and 1, respectively.

A family Δ of fuzzy sets on X is said sup-generating if $\forall \tilde{A} \in \mathcal{F}, \tilde{A} = \bigvee \{\delta \in \Delta \mid \delta \leq \tilde{A}\}$. We will consider in particular the family $\{\delta_{x,t}\}$ defined as $\delta_{x,t}(y) = t$ if $y = x$ and $\delta_{x,t}(y) = 0$ otherwise, which is sup-generating in the lattice $(\mathcal{F}, \leq)$.

As a metric on $\mathcal{F}$ we use: $d_\infty(\tilde{A}, \tilde{B}) = \sup_{x \in X} |\mu_A(x) - \mu_B(x)|$, and $(\mathcal{F}, d_\infty)$ is a metric space, inducing a definition of continuity.

Usual fuzzy connectivity – The first definition of fuzzy connectivity was proposed by Rosenfeld [8]. More precisely, a degree of connectivity between two points in a fuzzy set was defined, from which the connectivity of a fuzzy set was derived.

Definition 1. *[8] The degree of connectivity between two points x and y of X in a fuzzy set $\tilde{A}$ ($\tilde{A} \in \mathcal{F}$) is defined as:*

$$c_{\tilde{A}}^1(x, y) = \max_{\substack{l \in L_{x,y} \\ l = \{x_0 = x, x_1, \ldots, x_n = y\}}} \min_{0 \leq i \leq n} \mu_A(x_i)$$

where $L_{x,y}$ denotes the set of digitial paths from x to y, according to the underlying digital connectivity defined on X.

This degree of connectivity is symmetrical in x and y, weakly reflexive (i.e. $\forall (x, y) \in X^2, c_{\tilde{A}}^1(x, x) \geq c_{\tilde{A}}^1(x, y)$), and max-min transitive (i.e. $\forall (x, y, z)$

$\in X^3, c^1_{\tilde{A}}(x,z) \geq \min(c^1_{\tilde{A}}(x,y), c^1_{\tilde{A}}(y,z)))$. We have $c^1_{\tilde{A}}(x,x) = \mu_A(x)$ and $c^1_{\tilde{A}}(x,y)$
$\leq \min(\mu_A(x), \mu_A(y))$.

Based on this definition segmentation processes were designed in [9,10]. An affinity based on adjacency and grey level similarity was proposed, and the induced notion of connectivity was used to perform image segmentation, initialized with a set of seed points.

Definition 2. *[8] A fuzzy set is said connected if all its α-cuts are connected (in the sense of the connectivity on X).*

Proposition 1. *[8] A fuzzy set $\tilde{A}$ is connected iff $\forall (x,y) \in X^2, c^1_{\tilde{A}}(x,y) = \min(\mu_A(x), \mu_A(y))$.*

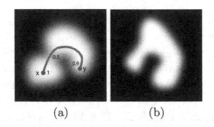

(a)   (b)

Fig. 1. Examples of non-connected (a) and connected (b) fuzzy sets according to Def. 2

These definitions are illustrated in Figure 1. One of the optimal paths between x and y (achieving the max-min criterion of the definition) is displayed in (a), and the minimal value on this path is 0.5, which provides the degree of connectivity between x and y. The fuzzy set in (a) is non-connected since $c^1_{\tilde{A}}(x,y) = 0.5$, which is stricly less than the membership degrees of x and y ($\mu_A(x) = 1$ and $\mu_A(y) = 0.9$). On the contrary, the fuzzy set in Figure 1(b) is connected.

Connection and hyperconnection – Definition 2 provides a crisp notion of the connectivity of a fuzzy set. However, if a set is fuzzy, it may be intuitively more satisfactory to consider that its connectivity is also a matter of degree. The notions of connection and hyperconnection [1,2,4] provide an appropriate framework to this aim.

Definition 3. *[2] Let $(L, \leq)$ be a lattice. A connected class, or connection, $\mathcal{C}$ is a family of elements of L such that:*

1. $0_L \in \mathcal{C}$,
2. $\mathcal{C}$ is sup-generating,
3. for any family $\{C_i\}$ of elements of $\mathcal{C}$ such that $\bigwedge_i C_i \neq 0_L$, then $\bigvee_i C_i \in \mathcal{C}$.

Let us first consider the lattice $(\mathcal{P}(X), \subseteq)$. On this lattice, we use the usual connection $\mathcal{C}_d$ induced by a digital connectivity c_d on X (in the sense of the graph of digital points). An element of $\mathcal{C}_d$ is then simply a subset A of X that is connected in the sense of c_d (i.e. $\forall (x,y) \in A^2, \exists x_0 = x, x_1, ..., x_n = y, \forall i < n, x_i \in A$, and $c_d(x_i, x_{i+1}) = 1$).

Now, on the lattice $(\mathcal{F}, \leq)$, let us consider the binary definition of connectivity in Definition 2, and the 1D examples in Figure 2. In (a), each fuzzy set is connected, and so is there union. However, in (b), the union is not connected, while each fuzzy set is connected and their intersection is not equal to $0_{\mathcal{F}}$. Therefore Definition 3 cannot account for this type of situation on the lattice of fuzzy sets. Dealing with such cases require to replace the infimum ($\wedge$) in condition 3 by another overlap mapping $\perp$ [2], leading to the notion of hyperconnection.

Definition 4. *[2,5] Let $(L, \leq)$ be a lattice. A hyperconnection $\mathcal{H}$ is a family of elements of L such that:*

1. *$0_L \in \mathcal{H}$,*
2. *$\mathcal{H}$ is sup-generating,*
3. *for any family $\{H_i\}$ of elements of $\mathcal{H}$ such that $\perp_i H_i \neq 0_L$, then $\bigvee_i H_i \in \mathcal{H}$.*

Note that it is sufficient to have $\Delta \subseteq \mathcal{H}$, for a sup-generating family Δ, in order to achieve condition 2.

On the lattice $(\mathcal{F}, \leq)$, the hyperconnection $\mathcal{H}^1$ containing the connected fuzzy sets according to Definition 2 is obtained for the following overlap mapping $\perp$ [5]:

$$\perp^1(\{\tilde{A}_i\}) = \begin{cases} 1 \text{ if } \forall \alpha \in [0,1], \bigcap_i \{(\mu_{A_i})_\alpha \mid (\mu_{A_i})_\alpha \neq \emptyset\} \neq \emptyset \\ 0 \text{ otherwise} \end{cases} \tag{1}$$

For the sake of simplicity, we denote the values taken by $\perp$ as 1 and 0 (instead of $1_{\mathcal{F}}$ and $0_{\mathcal{F}}$). It is easy to check that the union of connected fuzzy sets such that their non empty α-cuts intersect is connected in the sense of Definition 2.

This overlap mapping was extended in [5] to the following family:

$$\perp^1_\tau(\{\tilde{A}_i\}) = \begin{cases} 1 \text{ if } \forall \alpha \leq \tau, \bigcap_i \{(\mu_{A_i})_\alpha \mid (\mu_{A_i})_\alpha \neq \emptyset\} \neq \emptyset \\ 0 \text{ otherwise} \end{cases} \tag{2}$$

Let us define $\mathcal{H}^1_\tau = \{\tilde{A} \in \mathcal{F}, \forall \alpha \leq \tau, (\mu_A)_\alpha \in \mathcal{C}_d\}$.

Proposition 2. *[5] Each $\mathcal{H}^1_\tau$ is an hyperconnection, i.e. verifies all items of Definition 4, for the overlap mapping $\perp^1_\tau$. It contains in particular the sup-generating family $\Delta = \{\delta_{x,t}, x \in X, t \in [0,1]\}$. The family $\{\mathcal{H}^1_\tau, \tau \in [0,1]\}$ is decreasing with respect to τ: $\tau_1 \leq \tau_2 \Rightarrow \mathcal{H}^1_{\tau_2} \subseteq \mathcal{H}^1_{\tau_1}$.*

Now the connectivity of a fuzzy set can be defined as a degree, instead as a crisp notion, as follows: $c^1(\tilde{A}) = \sup\{\tau \in [0,1] \mid \tilde{A} \in \mathcal{H}^1_\tau\}$. This definition is equivalent to applying the extension principle [11] to the crisp connectivity: $c^1(\tilde{A}) = \sup\{\alpha \in [0,1] \mid (\mu_A)_\alpha \in \mathcal{C}_d\}$.

As an illustration, the fuzzy sets in Figure 2(c) and (d) have a degree of connectivity of 0.25 and 0.05, respectively. However, intuitively we would rather say that the example in (d) is more connected than the one in (c), which seems to have two very distinct parts. The degree of connectivity depends on the height of

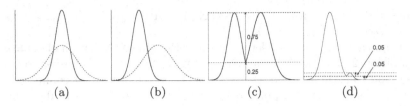

Fig. 2. Examples of fuzzy sets on $\mathbb{Z}$. (a) The union is connected in the sense of Definition 2, while in (b) it is not. The degree of connectivity of the fuzzy set in (c) is equal to 0.25, and in (d) to 0.05, although it seems to be more connected.

the lowest minimum or saddle point, and not on its depth. A small modification in (d) would make the fuzzy set fully connected, illustrating that this definition is not continuous.

The aim of this paper is to propose a new definition that overcomes these drawbacks.

3   A New Class of Connectivity

3.1   Connectivity Measure

In this section, we introduce a new measure of connectivity of a fuzzy set, with better properties than $c^1(\tilde{A})$. The proposed construction is based on the fact that, since it always holds that $c^1_{\tilde{A}}(x,y) \leq \min(\mu_A(x), \mu_A(y))$, then the condition for a fuzzy set to be connected, in the sense of Proposition 1, is equivalent to: $\forall (x,y) \in X^2$, $\min(\mu_A(x), \mu_A(y)) \leq c^1_{\tilde{A}}(x,y)$. We propose to replace the inequality by a degree of satisfaction of this inequality, based on Lukasiewicz' implication [12]: $\forall (a,b) \in [0,1]^2$, $\mu_{\leq}(a,b) = \min(1, 1 - a + b)$. Rewriting this expression for $a = \min(\mu_A(x), \mu_A(y))$ and $b = c^1_{\tilde{A}}(x,y)$ leads to the following definition.

Definition 5. *The connectivity degree between two points x and y in a fuzzy set $\tilde{A}$ is defined by:*

$$c^2_{\tilde{A}}(x,y) = \min(1, 1 - \min(\mu_A(x), \mu_A(y)) + c^1_{\tilde{A}}(x,y))$$
$$= 1 - \min(\mu_A(x), \mu_A(y)) + c^1_{\tilde{A}}(x,y). \quad (3)$$

This measure takes its values in $[0,1]$, it is symmetrical and reflexive ($c^2_{\tilde{A}}(x,x) = 1$). It is not transitive. From this degree of connectivity between two points we derive the following definition of the connectivity degree of a fuzzy set.

Definition 6. *The connectivity degree of a fuzzy set $\tilde{A}$ is defined as:* $c^2(\tilde{A}) = \min_{(x,y) \in X^2} c^2_{\tilde{A}}(x,y)$.

It is easy to show that, for given x and y, $c^1_{\tilde{A}}(x,y)$ and $c^2_{\tilde{A}}(x,y)$ are achieved for the same point on the same path from x to y, and that $c^2(\tilde{A})$ is achieved for

x such that $\mu_A(x) = \max_{x' \in X} \mu_A(x')$ (i.e. x is a global maximum), and for y belonging to a regional maximum (hence $c^2(\tilde{A}) = 1 - \mu_A(y) + c^1_{\tilde{A}}(x, y)$). Note that $c^1(\tilde{A})$ and $c^2(\tilde{A})$ are not achieved for the same points. Roughly speaking, the connectivity degree of a fuzzy set now depends on the depth of the deepest saddle point in the fuzzy set. On the examples illustrated in Figure 2, it can be observed that the fuzzy set in (c) is 0.25−connected $(1 - 0.75)$, while the fuzzy set in (d) is 0.95−connected. In the later case, if one of the modes is progressively shrinking to 0, the degree of connectivity will evolve smoothly towards 1. This is expressed formally by the following result, using as a distance between two function f_1 and f_2 from X^2 into $[0, 1]$: $d_\infty(f_1, f_2) = \sup_{(x,y) \in X^2} |f_1(x, y) - f_2(x, y)|$.

Proposition 3. *For fixed x and y, the mapping associating $\tilde{A}$ to $c^1_{\tilde{A}}(x, y)$ is continuous and Lipschitz, and the mapping associating $\tilde{A}$ to $c^2_{\tilde{A}}(x, y)$ is continuous and 2-Lipschitz. The mapping associating $\tilde{A}$ to $c^2(\tilde{A})$ is continuous and 2-Lipschitz.*

3.2 Link with a Hyperconnection

We propose a new overlap measure, considering that two fuzzy sets do not overlap if they "do not significantly overlap" (i.e. only low α-cuts can overlap), as follows:

$$\perp^2_\tau(\{\tilde{A}_i\}) = \begin{cases} 1 \text{ if } \forall \alpha \in [0, 1], \ \bigcap_i \{(\mu_{\gamma_{x_i}(\tilde{A}_i)})_\alpha \mid \alpha \le h_i - 1 + \tau\} \ne \emptyset \\ 0 \text{ otherwise} \end{cases} \quad (4)$$

where $h_i = \max_{x \in X} \mu_{A_i}$ (the height of $\tilde{A}_i$), x_i is a point of X such that $\mu_{A_i}(x_i) = h_i$, and $\gamma_{x_i}(\tilde{A}_i)$ denotes the geodesic reconstruction by dilation of $\tilde{A}_i$ from the marker δ_{x_i, h_i} (i.e. $\gamma_{x_i}(\tilde{A}_i) = (\delta_c(\delta_{x_i, h_i}) \wedge \mu_A)^\infty$, where δ_c denotes the elementary dilation on X, according to c_d).

Let us now define $\mathcal{H}^2_\tau$ as: $\mathcal{H}^2_\tau = \{\tilde{A} \in \mathcal{F} \mid c^2(\tilde{A}) \ge \tau\}$.

Proposition 4. $\mathcal{H}^2_\tau$ *defines a hyperconnection for the overlap mapping $\perp^2_\tau$.*

These definitions lead to connected components that are more interesting than using $\mathcal{H}^1_\tau$, as seen next.

3.3 Connected Components

In the general framework of connections, connected components of an element A of a lattice $(L, \le)$, relatively to a connection $\mathcal{C}$ on L, are the elements C_i of $\mathcal{C}$ such that: $C_i \le A$ and $\nexists C \in \mathcal{C}, C_i < C \le A$ (i.e. the largest elements of $\mathcal{C}$ that are smaller than A) [2].

This definition extends to hyperconnections [6]. Let $\mathcal{H}$ be a hyperconnection on L. The hyperconnected components of $A \in L$ are the elements H_i of $\mathcal{H}$ such that: $H_i \le A$ and $\nexists H \in \mathcal{H}, H_i < H \le A$. For any two hyperconnected components H_i and H_j of A, either $H_i = H_j$ or $H_i \perp H_j = 0$. Moreover, $\bigvee_i H_i = A$, where the supremum is taken over all connected components of A. If the overlap is taken as $\perp_\tau^2$, we will speak of $\tau-$hyperconnected component. In particular the $1-$hyperconnected components are exactly the reconstructions $\gamma_{x_i}(\tilde{A})$ where each x_i is a representative point of a regional maximum of μ_A.

These notions are illustrated in Figure 3, for the hyperconnection $\mathcal{H}_\tau^2$. Let $\tilde{A}$ be the fuzzy set in (a). It has four $1-$hyperconnected components, corresponding to each regional maximum of $\tilde{A}$, one of them being displayed in (b), two $0.5-$hyperconnected components (c and d), and one $0.1-$hyperconnected component, equal to $\tilde{A}$. The computation of the hyperconnected components will be explained in Section 3.4. The degree of connectivity of $\tilde{A}$ is $c^2(\tilde{A}) = 0.2$, hence $\tilde{A}$ is a connected component in the sense of $\mathcal{H}_\tau^2$ for $\tau \le 0.2$. If we denote by $\tilde{A}_1$ and $\tilde{A}_2$ the two $0.5-$hyperconnected components in (c) and (d), it is easy to check that $c^2(\tilde{A}_1) = c^2(\tilde{A}_2) = 0.5$, hence they are elements of $\mathcal{H}_{0.5}^2$ ($\tau = 0.5$). Let x_1 be the maximum of $\tilde{A}_1$ and x_2 the maximum of $\tilde{A}_2$. We have $h_1 = \mu_{A_1}(x_1) = h_2 = \mu_{A_2}(x_2) = 1$. The two reconstructions $\gamma_{x_1}(\tilde{A}_1)$ and $\gamma_{x_2}(\tilde{A}_2)$ overlap only until level $\alpha = 0.2$, which is less than $h_i - 1 + \tau = 0.5$. This shows that they actually do not overlap in the sense of $\perp_\tau^2$.

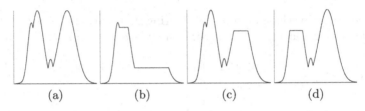

| (a) | (b) | (c) | (d) |

Fig. 3. (a) Fuzzy set (equal to its $\tau-$hyperconnected components for $\tau \le 0.2$). (b) One of the four $1-$hyperconnected components. (c, d) The two $0.5-$hyperconnected components.

3.4   Tree Representation

From an algorithmical point of view, the obtention of the hyperconnected components and their processing can benefit from an appropriate representation. Since the α-cuts are a core component of our definitions, we suggest to rely on the usual max-tree [3] representation of a function. From now on, we assume that the values of α are quantified, in a uniform way. For each level α of the quantification, nodes of a tree are associated with the connected components (in the sense of $\mathcal{C}_d$) of the α-cut of the considered fuzzy set. Edges are induced by the inclusion relation between connected components for two successive values of α. A fuzzy set $\tilde{A}$ is then bi-univoquely represented by a tree $T(\tilde{A})$, with:

- $\mathcal{V}$ the set of vertices of the tree (v denotes an element of $\mathcal{V}$ and $h(v)$ denotes its altitude, i.e. the value of α corresponding to this node),
- R the root of the tree,
- $\mathcal{L}$ the set of leaves,

- $\mathcal{E}$ the set of edges of the tree ($\mathcal{E} \subseteq \mathcal{V} \times \mathcal{V}$), defined from the inclusion relation,
- for $e = (v_1, v_2) \in \mathcal{E}$, $w(e) = |h(v_1) - h(v_2)|$.

There are several algorithms for computing the tree, a recent one being of quasi-linear complexity [13].

If it exists, the chain from v_1 to v_2 (a sub-tree of $T(\tilde{A})$) is denoted by $C_{T(\tilde{A})}(v_1, v_2)$. Its nodes are $v'_i, i = 0...n$ from v_1 to v_2 such that $\forall i, v'_i \in \mathcal{V}$, $v'_0 = v_1$, $v'_n = v_2$, $\forall i < n$, $(v'_i, v'_{i+1}) \in \mathcal{E}$, and $h(v'_i) - h(v'_{i+1})$ has the same sign as $h(v_1) - h(v_2)$ (i.e. $\{v'_i, i = 0...n\}$ is either a descending or an ascending path from v_1 to v_2, depending on the relative altitudes of v_1 and v_2). Its edges are $(v'_i, v'_{i+1}), i = 0...(n-1)$.

Let v_1 and v_2 two nodes such that $h(v_1) \leq h(v_2)$. We denote by $d(v_1, v_2)$ the length of the chain $C_{T(\tilde{A})}(v_1, v_2)$, expressed as the sum of $w(e)$, over all edges of this chain. If the chain does not exist (typically if v_1 and v_2 are on two different branches of the tree), then $d(v_1, v_2) = \infty$.

For any sub-tree G of $T(\tilde{A})$, we denote by $D_{T(\tilde{A})}(G, \nu)$ the dilation of G in $T(\tilde{A})$ of size ν ($\nu \in [0, 1]$), obtained by adding all ascending chains of length less or equal than ν issued from a node of G. A pseudo-erosion $E_{T(\tilde{A})}(G, \nu)$ is obtained by keeping all nodes v of G such that there exists at least one ascending chain of length ν issued from v. An important remark here is that E is not a true erosion (it does not commute with the infimum), and E and D are neither dual nor adjoint (even for the same ν), hence their composition does not have the usual property of an opening or a closing.

Proposition 5. *The set $\{\tilde{A}_i\}$ of $1-$hyperconnected components of $\tilde{A}$ is isomorphic to $\mathcal{L}$, and $T(\tilde{A}_i) = C_{T(\tilde{A})}(R, l_i)$, where l_i is the leaf associated with $\tilde{A}_i$.*

This result shows that it is possible to handle the $1-$hyperconnected components of a fuzzy set by processing the associated sub-tree.

Proposition 6. *If $G \subseteq T(\tilde{A})$ is a sub-tree representing a $\tau-$hyperconnected fuzzy subset of $\tilde{A}$, then $D_{T(\tilde{A})}(G, \nu)$ represents a $\max(0, \tau - \nu)-$hyperconnected fuzzy subset and $E_{T(\tilde{A})}(G, \nu)$ a $\min(1, \tau + \nu)-$hyperconnected fuzzy subset.*

Proposition 7. *The set of $\tau-$hyperconnected components of a fuzzy set $\tilde{A}$ is isomorphic to the set of leaves of $E_{T(\tilde{A})}(T(\tilde{A}), 1-\tau)$. A $\tau-$hyperconnected component of $\tilde{A}$ can then be obtained by a dilation of size $(1-\tau)$ of a $1-$hyperconnected component of $E_{T(\tilde{A})}(T(\tilde{A}), 1 - \tau)$.*

Figure 4 illustrates in (b) the component tree $T(\tilde{A})$ of the fuzzy set shown in (a). The $1-$hyperconnected components (c–f) correspond to each regional maximum of (a). The results of a pseudo-erosion of size 0.4 of $T(\tilde{A})$ and the dilation of size 0.4 of one of its connected components are shown in (g) and (h), respectively, providing exactly the sub-tree associated with one $0.6-$hyperconnected component of $\tilde{A}$. The corresponding $0.6-$hyperconnected component in the image is displayed in (i). Another $0.6-$hyperconnected component is shown in (j)

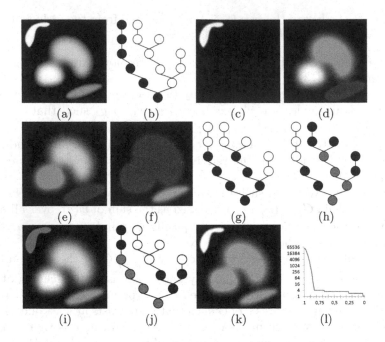

Fig. 4. (a) Fuzzy set. (b) Component tree (the α-cuts are quantizied with a step 0.2), with a chain from a leaf to the root shown in black. (c–f) $1-$hyperconnected components of the fuzzy set in (a). (g) Subtree corresponding to the pseudo-erosion of size 0.4 (in black). (h) A 0.6$-$hyperconnected component (in black and red) obtained by dilation of one connected component (in red) of the pseudo-erosion and the corresponding image (i). Another 0.6$-$hyperconnected component in the tree (j) and the corresponding image (k). (l) Number of τ-hyperconnected components of the noisy image as a function of τ.

(subtree) and (k) (image). If the fuzzy set in Figure 4 (a) is degraded by a Gaussian noise of variance 0.05, more than 20000 $1-$hyperconnected components are obtained. The evolution of the number of $\tau-$hyperconnected components as a function of τ is displayed in (l), showing a grouping effect.

3.5   Connected Filters

One of the main interests of the tree structure is that it allows finding efficiently the hyperconnected components, and therefore applying connected filters on the image. Let $f : \mathcal{F} \rightarrow \{0, 1\}$ be an increasing function defining a filtering criterion (e.g. on the size of the connected components). A connected filter according to criterion f is then defined as:

$$\xi(\tilde{A}) = \bigvee \{\tilde{h} \in \mathcal{H}_\tau^2 \mid \tilde{h} \leq \tilde{A} \text{ and } f(\tilde{h}) = 1\}. \tag{5}$$

This defines an increasing and idempotent operator and thus a morphological filter. In this particular form, it is moreover anti-extensive, and thus a morphological opening. Such a filter can be implemented in a very efficient way based on the tree structure.

4   Illustrative Example

In order to illustrate the proposed definitions, we consider the problem of segmenting brain structures in 3D MRI images. An axial slice is displayed in Figure 6 (a). Typically, we may want to segment connected objects such as the ventricular system or internal grey nuclei. As in our previous work [14,15], we rely on anatomical knowledge expressed as spatial relations between structures [16], and on grey level information. As illustrated in Figure 6, this knowledge is translated into fuzzy representations in the image space, that drive the segmentation and the recognition. Here we show how to include additional connectivity criteria in this procedure.

For each anatomical structure S_i, we define two fuzzy sets $\tilde{P}_i$ and $\tilde{N}_i$, corresponding to an over-estimation and an under-estimation of S_i, respectively: $\tilde{N}_i \leq S_i \leq \tilde{P}_i$. This idea is close to the concept of fuzzy rough sets. At the beginning of the procedure, $\tilde{N}_i$ is empty and $\tilde{P}_i$ is the whole space. Exploiting the available knowledge then allows reducing $\tilde{P}_i$ and extend $\tilde{N}_i$ so as to get as close as possible to the structure of interest. For instance for $S_i = S_{LV}$ being the lateral ventricle, we define $\tilde{P}_{Gl_{LV}}$ representing the knowledge on grey levels, so as to have $\tilde{S}_{LV} \leq \tilde{P}_{Gl_{LV}}$ (Figure 6 (b)). Once the brain has been segmented, it becomes possible to represent the central location of the ventricules inside the brain (Figure 6 (c)), so as to guarantee $\tilde{S}_{LV} \leq \tilde{P}_{Sp_{LV}}$. The conjunctive fusion of $\tilde{P}_{Sp_{LV}}$ and $\tilde{P}_{Gl_{LV}}$ is shown in Figure 6 (d), and provides an including fuzzy set $\tilde{P}_{LV}$. Although the over-estimation has been strongly reduced, it still exhibits several connected components. A connectivity contraint can now be introduced, via a connected filter based on a marker.

The criterion f used in the filter (Equation 5) relies on the inclusion of a marker $\tilde{N}$ in h and the inclusion of h in $\tilde{P}$. Here, for the first inclusion, we consider actually a degree of inclusion (as in Section 3.1), to achieve more robustness with respect to the position of the marker. The filter then writes:

$$\xi_{\tilde{N}}(\tilde{P}) = \bigvee \{\tilde{h} \in \mathcal{H}_\tau^2 \mid \max_{x \in X} \mu_h(x) \leq \mu_\leq(\tilde{N}, \tilde{h}) \text{ and } \tilde{h} \leq \tilde{P}\}. \tag{6}$$

Note that the criterion is not increasing in this case. However it is increasing if $\max_{x \in X} \mu_h(x)$ is constant. Equation 6 can thus be decomposed as a supremum over all possible values of this maximum (i.e. all levels of the quantification), and each term of this supremum can then be handled efficiently using the tree representation, as explained in Sections 3.4 and 3.5.

Proposition 8. *Let $\alpha = \max_{x \in X} \mu_N(x)$. The result of the connected filter defined in Equation 6 is $(\alpha - (1 - \tau))-hyperconnected$.*

A 1D example is shown in Figure 5, where a fuzzy set is progressively filtered by a marker getting larger and larger. Intuitively, hyperconnected components

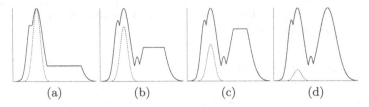

(a) (b) (c) (d)

Fig. 5. Progressive filtering of the fuzzy set of Figure 3 (a) by a marker of increasing size (in red). The result is displayed in blue.

verifying the inclusion constraint are kept, while the other ones are reduced to a level corresponding to the degree of satisfaction of the constraint.

We illustrate now the effect of this connected filter, applied to $\tilde{P}_{LV}$, based on a marker $\tilde{N}$ defined as a fuzzy set having a support reduced to one point centered in the right lateral ventricle, with a membership value taking values 1, 0.75, 0.5 and 0, respectively (Figure 6 (e–h)). A potential application of this approach is to perform a filter, preserving connectivity properties, and being more or less strong depending on the confidence we may have in the marker. This may lead to more robustness and can be used as a preliminary step in a segmentation process. This will be further explored in future work.

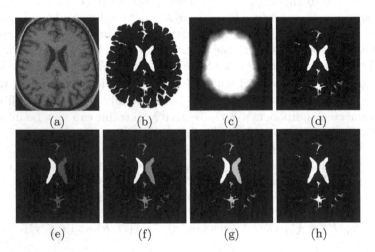

Fig. 6. (a) One axial slice of a 3D brain MRI. (b) Grey level information: $\tilde{P}_{Gl_{LV}}$. (c) Central location inside the brain: $\tilde{P}_{Sp_{LV}}$. (d) Conjunctive fusion. (e–h) Connected filter results using a marker centered in the right ventricle, with maximal value 1, 0.75, 0.5, 0, respectively.

5   Conclusion

In this paper we have introduced a new definition of fuzzy connectivity, based on the notion of hyperconnection, that overcomes some drawbacks of previous definitions, and that has in particular nice continuity properties. We have shown that a representation as a max-tree can lead to efficient extraction of hyperconnected components, and processing with connected filters. An illustrative example on a brain image has been shown. Future work aims at exploring other properties of the proposed definitions, and at developing a complete applicative framework for brain segmentation, including pathological cases.

References

1. Serra, J. (ed.): Image Analysis and Mathematical Morphology. Part II: Theoretical Advances. Academic Press, London (1988)
2. Serra, J.: Connectivity on Complete Lattices. Journal of Mathematical Imaging and Vision 9(3), 231–251 (1998)
3. Salembier, P., Oliveras, A., Garrido, L.: Antiextensive Connected Operators for Image and Sequence Processing. IEEE Transactions on Image Processing 7(4), 555–570 (1998)
4. Ronse, C.: Set-Theoretical Algebraic Approaches to Connectivity in Continuous or Digital Spaces. Journal of Mathematical Imaging and Vision 8(1), 41–58 (1998)
5. Braga-Neto, U., Goutsias, J.: A Theoretical Tour of Connectivity in Image Processing and Analysis. Journal of Mathematical Imaging and Vision 19(1), 5–31 (2003)
6. Braga-Neto, U., Goutsias, J.: Grayscale Level Connectivity: Theory and Applications. IEEE Transactions on Image Processing 13(12), 1567–1580 (2004)
7. Zadeh, L.A.: Fuzzy Sets. Information and Control 8, 338–353 (1965)
8. Rosenfeld, A.: Fuzzy Digital Topology. Information and Control 40, 76–87 (1979)
9. Udupa, J.K., Samarasekera, S.: Fuzzy Connectedness and Object Definition: Theory, Algorithms, and Applications in Image Segmentation. Graphical Models and Image Processing 58(3), 246–261 (1996)
10. Carvalho, B., Gau, C., Herman, G., Kong, T.: Algorithms for Fuzzy Segmentation. Pattern Analysis and Applications 2(1), 73–81 (1999)
11. Zadeh, L.A.: The Concept of a Linguistic Variable and its Application to Approximate Reasoning. Information Sciences 8, 199–249 (1975)
12. Dubois, D., Prade, H.: Fuzzy Sets and Systems: Theory and Applications. Academic Press, New-York (1980)
13. Najman, L., Couprie, M.: Building the Component Tree in Quasi-Linear Time. IEEE Transactions on Image Processing 15(11), 3531–3539 (2006)
14. Colliot, O., Camara, O., Bloch, I.: Integration of Fuzzy Spatial Relations in Deformable Models - Application to Brain MRI Segmentation. Pattern Recognition 39, 1401–1414 (2006)
15. Nempont, O., Atif, J., Angelini, E., Bloch, I.: Combining Radiometric and Spatial Structural Information in a New Metric for Minimal Surface Segmentation. In: Karssemeijer, N., Lelieveldt, B. (eds.) IPMI 2007. LNCS, vol. 4584, pp. 283–295. Springer, Heidelberg (2007)
16. Bloch, I.: Fuzzy Spatial Relationships for Image Processing and Interpretation: A Review. Image and Vision Computing 23(2), 89–110 (2005)

Directional Structures Detection Based on Morphological Line-Segment and Orientation Functions

Iván R. Terol-Villalobos[1], Luis A. Morales-Hernández[2],
and Gilberto Herrera-Ruiz[2]

[1] CIDETEQ,S.C., Parque Tecnológico Querétaro S/N, SanFandila-Pedro Escobedo,
76700, Querétaro Mexico
famter@ciateq.net.mx
[2] Facultad en Ingeniería, Universidad Autónoma de Querétaro, 76000, México

Abstract. In the present paper a morphological approach for segmenting directional structures is proposed. This approach is based on the concept of the line-segment and orientation functions. The line-segment function is computed from the supremum of directional erosions. This function contains the sizes of the longest lines that can be included in the structure. To determine the directions of the line segments, the orientation function which contains the angles of the line segments it is built when the line-segment function is computed. Next, by combining both functions, a weighted partition is built using the watershed transformation. Finally, the elements of the partition are merged according to some directional and size criteria for computing the desired segmentation of the image using a RAG structure.

1   Introduction

Anisotropic structures are frequently found in many classes of images (materials, biometry images, biology, ...), however, few works dealing with directional analysis in morphological image processing have been carried out [1,2,3,4,5,6] among others. It is maybe in the domain of fingerprint recognition, which is today the most widely used biometric features for personal identification, where the study of directional structures (called orientation-fields) based on directional structures detection is an active subject of research [7,8,9]. In fact, fingerprints can be considered as a structure composed by a set of line segments (see Fig. 1(a)). However, directional structure detection also plays a fundamental role in other domains [10,11,12]. Lee et al [10] propose a method based on oriented connectivity for segmenting solar loops, while Kass and Witkin [12] propose a method to analyze oriented patterns in wood grain. Given the interest in orientation pattern models for characterizing structures, this paper investigates the use of the mathematical morphology for modelling directional structures. As in the human vision, computer image processing of oriented image structures often requires a bank of directional filters or template masks, each of them sensitive to

D. Coeurjolly et al. (Eds.): DGCI 2008, LNCS 4992, pp. 458–469, 2008.
© Springer-Verlag Berlin Heidelberg 2008

a specific range of orientations [1]. Then, one investigates the use of an approach based on directional erosions. In the literature there exist several works to characterize directional structures based on the gradient computation that can be formalized in terms of mathematical morphology. See for example the works of [12,13,9]. The problem of the gradients is that they work at pixel scale, they are very sensitives to noise and a final stage to enhance directional structures is required. Then, the main idea in this paper is focused on another approach that permits to take into account the whole context of the structures contained in the image. A local approach using the concept of line-segment function combined with the watershed transformation is used. In our case, the line-segment function is computed from the supremum of directional erosions. This function contains the information of the longest line segments that can be placed inside the structure. In order to know their orientations, a second image is defined by observing the construction of the line-segment function and its evolution. This second image is computed by detecting the orientation of the supremum of directional erosions. These local descriptors, for the element size and orientation, enable the identification of the directional structures based on the watershed transformation. This paper is organized as follows. In Section 2, the concepts of morphological filter and directional morphology are presented. In Section 3 the notions of line-segment and orientation functions, derived from the supremum of directional erosions, are introduced. Also in the this section, an algorithm to compute the line-segment and the orientation function is analyzed. Next, in Section 4 an approach of working with directional morphology, the watershed transform and a region adjacency graph (RAG) for segmenting directional structures is proposed.

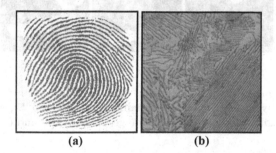

(a)   (b)

Fig. 1. (a) Fingerprint image, (b) Pearlitic phase image

2   Basic Morphological Directional Transformations

Mathematical morphology is based principally on so-called increasing transformations [14,15,16]. Between the different morphological tools, an interesting class of transformations are the directional ones. Morphological directional transformations are characterized by two parameters. A structuring element L depends on its length (size μ) and on the slope (angle α) of the element. Thus, the set of points of a line segment $L(\alpha, \mu)$ is computed by two sets of points for $\alpha \in [0, 90]$. The sets of points $\{(x_i, y_i)\}$ defined by the following expressions:

$$if \quad 0 \le \alpha \le 45 \quad then, \quad y_i = x_i tan(\alpha) \quad for \quad x_i = 0, 1, \cdots, \mu cos(\alpha)$$

$$if \quad 90 \ge \alpha > 45 \quad then, \quad x_i = y_i cot(\alpha) \quad for \quad y_i = 0, 1, \cdots, \mu cos(\alpha)$$

and the set of points $\{(-x_i, -y_i)\}$. This means, the structuring element is a symmetric set $L(\alpha, \mu) = \hat{L}(\alpha, \mu)$. Similar expressions can be described for $\alpha \in (90, 180]$. Then, the morphological opening and closing are given by:

$$\gamma_{L(\alpha,\mu)}(f) = \delta_{L(\alpha,\mu)}(\varepsilon_{L(\alpha,\mu)}(f)) \quad and \quad \varphi_{L(\alpha,\mu)}(f) = \varepsilon_{L(\alpha,\mu)}(\delta_{L(\alpha,\mu)}(f)) \quad (1)$$

where the morphological erosion and dilation are given by: $\varepsilon_{L(\alpha,\mu)}(f)(x) = \wedge\{f(y) : y \in L(\alpha,\mu)(x)\}$ and $\delta_{L(\alpha,\mu)}(f)(x) = \vee\{f(y) : y \in L(\alpha,\mu)(x)\}$.

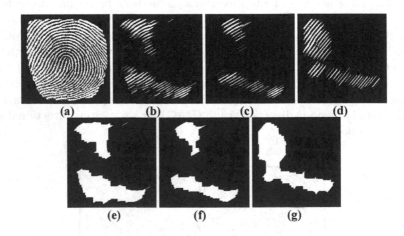

Fig. 2. Original image, (b), (c) Directional openings at direction 30 degrees with size 20 and 30, respectively, (d) Directional opening at direction 50 degrees size 30, (e), (f) and (g) Morphological closings size 10 of images in 2 (b), (c) and (d), respectively

3   Size and Orientation Codification Based on Directional Erosions (Line-Segment and Orientation Functions)

3.1   Line-Segment and Orientation Functions

In this section and in the following ones we will look for an approach to detect directional structures where the connectivity notion plays a fundamental role for segmenting them. In fact, it is well-known that the notion of connectivity is linked to the intuitive idea of segmentation task, where the objective is to split the connected components into a set of elementary shapes that will be processed separately. Then, the problem lies in determining what a connected component is for an image such as those illustrated in Fig. 1. One can take different ways for introducing such a concept. For example, in Fig. 2 some orientation structures are extracted using directional openings. Then, the directional structures are determined by a clustering process computed in this case by a morphological closing. Figures 2 (b) and (c) show the directional openings at direction 30 degrees with sizes of the structuring elements of 20 and 30, respectively. A hierarchy of structure sizes can be introduced by means of the granulometry. Thus, the directional granulometry and a connectivity introduced by means of closings for extracting some clustering enable us to compute the directional structures as illustrated in Figs. 2 (e) and (f). Nevertheless, the computing of the directional structures by this approach can become very complex. For instance, in Figs. 2 (d) and (g) the directional structures were determined for an angle of 50 degrees and a scale of 30. Some regions of the structures of the images in Figs. 2 (c) and (f) are the same than those of the images in Figs. 2 (d) and (g) (i.e., the intersection between these images is not empty). Given that deficiency, we look for another approach where the information of scales and directions of the structures of the image are easily accessible. Two functions that codify the size and the orientation are introduced below. The idea for codifying size structure comes from the notion of the distance function. Remember that the distance function $D_X(x)$ is a transformation that associates with each pixel x of a set X its distance from the background. Let us now to define a new function derived from the notion of distance function.

Definition 1. *The line-segment function $Dm_X(x)$ is a transformation that associates with each pixel x of a set X the length of the longest symmetrical line segment, centered at the origin, placed at point x and completely included in X.*

The goal of building this function consists in codifying the size information in such a way that local directional information can be accessed from each point of the function. This codification of the size information will be used to build a local approach for detecting directional structures on an image. This function, which we call line-segment function Dm_X, is computed by using the supremum of directional erosions. To stock the size information for all λ values, a gray-level image Dm_X is used. Let X be a given set, one begins with a small structuring element by taking into account all orientations to compute the set $Sup_{\alpha \in [0,180]}\{\varepsilon_{L(\lambda,\alpha)}(X)\}$. That means, one takes all points of the image that are not removed by at least one of the directional erosions. Then one increases Dm_X by one at all points x belonging to the set $Sup_{\alpha \in [0,180]}\{\varepsilon_{L(\lambda,\alpha)}(X)\}$, and one continues the procedure by increasing the size of the structuring element until the structure (the image) is completely removed. This means that the procedure continues until one has a λ_{max} value such that $Sup_{\alpha \in [0,180]}\{\varepsilon_{L(\lambda_{max},\alpha)}(X)\} = \emptyset$. Figures 3(b) and (c) show the output images computed from the original image in Fig. 3(a) for the size values 40 and 60, respectively. As expressed before, the gray-levels of the function Dm_X, are the sizes of the longest lines that can be included in the structure. Whereas, for the structures that can be considered as composed as a set of lines, as those in Figs. 1(a) and (b), we assume that the the maxima of the function Dm_X play a main role since they codifies the longest lines that take the whole context of the image. Thus, one knows the position of the largest structuring elements that can be included completely in the structure. However, the angles of these structuring elements (line segments) are not accessible from the image Dm_X. Then, let us introduce a second function associated to the line-segment function.

Definition 2. *The orientation function $Om_X(x)$ is a transformation that associates with each pixel x of a set X the angle of the longest symmetrical line segment, centered at the origin, placed at point x and completely included in X.*

Therefore, one stocks the directions of the line segments in a second image Om_X, called orientation function, when the line-segment function is computed. A real example (pearlitic phase micrograph) is shown in Fig. 4. The images in Fig. 4(b)-(c) illustrate the line-segment function Dm_X image and the image containing the orientation Om_X, respectively, computed from the binary image in Fig. 4(a). These functions can now be used for computing the line segments that characterize the structure. The line-segment function and its associated orientation image containing the angles serve to suggest a method for segmenting images with directional structures.

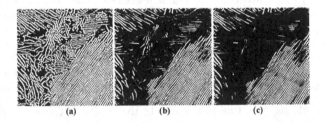

Fig. 3. (a)Original image, (b) and (c) $Sup_{\alpha \in [0,180]}\{\varepsilon_{L(\lambda,\alpha)}(X)\}$ with λ 40 and 60

3.2   Non-parametric Algorithm to Build the Line-Segment and Orientation Functions and the Limits of the Approach

Since the number of operations for computing the line-segment function and the orientation function is considerable, one can suppose that the method is very expensive in computation time. Moreover, the building of these functions seems to require a size parameter (the largest structuring element) and an angle step to compute them. Let us illustrate an algorithm to build these functions that does not require any parameter and that is not expensive in computation time. First, concerning the size parameter (largest structuring element), it was fixed to the size of the image diagonal that is the size of the largest structure that can be include in the image. Let S_x and S_y be the dimension of the images, horizontal and vertical axis, respectively. For example, for a VGA image 640 × 480, one has $L = \sqrt{S_x^2 + S_y^2} = 800$, then the largest structuring element has a size of

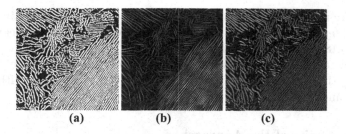

(a)   (b)   (c)

Fig. 4. (a) Original Binary image, (b) Line-segment function, (c) Orientation function

400 since one uses symmetrical structuring elements. It is clear that few cases of images, containing such structure characteristics, can be found in real images. Next, one requires to fixe the step in degrees to compute the line segments. In practice, a step between 5 and 10 degrees is sufficiently, but let us fix it to a smaller value (one degree) in order to show the limiting case in computation time. Then, one computes 180 structuring elements of size $L/2$, and they are stocked in a structure data (list of lists). Since symmetrical structuring elements are used, only half of the straight lines is stocked and centered at origin $(0,0)$. In fact, only the Freeman codes are stocked. Let $\{Ls_i\}$ with $i \in \{0,1,2,...,179\}$ be the lists containing the Freeman codes required to build a half of the structuring elements and let Ls_j be a given list. The structuring element is built using the list $Ls_j = \{c_k\}$ with $c_k \in \{0,1,2,3,...,7\}$ and its symmetrical data $\check{Ls}_j = \{\check{c}_k\}$ with $\check{c}_k = (c_k + 4) \bmod 8$. Consider the example in Fig. 5(a) where an erosion by a line segment is applied to the structure in gray color. The structuring element is obtained from the list $Ls_j = \{0,1,0,1,0,1,0,1,0\}$ (blue 0 and red 1) and $\check{Ls}_j = \{4,5,4,5,4,5,4,5,4\}$ (green 4 and yellow 5). Now, to compute the erosion at point (x,y) of an image f marked by a white dot, one begins by computing the smallest value between the points (x-1,y), (x,y) and (x+1,y). Then, the erosion size one is given by the infimum (the intersection for sets) $\varepsilon_1(f)(x,y) = f(x-1,y) \bigwedge f(x,y) \bigwedge f(x+1,y)$. Next, one computes the erosion size 2 with the following two points of the structuring element (x-2,y+1) and (x+2,y+1) and the erosion size one $\varepsilon_1(f)(x,y)$, thus, $\varepsilon_2(f)(x,y) = f(x-2,y-1) \bigwedge \varepsilon_1(f)(x,y) \bigwedge f(x+2,y+1)$. The procedure continues until the last pair of points of the structuring element is taken into account. In this example, one requires a longer structuring element to remove the point (x,y) of the image. Nevertheless, in the example in Fig. 5(b), when the third erosion is applied, the point is removed; i.e., $\varepsilon_3(f)(x,y) = f(x-3,y-2) \bigwedge \varepsilon_2(f)(x,y) \bigwedge f(x+3,y+3) = 0$, then, the procedure is stopped. This procedure is applied at each point of the image. It is clear that the fact of using the infimum (AND operation in a computer) to compute the erosion and to stop the procedure when it is no longer required, permits one to compute the erosion of the image faster. Then, instead of calculating the $Sup_{\alpha \in [0,180]}\{\varepsilon_{L(\lambda,\alpha)}\}$, one computes at each point x of the image, the longest structuring element that can not remove this point. Next, the length of this structuring element is used to affect the function Dm_X at point x. For instance, in the example in Fig. 1(b), an image of size 512×512 pixels, 5

seconds are required to compute the line-segment and orientation images using an angle step of one degree, whereas working with a step of 5 degrees in the interval [0,180] a second is only required. For the image in Fig. 1(a) (300x300 pixels) one requires less than two seconds using an angle step of one degree. The computer, that has been used for the experiments, is a laptop with 1.59 Ghz processor and 256 MB in RAM.

3.3   The Limits of the Approach

Let us take some geometrical examples to illustrate some limitations of the approach above proposed to extract the directional characteristics. The first example is the case of a rectangle of length l and wide h as shown in Fig.5 (c) and (d). The longest line segments that can be placed inside the rectangle are the diagonal lines of length $[l^2 + h^2]^{1/2}$ and their angle is given by $\alpha = \tan^{-1}(h/l)$. It is clear that larger is l and the smaller is h, then closer to 0 degrees is α ($l \to \infty$ or $h \to 0, \Rightarrow \alpha \to 0$). Consider an example where the rectangle has $l = 100$ pixels. Then, for $h = 20$, one has $\alpha = 11.5$ degrees, for $h = 10$, $\alpha = 5.7$ degrees, for $h = 5$, $\alpha = 2.8$ degrees. Another source of errors are the rectangle extremes, since at the limits the angle of the longest line (in this example) has an angle valu of 90 degrees. Moreover, the points remaining in the line that cuts the rectangle along the horizontal at $h/2$ crossing its center, can change between $\alpha = \tan^{-1}(h/l)$ and 90 degrees. Figures 5(c) and (d) show two examples in color representation. This drawback could be avoided by applying the supremum of directional openings, however, the computation time in determining the longest lines included in the structure will increase considerably. To illustrate that, remember the traditional geometrical interpretations of the erosion and the opening. In the erosion case one selects the center of the structuring element completely include in the structure, whereas in the opening case all the points hitting the structuring element are chosen. This means, when an opening is applied, the points of the image are analyzed several times and it will not possible to apply the algorithm above proposed. Now let us analyze another geometrical example given by a shape formed by two concentric circles of different radii (annulus or ring of wide h) as illustrated in Fig. 5(e). In the case of a disk, the maximum symmetrical distance remains at the middle point of the segment joining any two points of the circle (chord), the longest chord going through the center of the disk (diameter of the disk). In the case of a ring the longest line segment is tangent of the inner circle, and the smallest line is zero, a point on the outer circle. Between these two extremes there exit different chords all at the same angle. Then, when working with structures with a given curvature the error of the direction is smaller than the case of a straight right (or without error when the region of a shape is assimilable to a ring) as illustrated in Fig. 5 (f). However, the maxima of the line-segment function will remain close to the contour (inner contour) of the structure. The limits of the approach, shown in this section, are attenuated when the image is composed by thin structures as those in Fig. 1(b) or when the structures have a curvature as those in Fig. 1(a). To better illustrate the performance of the line-segment and the orientation functions several examples are shown in Fig. 6. Three examples of orientation detection in fingerprints are carried out. To illustrate the orientation detection the line segment and orientation functions were computed from the binary images obtained from the original images illustrated in Figs 6 (a), (b), and (c), then a line segment was placed at some maxima of the line segment function by taking into account the direction given by the orientation function as illustrated in Figs. 6 (d), (e) and (f). Only one point of each regional maximum was selected to place a line segment.

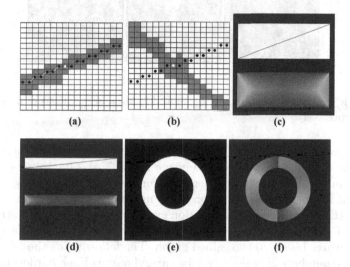

Fig. 5. a) and b) Directional erosions, c) and d) approach limits in rectangle structures, e) and f) approach limits in disk structures

4   Image Segmentation Using Directional Morphology and the Watershed Transformation

Image segmentation is one of the most interesting problems in image processing and analysis. The main goal in image segmentation consists in extracting the regions of greatest interest in the image [17,18,19]. In mathematical morphology, the watershed-plus-marker approach is the traditional image segmentation method [17]. Here, an alternative approach for segmenting images with directional structures is applied. Instead of looking for a set of markers signaling the regions, the watershed will be applied directly to obtain a fine partition. Then a systematic merging process will be applied to obtain the final segmentation. However, to carry out the merging process it is preferable to work with the catchment basins associated with the watershed image. Figure 7(a) shows the catchment basins, computed from the inverse line-segment function shows in Fig. 4(b) and weighted by the values of the angles of of the orientation function Om_X in Fig. 4(c). In order to take into account the neighborhood relationships a region

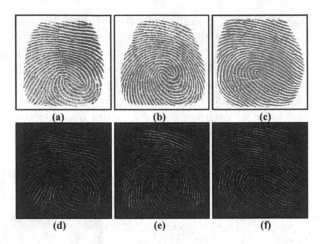

Fig. 6. a), b) and c) Original images, d), e) and f) Main orientations of images (a), (b) and (c), respectively

adjacency graph (RAG) must be computed. The RAG graph is constructed by using of the catchment basins of the image in Fig.7(a). One takes a point from each minimum of the inverse line-segment function for representing each catchment basin. The neighborhood graph of the maxima of the line-segment function Dm_X and the orientation Om_X function synthesize the directional structures of the image. Once the regions are codified on a graph, we can compute the directional structures based on the valued graph. The following method (see [20]) for reducing the numbers of regions may be carried out: a) Each border has assigned an orientation distance between the two regions it separates, b) The borders are sorted in increasing order, c) Two regions separated by the smallest distance are merged, d) The step (b) is repeated until the criterion can not be satisfied. We illustrate the method by identifying the adjacent regions with more-or-less similar orientation by considering the image in Fig. 1(b), a micrograph of the pearlite structure in steel. To achieve such a goal, one merges the vertices (catchment basins) with a difference of angles smaller than or equal to a given angle value $d(R_i, R_j) = |angle(R_i) - angle(R_j)| \leq \theta$. Figure 7(c) shows in color levels the orientation function of Fig. 4 (c), while Fig. 7 (d) shows the output image after the merging process using angle difference criteria θ of 20. The intensities of regions shown in Figs. 7 (b) and (d) were taken to be proportional to the mean value of the merged region angles. Once the grains of the perlitic structure are separated, it is now possible to compute some measures (for example, a granulometric study). It is clear that, when the regions of the image are codified under the form of a graph, many criteria can be easily introduced to segment the image. Figure 7 (e) shows this advantage of using a RAG structure for the merging process since the introduction of other criteria can improve the final segmentation. However, in this example, instead of computing the catchment basins on the whole image, one can do better by computing the weighted partition in a geodesic way. The mask, where the catchment basins transform will be

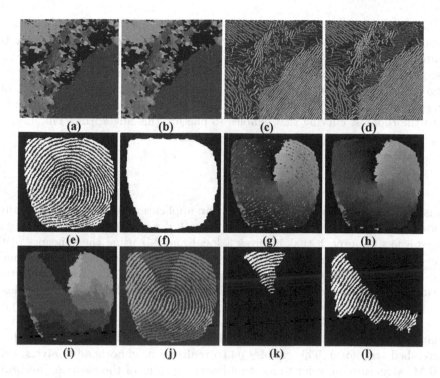

Fig. 7. a) Catchment basins of the partition weighted by the orientation (angles), b) Merged regions, c) and d) Orientation function and segmented image in color representation, e) Original binary image, f) Mask image, g) Catchment basins of the partition weighted by the orientation, h) Filtered partition by size criterion, i) Segmented partition, j) Color representation, k) and i) connected components

applied, is given by the image in Fig. 7 (f). This last image is the output image obtained by the morphological closing size 6 of the original image in Fig. 7 (e). Then, the image in Fig. 7(g) is computed by the catchment basins transform inside the mask. Since thin connections exist between the directional structures of the directional structures, one observes in Fig. 7 (g) small regions that are not representative of the structure (from a segmentation point of view). A size criterion was introduced to remove these regions as illustrated in Fig. 7 (h). Once the small regions are removed, two other criteria can be applied for obtaining the final segmentation. Let $\overrightarrow{\mu}_i$ and $\overrightarrow{\sigma}_i$ be the mean and the variance values in the region R_i. Then, two regions R_i and R_j can be merged if the orientation distance $d_{\overrightarrow{\mu}}(R_i, R_j) = |\overrightarrow{\mu}(i) - \overrightarrow{\mu}(j)| \leq \theta$. This means, after the merging process between regions is carried out, the mean value is used to describe the new region and a new variance value is computed. However, even if the mean difference criterion is satisfied, but one of the regions has a great variance ($\overrightarrow{\sigma}_i > \tau$ or $\overrightarrow{\sigma}_j > \tau$), these regions will not be merged. Figures 7 (i) and (j) show the segmented images, in gray and color levels, using a mean orientation difference criterion $\theta = 15$ degrees and using a variance criterion $\tau = 6.5$. In Section 3, we commented that the notion of connectivity is linked to the intuitive idea of segmentation. Under our approach the images in Figs. 7 (k) and (l) show two connected components of the original image in Fig.7. Compare these connected components with those of Figs. 2 (e), (f) and (g). From the point of view of fingerprint recognition, the connected components illustrate the existence of a singular point (core). The largest component in Fig. 7 (l) describes a separation with the topmost curving that enables to classify this fingerprint. The image in Fig. 7 (i) shows clearly the existence of the core.

5   Conclusion

This paper has shown the possibilities for application of morphological directional transformations to segment images with directional structures. One proposes a local approach that involves a local analysis using the concepts of the line-segment and orientation functions proposed in this paper. The maxima of the line-segment function were used for computing the loci of maximal structuring elements, and the orientation function was used to obtain the angles of the line segments. These pairs of local parameters enable us to produce a good description of the image by means of line segments. Then, a partition of the image may be computed by means of the catchment basins associated with the watershed transform. This enables us to realize a neighborhood analysis, using a RAG structure, in order to merge adjacent regions of the partition according to appropriate criteria, thus segmenting the images into connected components. The results based on the algorithms presented in this paper show the good performance of the approach.

Acknowledgements. The author Luis Morales acknowledges the government agency CONACyT for the financial support. The author I. Terol would like to thank Diego Rodrigo and Darío T.G. for their great encouragement. This work was funded by the government agency CONACyT, Mexico.

References

1. Soille, P., Talbot, H.: Directional morphological filtering. Trans. on Pattern Anal. Machine Intell. 23(11), 1313–1329 (2001)
2. Soille, P., Breen, E.J., Jones, R.: Recursive implementation of erosions and dilations along discrete lines at arbitrary angles. IEEE Trans. on Pattern Anal. Machine Intell. 18(5), 562–567 (1996)
3. Jeulin, D., Kurdy, M.: Directional mathematical morphology for oriented image restoration and segmentation. Acta Stereologica 11, 545–550 (1992)
4. Tuzikov, A., Soille, P., Jeulin, D., Vermeulen, P.: Extraction of grid patterns on stamped metal sheets using mathematical morphology. In: Proc. of International Conference on Pattern Recognition, vol. 1, pp. 425–428 (1992)
5. Oliveira, M.A., Leite, N.J.: Reconnection of fingerprint ridges based on morphological operators and multiscale directional information. In: Proc. of XVII Brazilian Symposium on Computer Graphics and Image Processing, pp. 122–129 (2004)

6. Morales-Hernández, L.A., Terol-Villalobos, I.R., Dominguez-González, A., Herrera-Ruiz, G.: Characterization of fingerprints using two new directional morphological approaches. In: Advances in Dynamics, Instrumentation and Control, pp. 325–334. World Scientific Publishing Co, Singapore (2007)
7. Cappelli, R., Lumini, A.: Fingerprint classification by directional image partitioning. IEEE Trans. on Pattern Anal. Machine Intell. 21(5), 402–421 (1999)
8. Park, C.H., Lee, J.J., Smith, M.J.T., Park, K.H.: Singular point detection by shape analysis of directional fields in fingerprints. Pattern Recognition 39, 839–855 (2006)
9. Li, J., Yau, W.Y., Wang, H.: Constrained nonlinear models of fingerprint orientations with prediction. Pattern Recognition 39, 102–114 (2006)
10. Lee, J.K., Newman, T.S., Gary, G.A.: Oriented connectivity-based method for segmenting solar loops. Pattern Recognition 39, 246–259 (2006)
11. Bahlmann, C.: Directional features in online handwriting recognition. Pattern Recognition 39, 115–125 (2006)
12. Kass, M., Witkin, A.: Analyzing oriented pattern. Computer Vision, Graphics, and Image Processing 37(3), 362–385 (1987)
13. Bazen, A.M., Gerez, S.H.: Systematic methods for the computation of the directional fields and singular points of fingerprints. IEEE-Trans. on Pattern Analysis and Machine Intelligence 24(7), 905–919 (2002)
14. Serra, J.: Image Analysis and Mathematical Morphology. Theoretical advances, vol. II. Academic Press, London (1988)
15. Heijmans, H.J.A.M.: Morphological Image Operators. Academic Press, New York (1994)
16. Soille, P.: Morphological Image Analysis: Principles and Applications, 2nd edn. Springer, Heidelberg, Berlin (2003)
17. Meyer, F., Beucher, S.: Morphological segmentation. J. Vis. Comm. Image Represent 1, 21–46 (1990)
18. Crespo, J., Schafer, R., Serra, J., Meyer, F., Gratin, C.: A flat zone approach: A general low-level region merging segmentation method. Signal Process 62, 37–60 (1997)
19. Salembier, Ph., Serra, J.: Morphological multiscale image segmentation. In: Proc. SPIE-Visual Communications and Image Processing, vol. 1818, pp. 620–631 (1882)
20. Shafarenko, L., Petrou, M., Kittler, J.: Automatic watershed segmentation of randomly textured color images. IEEE Trans. on Image Processing 6(11), 1530–1544 (1997)

Predicting Corresponding Region in a Third View Using Discrete Epipolar Lines

Hiroaki Natsumi[1], Akihiro Sugimoto[2,3], and Yukiko Kenmochi[3]

[1] Chiba University, Japan
[2] National Institute of Informatics, Japan
[3] Université Paris-Est, LABINFO-IGM, UMR CNRS 8049, A2SI-ESIEE, France
sugimoto@nii.ac.jp, y.kenmochi@esiee.fr

Abstract. The discrete epipolar line, a discrete version of the epipolar line, is recently proposed to give geometric relationships between pixels in two different views so that we can directly deal with pixels in digital images. A method is then proposed to determine the discrete epipolar line providing that fully calibrated images are available. This paper deals with weakly calibrated digital images and proposes a method for determining the discrete epipolar line using only weakly calibrated images. This paper also deepens the work further, presenting a method for identifying the corresponding region in a third view from a given pair of corresponding pixels in two views.

1 Introduction

Understanding the geometry of corresponding primitives across different views that arise from the perspective projection of 3D objects is fundamental for applications such as 3D reconstruction from stereo or motion, object recognition, image synthesis, image coding. In particular, the relationships between different views of a point in space have been deeply investigated and sufficient knowledge about epipolar geometry, more generally multi-view geometry, is already well established [3,8,9].

In the framework of multi-view geometry studied so far, points in images are assumed to be directly handled. In other words, digitization of image points is not concerned at all. In reality, however, we cannot deal with points themselves in digital images because digital images involve some digitization process. The smallest unit in digital images is not a point but a pixel. Therefore, even if geometric features are perfectly detected, corresponding points across two views, for example, do not necessarily satisfy the epipolar constraint because of digitization of images [2,11]. This problem cannot be overcome without paying attentions to pixels as the smallest unit of images.

Conventionally, digitization errors are always treated together with observation errors, and how uncertainty arising from errors propagates into the estimation of geometrical information such as epipolar lines is statistically analyzed, providing that knowledge of error statistics are known [7,13]. The employed statistical model for errors, however, is independent of the digitization scheme, and, moreover, the justification of the employed model is not sufficiently discussed.

D. Coeurjolly et al. (Eds.): DGCI 2008, LNCS 4992, pp. 470–481, 2008.

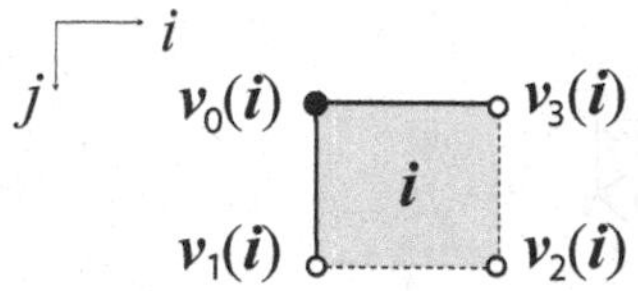

Fig. 1. The dominated region by pixel i and its four corner vertices

Hamanaka *et al.* [5], on the other hand, proposed the discrete epipolar geometry, i.e., a discrete version of the conventional epipolar geometry where the pixel is highlighted as the smallest unit of images. The discrete epipolar geometry aims at rebuilding the conventionally known geometrical relationships on points between multiple views in order to directly deal with pixels in images. With its help, images with different resolutions can be handled simultaneously, for example. Reconstruction ambiguity caused by image digitization alone can be also clarified. However, the proposed method for determining from a given pixel, its discrete epipolar line, is not practically useful. This is because it assumes that images are fully calibrated in advance. This indicates that not only intrinsic camera parameters but also extrinsic ones are required to identify the discrete epipolar line for a given pixel.

This paper advances the direction of directly dealing with pixels, proposing a method for determining the discrete epipolar line using only weakly calibrated images. This paper also deepens the work further, presenting a method for identifying the corresponding region in a third view from a given pair of corresponding pixels in two views where images are assumed to be weakly calibrated. In the both methods, we move a pixel across two views with making full use of the fundamental matrix relating the two views. This allows us to obtain the discrete epipolar line and the corresponding region with only two-dimensional computation. To show the advantage of our approach, we demonstrate some experiments using images with different resolutions, which cannot be observed without paying attentions to the size of the smallest unit of images.

2   Digitizing an Image into Pixels

The smallest unit of digital images is not a point but a pixel. This indicates that an image is not continuous but digitized. We thus introduce a digitization to an uncalibrated image to obtain a set of pixels as its representation.

For a given point with homogeneous coordinates $x = (x, y, 1)^\top$ in an uncalibrated image I, we define $i = \lfloor r_x x + \frac{1}{2} \rfloor$ and $j = \lfloor r_y y + \frac{1}{2} \rfloor$, where r_x, r_y are the resolutions of the x- and y-coordinate, respectively. $i = (i, j)^\top$ are the coordinates of the pixel representing x. Applying this digitization to all the points in I leads to the digitization D of I. We note that various digitization schemes exist to discuss geometric properties of digitized objects such as connectivities, bubble-freeness or topologies (see [1,10], for example); any digitization scheme causes no essential difference in subsequent discussion as far as any point in I corresponds to the unique pixel in D.

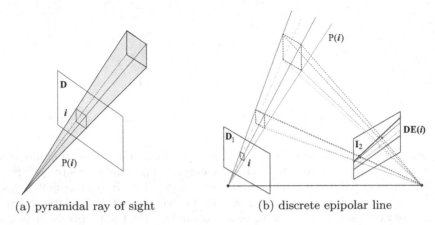

(a) pyramidal ray of sight   (b) discrete epipolar line

Fig. 2. Pixel-based geometry in the perspective projection

Conversely, the region dominated by a given pixel $i = (i, j)^\top$ is a rectangle and is easily computed. The four vertices determining the rectangle are called the *corner vertices* of i and denoted by $v_\kappa(i) \in I$ ($\kappa = 0, 1, 2, 3$).

$$v_0(i) = \left(\frac{1}{r_x}(i - \frac{1}{2}), \frac{1}{r_y}(j - \frac{1}{2}), 1\right), v_1(i) = \left(\frac{1}{r_x}(i - \frac{1}{2}), \frac{1}{r_y}(j + \frac{1}{2}), 1\right),$$

$$v_2(i) = \left(\frac{1}{r_x}(i + \frac{1}{2}), \frac{1}{r_y}(j + \frac{1}{2}), 1\right), v_3(i) = \left(\frac{1}{r_x}(i + \frac{1}{2}), \frac{1}{r_y}(j - \frac{1}{2}), 1\right).$$

We note that only $v_0(i)$ is included into the rectangle (Fig.1).

3   Determining Discrete Epipolar Line

3.1   Discrete Epipolar Line

The concept of discrete epipolar lines was introduced in [5]. The discrete epipolar line is geometrically defined as follows. A pixel in a calibrated image and the viewpoint define a quadrangular prism in space, called a pyramidal ray of sight for the pixel (Fig.2(a)). Projecting this pyramidal ray of sight onto the calibrated image observed from another viewpoint forms a discrete epipolar line[1] (Fig.2(b)).

As we can easily understand, the discrete epipolar line is bounded by two among four epipolar lines, each of which corresponds to one of the four corner vertices of the concerned pixel. The two epipolar lines that bound the discrete epipolar line are called *bounding epipolar lines* in this paper. Effectively identifying bounding epipolar lines under any viewpoint configuration thus becomes crucial in order to determine the discrete epipolar line.

---

[1] More precisely, the discrete epipolar line is the set of pixels intersecting with the image region of a pyramidal ray of sight. Namely, digitization of the image region in the second view is followed. The discrete epipolar line termed in this paper, however, is used without followed digitization. It is thus a continuous region represented by two inequalities. This causes no essential difference in discussion.

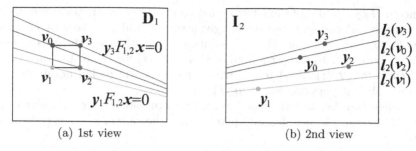

(a) 1st view (b) 2nd view

Fig. 3. A given pixel and epipolar lines of its four corner points

Hamanaka *et al.* [5] proposed to use the epipolar planes of four corner vertices and selected two among the four so that the pyramidal ray of sight is fully included in between the selected epipolar planes. This is because the boundary of the discrete epipolar line is identical with the projection of the selected epipolar planes onto the second image. As images are assumed to be calibrated, however, not only intrinsic camera parameters but also extrinsic ones are explicitly required to identify the discrete epipolar line for a given pixel. This prevents discrete epipolar lines from their practical usefulness.

3.2 Identifying the Boundaries Using the Fundamental Matrix

A pair of corresponding points x_1, x_2 in two different uncalibrated images are known to be related by fundamental matrix $F_{1,2}$:

$$x_2^\top F_{1,2}\, x_1 = 0. \tag{1}$$

All the information about camera parameters is aggregated into 3×3 matrix $F_{1,2}$. Eq.(1) indicates that from a given point x_1 in the first image, we can determine its corresponding epipolar line $x_1^\top F_{1,2}^\top x' = 0$ in the second image if two images are weakly calibrated, i.e., $F_{1,2}$ is known, where x' is the homogeneous coordinates of a point in the second image. In the below, we assume that two images are weakly calibrated.

An epipolar line divides an image into two parts. The projection of the pyramidal ray of sight is included in only one part if the line is a bounding epipolar line, while it overlaps with two parts otherwise. This property allows us to discriminate bounding epipolar lines from others. We investigate this not in the second image but in the first image. This is because the given pixel is nothing but the projection of the pyramidal ray of sight onto the first image. We note that epipolar lines are pair-wisely determined in the first and second images.

Let I_1, I_2 be given two images, and D_1 be the digitization of I_1 (Fig.3). For a given pixel $i \in D_1$, we have four corner points $v_\kappa(i) \in I_1$ ($\kappa = 0, 1, 2, 3$), each of which determines epipolar line $\ell_2(v_\kappa(i))$ in I_2. We then select any point $y_\kappa(\in I_2)$ on the epipolar line $\ell_2(v_\kappa(i))$. y_κ together with $F_{1,2}$ determines epipolar line $y_\kappa^\top F_{1,2}\, x = 0$ in I_1, where $x \in I_1$. We see that two epipolar lines $\ell_2(v_\kappa(i))$ and $y_\kappa^\top F_{1,2}\, x = 0$ are pair-wisely corresponding with each other, and, therefore, $y_\kappa^\top F_{1,2}\, x = 0$ is independent of the choice of y_κ.

Once $\boldsymbol{y}_\kappa^\top F_{1,2}\,\boldsymbol{x} = 0$ is obtained, it is easy to check the geometrical relationship between the region dominated by the given pixel i and two parts of $\boldsymbol{I}_1$ divided by $\boldsymbol{y}_\kappa^\top F_{1,2}\,\boldsymbol{x} = 0$. Since $\boldsymbol{y}_\kappa^\top F_{1,2}\,\boldsymbol{x} = 0$ always goes through the corner vertex $\boldsymbol{v}_\kappa(i)$, we have only to check the sign of $\boldsymbol{y}_\kappa^\top F_{1,2}\,\boldsymbol{v}_\mu(i)$ for the rest corner vertices $\boldsymbol{v}_\mu(i)$ $(\mu = 0,1,2,3; \mu \neq \kappa)$ to see whether or not the region dominated by i exists in only one part divided by $\boldsymbol{y}_\kappa^\top F_{1,2}\,\boldsymbol{x} = 0$.

The algorithm below computes the discrete epipolar line from a given pixel. We note that $\boldsymbol{v}_0(i)$ is included into pixel i; thus the case of $\kappa = 0$ is distinctively treated.

Input: pixel i in the first image and fundamental matrix F relating two images.
Output: discrete epipolar line **DE** for i in the second image.

Step 1 Let **DE**$:= \phi$.
Step 2 For $\kappa = 0,1,2,3$, compute corner vertex $\boldsymbol{v}_\kappa(i)$ and epipolar line $\ell_2(\boldsymbol{v}_\kappa(i))$.
Step 3 For $\kappa = 0,1,2,3$, do
    (1) let $W := \{0,1,2,3\} - \{\kappa\}$;
    (2) select a point $\boldsymbol{y}_\kappa$ on $\ell_2(\boldsymbol{v}_\kappa(i))$;
    (3) if $\boldsymbol{y}_\kappa^\top F \boldsymbol{v}_\mu \geq 0$ for all $\mu \in W$, then { if $\kappa = 0$, then put $\boldsymbol{y}_\kappa^\top F \boldsymbol{x} \geq 0$ into **DE**; else put $\boldsymbol{y}_\kappa^\top F \boldsymbol{x} > 0$ into **DE** };
    (4) else if $\boldsymbol{y}_\kappa^\top F \boldsymbol{v}_\mu \leq 0$ for all $\mu \in W$, then { if $\kappa = 0$ then put $\boldsymbol{y}_\kappa^\top F \boldsymbol{x} \leq 0$ into **DE**; else put $\boldsymbol{y}_\kappa^\top F \boldsymbol{x} < 0$ into **DE** }.

4   Predicting Corresponding Region in a Third View

We assume that we are given two corresponding pixels in two different views. The problem addressed here is to determine the positions of corresponding pixels in a third view.

In the conventional framework, we can easily predict the position of the corresponding point in a third view from a given corresponding pair of points in two views. Namely, we first determine epipolar lines in a third view from a given pair of points in the first and second views, and then identify the intersection point of the two epipolar lines in the third view. We remark that the intersection point is always uniquely determined. This relationship between corresponding points across three views is algebraically analyzed and obtained constraints are called the trifocal tensor in the literature [4,6,12].

Once we accept the fact that the smallest unit of images is a pixel, however, the problem becomes hard to analytically solve, as seen below. We therefore algorithmically solve the problem.

4.1   Corresponding Pixels in Two Views and a Third View

When two corresponding pixels i_1, i_2 are given in two views, two pyramidal rays of sight in space are determined. We denote them by $P(i_1)$ and $P(i_2)$. The region in space that forms the given pixels as its image is identical with the intersection of the two pyramidal rays of sight, which is denoted by $P(i_1) \cap P(i_2)$. When a third viewpoint comes in, $P(i_1) \cap P(i_2)$ is projected onto the third view and forms a region in the third view. This region corresponds to i_1, i_2 and is denoted by $R_3(i_1, i_2)$. We call $R_3(i_1, i_2)$ the *corresponding region* for i_1 and i_2. As we easily see, $R_3(i_1, i_2)$ is not a point but a polygon.

The definition of a pyramidal ray of sight allows us to see that $P(i_1)$ and $P(i_2)$ have four faces and that they are a convex polyhedron. Accordingly, $P(i_1) \cap P(i_2)$ is a convex polyhedron with at most eight faces. Depending on not only the positions of given i_1 and i_2 in two views but also the geometrical configuration of two viewpoints, the number of faces of $P(i_1) \cap P(i_2)$ changes. This indicates that the shape of $R_3(i_1, i_2)$ itself is not invariant against these factors. This kind of changes does never arise in the conventional framework. In contrast, convexity is preserved against any changes in positions of i_1 and i_2, and in geometrical configuration of two viewpoints because the perspective projection does not break convexity. These observations can be summarized as follows.

For given two pixels in two views, the corresponding region in a third view
- is a convex polygon under any configuration of viewpoints and any positions of the given pixels;
- has at most eight edges, and the number of edges depends on the geometrical configuration of viewpoints and also on the positions of the given pixels.

4.2   Identifying the Corresponding Region

Straightforwardly using discrete epipolar lines to predict a corresponding region in a third view does not effectively work. This can be easily understood from the fact that if we determine in a third view two discrete epipolar lines for a given pair of pixels and then compute the intersection of the two discrete epipolar lines, the obtained region is with at most four edges; this contradicts the property of the corresponding region addressed in the previous section.

In fact, two pyramidal rays of sight for given pixels carve each other, and, therefore, the corresponding region becomes strictly smaller than the simple intersection of the two discrete epipolar lines. For a given pair of corresponding pixels i_1 and i_2 in two views, let $\mathbf{DER}_3(i_1)$ and $\mathbf{DER}_3(i_2)$ be the regions satisfying the discrete epipolar line in a third view for i_1 and i_2, respectively. Then, $R_3(i_1, i_2) \subseteq \mathbf{DER}_3(i_1) \cap \mathbf{DER}_3(i_2)$.

To see how two pyramidal rays of sight carve each other, we focus on each ridge line of a pyramidal ray of sight and investigate how it intersects with the faces of the other pyramidal ray of sight in order to obtain all the vertices of convex polyhedron $P(i_1) \cap P(i_2)$. Investigating this in space is complicated whereas the projection onto each view makes this analysis even simpler.

Let us focus on the investigation about a corner point $v_\kappa(i_1)$ and pyramidal ray of sight $P(i_2)$ (Fig.4). We consider epipolar line $\ell_2(v_\kappa(i_1))$ for $v_\kappa(i_1)$ in the second view. We note that $\ell_2(v_\kappa(i_1))$ is the projection of the line in space going through the first viewpoint and $v_\kappa(i_1)$. The fact that this line in space intersects with $P(i_2)$ is identical with the fact that $\ell_2(v_\kappa(i_1))$ intersects with i_2. Furthermore, if it intersects in space, the projection of intersection points in space is identical with the intersection points of $\ell_2(v_\kappa(i_1))$ and i_2 in the second view. This allows us to identify the projection of intersection points in space onto the second view without directly dealing with 3D information.

Once we identify the projection of intersection points in space onto the second view, it is then straightforward to predict the position of the corresponding point

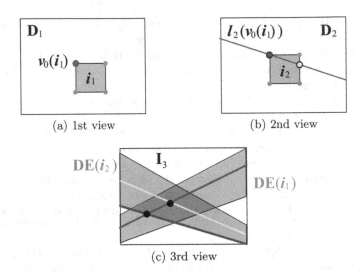

(a) 1st view   (b) 2nd view

(c) 3rd view

Fig. 4. Intersection points of two pyramidal rays of sight observed in three views

in a third view. Namely, using fundamental matrices relating the third view with the first and second views, we determine epipolar lines in the third view for $v_\kappa(i_1)$ and also those for the obtained intersection points, and then compute intersection points to identify their positions in the third view.

Iterating the above procedure with respect to the four corner points $v_\kappa(i_1)$ ($\kappa = 0, 1, 2, 3$) of i_1 and then exchanging the roles of two views in the procedure lead to identifying the positions of the vertices in the third view, i.e., the projection onto the third view of all the vertices of $P(i_1) \cap P(i_2)$. Computing the convex hull of the obtained vertices in the third view enables us to identify the corresponding region $R_3(i_1, i_2)$. We remark that $R_3(i_1, i_2)$ is a convex polygon whose vertices are a subset of the obtained vertices.

The algorithm below identifies the corresponding region in a third view from given two corresponding pixels in two views.

Input: corresponding pixels i_1 and i_2 in the first and second views and fundamental matrices relating two of three views: $F_{1,2}, F_{1,3}, F_{2,3}$.
Output: corresponding region $R_3(i_1, i_2)$ in the third view.

Step 1. Let $F_{2,1} := F_{1,2}^\top$ and $G := \phi$.
Step 2. For $(\alpha, \beta) \in \{(1, 2), (2, 1)\}$, do
    (1) for $\kappa = 0, 1, 2, 3$, do
        (I) for $\tau = 0, 1, 2, 3$, do
            (i) let $L_{\tau,\tau+1}^\beta$ be the line segment connecting $v_\tau(i_\beta)$ and $v_{\tau+1(\mathrm{mod}4)}(i_\beta)$;
            (ii) compute in view β the intersection point of line segment $L_{\tau,\tau+1}^\beta$ and epipolar line $v_\kappa^\top(i_\alpha)F_{\beta,\alpha}\,x = 0$; let the intersection point be t;
            (iii) compute the intersection of $v_\kappa(i_\alpha)^\top F_{\alpha,3}^\top x = 0$ and $t^\top F_{\beta,3}^\top x = 0$; put the intersection point into G.

Step 3. Compute the convex hull of G; let $R_3(i_1, i_2)$ be the region inside of the obtained convex hull.

5   Experiments

We demonstrate some experiments on predicting corresponding regions using weakly calibrated images with different resolutions. We remark that images with different resolutions are frequently obtained when they are captured using different digital cameras.

We set up experimental conditions as shown in Fig. 5 (a). Namely, for a randomly selected point X_0 in space, we first put two viewpoints, C_1 and C_2, so that they together with X_0 form the regular triangle in space, and obtained its two image points using camera parameters. We then put a third viewpoint C_{30} to form the regular tetrahedron together with X_0, C_1 and C_2, and computed fundamental matrices determined by these views. After digitizing the first and second views, we applied our method to identify the corresponding region in the third view. We also conducted experiments in the cases where the third viewpoint is changed to C_{31} and C_{32} so that it becomes close to the first viewpoint C_1 and the second viewpoint C_2, respectively. This is to evaluate how the corresponding region depends on the viewpoint configuration in space. In our experiments, we fixed the resolution of the first view, and changed that of the second view to see how the corresponding region in the third view depends on resolutions. The resolution ratio of the second view to the first view was set to be 2^k ($k = -3, -2, -1, 0, 1, 2, 3$). Under the same condition except for changing X_0 to X_1 (Fig. 5 (b)), we conducted the same experiments.

The results for point X_0 are shown in Fig. 6 while those for point X_1 are in Fig. 7. In these figures, (a), (c) and (e) illustrate the shape of the computed corresponding region, where the horizontal axis and the vertical axis correspond to the horizontal image coordinate and the vertical image coordinate, respectively. (b), (d), (f), on the other hand, show the number of edges and the area of the computed corresponding region, where the horizontal axis indicates the image resolution. We note that the vertical axis for the area of the computed corresponding region indicates the logarithm of the area to the base two.

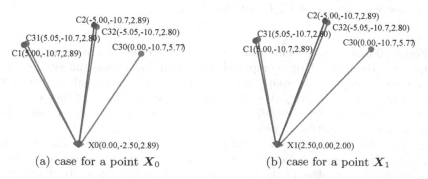

(a) case for a point X_0 (b) case for a point X_1

Fig. 5. Configuration of viewpoints C_1, C_2, C_{30}, C_{31} and C_{32}

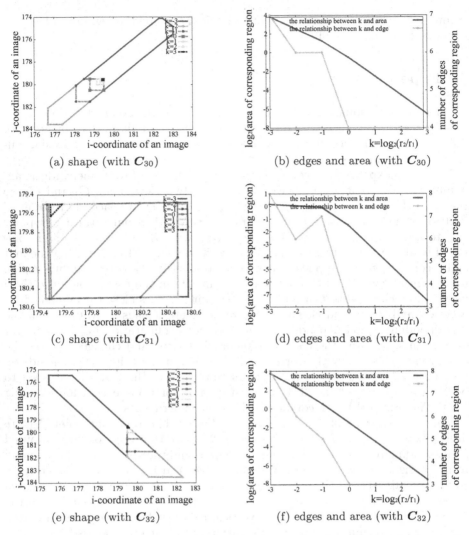

Fig. 6. Shape, edges and area of the corresponding region depending on image resolutions (case with viewpoints C_1, C_2 and point X_0)

Figures 6 and 7 show that in any case, as the resolution becomes higher the shape and the area of the computed corresponding region monotonically decrease. As we expect, the shape and the area tends to eventually converge to a point and zero, respectively. We also observe that the shape of the corresponding region itself changes depending on image resolutions even for the same point configuration. Furthermore, we observe some correlation exists between changes in number of edges and those in area; this observation is independent of the position of a third viewpoint. Namely, with a certain image resolution, both the area and the number of edges of the corresponding region significantly change at the same time and the changes depend on only the configuration of the first and the

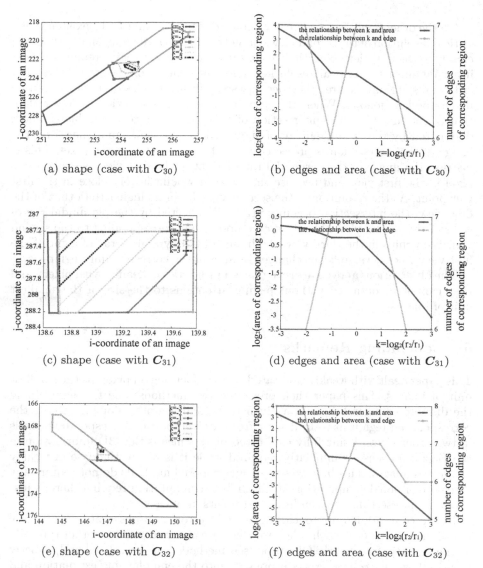

(a) shape (case with C_{30})

(b) edges and area (case with C_{30})

(c) shape (case with C_{31})

(d) edges and area (case with C_{31})

(e) shape (case with C_{32})

(f) edges and area (case with C_{32})

Fig. 7. Shape, edges and area of the corresponding region depending on image resolutions (case with viewpoints C_1, C_2 and point X_1)

second viewpoints. These observations come from the fact that the intersection of two pyramidal rays of sight varies with image resolutions, and in particular, the shape of the intersection drastically changes with a certain image resolution.

Figure 6 (a), (b), (e) and (f) indicate the existence of the cases in which the area of the corresponding region has the same changes in shape depending on image resolutions though their viewpoint configurations are different from each other. This implies that when a third viewpoint is far from the first viewpoint, changes in resolution of the second view are irrelevant to the viewpoint configuration. This observation is also supported by Fig. 7 (a), (b), (e) and (f).

On the other hand, we observe in (c) and (d) in Figs. 6 and 7, the existence of a critical resolution of the second view. Namely, when a third viewpoint is allocated closely to the first viewpoint, the area of the corresponding region remains almost invariant until the critical resolution, and once the second view achieves the critical resolution, the area then decreases significantly. This observation can be understood as follows. When the resolution of the second view is lower than that of the first view, the intersection of two pyramidal rays of sight is almost dominated by the pyramidal ray of sight of the first view, and, therefore, the image of the intersection is projected with the size of almost one pixel. This is because the image of the pyramidal ray of sight of the first view is exactly one pixel in the first view and because the third viewpoint is very close to the first viewpoint. As the resolution of the second view becomes higher than that of the first view, the intersection becomes gradually affected by the pyramidal ray of sight of the second view. The discrete epipolar line corresponding to the second view then chips in the third view the image of the pyramidal ray of sight of the first view to cause significant changes in area of the corresponding region.

The observations above are distinguished properties in dealing with pixels, and they cannot be obtained without paying attentions to the size of the smallest unit of images.

6   Concluding Remarks

This paper dealt with weakly calibrated images, focusing on pixels as the smallest unit of images. This paper then presented two methods: one for determining the discrete epipolar line from a given pixel, and the other for identifying the corresponding region in a third view from a given pair of corresponding pixels in two other views. Using fully calibrated images means that 3D information in the Euclidean sense is implicitly required while it is not in the case of weakly calibrated images. This indicates that our proposed methods do not require any three-dimensional computation at all; in fact, computation required here is just checking intersection between lines and points in images.

In the conventional approach, digitization errors and observation errors are not discriminated from each other and they are always treated together typically in the statistic framework. Our proposed methods, on the other hand, exactly identify how digitization errors propagate into the epipolar line estimation and how the ambiguity region looks like in a third image. Investigating practical superiority of our method to existing methods, in particular, the sub-pixel based method, is our next step. More detail analysis of the properties of the corresponding region from the theoretical point of view is also left for future work.

References

1. Andres, E.: Discrete Linear Objects in Dimension n:the Standard Model. Graphical Models 65, 92–111 (2003)
2. Blostein, S.D., Huang, T.S.: Error Analysis in Stereo Determination of 3-D Point Positions. IEEE Trans. on PAMI 9(6), 752–765 (1987)

3. Faugeras, O., Mourrain, B.B.: On the Geometry and Algebra of the Point and Line Correspondences between N images. In: Proc. of ICCV, pp. 951–956 (1995)
4. Faugeras, O., Robert, L.: What can Two Images Tell Us about a Third One? IJCV 18(1), 5–19 (1996)
5. Hamanaka, M., Kenmochi, Y., Sugimoto, A.: Discrete Epipolar Geometry. In: Andrès, É., Damiand, G., Lienhardt, P. (eds.) DGCI 2005. LNCS, vol. 3429, pp. 323–334. Springer, Heidelberg (2005)
6. Hartley, R.: Lines and Points in Three Views and the Trifocal Tensor. IJCV 22(2), 125–140 (1997)
7. Hartley, R., Sturm, P.: Triangulation. CVIU 68(2), 146–157 (1997)
8. Hartley, R., Zisserman, A.: Multiple View Geometry in Computer Vision. Cambridge University Press, Cambridge (2000)
9. Heyden, A.: A Common Framework for Multiple View Tensors. In: Burkhardt, H., Neumann, B. (eds.) ECCV 1998. LNCS, vol. 1407, pp. 3–19. Springer, Heidelberg (1998)
10. Montanari, U.: On Limit Properties in Digitization Schemes. J. of ACM 17(2), 348–360 (1970)
11. Rodriguez, J.J., Aggarwal, J.K.: Stochastic Analysis of Stereo Quantization Error. IEEE Trans. on PAMI 12(5), 467–470 (1990)
12. Shashua, A., Werman, M.: Trilinearity of Three Perspective Views and its Associated Tensor. In: Proc. of ICCV, pp. 920–925 (1995)
13. Stewenius, H., Schaffalitzky, F., Nister, D.: How Hard is 3-View Triangulation Really? In: Proc. of ICCV, pp. 686–693 (2005)

A Discrete Modelling of Soil Fragments Transport by Runoff

Gilles Valette[1,2], Stéphanie Prévost[1], Laurent Lucas[1], and Joël Léonard[2]

[1] CReSTIC/SIC/MADS, EA3804 University of Reims Champagne-Ardenne,
Reims, France
laurent.lucas@univ-reims.fr
[2] INRA UR1158 Agronomie Laon-Reims-Mons, France

Abstract. We aim to model and visualize the evolution of the surface structure of a cultivated soil surface during rainfall. In this paper, we briefly present our model, based on an Extended Cellular Automaton, and the different simulated processes. Among these processes, we focus on runoff which is of high relevance as it drives the evolution of the soil surface structure by transporting and depositing the detached fragments of soil and thus inducing an evolution in the granulometry of the surface material. We propose a simple algorithm to model, in a discrete way, runoff and also the transport and deposition of soil fragments according to their size. In that way we are able to derive information about the evolution of soil surface granulometry. A validation of the runoff model is proposed, based on the comparison of the results obtained with results from a numerical solution of the Saint Venant's equations. Although no validation was attempted for transport, simulations yielded visually promising results.

1 Introduction

The objective of our project, called SoDA (Soil Degradation Assessment), is to develop and validate a dynamic simulation of the evolution of the surface structure of a cultivated soil surface during a rainfall at the meter scale, keeping in mind a constant care for visualization. Major aspects of this evolution are the formation of soil crusts and the development of cracks [1, 2, 3]: presence of a crust due to rainfall can mechanically inhibit seedling emergence whereas presence of cracks can allow seedlings to break through the soil surface. Therefore, such a model, able to predict soil structure under different initial soil conditions and climatic scenarios, would be for example a useful tool to select adequate tillage and sowing practices.

Crusts form during rainfall because of the detachment of soil particles by raindrops and their subsequent redistribution, either by projection or transport in the flow [4]. The size of the fragments produced by the desaggregation processes varies with the initial size distribution of particles, the hydric state of the soil, the intensity of the stresses exerted by raindrops. The size of soil fragments determines if runoff will be able to transport them or not. It is thus very important to have and to keep some information about particle size, as some types of crusts —like sedimentary crusts— are precisely defined by the occurrence of

D. Coeurjolly et al. (Eds.): DGCI 2008, LNCS 4992, pp. 482–494, 2008.

sorting during sedimentation. This is one originality of our simulator to permit this tracking of the soil granulometry.

A primary version of our simulator has been presented in a previous article [5]. The aim of the present paper is to present our new model of runoff and transport of soil fragments, which is based on Cellular Automata rules which are simple interpretations of equations from hydraulics. The previous model used a single equilibrating law between two cells, thus without any physical principle, and was not parallelizable. We have validated this new model by comparing our results to the ones obtained using a numerical solution to the Saint Venant's equations [6]. Finally, we show that our model permits the simulation of the temporal and spatial dynamics of soil fragments sorting to produce some visually convincing results.

2   Related Work

We present in this section a brief review of models of transport of solid material by runoff found in the field of Computer Imagery and Soil Science followed by a few comments.

Musgrave et al. [7] proposed a new method for creating mountain fractal terrains with the use of height fields. They used a report algorithm to model runoff and an *ad hoc* model of hydraulic erosion. Beneš and Forsbach [8, 9] introduced a new data structure for visual simulation of 3D terrains. They divided the hydraulic erosion process into four independent steps that can be applied independently. The authors modeled the runoff with a report algorithm on a 8-neighbourhood, they used an empirical saturation level to calculate the amount of sediments in water, and they considered evaporation but not infiltration. Chiba et al. [10] presented a simple "quasi physically based" method for simulating the topography of eroded mountains based on velocity fields of water flow. In this method, the velocity fields of water flowing down the face of a mountain are calculated by simulation of the motion of "water particles", and these velocity fields are used for simulating erosion and sedimentation. Neidhold et al. [11] presented a real-time method that combined a non-expensive fluid simulation based on a Newtonian physics approach on a two dimensional grid storing acceleration, velocity and mass. They provided an erosion algorithm. Their method does not take into account infiltration, and it uses specific constants, whose value has to be arbitrarily given by the user. Beneš et al. [12] introduced a full 3D hydraulic erosion, by coupling Navier-Stokes equations with cohesive or cohesionless material transportation and solving them on a 3D grid. Recently, shallow water fluid models were used by Beneš [13] and Mei et al. [14] in order to provide real-time hydraulic erosion based on 2D regular grids.

In the field of Soil Science, prediction of soil erosion generally means prediction of soil loss rather than analysis of the evolution of the soil surface and its relief. Current models for soil erosion by water are WEPP (Water Erosion Prediction Project) [15] and EUROSEM (European Soil Erosion Model) [16]. Favis-Mortlock et al. [17] developed the "RillGrow". In this model emphasis is put on the evolution of the soil surface rather than on sediment exportation.

D'Ambrosio et al. [18,19] and Avolio et al. [20] suggested a Cellular Automata (CA) model for soil erosion by water. This model involves a larger number of states, including altitude, water depth, total head, vegetation density, infiltration, erosion, sediment transport and deposition. A similar CA approach, based on the model of *"precipitons"* introduced by Chase [21], is used by Luo et al. [22] for the WILSIM (Web-based Interactive Landform Simulation Model) project. In order to simulate landscape erosion and deposition, Haff [23] uses a CA model called *"waterbot"*. Servat [24] suggested a description of flows in terms of heterogeneous agents that interact in a continuous space. This research has led to the development of the RIVAGE simulator of runoff and infiltration, with some additional trials to incorporate erosion processes.

Most of the models presented share the same principles, whatever the objectives, the scale or the processes considered. For example, they almost have in common the principle of flow of water or material according to the greatest gradient, or the transport capacity concept. On that point, our simulator does not significantly differ from other models. However, Computer Imagery models aim visually plausible results and not simulation, thus they often make simplifications about the processes considered and they cannot apply in our simulator. For example, these models assume that the material is dissolved by running water. At the scale of our simulator, runoff is not capable of detachment: the fragments which are mobilized must exist and have been created by the raindrops energy. Moreover, these models are often based on some user-defined parameters and quantities which are not directly linked to the properties of soil or water. On the contrary, our model uses formulas and parameters which are in adequation with the current knowledge of soil scientists. Finally, our model considers explicitly a 3D space and the processes of splash and infiltration, which are rarely considered. It also permits the precise tracking of the granulometry of detached and transported particles of the soil (as soon as it has been modified). This is its main originality and it is an important issue as it allows to describe a spatial (vertical and horizontal) and temporal evolution in the state of the soil, and to establish a relationship between this evolution and the processes at work.

3   Our Model

The simulator is based on an Extended Cellular Automaton [18,20]. The patch of soil which the simulator has to manage is typically of 50cm x 50cm x 10cm, discretized into 2mm side cubic cells. Runoff can be considered as a centimetric scale process and thus, we could use a larger resolution, i.e., 1cm instead of 2mm, but we need a millimetric resolution to treat with a good accuracy the changes in the altitude due to the transport of the smallest fragments. The cells contain variable proportions of solid, water and air. In order to be able to keep an information on the granulometry of the soil, the solid phase can be constituted of "continuous" matter (i.e., not fragmented) or of discrete particles which are distributed into 6 classes, from 0 to 2mm. The rainfall events are also discretized by raindrops whose size and number vary through time. The 3D-grid containing the soil cells is completed with three 2D-grids: one to store the height of the terrain at the center of each surface cell, one for flow depth, one for the particles transported in the flow. The movement of water is due to runoff and infiltration. We don't take into account the effect of evaporation because the infiltration rate is always much greater than the evaporation rate, especially during a rainfall. In its main loop, our simulator considers successively the following processes:

- the raindrops generation, following a gamma distribution depending on the rain intensity;
- the soil detachment, which is proportional to the kinetic energy of each raindrop, and which is corrected with respect to the thickness of the surface water;
- the soil fragmentation, estimated from tests of structural stability;
- the projection by splash of the soil fragments, to a distance which is proportional to the momentum of each raindrop and which depends on the size of the fragments;
- infiltration, which is in the current version simply calculated with the physically based Green-Ampt method [25, 5], but will be replaced in a future work by a cellular automata infiltration model which will be coupled to the evolution of the soil surface structure;
- runoff, which is simulated using a report algorithm depending on the slope of the water surface. The altitude of the water surface is equal to the sum of flow depth and terrain height, and we use it to represent the hydraulic head, neglecting the kinetic component which is always very small in our study context.
- the transport and the deposition by runoff, with the hypothesis that runoff is only an agent of transport and does not play any role in the detachment of the fragments, which is true at the scale considered.

4   Runoff

Our modelling of the runoff process takes into account the flow depth and the altitude of the soil surface, the sum of these two quantities being used to define the total hydraulic head. For each iteration of our algorithm, for each cell, we have to determine how much water is transferred and where it is transferred. We have chosen to transfer water from one source cell to a unique target cell only, in the direction of the highest head gradient. This is a simplification of the reality, but as the gradient is changed when water has been transferred to a downstream cell, it remains possible that water on a cell which has only lower elevation neighbours be transferred all around during a few iterations. To compute the quantity of water which has to be transferred, we must consider four points: i) the discharge must be accurate; ii) the mass of water must be conserved; iii) the state of a cell established during the current iteration must not be necessary for establishing the state of another cell in the same iteration; iv) the algorithm has to converge to a stable state. The third point is necessary for two reasons: firstly because it is the correct behaviour in a CA model, secondly because this independance is required in order to get a parallelizable code.

4.1 Algorithm

In our runoff algorithm we treat all the surface cells from the lowest head cell to the highest head cell to calculate the quantity of water received by a cell and where it comes from. The pre-ordering of the cells allows to satisfy the constraint that a cell can transfer water only to the neighbouring cell which has the lowest hydraulic head. If we treated the cells receiving water in an arbitrary order, it would be possible that water be transmitted to a cell not being the lowest hydraulic head cell in the neighbourhood of the source cell (an example is shown Fig. 1). By using this pre-ordering and marking each cell which gives water, we insure that the current cell receives water only from unmarked neighbours. Instead of ordering the cells, the same result could be obtained by examining an extended neighbourhood of each cell, but the computional cost would be much higher. As we consider the 2D-space formed by the surface cells, the neighbourhood used can be the von Neumann neighbourhood (also called 4-neighbourhood) or the Moore neighbourhood (also called 8-neighbourhood) [19]. We have tried both these neighbourhoods and observed some differences (see Section 7). Notice that we use the standard CA behaviour, i.e., for one cell the transition function applies only to the adjacent cells of this cell. Thus, the size of the neighbourhood is limited. For the same reason, we do not have to consider a discrete distance computation, e.g., chamfer methods, because we only need the distance between two adjacent cells, which we consider as the Euclidean distance between their centers.

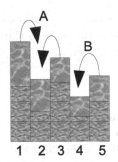

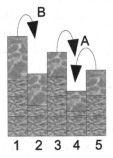

(a) Cells treatment following an arbitrary order.

(b) Cells treatment following the hydraulic head order.

Fig. 1. Why the cells must be ordered: in (a) an arbitrary order from cell 1 to cell 5 is considered, cell 2 is treated before cell 4, thus cell 3 gives water to cell 2 which is not the lowest hydraulic head cell in its neighbourhood, in (b) cells are ordered by their hydraulic head, cell 4 is now treated before cell 2, thus cell 3 gives water to cell 4, which is correct

In our method, boundaries are easy to treat. We distinguish three cases. If they are considered as walls, they are simply ignored and there is nothing to do. If the terrain is seen as a torus, a cell on the boundary can receive water from its opposite cells, so it suffices to add these cells in its neighbours. Finally, if boundaries are considered as holes, we treat them as "ghost" cells that are the lowest cells of the terrain. So we have to do a first pass in order to calculate the quantity of water these cells receive from their immediate neighbour, in the perpendicular direction relative to the boundary. This neighbour is marked, so it does not give any more water.

Finally, it is worth noticing that our algorithm is fully parallelizable, because the calculations are independant and exclusively established from the state obtained in the previous iteration. The ordering of the cells is not an issue for the majority of the cells, because it can be done locally, e.g., on the portion of the soil treated by a core. However, we have to treat differently the cells which are at the frontier of two cores by examining their extended neighbourhood, i.e., the neighbourhood of each (possible) emitting cell, in order to eliminate the cells which are adjacent to a cell with a lower hydraulic head.

4.2   Flow Discharge Calculation

The flow q from a source cell i to a target cell c is calculated using the Darcy-Weisbach equation [26]:

$$v = \sqrt{\frac{8\,g\,S(c,i)\,W_i}{f}} \;, \tag{1}$$

combined to the definition of the flow:

$$q = v\,W_i\,dx \;, \tag{2}$$

where v is flow velocity, $S(c,i)$ is the friction slope, which we assume to be the slope of altitude plus water depth, W_i is the water depth, g is the acceleration due to gravity, dx the spatial step, and f is the Darcy-Weisbach friction factor. f depends on soil roughness and strongly determines flow velocity; it is an init parameter of our simulator.

4.3   Equilibrating Flow Calculation

One main issue in the report algorithm is that the quantity of water transferred to a cell must not be higher than the difference in hydraulic head between the source cells and the target cell, because the target cell would then become a source cell at the next iteration, and so on, causing instability. For this reason we have to calculate a maximum quantity of water that a cell can receive from each neighbour. As a cell can receive water from up to 8 cells (or 4, depending on the chosen neighbourhood), we have to find a way to calculate this maximum quantity. It is worth noticing that to permit this equilibrating flow calculation, we choose to consider a cell receiving water from its neighbours, and not a cell transferring water. This calculation uses the following definitions:

- W_i^t is the depth of water on the cell i at iteration t;
- H_i is the height of the soil on the cell i, considered as constant;

- $\varphi_i^t = H_i + W_i^t$ is the hydraulic head of the cell i;
- $\delta_i = W_i^t - W_i^{t+1}$ is the emitted water from cell i (positive quantity);
- $\delta_c = W_c^{t+1} - W_c^t$ is the received water in cell c (positive quantity);
- A is the set of not marked neighbouring cells of central cell c;
- n is the cardinal of A, i.e., the number of cells which have to give water to c.

In order to avoid instability, the receiving cell c (i.e., the central cell of the neighbourhood) must always have a smaller hydraulic head than that of all the emitting cells i:

$$\forall i \in A \quad \varphi_c^t \leq \varphi_i^t . \tag{3}$$

Considering this condition at iteration $t+1$ and using the definitions of φ_i^{t+1}, φ_c^{t+1}, δ_i, δ_c it comes:

$$\forall i \in A \quad \varphi_c^t + \delta_c \leq \varphi_i^t - \delta_i . \tag{4}$$

The mass conservation implies that the quantity of water received should be equal to the quantity of water emitted:

$$\sum_{i \in A} \delta_i = \delta_c . \tag{5}$$

(4) and (5) give the condition for stability:

$$\forall i \in A \quad 2\delta_i + \sum_{j \in A - \{i\}} \delta_j \leq \varphi_i^t - \varphi_c^t . \tag{6}$$

If we consider the limit, i.e., the equivalence between these quantities, we get this matricial equation:

$$\begin{pmatrix} 2 & 1 & \dots & 1 \\ 1 & 2 & \dots & 1 \\ & & \dots & \\ 1 & 1 & \dots & 2 \end{pmatrix} \begin{pmatrix} \delta_1 \\ \delta_2 \\ \dots \\ \delta_n \end{pmatrix} = \begin{pmatrix} \varphi_1^t - \varphi_c^t \\ \varphi_2^t - \varphi_c^t \\ \dots \\ \varphi_n^t - \varphi_c^t \end{pmatrix} . \tag{7}$$

The first matrix can be easily inverted, thus we obtain the formulation of the solution:

$$\delta_i = \frac{n}{(n+1)}(\varphi_i^t - \varphi_c^t) - \sum_{j \in A - \{i\}} \frac{\varphi_j^t - \varphi_c^t}{(n+1)} . \tag{8}$$

In order to have a physically correct behaviour, we must insure that this value δ_i is always positive. Therefore, we eliminate from the neighbourhood A the cells for which this value is negative before recomputing all the values for this new neighbourhood. Finally, the transferred quantity of water is the minimum between δ_i and q, so the inequality (6) is always verified and the risk of instability is eliminated.

5 Mobilization and Deposition

Mobilization and deposition are based on a transport capacity approach. The transport capacity of the flow is defined by its maximal sediment concentration $\mathcal{C}$. The value of $\mathcal{C}$ depends on flow conditions. The quantity of matter $\mathcal{M}$ that a given flow can transport is then given by the product of the volume of water W and the maximal sediment concentration $\mathcal{C}$. Transport capacity equations of the form $C \propto (\tau - \tau_c)^b$, where τ is the shear stress, τ_c is the critical shear stress, above which significant transport occurs and whose values can be found in the soil science literature, and b a parameter often larger than one, have been shown to be generally appropriate [27]. We thus have chosen to use this kind of formulation to estimate transport capacity for the class p:

$$\mathcal{C}(p) = a_p \Big(\tau - \tau_c(p)\Big)^{b_p} , \tag{9}$$

and we used a sediment transport database [27] to derive parameters $\tau_c(p)$, a_p and b_p of (9) for each of the size classes considered. The shear stress is calculated from $S(i,j)$, the total head slope between the source cell i and the target cell j, and the water depth W_i according to the following equation [26]:

$$\tau = \rho_w \, g \, W_i \, S(i,j) . \tag{10}$$

The first step of our algorithm is the comparison between $\tau(i)$ and the threshold $\tau_c(p)$ to eliminate the classes of fragments for which no mobilization is possible: for these classes, we deposit all the fragments. From the other classes, we calculate the maximum sediment concentration $\mathcal{C}_{max}$ using (9) and the mean weighted diameter. This value permits us to obtain the maximal mass of sediment $\mathcal{M}_i$ we can have in the flow. If this mass is greater than the mass already present in the flow, we can take off some available fragments and put them in the flow. On the opposite, if this mass is less than the mass already present in the flow, deposition occurs, i.e., particles are added to the cell considered. To be able to mobilize the biggest fragments, we begin the loop starting from the coarsest particles, until we get the needed quantity. Finally, as the transformations are purely local, i.e., depending only on the state of the cell and made inside this cell, it is correct to apply them directly in a loop treating all cells in any order.

6 Transport

When a quantity of water moves because of runoff, it transports a proportional quantity of mobilized matter to the cell of destination. In this purpose, we calculate first the total mass of fragments mobilized in the source cell and deduce from this value the mass of fragments we have to move to the target cell. As we work with discrete number of fragments, in many cases it is impossible to move this exact quantity. Moreover, this quantity is often too small to move one of the biggest fragments. If this deficit was repeated too frequently, it would introduce a bias in the transport of fragments. In order to avoid this, we have chosen to randomly begin this movement from the biggest class or from the smallest class, with an equal probability. In that way, we insure first to sometimes move some big fragments, and second to compensate this excess of transported mass in a following iteration.

7   Results

7.1   Validation of the Runoff Algorithm

For validation purpose, we simulated rainfall over a 30 cm x 2 cm terrain having a constant slope of 5%. The rain intensity was 30mm/h and the duration of the rainfall was 30s. Infiltration was inhibited, the soil being considered as impermeable. The friction factor f was set to a value of 1. In Fig. 2(a) a plot of cumulated runoff at the outlet of the terrain versus cumulated rainfall shows that water conservation is satisfied. Fig. 2(b) shows that the discharge at the outlet of the terrain reaches rainfall intensity in about 15s, keeps this value as long as rainfall continues, and then decreases to zero in about 50s after rainfall has stopped. We have compared these results to results obtained using a numerical solution of the classical Saint Venant's equations [6]. Both results are in close agreement. Simulations done with both a 4-neighbourhood and an 8-neighbourhood gave similar results but the flow appears more regular with the 4-neighbourhood.

7.2   Visual Result of Soil Fragments Transport

In order to test the transport algorithm, we used a terrain combining 3 different slopes: 10% over the first 20cm, 40% over 20cm, and 5% over 30cm. We limited

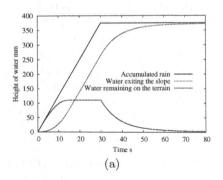
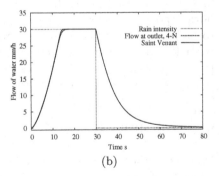

| (a) | (b) |

Fig. 2. (a) Verification of the conservation of the water during a simulation. (b) Comparison between the simulated discharge at the outlet of the terrain with a 4-neighbourhood and the curve given by a numerical solution of the Saint Venant's equations. The curve 8-neighbourhood is very close to the 4-neighbourhood but with only more oscillations on the flat part.

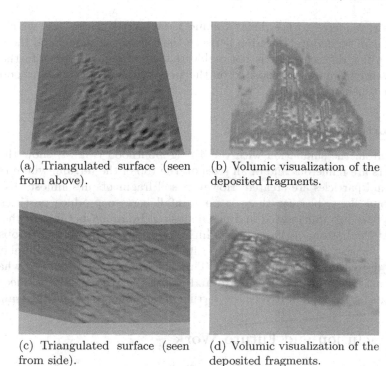

(a) Triangulated surface (seen from above).

(b) Volumic visualization of the deposited fragments.

(c) Triangulated surface (seen from side).

(d) Volumic visualization of the deposited fragments.

Fig. 3. Visual result of transport by runoff along a slope at its bottom: (a) and (b) with a 4-neighbourhood, (c) and (d) with an 8-neighbourhood

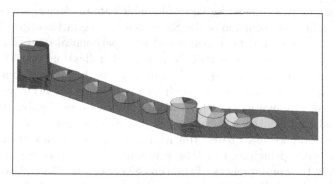

Fig. 4. Visualization of the granulometry: the height of a cylinder indicates the total mass of the fragments deposited. The levels of grey indicate the class of the fragments: white is the smallest size, black the biggest size.

rainfall to the first part of the terrain. Two simulations of 30 min duration, with a rainfall intensity of 30mm/h and without infiltration have been done: one with a 4-neighbourhood, the second with an 8-neighbourhood. Fig. 3 shows the results of these simulations, with both a triangulated surface, rendered by OpenGL with the Phong's illumination model, and a volumic visualization of the deposited fragments, rendered with a direct volume rendering algorithm [28] which takes advantage of the natural bijection between the cells of our model and the voxels. Among a few differences, it is noticeable that the fragments went further and that less fragments were deposited on the slope with the 4-neighbourhood than with the 8-neighbourhood.

7.3   Sorting During Transport

Results from the simulation with the 4-neighbourhood (Fig. 4) show that soil fragments are numerous and of all sizes on the top of the terrain, where rainfall occurs and particles are created. However, soil fragments are almost absent on the second slope, because there is no rainfall here to produce particles, and existing ones are transported and not deposited by runoff because of the steep slope. At the bottom of the terrain, which has a lower slope, it can be observed that the biggest fragments were deposited here, and that their number and size are decreasing with the distance. This behaviour is very close to what can be currently observed in the nature and shows that the simulator adequately represents, at least qualitatively, the sorting of the soil material during transport.

8   Conclusion and Future Work

We develop a simulator of the evolution of soil surface structure of cultivated soils under rainfall, at the meter scale. This is an important issue as it has both theoretical and practical interest. In this paper, we have detailed the discrete model of runoff and transport of soil fragments we have implemented in this simulator. Our runoff model was shown to give results which are in close agreement with results from a numerical solution of the Saint Venant's equations. Results from the transport model have not been compared to experimental ones, but they were very close to what can be observed, which is an indication that the simulator adequately represents the sorting of the soil material during transport by runoff. This is an important issue for us as this sorting is an essential process of the development of sedimentary crusts, and the prediction of the dynamics —both temporal and spatial— of soil crusting is the main objective of the simulator. As a future work, we have to establish and implement a model of infiltration based on the same principles than the runoff model. We also want to document more precisely the consequences of the choice between the 4-neighbourhood and the 8-neighbourhood. Finally, a real experiment of rain simulation will be realized in order to calibrate the parameters of the simulator and to validate all the simulated processes.

Acknowledgment

This work is part of the SoDA project, supported by the regions of Champagne-Ardenne and Picardie (information and multimedia material available at *www.soda-project.com*).

References

1. Hamblin, A.P.: The Influence of Soil Structure on Water Movement, Crop Root Growth, and Water Uptake. Advances in Agronomy 38, 95–158 (1985)
2. Baranowski, R., Bakowski, B.: The Influence of Differentiated Soil Structure on Temperature Dynamics. Roczniki Gleboznawcze 28(1), 37–44 (1977)
3. Valette, G., Prevost, S., Lucas, L.: Modeling and visualization of cracks in a desiccating soil. In: 11th International Fall Workshop Vision, Modeling, and Visualization, Aachen, Germany, pp. 177–184. Aka Verlag, Berlin (2006)
4. Le Bissonnais, Y., Bruand, A.: Crust Micromorphology and Runoff Generation on Silty Soil Materials During Different Seasons. Catena Supplement (24), 1–16 (1993)
5. Valette, G., Prévost, S., Lucas, L., Léonard, J.: SoDA Project: a Simulation of Soil Surface Degradation by Rainfall. Computers & Graphics 30(4), 494–506 (2006)
6. Zhang, W., Cundy, T.W.: Modeling of two-dimensional overland flow. Water Resources Research 25, 2019–2035 (1989)
7. Musgrave, F.K., Kolb, C.E., Mace, R.S.: The Synthesis and Rendering of Eroded Fractal Terrains. In: Proceedings of the 16Th Annual Conference on Computer Graphics and Interactive Techniques, pp. 41–50. ACM Press, New York (1989)
8. Beneš, B., Forsbach, R.: Layered Data Representation for Visual Simulation of Terrain Erosion. EEE Proceedings of SCCG 25(4), 80–86 (2001)
9. Beneš, B., Forsbach, R.: Visual Simulation of Hydraulic Erosion. Journal of Winter School of Computer Graphics 10, 79–94 (2002)
10. Chiba, N., Muraoka, K., Fujita, K.: An Erosion Model Based on Velocity Fields for the Visual Simulation of Mountain Scenery. The Journal of Visualization and Computer Animation 9, 185–194 (1998)
11. Neidhold, B., Wacker, M., Deussen, O.: Interactive Physically Based Fluid and Erosion Simulation. In: Galin, E., Poulin, P. (eds.) Eurographics Workshop on Natural Phenomena, pp. 25–32 (Dublin) (2005)
12. Beneš, B., Těšínský, V., Hornyš, J., Bhatia, S.: Hydraulic erosion. Computer Animation and Virtual Worlds 17(2), 99–108 (2006)
13. Beneš, B.: Hydraulic Erosion by Shallow Water Simulation. In: The 4th Workshop on Virtual Reality Interaction and Physical Simulation - Vriphys 2007 (to appear)
14. Mei, X., Decaudin, P., Hu, B.: Fast Hydraulic Erosion Simulation and Visualization on GPU. In: Pacific Graphics 2007 (to appear)
15. Lane, L., Nearing, M.: USDA - Water Erosion Prediction Project: Hillslope Model. Technical Report 2, USDA-ARS National Soil Erosion Research Laboratory, NSERL, West Lafayette, Indiana, USA (1989)
16. Morgan, R., Quinton, J., Smith, R., Govers, G., Poesen, J., Auerswald, K., Chisci, G., Torri, D., Styczen, M.: The European Soil Erosion Model (EUROSEM): a Dynamic Approach for Predicting Sediment Transport from Fields and Small Catchments. Earth Surface Processes and Landforms 23(6), 527–544 (1998)
17. Favis-Mortlock, D., Boardman, J., Parsons, A., Lascelles, B.: Emergence and Erosion: a Model for Rill Initiation and Development. Hydrological Processes 14(11-12), 2173–2205 (2000)
18. D'Ambrosio, D., Gregorio, S.D., Gabriele, S., Gaudio, R.: A Cellular Automata Model for Soil Erosion by Water. Physics and Chemistry of the Earth, Part B 26(1), 33–40 (2001)
19. D'Ambrosio, D., Gregorio, S.D., Iovine, G.: Simulating Debris Flows Through a Hexagonal Cellular Automata Model: Sciddica S3-hex. Natural Hazards and Earth System Sciences 3, 545–559 (2003)

20. Avolio, M.V., Crisci, G.M., D'Ambrosio, D., Di Gregorio, S., Iovine, G., Rongo, R., Spataro, W.: An Extended Notion of Cellular Automata for Surface Flows Modelling. WSEAS Transactions on Computers 2, 1080–1085 (2003)
21. Chase, C.G.: Fluvial Landsculpting and the Fractal Dimension of Topography. Geomorphology 5, 39–57 (1992)
22. Luo, W., Duffin, K.L., Peronja, E., Stravers, J.A., Henry, G.M.: A Web-Based Interactive Landform Simulation Model (WILSIM). Computers and Geosciences 30, 215–220 (2004)
23. Haff, P.K.: Waterbots In: Landscape Erosion and Evolution Modeling. In: Harmon, R.S., Doe III, W.W. (eds.), Kluwer Academic/Plenum Publishers (2001)
24. Servat, D., Léonard, J., Perrier, E., Treuil, J.: The RIVAGE-Project: a New Approach for Simulating Runoff Dynamics. In: Feyen, J., Wiyo, K. (eds.) Modelling of Transport Processes in Soils at Various Scales in Time and Space, Wageningen Pers, Wageningen, The Netherlands, pp. 592–601 (1999)
25. Green, W.H., Ampt, G.A.: Studies on Soil Physics Part I: The Flow of Air and Water Soils. Journal of Agricultural Science 4(1), 1–24 (1911)
26. Chow, V., Maidment, D., Mays, L.: Applied Hydrology. In: Water Resources and Environmental Engineering, McGraw-Hill, New-York (1988)
27. Govers, G.: Evaluation of transporting capacity formulae for overland flow. In: Parsons, A., Abrahams, A. (eds.) Overland flow. Hydraulics and erosion mechanics, pp. 243–273. UCL Press, London (1992)
28. Benassarou, A., Bittar, E., John, N., Lucas, L.: MC Slicing for Volume Rendering Applications. In: Sunderam, V.S., van Albada, G.D., Sloot, P.M.A., Dongarra, J. (eds.) ICCS 2005. LNCS, vol. 3515, pp. 314–321. Springer, Heidelberg (2005)

Optimal Difference Operator Selection

Peter Veelaert and Kristof Teelen

University College Ghent, Engineering Sciences - Ghent University Association,
Schoonmeersstraat 52, B9000 Ghent, Belgium
{Peter.Veelaert,Kristof.Teelen}@hogent.be

Abstract. Differential operators are essential in many image processing applications. Previous work has shown how to compute derivatives more accurately by examining the image locally, and by applying a difference operator which is optimal for each pixel neighborhood. The proposed technique avoids the explicit computation of fitting functions, and replaces the function fitting process by a function classification process. This paper introduces a new criterion to select the best function class and the best template size so that the optimal difference operator is applied to a given digitized function. An evaluation of the performance of the selection criterion for the computation of the Laplacian for digitized functions shows better results when compared to our previous method and the widely used Laplacian operator.

1 Introduction

Differential operators are essential in many applications in image processing and computer vision. For example, the Scale-Invariant Feature Transform (SIFT) [1] combines the Laplacian with Gaussian smoothing at different scales to extract keypoints from images. Different methods for discrete feature extraction and edge detection were proposed by Lindeberg [2], Lachaud *et al* [3], Gunn [4] and Demigny and Kamlé [5].

In fact, almost any algorithm that extracts interesting features from images, such as edges or remarkable points, relies in some way on first or higher order derivatives. Remarkably, the difference operators used to approximate a differential operator are often quite rudimentary. Furthermore, although difference operators are quite sensitive to noise, one usually assumes that Gaussian smoothing at different scales is the best way to remove noise. Previous work, however, has shown that it is possible to compute derivatives much more accurately by examining the image locally, and to apply to each pixel a difference operator which is optimal for that pixel neighborhood [6, 7, 8]. This approach gives more accurate results and needs less Gaussian blurring to obtain edges. The proposed technique avoids the explicit computation of fitting functions, a time consuming process, since it replaces the function fitting process by a function classification process.

The method proposed in this and previous papers depends heavily on the reduction of difference operators by Groebner bases [6, 7, 8]. One of the contributions of this work is that it relates difference operators to function approximation and feature detection, within the mathematical framework of ideals of difference operators. The application of Groebner bases to solve or reduce difference equations is not new, however. The

D. Coeurjolly et al. (Eds.): DGCI 2008, LNCS 4992, pp. 495–506, 2008.

pioneering work by Oberst on the use of Groebner bases to solve systems of partial difference equations has led to considerable interest in the field [9, 10]. Gerdt *et al* use Janet-like Groebner bases to generate simple difference schemes to compute solutions for partial differential equations that involve multiple functions [11].

In previous work, we have derived an optimal difference operator for a given function class and given template size. This paper focuses on the selection of the best class and best template size for a given digitized function. We will first extend previous work by showing how to compute an optimal difference operator for each function class starting from a template with a predefined shape. Incorporating known properties of the operator has the advantage that the optimization algorithm can be sped up. We give an overview of optimal operators for different function classes and template sizes. Next, we propose a method to select the most appropriate difference operator by a criterion which involves the local features of the image and the properties of the optimal difference operator. The proposed selection criterion is illustrated by the computation of a digitized difference operator for the Laplacian. We show how the accuracy of the computation of the Laplacian can benefit from locally applying the optimal version of the Laplacian difference operator.

In section 2, we present a mathematical framework to describe the classes of fitting functions. Next, we propose a general procedure for the computation of all difference operators of a predefined shape for a function class in section 3. The procedure for finding the optimal difference operator is also described in that section. In section 4, we define a criterion for the selection of the optimal operator in accordance with the local image properties. Finally, we conclude this paper in Section 5.

2   Function Classes

To find the derivative of a discrete function f, a standard approach is to approximate f by a continuous function $\tilde{g}$, before taking derivatives. This is a time consuming process, which involves unclear choices such as the size of the window in which the approximation takes place and the nature of the approximating function. The idea advocated in this and previous work is that it is not necessary to compute $\tilde{g}$ itself, but that it is sufficient to know by which class of continuous functions the digitized function can be approximated well, and to use a difference operator that is known to be optimal for this class. We introduce a criterion that will be used to select an optimal class of approximating functions, and an optimal window size.

We use a continuous real function $\tilde{g} : \mathbb{R}^m \to \mathbb{R}$ to approximate a digitized function $f : \mathbb{Z}^m \to \mathbb{Z}$. To approximate the value of a differential at a point x_0, it is sufficient to approximate f in a finite subset $D \subset \mathbb{Z}^m$ containing x_0. For all $x \in D$, $f(x) \simeq \tilde{g}(x)$ and $|f - \tilde{g}| < \epsilon$ is written as a shorthand for $|f(x) - \tilde{g}(x)| < \epsilon$.

The shift operator σ^j is defined by $\sigma^j f(x) = f(x+j)$, for $x, j \in \mathbb{Z}^m$. The functional composition of shift operators can be expressed as a multiplication of polynomials, i.e. $\sigma^j \sigma^k f = \sigma^{j+k} f$. A difference operator P can be represented as a polynomial in σ with non-negative, bounded exponents, that is $P = \sum p_j \sigma^j$, and $P \in \mathbb{R}[\sigma]$, the ring of polynomials in σ. We write $P\tilde{g} = 0$ as a shorthand for $\sum p_j \sigma^j \tilde{g}(x) = 0$, $x, j \in \mathbb{Z}^m$. If we write that $|Pf - P\tilde{g}| < \epsilon$, this means that $|Pf(x) - P\tilde{g}(x)| < \epsilon$ for all x for which

$Pf(x)$ is well defined, that is $(x + j) \in D$ for every non-vanishing coefficient p_j of the difference operator P. Since polynomial ideals are usually defined for polynomials with non-negative exponents, we will assume, without loss of generality, that D is a finite rectangular neighborhood containing only points with non-negative coordinates.

An important concept is the ideal $I =< P_1, P_2, \ldots >$ generated by a set of difference operators P_i. I consists of all operators $P = \sum_i S_i P_i$, where the S_i are arbitrary polynomials in σ. Often the ideal I will be represented by a Groebner basis [13].

Difference operators can be represented by templates. A two-dimensional difference operator $P = \sum p_j \sigma^j = \sum_{j_x, j_y} p_{j_x j_y} \sigma_x^{j_x} \sigma_y^{j_y}$, $j \in \mathbb{Z}^2$ is represented by a two-dimensional template:

$$\begin{array}{|c|c|c|}\hline p_{00} & p_{10} & p_{20} \\\hline p_{01} & p_{11} & \\\hline\end{array} \cdots \qquad (1)$$

We use the convention that the box at the upper left corner corresponds to p_{00}. Boxes with vanishing coefficients are either not drawn, or drawn as empty boxes.

Let L be the differential operator that must be replaced by a difference operator. The goal is to select a class G of fitting functions $\tilde{g}$, and a difference operator $Q = \sum q_j \sigma^j$ that satisfies $Q\tilde{g} = L\tilde{g}$ for every $\tilde{g} \in G$. If $\tilde{g}$ is an approximation for f such that $|f - \tilde{g}| < \epsilon$ in the domain D, then, because all operators are linear, we have

$$|Qf - L\tilde{g}| < \epsilon \sum |q_j|, \text{ in } D. \qquad (2)$$

Hence, the difference operator Q is a good approximation for the differential operator L, provided G contains at least one good approximation $\tilde{g}$ for f. It is important, however, to understand that the total error $\epsilon \sum |q_j|$ is due to two successive approximations. The error ϵ arises when f is replaced by some continuous fitting function $\tilde{g}$, a step which cannot be avoided if we want to compute differentials. The additional error $\sum |q_j|$ is due to the replacement of the differential operator L by a difference operator Q, which avoids the explicit computation of the continuous approximation. The above technique only works if two distinct requirements are satisfied. The approximation (2) is only valid provided $Q\tilde{g} = L\tilde{g}$ holds for the approximation function. On the other hand the approximation error ϵ should be as small as possible. The best way to satisfy both requirements is to define a class of fitting functions by a set of difference equations $P_i \tilde{g} = 0$. The difference operators P_i span an ideal $I =< P_1, P_2, \ldots >$, which means that the ideal I can be used to define the class of fitting functions.

Using an ideal of difference operators to define a class of fitting functions has two benefits. First, it becomes easy to verify whether the class of fitting functions contains at least one member $\tilde{g}$ such that $|f - \tilde{g}| \leq \epsilon$. In fact, if $|f - \tilde{g}| \leq \epsilon$, then the inequality $|f - \tilde{g}| \leq \epsilon$ implies $|Pf| \leq \epsilon|P|$ for each $P \in I$, since a difference operator is a linear operator and $P\tilde{g} = 0$. $|P| = \sum |p_j|$ where p_j are the coefficients of P. One can prove that the converse is also true [6], i.e. there is a finite set of difference operators P in I, such that $|f - \tilde{g}| \leq \epsilon$ holds, provided $|Pf| \leq \epsilon|P|$ for all P.

Features can be detected without error by verifying only a finite number of inequalities, when the solution space of the difference equations is a finite linear vector space. Assume that the solution set of the partial difference equations $P_1 g = 0, \ldots, P_n g = 0$

can be written as a linear vector space with $g_1, \ldots, g_l$ as a basis: $\alpha_1 g_1 + \cdots + \alpha_l g_l$. Let K_D be the set of all difference operators P_i of the form

$$
\begin{vmatrix}
g_1(x_1) & \ldots & g_l(x_1) & \sigma^{x_1} \\
\ldots & & & \\
g_1(x_{l+1}) & \ldots & g_l(x_{l+1}) & \sigma^{x_{l+1}}
\end{vmatrix}
\tag{3}
$$

with the points $x_j \in D$. The operators of K_D are written as determinantal expressions of the coefficients $g_j(x_j)$ and the shift operators σ^{x_j}. Since D is finite, K_D is finite. The operators P in K_D are all operators for which Pf is well defined. When we assume that D is rectangular, there is a unique point x_0 in D with smallest coordinates. Let $\sigma^{-x_0} K_D$ denote the polynomials of K_D multiplied by σ^{-x_0}. The polynomials $\sigma^{-x_0} K_D$ all have non-negative exponents, but some of the exponents will be zero. If D is chosen not too small then the polynomials in $\sigma^{-x_0} K_D$ will generate the entire ideal $I =< P_1, \ldots, P_n >$. In practice, for two-dimensional operators, it is usually sufficient that D has size at least $(a+2) \times (b+2)$ when the leading monomial of the fitting functions is $x^a y^b$, or smaller. In that case, for f to be in $G_{\epsilon;D}$, it is sufficient to verify $|Pf| \leq \epsilon |P|$ for all P in the finite set K_D [6]. In other words, it is sufficient to verify a finite set of inequalities to determine whether a discrete function f can be approximated well by at least one function in a given function class. Given a domain D, if there is a $\tilde{g} \in G$ such that $|f - \tilde{g}| \leq \epsilon$, we denote this as $f \in G_{\epsilon;D}$.

The second benefit is that it is possible to find an optimal difference operator $R_{G;D}$ for all $f \in G_{\epsilon;D}$, with arbitrary ϵ. Suppose we have at least one operator Q such that $Q\tilde{g} = L\tilde{g}$, then $R\tilde{g} = L\tilde{g}$ for any R for which $R - Q \in I$. Thus R can be chosen from an infinite set of difference operators. The optimization criterion is that $|R|$ should be as small possible.

The function classes used in this paper and their respective Groebner bases are enlisted in Table 1. We shall work with lattices of ideals or, equivalently, lattices of function classes. Suppose we have a set of function classes G_j and a set $\mathcal{I}$ of corresponding

Table 1. Functions and corresponding Groebner bases with lexicographic ordering $\Delta_x > \Delta_y$

$G_1 : \alpha_1$	$< \Delta_x, \Delta_y >$
$G_2 : \alpha_1 x + \alpha_2$	$< \Delta_x^2, \Delta_y >$
$G_3 : \alpha_1 y + \alpha_2$	$< \Delta_y^2, \Delta_x >$
$G_4 : \alpha_1(x + y) + \alpha_2$	$< \Delta_x - \Delta_y, \Delta_y^2 >$
$G_5 : \alpha_1 x + \alpha_2 y + \alpha_3$	$< \Delta_x^2, \Delta_x \Delta_y, \Delta_y^2 >$
$G_6 : \alpha_1 xy + \alpha_2 x + \alpha_3 y + \alpha_4$	$< \Delta_x^2, \Delta_y^2 >$
$G_7 : \alpha_1(x + y)^2 + \alpha_2(x + y) + \alpha_3$	$< \Delta_x - \Delta_y, \Delta_y^3 >$
$G_8 : \alpha_1(x + y)^2 + \alpha_2 x + \alpha_3 y + \alpha_4$	$< \Delta_x^2 - \Delta_y^2, \Delta_y(\Delta_x - \Delta_y), \Delta_y^3 >$
$G_9 : \alpha_1(x^2 + y^2) + \alpha_2 x + \alpha_3 y + \alpha_4$	$< \Delta_x^2 - \Delta_y^2, \Delta_x \Delta_y, \Delta_y^3 >$
$G_{10} : \alpha_1 x^2 + \alpha_2 y^2 + \alpha_3 x + \alpha_4 y + \alpha_5$	$< \Delta_x^3, \Delta_x \Delta_y, \Delta_y^3 >$
$G_{11} : \alpha_1(x^2 + y^2) + \alpha_2 xy + \alpha_3 x + \alpha_4 y + \alpha_5$	$< \Delta_x^2 - \Delta_y^2, \Delta_x \Delta_y^2, \Delta_y^3 >$
$G_{12} : \alpha_1 x^2 + \alpha_2 y^2 + \alpha_3 xy + \alpha_4 x + \alpha_5 y + \alpha_6$	$< \Delta_x^3, \Delta_x^2 \Delta_y, \Delta_x \Delta_y^2, \Delta_y^3 >$
$G_{13} : \alpha_1 x^3 + \alpha_2 x^2 y + \cdots + \alpha_n$	$< \Delta_x^4, \Delta_x^3 \Delta_y, \Delta_x^2 \Delta_y^2, \Delta_x \Delta_y^3, \Delta_y^4 >$
$G_{14} : \alpha_1 x^4 + \alpha_2 x^3 y + \cdots + \alpha_n$	$< \Delta_x^5, \Delta_x^4 \Delta_y, \Delta_x^3 \Delta_y^2, \Delta_x^2 \Delta_y^3, \Delta_x \Delta_y^4, \Delta_y^5 >$

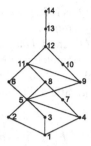

Fig. 1. Lattice of the operator ideals shown in Table 1

ideals I_j. Then $\mathcal{I}$ is a lattice provided for each pair of ideals I_k, I_l in $\mathcal{I}$ the intersection (or meet) of I_k and I_l, as well as their union (or join) are also included in $\mathcal{I}$. The ideals of Table 1 form in fact the lattice shown in the Figure 1.

3   Computing Optimal Operators

We will now go into more detail with regard to the optimization process for the optimal difference operator $R_{G;D}$, i.e. the operator for which $|R|$ is as small as possible. First, we give a general procedure for finding all operators with a predefined shape for a given function class. Then we can compute the optimal operator for each class by solving the optimization problem $\min(|R|)$.

3.1   Shaped Operators

Difference operators S with a predefined shape have the advantage that known properties of the operator can be taken into account so that the optimization algorithm is sped up. For example, if it is known that there exists a optimal symmetric operator, we can use this fact by starting with a predefined symmetrical shape, which reduces the optimization space considerably.

A general form for all operators of a predefined shape can be derived as follows. First compute a reduced operator modulo the function class for both the difference operator Q, with $Q\tilde{g} = L\tilde{g}$, and the operator S with the predefined shape. Set the reduced predefined operator equal to the reduced shaped operator and solve the resulting system of linear equations. When this system has no solution, there is no operator with the predefined form. When a solution is found, some variables are eliminated to obtain an expression in which the number of remaining variables equals the number of dimensions of the solution space. The resulting operator represents symbolically all correct operators of the predefined shape.

Example. We illustrate this procedure for the function class G_7, and the Laplacian $L = \frac{\partial^2}{\partial x^2} + \frac{\partial^2}{\partial y^2}$ as differential operator. For functions in G_7, the Laplacian L can be replaced by the difference operator $Q = 2\Delta_y^2 = 2\sigma_y^2 - 4\sigma_y + 2$. The difference operator Q is shifted such that the Laplacian is computed in the center of a 5x5 square:

$Q' = (2\sigma_y^2 - 4\sigma_y + 2)\sigma_x^2\sigma_y^2$. When Q' is reduced modulo the ideal $I_{G_7} =< (\sigma_x - 1) - (\sigma_y - 1), (\sigma_y - 1)^3 >$, we obtain Q' mod $I = 2 - 4\sigma_y + 2\sigma_y^2$.

Note that the reduced operators Q' mod I are in general not symmetric in σ_x, σ_y, because the Groebner basis of I is computed according to a predefined lexicographic ordering relation between the variables, $\sigma_x > \sigma_y$. Since the same Groebner basis is used to reduce the operator S, this will not affect the final result. We choose for example a 5x5 difference operator S of the form

$$
\begin{array}{ccccc}
a & b & c & b & -a \\
b & d & e & -d & b \\
c & e & f & e & c \\
b & -d & e & d & b \\
-a & b & c & b & a
\end{array}
\tag{4}
$$

with coefficient $b = 0$. The reduction of the operator S yields

$$
\begin{aligned}
S \quad \text{mod } I = {} & (16a + 20c + 4d + 14e + 3f) \\
& + (-32a - 48c - 8d - 36e - 8f)\sigma_y \\
& + (16a + 32c + 4d + 26e + 6f)\sigma_y^2.
\end{aligned}
\tag{5}
$$

We obtain 3 linear equations by setting Q' mod $I = S$ mod I, which enables us to eliminate 3 variables. The result is a general expression for all correct operators of the shape (4) in which the variables a, c, d can be chosen freely, and $b = 0$, $e = 1 - 8a - 4c - 2d$, $f = -4 + 32a + 12c + 8d$.

Several function classes will yield operators with the same form. For example, the classes $G_7, G_8, G_9, G_{10}, G_{11}, G_{12}$ all have the same symmetrical operator. This means that these function classes will also have the same optimal operator of that shape.

3.2 The Optimal Operator

An operator is optimal when the sum of the absolute values of its coefficients is minimal. So once the general form of an operator is known, an optimal operator can be found by solving a large set of linear programming problems. The cost of the operator in the previous example is $4|a|+4|c|+4|d|+4|1-8a-4c-2d|+|-4+32a+12c+8d|$. Although this expression is non-linear, the optimum can be found by solving 2^5 small linear programming problems. In each problem, we make an assumption for each absolute value about the sign of the term and replace it by a linear term. Since there are 5 terms in this example, there are 2^5 possible combinations. For our example, we obtain an optimal operator with cost $1/2$ and the template shown in Fig. 2(b).

Several function classes may have the same optimal operator. Figure 2 shows the optimal operators for some function classes and different window sizes. The costs for all functions classes and windows is given by Table 2.

In Table 2, the set of linear function classes comprises G_1, G_2, G_3, G_4 and G_5; the quadratic classes are G_9, G_{10}, G_{11} and G_{12}; the cubic class is G_{13}; the class of fourth order is G_{14}. Although G_6 is not a class of linear functions, with respect to the Laplacian it behaves as a linear function. We use the term quadratic symmetric to denote the set

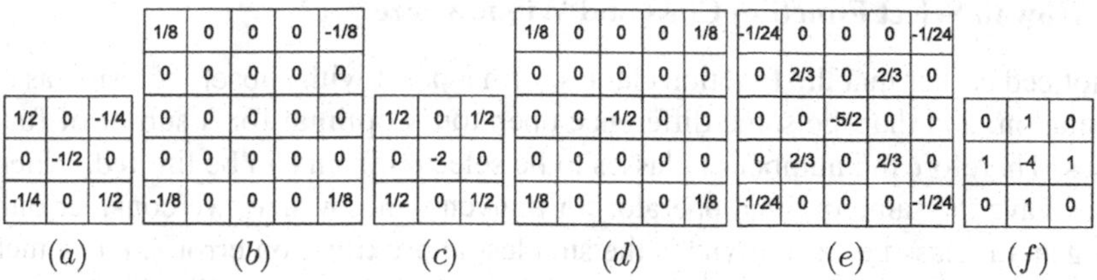

(a)

1/2	0	-1/4
0	-1/2	0
-1/4	0	1/2

(b)

1/8	0	0	0	-1/8
0	0	0	0	0
0	0	0	0	0
0	0	0	0	0
-1/8	0	0	0	1/8

(c)

1/2	0	1/2
0	-2	0
1/2	0	1/2

(d)

1/8	0	0	0	1/8
0	0	0	0	0
0	0	-1/2	0	0
0	0	0	0	0
1/8	0	0	0	1/8

(e)

-1/24	0	0	0	-1/24
0	2/3	0	2/3	0
0	0	-5/2	0	0
0	2/3	0	2/3	0
-1/24	0	0	0	-1/24

(f)

0	1	0
1	-4	1
0	1	0

Fig. 2. Optimal Laplacian operator templates for several sets of function classes and different window sizes: $\{G_7, G_8\}$(3x3) (a), $\{G_7, G_8\}$(5x5) (b), $\{G_9, G_{10}, G_{11}, G_{12}\}$(3x3) (c), $\{G_9, G_{10}, G_{11}, G_{12}, G_{13}\}$(5x5) (d), G_{14} (e), widely used template (f)

Table 2. Cost of optimal operators for different sets of function classes and window sizes

functions window	linear	quadratic symmetric	quadratic	cubic	4th order
3x3	0 (1)	2	4	n.e.	n.e.
5x5	0 (1/4)	1/2	1	1	16/3
7x7	0 (1/9)	2/9	4/9	4/9	9/5

of function classes $\{G_7, G_8\}$, which are symmetric in the second degree terms. There are no 3x3 operators that correctly compute the Laplacian for cubics and higher order functions.

For a given window size, the cost increases when the function class becomes more general. In fact, Table 1 shows that $I_{12} \subset I_{10}$, i.e. G_{12} contains G_{10} as a subset, and the functions in G_{10} have to satisfy the difference equations from I_{12} plus some extra difference equations. As a result, the class of operators from which an optimal operator has to be chosen is larger for I_{10} than for I_{12}. Second, for a given function class the cost decreases when the window size increases. In fact, the operators for a larger window contain the operators for a small window as a special case.

For linear functions, the optimal operator is the zero operator with cost equal to zero. This is correct since the Laplacian of a linear function (and of G_6) is zero, a value which can be estimated without error. This may give rise to some anomalies when we have to select an appropriate function class and window size. In the next section we discuss how these anomalies can be avoided by introducing an artificial cost for linear functions (shown between brackets in Table 2).

4 Optimal Operator Selection

We have now assigned an optimal operator to each function class for each window size. We now propose a selection criterion for the best operator with regard to the operator cost and the local image properties, i.e. the approximation cost for a certain function class and window size. Our selection method is then validated by experiments on artificial images of digitized surfaces.

4.1   How to Select Function Class and Window Size?

We noticed earlier that the function classes form a poset with subset inclusion as ordering relation, and that the same difference operator is optimal for a subset of function classes. Therefore the number of classes to be selected from can be limited. When two classes have the same optimal operator for a given window size, we consider only the most general class because it yields the smallest approximation error ϵ. ϵ is a measure for the local fit of the approximating continuous function to the digitized surface. A feature detection template P_i yields a cost ϵ_i when applied to the image part f:

$$\epsilon_i = \frac{|P_i f|}{|P_i|} \tag{6}$$

with $P_i \in K_D$ (3) for that function class. The fitting cost ϵ is then obtained as $\epsilon = \max_i(\epsilon_i)$ for a limited subset of feature detectors P_i (usually 10-15 suffices).

For practical reasons, we only consider windows of either 3x3 or 5x5 in our experiments. The following classes are available in our experiments:

- The function classes G_1, G_2, G_3, G_4, G_5 and G_6 form a subset of functions for which the optimal difference operator is the zero operator, for both the 3x3 and 5x5 window size. G_6 is the most general of these functions, and is therefore used as a representative for this set of classes.
- G_8 represents the set of quadratic function classes $\{G_7, G_8\}$. The optimal operators are shown in Fig. 2 for the 3x3 and 5x5 templates.
- Quadratic functions as in G_9, G_{10}, G_{11} and G_{12} are represented by G_{12}, with the operator shown in Fig. 2 as the optimal operator for the 3x3 templates. For 5x5 templates, the optimal operator generated for these classes is the same as that for class G_{13}. Therefore, we use G_{13} to generate the feature detection templates for this subset of function classes.

Up to now, we have an optimal operator for a given function class and a given window. When the window size increases, the cost goes down for a given function class. When the window size is fixed, the cost goes down when the function class becomes less general. Therefore, on the one hand it makes sense to use a large window with a more specialized function class, on the other hand this will increase the fitting cost. We now propose a criterion which seeks an optimum between these two choices: use the function class and window size for which $\epsilon|R|$ is minimal.

The product $\epsilon|R|$ represents the fitting cost ϵ as well as the operator cost $|R|$ so that the image surface must locally be well approximated by the function class and the error introduced by using a difference operator must be small. Unfortunately, this criterion does not work for zero operators as their operator cost $|R|$ is always zero, and therefore also the estimation error $\epsilon|R|$ is zero. Theoretically, this is correct because $\epsilon|R|$ represents the error with which we can estimate the value of $L\tilde{g}$. Since the Laplacian of a linear function is always zero, we can in fact estimate the value of the Laplacian without error. The fitting cost does not count in this case. One solution would be to define a more elaborate mixed cost $h(\epsilon, \epsilon|R|)$. We propose a simpler solution by introducing a non-zero cost for zero operators. By observing Table 2, one sees that the cost $|R|$ often goes down by a factor 2 when the function class gets more specialized. For example, in a 3x3 window, the cost is 4 for quadratic functions, and 2 for quadratic symmetric functions. We shall extrapolate this trend and use a cost for linear functions which is half the cost of symmetric quadratic functions, as shown between brackets in Table 2. Thus, quadratic or higher order functions will only be used if their fitting cost is at least less than one half the fitting cost of linear functions. By a fortunate coincidence, the cost used in a 3x3 window for linear functions is simply the fitting cost ϵ itself.

4.2 Evaluation

To verify whether this selection method yields good results, first the following question must be answered: Are the correct fitting classes chosen, i.e. is $\epsilon|R|$ locally minimal for the correct function class? Experimental data is obtained for images of digitized surfaces as shown in Fig. 3(a). The surfaces are composed by functions of different classes so that the exact value of the Laplacian is known beforehand for each pixel in the image. This allows for a comparison of the values for the Laplacian obtained by the proposed method, the hierarchical method discussed in [7] and the widely used operator as shown in Fig. 2(f).

We reconsider the hierarchical method [7], which determines the function class by descending a tree-structure in which three consecutive nodes respectively represent G_5, G_8 and G_{12}. Feature detection is performed by evaluating

$$\frac{|P_i f|}{|P_i|} \leq \epsilon_{th} \tag{7}$$

with ϵ_{th} a predefined threshold for a limited number of feature detection templates P_i (3). If this test succeeds at a node in the tree, the corresponding optimal difference operator is chosen. Although this method shows better results than the application of the classical Laplacian template, this approach still has some drawbacks. A first problem is the extension of the tree for templates of different sizes. Which size should be preferred in the hierarchical composition of the tree? Second, why should the first node where the test succeeds always be preferred? The second or third class could yield a better approximation for the digitized image surface. If we select an appropriate Laplacian operator by descending the tree, we notice that the zero-operator is often chosen, even for quadratic functions. In those cases, a linear function locally is a good enough approximation for the digitized image. Then the zero operator is applied, which does not correspond to the actual Laplacian value. A related problem is the choice of the threshold ϵ_{th}.

In the now proposed approach, the appropriate function class is chosen by the criterion $\min(\epsilon|R|)$ over all function classes. $\epsilon|R|$ is easily computed for different template sizes and discriminates well between different classes. Mostly the correct function class is chosen for each image part throughout the experiments. An example is shown in Fig. 3: the function class for which $\epsilon|R|$ is minimal is shown for each pixel. Fig. 3 shows that the use of larger feature templates proves its worth. A larger image region then influences the outcome of the Laplacian operator so that its result is more reliable. We notice however that in some cases the 'incorrect' class is chosen, e.g. the linear case is chosen for quadratic cases, when the image surface is locally represented by a rather flat

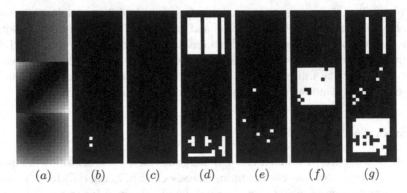

(a)   (b)   (c)   (d)   (e)   (f)   (g)

Fig. 3. Figure (a) shows a digitized image composed of three digitized surfaces: linear, quadratic symmetric and quadratic (from top to bottom). The next images respectively represent the function classes G_6(3x3) (b), G_8(3x3) (c), G_{12}(3x3) (d), G_6(5x5) (e), G_8(5x5) (f) and G_{13}(5x5) (g). When the criterion $\epsilon|R|$ is minimal for that function class and window size in a pixel, that pixel is indicated as white.

quadratic. Notice that only a limited subset of feature detectors P_i is used to compute ϵ. We must also notice that we give preference to larger windows and functions of higher order when $\epsilon|R|$ is minimal for multiple classes. This is e.g. the case for most pixels in the upper part of the image in Fig. 3, where $\epsilon|R| = 0$ for all 3x3 templates, so that the class with polynomials of the highest order (G_{12} in this case) is chosen.

4.3   Computation of the Difference Operator

It is not sufficient to select the correct function class. Does the optimal operator for a specific function class also yield a good estimation for the exact value of the difference operator? Is the estimated value considerably better than the value obtained by applying the widely used template for the difference operator?

When we compute the Laplacian for image surfaces composed of known digitized functions with the optimal operator for that function class, we can compare that result to the actual Laplacian value for the continuous functions. An example is shown in Fig. 4. We notice that the operators with a 5x5 template estimate the actual value more accurate than the value obtained by the corresponding 3x3 operator template. This is expected as the maximum error is $|R|$, which is smaller for larger templates. The experiments also show that in most cases the Laplacian is estimated much better than the expected maximum error of that operator. The mean and maximum error on the estimation of the actual value are respectively up to 8 and 2 times smaller than $|R|$ for the optimal difference operator when the class is known. The estimation is in almost all cases better than that of the default operator. The mean error on our estimation is much smaller (up to more than 10 times) than the mean error on the default kernel.

When we combine the class selection method and the application of the optimal difference operator for the selected class, we can evaluate the results of the proposed

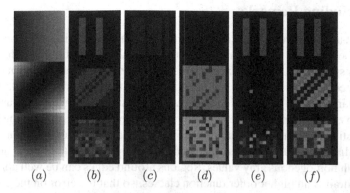

(a)   (b)   (c)   (d)   (e)   (f)

Fig. 4. An image (a) composed of three digitized surfaces: linear, quadratic symmetric and quadratic (from top to bottom). The other images indicate the absolute error on the estimation of the Laplacian in gray values, with black to white corresponding to the range [0,5]. In (b) the Laplacian is computed in each of the three image parts with respectively the optimal 3x3 operator for class G_6, G_8 and G_{12}. In (c) the Laplacian is computed with the optimal 5x5 operator for G_6, G_8 and G_{13} in the corresponding image parts. The error on the Laplacian is indicated when the operator is selected respectively by the hierarchical tree (d), by the criterion $\min(\epsilon|R|)$ (e) and for the default Laplacian kernel (f).

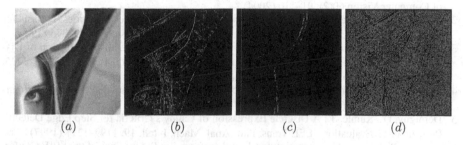

(a)   (b)   (c)   (d)

Fig. 5. Original image (a), edges detected as zero crossings after Laplacian computation on the unsmoothed image by the proposed method (b), by the hierarchical method (c) and by the default method (d)

method. When compared to the results of the hierarchical tree, we notice a considerable improvement, as can be observed for the example in Fig. 4. We notice that the zero operator is frequently chosen not only for linear, but also for quadratic surfaces as it appears first in the tree. This problem is solved by the proposed method. As already noted, the application of optimal operators of larger sizes brings increased accuracy. Even more, the experiments show that these cases are preferred over the less accurate smaller templates. It it also proven by the experiments that the proposed method always outperforms the default method. Both the mean and the maximum error of the estimated value for the Laplacian compared to the actual value is always greater in case of the default estimation method. The mean error is in the order of a few tenths for our method, while it attains mean error values greater than one for the default method.

5 Concluding Remarks

We present a procedure to replace a differential operator by a difference operator optimal for a specific class of functions. The criterion $\min(\epsilon|R|)$ determines which operator to use for a specific image part. Experiments on digitized functions show that the Laplacian is indeed better estimated by our method than by the widely used Laplacian kernel. A better estimation of the Laplacian offers a considerable advantage in feature detection applications, for example edge detection by looking for zero crossings of the Laplacian. A better estimate of the Laplacian results in a more accurate computation of the zero crossings. In particular, we expect a considerable improvement for large noisy image regions with homogenous gray values. Regions around edges can be well approximated by the quadratic and higher order function classes, so that the error on the computation of the Laplacian will decrease when our criterion is applied to select the best operator. In Fig. 5, we apply our method to compute the Laplacian for the unsmoothed Lena image. When we compare the edges obtained from zero crossings of the Laplacian, we notice that the proposed method gives better results than the hierarchical method, and much better than the widely applied method for the discrete Laplacian.

References

1. Lowe, D.G.: Distinctive image features from scale-invariant keypoints. International Journal of Computer Vision 60(2), 91–110 (2004)
2. Lindeberg, T.: Discrete Derivative Approximations with Scale-Space Properties: A Basis for Low-Level Feature Extraction. J. of Mathematical Imaging and Vision 3, 349–376 (1993)
3. Lachaud, J.O., Vialard, A., de Vieilleville, F.: Analysis and Comparative Evaluation of Discrete Tangent Estimators. In: Andrès, É., Damiand, G., Lienhardt, P. (eds.) DGCI 2005. LNCS, vol. 3429, pp. 240–251. Springer, Heidelberg (2005)
4. Gunn, S.: On the discrete representation of the Laplacian of Gaussian. Pattern Recognition 32, 1463–1472 (1999)
5. Demigny, D., Kamlé, T.: A Discrete Expression of Canny's Criteria for Step Edge Detector Performances Evaluation. IEEE Trans. Patt. Anal. Mach. Intell. 19, 1199–1211 (1997)
6. Veelaert, P.: Local feature detection for digital surfaces. In: Proceedings of the SPIE Conference on Vision geometry V, SPIE, vol. 2826, pp. 34–45 (1996)
7. Teelen, K., Veelaert, P.: Improving Difference Operators by Local Feature Detection. In: Kuba, A., Nyúl, L.G., Palágyi, K. (eds.) DGCI 2006. LNCS, vol. 4245, pp. 391–402. Springer, Heidelberg (2006)
8. Veelaert, P., Teelen, K.: Feature controlled adaptive difference operators. Discrete Applied Mathematics (preprint, submitted, 2007)
9. Oberst, U.: Multidimensional constant linear systems. Acta Appl. Math. 20, 1–175 (1990)
10. Oberst, U., Pauer, F.: The Constructive Solution of Linear Systems of Partial Difference and Differential Equations with Constant Coefficients. Multidim. Systems and Signal Processing 12, 253–308 (2001)
11. Gerdt, V., Blinkov, Y., Mozzhilkin, V.: Groebner Bases and Generation of Difference Schemes for Partial Differential Equations. Symmetry, Integrability and Geometry: Methods and Applications 2 (2006) (Paper 051, arXiv:math.RA/0605334)
12. Stoer, J., Witzgall, C.: Convexity and Optimization in Finite Dimensions I. Springer, Berlin (1970)
13. Cox, D., Little, J., O'Shea, D.: Ideals, Varieties and Algorithms: an Introduction to Computational Algebraic Geometry and Commutative Algebra. Springer, New York (1992)

First Results for 3D Image Segmentation with Topological Map*

Alexandre Dupas[1] and Guillaume Damiand[2]

[1] SIC-XLIM, Université de Poitiers, UMR CNRS 6172,
Bâtiment SP2MI, F-86962 Futuroscope Chasseneuil, France
`dupas@sic.univ-poitiers.fr`
[2] LaBRI, Université de Bordeaux 1, UMR CNRS 5800, F-33405 Talence, France
`damiand@labri.fr`

Abstract. This paper presents the first segmentation operation defined within the 3D topological map framework. Firstly we show how a traditional segmentation algorithm, found in the literature, can be transposed on a 3D image represented by a topological map. We show the consistency of the results despite of the modifications made to the segmentation algorithm and we study the complexity of the operation. Lastly, we present some experimental results made on 3D medical images. These results show the process duration of this method and validate the interest to use 3D topological map in the context of image processing.

Keywords: Topological model, 3D Image segmentation, Intervoxel boundaries, Combinatorial maps.

1 Introduction

Segmentation of 3D images is a great challenge in many fields as for example in the analysis of medical images. The segmentation refers to the process of partitioning an image into regions which are homogeneous to a criterion. This kind of approach, called *region-based* segmentation, requires a representation of regions in the image.

There are many works that have studied the definition of such a structure to represent images. Topological data structures describe the image as a set of elements and their neighborhood relations. The most famous example is the Region Adjacency Graph (RAG) [1] which represents each region by a vertex, and where neighboring regions are connected by an edge. But the RAG suffers from several drawbacks as it does not represent multiple adjacency or makes no difference between inclusion and adjacency relations. To solve these issues the RAG model has been extended, for instance in dual-graph structure to represent 2D images [2] or in topological maps [3,4,5,6]. These last have already been used in segmentation of 2D images [7,8] and a previous work has defined two operations on 3D topological maps needed to achieve image segmentation [9] but without using it in a segmentation process.

---

* Partially supported by the ANR program ANR-06-MDCA-008-05/FOGRIMMI.

D. Coeurjolly et al. (Eds.): DGCI 2008, LNCS 4992, pp. 507–518, 2008.

Our general objective is to develop segmentation operations which profit from all the information stored by topological maps. As a first step toward this goal, we show in this paper how topological maps may be used as a 3D image representation model in a traditional segmentation process. In our knowledge, this is the first time that 3D image segmentation is achieved by using combinatorial maps.

In the literature, we found a region-based segmentation method, proposed by P. F. Felzenszwalb and D. P. Huttenlocher in [10], which seems to provide interesting results. This method merges neighboring homogeneous regions in order to produce another homogeneous region. This approach of the segmentation is called *bottom-up*. As a variant of the *split-and-merge* methods, it consists in the merging of small regions into bigger ones. In the original work, the segmentation using a local criterion is defined on 2D graph representation of the image, but was extended to different models and in particular to a hierarchical combinatorial map representation of 2D image in [8].

In this paper we present how this segmentation technique can be transposed to the topological map framework. The implementation of the homogeneity criterion, its computation and the algorithm used are also provided. We show that the results obtained with this method are similar to those that could be obtained with the original approach. Moreover the processing time is suitable for 3D image segmentation as it allows to segment real medical images. At this time, using topological map does not improve original methods. The next step of this work is to mix this approach with topological criteria to improve the segmentation results and show the contribution of topological maps to the 3D image processing.

We first present in Sect. 2 topological maps, which are combinatorial maps verifying specific properties, used to represent 3D images. We also introduce a topological map manipulation operation used in this work : the region merging. Then, Sect. 3 details the criterion and presents the segmentation algorithm. We also explain why despite the differences between this method and the classical one, the results are similar. In Sect. 4 we present the complexity analysis of the segmentation operation and we give some experimental results on 3D medical images. Lastly, we conclude and give some perspectives in Sect. 5.

2   Recalls on 3D Topological Maps

A 3D topological map is an extension of a combinatorial map used to represent a 3D image partition. Let us recall the notions of combinatorial maps, 3D images, intervoxel elements and topological maps that are used in this work.

2.1   Combinatorial Map

A combinatorial map is a mathematical model describing the subdivision of a space, based on planar maps. A combinatorial map encodes all the cells of the subdivision and all the incidence and adjacency relations between the different cells, and so describe the topology of this space.

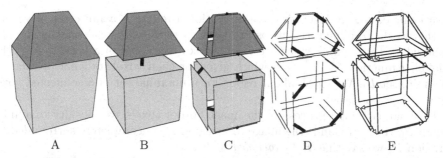

Fig. 1. The successive decompositions of an object to obtain the corresponding 3-map. (A) A 3D object. (B) Disjointed volumes. (C) Disjointed faces. (D) Disjointed edges. (E) Corresponding combinatorial map.

A combinatorial map can be obtained intuitively by successive decompositions as we can see in Fig. 1. To describe the 3D object shown in Fig. 1 A, we first decompose the volumes of this object (Fig. 1 B) then the faces of these volumes (Fig. 1 C) and finally the edges of these faces (Fig. 1 D). At each step, we keep the adjacency relations between the decomposed cells (drawn by black segments but only partially drawn for the last step). The obtained elements after the last decomposition are called *darts* and are the only basic elements used in the definition of the combinatorial maps. In order to obtain the map, we report each adjacency relation onto darts. We call β_i the relation between two darts which describes an adjacency between two i-dimensional cells. Figure 1 E presents the combinatorial map corresponding to object shown in Fig. 1 A.

Let us see now the formal definition of 3D combinatorial maps:

Definition 1 (3D combinatorial map). *A 3D combinatorial map, (or 3-map) is a 4-tuple $M = (D, \beta_1, \beta_2, \beta_3)$ where:*

1. *D is a finite set of darts;*
2. *β_1 is a permutation[1] on D;*
3. *β_2 and β_3 are two involutions[2] on D;*
4. *$\beta_1 \circ \beta_3$ is an involution[3] on D.*

The different constraints of the 3-map definition (β_1 is a *permutation*, other β_i are *involutions* and $\beta_1 \circ \beta_3$ is an involution) ensures the topological validity of described objects. For example, intuitively the last constraint says that two volumes can not be partially adjacent. If two volumes are adjacent for a face, they must be adjacent for all the edges of the face. See [11] for more details on maps and comparison with other combinatorial models.

2.2   3D Images and Intervoxel Elements

Let us now recall some usual notions about images and intervoxels elements. A voxel is a point of discrete space $\mathbb{Z}^3$ associated with a value which could

---

[1] A *permutation* on a set S is a one to one mapping from S onto S.
[2] An *involution* f on a set S is a one to one mapping from S onto S such that $f = f^{-1}$.
[3] $\beta_1 \circ \beta_3$ is the composition of both permutations: $(\beta_1 \circ \beta_3)(x) = \beta_1(\beta_3(x))$.

be a color or a gray level. A three dimensional image is a finite set of voxels. In this work, combinatorial maps are used to represent voxel sets having the same labeled value and which are 6-connected. The label of a voxel is given by a labeled function $l : \mathbb{Z}^3 \rightarrow L$ which gives for each voxel its label (a value in the finite set L). We speak about region for a maximal set of 6-connected voxel having the same label.

To avoid particular process for the image border voxels, we consider an infinite region R_0 that surrounds the image. If a region R_j is completely surrounded by a region R_i we say that R_j is *included* in R_i.

In the intervoxel framework [12], an image is considered as a subdivision of a 3-dimensional space in a set of cells: voxels are the 3-cells, surfels the 2-cells between two 3-cells, linels the 1-cells between two 2-cells and pointels the 0-cells between two 1-cells.

2.3   Topological Map

The topological map is a data structure used to represent the subdivision of an image into regions. It is composed of three parts:

- a minimal combinatorial map representing the topology of the image;
- an intervoxel matrix used to retrieve geometrical information associated to the combinatorial map. The intervoxel matrix is called the *embedding* of the combinatorial map;
- an inclusion tree of regions.

Figure 2 presents an example of topological map. The 3D image, divided into three regions plus the infinite region R_0 (Fig. 2 A), is represented by the topological map which is divided in three parts labeled B, C and D. The minimal combinatorial map extracted from this image is displayed in Fig. 2 B. The embedding of the map is represented in Fig. 2 C and the inclusion tree of regions in Fig. 2 D.

The combinatorial map allows the representation of all the incidence and adjacency relations between cells of the subdivision. In the topological map framework, we use the combinatorial map as a topological representation of the partition of an image in regions. Each face of the topological map is separating two adjacent regions and two adjacent faces do not separate the same two regions. With these rules, we ensure the minimality (in number of cells) of the topological map (see [13,9] for more details on topological maps).

The intervoxel matrix is the embedding of the combinatorial map. Each cell of the map is associated with intervoxel elements representing geometrical information of the cell. A face in the combinatorial map is embedded by a set of surfels separating voxels of the two incident regions. The edges, which are the border of faces, are represented by a set of linels. The vertices, which are the border of edges, are embedded by pointels. Thus the intervoxel matrix allows to retrieve the geometry of the labeled image represented by the combinatorial map.

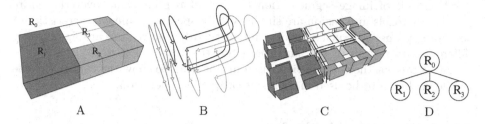

Fig. 2. The different parts of the topological map used to represent an image. (A) 3D image. (B) Minimal combinatorial map. (C) Intervoxel matrix (embedding). (D) Inclusion tree of regions.

The inclusion tree of regions represents the inclusion relations. Each region in the topological map is associated to a node in the inclusion tree. The nodes are linked together by the inclusion relation defined in Sect. 2.2. To link this tree with the combinatorial map, each dart d of the map knows its belonging region (called $region(d)$). Each region R knows one of its dart called *representative dart* (called $rep(R)$).

The topological map can be modified by using an operation of regions merging [14]. This process, called *global merge*, allows to merge any number of sets of connected regions. In the resulting topological map, each set of connected regions is represented by only one region.

In order to handle such sets, we use a *disjoint-set forest* [15] of regions. We use the *union-find* trees to represent the disjoint-sets. The two possible operations on sets are the union of two sets, and find the corresponding set of a particular element. In [16], R. Tarjan shows that union and find on disjoint-set represented by trees can be considered as constant time operations.

The principle of the algorithm of global merging is to remove all existing faces between regions of a same set: we remove the faces in the combinatorial map and their corresponding embedding. The next step is to simplify incident cells to respect the minimality of the topological map. This process is also performed both in the combinatorial map and in the embedding. The last step of the global merging process is to rebuild the inclusion tree by using the list of the remaining regions and the combinatorial map already built.

The complexity of this operation is $O(|D| + |S|)$ where $|D|$ is the total number of darts in the combinatorial map and $|S|$ is the number of surfels of the removed faces in the embedding (see [14] for more details).

3 Operation of Segmentation

The segmentation is an image processing operation which leads to partition an image into multiple regions. The goal of segmentation is to simplify the representation of an image into something that is more meaningful and easier to analyze. Image segmentation is typically used to locate objects in images.

The result of image segmentation is a set of regions that cover the entire image. All voxels in a region are similar with respect to some characteristics or computed properties, such as color or intensity. Adjacent regions are significantly different with respect to the same characteristics.

Let us present the characteristic used in this work to show the capacity of the topological map to be used in an image processing operation.

3.1   Criterion Used in the Segmentation Process

This method uses a local criterion proposed by P. F. Felzenszwalb and D. P. Huttenlocher in [10]. This segmentation criterion is based on intensity differences between neighboring pixels in 2D image. It is a region based segmentation using a characteristic onto regions and another characteristic between neighboring regions to produce a well-segmented (i.e. not over-segmented neither under-segmented) partition of the image.

We recall the principle of this segmentation. Each couple of adjacent regions is characterized by an external variation ($Ext(R_i, R_j)$). This value is the smallest intensity difference between neighboring voxels, one belonging to R_i and the other one belonging to R_j.

Each region is characterized by an internal variation ($Int(R)$): this value depends on voxels that are contained in the region. If a region contains voxels of the same intensity, its internal variation is 0. The authors in [10] prove that the new internal variation of two neighboring regions merged is equal to the external variation between the two regions. It allows the incremental computing of the internal variation of regions during the segmentation process.

We say that two regions are similar, and should be merged into one region, when the external variation between the regions is smaller than their minimum internal variation:

$$Ext(R_i, R_j) \leq MInt(R_i, R_j)$$

Where the minimum internal variation $MInt$ is

$$MInt(R_i, R_j) = min(Int(R_i) + \tau(R_i), Int(R_j) + \tau(R_j))$$

The threshold function τ controls the degree to which the external variation can actually be larger than the internal variations, and still have the regions be considered similar. We use, in this work, the same function as in the original work which depends on the size of the region, $\tau(R) = k/|R|$ where $|R|$ is the size (in number of voxels) of the region R and k is a constant defined by the user depending on the considered image.

To use this process with the topological map, the notion of external variation is transposed on faces. We define the external difference of a face F to be the smallest intensity difference between neighboring voxels across the face F (i.e. one voxel is on a side of the face, and its neighboring voxel is on the other side).

This value is related to the external variation of regions by the following formula. Let be two regions R_i and R_j witch are adjacent by k faces F_p, $p \in [1..k]$.

The external variation $Ext(R_i, R_j)$ is equal to the minimum external variation of the faces F_p with $p \in [1..k]$. This gives the following formula:

$$Ext(R_i, R_j) = min(Ext(F_p), p \in [1..k])$$

For performance purposes, we store these values into the topological map. Each dart of the topological map knows both its belonging region and its belonging face. With the help of this structure, we store on each face its external variation and on each region its internal variation. These values are computed during the creation of the topological map and each operation applied to the map updates these values accordingly. So for each region and face of the topological map, we retrieve the corresponding variations in a constant time.

3.2 Algorithm

Algorithm 1 shows the segmentation process using a threshold function to produce the optimal[4] segmentation of the 3D image represented by the given topological map.

This process is divided in two steps. First, we build a symbolic merging of the regions. This step corresponds to the choice of the regions to merge in the topological map. Then, we use the merging operation recalled in Sect. 2.3 to convert the symbolic merging into an effective one.

The first step uses a disjoint-set forest of regions to efficiently represent the symbolic merging of regions. Before the segmentation, each region belongs to its own set. To merge symbolically two regions, we merge the two disjoint-set representing these regions. We compute the new characteristics of the resulting region and we set them to the head of the disjoint-set (which represents the resulting region).

To initialize the process, we build a list of faces sorted by increasing external variation. The first face in the list has the lowest external variation. In fact, each face his represented by one of its darts. To build the list, we run through each dart of the topological map. If its incident face is not marked, we add the current dart to the list L and we mark the corresponding face. When all the darts have been processed, the list contains one dart for each face of the image. Then the list is sorted by increasing order of the external variation of the incident face, by using a classical sort algorithm.

For each dart, we first check if the two incident regions around the incident face are not already merged. This is performed by checking if the belonging region of the current dart d does not belong to the same disjoint-set as the belonging region of the dart $\beta_3(d)$ (i.e. the dart which belongs to the second region incident to the face). Then we compute the $MInt$ value for the two incident regions and if this value is lesser or equal to the external variation of the current face, we symbolically merge the two regions.

When all faces have been processed, the segmentation algorithm uses the merging operation to produce the topological map corresponding to the resulting regions.

---

[4] According to P. F. Felzenszwalb and D. P. Huttenlocher in [10].

Algorithm 1. Segmentation

Input: Topological map M, Threshold function τ
Result: M is modified and represents the segmentation of the initial map

$L \leftarrow$ build the sorted list of faces;
while $L \neq \emptyset$ **do**
    $F \leftarrow L.pop()$;
    $R_1 \leftarrow region(F), R_2 \leftarrow region(\beta_3(F))$;
    if $R_1 \neq R_2$ **then**
        if $Ext(F) \leq MInt(R_1, R_2)$ **then**
            $R \leftarrow$ union of R_1 and R_2 in the disjoint-set forest;
            $Int(R) \leftarrow Ext(F)$;

Apply the global operation of region merging;

The main difference between this algorithm and the graph-based process described in [10] is the usage of the external variation onto each face instead of the external variation between couple of regions. This modification is necessary since the topological map represents multi-adjacency while RAG only represents simple adjacency. We have to prove that both approaches are equivalent.

Property 1 ensures that two regions are merged if and only if they respect the original criterion equation. This property says that two regions R_i and R_j may be merged together if and only if they merge during the process of the first face which is incident to both regions in the increasing order of their external variation (this is due to the fact that the external variation of this first face is equal to the external variation between the two regions).

Property 1 (Validity). If two regions R_i and R_j are not merged when the algorithm process the first separating face, then they will never merge during the algorithm.

Proof. We consider two regions R_1 and R_2 in the topological map that are multi-adjacent. Let be F_p, $p \in [1..k]$, the different faces incident to both R_1 and R_2. We suppose that $Ext(F_1) \leq Ext(F_2) \leq \ldots \leq Ext(F_k)$. We know that the external variation of the two regions is equal to the external variation of the face F_1.

Algorithm 1 processes the faces in increasing external variation order. The first face considered to merge R_1 and R_2 is thus F_1.

Suppose $Ext(F_1) > MInt(R_1, R_2)$ then the two regions R_1 and R_2 are not merged according to the criterion. Given this hypothesis, we want to know if it is possible for the two regions to merge when considering the other faces. There are two cases to consider.

If $MInt(R_1, R_2)$ does not change, the process of the other faces can not leads to the merging of the two regions since they have a greater external variation.

If $MInt(R_1, R_2)$ increases its value between the treatment of two faces F_i and F_j ($i \geq 1$ and $j > i$) we have to show that this modification will not alter the result of the algorithm.

Let suppose $Int(R_1) + \tau(R_1) < Int(R_2) + \tau(R_2)$ and thus $MInt(R_1, R_2) = Int(R_1) + \tau(R_1)$ (this supposition can be made without lost of generality, eventually by renaming the two regions). So $MInt(R_1, R_2)$ increases only if $Int(R_1) + \tau(R_1)$ increases. This is only possible if R_1 is merged with another neighboring region which will be called R_3.

To merge R_1 and R_3, we need a face F so that $Ext(F) \leq MInt(R_1, R_3)$. Since the list is sorted $Ext(F_1) \leq Ext(F)$, and by definition of $MInt$, $MInt(R_1, R_3) \leq Int(R_1) + \tau(R_1)$. Our starting hypothesis gives that $Ext(F_1) > Int(R_1) + \tau(R_1)$. When combining these assertions, we have: $Ext(F) \geq Ext(F_1) > Int(R_1) + \tau(R_1) \geq MInt(R_1, R_3)$. Thus $Ext(F_1) > MInt(R_1, R_3)$, the region R_1 and R_3 will not be merged and $Int(R_1) + \tau(R_1)$ will not increase. This property shows that R_1 will never be merged during the algorithm. So $MInt(R_1, R_2)$ does not increase and each face F_p, $p \in [2..k]$, $Ext(F_p) > MInt(R_1, R_2)$. The region R_1 and R_2 will never merge.   □

Thus Prop. 1 is verified. Algorithm 1 gives the same result as the one proposed in [17].

4   Results and Analysis

Let us talk about the complexity of the segmentation. Let be $|F|$ the number of faces of the topological map and $|D|$ the number of darts of the topological map. We know that $|D| > |F|$ since there is at least one dart for each face of the map.

The initialization of the list of faces needs to run through all the darts. The marking operation as well as the checking of a mark are constant time operations. We perform $|F|$ insertions in the list and then we sort it which leads to a complexity for this operation in $O(|D| + |F| * \log(|F|))$. The union and find operations on disjoint-set are quasi-linear according to [16] and thus we consider it linear for our usage.

Then each face is considered in a loop executed $|F|$ times. The extraction of the list is a constant time operation as well as the retrieval of the belonging region of a dart and the β_3 operation. With the previous hypothesis on union-find trees, checking if the two regions are not merged is a constant time operation. The computation of the equation is also a constant time operation because we store the needed values on regions and faces. The access to these values is a constant time operation. The symbolic merging corresponds to the union operation on disjoint-sets and thus we consider it constant. So this part is performed in $O(|F|)$.

The last step of this algorithm is the application of the global merging operation on the topological map. We use the complexity given in Sect. 2.3 which is $O(|D| + |S|)$ where $|S|$ is the number of surfels of deleted faces.

This gives the complexity for the whole segmentation algorithm to $O(|D| + |F| * \log(|F|) + |S|)$. The operation depends on the number of darts of the whole topological map, the number of faces of the topological map and the number of surfels belonging to deleted faces.

We have written an application which allows to segment an image in the 3D topological map framework. We have measured several values during the segmentation of three images (which are medical brain images). In order to decrease the memory needed by the representation of the image, we pre-segment the image during the extraction of its topological map. Table 1 shows the initial and final number of regions for each of the 3 images so as it gives an idea of the effectiveness of the segmentation process. We have also collected the processing time of the three steps of the segmentation.

Table 1. Image information and processing time for the three steps of the segmentation algorithm (segmentation threshold $k = 5000$ and pre-segmentation $p = 3$)

Image	Img1	Img2	Img3
Size	256x256x44	256x256x111	256x256x124
Initial Regions	147924	431486	310421
Remaining Regions	10121	30179	22523
List initialization	2.93s	8.72s	6.25s
Symbolic merge	0.46s	1.30s	0.97s
Global merge	5.50s	14.99s	11.69s
Total	8.89s	25.10s	18.92s

The fastest step is the symbolic merging since it only manipulates regions by using union-find trees and the sorted list of faces. The list initialization is a slow step as it depends on the topological map size. The slowest part is, as expected, the region merging operation since it heavily depends on the number of darts in the topological map. The total processing time is nevertheless suitable for a use of the algorithm in the image analysis framework.

We present a segmentation result obtained on Img1. Figure 3 shows one slice of the medical image on left side. The image on the right represents the labeled view of the slice after the segmentation process. This view allows to identify all the resulting regions of the image since they have different colors. We could observe that most of the brain is represented by only one region whereas the skull is composed of many regions. The segmentation needs to be tuned in order

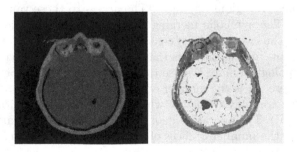

Fig. 3. A slice of a 3D brain image (256 x 256 x 44 gray image) and the segmentation result produced by the algorithm on the same slice ($k = 5000$, $p = 3$)

to produce more accurate results depending on the problem. For example, a low segmentation threshold leads to an over-segmentation which could be used in an image processing chain as the initial input.

5 Conclusion

In this paper, we have presented the implementation of an existing bottom-up segmentation process in the 3D topological map. This work shows that the topological map is a suitable data structure to represent 3D images within the image processing framework.

We have detailed the homogeneity criterion used in the segmentation operation. It relies on intensity differences between neighboring voxels of the image. The aim of the segmentation process is to provide a set of homogeneous regions according to the criterion such as the merge of two regions produces another region which is not homogeneous.

The segmentation algorithm is divided in two main parts. Firstly we aim to produce a symbolic segmentation of the image, by handling its regions on a high level, and computing the criterion step by step during the merging process. Secondly, we use the existing operation of region merging in order to reflect the symbolic changes in the topological map.

We then proved that the modifications made to the initial segmentation algorithm do not modify its initial properties. We have analyzed the complexity of the whole segmentation process. It depends on the number of darts $|D|$, the number of faces $|F|$ and the number of surfels $|S|$ of the topological map. The complexity is given by the relation $O(|D| + |F| * \log(|F|) + |S|)$.

We have made some experiments on the segmentation process. Obtained processing times prove that the most expensive part of the algorithm is the region merging one and show that 3D topological maps can be used in real image processing applications.

The following step is to improve the segmentation process by taking advantage of the topological and geometrical information stored by the 3D topological map. We intend to use these information in addition with the existing criterion to obtain results that not only depend on the homogeneous criterion but also take into account the shape of regions as well as the adjacency relations between regions. For example one topological criterion may disallow the creation of double torus region if no such object could exist in the image. This future work will show the advantage of using the topological map for 3D image segmentation.

References

1. Rosenfeld, A.: Adjacency in digital pictures. Information and Control 26, 24–33 (1974)
2. Kropatsch, W.G., Macho, H.: Finding the structure of connected components using dual irregular pyramids. In: Discrete Geometry for Computer Imagery, pp. 147–158 (September 1995) (invited lecture)

3. Bertrand, Y., Damiand, G., Fiorio, C.: Topological map: Minimal encoding of 3d segmented images. In: Workshop on Graph-Based Representations in Pattern Recognition, Ischia, Italy, IAPR-TC15, pp. 64–73 (May 2001)
4. Braquelaire, J.P., Domenger, J.P.: Representation of segmented images with discrete geometric maps. Image and Vision Computing 17(10), 715–735 (1999)
5. Damiand, G., Bertrand, Y., Fiorio, C.: Topological model for two-dimensional image representation: definition and optimal extraction algorithm. Computer Vision and Image Understanding 93(2), 111–154 (2004)
6. Fiorio, C.: A topologically consistent representation for image analysis: the frontiers topological graph. In: Miguet, S., Ubéda, S., Montanvert, A. (eds.) DGCI 1996. LNCS, vol. 1176, pp. 151–162. Springer, Heidelberg (1996)
7. Braquelaire, J.P., Brun, L.: Image segmentation with topological maps and interpixel representation. Journal of Visual Communication and Image Representation 9(1), 62–79 (1998)
8. Haxhimusa, Y., Ion, A., Kropatsch, W.G., Brun, L.: Hierarchical image partitioning using combinatorial maps. In: Hanbury, A., Bischof, H. (eds.) 10th Computer Vision Winter Workshop, pp. 43–52 (February 2005)
9. Damiand, G., Resch, P.: Split and merge algorithms defined on topological maps for 3d image segmentation. Graphical Models 65(1-3), 149–167 (2003)
10. Felzenszwalb, P.F., Huttenlocher, D.P.: Image segmentation using local variation. In: Computer Vision and Pattern Recognition, 1998. Proceedings. IEEE Computer Society Conference on, June 1998, pp. 98–104 (1998)
11. Lienhardt, P.: Topological models for boundary representation: a comparison with n-dimensional generalized maps. Computer-Aided Design 23, 59–82 (1991)
12. Khalimsky, E., Kopperman, R., Meyer, P.R.: Boundaries in digital planes. Journal of Applied Mathematics and Stochastic Analysis 3(1), 27–55 (1990)
13. Damiand, G.: Définition et étude d'un modèle topologique minimal de représentation d'images 2d et 3d. Thèse de doctorat, Université Montpellier II (Décembre 2001)
14. Dupas, A., Damiand, G.: Comparison of local and global region merging in the topological map. In: Brimkov, V.E., et al. (eds.) IWCIA 2008, vol. 4958, pp. 420–431. Springer, Heidelberg (2008)
15. Cormen, T.H., Leiserson, C.E., Rivest, R.: Introduction to Algorithms. MIT Press, Cambridge (1990)
16. Tarjan, R.: Efficiency of a good but not linear set union algorithm. Journal of the ACM 22, 215–225 (1975)
17. Felzenszwalb, P.F., Huttenlocher, D.P.: Efficient graph-based image segmentation. International Journal of Computer Vision 59(2), 167–181 (2004)

Adaptive Morphological Filtering Using Similarities Based on Geodesic Time

Jacopo Grazzini and Pierre Soille

Spatial Data Infrastructures Unit
Institute for Environment and Sustainability
Joint Research Centre - European Commission
TP 262 - via E.Fermi, 2749 - 21027 Ispra (VA), Italy
{Jacopo.Grazzini,Pierre.Soille}@jrc.it

Abstract. In this paper, we introduce a novel image-dependent filtering approach derived from concepts known in mathematical morphology. Like other adaptive methods, it assumes that the local neighbourhood of a pixel contains the essential process required for the estimation of local properties. Indeed, it performs a local weighted averaging by combining both spatial and tonal information in a single similarity measure based on the local calculation of discrete geodesic time functions. Therefore, the proposed approach does not require the definition of any initial spatial window but determines adaptively, directly from the input data, the neighbouring sample points and the associated weights. The resulting adaptive filters are consistent with the content of the image and, therefore, they are particularly designed for the purpose of denoising and smoothing of digital images.

1 Introduction

The dictionary of computer vision [1] states that adaptivity is the property of an algorithm to adjust its parameters to the data at hand in order to optimise performance. In this context, adaptive filtering, in which the parameters of the filter vary over different parts of the data, is a suitable approach for advanced image processing [2]. Indeed, the need for an adaptive approach to cope with inhomogeneities in images is well recognized [3,4]: adaptive filtering is a class of typical nonlinear smoothing techniques that has been applied to many computer vision tasks [5,6,7,8].

Strong relations have been established between a number of widely-used adaptive filters for digital image processing [9]. It has been shown that images filtered using adaptive neighbourhoods are superior to those filtered using fixed neighbourhoods, as the adaptive-neighbourhood techniques tune themselves to the contextual details in the image [4]. Indeed, the use of spatially adaptive approaches implies that operators must vary over the whole image, taking into account the local image context. It assumes intrinsically that the neighbourhood of a pixel contains the essential process required for local estimation of its true intensity value. Adaptive operators can be subdivided in two main classes: the

D. Coeurjolly et al. (Eds.): DGCI 2008, LNCS 4992, pp. 519–528, 2008.

adaptive-weighted operators and the spatially-adaptive operators [8,10], where the adaptive concept results respectively from the adjustment of the weights upon the operational window and from the spatial adjustment of the window. The operator of the first class typically involves convolution between the image and a fixed-size sliding window positioned over each pixel whose coefficients are dependent on the image statistics under the sliding window. The adaptive neighbourhood of the second class surrounds the central pixel to be filtered, but its shape and the area it covers are dependent on the local characteristics of the image rather than being arbitrarily defined. Besides, some recently developed techniques propose the use of a non-local approach for filtering, where the neighbourhoods used in the estimation are not necessarly spatially connected to the central pixel [11].

In the context of mathematical morphology (MM), the adaptive paradigm proposes solutions to the limitations of classical morphological transformations in the absence of any prior knowledge about the analysed images. By adopting an adaptive strategy, one addresses the lack of flexibility of morphological operators based on the difficult choice of a structuring element [12]. Several morphological techniques have been developed to take into account the local features of the image. In [13], adaptive-neighbourhood sequential filters are proposed, where neighbourhood of increasing size are calculated for each pixel using a similarity criterion and a region-growing approach. In [10], spatially adaptive operators using operational windows, whose extent is defined around each pixel by the connected regions for a selected criterion mapping (*e.g.*, luminance, local contrast, or local curvature), are built. In [14], similar neighbourhoods are calculated by introducing at every image pixel a distance defined between the values of the image only. Pixels reached within this adaptive neighbourhood are referred to as an *amoeba*; the mean or rank statistics are then computed within the amoeba to filter the image. Another approach to investigate the morphology of the image structures is to define them as connected components of pixels satisfying some properties [15,16,17].

In this paper, we propose an image-dependent filtering approach also aiming at extending the well-defined concepts of MM in order to consider radiometric, geometrical and morphological characteristics of the images [8,10,18]. The main idea is to associate with each pixel of the input image a weighted convolution of sample points within an adaptive neighbourhood, where the weights depend not only on the points location but also on their greylevel distance to the central pixel. For that purpose, we define some local similarity measure of the twofold spatial and tonal information based on the local estimation of pairwise discrete *geodesic time functions* in the local neighbourhood of the central pixel. This approach can be seen as an extension of [19], where a geodesic dilation was obtained by thresholding the generalised geodesic time function computed within the mask of greylevels. It can also be related to the concept of amoeba. Similarly, the method we introduce does not require the definition of any initial spatial window as it determines the neighbouring sample points directly. Besides, it is also able to estimate adaptively the associated weights from the input data.

The rest of the paper is organised as follows. In the next section, we recall the fundamental notions of discrete geodesic path and geodesic time known in MM, and the way adaptive neighbourhoods can be defined. In section 3, we propose two different filtering approaches based on the estimation of local geodesic time. We also precise the filtering strategy for the case of multispectral images. The conclusion and a description of future foreseen developments are presented in section 4.

2   Geodesic Time and Adaptive Neighbourhoods

2.1   Discrete Geodesic Time

Geodesic transforms are classical operators in discrete image analysis [20,21,22]. The *geodesic distance* between two pixels of a discrete connected set (typically, a binary image) is defined as the length of the shortest path(s) linking these points and remaining in the set (the so-called *geodesic paths*) [12,22]. This idea can been generalised to greylevel images using the *geodesic time on geodesic mask* [19,23]. The geodesic mask image is then treated as a 'height map', *i.e.* a surface embedded in a 3D space, with the third coordinate given by the greylevel values.

Formally, we define a discrete path $\mathcal{P}$ of length $l-1$ going from $\mathbf{p}$ to $\mathbf{q}$ as a l-tuple $(\mathbf{x}_1, \ldots, \mathbf{x}_l)$ of pixels such that $\mathbf{x}_1 = \mathbf{p}, \mathbf{x}_l = \mathbf{q}$, and $(\mathbf{x}_{i-1}, \mathbf{x}_i)$ defines adjacent pixels for all $i \in [2, l]$. Introducing the greylevel geodesic mask g, the time $\tau_g(\mathcal{P})$ necessary to cover $\mathcal{P}$ represents the sum of the greylevel values of the pixels along $\mathcal{P}$ [19]. This assumes the cost c_i of travelling from a pixel $\mathbf{x}_i$ to an adjacent pixel $\mathbf{x}_{i+1}$ to be:

$$c_i = \frac{1}{2}(g(\mathbf{x}_i) + g(\mathbf{x}_{i+1})) \cdot |\mathbf{x}_i - \mathbf{x}_{i+1}| \tag{1}$$

Here, the spatial distance $|\mathbf{x}_i - \mathbf{x}_{i+1}|$ refers to the elementary step in the image graph: it is chosen as either the Euclidean distance or the optimal Chamfer propagating weights in a binary 3×3 mask [24]. Intuitively, the lower the intensity value in g, the faster the propagation. The geodesic time $\tau_g(\mathbf{p}, \mathbf{q})$ separating two points $\mathbf{p}$ and $\mathbf{q}$ is then the smallest amount of time allowing to link $\mathbf{p}$ to $\mathbf{q}$ in g, *i.e.* it consists in finding the path with the lowest sum of greylevel values along all possible discrete paths linking $\mathbf{p}$ to $\mathbf{q}$ [19]:

$$\tau_g(\mathbf{p}, \mathbf{q}) = \min\{\tau_g(\mathcal{P}) \mid \mathcal{P} \text{ is a path linking } \mathbf{p} \text{ to } \mathbf{q}\}.$$

This concept is closely related to the notion of grey weighted distance transform defined in [25] and to the continuous framework of the minimal path approach of [26]. It leads to efficient algorithms because classical shortest path algorithms can be applied such as the Dijkstra's graph search algorithm. Typically, algorithms based on priority queue data structures [12,27] enable the implementation of the local geodesic time with a computational complexity of $O(n \log n)$, where n is the number of pixels in the considered spatial domain [22,27]. They

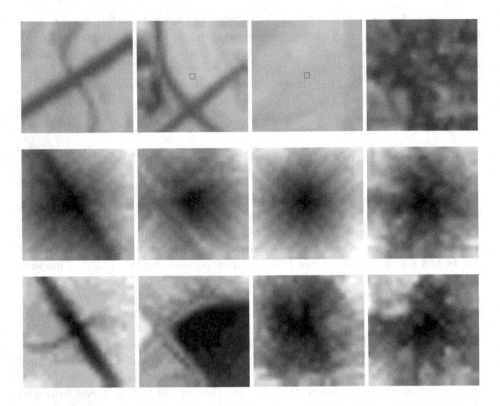

Fig. 1. Examples of adaptive geodesic neighbourhoods. Top: excerpt of a satellite image (21×21 pixels) with a marker pixel located in its center (red square); middle: geodesic time computed on the magnitude of the image gradient; bottom: *ibid.* on the image variations. From left to right, the marker pixel is *resp.* situated on a thin linear structure, near to a strong discontinuity, on a homogeneous textured area and in a noisy region. The estimated time was discretised in the range [1,12] in order to visualise the different geodesic levels.

take advantage of the fact that images are finite and their intensity values - in the monospectral case - are totally ordered. Therefore, they guarantee that pixels that effectively contribute to the output are processed only once.

2.2 Adaptive Neighbourhoods

The local estimation of discrete geodesic time at every pixel location enables to combine both spatial and tonal information into the definition of new adaptive neighbourhoods. Given an input image f, we are able to derive local adaptive neighbourhoods in f using the previous approach. Namely, by considering at every single pixel location $\mathbf{x} \in f$ the geodesic time $\tau_g(\mathbf{x}, \cdot)$ estimated from $\mathbf{x}$ ($\mathbf{x}$ is then said to be a *marker*) over the magnitude of the spatial gradient $|\nabla f|$, new neighbourhoods accounting for local image variability can be built. The

underlying idea is that the geodesic paths associated to this new measure -
instead of the standard geometric distance - define the intrinsic neighbourhood
relationship between the sample points when the 2D image is projected onto the
3D spatial-tonal domain. Indeed, the geodesic time expressed with Eq. (1) and
the geodesic mask $g = |\nabla f|$ is going through the lowest values of the spatial
gradient.

Similarly, we can define a geodesic measure that accounts for both the distance
between pixels and the roughness of the 'height map', $i.e.$ a measure of the
shortest path drawn on the projection of the 2D image onto the spatial-tonal
domain. For this purpose, we redefine the cost c_i of crossing pixels as:

$$c_i = \frac{1}{2}|g(\mathbf{x}_i) - g(\mathbf{x}_{i+1})| + |\mathbf{x}_i - \mathbf{x}_{i+1}| \qquad (2)$$

with the distance $|\mathbf{x}_i - \mathbf{x}_{i+1}|$ as before and the geodesic mask set to $g = f$.
This definition is equivalent to the weighted distance on curves space of [23].
Its intuitive interpretation is that it represents the minimal amount of ascents
and descents to be travelled to reach a neighbouring pixel. Indeed, the geodesic
time $\tau_g(\mathbf{x}, \cdot)$ estimated with Eq. (2) minimises the changes in greylevel values.
Note that in practise the geodesic mask is in fact constantly updated through
the propagation of the geodesic time [23,27]. The neighbourhoods defined this
way coincide with the kernels defined by the morphological amoebas of [14].

We can observe the way neighbourhoods adapt to the image context on the
geodesic levels in Fig. 1. In particular, on the second column, the pixels across
the road are not included in the closest neighbourhood of the marker as the
cost of crossing the road is high when considering either the spatial gradient
or the image variations. Thus, by considering such neighourhoods in filtering
applications, the contributions of pixels from the other side of the edge will
be suppressed. In other words, if a pixel is located near an edge, then pixels
on the same side of the edge will have much stronger influence in the filtering:
greylevel values from across a sharp feature are given less influence because they
are penalised by the geodesic time functions. This is a desirable property of a
filtering approach.

3   Adaptive Geodesic Filtering

3.1   Similarity Measures and Filtering Kernels

The most common strategy encountered in adaptive filtering consists in building
local kernel functions over image regions according to their contents [2,9]: around
a pixel $\mathbf{x}$ to be updated, one defines a kernel $\mathcal{K}$ with proper weights depending on
the actual image variability in the neighbourhood of $\mathbf{x}$; filtering is then performed
through a weighted average of local samples with $\mathcal{K}$. A critical issue is then how
to account for image variability for generating the weights of the kernel.

A possible approach consists in adapting the local effect of the filter by using
both the location of the nearby samples and their intensity values within the

neighbourhood defined by the kernel $\mathcal{K}$. Accordingly, we propose to use the similarity measure based on the local geodesic time values for defining the weighs of $\mathcal{K}$. Precisely, the local similarity is measured in the neighbourhood of the marker $\mathbf{x}$ as a (monotonically) decreasing function Ψ of the geodesic time $\tau_g(\mathbf{x}, \cdot)$ over the geodesic mask g: $\mathcal{K}(\mathbf{x}, \cdot) = \Psi(\tau_g(\mathbf{x}, \cdot))$. This way, the filtering is performed along the geodesic paths going through $\mathbf{x}$ as defined by the selected geodesic function (and depending on the geodesic mask). A standard Gaussian can be used for the function Ψ, but other functions are not excluded. By introducing such a tonal weight in the kernel function, the mixing of different intensity populations is prevented and the effective sampling procedure is adapted locally to the image features such as edges. Employing Eq. (1) with $g = |\nabla f|$, the time necessary to travel between two pixels separated by high gradient values is higher than the time necessary to travel between two pixels separated by low gradient values. Therefore, higher weights are assigned to the nearby sample pixels that involve low gradient values along the minimal geodesic paths from $\mathbf{x}$, as compared to samples that are either far away from $\mathbf{x}$ or separated by high gradient values. Using Eq. (2) with $g = f$, higher weights are assigned to the nearby sample pixels which are linked to $\mathbf{x}$ and with similar greylevel values. Note moreover the difference in this latter case with the amoeba filtering [14]: not only the domain defined by the geodesic neighbourhoods is used in the filtering procedure, but also the geodesic time values themselves as they define the weights of the samples in the kernel, unlike the amoebas which consider rank statistics instead. Indeed, in the proposed filtering techniques, the adaptive concept results from both the spatial adjustment of the operational kernel and the adjustment of the weights upon the kernel.

In practice, due to memory and computational limitations, the support of $\mathcal{K}$ is limited to a fixed size, *i.e.* sample pixels that are further away (in the spatial domain) than a given distance ω to the central pixel are not considered. Running the filtering procedure that calculates the local geodesic time from each single pixel in the image finally results in a total complexity of $O(n \cdot \Omega^2 \log \Omega^2)$ where n is the number of pixels of the input image and $\Omega = 2\omega + 1$.

3.2   Application to Multispectral Image Smoothing

As a special case of filtering, image smoothing is a common preprocessing stage used to improve the visual appearance in an image, and to simplify subsequent image processing stages such as feature extraction, image segmentation or motion estimation [2]. Traditionally, the problem of image smoothing is to reduce undesirable distortions - due to the presence of noise or the quality of the image acquisition process - while preserving important features such as homogeneous regions, discontinuities, edges and textures [3,9,28].

We consider the geodesic filters introduced in section 3 for edge-preserving smoothing of multispectral images. When dealing with multivalued image, the previous approach needs however to be adapted. In particular, the calculation of the gradient magnitude used by the first proposed filter needs to be attentively considered in order to take into account the actual multispectral information. A

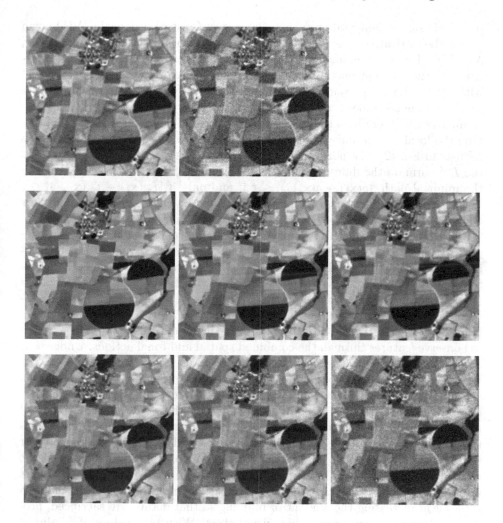

Fig. 2. Outputs of the proposed filtering approaches compared to the amoeba filtering of [14]. Results for a multispectral satellite image (top left) and its noisy version (Gaussian noise added, PSNR = 22.43 dB top right) are displayed on the second and third lines *resp.* From left to right: geodesic time computed on the magnitude of the image gradient, geodesic time computed on the image variations and morphological amoeba with median statistics. The parameters for the geodesic filters were set to $\alpha = 10$ (second line) and $\alpha = 7$ (third line); the window size was set to $\omega = 7$. For comparison, the spatial extent of the amoeba kernels was similarly limited to a window of 7×7 pixels. PSNR values when filtering the noisy image: PSNR = 30.86 dB and 28.95 dB for the geodesic filters, and 27.57 dB for the amoeba filter.

way to estimate it for a multichannel image f with components $f_m, m = 1, \ldots, M$ is by means of the eigenvalue analysis of the image squared differential expressed by the 2×2 matrix: $\left[\sum_m (\frac{\partial f_m}{\partial x})^2, \sum_m \frac{\partial f_m}{\partial x} \frac{\partial f_m}{\partial y}; \sum_m \frac{\partial f_m}{\partial x} \frac{\partial f_m}{\partial y}, \sum_m (\frac{\partial f_m}{\partial y})^2 \right]$, the so-called first fundamental form [29]. The largest (postive) eigenvalue λ is known to be the derivative energy in the most prominent direction (in particular, $\lambda = |\nabla f|^2$ for greylevel images with $M = 1$) and is consequently a natural estimate for the gradient magnitude of the image. Therefore, we can apply Eq. (1) with $g(\mathbf{x}) = \lambda(\mathbf{x})$. Approaches based on gradient information are however quite sensitive to noise. Indeed, noise corruption can generate discontinuities that are confused with other discontinuities that correspond to important features. Therefore, the local discontinuity measure based on the spatial gradient is not always robust. In Eq. (2), the norm must be understood as a multispectral norm, $e.g$ the L^∞ norm on the different channels; in such case, we have, when estimating the minimal path, $|g(\mathbf{x}_i) - g(\mathbf{x}_{i+1})| \leq t$ if and only if $|f_m(\mathbf{x}_i) - f_m(\mathbf{x}_{i+1})| \leq t$ for all $m = 1, \ldots, M$; thus, the approach adopted by the second proposed filter depends on the dimension of the tonal space, which may increase its complexity.

Both filters were efficiently implemented for processing digital images using priority queues. They result in visually satisfying smoothed versions of the original images, blurring small discontinuities and sharpening edges (Fig. 2, bottom lines, first two columns). When compared with the morphological amoebas (Fig. 2, bottom lines, last column), they take advantage of the fact that the estimated geodesic time to the central pixel is also taken into account as a weighting function in the filtering procedure. Indeed, the generic filtering approach enables to conserve features through the combined spatial and tonal actions. Under that aspect, this approach can be linked to the *bilateral filtering* algorithm of [28]. Using a twofold similarity measure like the geodesic time, defined on either the spatial gradient or the image variation, enables moreover to account for the correlations between the positions of the pixels and their values, while bilateral filtering is breaking this correlation. Close inspection of the images also shows that the method is able to enhance texture regions. The degree of smoothing or sharpening can be further adjusted introducing a control parameter. Precisely, multiplying the cost c_i of crossing pixels by a given fixed value, say α, results in amplifying or attenuating the local contrast in parts of an image. Small α values lead increasing the amount of blurring so that details are sacrificed, producing the well-known cartoon-like visual effect. With high values of α, almost all contrasts are preserved and filtering has very little effect on the image. α will control the relative influence of tone and space in the calculation of the similarity measure of neighbour pixels. Other experiments have been conducted on benchmark natural images and compared with the outputs of some standard filtering techniques in [30].

4   Conclusion

In this paper, we introduce new morphological adaptive operators for filtering. The adaptive concept results from both the spatial adjustment of the operators and the adjustment of the weights upon them. The basic idea is similar to that of spatial-tonal filtering approaches, which consist in employing both geometric and intensity closeness of neighbouring pixels. The originality of our approach lies in the definition of a new similarity measure combining both spatial and tonal information and based on the local estimation of some geodesic time functions. The proposed approach could be used as a preprocessing stage in feature extraction and/or image classification. Indeed, by blurring small discontinuities and sharpening edges, the image structures are not geometrically damaged, what might be fatal for further processing like classification or segmentation. Like other spatial-tonal based techniques, the degree of smoothing in the image can also be tuned. Further experiments should be led on strongly textured images and compared with other recently developed techniques.

Current research is geared towards improving and extending the present work. Improvements regard mainly the selection of the different parameters involved in the filtering strategy. A specific issue regards the spatial extent of the window used for estimating the local geodesic time functions (represented by ω herein). An alternative approach to the one adopted in this paper would be to limit the weighting average to the sample pixels reached from the central pixel with a time inferior to a given threshold value. This would simply be achieved by checking if the time is less than this threshold when we extract a new element from the priority queue and stopping the propagation process in such case. The role of the parameter α should also be studied more in depth.

The proposed approach is of particular interest for filtering data for which the discrete framework should be assumed, in order to avoid creating spurious artifacts through diffusion-like processes. We foresee further applications in the fields of remote sensing and medical imaging.

References

1. Fisher, R., Dawson-Howe, K., Fitzgibbon, A., Robertson, C., Trucco, E.: Dictionary of Computer Vision and Image Processing. Wiley, Chichester (2005)
2. Jähne, B.: Digital Image Processing: Concepts, Algorithms and Scientific Applications, 4th edn. Springer, Heidelberg (1997)
3. Saint-Marc, P., Chen, J., Medioni, G.: Adaptive smoothing: A general tool for early vision. IEEE Trans. Patt. Ana. Mach. Intel. 13, 514–529 (1991)
4. Paranjape, R., Rangayyan, R., Morrow, W.: Adaptive neighborhood mean and median image filtering. J. Elec. Im. 3, 360–367 (1994)
5. Pitas, I., Venetsanopoulos, A.: Nonlinear Digital Filters: Principles and Applications. Kluwer Academic Publishers, Norwell, USA (1990)
6. Nitzberg, M., Shiota, T.: Nonlinear image filtering with edge and corner enhancement. IEEE Trans. Patt. Ana. Mac. Intel. 14(8), 826–833 (1992)
7. Cheng, F., Venetsanopoulos, A.: Adaptive morphological operators, fast algorithms and their applications. Patt. Recog. 33, 917–933 (2000)
8. Debayle, J., Gavet, Y., Pinoli, J.C.: General adaptive neighborhood image restoration, enhancement and segmentation. In: Campilho, A., Kamel, M. (eds.) ICIAR 2006. LNCS, vol. 4141, pp. 29–40. Springer, Heidelberg (2006)
9. Mrázek, P., Weickert, J.J., Bruhn, A.: On robust estimation and smoothing with spatial and tonal kernels. In: Geometric Properties for Incomplete Data, pp. 335–352. Springer, Heidelberg (2006)
10. Debayle, J., Pinoli, J.C.: General adaptive neighborhood image processing. J. Math. Im. Vis. 25(2), 245–284 (2006)

11. Buades, A., Coll, B., Morel, J.M.: Neighborhood filters and PDE's. Num. Math. 105(1), 1–34 (2006)
12. Soille, P.: Morphological Image Analysis: Principles and Applications, 2nd edn. Springer, Heidelberg (2004)
13. Braga-Neto, U.: Alternating sequential filters by adaptive-neighborhood structuring functions. In: Mathematical Morphology and its Applications to Image and Signal Processing, pp. 139–146. Kluwer Academic Publishers, Dordrecht (1996)
14. Lerallut, R., Decencière, E., Meyer, F.: Image filtering using morphological amoebas. In: Proc. of ISMM. CIV, pp. 13–22. Springer, Heidelberg (2005)
15. Meyer, F., Maragos, P.: Nonlinear scale-space representation with morphological levelings. J. Vis. Comm. Im. Repres. 11(3), 245–265 (2000)
16. Soille, P.: Constrained connectivity for hierarchical image decomposition and partitioning. IEEE Trans. Patt. Ana. Mac. Intel (2008) (available online since October 2007)
17. Soille, P., Grazzini, J.: Advances in constrained connectivity. In: Proc. of DGCI. LNCS, vol. 4992, pp. 423–433. Springer, Heidelberg (2008)
18. Lavialle, O., Delord, D., Baylou, P.: Adaptive morphology applied to grey level object transformation. In: Proc. of ESPC, pp. 231–234 (2000)
19. Soille, P.: Generalized geodesy via geodesic time. Patt. Recog. Lett. 15(12), 1235–1240 (1994)
20. Lantuéjoul, C., Maisonneuve, F.: Geodesic methods in image analysis. Patt. Recog. 17, 177–187 (1984)
21. Verwer, B., Verbeek, P., Dekker, S.: An efficient uniform cost algorithm applied to distance transforms. IEEE Trans. Patt. Ana. Mach. Intel. 11(4), 425–429 (1989)
22. Coeurjolly, D., Miguet, D., Tougne, L.: 2D and 3D visibility in discrete geometry: an application to discrete geodesic paths. Patt. Recog. Lett. 25(5), 561–570 (2004)
23. Ikonen, L., Toivanen, P.: Distance and nearest neighbor transforms on gray-level surfaces. Patt. Recog. Lett. 28, 604–612 (2007)
24. Borgefors, G.: Distance transformations in digital images. Comp. Vis. Graph. Im. Proc. 34, 344–371 (1986)
25. Levi, G., Montanari, U.: A grey-weighted skeleton. Inform. Cont. 17, 62–91 (1970)
26. Cohen, L., Kimmel, R.: Global minimum for active contour models: a minimal path approach. Int. J. Comp. Vis. 24, 57–78 (1997)
27. Ikonen, L.: Pixel queue algorithm for geodesic distance transforms. In: Andrès, É., Damiand, G., Lienhardt, P. (eds.) DGCI 2005. LNCS, vol. 3429, pp. 228–239. Springer, Heidelberg (2005)
28. Tomasi, C., Manduchi, R.: Bilateral filtering for gray and color images. In: Proc. of ICCV, pp. 839–846 (1998)
29. Di Zenzo, S.: A note on the gradient of a multi-image. Comp. Vis. Graph Im. Proc. 33, 116–125 (1986)
30. Grazzini, J., Soille, P.: Edge-preserving smoothing of natural images based on geodesic time functions. In: Proc. of VISAPP, pp. 20–27 (2008)

Book Scanner Dewarping with Weak 3d Measurements and a Simplified Surface Model

Erik Lilienblum and Bernd Michaelis

Institute for Electronics, Signal Processing and Communications
Otto-von-Guericke University Magdeburg
P.O.Box 4120, 39106 Magdeburg, Germany
{erik.lilienblum,bernd.michaelis}@ovgu.de

Abstract. For book scanner technologies projective distortions are the main problem. In general, the use of 3d measurements of a warped surface is the best way to remove the projective distortions. But if the quality of the 3d measurements is very low, it is difficult to get satisfying dewarping results. In our paper we present a new technique handling this problem by introducing a simplified surface model. We use this model as a basis to compute a linear approximation parallel to the geometrical position of the book crease. The resulting method leads to a robust and fast computation. It provides us with a reliable dewarping output even for weak measurements given by a light sectioning method of top view scanners.

1 Introduction

Digital archiving of hard-back literature increasingly becomes an essential part of the work of libraries and museums. Although for this purpose the modern computer technology already accomplishes major premises, the status quo of the scanner technology is not satisfying. In particular it is hardly possible to get distortion-free copies from thick books without damaging them. This currently represents a large problem both for high quality digitalisation of valuable historical books in museums and for simple copying of common books in public libraries. Besides, the image distortions bring a further problem with itself. A full automated character recognition (e.g. OCR) in the area of book crease is often impossible.

In the last years a special kind of book scanner, which is called top view scanner, became generally accepted on the market. To use the books in their natural way those scanners capture a copy of the page from above and from a certain distance. Through the use of a line camera with an appropriate stripe lighting it is possible to receive an evenly sharp and well illuminated two-dimensional image. However, we inevitably get a distorted copy in consequence of the projective geometry of the scanner and the warped surface of the page. An example can be seen on the upper left in figure 1.

Removing the projective distortions supported by 3d measurements of the page surface is a well known idea which we also use in our book scanner application. For the 3d surface measurement we extend the scanner construction by

D. Coeurjolly et al. (Eds.): DGCI 2008, LNCS 4992, pp. 529–540, 2008.

original scanner image:

original surface measurement:

dewarped output image:

approximated book surface:

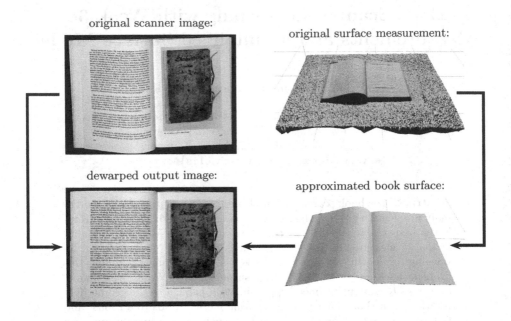

Fig. 1. Dewarping concept with 3d surface measurements

an additional matrix camera and apply a special kind of light sectioning. The result is represented on the upper right in figure 1. For homogeneous textures on the page this method provides sufficient accuracy. In contrast, if there are figures with very dark parts on the page, we get inapplicable measurements in these regions. Independent from the applied method this is a common problem for optical measurement systems.

However, to use the surface reconstruction despite their low quality in regions of weak measurements we have to approximate the book surface. The essence of our proposed method is to get a save and coherent surface reconstruction of the page, which is widely independent from the original measurement quality. The result of our approximation is represented down on the right in figure 1. By combining our approximated surface reconstruction and the original scanner image we can calculate a nearly distortion-free copy of the page. An example of an output result is shown down on the left in figure 1.

2   Related Work

Removing the projective distortion in a two-dimensional image of a three-dimensional warped surface is an abstract mathematical problem. A possibility without using additional information or hardware support offers the "shape from shading" principle. In [1,2] the shape of the surface is drawn directly from the distribution of brightness. Because of strict conditions the exactness and reliability of this approach are not particularly high.

Another approach without an extension of the scanner hardware is to recognize the lines of text in a page [3,4,5]. Supposing that the lines must be straight, the original scanner image can be corrected accordingly. But this method works only with text pages. Pages with figures or insular text blocks can not be dewarped.

A quite similar technique is used in systems recognizing the margins. One approach is given by Rebmann et al. in [6]. Another approach and an example of its application can be found in a software tool which is based on the patent [7]. The patent describes a dewarping method disposing the geometrical position of the lower page margin. In most cases the software solution works well, but there are problems, if the book is not aligned well or if there are notes, which exceed the lower paper delimitation. These problems were a main motivation for us to develop an alternative solution.

Book scanner systems with additional hardware support are mostly based on the construction of a vision system and the calculation of a 3d surface reconstruction of the page. There are constructions with one matrix camera and one projector in [8], two matrix cameras in [9] or laser triangulation in [10]. These techniques are often costly concerning the additional hardware and provide low resolutions in the copy of documents due to the exclusive use of matrix cameras. Additionally, there are a lot of other techniques to get a surface reconstruction, which we do not describe here. A good survey is given in [11].

The surface reconstruction of a book page is the basis to solve the dewarping problem. The success of an approach depends more on the quality of the 3d data than on the quality of the dewarping algorithm. For exact measurements the approach from Brown [12] should provide a good dewarping result even for rigid surface deformations. But computational costs of that approach are very high and it is difficult to bear measurement errors. For this reason we prefer the works from Liang [13,14]. As in our own approach it integrates the mathematical model of a warped page.

3 Weak Surface Measurements

As described in [15] and patented in [16] our approach to calculate the surface reconstruction is based on a further development of classical light sectioning [17]. According to this well known principle the position of the light stripe in an image of the matrix camera depends on the surface shape of the page we are scanning. Thereby from the captured image sequence we can infer the form of the whole surface. Figure 2 shows a schematic representation of our setup.

In classical light sectioning one determines in an image of the matrix camera the local pixel position of a projected light stripe. Usually this only works if the light stripe is very small, e.g. a projected laser line. Because the stripe lighting of a top view scanner generates a broad light stripe we have to apply an alternative technique. For a pixel of the matrix camera we determine from the image sequence the point in time, on which the light stripe was passing the pixel. Independent from the width of the stripe lighting we can calculate the point in

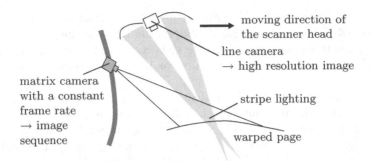

Fig. 2. Schematic representation of our setup

time with comparative high accuracy. By triangulation of the optic beam of the pixel and the plane of the stripe lighting we get the position of a surface point.

Evaluating the measurements of our method partially unveils some notable deviations between measured values and actual values. These measurement errors are not due to incorrectly computed time values or errors in the calibration of the system. They are caused by inhomogeneous paper properties and reflections. Especially at very dark textures the light of the surrounding areas has a large influence. That leads to weak measurements which are shown in figure 3.

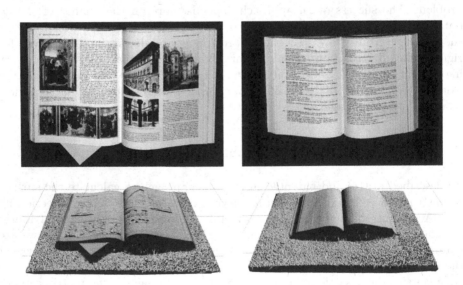

Fig. 3. Weak 3d measurements (left) and good 3d measurements (right)

The rough 3d-surface of the left example in figure 3 is caused by errors of weak measurements. Especially at the dark parts in the illustration of the page we get notable measurement deviations, which we cannot level out by local operations. On the right side of figure 3 we see an example with a proper 3d surface measurement.

Weak surface measurements are not only a problem of our time based light sectioning method. The quality of all approaches mentioned in section 2 depend on the page texture. Generally, we can state that the more we invest in the scanner hardware and the calculation time the fewer weak measurements we get. Both a better scanner hardware and a higher calculation time cause costs, which we can avoid by developing an efficient dewarping method working robust with weak measurements.

4 The Book Model Calculation

4.1 Assumptions

The approximation of a surface can be computed under very different points of view. If the quality of the given measurements is low, information about the surface characteristics like smoothness or boundary conditions are very important. Especially the kind of the approximation method should depend on the purpose of the expected approximation result.

In case of book surface approximation we know that the page surface is very smooth, normally. Therefore we can level out local measuring noise by simply choosing a large approximation area. But this procedure is not sufficient in case of weak or missing measurements, which are often locally connected. Here we need more stringent conditions, which we define for our special approximation problem through a geometric book model. The price we have to pay for introducing a stringent book model is the inability to dewarp rigid deformed books. But the advantage is that we can develop an algorithm which is fast, robust, and applicable for nearly all other kinds of books.

The definition of the book model is based on the following main assumptions:

1. There is a book crease which is detectable in the 3d measurement set.
2. Parallel to the book crease there is only a very small surface warping.

Clearly, both assumptions do not claim completeness for all kinds of books. But this is rarely a problem. A book without a crease is very unusual and we can exclude such cases for our proposed method by an exception rule. The second point, the warping direction, is more difficult because for instance a dog-ear cannot be mapped by our model. But generally dog-ears should not possess a main meaning for book scanners.

4.2 Improving the Crease Detection

The most important characteristics of our book model is the geometric location of the book crease. It subdivides the book surface into two parts and gives the orientation of the pages. The easiest way to detect the book crease is searching a minimum line in the middle of the 3d measurement set.

In case of weak measurements this procedure provides us only with a first approximation. To improve this result we have to analyse the surface measurements in the neighbourhood of the book crease. Therefore we calculate a linear approximation of the surface through fitting a plane on the left and the right side of the crease, respectively. The planes which we denote by E_l and E_r are schematically shown in figure 4.

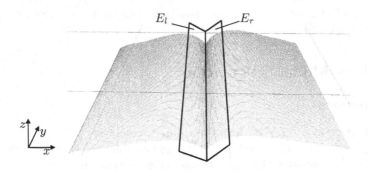

Fig. 4. Approximation planes left and right from the book crease

Let be given the first calculation of the geometric location of the book crease, which we get by searching the minimum line. Our improvement refers to the direction of the book crease in the telecentric view to the xy-level. In this case we can set $z = 0$ and describe the geometric location of the crease in the xy-level with the normal form for straight lines

$$ax + by + c = 0 . \tag{1}$$

We now compute the approximation planes E_l and E_r given through

$$\begin{aligned} E_l : & \quad A_l x + B_l y + C_l z + D_l = 0 , \\ E_r : & \quad A_r x + B_r y + C_r z + D_r = 0 . \end{aligned} \tag{2}$$

Then we form an intersection between the planes and project the result into the xy-level. Thus we get in the xy-level an improvement of equation 1 with

$$\begin{aligned} a &= A_r - A_l , \\ b &= B_r - B_l . \end{aligned} \tag{3}$$

The parameter c of equation 1 cannot be improved by this method because the intersection of the approximation planes can induce a displacement. After correcting a and b parameter c has to be recomputed by searching the minimum.

The exact alignment of the book crease has a direct effect on the alignment of the output result. An example of the consequence of a displaced book crease is given in figure 5. Here the difference between the uncorrected book crease and the corrected book crease is about 2 degrees. That sounds not much, but it is important for the visual overall impression of the dewarped output. With the described method we typically gain error corrections within 2 and -2 degrees.

ARCHITECTURE IN ITALY

Palazzo Medici, which set a pattern for Florentine town-houses, is sober and severe, even a little forbidding. On the exterior, windows are simple and regular and the design makes its effect largely by very carefully adjusted

ARCHITECTURE IN ITALY

Palazzo Medici, which set a pattern for Florentine town-houses, is sober and severe, even a little forbidding. On the exterior, windows are simple and regular and the design makes its effect largely by very carefully adjusted

Fig. 5. Uncorrected and corrected alignment of the book crease

4.3 Weighted Linear Approximation

The second point of our book model definition gives us the possibility to level out the weak measurements by linear approximation patches over the complete vertical page length. Because this computation only leads to an error reduction similar to forming an average we have to introduce a quality estimation of the used measurements. Because in our light sectioning method dark surface areas are influenced by the surrounding light, the brightness of a measured point can be a good value for this purpose. This certainly is a compromise but it is also an important condition to develop a robust and fast method based on a linear approximation of weighted measurements.

The set of weighted measurements that goes into one linear approximation we define by

$$\Gamma(d_1, d_2) = \{(p, \nu(p)^2) \,|\, p \in M, d_1 \le \kappa(p) \le d_2\}, \tag{4}$$

whereby $\kappa(p)$ is the Euclidean distance between a point p and the detected book crease relating to the xy-level, d_1 and d_2 are constraints to set the boundaries of the approximation area, $M \subset \mathbb{R}^3$ is the set of measured surface points, and $\nu(p)$ is the brightness of a measured point.

The actual operation to fit a plane into a set of weighted points is a standard problem. We do not describe it here. We denote that fitting operation with P, whereby the result of the operation $P : M \to \mathbb{R}^3$ is defined by an infinite set of 3d points forming a plane. We now can form a straight line defined by the set

$$S(i, g) = \{p \,|\, p \in P(\Gamma((i-1)r, (i+1)r)) \text{ and } \kappa(p) = ir\}, \tag{5}$$

whereby $i \in \mathbb{G}$ is an integer value we call grid coordinate and $r \in \mathbb{R}$ is a real value we call grid distance. Through a computation of these straight lines for all i we get an equidistant grid of approximation values along the whole book surface. An example can be seen schematically in figure 6.

The new format of our 3d approximation values provides us with good properties for the further processing. We are easily able to detect the page borders by analysing the difference between the original measurements and the linear approximation. Such a 3d supported border recognition is much more stable than

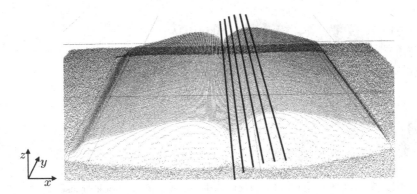

Fig. 6. Linear approximation with straight lines over the vertical book length

a brightness supported border recognition. The result is convincing even if there are notes going beyond the page border.

4.4 Dewarping and Output

After applying the weighted linear approximation of the page surface we have got an aligned equidistant measurement grid. To receive a dewarped output result we have to develop the warped surface into a plane. For this our data structure of straight lines is a good basis.

We start the dewarping algorithm from the book crease toward the outer delimitation of the left and right page, respectively. In each step we fold the 3d distance between two neighbouring approximation lines into the xy-level. Thereby we get a new line in a 2d output coordinate system. The height information of each line is saved in a relation to the original line.

In figure 7.a we see the origin in a telecentric view toward the xy-level of our xyz-coordinate system. It schematically diagrammes the detected book crease and the first 5 approximation lines to the right in its original geometrical direction. In figure 7.b we see the approximation lines in a transformed $x'y'z$ coordinate system aligned to the book crease. The $x'y'$-level of the transformed $x'y'z$ coordinate system is equal to the 2d output uv-coordinate system.

In the uv coordinate system we fold the approximation lines into the plane which is shown in figure 7.c. Through the transformation of the surface into a plane we loose the equidistant grid structure and the straight lines are not longer aligned. As it is shown in figure 7.d we renew the grid structure through the computation of equidistant supporting points by a linear interpolation. Thereby we save the relation to the original 3d points. As a result we obtain at the uv coordinate system an equidistant set of supporting points with a relation to 3d points of the approximated book surface. This data structure is a direct basis to generate a dewarped output image of the page.

The calibration of the complete book scanner system provides us with a mapping from 3d points into the original scanner image. From this mapping we

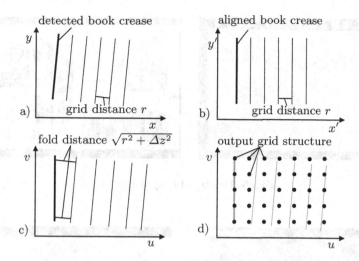

Fig. 7. Dewarping steps after the linear approximation

obtain for any point in the 3d space a subpixel coordinate in the scanner image with high accuracy. To generate a dewarped image we only have to output the subpixel values, which we get by mapping the supporting points. We get a defined high resolution equal to the original scanner image by bilinear interpolation between the supporting points related to our equidistant grid.

5 Results

It is difficult to evaluate the results of our method by objective mathematical calculations. On one hand the evaluation depends much on the quality and characteristics of the input book. On the other hand there is no standard evaluation method to other dewarping algorithms. So we present here some examples with different book characteristics, which we analyse concerning the dewarping quality of the output.

At first we show in figure 8 the dewarped output of the two examples from figure 3 in section 3. It gives a general impression of the ability of our dewarping method. A more detailed illustration is already shown on the left in figure 5. The text lines are nearly straight and well aligned. There are no problems with the illustrations even if they include deep black parts. Additionally, the characters situated near the book crease appear in there original size. We still get sufficient results even if the book crease is very deep like in the example of figure 9.

The examples are only a small selection of our test scans with many different books. If near the book crease the brightness is high enough we detect the correct geometrical location with nearly 100%. The detection only fails if there is no book crease at all or the geometric form of the book crease is very flat. In this case we cannot apply the method presented in this paper.

Fig. 8. Dewarped output results

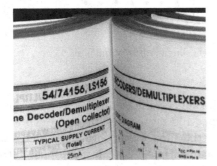 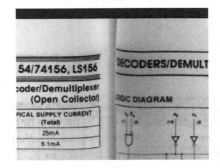

Fig. 9. Character dewarping near a deep book crease

If we detect a book crease the method works robust and fast. On a standard PC with 3.4 GHz the computation time for our method is below 1s. Within this period we have to calculate up to 640×448 3d points according to the resolution of the matrix camera. Without dewarping the pure scanning time of our book scanner with a resolution of 300 dpi is about 1s. That means we receive a total time of 2s for generating a nearly distortion-free book copy. This is a very good value for a publicly used book scanner system.

The robustness of the method depends on the distribution of the brightness values along the book surface. In most cases we get the expected completely dewarped output results. Only if there are deep black values over the whole vertical length of the page our method yields incorrect output results, e.g. a wrong outer delimitation of the page or additional distortions not coming from the original scanner image. An examples is given in figure 10. While on the left page the dewarping is nearly correct, we see on the right page deviations from the expected output, which are due to weak measurements over the vertical length of the page. We can stop such incorrect outputs through defining some consistency rules relating the brightness values, but we cannot sufficiently correct them. The analysis and improvement of this infrequent problem will be one of the next steps our research work.

Fig. 10. Example of incorrect output due to low vertical brightness values

6 Conclusion and Further Work

We have presented an improved method to generate nearly distortion-free copies of bounded books. As input we use the original images of a top view scanner and a set of 3d surface measurements we get from a special light sectioning. Because the light sectioning only provides with weak 3d measurements we introduced a simplified surface model of bounded books and developed a new dewarping method based on a weighted linear approximation along the vertical book length. Thereby we additionally described a technique to detect the geometrical location of the book crease. For the common application of book scanning our method yields well dewarped and aligned output results. There are only very few cases for which our method is not suitable.

If the technique will be implemented in a book scanner product we hope for extensive feed back from the users regarding the quality of our method. Furthermore we will do some impartial investigations about the OCR recognition. Depending on the geometric form we presume a significant improvement of the OCR recognition near the book crease.

To further the improvement of our dewarping system we have to upgrade the quality of the 3d measurement. One way is to support the light sectioning by an additional area correlation method to reduce the weak measurements. The scanner hardware already fulfilled all conditions for this approach. Another idea is to use a second matrix camera. This brings less weak measurements because we can use pure stereo photogrammetry instead of light sectioning. But it increases the cost of the book scanner system. And it was a major concern designing the book scanner with a very good cost-benefit ratio.

Acknowledgment

This work was supported by AiF/Germany grants (FKZ: KF0056101SS4). We thank the company Chromasens for providing us with the book scanner for research purposes. And special thanks to Roger Klein.

References

1. Wada, T., Ukida, H., Matsuyama, T.: Shape from shading with interreflections under a proximal light source: Distorsion-free copying of an unfolded book. International Journal of Computer Vision 24(2), 125–135 (1997)

2. Tan, C.L., Zhang, L., Zhang, Z., Xia, T.: Restoring warped document images through 3d shape modeling. IEEE Transactions on Pattern Analysis and Machine Intelligence 28(2), 195–208 (2006)
3. Wu, C., Agam, G.: Document image de-warping for text/graphics recognition. In: Proc. of Joint IAPR and SPR, pp. 348–357 (2002)
4. Zhang, Z., Tan, C.L.: Correcting document image warping based on regression of curved text lines. In: Proc. of 7th International Conference on Document Analysis and Recognition, vol. 1, pp. 589–595 (2003)
5. Ulges, A., Lampert, C.H., Breuel, T.M.: Document image dewarping using robust estimation of curled text lines. In: Proc. of 8th International Conference on Document Analysis and Recognition, vol. 2, pp. 1001–1005. IEEE Computer Society, Washington, DC (2005)
6. Rebmann, R., Michaelis, B., Krell, G., Seiffert, U., Püschel, F.: Improving image processing systems by artificial neural networks. In: Dengel, A., Junker, M., Weisbecker, A. (eds.) Reading and Learning. LNCS, vol. 2956, pp. 37–64. Springer, Heidelberg (2004)
7. Frei, B.: Method and device for the correction of a scanned image. Patent US020050053304A1 (2002)
8. Donescu, A., Bouju, A., Quillet, V.: Former books digital processing: image warping. In: Proc. Workshop on Document Image Analysis, pp. 5–9 (1997)
9. Yamashita, A., Kawarago, A., Kaneko, T., Miura, K.: Shape reconstruction and image restoration for non-flat surfaces of documents with a stereo vision system. In: Proc. of 17th International Conference on Pattern Recognition, vol. 1, pp. 486–489. IEEE Computer Society, Washington, DC (2004)
10. Chu, K.B., Zhang, L., Zhang, Y., Tan, C.L.: A fast and stable approach for restoration of warped document images. In: Proceedings of the Eighth International Conference on Document Analysis and Recognition, pp. 384–388. IEEE Computer Society, Washington, DC (2005)
11. Klette, R., Koschan, A., Schlüns, K.: Copmuter Vision: Three-Dimensional Data from Images. Springer, Heidelberg (1998)
12. Brown, M., Seales, W.: Image restoration of arbitrarily warped documents. IEEE Transactions on Pattern Analysis and Machine Intelligence 26(10), 1295–1306 (2004)
13. Liang, J., DeMenthon, D., Doermann, D.: Flattening curved documents in images. In: Proceedings of the 2005 IEEE Computer Society Conference on Computer Vision and Pattern Recognition, vol. 2, pp. 338–345 (2005)
14. Liang, J., DeMenthon, D., Doermann, D.: Geometric rectification of camera-captured document images. IEEE Transactions on Pattern Analysis and Machine Intelligence (Preprint, 2007)
15. Lilienblum, E., Michaelis, B.: Digitalisation of warped documents supported by 3d-surface reconstruction. In: The 5th International Conference on Computer Vision Systems Conference Paper (2007), doi:10.2390/biecoll-icvs2007-69
16. Lilienblum, E., Michaelis, B., Schnitzlein, M.: Verfahren zum Entzerren einer mittels einer Scanneinrichtung eingescannten zweidimensionalen Abbildung einer nicht ebenflächigen Vorlage. German patent, No. 102006032533 (2006)
17. Shirai, Y.: Recognition of polyhedrons with a range finder. Pattern Recognition 4, 243–250 (1972)

3D Image Topological Structuring with an Oriented Boundary Graph for Split and Merge Segmentation

Fabien Baldacci, Achille Braquelaire, Pascal Desbarats,
and Jean-Philippe Domenger

LaBRI, Université Bordeaux 1, 351 cours de la liberation F-33405
Talence cedex France
{baldacci,braquelaire,desbarats,domenger}@labri.fr

Abstract. In this paper, we present a new representation model for the topology and the geometry of a 3D segmented image. This model has been designed to provide main features and operations required by a 3D image segmentation library. It is mainly devoted to region based segmentation methods such as split and merge algorithms but is also convenient for contour based approaches. The model has been fully implemented and tested both on synthetic and real 3D images.

Keywords: 3D Images Segmentation. Topological graph. Image representation.

1 Introduction

Split and merge segmentation [1] basically consists in building and refining a partition of an image. The partition elements are called regions. The split operation consists in dividing a domain into regions, and the merge operation consists in fusing two or more adjacent regions. Split and merge algorithms consist in alternatively splitting and merging regions according to one or more criteria. Those methods have been extensively studied for 2D images and can be transposed to 3D images. Related algorithms require the extraction of some features to drive the segmentation process, such as the domain corresponding to a region (to compute geometrical features, region mean, region variance, etc.), the set of regions being adjacent to another one, or the set of regions included into another one. Boundary of a given region or boundary shared by two regions are also usually needed.

Efficiency of features extraction is a critical aspect of the development of 3D image segmentation methods. Feature extraction can be improved by structuring the segmented image. This structuring must represent both the topology and the geometry of the partition of a 3D image into 3D regions [2]. Several topological models have been proposed in 2D. One popular model is the model of *Region Adjacency Graph* (RAG) [3] which is simple to implement but does not capture the whole topology of the segmented image, such as the multiple adjacency of regions relation (related to the number of disconnected surfaces

D. Coeurjolly et al. (Eds.): DGCI 2008, LNCS 4992, pp. 541–552, 2008.

shared by two regions). Other popular and more sophisticated models are the models based on combinatorial maps [4,5]. Combinatorial maps can be defined in 3D [6] and two models using 3D combinatorial maps have been proposed: the *Hierarchical Local Embedding* (HLE) [7,8] and the *Geometrical Embedding* (GE) [9]. Both these models use 3D combinatorial maps associated with inclusion relations or inclusion tree, and an intervoxel representation of the geometry of regions boundary [10,11]. Intervoxel representation of geometry [12] lays on the cellular decomposition of 3D discrete space [13,14] in which elements are voxels (elements of dimension 3), surfels (intersection of two voxels), linels (intersection of two surfels) and pointels (intersection of two linels) (see Fig. 1).

Nevertheless, the development of these models for 3D images raises some unstraightforward problems. For instance, a GE representation is not minimal in term of topological elements, and some extra elements need to be added with some voxel configurations. A HLE representation uses fictive topological elements that cannot be retrieved in the geometrical level (this leads to some constraints on the update operations of the representation), and local geometry representation which may be inefficient for split and merge methods. These models have been compared in order to be merged [15], but the resulting model still uses fictive elements and related drawbacks remains the same. Furthermore, combinatorial maps require to decompose region boundary into surface elements homeomorphic to a topological disc, which need extra processing on surfaces resulting from a segmentation.

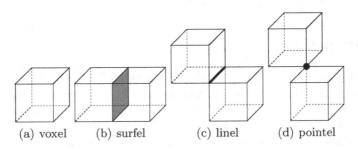

(a) voxel   (b) surfel   (c) linel   (d) pointel

Fig. 1. Elements of the cellular decomposition of 3D discrete space

Models based on maps encode all topological features of a segmented image like regions neighborhood, regions inclusion, and higher level topological features such as Euler characteristic — and thus the number of regions' holes — and Betti numbers [16]. However many segmentation algorithms do not require such high level topological features and it is relevant to consider alternative topological models, less powerful than maps but powerful enough to implement most of split and merge methods. Thus we propose in this work a new model for structuring 3D segmented image which is lighter than models based on combinatorial maps, compatible with them (it is still possible to extract the combinatorial map of some regions of interest) and which avoids their drawbacks (the compromise is that high level topological features require specific processing to be extracted from our model).

To sum up, the goal of this work is to develop a model which is on one hand not space consuming and more suitable than other models when considering large 3D images, and on the other hand which allows efficient implementation of 3D split and merge algorithms. We retained the following approach:

1. Identify the main geometrical and topological features and the representation updates required by region based segmentation algorithms.
2. Define a minimal set of functions satisfying these requirements (this set of functions can be seen as the API of a segmentation library).
3. Build a model with a minimum memory cost that allows efficient computation of these functions.

Those requirements and functions are listed in Section 2. The model is presented in Section 3, the related functions are detailed in Section 4. Finally examples and results are given in Section 5.

2   Requirements for a Segmentation Environment

A partition P of an image is a region set such as: $P = \{\{r_1...r_n\} \mid \cap_{i \neq j} r_i r_j = \emptyset, \cup_{1 \leq i \leq n} r_i = P\}$; a region is a set of 6-connected voxels. A segmented image is a partition of the image satisfying criteria.

The boundary of a region r_i is the set of surfels that separates r_i from other regions, it is denoted by ∂r_i. A region boundary consists of the surfels of the outside boundary (the shell of the region) and of the surfels of the region cavities. The common boundary of two adjacent regions r_i and r_j is decomposed into a set of connected surfaces $\{S_{ij}^1...S_{ij}^n\}$, such as each connected surface S_{ij}^k is the k-nth maximum set of connected surfels $s \in \partial r_i \cap \partial r_j$, called S-set. These surfaces are not necessarily simple (a simple surface is a surfce whithout hole). need any other processing for our representation.

Update primitives required by region based segmentation methods are basically the construction of the representation of a partitioned region (resulting from the splitting step of a split and merge method) and the merging of two or more adjacent regions. The first one requires the specification of a region partition. We chose to specify a region partition by using a function $Oracle(v_i, v_j)$ which indicates whether two adjacent voxels v_i and v_j belong or not to the same region. This function provides an abstract representation of a region (or an image) partition which is not dependent on the concrete representation of the segmented image (label array, boundary representation, etc.).

Let us now give a short description of the functions we have retained for a 3D image segmentation environment. The list of these functions is derived from an analogous work for 2D images [17]. The reader can refer to this work to find justifications of the choice of the selected function.

Updating functions. The two following functions respectively build the representation of a partitioned region and merge two adjacent regions.

Split$(r, Oracle)$: returns $\{r_i | \cup_i r_i = r\}$. This function refines the partition of the image by dividing r_i, according to the $Oracle$ function which specifies the partition.

Merge(r_i, r_j): returns $\{r | r = r_i \cup r_j\}$. This function merges two adjacent regions r_i and r_j.

Geometrical functions. The following functions extract features from the geometrical representation.

RegionDomain(r_i): returns $\{v | v \in r_i\}$. This function returns a list of pairs of voxels defining a set of lines covering the region r_i: for each pair $\{v_i(x_i, y, z), v_j(x_j, y, z)\}$, voxels $\{v_k(x_k, y, z) | x_i < x_k < x_j\} \in r_i$.

RegionBoundary(r_i): returns $\{s | s \in \partial r_i\}$. This function returns a list of closed surfaces. The number of returned surfaces is equal to the number of cavities of the region, plus one for the outside boundary.

CommonBoundary(r_i, r_j): returns $\{s | s \in \partial r_i \cap \partial r_j\}$. This function returns a set of non simple surfaces $\{S_1 ... S_n\}$, such as $\forall s \in S_i, s' \in S_j, i \neq j, \forall surfel$ $chain\{s...s'\}, \exists s'' \in \{s...s'\}, s'' \notin \partial r_i \cap \partial r_j$.

RegionAt(**voxel v**): returns $r | v \in r$.

Topological functions. The following functions extract topological features: neighborhood and inclusion relations. A region r_i is said to be included in another region r_j if each possible voxel chain (sequence of connected voxels) from any voxel of r_i to the outside of the image contains at least one voxel of r_j. The nearest region in which r_i is included is denoted by $Parent(r_i)$.

IncludedRegions(r_i): returns $\{r | r_i = Parent(r)\}$.

Neighbors(r_i): returns $\{r_j | \partial r_i \cap \partial r_j \neq \emptyset\}$.

OppositeRegion(**Surface** S_{ij}^k, **Region** r_i): returns r_j.

Parent(r_i): returns $r | r = Parent(r_i)$.

This list of *required functions* can be seen as the basic API of a 3D image segmentation library. In the following section we describe the model we retained as a framework to implement these functions.

3   The Model

The model described in this section is based on the *S-sets* induced by a partition. It is made of two representation levels, a topological one and a geometrical one. These levels are linked together by associating an oriented surfel with each *S-set*. The geometrical level must at least encode the boundary surfels in order to be able to represent *S-sets*. It must also encode the edges of *S-set*, denoted by ∂S, which are sequences of linels. The encoding of such linels is necessary to traverse the boundary elements composing surface boundaries. We chose to use a boundary image [15] (also called inter-voxel matrix) to encode the geometrical elements. This matrix encodes the presence of surfels, linels and pointels corresponding to the image with seven bits per voxel (pointels are not used by our model, but are necessary to the embedding of a combinatorial map). Another advantage of the boundary image is to provide a global representation of the geometry which allows the access to the topological representation from a geometrical element. Moreover such models are more convenient for implementing contour based methods.

From the required functions list of the previous section we know we have to provide an explicit encoding of two topological features: the region adjacency relation and the inclusion relation. Coding region adjacency is achieved by using an adjacency graph that we call *oriented boundary graph* (in short *OBG*). This adjacency graph, which is a multigraph, needs one edge per *S-set* to encode multiple adjacency. More precisely a graph $OBG(E, V)$ is defined on:

- a set of vertices $V = \{r_1, ... r_n\}$;
- a set of edges $E = \{e(r_i, r_j)^k | \exists S_{ij}^k \text{ with } \cup_k S_{ij}^k = \partial r_i \cap \partial r_j\}$.

This graph provides the **Neighbors** function simply by traversing all the incident edges of the given region. Since the *OBG* is a connected graph, it is possible to compute the inclusion relation without adding an extra structure such as an inclusion tree. The algorithm 1 is used to compute the **IncludedRegion** function. It returns the list of regions included in a given one. In this algorithm, the *ParentSurface* is a *S-set* belonging to the outside boundary of the region. Then all *S-sets* of the outside boundary have to be marked, then unmarked *S-sets* correspond to region cavities. This algorithm needs to encode the *S-set* adjacency relation. This relation explicits a relation implicitly encodede in the *OBG* definition, and is encoded by linking each edge $e(r_i, r_j)^p$ with $e(r_l, r_m)^q$ if $\partial S_{ij}^p \cap \partial S_{lm}^q \neq \emptyset$. This link is done by a representative linel pointing to all edges corresponding to *S-set* adjacent to it. Note that if the inclusion relation has to be computed many times, it is possible to perform a lazy evaluation by storing the ParentSurface in order to avoid the breadth-first search in the *OBG*. The orientation of the *OBG* will be precised in the next paragraph.

The boundary image and the *OBG* allows to efficiently retrieve all features needed by the required functions list. Both structures now have to be linked

---

Algorithm 1. Inclusion

Require: An *OBG* and a region r_i
Returns: List of regions included in r_i
  1: ParentSurface $\leftarrow$ BreadthFirstSearchUntil(OBG, r_i)
  2: Mark(ParentSurface)
  3: **for all** Surface $\in \partial r_i$ and connected to ParentSurface **do**
  4:     Mark(Surface)
  5: **end for**
  6: **for all** Surface $\in \partial r_i$ **do**
  7:     **if** IsUnmarked(Surface) **then**
  8:         AddToList(ResultList, OppositeRegion(Surface, r_i))
  9:     **end if**
10: **end for**

---

together in order to provide efficient updates of the representation (**Split** and **Merge** functions). It is done by selecting a representative surfel for each *S-set*. This surfel is associated in the geometrical level with its corresponding edge in the topological level. It is thus possible to go from an element of the boundary image to the associated element of the *OBG*. Representative surfels are also stored in the *OBG* such that with each edge $e(r_i, r_j)^k$ is associated one surfel of its corresponding *S-set* S_{ij}^k. This allows to retrieve all surfaces from the *OBG*. Finding regions is done by orienting edges. Edges are oriented depending on their geometrical embedding. For each type of surfel s it is possible to define a positive voxel $Positive(s)$ and a negative one $Negative(s)$ according to the image orthonormal coordinates system. Those voxels belong to the two different regions the surface is separating. Edges are oriented such that with each edge $e(r_i, r_j)$ a surfel $s|Positive(s) \in r_i, Negative(s) \in r_j$ is associated. This orientation is not used for traversing the *OBG*.

An example is shown on Fig. 2. The region r_0 is the outside of the image, called the infinite region. There are four other regions, such as r_2, r_3 and r_4 are included in r_1. Regions are linked in the *OBG* by edges representing *S-sets*. Region r_1 is separated from region r_0 by one *S-set* which is the outside boundary of r_1. Region r_1 has two cavities: the connected component made of r_2 and r_3, and the one composed of r_4. The outside boundary of r_2 (as the outside boundary of r_3) is cut into two *S-set*: one common with r_1 and the other one common with r_3. Each edge is linked to the representative surfel of its corresponding *S-set*, and adjacent surfaces are linked together by a linel. Black side of a representative surfel s corresponds to their $Negative(s)$, and grey side corresponds to their $Positive(s)$.

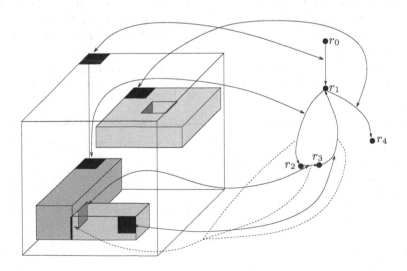

Fig. 2. Example of image with the corresponding representation with our model

4   Updating and Feature Extraction Functions

The construction of the representation of a partitioned region is done from the region domain and the *Oracle* function representing the segmentation. The region may be the whole image, for instance, at the first step of a split and merge process. The construction is based on a temporary labelling of each elements: voxels, surfels and linels, such as all the voxels (resp. surfels) of a region (resp. a *S-set*) have the same label, and two different regions (resp *S-set*) have different labels. Each voxel label is associated with one region in the *OBG*, and each surfel label is associated with one edge. Since traversing is done by a scan-line algorithm, a region can temporarily have different labels. When such a case is detected, the two related vertices of the *OBG* are merged and an indirection table is updated in order not to relabelling voxels.

The split function is described in Algorithm 2. Each voxel of the domain is compared with its labelled neighbors, and then is labelled with possibly the appropriate treatment (creating a node or merging two ones). The way the domain is traversed (by scan-line) allows to determine which neighbors are already labelled without any test. In the same time surfels are detected. When all the voxels of a 2x2x1 cuboid (Figure 3) are labelled, all the surfels contained in the cuboid have been detected and it is possible to decide whether a linel is present. Surfels and linels have to be labelled in the same way voxels are.

Construction and split is done by traversing all the voxels of the image once, and since there is no constraint on the configuration of treated voxels for treating a new one, it is possible to parallelize this operation [18].

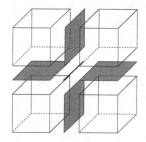

Fig. 3. A 2x2x1 voxel cuboid and its contained surfels and linel

Merging two regions r_i and r_j is done by deleting all the *S-sets* S_{ij}^k from the boundary image and their corresponding edges $e(r_i, r_j)^k$ from the *OBG*. Some *S-sets* may have to be merged. Such *S-sets* are detected by checking all the encountered linels: if they bound only two *S-sets*, then those *S-sets* have to be merged (*S-sets* to be merged are retrieved directly from the representative linel). The merging of two regions is described by Algorithm 3.

CommonBoundary of two regions can be retrieved since each edge is pointing to a representative surfel from which the *S-set* can be traversed by looking

---

Algorithm 2. Split Algorithm

Require: An *OBG*, A domain D and an *Oracle* function
1: **for all** Voxel $v \in D$ **do**
2:   **for all** labelled voxel v_i neighbors of v **do**
3:     **if** $Oracle(v_i, v)$ **then**
4:       SaveLabel(list, v_i)
5:     **else**
6:       addSurfelBetween(v_i, v)
7:     **end if**
8:   **end for**
9:   **if** NumberOfLabel(list) > 1 **then**
10:     MergeRegion(list)
11:   **else**
12:     **if** NumberOfLabel(list) $== 0$ **then**
13:       CreateNewRegion(NewLabel)
14:     **end if**
15:   **end if**
16:   **for all** 4 labelled voxel cuboid containing v **do**
17:     **if** more than 2 surfels are present **then**
18:       AddLinel(x, y, z)
19:     **end if**
20:     LabelAddedLinels()
21:     LabelSurfels()
22:   **end for**
23: **end for**

Algorithm 3. Merge Algorithm

Require: 2 regions r_1 and r_2 to merge
Returns: The new region $r = r_1 \cup r_2$
1: **for** each surface of r_1 **do**
2:   **if** it separates r_1 from r_2 **then**
3:     linellist $\leftarrow$ deleteSurfaceFromGeometry()
4:     DeleteSurfaceFromTopology()
5:   **end if**
6: **end for**
7: **for** each linel of linellist **do**
8:   **if** its degree is 2 **then**
9:     DeleteLinelFromGeometry()
10:     **if** it is a representative linel **then**
11:       MergeSurfaces(linel)
12:     **end if**
13:   **else**
14:     **if** its degree is less than 2 **then**
15:       DeleteLinelFromGeometry()
16:     **end if**
17:   **end if**
18: **end for**

---

for connected surfels up to encountering linels. Adjacent *S-sets* can be traversed together (to compute the **RegionBoundary**) without stopping on linels by using the surfel orientation in order to stay in a same region during the traversing. **RegionDomain** can be built from the boundary by sorting each encountered surfels of same type, each surfel pair of the results bounds a voxel line of the region.

Some regions features are additive for the split and the merge operation: size (in number of voxels), sum of values, sum of squares values, etc. It is possible to save those values with each region to compute features extraction such as mean and variance of regions. Furthermore, the model allows the use of active contours, useful for example to smooth the surface of a region.

5   An Example of Segmentation Algorithm

In this section we present an example of segmentation algorithm (Algorithm 4) using functions of the proposed framework. Note that this algorithm is not supposed to give accurate results but to show how the proposed model can be used to specify and implement a split and merge algorithm based on topological and geometrical features.

This algorithm first creates the partition of an image by classifying the voxels into classes according to given thresholds (on the examples we use five thresholds

---

Algorithm 4. Simple Segmentation Algorithm

Require: An image I, a size minsize and a value minmean
1: **Split**(I, Classify($\{s_i\}$))
2: **for all** region r **do**
3:    **if** Size(r) < minsize **then**
4:       **if** **Parent**(r) $\in$ **Neighbors**(r) **then**
5:          **Merge**(r, **Parent**(r))
6:       **end if**
7:    **end if**
8: **end for**
9: **for all** region r **do**
10:    **if** **Parent**(r) $\in$ **Neighbors**(r) **then**
11:       **if** abs(Mean(Parent(r)) - Mean(r)) < minmean **then**
12:          **Merge**(r, Parent(r))
13:       **end if**
14:    **end if**
15: **end for**
16: **for all** region r_i **do**
17:    **for all** region r_j $\in$ **Neighbors**(r_i) **do**
18:       **if** abs(Mean(r_i) - Mean(r_j)) < minmean **then**
19:          **Merge**(r_i, r_j)
20:       **end if**
21:    **end for**
22: **end for**

---

and six classes). Then it uses three criteria for merging the regions: first it merges small isolated regions with their parent (this corresponds to noise removing). Then it applies two kind of merges on adjacent regions that are similar enough according to segmentation criteria (mean difference, inclusion relation).

We have tested our representation both on synthetic and real images with the segmentation algorithm 4. Figure 4 shows the obtained result on synthetic noisy image. Figure 4(a) represent the obtained result after the split operation, and Figure 4(b) the obtained result at the end of the algorithm.

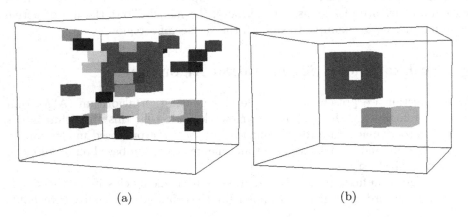

(a)   (b)

Fig. 4. Example of segmentation on a noised synthetic image

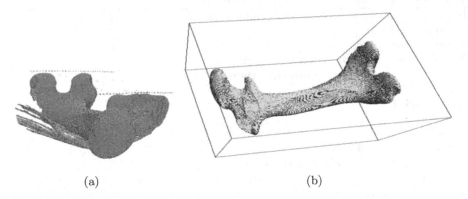

(a)   (b)

Fig. 5. Example of segmentation on a real medical image of size 512x512x475 voxels

Figure 5 shows the result obtained on a real medical image (representing a femur) sizing 512x512x475 voxels. We obtain from the split operation more than 200000 regions. Figure 5(a) is a capture during the segmentation process and it contains a lot of insignificant regions. Merges operations lead to the Figure 5(b) which is composed of 8 regions (outside of the image, the bone, the background and some regions inside the bone). This example shows the possibility of using our model to represent and update a partition with large size images.

6   Conclusion

The proposed model has been defined according to a set of topological and geometrical functions required by a segmentation environment. On one hand it is simpler than models based on combinatorial maps, and on the other hand it is convenient for implementing a library for 3D image segmentation. This model does not explicitly encode the whole topology of a segmented image as combinatorial maps do but it avoids the drawbacks of map based models. We chose to encode less information, in order to have a lighter and a more efficient model for all segmentation applications which do not use high level topological features of the regions. Nevertheless we think that both this model and map based models are interesting for image segmentation, depending on the kind of segmentation methods to implement. Thus the model has been designed in order to allow local extraction of a combinatorial map when it appears necessary to explicit the encoding of higher topological features. Furthermore it could be possible to use contour based methods in order to refine the result of a region based segmentation.

The model have been implemented and is fully functional with real images. The first version was a prototype implemented in Python language in order to experimentally validate the model, that is the reason why time an memory consumption cannot be given. A second optimised version in C language is currently under development. This development is achieved for the construction of the model, and for a 512x512x475 image with more than 1 million regions, it takes about 30 seconds to build the model on a 2,33Ghz intel Xeon machine.

Future works consist in studying the extraction of combinatorial maps from our model, in order to use high level topological features on some regions, and then to update our model after treatments. A fast surface traversing algorithm optimising the representative surfel search may be developed using two ideas: placing the representative surfels with priority to some directions, or saving a distance map with the boundary image. A further work will consist in parallelising most of the operations (especially the split one, for which a first result is presented in [18]).

References

1. Horowitz, S., Pavlidis, T.: Picture segmentation by a directed split and merge procedure. In: ICPR 1974, pp. 424–433 (1974)
2. Braquelaire, J.P., Brun, L.: Image segmentation with topological maps and interpixel representation. Journal of Visual Communication and Image Representation 9(1), 62–79 (1998)
3. Rosenfeld, A.: Adjacency in digital pictures. InfoControl 26 (1974)
4. Braquelaire, J.-P., Domenger, J.P.: Representation of segmented images with discrete geometric maps. Image Vision Comput. 17(10), 715–735 (1999)
5. Braquelaire, A., Desbarats, P., Domenger, J.P.: 3d split and merge with 3-maps. In: 3rd IAPR-TC-15 Workshop on Graph-based representation. CUEN, pp. 32–43 (2001) ISBN 887146579-2

6. Lienhardt, P.: Topological models for boundary representation: a comparison with n-dimensional generalized maps. Comput. Aided Des. 23(1), 59–82 (1991)
7. Bertrand, Y., Damiand, G., Fiorio, C.: Topological encoding of 3d segmented images. In: Nyström, I., Sanniti di Baja, G., Borgefors, G. (eds.) DGCI 2000. LNCS, vol. 1953, pp. 311–324. Springer, Heidelberg (2000)
8. Damiand, G.: Définition et étude d'un modèle topologique minimal de représentation d'images 2d et 3d. PhD thesis, PhD Thesis, Montpellier II University (2001)
9. Desbarats, P.: Structuration d'images segmentées 3D discrètes. PhD thesis, PhD Thesis, Bordeaux I University (2001)
10. Braquelaire, A., Desbarats, P., Domenger, J.P., Wütrich, C.: A topological structuring for aggregates of 3d discrete objects. In: 2nd IAPR-TC-15 Workshop on Graph-based representation, pp. 193–202. OCG (1999) ISBN 3-8580-126-2
11. Damiand, G., Resch, P.: Topological map based algorithms for 3d image segmentation. In: Braquelaire, A., Lachaud, J.-O., Vialard, A. (eds.) DGCI 2002. LNCS, vol. 2301, pp. 220–231. Springer, Heidelberg (2002)
12. Brice, C.R., L., F.C.: Scene analysis using regions. Artif. Intell. 1(3), 205–226 (1970)
13. Kovalevsky, V.: Finite topology as applied to image analysis. CVGIP 46(2), 141–161 (1989)
14. Kovalevsky, V.: Multidimensional cell lists for investigating 3-manifolds. Discrete Appl. Math. 125(1), 25–43 (2003)
15. Braquelaire, A., Damiand, G., Domenger, J.P., Vidil, F.: Comparison and convergence of two topological models for 3d image segmentation. In: Hancock, E.R., Vento, M. (eds.) GbRPR 2003. LNCS, vol. 2726, pp. 59–70. Springer, Heidelberg (2003)
16. Desbarats, P., Domenger, J.P.: Retrieving and using topological characteristics from 3D discrete images. In: Proceedings of the 7th Computer Vision Winter Workshop, PRIP-TR-72, pp. 130–139 (2002)
17. Braquelaire, A.: Representing and segmenting 2d images by means of planar maps with discrete embeddings: From model to applications. In: Brun, L., Vento, M. (eds.) GbRPR 2005. LNCS, vol. 3434, pp. 92–121. Springer, Heidelberg (2005)
18. Baldacci, F., Desbarats, P.: Parallel 3d split and merge segmentation with oriented boundary graph. In: Proceedings of The 16th International Conference in Central Europe on Computer Graphics, Visualization and Computer Vision (accepted, 2008)

Author Index

Lecture Notes in Computer Science

Sublibrary 6: Image Processing, Computer Vision, Pattern Recognition, and Graphics

For information about Vols. 1– 2396
please contact your bookseller or Springer

Vol. 4417: A. Kerren, A. Ebert, J. Meyer (Eds.), Human-Centered Visualization Environments. XIX, 403 pages. 2007.

Vol. 4391: Y. Stylianou, M. Faundez-Zanuy, A. Esposito (Eds.), Progress in Nonlinear Speech Processing. XII, 269 pages. 2007.

Vol. 4370: P.P. Lévy, B. Le Grand, F. Poulet, M. Soto, L. Darago, L. Toubiana, J.-F. Vibert (Eds.), Pixelization Paradigm. XV, 279 pages. 2007.

Vol. 4358: R. Vidal, A. Heyden, Y. Ma (Eds.), Dynamical Vision. IX, 329 pages. 2007.

Vol. 4338: P.K. Kalra, S. Peleg (Eds.), Computer Vision, Graphics and Image Processing. XV, 965 pages. 2006.

Vol. 4319: L.-W. Chang, W.-N. Lie (Eds.), Advances in Image and Video Technology. XXVI, 1347 pages. 2006.

Vol. 4292: G. Bebis, R. Boyle, B. Parvin, D. Koracin, P. Remagnino, A. Nefian, G. Meenakshisundaram, V. Pascucci, J. Zara, J. Molineros, H. Theisel, T. Malzbender (Eds.), Advances in Visual Computing, Part II. XXXII, 906 pages. 2006.

Vol. 4291: G. Bebis, R. Boyle, B. Parvin, D. Koracin, P. Remagnino, A. Nefian, G. Meenakshisundaram, V. Pascucci, J. Zara, J. Molineros, H. Theisel, T. Malzbender (Eds.), Advances in Visual Computing, Part I. XXXI, 916 pages. 2006.

Vol. 4245: A. Kuba, L.G. Nyúl, K. Palágyi (Eds.), Discrete Geometry for Computer Imagery. XIII, 688 pages. 2006.

Vol. 4241: R.R. Beichel, M. Sonka (Eds.), Computer Vision Approaches to Medical Image Analysis. XI, 262 pages. 2006.

Vol. 4225: J.F. Martínez-Trinidad, J.A. Carrasco Ochoa, J. Kittler (Eds.), Progress in Pattern Recognition, Image Analysis and Applications. XIX, 995 pages. 2006.

Vol. 4191: R. Larsen, M. Nielsen, J. Sporring (Eds.), Medical Image Computing and Computer-Assisted Intervention – MICCAI 2006, Part II. XXXVIII, 981 pages. 2006.

Vol. 4190: R. Larsen, M. Nielsen, J. Sporring (Eds.), Medical Image Computing and Computer-Assisted Intervention – MICCAI 2006, Part I. XXXVVIII, 949 pages. 2006.

Vol. 4179: J. Blanc-Talon, W. Philips, D. Popescu, P. Scheunders (Eds.), Advanced Concepts for Intelligent Vision Systems. XXIV, 1224 pages. 2006.

Vol. 4174: K. Franke, K.-R. Müller, B. Nickolay, R. Schäfer (Eds.), Pattern Recognition. XX, 773 pages. 2006.

Vol. 4170: J. Ponce, M. Hebert, C. Schmid, A. Zisserman (Eds.), Toward Category-Level Object Recognition. XI, 618 pages. 2006.

Vol. 4153: N. Zheng, X. Jiang, X. Lan (Eds.), Advances in Machine Vision, Image Processing, and Pattern Analysis. XIII, 506 pages. 2006.

Vol. 4142: A. Campilho, M. Kamel (Eds.), Image Analysis and Recognition, Part II. XXVII, 923 pages. 2006.

Vol. 4141: A. Campilho, M. Kamel (Eds.), Image Analysis and Recognition, Part I. XXVIII, 939 pages. 2006.

Vol. 4122: R. Stiefelhagen, J.S. Garofolo (Eds.), Multimodal Technologies for Perception of Humans. XII, 360 pages. 2007.

Vol. 4109: D.-Y. Yeung, J.T. Kwok, A. Fred, F. Roli, D. de Ridder (Eds.), Structural, Syntactic, and Statistical Pattern Recognition. XXI, 939 pages. 2006.

Vol. 4091: G.-Z. Yang, T. Jiang, D. Shen, L. Gu, J. Yang (Eds.), Medical Imaging and Augmented Reality. XIII, 399 pages. 2006.

Vol. 4073: A. Butz, B. Fisher, A. Krüger, P. Olivier (Eds.), Smart Graphics. XI, 263 pages. 2006.

Vol. 4069: F.J. Perales, R.B. Fisher (Eds.), Articulated Motion and Deformable Objects. XV, 526 pages. 2006.

Vol. 4057: J.P.W. Pluim, B. Likar, F.A. Gerritsen (Eds.), Biomedical Image Registration. XII, 324 pages. 2006.

Vol. 4046: S.M. Astley, M. Brady, C. Rose, R. Zwiggelaar (Eds.), Digital Mammography. XVI, 654 pages. 2006.

Vol. 4040: R. Reulke, U. Eckardt, B. Flach, U. Knauer, K. Polthier (Eds.), Combinatorial Image Analysis. XII, 482 pages. 2006.

Vol. 4035: T. Nishita, Q. Peng, H.-P. Seidel (Eds.), Advances in Computer Graphics. XX, 771 pages. 2006.

Vol. 3979: T.S. Huang, N. Sebe, M. Lew, V. Pavlović, M. Kölsch, A. Galata, B. Kisačanin (Eds.), Computer Vision in Human-Computer Interaction. XII, 121 pages. 2006.

Vol. 3954: A. Leonardis, H. Bischof, A. Pinz (Eds.), Computer Vision – ECCV 2006, Part IV. XVII, 613 pages. 2006.

Vol. 3953: A. Leonardis, H. Bischof, A. Pinz (Eds.), Computer Vision – ECCV 2006, Part III. XVII, 649 pages. 2006.

Vol. 3952: A. Leonardis, H. Bischof, A. Pinz (Eds.), Computer Vision – ECCV 2006, Part II. XVII, 661 pages. 2006.

Vol. 3951: A. Leonardis, H. Bischof, A. Pinz (Eds.), Computer Vision – ECCV 2006, Part I. XXXV, 639 pages. 2006.

Vol. 3948: H.I. Christensen, H.-H. Nagel (Eds.), Cognitive Vision Systems. VIII, 367 pages. 2006.

Vol. 3926: W. Liu, J. Lladós (Eds.), Graphics Recognition. XII, 428 pages. 2006.

Vol. 3872: H. Bunke, A.L. Spitz (Eds.), Document Analysis Systems VII. XIII, 630 pages. 2006.

Vol. 3852: P.J. Narayanan, S.K. Nayar, H.-Y. Shum (Eds.), Computer Vision – ACCV 2006, Part II. XXXI, 977 pages. 2006.

Vol. 3851: P.J. Narayanan, S.K. Nayar, H.-Y. Shum (Eds.), Computer Vision – ACCV 2006, Part I. XXXI, 973 pages. 2006.

Vol. 3832: D. Zhang, A.K. Jain (Eds.), Advances in Biometrics. XX, 796 pages. 2005.

Vol. 3736: S. Bres, R. Laurini (Eds.), Visual Information and Information Systems. XI, 291 pages. 2006.

Vol. 3667: W.J. MacLean (Ed.), Spatial Coherence for Visual Motion Analysis. IX, 141 pages. 2006.

Vol. 3417: B. Jähne, R. Mester, E. Barth, H. Scharr (Eds.), Complex Motion. X, 235 pages. 2007.